Methods in Enzymology

Volume XV
STEROIDS AND TERPENOIDS

METHODS IN ENZYMOLOGY

EDITORS-IN-CHIEF

Sidney P. Colowick   Nathan O. Kaplan

Methods in Enzymology

Volume XV

Steroids and Terpenoids

EDITED BY

Raymond B. Clayton

DEPARTMENT OF PSYCHIATRY
STANFORD UNIVERSITY SCHOOL OF MEDICINE
STANFORD UNIVERSITY MEDICAL CENTER
STANFORD, CALIFORNIA

1969

ACADEMIC PRESS   New York and London

ACADEMIC PRESS, INC.
111 Fifth Avenue, New York, New York 10003

United Kingdom Edition published by
ACADEMIC PRESS, INC. (LONDON) LTD.
Berkeley Square House, London W.1

LIBRARY OF CONGRESS CATALOG CARD NUMBER: 54-9110

PRINTED IN THE UNITED STATES OF AMERICA

Contributors to Volume XV

Article numbers are shown in parentheses following the names of contributors. Affiliations listed are current.

B. L. ARCHER (16), *Natural Rubber Producers' Research Association, Welwyn Garden City, Herts, England*

JOEL AVIGAN (21), *Molecular Disease Branch, National Heart Institute, National Institutes of Health, Bethesda, Maryland*

OLLE BERSÉUS (25), *Kemiska Institutionen, Karolinska Institutet, Stockholm, Sweden*

HEINZ BREUER (37), *Institut für Klinische Biochemie, Universität Bonn, Bonn-Venusberg, West Germany*

N. K. CHAUDHURI (8), *The Worcester Foundation for Experimental Biology, Shrewsbury, Massachusetts*

C. O. CHICHESTER (14), *Department of Food Science and Technology, University of California, Davis, California*

STANLEY J. CLARK (2), *Department of Biochemistry, Boston University, School of Medicine, Boston, Massachusetts*

E. G. COCKBAIN (16), *Natural Rubber Producers' Research Association, Welwyn Garden City, Herts, England*

MICHAEL COLVIN (31), *Department of Pharmacology and Experimental Therapeutics, Johns Hopkins University, School of Medicine, Baltimore, Maryland*

R. H. CORNFORTH (11), *Milstead Laboratory of Chemical Enzymology, Sittingbourne, Kent, England*

HENRY DANIELSSON (25), *Kemiska Institutionen, Karolinska Institutet, Stockholm, Sweden*

SAMUEL JAMES DAVIDSON (33), *Department of Physiology, Tufts University, School of Medicine, Boston, Massachusetts*

PETER D. G. DEAN (19), *Biochemistry Department, The University of Liverpool, Liverpool, England*

MARY E. DEMPSEY (20), *Department of Biochemistry, University of Minnesota, Minneapolis, Minnesota*

CHARLES H. DOERING (28), *Department of Psychiatry, Stanford University School of Medicine, Stanford Medical School, Stanford, California*

PERCY E. DUNAGIN, JR. (7), *Gorgas Hospital, Balboa Heights, Canal Zone*

KURT EINARSSON (25), *Kemiska Institutionen, Karolinska Institutet, Stockholm, Sweden*

P. ENEROTH (5), *Kemiska Institutionen, Karolinska Institutet, Stockholm, Sweden*

ENRICO FORCHIELLI (27), *Institute of Hormone Biology, Syntex Research Center, Palo Alto, California*

CHARLES S. FURFINE (34), *Department of Chemistry, Georgetown University, Washington, D.C.*

JOHN GLOMSET (24), *Department of Medicine, Division of Endocrinology, University of Washington, Seattle, Washington*

DEWITT S. GOODMAN (15, 22), *Department of Medicine, College of Physicians and Surgeons, Columbia University, New York, New York*

MARCEL GUT (8), *The Worcester Foundation for Experimental Biology, Shrewsbury, Massachusetts*

R. P. HANZLIK (9), *Department of Chemistry, Stanford University, Stanford, California*

ERICH HECKER (38), *Biochemisches Institut, Deutsches Krebsforschungszentrum, Heidelberg, Germany*

NOBUO IKEKAWA (3), *Laboratory of Chemistry for Natural Products, Tokyo Institute of Technology, Tokyo, Japan*

JOSEPH JARABAK (39), *Laboratory of Pharmacology, Baltimore Cancer Re-*

search Center (*National Cancer Institute*), *United States Public Health Service Hospital, Baltimore, Maryland*

REBECCA JARABAK (31), *Department of Pharmacology and Experimental Therapeutics, Johns Hopkins University, School of Medicine, Baltimore, Maryland*

R. CLIFTON JENKINS (10, 36), *Department of Psychiatry, Stanford University School of Medicine, Stanford Medical Center, Stanford, California*

FIROZE B. JUNGALWALA (13), *Department of Biochemistry, Agricultural Research Council, Institute of Animal Physiology, Babraham, Cambridge, England*

ANDREW A. KANDUTSCH (18), *The Jackson Laboratory, Bar Harbor, Maine*

RUDOLF KNUPPEN (37), *Institut für Klinische Biochemie, Universität Bonn, Bonn-Venusberg, West Germany*

SAMUEL S. KOIDE (32), *Population Council, The Rockefeller University, New York, New York*

BELISÁRIO P. LISBOA (1), *University of Stockholm, The Wenner-Gren Institute, Stockholm, Sweden*

FRIEDRICH MARKS (38), *Department of Pharmacology, Baylor University College of Medicine, Texas Medical Center, Houston, Texas*

ROBERT A. MEIGS (40), *Department of Reproductive Biology, Case Western Reserve University, School of Medicine, Cleveland, Ohio*

CARL MONDER (34), *Research Institute for Skeletomuscular Diseases of the Hospital for Joint Diseases, The Dr. Harry Britenstool Memorial, New York, New York*

SURESH H. MOOLGAVKAR (31), *Department of Pharmacology and Experimental Therapeutics, Johns Hopkins University, School of Medicine, Baltimore, Maryland*

R. G. NADEAU (9), *Hercules Inc., Research Center, Wilmington, Delaware*

SHAKUNTHALA NARASIMHULU (29), *Harrison Department of Surgical Research,*

University of Pennsylvania, School of Medicine, Philadelphia, Pennsylvania

JAMES ALLEN OLSON (7, 15), *University of Bangkok, Bangkok, Thailand*

ROBERT H. PALMER (6), *Department of Medicine, The University of Chicago, Chicago, Illinois*

G. POPJÁK (11, 12), *Department of Biological Chemistry, University of California, Los Angeles, California*

JOHN W. PORTER (13), *Department of Physiological Chemistry, University of Wisconsin, Madison, Wisconsin*

M. A. RAHIM (35), *Department of Biochemistry, University of Dacca, Dacca, East Pakistan*

OTTO ROSENTHAL (29), *Harrison Department of Surgical Research, University of Pennsylvania, School of Medicine, Philadelphia, Pennsylvania*

KENNETH J. RYAN (40), *Department of Reproductive Biology, Case Western Reserve University, School of Medicine, Cleveland, Ohio*

EUGENE C. SANDBERG (10, 36), *Department of Obstetrics and Gynecology, Stanford University School of Medicine, Stanford Medical Center, Stanford, California*

TOKUICHIRO SEKI (4), *Department of Genetics, Medical School, Osaka University, Osaka, Japan*

C. J. SIH (35), *The University of Wisconsin, School of Pharmacy, Madison, Wisconsin*

J. SJÖVALL (5), *Kemiska Institutionen Karolinska Institutet, Stockholm, Sweden*

EZRA STAPLE* (26), *Department of Biochemistry, University of Pennsylvania, School of Medicine, Philadelphia, Pennsylvania*

DANIEL STEINBERG (21), *Department of Medicine, University of California at San Diego, La Jolla, California*

PAUL TALALAY (31), *Department of Pharmacology and Experimental Therapeutics, Johns Hopkins University,*

* Deceased.

School of Medicine, Baltimore, Maryland

C. R. TREADWELL (23), *Department of Biochemistry, The George Washington University, School of Medicine, Washington, D.C.*

CHRISTEN D. UPPER (17), *Department of Plant Pathology, University of Wisconsin, Madison, Wisconsin*

GEORGE V. VAHOUNY (23), *Department of Biochemistry, The George Washington University, School of Medicine, Washington, D.C.*

CHARLES A. WEST (17), *Chemistry Department, University of California, Los Angeles, California*

ULRICH WESTPHAL (41), *Biochemistry Department, University of Louisville, School of Medicine, Louisville, Kentucky*

WALTER G. WIEST (30), *Department of Obstetrics and Gynecology, Washington University, School of Medicine, St. Louis, Missouri*

HERBERT H. WOTIZ (2), *Department of Biochemistry, Boston University School of Medicine, Boston, Massachusetts*

HENRY YOKOYAMA (14), *United States Department of Agriculture Fruit and Vegetable Chemistry Laboratory, Pasadena, California*

Preface

This volume of "Methods in Enzymology" is a fission product of an undertaking originally planned jointly with Dr. J. M. Lowenstein in which lipids and steroids were to be covered in one book. Progress in these areas during the past few years has been so great that the need for separate treatment soon became apparent, for the bulk of material relating to steroids and terpenoids, on the one hand, and to lipids, on the other, clearly exceeded the limits that could be satisfactorily encompassed by any single volume.

This resulting volume, devoted to methodology in the steroid and terpenoid fields, is divided into four main sections that deal, respectively, with special analytical methods, methods of synthesis of labeled substrates, enzyme preparation methods, and steroid-binding protein methods as exemplified by corticosteroid binding globulin (C.B.G.). As in the other volumes of this publication, the aim has been to include details of new procedures or improvements in existing methods that have become available since the last contributions in the area of steroid biochemistry were made to the treatise. Thus, in the analytical section, thin-layer chromatography and a number of applications of gas–liquid chromatography receive particular attention, and, with regard to the latter, methods for the isolation of suitably purified extracts for analysis by the highly sensitive electron-capture technique are emphasized.

In the section devoted to methods of synthesis an extensive survey of chemical techniques for the labeling of steroids is given that represents the broad advances which have taken place since they were last summarized in this treatise (Volume IV, 1957). The outstanding progress in synthetic labeling methods for mevalonic acid and terpenoid pyrophosphate intermediates that has been achieved since 1960 is also represented in a separate chapter. This work, which underlies so many important recent advances in the understanding of biosynthetic mechanisms, will no doubt continue to be a keystone in the edifice of terpenoid and steroid biochemistry for many years.

In the coverage of enzyme preparative methods and assay techniques, the data presented in earlier volumes are now supplemented with more recent information in areas in which notable progress has been made, as, for example, in the analysis of steroid oxygenase systems and the characterization of several esterases and specific dehydrogenases. To some extent this volume departs from the general approach in other volumes in that several chapters deal extensively with enzyme systems that are obviously complex, such as, for example, a number of microsomal systems of sterol and steroid metabolism. Since so much of steroid enzymology is still at this stage of development, it seemed appropriate to summarize the present status of work in these areas in the hope that the ready availability of the related methodology would help to promote further work toward their clarification.

The final chapter, devoted primarily to corticosteroid binding protein, summarizes methods that are useful in a number of areas of growing importance in hormone assay studies and in the characterization of specific binding proteins possibly involved in the mechanism of hormone action.

The understanding cooperation of the authors whose devotion to the consolidation and advancement of steroid and terpene biochemistry that has made this book possible has been most heartening and deserves the sincere thanks of all practitioners in the field. The expert attention of Dr. Charles H. Doering and his assistant Mrs. Joanne Richards to the indexing of the book and their help in proof reading are also most gratefully acknowledged. Indexing was greatly facilitated by the use of the Advanced Computer for Medical Research (ACME) at the Stanford University School of Medicine, supported by U.S.P.H.S. Grant FR-00311. We would like to thank Mr. Robert L. Bassett for his generous advice in computer programming. We sincerely appreciate the cooperation of Academic Press in the production of this work.

RAYMOND B. CLAYTON

June, 1969

Table of Contents

Section I. Newer Analytical Techniques

Section II. Special Synthetic Methods

Section III. Enzyme Systems of Terpenoid and Sterol Biosynthesis and Sterol Ester Formation

Section IV. Enzyme Systems of Bile Acid Formation

Section V. Enzyme Systems of Steroid Hormone Metabolism

Section VI. Corticosteroid-Binding Globulin and Other Steroid-Binding Serum Proteins

METHODS IN ENZYMOLOGY

EDITED BY

Sidney P. Colowick and Nathan O. Kaplan

VANDERBILT UNIVERSITY
SCHOOL OF MEDICINE
NASHVILLE, TENNESSEE

DEPARTMENT OF CHEMISTRY
UNIVERSITY OF CALIFORNIA
AT SAN DIEGO
LA JOLLA, CALIFORNIA

METHODS IN ENZYMOLOGY

EDITORS-IN-CHIEF

Sidney P. Colowick   Nathan O. Kaplan

VOLUME VIII. Complex Carbohydrates
Edited by ELIZABETH F. NEUFELD AND VICTOR GINSBURG

VOLUME IX. Carbohydrate Metabolism
Edited by WILLIS A. WOOD

VOLUME X. Oxidation and Phosphorylation
Edited by RONALD W. ESTABROOK AND MAYNARD E. PULLMAN

VOLUME XI. Enzyme Structure
Edited by C. H. W. HIRS

VOLUME XII. Nucleic Acids (in two volumes)
Edited by LAWRENCE GROSSMAN AND KIVIE MOLDAVE

VOLUME XIII. Citric Acid Cycle
Edited by J. M. LOWENSTEIN

VOLUME XIV. Lipids
Edited by J. M. LOWENSTEIN

VOLUME XV. Steroids and Terpenoids
Edited by RAYMOND B. CLAYTON

In Preparation

Fast Reactions
Edited by KENNETH KUSTIN

Metabolism of Amino Acids and Amines
Edited By HERBERT TABOR AND CELIA WHITE TABOR

Vitamins and Coenzymes
Edited by DONALD B. MCCORMICK AND LEMUEL D. WRIGHT

Proteolytic Enzymes
Edited by GERTRUDE E. PERLMANN AND LASZLO LORAND

Methods in Enzymology

Volume XV
STEROIDS AND TERPENOIDS

Section I

Newer Analytical Techniques

[1] Thin-Layer Chromatography of Steroids, Sterols, and Related Compounds

By BELISÁRIO P. LISBOA

I. Introduction

As one of the most utilized analytical procedures for the investigation of steroids, thin-layer chromatography (TLC) is the subject of several recent reviews.[1-5] The simple apparatus and techniques required for the development of thin-layer chromatograms, the high-power of resolution of TLC, and the possibility of employing a large number of reactions of great specificity and high sensitivity for the visualization of the analyzed compounds are major reasons for the acceptance that TLC has found as a separation procedure.

[1] J. M. Bobbit, "Thin-Layer Chromatography." Reinhold, New York, 1963.
[2] K. Randerath, "Thin-Layer Chromatography." Verlag Chemie, Weinheim, 1966.
[3] J. G. Kirchner, "Technique of Organic Chemistry; Thin Layer Chromatography" (E. S. Perry and A. Weissberger, eds.), Vol. XII. Wiley (Interscience), New York, 1966.
[4] E. Heftmann, *Chromatog. Rev.* **7**, 179 (1965).
[5] A. A. Akhrem and A. I. Kuznetsova, "Thin-Layer Chromatography. A Practical Laboratory Handbook." Davey, New York, 1965.

Although TLC is employed in quantification procedures, it is even more widely used for preliminary separations and for purposes of identification. For preliminary separations, TLC is often used in conjunction with other methods of fractionation, such as column-partition chromatography and gas-liquid chromatography.

This review deals with practical considerations concerning the various techniques which may be applicable to the separation of individual steroids by TLC.

II. Practical Considerations

Factors that must be taken into consideration when TLC is chosen as the method of fractionating mixtures of steroids include the adsorbent, the preparation of the layer, the type of solvent system to be employed for development of the chromatoplates, and the technique to be used.

A. Adsorbents

Although all the adsorbents used for column chromatography could be employed in the preparation of thin-layer plates, silica gel, aluminum oxide, kieselguhr, and cellulose powder are the most utilized. Magnesium silicate, hydroxylapatite, and talc also have been reported to be useful adsorbents. Whereas aluminum oxide is used in adsorption chromatography and kieselguhr is most generally used as support for the stationary phase in partition chromatography or as support for the nonpolar phase in reversed-phase-TLC, silica gel is used in all these fractionating procedures.

These different adsorbents can be used in pure form or in mixtures with binders, such as rice starch, polyvinyl alcohol, or gypsum (calcined calcium sulfate). There are different kinds of silica gel and aluminum oxide adsorbents, containing amounts of calcium sulfate ranging between 5 and 13%. Silica gel G, Merck (13%), Alumina G, Merck (13%), DS-5, Camag (5%), silica gel D-5, Fluka (5%), and silica gel, Serva (about 10%) are examples of adsorbents containing gypsum.

Adsorbents with or without gypsum are available which contain small amounts of agents, such as Rhodamine 6 G, sodium fluorescein, or zinc sulfate activated with manganese, these fluoresce under ultraviolet (UV) light and thus facilitate the vizualization of UV-absorbing materials undergoing chromatography.

Also mixtures of two different adsorbents have been employed for the separation of steroids, e.g., aluminum oxide G–silica gel G (1:19)[6] and silica gel G–kieselguhr G (1:1).[7] An alkaline mixture containing silica

[6] C. Dumazert, C. Ghiglione, and T. Pugnet, *Ann. Pharm. Franç.* 21, 227 (1963).
[7] R. D. Bennett and E. Heftmann, *J. Chromatog.* 9, 359 (1962).

gel, magnesium oxide, and sodium sulfate (84:15:1), to which some calcium sulfate is added (Anasil B, Analabs Inc., USA), has been used for separation of methylcholanoates.[8]

Powdered cellulose and adsorbents with rice starch as binder do not permit the use of strong acids for the detection of the spots.

B. Preparation of the Layer

The preparation of the support layer depends on the technique to be employed. According to the consistency of the supporting media on the glass surface, one can utilize two different thin-layer chromatographic techniques. One of them, so-called *spread* or *nonbound*-TLC, employs a loose layer of an adsorbent without binder uniformly spread on the surface of a glass plate. The other procedure, utilized by Ismailov and Shraiber[9] in 1938 to make microslides in their pioneer work, involves adherent layers obtained by using a suspension of the adsorbent (with or without binders), which are allowed to dry on the glass surface. This procedure is called surface chromatography,[10] *bound*-TLC, or simply TLC.

1. *Spread TLC*

The layers, 0.4–0.6 cm thick, can be prepared by very simple devices[11–14] and can be employed for both ascending and descending chromatography (see, Mistryulov[11] and Lábler[12]). They are very sensitive to shaking, and the angle of inclination of the chromatoplate with the solvent surface cannot be greater than 40 degrees. Also the spots cannot be visualized by direct spraying of the layer, but only by exposure to vapors or indirect dispersion. One method for application of detecting reagents through saturation of a closed atmosphere without air turbulence has been suggested by Mistryulov.[15]

Since the loose layers for spread TLC are obtained from dry powders, the activity of the aluminum oxide used can be determined before the preparation of the plates; a change in activity of the adsorbent does not take place during the preparation procedure.

[8] A. F. Hofmann, *in* "New Biochemical Separations" (A. T. James and L. J. Morris, eds.), p. 262. Van Nostrand, Princeton, New Jersey, 1964.

[9] N. A. Ismailov and M. S. Schraiber *Farmatsiya (Sofia)* **3**, 1 (1938).

[10] J. E. Meinhard and N. F. Hall, *Anal. Chem.* **21**, 185 (1949).

[11] E. A. Mistryulov, *J. Chromatog.* **9**, 311 (1962).

[12] L. Lábler, *in* "Thin-Layer Chromatography" (G. B. Martini-Bettòlo, ed.), p. 32. Elsevier, Amsterdam, 1964.

[13] M. Mottier and M. Potterat, *Anal. Chim. Acta* **13**, 46 (1955).

[14] M. Mottier, *Mitt. Gebiete Lebensm. Hyg.* **49**, 454 (1958).

[15] E. A. Mistryulov, *Collection Czech. Chem. Commun.* **26**, 2071 (1961).

2. Bound TLC

Chromatoplates with bound layers up to 2 mm in thickness can be prepared by means of mechanical applicators. Glass or aluminum plates of different sizes, such as 20 × 40, 20 × 20, 20 × 10, and 20 × 5 cm can be used as supports. Most commonly employed are 250 μ-thick adsorbent layers on 20 × 20 cm plates. Ribbed glass plates (squared or round ridge plates) permitting the preparation of layers of uniform thickness (0.35 mm) without use of expensive apparatus have been developed by Gamp et al.;[16] however, these plates cannot be used if bidimensional developments are desirable. Precoated plates (Merck AG, Darmstadt, Germany) and flexible polyester sheets[17,18] (Distillation Products Industries Eastman Kodak Co., Rochester, New York) are also available for the more commonly employed adsorbents.

For the preparation of the layers, a slurry containing the adsorbent is prepared according to the recommendation of the manufacturer. For instance, for silica gel G (Merck) the ratio of adsorbent to water is 1:2 (w/v); devices are also indicated for the time that the slurries should be shaken. The plates are dried at room temperature and then activated by heating at 100–105° for 30 minutes. Special activation times are required if aluminum oxide is used as adsorbent, depending on the degree of activity required.

3. Partition TLC

Although most of the investigations using TLC have employed pure systems or multicomponent uniphasic mixtures which act principally, if not exclusively, by elution of the substances adsorbed on solid supporting media, partition chromatography also has been successfully used on chromatoplates.

For this purpose the liquid phase which is employed as suitable support, such as ethylene glycol or formamide, must be mixed with the adsorbent by dipping, by spraying, or by development of the chromatoplate with a solvent mixture containing the stationary phase.

A special example of partition TLC is reverse-phase partition TLC, in which a nonpolar stationary phase is obtained by impregnation with such lipophilic substances as paraffin, undecane, or tetradecane.

a. *Impregnation with Ethylene Glycol.*[19] Coated plates are developed in a tank containing a mixture of ethylene glycol–methanol (30:70) until

[16] A. Gamp, P. Studer, H. Linde, and K. Meyer, *Experientia* 18, 292 (1962).

[17] A. Lestienne, E. P. Przybylowicz, W. J. Staudenmayer, E. S. Perry, A. D. Baitsholts, and T. N. Tischer, *3rd Intern. Symp. Chromatog. Brussels, 1964*, p. 233. Soc. Belge Sci. Pharmaceutiques, Brussels, 1964.

[18] R. O. Quesenberry, E. M. Donaldson, and F. Ungar, *Steroids* 6, 167 (1965).

[19] E. Chang, *Steroids* 4, 237 (1964).

the solvent has ascended to a height of 15 cm (1 hour development); the methanol is removed by allowing the plates to dry for 30 minutes at room temperature. To avoid disturbing the adsorbed stationary phase on the starting line, the steroids should be applied in benzene solutions.

b. *Impregnation with Formamide.* The plates are put vertically, to a depth of 1 cm, into a 40% solution of formamide in acetone. Development proceeds until the solvent has covered all the adsorbent surface. To obtain a uniform formamide-containing adsorbent layer, the plates are allowed to remain for 24 hours over a formamide–acetone solution.[20] Impregnated formamide–talc layers have been obtained by adding a mixture of form-amide–acetone (3:7) directly to the slurry (talc–ethanol–formamide-mixture 4:7:1, w/v/v) and shaking for 0.5 minute;[21] the acetone and ethanol are eliminated from the plates by putting them at 40° for 10 minutes under an air stream.

c. *Impregnation with Undecane.* Chromatoplates are prepared by dipping[22, 23] or by developing[24] them in 10–15% *n*-undecane in petroleum ether (or in the volatile system used); the petroleum ether is evaporated in air. The time required for drying affects the concentration of undecane in the layer, since the undecane also evaporates. After 1 hour of drying, the concentration is 0.09 g of undecane per gram of adsorbent. An increase of the impregnation is obtained when tetradecane is added to the petro-leum ether solution (undecane–tetradecane–petroleum ether 12:3:85), and the drying time is reduced to 30 minutes. These conditions give 0.25 g of stationary phase per gram of adsorbent.[23]

d. *Impregnation with Other Stationary Phases.* Impregnation with paraffin is obtained by dipping the plate in a 0.5% solution of liquid paraffin in ether.[22, 25] Impregnation with 2-phenoxyethanol or 2-methoxy-ethanol is obtained by dipping the plates in a solution of these sub-stances in acetone.[22]

4. *Special Impregnation Techniques*

The formation of complexes between silver nitrate and unsaturated sterols,[26] or between steroids with *cis*-glycol structures and borate ions,[27, 27a]

[20] L. Göldel, W. Zimmermann, and D. Lommer, *Z. Physiol. Chem.* 333, 35 (1963).
[21] J. Zurkowska and A. Ozarowski, *Planta Med.* 12, 222 (1964).
[22] D. I. Cargill, *Analyst* 87, 865 (1962).
[23] J. W. Copius-Peereboom, "Chromatographic Sterol Analysis as Applied to the Investigation of Milk Fat and Other Oils and Fats." PUDOC (Centrum voor Landbouwpublikaties en Landbouwdocumentatie), Wageningen, 1963.
[24] A. Truswell and W. C. Mitchell, *J. Lipid Res.* 6, 438 (1965).
[25] Č. Michalec, M. Šulc, and J. Měštan, *Nature* 193, 63 (1962).
[26] B. de Vries and G. Jurriens, *Fette Seifen Anstrichmittel* 65, 725 (1963).
[27] B. P. Lisboa, *Excerpta Med. Intern. Congr. Ser.* 111, 115 (1966).
[27a] B. P. Lisboa, *Pharm. Biologiste* 5, 463 (1968).

has been utilized for the resolution by TLC of some pairs of substances of similar properties. For this purpose, both silica gel and aluminum oxide layers have been impregnated with complex-forming substances, such as silver nitrate, boric acid, and sodium borate.[28] The incorporation of these substances in the adsorbent layer can be carried out by spraying, but the best procedure for obtaining a uniform and reproducible level of impregnation is to use a solution of the complexing agent instead of water for the preparation of the chromatoplates.

a. *Impregnation with Silver Nitrate.*[29] Silica gel G plates impregnated with silver nitrate are prepared by thoroughly mixing 40 g of silica gel G (Merck) with 100 ml of a 10% aqueous solution of silver nitrate. The plates are dried in air, stored away from light, and activated just before use by heating for 20–30 minutes at 95°. Impregnated plates are now available commercially, (Applied Science Laboratories, State College, Pennsylvania; Analtech. Inc., USA).

b. *Impregnation with Boric Acid or Sodium Borate.* Reichelt and Pitra[30] make a homogeneous mixture of one part of silica gel and four parts of a 1.25% solution of borax. This is allowed to dry, and, after activation, can be used for the preparation of the plates in the conventional manner. It seems more practical to make the plates by mixing the adsorbent with a 10% solution of boric acid or sodium borate instead of water.[31]

5. Preparative Layers

Layers with a thickness up to 2 mm have been used for the separation of substances on a preparative scale, often by using one-dimensional multiple developments. Halpaap[32, 33] has preferred to use plates that are 1 m wide and 20 cm high; however, for their development special tanks are required, in which 5 plates may be run together.

Preparative layers require the use of highly active gypsum-free supports, to which stabilizers of great binding power have been added, in order to permit the adhesivity of thick layers on the plates. To avoid the use of chemical reactions for the developments of the spots or bands, fluorescence indicators e.g., 3,5-dihydroxybenzpyrene-8,10-disulfonic acid, disodium salt (Bayer, Leverkusen, Germany),[34] are also added to the

[28] L. J. Morris, *in* "New Biochemical Separations" (A. T. James and L. J. Morris, eds.), p. 296. Van Nostrand, Princeton, New Jersey, 1964.

[29] B. P. Lisboa and R. F. Palmer, *Anal. Biochem.* **20**, 77 (1967).

[30] J. Reichelt and J. Pitra, *Czech. Farmacie* **12**, 416 (1963).

[31] B. P. Lisboa, A. I. Palomino, and G. Zucconi, unpublished results, 1968.

[32] H. Halpaap, *Chem. Ing. Tech.* **35**, 488 (1963).

[33] H. Halpaap, *Chemiker Ztg.* **89**, 835 (1965).

[34] R. Tschesche, G. Biernoth, and G. Wulff, *J. Chromatog.* **12**, 342 (1963).

slurry. Special gypsum-free silica gel with added fluorescence indicators, are commercially available (Silica gel HF_{254}, and Silica gel $HF_{254 + 366}$, Merck AG, Darmstadt, Germany). Although layers 2 mm in thickness are optimal, layers up to 10 mm thick can be used.

Milligram amounts of steroids can be completely separated by the use of layers 1.5–2.0 mm thick. Tschesche *et al.* (see footnote 53) have obtained, for instance, by chromatography of a mixture of cholestane-3β, 5α,6β-triol-3,6-diacetate, cholestane-3β,5α,6β-triol-3-methyl ether-6-acetate, and cholestane-3β,5α,6β-triol-3-methyl ether (30 mg each) on a 1.5–2.0 mm thick silica gel layer on conventional 20 $\times$ 20 cm plates, 29, 29, and 28 mg of pure steroid, respectively. The capacity of separation of two steroids on thick layers decreases with increase in their polarity (see footnote 53).

6. *Gradient Layers*

In this procedure, mixtures of two supports are simultaneously employed in different proportions[35] using special devices such as the GM-applicator (Desaga, Germany). In the adsorbent layer, there is a continuous transition between two different adsorbents.

No use of this technique in the study of steroids has been reported. It could be useful for preliminary investigations, when one wishes to know the most appropriate ratio of a mixture of adsorbents for achieving a maximal resolution of two compounds.

Layers with gradient activation have been also prepared by successive immersion of prepared thin-layer plates in an acetone–water mixture containing 5–30% water.[36] The plate is immersed at 5-second intervals, and the depth of immersion is reduced some millimeters each time. The gradient activation thus obtained can be used either along the starting line or in the direction of the run. Separation of cholesterol, cholesterol stearate, and squalene on such kieselgel G layers has been reported.

7. *Layers with Shaped Areas*

Shaped layers have been employed to obtain a better resolution of closely related substances principally if small amounts of one of them is in the presence of large amounts of the other. Since the solvent system spreads over a wide angle, the substances are resolved with the formation of narrow slightly curved bands without overlap. Depending on the shape of the area, the technique is called wedge-shape (or wedge-tip), wedge-strip, or circular technique.

[35] E. Stahl, *Angew. Chem. Intern. Ed.* **3**, 784 (1964).
[36] C. G. Honegger, *Helv. Chim. Acta* **47**, 2384 (1964).

The circular technique employed for paper chromatography, and called radial chromatography,[37] has been utilized on unbound aluminum oxide layers by Wirz.[38]

Wedge-tip chromatograms, suggested by Prey et al.,[39] have been applied in the partition TLC of saponins and triterpenoid glycosides by Tschesche et al.[40] The adsorbent is scraped off to form a V-shaped wedge strip, in the thin edge (less than 1 cm wide) of which the mixture is spotted.

The Matthias[41] wedge-strip paper chromatography has been applied to reversed-phase TLC by Copius-Peereboom and Beekes[42] and is the most widely employed shaped-area layer technique in the steroid field. Hexagonal holes (23 × 20 mm) are brushed out of the impregnated adsorbent layer by using an appropriate template, and the substances are spotted in the middle of the 8 mm wide "bridges" situated between each of the hexagonal areas, 4 cm from the bottom of the plate.

C. Solvents; Adsorption and Partition Chromatography

The choice of a solvent or mixture of solvents depends on the polarity of the substances submitted to chromatography, and therefore on the type of chromatography utilized: adsorption, normal partition, or reversed-phase partition chromatography. Since the best separations are those obtained when the R_f values are between 0.15 and 0.70, solvents or solvent systems should be used which allow steroid migration in this range of values.

Eluotropic series of solvents have been proposed by several authors, based on the ability of different solvents to elute polar compounds[43] from aluminum oxide or based on their dielectric constants.[44]

A microcircular technique has been employed by Ismailov and Schraiber[9] to test the adsorptivity of different supports, such as magnesium oxide, calcium oxide, aluminum oxide, and talc. The separation of the constituents of the galenical preparations, e.g., Digitalis purpurea on 2 mm-thick bound layers prepared on microslides was observed under UV light (ultrachromatography).

[37] J. G. Marchal and T. Mittwer, Compt. Rend. Soc. Biol. 145, 417 (1951).
[38] P. Wirz, Mitt. Gebiete Lebensm. Hyg. 49, 454 (1958).
[39] V. Prey, H. Berbalk, and M. Kausz, Mikrochim. Acta, p. 968 (1961).
[40] R. Tschesche, I. Duphorn, and G. Snatzke, in "New Biochemical Separations" (A. T. James and L. J. Morris, eds.), p. 248. Van Nostrand, Princeton, New Jersey, 1964.
[41] W. Matthias, Naturwissenschaften 41, 17 (1954).
[42] J. W. Copius-Peereboom and H. W. Beekes, J. Chromatog. 9, 316 (1962).
[43] W. Trappe, Biochim. Z. 305, 150 (1940).
[44] J. Jaques and P. J. Mathiew, Bull. Soc. Chim. France p. 94 (1946).

Stahl[45] has adapted this technique to assay the elutive power of different solvents on the same adsorbent. Several spots of the substance are applied on a thin layer, and different solvents are tested by putting them individually in the center of these spots by means of a capillary tube. The area of the circle formed by the spreading of the substance gives an indication of the elutive power of the solvent.

All the solvents used for chromatography should be of high purity and freshly distilled to remove peroxides. It has been reported that ethyl acetate, one of the most widely employed solvents in TLC of steroids, could cause extensive side-chain destruction of corticosteroids, if not freshly distilled.[46]

If single solvents do not allow a satisfactory resolution, the use of simple mixtures is recommended. In the chromatography of steroids and related compounds, most of the mixtures are two-component systems in which more polar solvents, such as ethyl acetate, acetone, or alcohols, are added to solvents with low dielectric constants, such as hydrocarbons. Particularly useful are systems of the type cyclohexane (or n-hexane)–ethyl acetate, chloroform–ethanol (or methanol), benzene–ethanol (or methanol), and chloroform (or benzene)–acetone.

Addition of water, acetic acid, or both to a system increases the participation of the partition mechanism in the chromatography and therefore is very useful for the separation of more hydrophilic steroids, or of these steroids that show tailing with the use of adsorption chromatography. Systems of this type are used, for instance, for the fractionation of bile acids.[47] Partition chromatography on thin layer can be carried out by employing one-phase mixtures containing a small amount of the stationary phase, such as ethylene glycol monomethyl ether or formamide. n-Butanol–methanol–formamide (17:2:1) employed for the separation of *Strophanthus* glycosides by Lukas[48] and chloroform–ethyleneglycol monomethyl ether (9:1) used for separation of phytosterols from *Holarrhena antidysenterica* (Wall) by Tschesche et al.,[49] are examples of one-phase systems for partition TLC.

The best separations of strongly polar substances such as hydrophilic steroids and steroid conjugates, have been obtained by using normal partition chromatography in which the impregnation of the adsorbent can be carried out in the same manner as with paper chromatography.[50]

[45] E. Stahl, *Chemiker Ztg.* 82, 323 (1958).
[46] S. Burstein and H. L. Kimball, *Steroids* 2, 209 (1963).
[47] H. Gänshirt, F. W. Koss, and K. Morianz, *Arzneimittel Forsch.* 10, 943 (1960).
[48] G. Lukas, *Sci. Pharm.* 30, 47 (1962).
[49] R. Tschesche, I. Mörner, and G. Snatzke, *Ann. Chem.* 670, 103 (1963).
[50] J. Vaedtke and A. Gajewska, *J. Chromatog.* 9, 345 (1962).

By using two-dimensional chromatography on formamide-impregnated plates, it is possible to utilize layers of different activity.[20] Between the two developments, the degree of impregnation of the stationary phase is altered by treatment of the formamide-impregnated layer with water vapor or by removal of part of the formamide by means of a hot air stream. A temporary impregnation with undecane is also possible, since undecane can be removed by heating. After partition chromatography the stationary phase is best removed to prevent interference with the various detecting reagents.

For the resolution of weakly polar steroids such as sterol acetates, reversed-phase partition chromatography is the method of choice.[51] Higher paraffins, undecane, tetradecane, and decalin are among the nonpolar solvents most commonly employed for the stationary phase. The more polar phase (mobile phase) might be saturated with the non-polar one; but there are different views regarding the optimal degree of saturation of the mobile phase. Kaufmann and Makus[52] and Tschesche et al.[53] prefer a partial saturation of 80% to prevent the nonpolar phase from separating out during development, whereas Copius-Peereboom[23] considers complete saturation essential in order to avoid the formation of secondary fronts. A great number of solvents, alone or in mixtures, have been used to develop impregnated kieselguhr plates. Examples are acetone, acetonitrile, acetic acid, and methanol.

Sometimes substances have been added to systems of solvents for different reasons. Mistryulov[15] has used small amounts of decalin in his systems to prevent depletion of the unbound aluminum oxide layer. Tschesche et al. have added to the developing mixtures pyridine[54] or diethylamine[40] to prevent tailing. Bromination has been carried out during the development of the chromatogram by adding 0.5% bromine to the solvent system.[55]

Multiple one-dimensional TLC, in which the chromatoplate is submitted to several runs in one or different systems is a commonly employed procedure. One special case of this technique is stepwise development-TLC[56] (see Section D,2 below).

During adsorption TLC, a frontal analysis of the multicomponent

[51] J. W. Copius-Peereboom and H. W. Beekes, *J. Chromatog.* **17**, 99 (1965).
[52] H. P. Kaufmann and Z. Makus, *Fette Seifen Anstrichmittel* **62**, 1014 (1960).
[53] R. Tschesche, G. Wulff, and K. H. Richert, *in* "New Biochemical Separations" (A. T. James and L. J. Morris, eds.), p. 198. Van Nostrand, Princeton, New Jersey, 1964.
[54] R. Tschesche, F. Lampert, and G. Snatzke, *J. Chromatog.* **5**, 217 (1961).
[55] H. P. Kaufmann and T. H. Khoe, *Fette Seifen Anstrichmittel* **64**, 81 (1962).
[56] E. Stahl and U. Kaltenbach, *J. Chromatog.* **5**, 458 (1961).

system[57] has been reported, resulting in different adsorption affinities of the solvents for a given adsorbent. At the immersion line of the plate, the solvent system is demixed and each solvent runs independently of the other; the less polar component (e.g., hydrocarbon, chloroform) penetrates into the adsorbent more rapidly than the more polar one (e.g., alcohols, esters). As a consequence of frontal analysis, internal secondary fronts are formed, visible after the plate has been sprayed with some acid-containing reagents. It seems that as a consequence the more polar steroids are chromatographed by a partition procedure whereas the nonpolar substances are submitted to adsorption chromatography.

The use of systems in which a partition mechanism in the chromatography is more accentuated, as when acetic acid, methanol, or water are added, reduces the formation of tailing and leads to more constant R_{Mg} values.

Gradient elution techniques have been employed also on thin-layer chromatography,[58, 59] but their application to the steroid field has been restricted to examples of separations of lipid mixtures containing cholesterol.[60, 61] In this technique, the less polar solvent changes during the development of the chromatogram by gradual addition of one or more polar components to the system.

D. Techniques of Development

A great number of procedures have been employed to carry out TLC. Ascending, horizontal, or descending chromatography have been performed using one run or multiple runs, developed with one or more solvent systems, in the same direction or in two dimensions.

1. Ascending Thin-Layer Chromatography

Ascending TLC is the most commonly employed procedure. The plate is developed by putting it in a chamber that has been lined with filter paper and filled to a depth of 0.8 cm (immersion line) with the solvent system used. In ascending one-dimensional TLC the development is carried out until the solvent system reaches a distance of 10–15 cm (the so-called "front") from the starting line.

In order to obtain reproducible results with this technique, it is

[57] G. P. Pataki, Dunnschichtchromatographie. Inaugural Dissertation, Philosophisch-Naturwiss. Fakultät, Univ. of Basel, Basel, Switzerland, 1962.
[58] S. M. Rybicka, Chem. & Ind. p. 308 (1962).
[59] S. M. Rybicka, Chem. & Ind. p. 1947 (1962).
[60] A. Niederwieser, J. Chromatog. 21, 326 (1966).
[61] A. Niederwieser and C. G. Honegger, Advan. Chromatog. 2, 123 (1966).

necessary to work under controlled experimental conditions,[62-64] regulating the following factors, which influence the mobility of the substance and therefore of the R_f values: (a) tank size and saturation of the chamber-atmosphere; (b) layer thickness; (c) temperature of development; (d) quantity of solvent system at the bottom of the tank and therefore the immersion line; (e) position of the starting line; (f) running distance of the front (distance starting line/front); (g) angle of the plate with the solvent surface; (h) amount of the substance deposited on the starting line. Only those results that have been obtained under the same conditions can be compared. Therefore, most of the R_f values quoted in the literature must be interpreted more as relative values, permitting the grouping of substances on the basis of their polarity sequence in the system employed, than as absolute values.

In order to obtain satisfactory and more uniform vapor saturation of the chamber, Fauconnet and Waldesbühl,[65] in their TLC of cardenolides, first equilibrate the tank containing the solvent system and an adsorbent-covered thin-layer plate for 1 day. Then they introduce into the chamber the second thin-layer plate, on which the cardenolides have been deposited, so that the plates are face to face with a space of 2–5 mm between the layers. Because the first chromatoplate was totally covered with the solvent system when the second plate was run, the thin atmosphere-space between the layers is and remains uniformly saturated during the development.

For both better saturation and a reduced amount of the solvent required for the development, the so-called "sandwich" chambers, S-chambers, have been used (Desaga Co., Germany; Camag Chemie-Erzeugnisse AG, Switzerland). The device used by Fauconnet and Waldesbühl can also be employed in S-chambers to obtain a uniform saturation of the narrow chamber atmosphere: Jänchen[66] uses as a second wall a prepared thin-layer plate with the adsorbent layer soaked with the solvent to be used.

2. Multiple One-Dimensional TLC

When components of very similar polarities are to be resolved, single runs are sometimes not sufficient, and multiple runs or continuous development are necessary.

Multiple one-dimensional TLC is carried out by developing the chromatoplate in the same direction in one or several solvent systems.

[62] M. S. J. Dallas, *J. Chromatog.* 17, 267 (1965).
[63] B. P. Lisboa, *Steroids* 7, 41 (1966).
[64] C. G. Honegger, *Helv. Chim. Acta* 46, 1772 (1963).
[65] L. Fauconnet and M. Waldesbühl, *Pharm. Acta Helv.* 38, 423 (1963).
[66] D. Jänchen, *J. Chromatog.* 14, 261 (1964).

This chromatographic procedure proposed by Jeanes *et al.*[67] has been shown to be very effective for the separation of polar steroids[68-70] as well as for the resolution of weakly polar steroids.[29] The multiple runs, by increasing the total distance of development, amplify the number of theoretical plates and make possible better separations. However, the time required to carry out several runs makes this technique unsuitable for the separation of unstable substances which can undergo transformations in presence of the adsorbents, such as 2-hydroxylated estrogens and some corticosteroids.

During multiple runs the solvent front is allowed to rise to the top of the plate, a distance of 17–17.5 cm from the starting line, under the same conditions of saturation as those employed for single runs. In order to work with optimal saturation conditions it is necessary to equilibrate at least two tanks with the system employed and allow an equilibration period of at least 90 minutes after the preceding run before reusing a tank.

The separation of two substances increases with the number of runs until a maximum is reached. Because the length of the adsorbent is limited, the resolution obtained for two substances decreases on approaching the top of the plate. The maximum separation is obtained with an average R_f value (R_{Pmax}) corresponding to 0.632 times the length of the support.[71, 72] This means that polar solvent systems must be employed on multiple TLC to resolve closely related compounds, i.e., systems in which the R_f values obtained for these compounds are low. If the R_f values obtained for an unresolved pair of steroids in a system is greater than 0.40, it is unlikely that they can be separated by multiple developments in that system.[68]

Thoma has calculated the minimum number of developments required to separate two solutes based on their R_f values on single chromatograms. Tables with these data are available.[71] The degree of resolution, DR, obtained after n runs is defined by the formula DR $= \Delta p/L$. Δp is the distance separating the two solutes, and L is the length of the support.

The mobility value obtained for a steroid after multiple runs is expressed in centimeters and defines the distance between the maximal concentration of the solute and the starting line. The steroid distance d can be calculated by the formula suggested by Prusíková:[73]

[67] A. Jeanes, C. S. Wise, and R. S. Dimler, *Anal. Chem.* **23,** 415 (1951).
[68] N. Zöllner and G. Wolfram, *Klin. Wochschr.* **40,** 1098 (1962).
[69] B. P. Lisboa, *Acta Endocrinol.* **43,** 47 (1963).
[70] B. P. Lisboa, *J. Chromatog.* **19,** 333 (1965).
[71] J. A. Thoma, *J. Chromatog.* **12,** 441 (1963).
[72] J. A. Thoma, *Anal. Chem.* **35,** 214 (1963).
[73] M. Prusíková, *Experientia* **15,** 460 (1959).

$$d = aR_{f_1}(1 - R_{f_2}) + bR_{f_2}$$

where R_{f1} and R_{f2} are the R_f values of the substance in the two different solvent systems and a and b are the distances between the front and the starting line in the systems 1 and 2, respectively.

The stepwise elution technique (*Stufentechnik*)[56, 74] is a special case of multiple one-dimensional TLC in which there are two developments achieved by using different systems, one polar and the other less polar. During the first development, the polar system resolves the polar substances and the less polar solutes remain on the front, which is allowed to rise only halfway up the layer. In the second run, the solvent system covers all the chromatoplate and resolves the nonpolar compounds in the top half of the plate. This technique is useful in the separation of mixtures of glycosides and aglycones; the more polar glycosides are separated in the lower half of the layer by means of a system of great eluotropic action, whereas the aglycones are resolved by less polar solvents on the top half of the plate.[56, 75]

3. Ascending TLC with Continuous Development

Continuous development on ascending TLC has been achieved by the use of two different techniques. The first, developed by Zöllner and Wolfram,[68] is based on the evaporation of the solvent on the upper portion of the plate by using a partially open tank[68] or by leaving the upper 2 cm of the layer above the top of the tank.[76, 77] The mobility of the compounds in this continuous-flow arrangement are not reproducible because of the differential evaporation of the solvents, especially if the solvents are very different in their volatility. The application of this method to steroidal sapogenins, sterols, and C_{21}-steroids by Bennett and Heftmann[77] has permitted the successful separation of some pairs of compounds such as diosgenin–tigogenin and 5α-pregnanolone–Δ^5-pregnenolone, which are otherwise difficult to resolve.

In the other device for overflow TLC, the solvent system is soaked up in the upper part of the plate. Using a thick cotton layer pressed between filter paper and fixed on the upper quarter of a ribbed chromatoplate, Lewbart *et al.*[78] obtained a perfect separation of sarmentogenin and 3-episarmentogenin, among other cardenolides, after development for 4 hours with ethyl acetate. A metal trough containing dry adsorbent and fixed in the upper region of the plate can be used to soak up the solvents.

[74] E. Stahl, *Arch. Pharm.* **292/64**, 411 (1959).

[75] M. Ishikawa and T. Miyasaka, *Shika Zairyo Kenkyusho Hokoku* **2**, 397 (1962).

[76] T. M. Lees, M. J. Lynch, and F. R. Mosher, *J. Chromatog.* **18**, 595 (1965).

[77] R. D. Bennett and E. Heftmann, *J. Chromatog.* **21**, 488 (1966).

[78] M. L. Lewbart, W. Wehrli, and T. Reichstein, *Helv. Chim. Acta* **46**, 505 (1963).

This device was used by Bennett and Heftmann[79] to resolve sterol acetates and by Heusser[80] to separate *Digitalis* glycosides. Similar techniques also have been employed for overflow TLC on unbound aluminum oxide layers by Lábler[12] (mound of adsorbent procedure) and Schwarz (ascending-descending technique).[81]

4. Multiple Bidimensional TLC

Ascending TLC using bidimensional developments has been used to separate steroidal compounds.[82, 83] For the execution of this technique the extracts or mixtures of steroids are spotted in one of the corners of a square plate, 2.5 cm from the bottom and at least 1.5 cm from the right edge. A mixture of steroid standards is spotted at the other corner, in a comparable position. The chromatoplate is then developed in the first system, removed, dried, and placed in the second tank, to run in a direction perpendicular to the first chromatogram. Before this second run is performed, a mixture of steroid standards is spotted in a third corner for development parallel to the substances resolved in the first chromatogram.

Good separations are obtained by bidimensional TLC if solvent systems are used in which the substances to be resolved show an inversion in their sequence of polarity.[83, 84] One particular advantage of two-dimensional developments is the possibility of developing a mixture containing substances of very great difference in their polarity by a combination of adsorption and partition TLC. This was shown by Görlich[85] in the separation of steroid glycosides of *Nerium oleander* L. using the systems ethyl acetate–chloroform 9:1 (adsorption TLC) and ethyl methyl ketone–toluene–water–acetic acid–methanol 40:5:3:1:2.5 v/v (partition TLC). This technique can be utilized in partition TLC by using different degrees of impregnation—for instance, for the separation of polar corticosteroids on formamide-impregnated silica gel plates.[20]

5. Horizontal Development TLC

The horizontal TLC can be developed using single or overrun techniques by means of very simple devices.[15, 76, 86] One of them[76] consists of

[79] R. D. Bennett and E. Heftmann, *J. Chromatog.* **12**, 245 (1963).

[80] D. Heusser, *Planta Med.* **12**, 237 (1964).

[81] V. Schwarz, communication to L. Lábler, *in* "Thin-Layer Chromatography" (G. B. Martini-Bettòlo, ed.), p. 32. Elsevier, Amsterdam, 1964.

[82] H. P. Kaufmann, Z. Makus, and F. Deicke, *Fette Seifen Anstrichmittel* **63**, 235 (1961).

[83] B. P. Lisboa and E. Diczfalusy, *Acta Endocrinol.* **40**, 60 (1962).

[84] B. P. Lisboa and E. Diczfalusy, *Acta Endocrinol.* **43**, 545 (1962).

[85] B. Görlich, *Planta Med.* **9**, 237 (1961).

[86] M. Brenner and A. Neiderwieser, *Experientia* **17**, 237 (1961).

a flat rectangular tank containing the solvent system and a rolled paper wick, arranged to contact the layer just before the starting line. The plate, which is placed inverted over the tank, has three of the four edges scraped, in order to permit a perfect adhesion to the tank walls and to guide the migration of the solvent in the proper direction; the fourth unscraped edge extends 2 cm outside of the tank, permitting the evaporation of the solvent, if overrun chromatograms are developed. Using this arrangement for continuous-flow development, Lees et al.[76] have obtained a complete separation of cholesterol and desmosterol acetates on silica gel G in only 1 hour, using benzene–hexane 1:3.

In the variant of Brenner and Neiderwieser[86] (commercially available BN-chamber, Desaga, Germany), the adsorbent layer faces upward, and, except at the end opposite to the starting line (last 1.5 cm at the left edge), it is protected by a cover plate which does not touch the layer. The solvent is allowed to ascend to the layer by means of a filter paper wick to the starting edge, runs through the chromatoplate in the direction of development, and evaporates when it reaches the exposed edge opposite the starting edge. This procedure permits a complete separation on silica gel G of the closely related acetates of adynerigenin (3β-hydroxy-8,14β-epoxy-5β-card-20:22-enolide) and 8,14-epiadynerigenin after 7 hours' overrun on ethyl acetate–cyclohexane 1:2.[87]

6. Descending TLC

Procedures for descending TLC have been reported by Zöllner and Wolfram[68] and by Göldel et al.[20] In the Göldel procedure, also used for overrun TLC, a glass trough containing the solvent system similar to those used in paper chromatography, is fixed on the upper part of the tank, at the same height as the chromatoplate. One end of a paper filter strip as large as the chromatoplate is immersed in the trough. The other end is rolled on a glass rod to form a wick which touches the adsorbent layer just above the starting line. A uniform flow of solvent rises to the adsorbent layer transferred by this rolled paper wick, runs downward through the adsorbent layer, and is taken up by loose adsorbent in which the lower end of the chromatoplate is placed. By using this procedure in two dimensions, Göldel et al.[20] have obtained satisfactory results in the separation of C_{21}-steroids of very different polarities.

Descending TLC, employing a plate angle approximately 10 degrees to the horizontal, has been used for unbound aluminum oxide[11] as well as for bound layers. In such a technique, the solvent system reaches the

[87] P. S. Janiak, E. K. Weiss, J. von Euw, and T. Reichstein, Helv. Chim. Acta **46**, 374 (1963).

support layer by means of a strip of filter paper, as in the procedure described for horizontal TLC.

Satisfactory separations of C_{19}-steroids on silica gel G by means of this technique have been reported by Reisert and Schumacher.[88] A complete resolution of androsterone, etiocholanolone, and dehydroepiandrosterone is obtained after 5.5 hours' overrun chromatography in the system chloroform–ethanol 100:1.5.

7. Low-Temperature TLC

TLC developed at low temperature allows the application of this technique to the separation of volatile compounds and compounds that undergo decomposition on the adsorbent surface at room temperature. Also, the possibility of separating groups of substances that are difficult to separate is increased by the use of new systems containing solvents with low boiling points.

Stahl[35] has investigated this type of TLC, using a special thermostatically controlled vessel (Desaga, Heidelberg, Germany), in which the layer can be preequilibrated, and he could show that, in some cases, low-temperature TLC is able to resolve pairs of substances that are inseparable at room temperature.

Truswell and Mitchell[24] were the first to apply this technique to the separation of steroids; a mixture of cholest-7-en-3β-ol and 5α-cholestan-3β-ol has been completely resolved by reversed-phase TLC on n-undecane-impregnated silica gel plates after 3–3.5 hours of development in the system acetic acid–acetonitrile 1:1, 70% saturated with n-undecane, at −15 to −20°. Lisboa[63] has also employed low-temperature TLC for the separation of polar corticosteroids on silica gel G (chloroform–ethanol 9:1); the separation factor of some of the investigated polar Δ^4-3-oxopregnane steroids was increased by development at 4°, compared to the results obtained at room temperature.

III. Detection of the Steroids on TLC

The color reactions developed *in situ* for substances submitted to TLC not only have the capacity to localize the maximal concentration of the solute, but, much more, to corroborate the identification of the substance through the specific reactions of their different functional groups.

For the simple localization of the solute, a high degree of sensitivity is desirable, but for the characterization of the substance, reactions of definite specificity are required. Besides the so-called "universal" or general reagents, there are a great number of group-specific reactions which have been developed for a large number of steroids. However, it

[88] P. M. Reisert and D. Schumacher, *Experientia* **19**, 84 (1963).

TABLE I

PROCEDURES UTILIZED FOR THE VISUALIZATION OR CHARACTERIZATION OF THE DIFFERENT STEROIDAL COMPOUNDS OR SOME OF THEIR STRUCTURAL GROUPS

Compounds or groups	Procedures[a]
Hormonal steroids	
α,β-Unsaturated steroids	A; B1(h,k); B3(c); C2(b,c,e,g,h); E(a)
C_{21}-Side-chain identification	B2(d); C1(c,d); C2(a–f,i); C3(a)
Ketonic groups	C1(a,b); C3(b)
α,β-Diketones	C2(i,j)
Δ^5-3-Hydroxysteroids	B1(d,g,r); B3(a); B5(a)
Δ^5-7-Hydroxysteroids	B1(d,e,g,j); B2(o); B5(a); F(a)
Δ^{16}-C_{19}-Steroids	B2(i); B3(g)
4-Hydroxy-Δ^4-group	C2(j)
Isolated double bonds	E(c)
Phenolic steroids	B1(l); B2(e); B3(a,b); B5(a,c); D(a–f); E(a–c)
Ketolic estrogens	C2(i)
Corticosteroids (nonspecific reactions)	B4(d); B5(e); B6(b); F(g)
C_{18}-, C_{19}-, and C_{21}-steroids (general)	B1(a–d,f,g); B2(a–e,j); B6(a)
Sterols	
C_{25}- to C_{29}-sterols	B1(a,e,f,i,q); B2(c,j); B3(d,f); B4(a–e,e); B5(a,c,d); F(i)
Trimethylsterols	B1(a,f)
Sterol bromides	B5(f)
Cholesterol and esters	B1(a,f,i,o,p); B2(h,j); B5(a); F(g)
Bile acids and alcohols	B1(a,g,m); B2(a); F(h)
Cardenolides and bufogenins	B1(a,f); B5(a); C4(a–e); F(c,d)
Cardiac glycosides	B1(a,b,d,n,o); B2(g); B3(e); B5(a,b); C4(a,c); F(b–e)
Sapogenins	B1(a,f); B2(e,f)
Saponins	B1(a); B3(e); B5(e)
Steroid alkaloids	B3(c)
Triterpenes	B1(a,f); B5(a,c); F(f)

[a] The capital letters and arabic numbers refer to subsections of Section III, Detection of Steroids on TLC. Lower case letters denote the specific reactions, described in those subsections, which are particularly valuable in the characterization of the structural features indicated at the left of the table.

must be pointed out that the general reagents can also be specific, depending, for instance, on the temperature and duration of heating.

In some cases the presence of certain structural groups in the molecule allows the detection of the substance by the absorption of shortwave or longwave ultraviolet light. This is possible on TLC if fluorescence indicators are added to the layer during their preparation or sprayed on after development. Sodium fluorescein,[69] 2',7'-dichlorofluorescein,[89] morin (2',3, 4',5,7-pentahydroxyflavone),[90] and pyrene derivatives[34] are widely used. The localization of the substance by this visualization procedure does not modify the structure, and therefore it is very useful for following the migration of steroids containing α,β-unsaturated groups during multiple runs for isolation purposes.

One of the great advantages of the use of TLC over paper chromatography is the possibility of employing, in the visualization of the substances, corrosive and strong dehydrating reagents with a sensitivity down to 0.005 μg per spot. In general, all the reagents employed on paper chromatography can be used also on TLC, if two exceptions are made: the "soda fluorescence" reaction of Bush[91] and the *tert*-butanolic sodium butoxide reaction of Abelson and Bondy,[92] both characteristic of Δ^4-3-oxo groups.

The presence of impurities sometimes causes difficulties in the application of color reactions for the *in situ* characterization of a steroid, and thus a preliminary purification sometimes is necessary.

Together with the formation of derivatives *in situ* or prior to the development of the chromatogram, and the chromatographic mobilities of the steroid in different systems, the color reactions give a good approach to the characterization of a substance.

The exact preparation and utilization of the reagents employed for the visualization of the steroids and derivatives, the factors influencing the reaction development, their sensitivity and specificity, are treated below, according to the reacting group.

Table I summarizes the different color reactions that can be employed for each class of steroid.

A. UV-Absorbing Steroids

Steroids chromatographed on supports containing fluorescent chemicals or sprayed with them after development, as described above, can be

[89] H. K. Mangold, *J. Am. Oil Chemists' Soc.* **38**, 708 (1961).
[90] V. Černý, J. Joska, and L. Lábler, *Collection Czech. Chem. Commun.* **26**, 1658 (1961).
[91] I. E. Bush, *Biochem. J.* **50**, 370 (1952).
[92] D. Abelson and P. K. Bondy, *Anal. Chem.* **28**, 1922 (1956).

visualized under a short- or longwave UV lamp if they possess a functional UV-absorbing group in their molecule.

α,β-Unsaturated ketosteroids, such as steroids containing a Δ^4-3-oxo, Δ^{16}-20-oxo, and Δ^5-7-oxo structure, together with 6-oxo estrogens can be visualized by using a low-pressure mercury resonance arc lamp (principal emission at 253.7 mμ). Also, other estrogens and their methyl ether derivatives which exhibit an absorption maximum at 280 mμ can be detected on fluorescein-containing layers under this UV source.[84]

Under a longwave ultraviolet lamp, it is possible to detect steroidal compounds containing an α,β-unsaturated lactone (butenolide) or double unsaturated lactone attached at the position C-17. Examples of this group are the cardenolide glycosides, the bufadienolide glycosides, and their aglycones.

B. General and Unspecific Reagents

1. Universal Acidic Reagents and Other Acids Used as Reagents

Universal reagents (strong acid reagents) are strongly dehydrating acids, alone or in mixtures, that react with all or most of the steroids. Sometimes the reaction at elevated temperatures consists of a carbonization of the chromatographed substances. These reagents usually cannot be employed when powdered cellulose or starch-bound adsorbents are used.

a. Sulfuric Acid. This acid usually has been used diluted (2–50%) in ethanol or water. Colors developed by thirty-two estrogens[84] and thirteen corticosteroids,[93] in daylight and UV light after spraying with a 2% H_2SO_4 solution on H_2O–ethanol (1:1)[83] (95–100° for 20 minutes), have been reported. The color reactions have been recorded[94] for 148 steroids (C_{18}-, C_{19}-, and C_{21}-steroids, cardenolides, bile acids, sterols, sapogenins, and alkaloids) spotted, but not developed, on TLC plates after spraying with a 50% aqueous solution of H_2SO_4 and heating on a hot plate. The lower limit of detection was between 0.01 and 0.05 μg.

To detect cardenolides,[95] 10% H_2SO_4–ethanol has been employed; a 50% solution (10 minutes at 160°) has been employed for cholesterol and cholesterol esters,[96] sterols, acetates, and propionates,[97] Δ^4-3-oxo-steroids,[98]

[93] B. P. Lisboa, J. Chromatog. 16, 136 (1964).
[94] E. Heftmann, S.-T. Ko, and R. D. Bennett, J. Chromatog. 21, 490 (1966).
[95] J. Binkert, E. Angliker, and A. von Wartburg, Helv. Chim. Acta 45, 2122 (1962).
[96] N. Zöllner and G. Wolfram, Klin. Wochschr. 40, 1101 (1962).
[97] J. R. Claude, J. Chromatog. 17, 596 (1965).
[98] C. Tamm, A. Gutler, G. Juhasz, E. Weiss-Berg, and W. Zürcher, Helv. Chim. Acta 46, 889 (1963).

cardenolides,[99] and sterols (also on silver-nitrate impregnated layers);[100] and an 80% solution (70° for 5 minutes) has been used to detect Δ^7- and Δ^8-cholestenediols.[101]

Concentrated sulfuric acid was employed for *Strophanthus* glycosides[47] (100° for 2 minutes, sensitivity 0.2 μg; brown to black spots), stero-bile acids and bile sterols[102] (130°, yellow to black spots), bile acids,[103] C_{18}-, C_{19}-, and C_{21}-steroids[104] (100° for 10–15 minutes), and sapogenins.[105] The colors in daylight and UV light of 62 sapogenins and androstane, pregnane, and cholestane steroids, obtained on unbound aluminum oxide layers, are published by Černý *et al.*[90]

b. p-Toluenesulfonic Acid. Spraying with a 20% ethanolic solution (95–100° for 20 minutes) has been used to detect estrogens,[84] etienic acids,[106,107] sapogenins,[108] cardenolide acetates,[87] and *Digitalis* glycosides.[109] The colors for 32 steroid estrogens in daylight and UV light (sensitivity: 2–4 μg/spot) have been recorded;[84] characteristic pink fluorescence on UV was found for 2-methoxy-3-hydroxyestrogens; 6-oxo and 7-oxo estrogens react giving a gray color. This reaction can be applied to formamide-impregnated layers.[109] A 50% aqueous solution (140° for several minutes) was used for $9\beta,10\alpha$-steroids of the C_{19} and C_{21} series.[110] Concentrated aqueous solution of *p*-toluenesulfonic acid (100% w/v) has been used for different classes of steroids[111] (heating: 10 minutes at 100°; sensitivity 2 μg/cm^2). Several colors are useful to distinguish steroids with similar polarities, such as 5α-androstane-3,17-dione (pale yellow), androst-4-ene-3,17-dione (green) and androsta-1,4-diene-3,17-dione (orange red); testosterone (green blue) and 19-nortestosterone (orange); and 11α-hydroxyprogesterone (yellow) and 12α-hydroxyprogesterone (purple).

c. Liebermann[112]–Burchard[113] Reaction. The plates were sprayed with

[99] E. Weiss-Berg and C. Tamm, *Helv. Chim. Acta* **46**, 2435 (1963).
[100] J. R. Claude and J. L. Beaumont, *Ann. Biol. Clin. (Paris)* **22**, 815 (1964).
[101] M. Slaytor and K. Bloch, *J. Biol. Chem.* **240**, 4598 (1965).
[102] T. Kazuno and T. Hoshita, *Steroids* **3**, 55 (1964).
[103] P. Eneroth, *J. Lipid Res.* **4**, 11 (1963).
[104] M. Takeuchi, *Chem. Pharm. Bull. (Tokyo)* **11**, 1183 (1963).
[105] N. Matsumoto, *Chem. Pharm. Bull. (Tokyo)* **11**, 1189 (1963).
[106] J. von Euw and T. Reichstein, *Helv. Chim. Acta* **46**, 142 (1963).
[107] H. H. Sauer, E. Weiss, and T. Reichstein, *Helv. Chim. Acta* **49**, 1632 (1966).
[108] J. von Euw and T. Reichstein, *Helv. Chim. Acta* **49**, 1468 (1966).
[109] D. Sonanini, *Pharm. Acta Helv.* **39**, 673 (1964).
[110] G. Saucy, H. Els, F. Miksch, and A. Fürst, *Helv. Chim. Acta* **49**, 1529 (1966).
[111] V. Vlasinich and J. B. Jones, *Steroids* **3**, 707 (1964).
[112] C. Liebermann, *Ber. Deut. Chem. Ges.* **18**, 1803 (1885).
[113] H. Burchard, Beiträge zur Kenntnis des Cholesterins. Inaugural Dissertation, Universitätsbuchdruckerei, Adler's Erben, Rostock, 1889.

a reagent consisting of 1 ml of sulfuric acid, 20 ml of acetic acid anhydride, and 50 ml of chloroform[114] and heated at 85–90° for 15 minutes. Colors for 32 estrogens[84] and 29 α,β-unsaturated pregnene steroids[93] in daylight and UV light have been investigated (sensitivity: 2–5 μg). Estradiol-17β (ochre; UV: gray yellow) and its 17α-epimer (purple; UV: yellow orange) can be differentiated by this reaction.

Successive spraying with acetic acid anhydride and sulfuric acid (100° for 10–15 minutes) has been employed to detect estrogens, and α,β-unsaturated C_{19}- and C_{21}-steroids.[104] C_{19}-, C_{18}-, C_{21}-, and C_{27}-steroids (including sapogenins) have been visualized on starch–silica gel layers[115] by means of a mixture of acetic anhydride and sulfuric acid (4:1).

Most of the sterols develop a brown spot (110° for 3–8 minutes) with the Liebermann–Burchard reaction.[116] The enols develop a blue color before heating.

d. Phosphoric Acid. Diluted phosphoric acid (20% ethanol) reacts with estrogens[84] after 20 minutes of heating at 100° to give colors in daylight and UV light with a sensitivity of 2–4 μg per spot; colors are given for 32 estrogens. A 10% solution (125° for 12 minutes at least) gives, if observed at longwave UV, a blue fluorescence with cardenolide glycosides of the gitoxigenin, digoxigenin, and gitaloxigenin series,[117] similar to those found by using other organic acids (as formic, tartaric, and trichloroacetic acids). For steroid oximes,[118] a 50% aqueous solution (110° for 20 minutes) was employed.

Spraying with concentrated phosphoric acid (85%) followed by heating at 90–95° for 15 minutes has been employed to detect Δ^5-3-hydroxysteroids of the cholestane, androstane, and pregnane series and estrogens.[70] Colors for 20 Δ^5-3-hydroxysteroids and 16 estrogenic steroids in daylight and UV light, some of them highly specific, have been reported. 7α- and 11β-Hydroxyestradiol present a green fluorescence by shortwave UV, and 3β-hydroxysteroids with a 7-hydroxy-Δ^5 structure or $\Delta^{5,7}$-diene structure react immediately, without heating, to give a blue, blue-gray, or blue-violet color. The presence of a 7-oxo group in neutral steroids inhibits this reaction, and a yellow color can be obtained only after prolonged heating, if at all. The application of this reaction to 20 Δ^4-3-oxo-C_{21}-steroids showed valuable fluorescence by shortwave UV—for instance, a pink color for 16α,21- and 17α,21-dihydroxypregn-4-ene-3,20-diones.[93]

For the detection of several C_{21}-steroids on unbound aluminum oxide

[114] R. Neher and A. Wettstein, *Helv. Chim. Acta* **34**, 2278 (1951).
[115] L. L. Smith and T. Foell, *J. Chromatog.* **9**, 339 (1962).
[116] J. R. Claude, *J. Chromatog.* **23**, 267 (1966).
[117] L. Fauconnet and M. Waldesbühl, *Pharm. Acta Helv.* **38**, 423 (1963).
[118] G. Göndös, B. Matkovics, and Ö. Kovàcs, *Mikrochem J.* **8**, 415 (1964).

layers, a 70% solution of phosphoric acid has been used;[119] it is reported that pregnanetriol may be determined even after spraying, without any interference of the spray-reagent in the subsequent determination.

e. Winogradow Reaction.[120] The plates are sprayed with a 90% solution of trichloroacetic acid in water (or hydrochloric acid)[121] and heated at 100° for 15 minutes. Allyl alcohols[70] develop colors immediately after spraying: 7-hydroxy-Δ^5-steroids show a blue color (after heating: green, which changes to carmine brown), whereas the color developed by 3-hydroxy-Δ^4-steroids is pink. 7-Dehydrocholesterol reacts in the cold (green gray), but cholesterol, lanosterol and cholesta-3,5-dien-7-one are visualized only after heating.[70] 17α-Estradiol gives a vivid pink color with this reagent, whereas its 17β-epimer does not react.

Strophanthus glycosides show a yellow fluorescence in UV light after they have been sprayed with a 25% solution of trichloroacetic acid in chloroform and heated for a short period (2 minutes at 100°).[48]

f. Chlorosulfonic Acid–Acetic Acid Reagent.[122] This reagent has been employed in the proportions 1:2; plates are heated before spraying, and in some cases again (3–5 minutes at 110–130°). Colors are observed in daylight and UV light, and some steroids, such as tigogenin and phytosterols,[122] can be visualized without supplementary heating.

This reaction is utilized to detect sapogenins,[123,124] cardenolides,[122] phytosterols,[49] neutral[125,126] and acid[54] triterpenes, trimethylsterols,[54] abeo- and diabeosteroids,[127] C_{27}- and C_{28}-sterols,[128] and C_{18}-, C_{19}-, and C_{21}-steroids.[104] The colors obtained for 20 sapogenins[105] with this reaction were, except that of nogiragenin (yellow green), yellow, orange, or red in daylight. Under an ultraviolet lamp a specific green fluorescence was noted for nogiragenin, yonogenin, and gitogenin.

Different colors were obtained with this reaction[23] for the acetates of cholesterol (strong violet), dihydrocholesterol (white), and 7-dehydrocholesterol (gray brown); a differentiation can also be established between the acetates of Δ^7-ergostenol (yellow green) and 5-dihydroergosterol (gray brown).

[119] L. Stárka and J. Maliková, *J. Endocrinol.* **22**, 215 (1961).
[120] K. Winogradow, personal communication to E. Drechsel, *Zentr. Physiol.* **9**, 361 (1897).
[121] E. Hirschsohn, *Pharm. Zentralhalle* **43**, 357 (1902).
[122] R. Tschesche, W. Freytag, and G. Snatzke, *Chem. Ber.* **92**, 3053 (1959).
[123] R. Tschesche and G. Wulff, *Chem. Ber.* **94**, 2019 (1961).
[124] H. Sander, *Naturwissenschaften* **48**, 303 (1961).
[125] R. Tschesche and F. Ziegler, *Ann. Chem.* **674**, 185 (1964).
[126] R. Tschesche, B. T. Tjoa, and G. Wulff, *Ann. Chem.* **696**, 160 (1966).
[127] G. Snatzke and A. Nisar, *Ann. Chem.* **683**, 159 (1965).
[128] R. Tschesche and G. Snatzke, *Ann. Chem.* **636**, 105 (1960).

g. Phosphomolybdic Acid.[129] This has been employed in 5–20% ethanolic solutions to detect a wide range of steroids. The intensity of colors produced is highly influenced by the temperature of heating[130] and by the support used. On Florisil layers it could not be used unless the plate was also sprayed with dilute sulfuric acid.[131] For use on neutral silica gel layers containing sodium hydroxide, it is necessary to add hydrochloric acid to the reagent in order to achieve the sensitivity found on normal layers.[115]

A 10% solution (15 minutes at 90° or 5 minutes at 100° heating) was employed to detect Δ^4-3-oxo-C_{21}-steroids,[93] Δ^5-3-hydroxysteroids,[70] Δ^4-3-oxo-C_{19}-steroids,[132] saturated C_{19}- and C_{21}-steroids,[132] estrogens,[132] bile acids,[133] sterols,[116, 134] etienic acid methyl esters,[134] cholesterol and its esters[82] with a sensitivity between 2 and 5 μg per spot. This reaction is very sensitive on rice starch-bound TLC[115] for the detection of C_{18}-, C_{19}-, C_{21}-, and C_{27}-steroids, but the plates should not be heated for more than 10 minutes at 100°. Most of the steroids develop a molybdenum blue color, sometimes immediately after spraying (7-hydroxy-Δ^5-steroids); 17α,20,21-trihydroxysteroids develop a blue-violet color,[93] and cholesterol or its esters a red spot.[135]

The blue-gray colors developed by steroids with a keto group are less intense than those of alcoholic sterols and change to brown or violet red after 24 hours in the dark.[116] The original colors reappear after exposure to UV light. Intense colored background can be eliminated if the plate is sprayed twice with an ethanolic alkali solution (10%).[115]

A 20% phosphomolybdic acid solution was employed to visualize sterols and their acetates (5–10° minutes at 90°),[42] and a 5% solution to detect cholesterol and its esters (10 minutes at 100°).[82] For the visualization of bile acids, Hofmann[136] prefers to heat the plate first at 160° before using this reaction.

h. Arsenomolybdic Acid.[137] This is a more sensitive reagent than phosphomolybdic acid for detecting reducing compounds such as Δ^4-3-

[129] D. Kritchevsky and M. R. Kirk, *Arch. Biochem. Biophys.* **35**, 346 (1962).
[130] A. F. Hofmann, *J. Lipid Res.* **3**, 391 (1962).
[131] A. F. Hofmann, *in* "Biochemical Problems of Lipids" (A. C. Frazer, ed.), BBA Library, Vol. 1, p. 1. Elsevier, Amsterdam, 1963.
[132] F. A. Vandenheuvel, G. J. Hinderks, J. C. Nixon, and W. G. Layng, *J. Am. Oil Chemists' Soc.* **42**, 283 (1965).
[133] G. Beisenherz, F. W. Koss, and U. Chuchra, Festschrift "Walter Graubner," p. 31. Boehringer, Ingelheim a. Rhein, 1961.
[134] M. Barbier, H. Jäger, H. Tobias, and E. Wyss, *Helv. Chim. Acta* **42**, 2440 (1959).
[135] H. Jatzkewitz and E. Mehl, *Z. Physiol. Chem.* **320**, 251 (1960).
[136] A. F. Hofmann, *Anal. Biochem.* **3**, 145 (1962).
[137] V. Schwarz, *Nature* **169**, 506 (1952).

oxo-C_{21}-steroids[93] (blue color, sensitivity 0.5 μg per spot). The reagent[138] is prepared by dissolving 25 g of ammonium molybdate in 450 ml of water, adding to it 21 ml of concentrated sulfuric acid and 25 ml of a 12% aqueous solution of $Na_2HAsO_7 \cdot 7 H_2O$, and filtering the mixture after 2 days at 37°. After spraying, the TLC plates are heated for 5–10 minutes at 100°.

 i. Phosphotungstic Acid. This acid is employed to detect cholesterol and its esters[25, 82] and other sterols.[23] A 10% alcoholic solution of the acid[25] is used; followed by heating for 5–10 minutes at 100°.

 j. Hammarsten[139]–Yamasaki[140] Reaction. The plates are intensively sprayed with concentrated hydrochloric acid and heated at 80–100° for 5 minutes. 3,7-Dihydroxy-Δ^5-steroids develop a blue color, whereas other 3-hydroxy-Δ^5-steroids give no color. Even other allyl alcohols such as 3-hydroxy-Δ^4-steroids remain colorless.[70]

 k. Heard and Sobel Reaction.[141] Equal parts of Folin–Ciocalteau reagent[142] and glacial acetic acid are mixed before spraying. The Folin–Ciocalteau reagent is available commercially. After heating for 30 minutes (95–100°), α,β-unsaturated ketones[69] show a molybdenum blue color (sensitivity: 5 μg per spot). Also steroids with a primary or cyclic secondary α-ketol group, and α,β-diketones give a strong reaction.[69]

 l. Vanadic Acid–Sulfuric Acid.[143] The layer is first sprayed with a saturated aqueous solution of ammonium metavanadate, then dried at 80° and sprayed again with 1 N sulfuric acid. Estrogens and neutral steroids (e.g., progesterone, deoxycorticosterone, etiocholanolone)[84] give a nonspecific green-gray color on a yellow background.

 m. Phosphomolybdic–Acetic–Sulfuric Acid Mixture (5g:100ml:5ml). The mixture has been shown to be much better than phosphomolybdic acid alone for visualizing bile acids.[144] Even keto acids, e.g., 3α-hydroxy-7,12-dioxocholanoic acid, can be detected if they are previously reduced *in situ* by means of a 5% methanolic (80% methanol–water) solution of sodium borohydride.

 n. Perchloric Acid. The spray solution used to detect cardiac glycosides[145] is prepared by adding 15 ml of 70% perchloric acid to 100 ml of

[138] N. Nelson, *J. Biol. Chem.* **153**, 375 (1944).
[139] O. Hammarsten, *Z. Physiol. Chem.* **61**, 495 (1909).
[140] K. Yamasaki, *Biochemistry (Tokyo)* **18**, 311 (1933).
[141] R. D. H. Heard and H. Sobel, *J. Biol. Chem.* **165**, 687 (1946).
[142] O. Folin and V. Ciocalteau, *J. Biol. Chem.* **73**, 627 (1927).
[143] K. F. Mandelin, *Pharm. Z. Russland* **22**, 345 (1884). Quoted by H. D. Gibbs, *Chem. Rev.* **3**, 291 (1926).
[144] T. Usui, *J. Biochem. (Tokyo)* **54**, 283 (1963).
[145] E. Johnston and A. L. Jacobs, *J. Pharm. Sci.* **55**, 531 (1966).

water. After spraying, the plates are heated for a few minutes at 100°
and fluorescence is observed with a longwave UV lamp. Also a 2% solution has been used to detect C_{19}- and C_{21}-steroids.[146] It has been used for
their semiquantitative determination by comparing the spots with known
amounts of standards.

   o. Lifschütz Reaction.[147] A sulfuric acid–acetic acid mixture (1:1)
has been employed to detect cholesterol and cholesterol esters[135] (3–15
minutes at 90°). The sterols develop a red color, with a sensitivity of
2–4 μg per spot. It also has been used to detect steroid glycosides.[148]

   p. Sulfuric Acid–Nitric Acid. This has been used for the detection of
lipids[131] (including bile acids and cholesterol) in the proportions: sulfuric
acid–nitric acid–water 4:3:3 (v/v/v).

   q. Chromsulfuric Acid Reagent. This is a saturated solution of
chromic acid in concentrated sulfuric acid;[89] Δ^5- and Δ^7-sterols[148] may be
detected on hot TL plates with this reagent.

   r. Picric Acid–Perchloric Acid.[149] The reagent is prepared by dissolving 100 ml of picric acid in 36 ml of glacial acetic acid and adding 6 ml of
70% perchloric acid.

   Hot TL plates are sprayed with the reagent (in a hood). Most of the
3-hydroxy-Δ^5-steroids investigated[149] show a red or lavender color; 7-oxo
substitution inhibits the color development.

   The reaction of Δ^5-3β-hydroxysteroids with the picric acid–perchloric
acid reagent is very similar to that of ammonium molybdate–perchloric
acid (see below). It must be noted that perchloric acid alone develops
pink or purple halochromic compounds with steroids containing a conjugated double bond system in rings A and B[150] and with steroids that
form such a system when treated with an acid.

2. Reagents Containing an Aromatic Aldehyde

   a. Reaction of Ekkert.[151] The reagent consists of anisaldehyde added
to a strong acid. The reagent according to Miescher[152] consists of 1%
(w/v) anisaldehyde (*p*-methoxybenzaldehyde) solution in glacial acetic
acid–sulfuric acid (98:2). The plates are sprayed and heated at 85°
(estrogens), 90° (sterols), or 100° for 3–15 minutes. The color sequence
is particularly useful for the characterization of some structures of the

[146] H. Metz, *Naturwissenschaften* **48**, 569 (1961).
[147] J. Lifschütz, *Z. Physiol. Chem.* **53**, 140 (1907).
[148] W. Sucrow, *Chem. Ber.* **99**, 2765 (1966).
[149] W. R. Eberlein, *J. Clin. Endocrinol. Metab.* **25**, 288 (1965).
[150] W. Lang, R. G. Folzenlogen, and D. G. Kolp, *J. Am. Chem. Soc.* **71**, 1733 (1949).
[151] L. Ekkert, *Pharm. Zentralhalle* **69**, 97 (1928).
[152] K. Miescher, *Helv. Chim. Acta* **29**, 743 (1946).

steroid molecule. This is the most widely used reaction with an aromatic aldehyde and has a sensitivity down to 2 μg per spot.

Colors in daylight have been investigated for 32 estrogens,[84] 80 C_{19}-steroids,[70, 153, 154] 54 saturated C_{21}-steroids,[155] 27 formaldehydogenic steroids,[63] 39 Δ^4-3-oxo-21-deoxypregnenesteroids,[156] 22 weakly polar steroids,[29] 11 sterols,[23] and several 7-oxygenated steroids (estrogens and Δ^5-3β-hydroxysteroids).[157, 158] In this reaction the colors obtained are dependent upon temperature, duration of heating, and intensity of spraying. A characteristic green color was found for estradiol-17β and its 7α-hydroxy, 11β-hydroxy, and 15α-hydroxy derivatives.[84, 159] A differentiation between testosterone (color sequence: yellow-brown-blue) and epi-testosterone (blue spot after 3 minutes)[154] is possible. This color reaction permits one also, by the color sequence developed during gradual heating, to recognize the structures Δ^4-3-oxo-, 5$\alpha(H)$-3-oxo-, and 5$\beta(H)$-3-oxo in the androstane steroids[160] oxygenated at positions 3 and 17. The development of color in the anisaldehyde reaction is inhibited by the presence of a 7-oxo group[154] on neutral steroids and 6-oxo groups on estrogens.[84]

The application of the anisaldehyde reaction to 24 bile acids[161] (sensitivity: 1 μg per spot; colors with visible and longwave UV) gives some UV fluorescences of special interest: green (blue, visible light) was observed for 6α,6β,7α- and 7β-hydroxy derivatives of lithocholic acid; orange for 12α-hydroxy-, 3α,12α-dihydroxy-, and 3α,7α,12α-trihydroxycholanoic acids; the four isomeric 6,7-dihydroxy derivatives of lithocholic acid show yellow fluorescence; and the 3-oxo-, 3,7-dioxo-, 3,12-dioxo-, and 3,7,12-trioxocholanoic acids are pink. Sometimes the differences observed in daylight are very useful. For instance, for the hyocholic isomers: olive (3α,6α,7α- and 3α,6β,7α-trihydroxycholanoic acid), lavender (3α,6α,7β-trihydroxycholanoic acid), and pink (3α,6β,7β-trihydroxycholanoic acid). The colors obtained for chenodeoxycholic (blue; UV: green) and deoxycholic (brown; UV: orange) acids are useful for the identification of these two acids of similar chromatographic mobilities.

b. Reaction of Inouye–Ito.[162] Plates are sprayed with a 1% (w/v)

[153] B. P. Lisboa, *J. Chromatog.* **13**, 391 (1964).
[154] B. P. Lisboa, *J. Chromatog.* **19**, 81 (1965).
[155] B. P. Lisboa, *Steroids* **6**, 605 (1965).
[156] B. P. Lisboa, *Steroids* **8**, 319 (1966).
[157] B. P. Lisboa, R. Knuppen, and H. Breuer, *Biochim. Biophys. Acta* **97**, 557 (1965).
[158] L. Cédard, B. Fillmann, R. Knuppen, B. P. Lisboa, and H. Breuer, *Z. Physiol. Chem.* **338**, 89 (1964).
[159] B. P. Lisboa, U. Goebelsmann, and E. Diczfalusy, *Acta Endocrinol.* **54**, 467 (1967).
[160] B. P. Lisboa and H. Breuer, *Gen. Comp. Endocrinol.* **6**, 114 (1966).
[161] D. Kritchevsky, D. S. Martak, and G. H. Rothblat, *Anal. Biochem.* **5**, 388 (1963).
[162] K. Inouye and H. Ito, *Z. Physiol. Chem.* **57**, 313 (1908).

vanillin (4-hydroxy-3-methoxybenzaldehyde) solution in glacial acetic acid–sulfuric acid 98:2, and heated at 95–100°. The sequence of color obtained during heating for 5–15 minutes has been observed for 32 estrogens[84] and 37 Δ^4-3-oxo-C_{21}-steroids;[93] the colors are less specific than those obtained with the anisaldehyde reaction.

c. *Reaction of Matthews*.[163] The reagent, a 0.5% solution of vanillin in sulfuric acid–ethanol 4:1, is freshly prepared each day. The plates are sprayed and the colors observed before and after heating (5 minutes at 100°). The reaction is used for C_{18}-, C_{19}-, C_{21}-, and C_{27}-steroids[154, 158, 163] with a sensitivity of 5 $\mu g/cm^2$. Androstan-3-one develops a yellow spot, whereas androstan-17-one develops a reddish purple spot.[163]

This reaction is useful for 7-oxygenated Δ^5-steroids:[70, 158] Δ^5-3β-hydroxysteroids with a 7-hydroxyl group develops blue; and the 7-oxo-Δ^5-3β-hydroxysteroids, pink. Dehydroepiandrosterone gives a pink spot.

d. *Reaction of Chabrol*.[164] This is a vanillin–phosphoric acid reaction. (i) Plates are sprayed with a 2% solution of vanillin in concentrated phosphoric acid (85%),[165] heated at 90–95°, and observed after 15 and 30 minutes. This reaction permits the detection of 17α,20,21-trihydroxysteroids and 17α-hydroxy-20-oxo-21-deoxysteroids with a sensitivity of 1 μg per spot.[93, 155] A specific orange color is observed with the latter group of steroids. (ii) Spray with a 5% solution of vanillin in ethanolic phosphoric acid (50%)[166] and heat at 100° for 2–10 minutes. Spiranosteroids[167] give yellow spots (after 2 minutes) which change after continued heating to different colors. (iii) Spray with a 1% solution of vanillin in 50% phosphoric acid and heat (120° for 10–20 minutes). This spray has been used to detect steroids[146] and sapogenins.[124, 168]

Vanillin–phosphoric acid has been used also for detecting alkaloid steroids,[168] e.g., soladulcidin (blue-gray color).

e. *Komarowsky's Reagent*.[169, 170] The spray reagent is made[171] by using one part of a 50% (v/v) aqueous solution of sulfuric acid and 10 parts of a 2% methanolic solution of *p*-hydroxybenzaldehyde. The plate is sprayed, then heated for 3–4 minutes at 105° (or 10 minutes at 60°). Yellow spots (white background) in daylight, which after heating turn

[163] J. S. Matthews, *Biochim. Biophys. Acta* 69, 163 (1963).
[164] E. Chabrol, R. Charonnat, J. Cottet, and P. Blonde, *Compt. Rend. Soc. Biol.* 115, 834 (1934).
[165] W. J. McAleer and M. A. Kozlowski, *Arch. Biochem. Biophys.* 62, 196 (1956).
[166] W. J. McAleer and M. A. Kozlowski, *Arch. Biochim. Biophys.* 66, 120 (1957).
[167] R. Tschesche, H. Schwarz, and G. Snatzke, *Chem. Ber.* 94, 1699 (1961).
[168] H. Sander, M. Alkemeyer, and R. Hänsel, *Arch. Pharm.* 295, 6 (1962).
[169] A. Komarowsky, *Chem. Ztg.* 27, 807 (1903).
[170] A. Komarowsky, *Chem. Ztg.* 27, 1086 (1903).
[171] P. J. Stevens, *J. Chromatog.* 14, 269 (1964).

to pink, were found for steroidal sapogenins (unsubstituted at position C-23) and 3-oxocorticosteroids (unsubstituted at C-2);[171] the sensitivity was down to 0.1 μg per spot. $\Delta^{1,4}$-3-oxosteroids (such as prednisone) do not react. This reagent gives valuable colors with estrogens.[172]

f. Okanishi Reaction.[173] The plate is sprayed first with a 1% ethanolic solution of cinnamic aldehyde, dried at 70° for 3 minutes and sprayed again with a solution of antimony–trichloride in nitrobenzene 5:1 (w/v). Sapogenins develop an orange-yellow color, with 2–3 μg sensitivity.[167]

g. Sjöholm Reagent.[174] This consists of a solution of 1% anisaldehyde and 7% perchloric acid in acetone–water 1:4. The plates are sprayed and colors are observed in visible light and the fluorescence in longwave UV light after 11–15 minutes of heating at 75–80°. With visible light digitoxigenin derivatives exhibit a blue, blue-gray, or blue-green color and have a typical violet fluorescence; gitoxigenin glycosides give red, red-violet, or brown; digoxigenin glycosides, blue to violet; and gitaloxigenin glycosides, red-blue or red-violet spots.

h. Zöllner Reaction.[175] This is a sulfo-phospho-vanillin reaction. Plates are sprayed first with dilute sulfuric acid, heated 10 minutes at 100°, and sprayed again with a mixture of phosphoric acid (4 parts) and 0.6% aqueous vanillin (1 part). Cholesterol and cholesterol esters develop a blue color.[96]

i. Brooksbank–Haslewood Reaction.[176] This reaction has been used to detect 16-dehydrosteroids and their acetates.[177] Equal volumes of 0.5% (w/v) resorcylaldehyde in glacial acetic acid and 5% (v/v) concentrated sulfuric acid in glacial acetic acid are mixed prior to spraying. The plates are heated at 90–100° for 5 minutes.

16-Dehydro-C_{19}-steroids develop[177,178] mauve (5α- and 5β-androst-16-en-3α-ols, 5α-androst-16-en-3β-ol and their acetates; androsta-5,16-dien-3β-ol acetate), purple (5α- and 5β-androst-16-en-3-ones), or blue (androsta-5,16-dien-3β-ol) colors. Estratetraen-3-ol and its acetate give red spots. Color formation is inhibited by traces of nonvolatile stationary phases.

j. Other Aromatic Aldehydes. p-Dimethylaminobenzaldehyde,[152] benzaldehyde,[169] and salicylaldehyde[169] have also been used as 1% solutions

[172] G. Zucconi, B. P. Lisboa, E. Simonitsch, L. Roth, A. A. Hagen, and E. Diczfalusy, *Acta Endocrinol.* **53**, 413 (1967).

[173] T. Okanishi, A. Akahori, and F. Yasuda, *Ann. Rept. Shionogi Res. Lab. Osaka* **8**, 927 (1958).

[174] I. Sjöholm, *Svensk Farm. Tidskr.* **66**, 321 (1962).

[175] N. Zöllner and K. Kirsch, *Z. Ges. Exptl. Med.* **135**, 545 (1962).

[176] B. W. L. Brooksbank and G. A. D. Haslewood, *Biochem. J.* **80**, 488 (1961).

[177] D. B. Gower, *J. Chromatog.* **14**, 424 (1964).

[178] B. W. L. Brooksbank and D. B. Gower, *Steroids* **4**, 787 (1964).

in acetic–sulfuric acid (98:2), to detect several steroids[84] (heating: 5–10 minutes at 95–100°; sensitivity: 3–6 μg per spot). Colors in visible light and fluorescence in UV light have been recorded for 32 estrogens[84] and 37 Δ^4-3-oxo-C_{21}-steroids[93] by the use of these three aromatic aldehydes. Useful fluorescence reactions are obtained with benzaldehyde and salicylaldehyde.

The salicylaldehyde reaction[161] gives different colors with the common bile acids: pink (lithocholic), brown (cholic), tan (deoxycholic), and purple (hyodeoxycholic and chenodeoxycholic). With these bile acids, the furfuraldehyde–sulfuric acid reagent (Pettenkofer reaction)[179] gives a blue color.[161]

Copius-Peereboom[23] has detected sterols and sterol acetates by spraying them first with pure salicylaldehyde, heating for 5 minutes (80°), then spraying with 1 N sulfuric acid and heating at 90° for 10 minutes. Vivid colors of relatively good specificity were developed: violet (cholesterol, stigmasterol, and β-sitosterol acetates), purple (brassicasterol acetate and Δ^7-cholestenol), gray blue (ergosterol and 7-dehydrocholesterol acetates), orange brown (lanosterol and 5-dihydroergosterol acetates), yellow green (7-ergostenol acetate), gray brown (vitamin D_2), and white (dihydrocholesterol acetate).

3. Acids with Salts

a. Ammonium Molybdate–Perchloric Acid Reagents.[180] Spray first with water-saturated phenol, then heat at 70° for drying. After cooling, spray with a reagent containing 1 g of ammonium molybdate and 2.5 ml of perchloric acid (60%) in 100 ml of 0.1 N HCl hydrochloric acid and heat again at 80° for 10–15 minutes. Δ^5-3β-Hydroxysteroids[70] develop rose, lilac, carmine, or violet spots, and Δ^5-pregnene-3α,16α,20α-(or 20β) triols develop brown spots whereas steroids with 7-hydroxy-Δ^5-groups give a blue color; a 7-oxo-group inhibits the reaction. Estrogens[70] react with this reagent to give different colors—mostly blue, but also lilac and rose.

b. Tschugajeff Reaction.[181] Thin-layer plates are sprayed[84] with a 60% solution of zinc chloride in glacial acetic acid (w/v) and heated at 90° for 10 minutes (daylight and UV light colors). Most of the 32 estrogens tested have shown in UV light a yellow or orange fluorescence (sensitivity 2–5 μg per spot).

[179] M. Pettenkofer, *Ann. Chem. Pharm.* **52/88,** 90 (1844).
[180] R. F. Witter and S. Stone, *Anal. Chem.* **29,** 156 (1957).
[181] L. A. Tschugajeff, *Russ. Arch. Pathol.* **9,** 289 (1900).

c. *Dragendorff's Reagent.*[182] This reagent is employed in two different modifications: (i) Reagent according to Lisboa:[153] 10 ml of a 0.3% bismuth subnitrate solution in 50% (v/v) sulfuric acid is added with constant stirring to 30 ml of a 10% solution of potassium iodide in 70% (v/v) ethanol. Prepare each second day and store at +4°. Δ^4-3-Oxo-C_{21}-steroids[93, 153] appear as orange spots; several other saturated steroids give a yellow or yellow-orange color. (ii) Reagent according to Munier and Macheboeuf:[183] Mix equal parts of (I) an aqueous solution of 2.15 g of bismuth subnitrate and 25 g of tartaric acid in 100 ml of water, and (II) a 2% aqueous solution of potassium iodide. Dilute 50 ml of the mixture to 550 ml with a 20% aqueous tartaric acid solution. This reagent reacts with steroidal alkaloids to give yellow spots in daylight.[90]

d. *Potassium Dichromate–Sulfuric Acid.*[184] Slowly add 70 ml of concentrated sulfuric acid (98%) to 30 ml of an aqueous saturated potassium dichromate solution.[185] Colors for various sterols and epiandrosterone[185] were followed between 18° and 92° by gradual rise of temperature. Under these conditions valuable changes of color were observed; for instance, stigmasterol and sitosterol give different colors at 76° (temperature reached after 20 minutes). This reaction has been employed, on silver nitrate-impregnated silica gel layers, to detect sterols.[186]

e. *Sonnenschein's Reagent.* (i) A 1% spraying solution of cerium-(IV)–ammonium nitrate in 50% sulfuric acid[187] has been used to detect neutral and phenolic C_{18}-steroids[188] and scillarenin glycosides[189] (scillaroside: yellow; scillaglaucoside: greenish; proscillaridin A and scillaren A: blue). (ii) Solanum glycosides were detected[190] by spraying with a saturated cerium(IV)–sulfate solution in 70% sulfuric acid and heating at 120°.

f. *Ferric Chloride–Sulfuric Acetic Acid.* Spray with a mixture of 10% ferric chloride in acetic acid (0.2 ml), acetic acid (30 ml), and sulfuric acid (20 ml); heat for 10 minutes at 110°. Sterols[116] of similar polarity give different colors, e.g., Δ^5-cholestene-3β,4β-diol (blue violet) and 3β-

[182] V. Pelcová, personal communication to O. Siblíková, quoted by I. M. Hais and K. Macek (eds.) "Handbuch der Papierchromatographie," Vol. 1, p. 349. Fischer, Jena, 1958.
[183] R. Munier and M. Macheboeuf, *Bull. Soc. Chim. Biol.* 31, 1144 (1949).
[184] E. Haaki and T. Nikkari, *Acta Chem. Scand.* 17, 536 (1963).
[185] I. S. Shepherd, L. F. Ross, and I. D. Morton, *Chem. & Ind.,* p. 1706 (1966).
[186] E. Haaki and T. Nikkari, *Acta Chem. Scand.* 17, 338 (1963).
[187] O. E. Schultz and D. Strauss, *Arzneimittel-Forsch.* 5, 342 (1955).
[188] A. von Wartburg, *Helv. Chim. Acta* 46, 591 (1963).
[189] A. von Wartburg, *Helv. Chim. Acta* 47, 1228 (1964).
[190] K. Schreiber and H. Rönsch, *Tetrahedron Letters* p. 329 (1963).

hydroxy-5α-cholestan-6-one (brown orange). Steroids with an enol struc-ture [Δ⁵-cholestene-3β,4β-diol, Δ⁷-cholestene-3β,7β (or 7α)-diols] develop a blue color without heating.

Ferric chloride in butanol–sulfuric acid has been also suggested to localize bile acids.[191]

g. *Uranyl Nitrate–Sulfuric Acid*.[192] Spray with a 5% (w/v) solution of uranyl nitrate in 10% (v/v) aqueous sulfuric acid and heat at 110° for 6–7 minutes. This reagent gives specific colors with several Δ¹⁶-steroids:[177] purple (5α-androst-16-en-3β-ol, androsta-5,16-dien-3β-ol, 5α-androst-16-en-3β-ol acetate); red (estratetraen-3-ol and its acetate); brown (andro-sta-5,16-dien-3β-ol acetate); pink (5α-androst-16-en-3α-ol acetate); bluish gray (5β-androst-16-en-3α-ol and its acetate); gray (5α-androst-16-en-3α-ol).

4. Acid–Phenol Reagents

a. *Thymol–Sulfuric Acid Reaction*.[23] First spray with a 20% solution of thymol in 96% ethanol, heat 10 minutes at 80°, and spray again with 1 N sulfuric acid; then heat 10 minutes at 90°. Steroids difficult to separate can be differentiated by using this reaction; e.g., the sterol acetates of cholesterol (strong violet), dihydrocholesterol (white), 7-de-hydrocholesterol (gray brown), lanosterol (orange brown), Δ⁷-ergostenol (yellow green), and 5-dihydroergosterol (gray green). For other sterols, see Copius-Peereboom.[23]

b. *β-Naphthol–Sulfuric Acid*.[23] Spray with a 0.2% solution of β-naphthol in 4 N sulfuric acid and heat for 10 minutes at 90°. Most of the sterol acetates give a bluish or blue color. This reaction is useful in the characterization of the acetates of cholesterol (strong blue), stigmasterol (strong violet), and β-sitosterol (strong purple). For the colors developed by other sterols and their acetates, see Copius-Peereboom.[23]

c. *Resorcinol–Sulfuric Acid*.[23] Spray with a 20% resorcinol solution on 96% ethanol (heat 5 minutes at 80°) followed by 1 N sulfuric acid; heat again 10 minutes at 90°. This reaction is valuable for sterols and their acetates. Sterol acetates gave various colors: e.g., cholesterol (faint blue), 7-dehydrocholesterol (purple), ergosterol (faint green), Δ⁷-ergos-tenol (yellow brown), and 5-dihydroergosterol (purple brown).

d. *Guaiacol–Sulfuric Acid*.[193] Spray with a 0.5% guaiacolsulfonate solution in 50% sulfuric acid and heat at 110° until colors appear. This reaction is used for corticosteroids.

[191] W. L. Anthony and W. T. Beher, *J. Chromatog.* **13**, 567 (1964).
[192] R. J. Bridgwater, private communication to D. B. Gower, *J. Chromatog.* **14**, 424 (1964).
[193] C. Monder, *Biochem. J.* **90**, 522 (1964).

e. Naphthoquinone–Perchloric Acid.[194] This reagent allows the differentiation of stigmasterol and cholesterol. Spray with a 0.1% solution of 1,2-naphthoquinonesulfonic acid in ethanol–60% perchloric acid–40% formaldehyde–water 2:1:0.1:0.9 (v/v) mixture. Heat at 70–80° for 15–60 minutes, and observe the development of colors during the heating. Stigmasterol changes from rose to blue after 30 minutes of heating whereas cholesterol turns to blue only after 45 minutes.

5. Metal Salts

a. Carr and Price's Reaction.[195] This is perhaps the most commonly employed reaction for all classes of steroids, including estrogens,[196, 197] sterols,[100, 116, 134, 198] norsteroids,[199] cholesterol esters,[25, 198] cardenolides,[85, 99, 200] bufadienolides,[201, 202] triterpenes,[125] and glycosides of *Strophanthus,*[48, 203] *Digitalis,*[56, 204] *Oleander,*[85, 200] *Scilla,*[189, 200] and *Convallaria.*[200]

(i) Spray with a saturated solution of antimony trichloride in chloroform (20% w/v) and heat at 85° for 15 minutes (or at 100° for 8–10 minutes). The reaction gives several colors in daylight and UV light, with a sensitivity of 3–6 μg per spot with 32 estrogens[84] and 20 Δ^5-3β-hydroxysteroids.[70] It is less sensitive for 6-oxo estrogens and negative for 7-oxo-Δ^5-steroids. Allyl alcohols react immediately: blue (7-hydroxy-Δ^5-steroids[70] and 4β-hydroxy-Δ^5-steroids[116]), pink (3-hydroxy-Δ^4-C_{19}-steroids),[70] or grass-green (17α-ethynyl-3α,10α,17β-trihydroxyestr-4-ene).[199] Less-sensitive and nonspecific colors are given by Δ^4-3-oxo-C_{21}-steroids;[93] 20-dihydro compound S and compound E of Reichstein develop red-orange fluorescence with UV light. The change in color after 10 minutes of heating and after 24 hours at room temperature has been recorded for 38 norsteroids.[199]

The limit of detection depends on the steroid, and in some cases it is as low as 0.1 μg per spot (*Strophanthus* glycosides).[203] Some colors have a certain degree of specificity. Although most of the sterols give violet

[194] J. Richter, *J. Chromatog.* **18**, 164 (1965).
[195] F. M. Carr and E. A. Price, *Biochem. J.* **20**, 497 (1926).
[196] M. Barbier and S. I. Zavialov, *Izvest. Akad. Nauk SSSR Otd. Khim. Nauk,* p. 1309 (1960).
[197] H. Struck, *Mikrochim. Acta,* p. 634 (1961).
[198] M. J. D. Van Dam, G. J. de Kleuver, and J. G. de Heus, *J. Chromatog.* **4**, 26 (1960).
[199] T. Golab and D. S. Layne, *J. Chromatog.* **9**, 321 (1962).
[200] B. Görlich, *Arzneimittel-Forsch.* **15**, 493 (1965).
[201] R. Zelnik, L. M. Ziti, and C. V. Guimarães, *J. Chromatog.* **15**, 9 (1964).
[202] M. Schüpbach and C. Tamm, *Helv. Chim. Acta* **47**, 2217 (1964).
[203] G. L. Corona and M. Raiteri, *J. Chromatog.* **19**, 435 (1965).
[204] H. Lichti, M. Kuhn, and A. von Wartburg, *Helv. Chim. Acta* **45**, 868 (1962).

spots (8 minutes at 110°),[100] 7α-hydroxy-cholesterol is green and latho-sterol is yellow orange; a differentiation between 7-dehydrocholesterol (rose spot which changes to violet after heating) and lathosterol (brown after heating) is possible.[116] Bufadienolides[201] give yellow, orange, or pink colors after only 2 minutes of heating.

(ii) For detecting cardenolides, spray with a saturated solution of antimony trichloride in chloroform–acetic acid anhydride (5:1)[205] and heat at 130° for 5 minutes.

(iii) Spraying with a solution of acetic acid 1:1 (w/v) and heating for 5 minutes (95°) is employed to detect β-sitosterol[206] and steroid-oximes.[118] Also 50% antimony chloride solutions are used to detect cardenolides.[109]

(iv) A mixture of 50% antimony chloride solution in acetic anhydride and 50% sulfuric acid (1:2) has been used to detect brassicasterol and other C_{28}-sterols in bromine systems.[51]

b. *Antimony Trichloride–Thionyl Chloride.*[207] Plates are sprayed with a saturated solution of antimony trichloride in chloroform–thionyl chloride 10:1. It is reported[208] that more intense colors can be obtained than with antimony trichloride alone. This reaction has been used for several steroids, including Δ^4-3-oxosteroids and glycosides.[208]

c. *Antimony Pentachloride.* This is used in chloroform solutions (20–40%), especially for ketonic triterpenes,[40] but also for estrogens[197] and Δ^7- or Δ^8-cholestenediols.[101] Triterpenes[209] give generally a rose, red, or violet color; zeorininone reacts brown, and 22-deoxyzeorin, blue violet.

d. *Bismuth(III) Chloride.*[23] Spray with a 33% solution of the salt in 96% ethanol and heat at 90° for 10–15 minutes. The reaction has been used for sterols, which give gray (ergosterol and 7-dehydrocholesterol acetates), strong violet (acetates of cholesterol, stigmasterol, and β-sito-sterol), or yellow brown (vitamin D_2, Δ^7-cholestenol, and the acetates of lanosterol, 5-dihydroergosterol, and Δ^7-ergostenol). Dihydrocholesterol acetate does not react.

This reaction permits the differentiation of cholesterol and brassi-casterol. Both substances initially react with an orange-brown color, but cholesterol acetate changes to strong violet whereas brassicasterol acetate changes to a faint blue gray.

[205] J. Reichelt and J. Pitra, *Collection Czech. Chem. Commun.* **27**, 1709 (1962).
[206] K. Schreiber, G. Osske, and G. Sembdener, *Experientia* **17**, 463 (1961).
[207] J. Kučera, Ž. Procházka, and K. Vereš, *Chem. Listy* **51**, 97 (1957).
[208] S. Heřmánek, V. Schwarz, and Z. Čekan, *Collection Czech. Chem. Commun.* **26**, 1669 (1961).
[209] S. Huneck, *J. Chromatog.* **7**, 561 (1962).

e. Zinc Chloride.[171] Spray with a 30% methanolic salt solution and heat (60 minutes at 110°); after removal from the oven, the layer has to be protected against the humidity of the atmosphere with a second plate in order to avoid a change of the colors. The spots are observed under longwave UV. The reaction is suitable for the localization of C_{27}-, C_{21}-, and 16β-methyl-C_{21}-steroids,[171, 210] and the limits of detection are 0.1 μg. All steroidal saponins including the 3-deoxy derivatives react. In C_{21}-steroids, the 3-, 11-hydroxyl, and Δ^4-3-ketone groups are reactive.

f. Cadmium Chloride.[23] Spray the layer with a 50% solution of cadmium chloride in aqueous ethanol (50%) and heat for 15 minutes at 90°. Observe the fluorescence in UV light. This reaction is used to detect sterol bromides.

6. Iodine Reagent

a. Iodine. The vapor form is used by placing the plate in a closed atmosphere together with iodine crystals. Alternatively the layer may be sprayed with a saturated solution of iodine in hexane. Brown spots are developed by different classes of steroids: cholestane steroids,[25, 211, 212] triterpenes,[209] bufadienolides,[213] Δ^4-3-oxo-C_{19}- and Δ^4-3-oxo-C_{21}-steroids,[93] various C_{18}-, C_{19}-, and C_{21}-steroids.[214]

The detection is an adsorption phenomenon without decomposition of the steroid;[214] therefore the reaction is useful when subsequent chromatography, elution, or chemical reactions are carried out. It cannot be employed on silver nitrate-impregnated layers.[215]

b. Milius Reaction.[216] The plate is sprayed with a 0.3% iodine solution in a 0.5% aqueous potassium iodide solution.[217] The colors are observed, and the plate is sprayed again with ether.[218] This reaction was investigated for 37 Δ^4-3-oxo-C_{21}-steroids.[93] For most of these, the colors obtained were yellow or blue; blue spots were given by 11α-hydroxyprogesterone, 11α-epicorticosterone and cortisone (sensitivity 8–10 μg), 17α-hydroxyprogesterone (15 μg), and cortexone (75 μg). After the ether spray, in some cases the yellow color changes to blue or brown.

[210] P. J. Stevens, *Proc. Assoc. Clin. Biochemists* **2**, 156 (1963).
[211] C. Tamm, *Helv. Chim. Acta* **43**, 1700 (1960).
[212] O. Berséus, H. Danielsson, and A. Kallner, *J. Biol. Chem.* **240**, 2397 (1965).
[213] M. Schüpbach and C. Tamm, *Helv. Chim. Acta* **47**, 2226 (1964).
[214] J. S. Matthews, V. A. L. Pereda, and P. A. Aguilera, *J. Chromatog.* **9**, 331 (1962).
[215] L. J. Morris, *J. Chromatog.* **12**, 321 (1963).
[216] F. Mylius, *Z. Physiol. Chem.* **11**, 306 (1887).
[217] R. B. Burton, A. Zaffaroni, and E. H. Keutmann, *J. Biol. Chem.* **188**, 763 (1951).
[218] W. J. McAleer and M. A. Kozlowski, *Arch. Biochem. Biophys.* **66**, 125 (1957).

C. Reactions Involving the Presence of Oxo Groups

1. *Hydrazone Formation*

a. *Gornall–McDonald Reaction.*[219] This reaction employs 2,4-dinitrophenylhydrazine (2,4-DNP).

(i) Spray with 0.1% 2,4-DNP in a 10% ethanolic solution of HCl (v/v) and leave at room temperature. The reaction has been used for estrogens,[84, 220] Δ^4-3-oxo-C_{21}-steroids,[93] and etienic acid methylethers.[134] A yellow color is seen with ketonic estrogens and 20,21-ketolic steroids, orange-yellow with 20-oxo-C_{21}-steroids and 17α,21-dihydroxy-20-oxo-steroids, and orange-red colors with Δ^4-3-oxo groups. The sensitivity is 2 μg per spot. 2,3-Dihydroxy estrogens[84] give a yellow-brown color which turns to brown-violet after exposure to ammonia vapors.

(ii) Spray with a 5% (w/v) solution of 2,4-DNP in an 8:1 (v/v) methanol–sulfuric acid mixture (Brady's reagent) for the detection of ketonic sterols.[221]

(iii) A solution of 2,4-DNP in aqueous acetic acid was used to detect triterpenes with ketonic (e.g., friedelin) or aldo groups (e.g., diacetyl lupenol).[209]

b. *Isonicotinic Acid Hydrazide*[222] *(INH)*. Hydrazone formation with INH permits the differentiation of cross-conjugated dienones, heteroannular dienones and α,β-unsaturated ketones.[154, 223] The reaction can be carried out on starch–silica gel plates[115] and after partition TLC on formamide- or propyleneglycol-impregnated Celite layers.[50]

The spray consists of a 1% solution of INH in 1% ethanolic glacial acetic acid. Δ^4-3-Oxo-C_{21}-steroids[69, 156] and Δ^4-3-oxo-C_{17}-steroids[93, 154] react to give yellow spots (yellow fluorescence with UV light) after some minutes.

Greenish yellow spots appear immediately on reaction with $\Delta^{4,6}$-3-oxo-steroids; $\Delta^{1,4}$-3-oxosteroids and $\Delta^{1,4,6}$-3-oxosteroids react only after 16–24 hours or after heating at 30–60° for 30 minutes.[154]

c. *Porter and Silber Reaction*[224] *(Phenylhydrazine–H_2SO_4)*. Spray reagent: 65 ml of phenylhydrazine is dissolved in 10 ml of a sulfuric acid reagent (310 ml sulfuric acid with 190 ml water) and 50 ml of ethanol. Before use, dilute with water 1:1. Spray and leave at room temperature

[219] A. G. Gornall and M. P. McDonald, *J. Biol. Chem.* **201**, 279 (1953).

[220] E. Diczfalusy, C. Franksson, B. P. Lisboa, and B. Martinsen, *Acta Endocrinol.* **40**, 537 (1962).

[221] P. D. G. Dean and M. W. Whitehouse, *Biochem. J.* **98**, 410 (1966).

[222] A. Ercoli, L. Giuseppe, and P. de Ruggieri, *Farm. Sci. Tec. (Pavia)* **7**, 170 (1952).

[223] L. L. Smith and T. Foell, *Anal. Chem.* **31**, 102 (1959).

[224] R. H. Silber and C. C. Porter, *J. Biol. Chem.* **210**, 923 (1954).

for 1 hour or heat for some minutes at 60°. For 17α,21-dihydroxy-20-oxosteroids, the sensitivity is as high as 1 μg per spot; 16α-substituted 17α,21-dihydroxy-20-oxosteroids do not react.

This reaction is positive[93] for 21-aldo-20-ketosteroids, 21-hydroxy-20-oxo-16-dehydrosteroids and 16,21-dihydroxy-20-oxosteroids (after some hours). Other ketosteroids react weakly, and only after prolonged heating, probably by the formation of a 3-phenylhydrazone.

d. *Lewbart and Mattox Reaction.*[225] This reaction is used to oxidize 17-deoxy-α-ketolic steroids to the corresponding glyoxal. 20,21-Glyoxal steroids give an almost instantaneous Porter–Silber reaction. Cortico-steroids with 17-deoxy-α-ketolic structures[93] are sprayed with a 0.01 M methanolic cupric acetate solution, left at room temperature overnight, and subsequently submitted to the Porter-Silber reaction as described above.

2. Reactions for α,β-Ketols, α,β-Diketones, and α,β-Unsaturated Ketones

a. *Schwartz*[226]*–Pan*[227] *Reaction for Formaldehydogenic Steroids.* Spray[93] with a 1% solution of potassium periodate in ethanol (70% v/v), and leave at room temperature. Before it has dried completely (after 10 minutes), spray again with a methanolic solution containing 15% ammonium acetate, 1% acetic acid, and 1% 2,4-pentanedione. After 15–20 minutes at room temperature observe the formaldehydogenic steroids in daylight (yellow spots) and in UV light (yellow green spots). The sensitivity is 1–2 μg per spot.

b. *Blue Tetrazolium (BT)* (*3,3′-Dimethoxy-4,4′-biphenylene)-bis-2,5-diphenyl-2H-tetrazolium Chloride).* BT is used for the detection of α-ketolic groups in Δ⁴-3-oxo- and tetrahydrocorticosteroids on silica gel layers[228-230] and on polyamide layers.[231]

(i) Spray[228] the plates with a mixture of 13 ml of methanol, 2 ml of a 0.5% methanolic BT solution, 2 ml of concentrated ammonium hydroxide and 3 ml of 6 N NaOH and leave for 1–2 hours in darkness. When the surface layer is only lightly sprayed it has been reported that the visu-alized spots can be scraped for subsequent steroid determination using a BT reaction.[228]

(ii) Mix before use equal parts of a 0.2% methanolic BT and a 10%

[225] M. L. Lewbart and V. R. Mattox, *Anal. Chem.* **33,** 559 (1961).
[226] D. P. Schwartz, *Anal. Chem.* **30,** 1855 (1958).
[227] S. C. Pan, *J. Chromatog.* **9,** 81 (1962).
[228] O. Nishikaze, R. Abraham, and H. Staudinger, *J. Biochem. (Tokyo)* **54,** 427 (1963).
[229] O. Adamec, J. Matis, and M. Galvánek, *Lancet* **I,** 81 (1962).
[230] O. Nishikase and H. Staudinger, *Klin. Wochschr.* **40,** 1014 (1962).
[231] O. Freimuth, B. Zawta, and M. Büchner, *Acta Biol. Med. Ger.* **13,** 624 (1964).

sodium hydroxide solution (in 60% aqueous methanol).[231] On polyamide layers, the sensitivity is 1 $\mu g/cm^2$.

    c. *2,3,5-Triphenyltetrazolium Chloride (TTZ)*. TTZ is used to detect α,β-unsaturated ketones and α,β-ketols[50, 93, 146,] on silica gel layers,[93, 146] on formamide- or propylene glycol-impregnated Celite layers,[50] and on polyamide layers.[231] (i) Reagent:[232] Mix before use 10 ml of concentrated potassium hydroxide (10 g/10 ml water) and 100 ml of 0.5% TTZ in ethanol; spray the plate and leave at room temperature for 5 minutes. α,β-Ketolic steroids form red formazans during this period. For detection of Δ^4-3-oxosteroids, heat the plate at 80° until the spots are visible.

    (ii) Also used are mixtures (1:1) of a 4% solution of TTZ (methanol) and 4% (methanolic)[146] or 10% (in 60% aqueous methanol)[231] sodium hydroxide solution.

    d. *2,5-Diphenyl-3(4-styrylphenyl)tetrazolium Chloride*[210] *(TPTZ)*. α,β-Ketols react with strong purple colors. Steroids with a Δ^4-3-oxo group alone are negative.

    (i) Reagent for aluminum oxide layers:[171] Mix and spray immediately 1 volume of 1% TPTZ in methanol and 10 volumes of a 3% solution of sodium hydroxide in water. Observe the color in daylight.

    (ii) Reagent for silica gel G layers: Use an 8% solution of sodium hydroxide (2 N NaOH) instead of 3%.

    e. *Potassium Ferricyanide Reaction.*[233] Spray first with a 0.1% solution of potassium ferricyanide in 0.25% sodium carbonate and heat for 30 minutes at 80°; spray again with a mixture of 100 ml of 0.2% solution of ferric ammonium sulfate and 5 ml of concentrated phosphoric acid (85%).

    A Prussian blue color immediately appears with Δ^4-3-oxo-C_{21}-steroids[93] (2–5 μg per spot). 17,21-Dihydroxy-20-ketones and 20,21-ketols react more strongly than other corticosteroids, and the reaction is more sensitive for 17-hydroxy-C_{21}-steroids than for 17-deoxy-C_{21}-steroids.

    f. *Tollens Reagent*. This is used for α,β-ketolic steroids[93] (sensitivity on silica gel layers 10–15 μg per spot). Spray reagent:[234] mix just before use 10 ml of 0.1 N silver nitrate, 10 drops of concentrated ammonium hydroxide, and 5 ml of 10% sodium hydroxide.

    g. *Bodánsky–Kollonitsch Reaction.*[235] This reaction is useful for α,β-unsaturated oxosteroids.[69, 93, 154] Spray with water-saturated n-butanol

[232] G. M. Shull, J. L. Sardinas, and R. C. Nubel, *Arch. Biochem. Biophys.* **37**, 186 (1952).

[233] N. R. Stephenson, *Can. J. Biochem. Physiol.* **37**, 391 (1959).

[234] A. Zaffaroni, R. B. Burton, and E. H. Keutmann, *Science* **111**, 6 (1950).

[235] A. Bodánsky and J. Kollonitsch, *Nature* **175**, 729 (1965).

containing 0.9% p-phenylenediamine and 1.6% phthalic acid and heat at 100–110° for 5–10 minutes.

Δ^4-3-Oxo- and Δ^1-3-oxosteroids show a yellow-brown to yellow-olive color; heteroannular conjugated dienones give orange to orange-brown spots, and cross-conjugated structures (as in $\Delta^{1,4}$-3-oxosteroids) are negative.

h. Alkaline Potassium Permanganate Reaction.[217] Spray with a 0.2% solution of potassium permanganate in 5% sodium carbonate and observe the colors immediately and after waiting overnight.

Δ^4-3-Oxosteroids[93] give immediate olive-yellow spots on a violet background; these turn overnight to olive on a pale yellow background. Sensitivity: 2 μg per spot.

i. p-Aminodiethylaniline Sulfur Dioxide. This is used for 16-oxo-estrone,[84] ketolic estrogens,[84] and $17\alpha,21$-dihydroxy-20-oxo-C_{21}-steroids.[93] Spray with a 0.5% (w/v) solution in a 5% aqueous solution of sodium bicarbonate (w/v). 16-Oxoestrone reacts immediately (sensitivity: 5 μg per spot). Ketolic estrogens and α,β-unsaturated steroids develop orange spots overnight at room temperature. $17\alpha,21$-Dihydroxy-20-oxosteroids (2–5 μg) give red-orange, and $16\alpha,17\alpha,21$-trihydroxy-20-oxosteroids give orange or yellow-orange, spots.

j. Ishidate Reaction.[236] The reaction is used for 16-oxoestrone[84] and some neutral steroids.[154] Spray with a 20% (w/v) hydroxylamine hydrochloride solution in a 20% (w/v) sodium acetate solution and heat at 90° for 15 minutes; spray again with a 5% (w/v) solution of cupric acetate.

16-Oxoestrone (5 μg per spot) reacts to give a yellow-orange color; 4-hydroxy-Δ^4-3-oxo-steroids and 2α-hydroxytestosterone give yellow-olive spots on a blue background.

3. Reagents for Methyl Ketones and Ketones with Unsubstituted α-Methylene Groups

a. Nitroprusside Reaction.[237] The reaction is used for 20-oxo-21-deoxy-C_{21}-steroids.[93] Methanolic nitroprusside paste reagent,[238] consists of 6 parts of sodium nitroprusside, 100 parts of anhydrous sodium carbonate, and 100 parts of ammonium acetate mixed with methanol to form a paste. The plate is covered with a filter paper, and the paper is then covered with a thin layer of the paste. The spots are observed through the uncovered surface of the plate. After 60 minutes at room temperature,

[236] M. Ishidate, *Mikrochim. Acta* 3, 284 (1938).
[237] F. Feigl, "Spot Tests in Organic Analysis," 5th ed. Elsevier, Amsterdam, 1956.
[238] S. C. Pan, *J. Chromatog.* 8, 449 (1962).

20-oxo-21-deoxysteroids give a bright violet spot. Δ^{16}-Progesterone does not react.

   b. *Zimmermann Reaction.*[239] The reaction is useful for ketonic groups with an α-unsubstituted methylene group. Spray[84] with a mixture (1:1), prepared just before use, of 2% ethanolic *m*-dinitrobenzene and 1.25 N ethanolic potassium hydroxide, and dry with a stream of hot air. The reaction has been used to detect 3- and 17-oxo-C_{19}-steroids,[70, 153, 154] 17- and 16-oxo estrogens,[84] and corticosteroids.[93] It can be applied to rice starch–silica gel plates.[115] It is employed for butenolides (as the Raymond reagent). It is useful for the differentiation of 16-ketolic estrogens: 16-ketoestradiol is positive, whereas 16α- and 16β-hydroxyestrone are negative: 16-oxoestrone is also positive. 3,20-Dioxo-21-deoxy-Δ^4-C_{21}-steroids give a blue-gray color; C_{21}-steroids with ketolic and dihydroxy-ketonic side chains are pink.[93] 4-Hydroxy-Δ^4-3-oxosteroids develop a red-brown or lilac color.[154]

   This reaction can be used for alcoholic steroids, if they are previously converted *in situ* to ketosteroids.[238, 240] Spray[153] first with a 0.25% chromic acid anhydride solution in glacial acetic acid, heat for 15 minutes at 90–95°, and submit the plate to the Zimmermann reaction as described before.

4. Reactions for α,β-Unsaturated Ring Lactones

   a. *Kedde Reagent.*[241]

   (i) Mix before use[242] a 2% methanolic solution of 3,5-dinitrobenzoic acid and 2 N aqueous potassium hydroxide. This has been used to detect *Strophanthus* glycosides,[48] *Digitalis* glycosides,[21] cardenolides,[78, 95] and butenolides.[78] *Strophanthus* glycosides[48] give yellow spots on a violet background (0.3 μg per spot). Steroids with a butenolide ring react with a blue color (5 μg per spot).

   (ii) Use a 5.7% methanolic potassium hydroxide solution[78] instead of the aqueous solution used in (i). This reagent is more sensitive than the preceding one, gives spots with more stable colors, and can be used also for benzoyl derivatives of cardenolides.

   The Kedde reaction can be used after *p*-toluenesulfonic acid[78] or vanillin–perchloric acid spray.[78] This becomes useful when the latter reagent is utilized to identify glucosides with a 2-deoxy sugar.

   b. *Lichti Reaction.*[243] Spray[95] with a saturated solution of 2,4,2',4'-

[239] W. Zimmermann, *Z. Physiol. Chem.* **233,** 257 (1935).
[240] D. Kupfer, E. Forchielli, M. Stylianou, and R. I. Dorfman, *J. Chromatog.* **4,** 449 (1962).
[241] D. L. Kedde, Inaugural Dissertation, Univ. of Leyden, 1946.
[242] E. Schenker, A. Hunger, and T. Reichstein, *Helv. Chim. Acta* **37,** 680 (1954).

tetranitrodiphenyl in benzene (ca. 5%); dry 10 minutes at room temperature and spray again with a 10% methanolic potassium hydroxide solution. The method is useful for the detection of cardenolides[95] and C_{18} neutral steroids.[188]

c. Raymond Reagent.[244] The reagent is used for cardenolide glycosides; it is an alkaline *m*-dinitrobenzene reagent[244, 245] (see Zimmermann Reaction).

D. Reactions for Phenolic Compounds

a. Folin-Ciocalteau Reagent.[142] The reagent is obtainable commercially (Merck AG, Darmstadt). Spray the plates, after they have been exposed to an ammonia-saturated atmosphere, with Folin-Ciocalteau reagent diluted with water (1:4).[83] Estrogens develop a blue color.

b. Diazo Reaction.[246, 247] The reaction is carried out with stabilized diazo salts: (i) fast black salt K (4-amino-2,5-dimethoxy-4'-nitroazobenzene); (ii) fast blue VB-salt (4-amino-4'-methoxydiphenylamine); and (iii) fast red salt GG (*p*-nitroaniline).

First spray the plate with a 0.05% aqueous solution of the stabilized salt and leave at room temperature for 30 minutes (or heat for 5 minutes at 80°), then spray with a 15% aqueous sodium carbonate solution. Estrogens[84] (3–5 μg per spot) develop a purple spot on a pale lilac background (i), an orange color on pink background (ii), or a red-brown color on a white background (iii).

c. Boute Reaction.[248] The reaction is employed for estrogens and their Girard's hydrazones[83] (yellow color); 2-hydroxy- and 2-methoxy derivatives give a buff color.[84] Expose the plate to ammonia vapors, followed by nitrogen dioxide vapors obtained by the reaction between metallic copper and nitric acid.

d. Ilinski–von Knorre Reaction.[249] This is used for estrogens[84] (sensitivity 4–5 μg per spot; yellow-brown color on lilac background). Spray with a 5% solution of sodium cobalt nitrate in 0.2 *N* acetic acid[250] and dry by heating at 100°. The reaction is specific for phenolic groups with both *ortho* positions unsubstituted (capable of forming 1-nitroso-2-hydroxyphenols).

[243] H. Lichti, personal communication to R. Mauli, C. Tamm, and T. Reichstein, *Helv. Chim. Acta* **40**, 284 (1957).
[244] W. D. Raymond, *Analyst* **63**, 478 (1938).
[245] R. Marthoud, Inaugural Dissertation, Univ. of Lyon, 1935.
[246] W. Marx and H. Sobotka, *J. Biol. Chem.* **124**, 693 (1938).
[247] I. A. Pearl and P. F. McCoy, *Anal. Chem.* **32**, 1407 (1960).
[248] J. Boute, *Ann. Endocrinol. (Paris)* **14**, 518 (1953).
[249] M. Ilinski and G. von Knorre, *Ber. Deut. Chem. Ges.* **18**, 699 (1889).
[250] F. Feigl, *Anal. Chem.* **27**, 1315 (1955).

e. Phloroglucinol–Sodium Hydroxide.[251] Spray with a 0.1% solution of phloroglucinol in 1 N sodium hydroxide and dry at 80°. The reagent is useful for the detection of 2,3-dihydroxyestrogens:[84] 7-oxoestrone also reacts (orange brown to lilac brown).

f. Ferricyanide–Ferric Chloride.[252] Spray with a mixture of equal parts of 1% aqueous ferric chloride and 1% aqueous potassium ferricyanide.[253] Estrogens[84] give a Turnbull blue color (1–2 μg per spot), and 2,3-dihydroxyestrogens give a blue-violet spots (0.2 μg per spot). The reaction is less sensitive (10 μg per spot) for α-ketolic and α,β-unsaturated ketosteroids (blue spots). It can be used on rice starch-silica gel layers.[115]

E. Reactions for Unsaturated Bonds

a. Ostromisslensky[254]*–Werner*[255] *Reaction.* This involves complex formation with tetranitromethane. Spray with 5% (v/v) tetranitromethane in ethyl acetate and expose overnight in a sealed tank to an atmosphere saturated with tetranitromethane vapors. Estrogens[84] give an olive-yellow to yellow-orange color (5–8 μg per spot); α,β-unsaturated steroids[84, 93] give a pale lemon (10–15 μg per spot). Estrogen 3-methyl ethers remain colorless.

b. Tortelli–Jaffé[256] *Reaction.* This is useful for characterizing ditertiary double bonds at C_{13-14}, C_{8-9} or C_{5-10}. Spray[257] with glacial acetic acid–chloroform 1:1, then with a 5% bromine solution in chloroform (v/v). Estrogens[83] give a yellow-green color (15 μg per spot).

c. Osmium Tetroxide. This is useful for steroids with isolated double bonds.[84, 93, 96, 153] The TL plate is exposed to vapors in well-sealed tanks. Steroids with isolated double bonds give a brown-violet to violet-gray color after 5–10 minutes; 2,3-dihydroxyestrogens[84] react immediately.

F. Other Visualization Procedures

a. Dimethylphenylenediamine Reagent.[258] Spray with a 1% solution of N,N-dimethyl-p-phenylenediammonium chloride in 50% methanol, to which 1 ml of acetic acid is added. 7-Hydroxycholesterol develops a purple-red color.

b. Sulfuric Acid–Hypochlorite Solution.[117] Mix 3 ml of Javel water

[251] E. Egrive, *Z. Anal. Chem.* **125**, 241 (1943).
[252] P. Brouardel and E. Boutmy, *Compt. Rend. Acad. Sci.* **92**, 1056 (1881).
[253] G. M. Barton, R. S. Evans, and J. A. F. Gardner, *Nature* **170**, 249 (1952).
[254] I. Ostromisslensky, *J. Russ. Phys. Chem. Ges.* **12**, 731 (1909).
[255] A. Werner, *Ber. Deut. Chem. Ges.* **72**, 4324 (1909).
[256] M. Tortelli and E. Jaffé, *Chem. Ztg.* **39**, 14 (1915).
[257] L. R. Axelrod and J. E. Pulliam, *Anal. Chem.* **32**, 1200 (1960).
[258] L. Acker and H. Greve, *Fette Seifen Anstrichmittel* **65**, 1009 (1963).

(sodium hypochlorite solution with ca. 10% active chlorine) and 10 ml of sulfuric acid. Spray the layer and heat at 125° for 10–15 minutes. Cardenolide glycosides give highly sensitive colors: glycosides of the digitoxigenin, gitoxigenin, and gitaloxigenin series develop rose to brown spots; those of the digoxigenin series, green or green-blue spots; and those of the diginatigenin series, gray-blue spots. The sensitivity is down to 0.002 μg per spot for digoxin and 0.04 for the other *Digitalis* glycosides.

    c. Trichloroacetic Acid–Sodium Sulfaminochloride Reaction.[109] Mix just before use 15 parts of a 25% ethanolic trichloroacetic acid solution and 1 part of freshly prepared 3% aqueous sodium sulfaminochloride. After spraying, heat the plates for 5 minutes at 110–120°. Specific fluorescence in UV light is developed for the different *Digitalis* cardenolides: yellow to brown-red for digitoxigenin, light blue for gitoxigenin and its glycosides, and steel blue for digoxin.

    d. Trichloroacetic Acid–Chloramine Reaction.[259] Mix just before use[260] 15 parts of a 25% ethanolic trichloroacetic acid solution and 1 part of a 3% chloramine solution. The plates are sprayed with the reagent and briefly heated. Used for the detection of *Strophanthus* glycosides,[48, 56] *Digitalis* glycosides,[56] and cardenolides.[99]

    e. Phosphoric Acid–Bromine.[117] The plates are sprayed with a 10% phosphoric acid solution, heated at 125° for 12 minutes, then sprayed again, before cooling, with a mixture containing 2 ml of a saturated aqueous solution of potassium bromide, 2 ml of a saturated aqueous solution of potassium bromate, and 2 ml of concentrated hydrochloric acid (25%). The fluorescence is observed in UV light as developed for *Digitalis* glycosides. This reaction allows the differentiation of cardenolide glycosides of the digitoxigenin series (orange) from those of gitoxigenin, digoxigenin and diginatigenin series (blue). The sensitivity is high, between 0.001 (genins B and C) and 0.01 μg per spot (lanatoside A).

    f. Rhodamines. Rhodamine 6G and rhodamine B are used for the detection of various steroids (e.g., cholesterol and esters[96]) and terpenes,[40, 209] principally for substances without reactive groups, such as diacetylbetulin.[209] The plates are sprayed with a 0.05% aqueous solution of rhodamine and observed under UV light.

    g. Bromothymol Blue. Spray with a 0.04% solution of bromothymol blue in 0.01 N sodium hydroxide. This is used for the detection of cholesterol,[96] cholesterol esters,[96] and corticosteroids.[261]

    h. Water. Spray the plate with water and observe the surface in

[259] K. B. Jensen, *Acta Pharmacol. Toxicol.* **9**, 99 (1953).
[260] F. Kaiser, *Chem. Ber.* **88**, 556 (1955).
[261] R. D. Bennett and E. Heftmann, *J. Chromatog.* **9**, 348 (1962).

reflected light: lipophilic substances appear as opaque spots. Water is used principally for the detection of bile acids,[47, 133] but also for bufadienolides[201] and pregnanediols.

i. Protoporphyrin and Silicotungstic Acid. These acids have been reported as reagents for sterols on TLC.[23]

G. Methods for Radioactive Steroids

1. *Autoradiograms*

After fractionation of radioactive substances by TLC, the resolved spots can be visualized by direct exposure of X-ray films (e.g., No Screen Medical X-Ray Safety Film, Eastman Kodak Corp., Rochester, New York) during a period depending on the labeled sample examined (nature of radiation, activity of the labeled substance and effect desired).

For samples labeled with higher energy β-emitters as ^{14}C an exposure of 1 to 8 days is necessary; the blackening of the photographic emulsion requires an exposure to 1 to 10 million β-particles per cm².[261a] A method using a direct exposure under the same conditions is not suitable for tritiated compounds. While the range of the β-particle emitted by ^{3}H-compounds in silica gel is less than 13 μ[261b] on conventional layers only 5% of the radiation contacts the X-ray film; furthermore, the β-radiation range in the film is 1 μ, whereas the distance between the silver halide crystals in the emulsion is 1 to 12 μ[261c] Luthi and Waser[261b, 261d] have modified the autoradiographic technique by conversion of a β-radiation effect to a fluorescence effect (fluorography) in order to detect ^{3}H-labeled compounds. By adding a scintillator such as anthracene to the silica gel layer, the β-radiation-induced fluorescence led to a blackening effect 100 times that of the nontreated thin layers. As little as 0.1 μC ^{3}H per spot can be detected after 1 day's exposure. Lower temperature will enhance this effect (thirty times more intense at $-70°$ than at $+4°$).[261b] Furthermore, the mobilities of several steroids and bile acids on anthracene-silica gel layers did not differ from those on untreated adsorbents.

2. *Evaluation by Means of Counters*

This can be achieved on the layer by direct counting with a gas-flow counter (strip scanner), or after the substances have been scraped off. In

[261a] R. H. Herz, *Nucleonics* **9**, 24 (1951), quoted by H. K. Mangold, *in* "Thin-layer Chromatography" (E. Stahl, ed.), p. 58. Academic Press, New York, 1965.
[261b] U. Luthi and P. G. Waser, *in* "Advances in Tracer Methodology" (S. Rothchild, ed.), Vol. III, p. 149. Plenum Press, New York, 1966.
[261c] J. H. Handloser, *in* "Advances in Tracer Methodology" (S. Rothchild, ed.), Vol. I, p. 201. Plenum Press, New York, 1963.
[261d] U. Luthi and P. G. Waser, *Nature* **205**, 1190 (1965).

the first case the plates can be automatically measured by running the layer continuously through a gas-flow counter (e.g. Dünnschicht-Scanner II LB 2722, Dr. Berthold Lab., Wildbad, Germany, for chromatoplates up to 20×40 cm; Packard Model 7201 Radiochromatogram Scanner, Packard Instr. Co. Inc., Downers Grove, Illinois, for 5×10 cm chromatoplates); for each chromatogram an adequate slit width and rate of scanning should be chosen. With Dr. Berthold-Scanner II it is possible to detect a spot with an activity of only 10^{-3} μC ^{14}C or 10^{-2} μC ^{3}H (yield: 15–30% for ^{14}C, 0.7–3% ^{3}H, depending on the thickness of the adsorbent). The sensitivity of this procedure makes it inadequate for measuring tritium-labeled compounds of low activity.

Measurement with Scintillation Counting Apparatus. After detection of the spots by means of nondestructive procedures such as the use of iodine, or by the location of standards chromatographed simultaneously at both sides of the radioactive material, the strips containing the radioactive substance are uniformly scraped (0.125 or 0.250 cm-wide bands) and directly mixed with the scintillation liquids, without previous extraction.[262] During the spraying of the simultaneously chromatographed standards, the region to be counted must be carefully protected to avoid both chemical and color quenching;[262a] thus some acid-containing reagents give strong quenching, even in small amounts. As scintillation medium for direct counting a toluene–ethanol solution (98:2) containing 0.4% of 2,5-diphenyloxazole (PPO) and 0.005% of 1,4-bis-2-(4-methyl-5-phenyl-oxazolyl)benzene (POPOP) is used.

Rivlin and Wilson[263] have developed a procedure to recover radioactive material from the liquid scintillation medium. The sample containing the radioactive steroid and PPO solution is twice chromatographed in silica gel G with ethyl acetate as developer: the large amount of PPO occupies a 2-cm wide band at the farther side of the layer, completely separated from the small amounts of radioactive material. The mobility of the steroid is not appreciably changed in spite of the large amount of PPO.

IV. Formation of Derivatives

The formation of steroid derivatives is investigated for various reasons: in order to improve the separation of substances with similar chromatographic mobilities and separate classes of compounds with different chemical functions, or to characterize an already isolated steroid.

[262] J.-Å. Gustafsson and B. P. Lisboa, *Steroids* 11, 555 (1968).
[262a] C. T. Peng, *in* "Advances in Tracer Methodology" (S. Rothchild, ed.), Vol. III. p. 81. Plenum Press, New York, 1966.
[263] R. S. Rivlin and H. Wilson, *Anal. Biochem.* 5, 267 (1963).

The preparation of steroid derivatives can be achieved before their application to the adsorbent layer or on the plate. In the latter case, formation of the derivative takes place *in situ*, either before the development on the starting point, or after the development. In some cases the derivative is formed during the run, with substances added to the solvent system, as in the so-called "bromine system."

A. Preparation of Derivatives before Their Submission to TLC

The derivatives may be prepared in stoppered test tubes, then extracted with a suitable organic solvent and spotted on the plate, or they may be prepared in capillary tubes;[264, 265] in the latter case, the derivative is applied on the starting point with the same capillary in which it has been prepared. The most usual derivatives and reactions follow.

a. Acetates. An amount of 1–20 μg of steroid is mixed in a ground-glass stoppered tube with pyridine and acetic acid anhydride, 1:1 (0.2 to 0.5 ml of each), under nitrogen atmosphere and left overnight. After evaporation the acetates are directly spotted on the layer. Under these mild conditions 11β-hydroxyl groups (in steroids with angular methyl group at C-10) and tertiary hydroxyl groups are not acetylated. This allows the separation of steroid pairs of similar polarities after acetylation, when one of them contains a hindered hydroxyl group, e.g., 20β-hydroxypregn-4-en-3-one/17α-hydroxyprogesterone.[63]

Generally the formation of acetates does not serve to separate isomers. However, several isomeric sapogenin acetates and cardenolide acetates show on silica gel G different mobilities that are of practical use. The following $R_f \times 100$ values have been found for tigogenin/neotigogenin (64/57), gitogenin/25β-gitogenin (49/44), and digalogenin/25β-digalogenin (46/40) in the system chloroform–acetone (96:4);[266] 25α-digitogenin/25β-digitogenin (59/52) in the system chloroform–acetone (92:8);[266] and for tigogenin/smilagenin/neotigogenin (32/26/23) in the system chloroform–toluene (9:1).[267] Also in the cardenolide series Tschesche *et al.*[122] have achieved the separation of acetate isomers, including those of uzarigenin, allouzarigenin, and 3-epiuzarigenin (see under cardenolides).

b. Propionates.[97] Add to the dry steroid extract 0.5 ml of propionyl chloride and heat several minutes to dissolve it. After 10 minutes the propionate can be extracted with hexane; wash successively with water, 10% sodium bicarbonate, and water again, and apply to the layers.

[264] C. Mathis and G. Ourisson, *J. Chromatog.* **12**, 94 (1963).
[265] C. Mathis, *Ann. Pharm. Franc.* **23**, 331 (1965).
[266] R. Tschesche, G. Wulff, and G. Balle, *Tetrahedron* **18**, 959 (1962).
[267] R. D. Bennett and E. Heftmann, *J. Chromatog.* **9**, 353 (1962).

Sterol propionates present greater chromatographic mobilities and resolution on silver nitrate-impregnated silica gel layers than the corresponding acetates, if the system hexane-benzene 5:1 is used.

c. Benzoates.[268] Dissolve the steroid in 0.2 ml of dry pyridine and cool on ice; add 0.1 ml of cold benzyl chloride and leave at room temperature overnight. Add 2 ml of ice water and after 4 hours extract with methylene chloride. The organic phase is washed twice with 3 N hydrochloric acid, 1 N sodium carbonate, and water and then spotted on the layers. Benzoates are much less polar than acetates: in the system cyclohexane–benzene 4:1 on silica gel G, the acetate and benzoate esters of cholesterol have R_f values of 0.17 and 0.44, respectively.[198]

d. Trifluoroacetates.[267] Dissolve the steroid (20 μg) in 0.2 ml of hexane and add 0.002 ml of trifluoroacetic acid anhydride. Shake for 1 minute, then add 1 ml of 2 N aqueous sodium carbonate and shake again; the organic phase can be transferred to the plates. Separation of sapogenin trifluoroacetates on silica gel G is possible by the use of chloroform-toluene 9:1; noteworthy is the separation obtained for the trifluoroacetates of tigogenin (5α,25α-spirostan-3β-ol) and neotigogenin (5α,25β-spirostan-3β-ol) or sarsasapogenin (5β,25β-spirostan-3β-ol) and smilagenin (5β,25α-spirostan-3β-ol) in this system.

e. Trimethylsilyl Ethers[269] (*TMeSi-ethers*). Dissolve the steroid on 1 ml of anhydrous pyridine and add 0.1 ml of hexamethyldisilazane and 0.03 ml of trimethylchlorosilane; the reaction is complete after 10 minutes. For steric reasons the reaction for 11β-hydroxyl groups must be carried out overnight at room temperature; under these conditions a 17α-hydroxyl group in the C_{21} series remains unreacted.[270] Partial ether formation with bile acids[271] for 3α,3β,6α(equatorial), 7β(equatorial), and 12β (equatorial) hydroxyl groups: heat the steroid for 3 hours at 50° with 0.1 ml of hexamethyldisilazane in dry acetone (1:2) or in dimethyl formamide (1:2).

TLC of trimethylsilyl ethers of steroids and sterols has been investigated by Luukkainen and Adlercreutz,[272] Brooks and Carrie,[273] Brooks *et al.*,[270] and Lindgren and Svahn.[274]

Brooks *et al.*[270] have obtained the separation of androsterone, etio-

[268] R. V. Brooks, W. Kline, and E. Miller, *Biochim. J.* **54**, 212 (1953).

[269] M. Makita and W. W. Wells, *Anal. Biochem.* **5**, 523 (1963).

[270] C. J. W. Brooks, E. Chambaz, and E. C. Horning, *Anal. Biochem.* **19**, 234 (1967).

[271] J. Sjövall, *in* "Biomedical Applications of Gas Chromatography" (H. A. Szymanski, ed.), p. 151. Plenum Press, New York, 1964.

[272] T. Luukkainen and H. Adlercreutz, *Biochim. Biophys. Acta* **107**, 579 (1965).

[273] C. J. W. Brooks and J. G. Carrie, *Biochem. J.* **99**, 47P (1966).

[274] B. O. Lindgren and C. M. Svahn, *Acta Chem. Scand.* **20**, 1763 (1966).

cholanolone, and dehydroepiandrosterone TMeSi-ethers on silica gel–10% calcium sulfate layers (Adsorbosil-3, Applied Science Laboratories, State College, Pennsylvania) after development with cyclohexane–ethyl acetate 9:1 ($R_f \times 100$ values are respectively, 49, 45, and 41) or with benzene–chloroform 8:2 ($R_f \times 100$ values: 32, 16, and 29).

f. *Isonicotinic Acid Hydrazones.*[69] Dissolve 1–10 μg of steroid in 0.5 ml of a 0.1% isonicotinic acid hydrazide (INH) solution in a 1% ethanolic glacial acetic acid solution and leave for 120 minutes at 37° under nitrogen atmosphere. After evaporation the hydrazones are submitted to TLC, benzene–methanol 75:25 being used as the developing system.

INH reacts much more slowly with unsaturated 16-dehydro-20-ketones than with unconjugated ketones or Δ^4-3-ketones; this different rate of hydrazone formation can be used to characterize 16-dehydroprogesterone or 16-dehydropregnenolone in the presence of progesterone or pregnenolone.[156]

g. *Girard's Reagent T-Hydrazones.* The steroid[83] is dissolved in 2 ml of a 0.1% Girard's reagent T (trimethylacetohydrazide) ammonium chloride in a 10% (v/v) methanolic glacial acetic acid solution, and left for 18 hours at 37° under nitrogen atmosphere. After evaporation the hydrazones are submitted to TLC, commonly in the solvent system developed for this purpose by Zaffaroni *et al.*,[275] which consists of the organic phase obtained after the separation of a mixture of *n*-butanol, *tert*-butanol, and water (1:1:1).

Steroids with a cross-conjugated trienone or dienone, such as 17β-hydroxyandrosta-1,4,6-trien-3-one and 17β-hydroxyandrosta-1,4-dien-3-one do not react with Girard's reagent, in contrast to those steroids with an unsaturated ketone (such as testosterone) or heteroannular dienone (such as 6-dehydrotestosterone). By formation of Girard hydrazones, these two groups of closely related steroids can be separated easily.[154]

h. *2,4-Dinitrophenylhydrazones (DNPH).*[276] Add to 0.1 μmole of the steroid 0.05 ml of a 0.2% ethyl acetate solution of 2,4-dinitrophenylhydrazine and evaporate until dry at 40° under nitrogen; add 1 ml of a 0.03% solution of trichloroacetic acid in benzene, and after one-half hour evaporate again. The dry hydrazones are directly spotted onto the layer and developed using the systems given in the Table XXIX (see under C_{19}-steroids); the data presented in this table show that several monohydroxymonoketonic C_{19}-steroids present different chromatographic mobilities as DNPH-derivatives.

i. *Reduction with Potassium Borohydride.* Steroid, 1–20 μg, is in-

[275] A. Zaffaroni, R. B. Burton, and E. H. Keutmann, *J. Biol. Chem.* **177**, 109 (1949).
[276] L. Treiber and G. W. Oertel, *Z. Klin. Chem. Klin. Biochem.* **5**, 83 (1967).

cubated in aqueous methanol with 5 mg/ml potassium borohydride for 1 hour. The methanol is evaporated and the water is made up to 5 ml, treated with some drops of hydrochloric acid to destroy the unreacted borohydride, and extracted with ether. The organic phase containing the reduced steroids is evaporated, redissolved in small amounts of methanol, and submitted to TLC.

The identification of 16α-hydroxyestrone in mixtures with 16β-hydroxyestrone and 16-oxo-17β-estradiol can be accomplished after the treatment of the mixture by sodium borohydride: 16α-hydroxyestrone gives estriol after reduction whereas the two other isomers mainly give 16-epiestriol, completely separable in several chromatographic systems.[83]

j. Oxidation with Chromium Trioxide. To 5–10 μg of dry steroid, add 0.5 ml of a 0.25% solution of chromic acid anhydride in glacial acetic acid. After 2–4 hours at room temperature, add 2 ml of water and extract twice with ether; wash with water, evaporate, and apply to the layer.

The A/B ring junction in the 3,17-dihydroxy- and 3,17-monohydroxymonoketonic C_{19}-steroids can be easily characterized after oxidation with chromium trioxide; thus the resulting androst-4-enedione, androstanedione, and etiocholanolone are completely separated after one-dimensional multiple TLC in *n*-hexane–ethyl acetate 75:25 (silica gel G, mobility, respectively, 3.2, 5.2, and 5.7 cm) and show different color development with the anisaldehyde–sulfuric acid reaction[160] (see Section II,A,2,a).

k. Oxidation of Allyl Alcohols. Specific oxidation of allyl alcohols to α,β-unsaturated ketones can be accomplished with 2,3-dichloro-5,6-dicyanobenzoquinone (DDQ), according to Burn *et al.*[277] Microamounts of steroids are left overnight at room temperature with 2–4 ml of a 0.3% dioxane solution of DDQ. The solution is then filtered, dried, redissolved in ethyl acetate, and washed repeatedly with one-third of its volume of 0.5 N sodium hydroxide and thereafter water until a neutral reaction is obtained. Aliquots of the oxidized steroid in ethyl acetate are spotted directly on the layer.

The oxidation of androst-4-enediols to 17-hydroxyandrost-4-en-3-one or 3-hydroxyandrost-4-en-17-ones to androst-4-ene-3,17-dione with DDQ gives a good approach for their characterization in mixtures with the closely related androstanediols or androstanolones, respectively.

l. Oxidation with Periodic Acid.[278] Add to 0.5 ml of a methanolic solution containing 10–50 μg of steroid, 0.5 ml of a 0.02 M solution of periodic acid ($HIO_4 \cdot 2\,H_2O$) in a 0.3 N sulfuric acid solution. After 18 hours in the

[277] D. Burn, V. Petrow, and G. O. Werton, *Tetrahedron Letters* p. 14 (1960).
[278] S. A. Simpson, J. F. Tait, A. Wettstein, R. Neher, J. von Euw, O. Schindler, and T. Reichstein, *Helv. Chim. Acta* 37, 1163 (1954).

dark at room temperature, add 4 ml of water, adjust the pH to 5 and extract the oxidation products with ethyl acetate.

The separation of the Δ^4-3-oxoetienic acids resulting from the periodate oxidation of formaldehydogenic C_{21}-steroids on silica gel G layers has been investigated by Lisboa[63] on water saturated n-butanol and by Duvivier[279] in the systems dichloromethane–benzene–methanol 6:2:2 and chloroform–methanol–ammonium hydroxide type.

The calculation of ΔR_{Mr} values for the changed side chain of C_{21}-steroids gives valuable information on the side-chain structure of the corticosteroids, as Lisboa[63, 155, 280] has shown. 21-Unsubstituted steroids with a 20-keto group or a 17α,20-ketolic structure do not change their R_f values ($\Delta R_{Mr} = 0$) after periodate treatment, whereas 17α,20-dihydroxy-steroids (I) or 17α,20,21-trihydroxysteroids (II) give 17-ketosteroids and 20,21-ketolic steroids (III) or 17α,21-dihydroxy-20-ketonic steroids (IV) give etienic acids on oxidation. The ΔR_{Mr}-values calculated for these conversions on silica gel layers in the systems (a) chloroform–ethanol 9:1 and (b) benzene–ethanol 8:2 give a complete characterization of these four types of side chains:

system (a): (I) $= -0.48 + 0.10$, (II) $= -1.13 + 0.17$,
            (III) $= +0.70 + 0.16$ and (IV) $= +1.47 + 0.35$
system (b): (I) $= -0.21 + 0.05$, (II) $= -0.66 + 0.12$,
            (III) $= +0.36 + 0.20$ and (IV) $= +1.32 + 0.17$

m. Epoxidation. (i)[265] The steroid is left to react with a 1% solution of p-nitroperbenzoic acid in ether for 3–6 hours, in capillary tubes or in glass-tubes. (ii)[281] Add to a steroid solution in 1–2 ml of chloroform a 5–20-fold molar excess of m-chloroperbenzoic acid; after 30 minutes at room temperature add 4 ml of ether and wash successively with 10% sodium bicarbonate, water, and saturated sodium chloride solution. After evaporation of the organic phase, redissolve the steroid epoxide in benzene.

Epoxide formation permits a very useful separation between Δ^5-3β-steroids and their closely related 3β-hydroxy-5α-analog, for instance, cholesterol from cholestanol and stigmasterol from stigmastanol. With the system benzene–butyl acetate–butanone 75:25:10[281] on silica gel layers, cholestanol ($R_S = 0.94$; S = cholesterol, $R_f = 0.48$) and stigmastanol ($R_S = 1.02$) can be completely separated from the much more polar epoxides of cholesterol ($R_S = 0.23$) and stigmasterol ($R_S = 0.46$).

[279] J. Duvivier, *3rd Intern. Symp. Chromatog. Brussels, 1964* p. 175. Soc. Belge des Sciences Pharmaceutiques, Brussels, 1964.
[280] B. P. Lisboa, *3rd Intern. Symp. Chromatog. Brussels, 1964* p. 33. Soc. Belge des Sciences Pharmaceutiques, Brussels, 1964.
[281] D. L. Azarnoff and D. R. Tucker, *Biochim. Biophys. Acta* **70**, 589 (1963).

n. Specific Oxidations. At the positions C-17 and C-20, these oxidations can be accomplished by use of enzymatic preparations, such as 17-hydroxysteroid oxidoreductase crude preparation of Sigma Co., USA, and crystalline 20β-hydroxysteroid oxidoreductase[282] of C. F. Boehringer & Söhne GmbH, Mannheim, Germany.

The transformation of the steroid molecule by treatment with these two enzyme preparations in association with the chromic acid oxidation gives useful information for the differentiation of 17-oxygenated steroids of the C_{18}- and C_{19}-series from 20-oxygenated C_{21}-steroids.[156]

B. Derivative Formation on the Layers prior to or during Development

A great number of chemical reactions, such as oxidation, dehydration, formation of phenylhydrazones and semicarbazones, can be accomplished on the layers, by spotting the reagents to the applied substances on the starting points.[283] Cargill[22] has obtained bromination of steroids with the same procedure, for the separation of cholesterol and dihydrocholesterol on benzene–ethyl acetate 2:1 mixture.

Also it is possible to carry out the formation of dibromo derivatives during the development by adding the reagent (0.5% bromine) to the mobile phase.[284] This method has been utilized by Copius-Peereboom and Beekes[51] to separate the acetates of cholesterol or stigmasterol from that of brassicasterol, using the system undecane–acetic acid–acetonitrile 1:3. Comparison of results in four procedures using the bromine system, untreated reversed-phase TLC, silver nitrate-impregnated TLC, and gas chromatography has shown the method to be very useful for the identification of different phytosterols, e.g., those from rapeseed oil and coconut fat.[285]

In the techniques described for the formation of π-complexes between unsaturated bonds and silver nitrate[28] or steroid glycol complexes with boric acid and cupric ions,[286] the impregnation with the reagent is done during the preparation of the layer or by spraying, and all the layer surface is treated. Becker[287] has described a procedure for simultaneous formation and separation of derivatives on paper chromatography (elatography) in which only a section of the support is treated by the reagent; this section can include the starting line itself or consist of the

[282] H. D. Henning and J. Zander, *Z. Physiol. Chem.* **330**, 31 (1962).

[283] J. M. Miller and J. G. Kirchner, *Anal. Chem.* **25**, 1107 (1953).

[284] H. P. Kaufmann, Z. Makus, and T. H. Khoe, *Fette Seifen Anstrichmittel* **64**, 1 (1962).

[285] J. W. Copius-Peereboom, *J. Gas Chromatog.* p. 325 (1965).

[286] B. P. Lisboa, *4th Intern. Symp. Chromatog. Electrophoresis, Brussels, 1966,* p. 88. Presses Academiques Européennes, Brussels, 1968.

[287] A. Becker, *Z. Anal. Chem.* **174**, 161 (1960).

region just after it, and functions in this case as a barrier to the steroid. The compounds are left to react under the optimal reaction conditions for a long time if they are deposited on the treated region, or for some minutes only if, during the run, they migrate through this barrier. The experimental conditions for the elatography of steroid hydrazones with Girard's reagent T have been investigated by Lisboa;[288] when this technique is used, only one extraction or transfer of samples is necessary.

C. Microchemical Reactions on the Layers after Development

Sometimes the conversion of the steroids after chromatography is necessary in order to use specific color reactions for their visualization or characterization. Particularly useful are reactions a–d below.

a. Conversion of alcoholic steroids to ketosteroids prior to the application of the Zimmermann reaction (see Section III,C,3,b).

b. Oxidation of 17-deoxy-α-ketolic steroids to the corresponding glyoxals, which gives a positive Porter-Silber reaction (see Lewbart and Mattox reaction, Section III,C,1,d).

c. Reduction of bile acids, e.g., reductodehydrocholic acid (3α-hydroxy-7,12-dioxocholanoic acid), to bile alcohols by spraying them with a 5% solution of sodium borohydride in methanol (85%), before the application of the phosphomolybdic–sulfuric acid mixture.[144]

d. Hydrolysis of conjugates. Steroid conjugates can be hydrolyzed after spraying of the plates with concentrated hydrochloric acid solution followed by heating for 15 minutes at 120° in an oven saturated with hydrochloric acid vapors. After cooling, the plate can be submitted to further reactions, for characterization of both moieties of the conjugate molecule, such as the naphthoresorcinol of reactions for glucuronic acid, or the Folin-Ciocalteau reaction for phenolic groups.

V. Applications of TLC Techniques to Specific Types of Steroids

1. *Sterols*

Barbier *et al.*,[134] Sigg and Tamm,[289] and Tamm[211] first used TLC on silica gel to separate less polar sterols. $R_f \times 100$ values for some cholestanones in benzene[289] are: 2β-methoxy-4,4-dimethylcholestan-3-one, 10; 2α-hydroxycholestan-1-one, 16; 2α-methoxy-4,4-dimethylcholestan-3-one, 37; 4,4'-dimethylcholestane-2,3-dione, 28; and cholestan-1-one, 54. Barbier *et al.*[134] have obtained the separation, on silica gel G-layers, of coprostan-3-one (0.71), cholestan-3-one (0.86), and cholest-4-en-3-one (0.63), using ethyl acetate–cyclohexane 15:85.

[288] B. P. Lisboa, *J. Chromatog.* **24**, 475 (1966).
[289] H. P. Sigg and C. Tamm, *Helv. Chim. Acta* **43**, 1402 (1960).

Because of the existence of a great number of isomeric unsaturated compounds and epimers, separations of the sterols are often difficult and sometimes are possible only by the formation of derivatives. In Tables II and III are summarized the chromatographic mobilities obtained for a large number of sterols on TLC.

The order of migration of the 3-hydroxy-C_{27}-sterols on adsorption TLC can be established as

$$3\alpha(5\alpha) \geqslant 3\alpha(5\beta) > 3\beta(5\beta) > 3\beta\text{-}\Delta^5 > 3\beta(5\alpha)$$

Nambara et al.[290] have developed a large number of systems for the TLC on silica gel of cholestanol, epicholestanol, and cholesterol as free compounds ($R_f \times 100$ values on benzene–acetone 10:1 are, respectively, 33, 50, and 37) or as dinitrobenzoates. Černý et al.[90] using unbound aluminum oxide (grade V) layers have resolved the four cholestanol isomers ($R_f \times 100$ values: cholestanol, 50; epicholestanol, 77; coprostanol, 65; and epicoprostanol, 69; value for cholesterol, 50).

The optimal TLC conditions for the separation of these steroids as well as for the pair cholestanone/coprostanone are those of Cargill[22] (see Table IV) using partition chromatography.

Some chromatographic mobilities for 7-oxygenated derivatives of cholesterol are given in Table II. Other systems are of interest: 7β-hydroxycholesterol[291] presents $R_s \times 100$ values (S = cholesterol) of 64, 52, and 23 in the systems benzene–ethyl acetate–acetic acid 7:3:1, benzene–dioxane–acetic acid 75:20:2, and benzene–ethyl acetate 2:1, respectively. Cholesterol and 7-oxocholesterol[114] can be separated on silica gel layers (n-butyl acetate; $R_f \times 100$ values, 69 and 54) or aluminum oxide layers (ethyl acetate; $R_f \times 100$ values, 67 and 36).

In their studies on the elucidation of the intermediates in the conversion of cholesterol to cholic acid, Danielsson and Einarsson[292] obtained the separation of several 7-oxygenated sterols on silica gel layers with the system benzene–ethyl acetate 3:7, with the following sequence of mobilities (starting line = 2 cm, front = 17.5 cm; $R_s \times 100$ values S = 5β-cholestane-3α,7α,12α-triol): cholesterol (400) > 7α-hydroxycholest-4-en-3-one (345) > 7α,12α-dihydroxy-5β-cholestan-3-one (295) > cholest-5-ene-3β,7β-diol (245) > cholest-5-ene-3β,7α-diol (205) > 7α,-12α-dihydroxycholest-4-en-3-one (172) > 5β-cholestane-3α,7α,12α-triol (100). Further separation between 7α,12α-dihydroxycholest-4-en-3-one and cholest-5-ene-3β,7α-diol has been achieved after 2 hours of ascending TLC on silica gel G with benzene–ethyl acetate–trimethylpentane 3:7:3

[290] T. Nambara, R. Imai, and S. Sakurai, Yakugaku Zasshi 84, 680 (1964).
[291] C. A. Anderson, Australian J. Chem. 17, 949 (1964).
[292] H. Danielsson and K. Einarsson, J. Biol. Chem. 241, 1449 (1966).

TABLE II

CHROMATOGRAPHIC MOBILITIES (R_f OR $R_B \times 100$ VALUES) REPORTED ON TLC FOR STEROLS OF THE NORCHOLESTANE, CHOLESTANE, ERGOSTANE, AND STIGMASTANE SERIES

Sterols	Chromatographic systems[b]														
	Ia	Ib	IIa	IIb	III	IV	Va	Vb	Vc	Vd	VI	VII	VIIIa	VIIIb	VIIIc
A. C_{26}-Sterols (norcholestanes)															
25-Keto-27-norcholesterol (1)[a]	—	—	63	24	—	—	—	—	—	—	—	—	—	—	—
21-Norcholesterol	—	—	—	—	—	—	—	—	—	—	—	—	—	—	—
B. C_{27}-Sterols (cholestanes)															
Cholestane	94	—	—	—	—	72	93	100	102	—	27	94	101	100	91
Cholestanol (dihydrocholesterol) (2)	118	—	—	—	—	—	146	126	108	—	—	—	—	—	—
Epicholestanol (3)	110	—	—	—	—	—	131	120	104	—	—	—	—	—	—
Coprostanol (4)	116	—	—	—	—	—	—	—	—	—	40	—	—	—	—
Epicoprostanol (5)	138	—	—	—	—	—	—	—	—	—	42	—	—	—	—
Cholestanone (6)	141	255	—	—	—	—	—	—	—	—	—	—	—	—	—
Coprostanone (7)	123	296	—	—	—	—	—	—	—	—	—	—	—	—	—
Cholesterol dibromide (8)	—	—	—	—	—	—	—	—	—	—	—	—	—	—	—
Cholest-16-ene	—	—	—	—	—	66	—	—	—	163	—	—	—	—	—
Cholesta-3,5-diene	—	—	—	—	—	—	258	153	110	—	50	—	—	—	—
Cholest-4-en-3-one	—	—	—	—	—	—	160	136	109	—	—	—	—	—	—
Cholest-5-en-3-one	—	—	—	—	—	—	138	125	117	—	—	—	—	—	—
Cholesta-1,4-dien-3-one	—	—	—	—	—	—	151	132	108	—	—	—	—	—	—
Cholesta-4,6-dien-3-one	—	—	—	—	—	—	216	146	109	122	—	—	—	—	—
Cholesta-3,5-dien-7-one	—	—	—	—	—	—	160	135	109	—	—	—	—	—	—
Cholest-4-ene-3,6-dione	—	—	—	—	—	—	—	—	—	—	—	—	—	—	—
Cholesterol (9)	100	100	72	40	64	—	100	100	100	100	30	100	100	100	100
Lathosterol (10)	—	—	—	—	—	—	100	100	100	—	26	—	94	97	93
Epicholesterol (11)	—	—	—	—	—	—	—	—	—	—	—	123	118	121	123
Allocholesterol (12)	—	—	—	—	—	—	—	—	—	—	—	—	—	—	—

(Continued)

7-Dehydrocholesterol (13)	—	—	—	—	100	100	100	—	28	93	93	93	100
Desmosterol (14)	—	—	—	—	—	—	—	—	—	—	—	—	—
Zymosterol (15)	—	—	—	—	—	—	—	—	—	102	93	100	92
4β-Hydroxycholesterol	—	21	—	—	23	42	69	—	—	—	—	—	—
7α-Hydroxycholesterol	—	5	37	—	10	28	46	11	—	—	—	—	—
7β-Hydroxycholesterol	—	—	43	—	11	33	55	11	—	—	—	—	—
20α-Hydroxycholesterol	—	—	—	—	—	—	—	—	—	—	—	—	—
25-Hydroxycholesterol	—	41	16	—	24	54	82	34	—	—	—	—	—
26-Hydroxycholesterol	—	—	—	—	—	—	—	—	—	—	—	—	—
20,22-Dihydroxycholesterol	—	—	—	—	—	—	—	—	—	—	—	—	—
7-Ketocholesterol	—	—	—	—	232	153	110	187	—	—	—	—	—
6-Ketocholestanol	—	—	—	—	25	54	73	—	—	—	—	—	—
6-Ketolathosterol	—	—	—	—	—	—	—	—	—	—	—	—	—
3β,5α-Dihydroxycholestan-6-one	—	—	—	—	—	5	17	—	—	—	—	—	—
6β-Hydroxylathosterol	—	—	—	—	—	—	—	—	—	—	—	—	—
6α-Hydroxylathosterol	—	—	—	—	—	—	—	—	—	—	—	—	—
Cholestane-3β,5α,6β-triol	—	—	—	—	—	5	17	—	—	—	—	—	—
C. 24-Methylcholestanes (ergostanes)													
Campesterol (16)	—	—	—	—	—	—	—	—	—	—	—	—	—
Campestanol (17)	—	—	—	—	—	—	—	—	—	—	—	—	—
Ergosterol (18)	—	—	—	—	—	—	—	—	—	89	91	99	93
Pyrocalciferol (10α-ergosterol)	—	—	—	—	—	—	—	—	—	135	128	113	130
Isopyrocalciferol (9β-ergosterol)	—	—	—	—	—	—	—	—	—	92	91	99	98
Ergocalciferol (vitamin D$_2$) (19)	—	—	—	—	—	—	—	—	—	101	102	116	111
D. 24b-Ethylcholestanes (stigmastanes)													
Stigmastanol (β-sitostanol) (20)	—	—	—	—	—	—	—	—	30	100	100	101	100
β-Sitosterol (21)	—	—	—	—	—	—	—	—	—	—	—	—	—
Stigmasterol (22)	—	—	—	—	—	—	—	—	31	100	100	101	100
Stigmasta-5,25-dien-3β-ol	—	—	—	—	—	—	—	—	—	—	—	—	—
Stigmast-5-ene-3β,25-diol	—	—	—	—	—	—	—	—	—	—	—	—	—
3β-Hydroxy-27-norstigmast-5-en-25-one	—	—	—	—	—	—	—	—	—	—	—	—	—

(Footnotes appear on p. 60.)

TABLE II (Continued)

Sterols	Chromatographic systems[b]											
	IXa	IXb	X	XI	XII	XIII	XIV	XV	XVI	XVIIa	XVIIb	XVIIc
A. C$_{26}$-Sterols (norcholestanes)												
25-Keto-27-norcholesterol (1)[a]	—	—	—	—	—	—	—	—	68	—	—	—
21-Norcholesterol	—	—	52	—	—	—	—	—	—	—	—	—
B. C$_{27}$-Sterols (cholestanes)												
Cholestane	100	—	—	—	—	—	—	—	—	—	—	—
Cholestanol (dihydrocholesterol) (2)	—	91	51	25	39	—	—	—	—	—	—	—
Epicholestanol (3)	—	—	—	—	58	—	—	—	—	—	—	—
Coprostanol (4)	69	—	69	35	51	—	—	—	—	—	—	—
Epicoprostanol (5)	—	—	—	—	51	—	—	—	—	—	—	—
Cholestanone (6)	—	—	—	—	—	—	—	—	—	—	—	—
Coprostanone (7)	—	—	—	—	—	—	—	—	—	—	—	—
Cholesterol dibromide (8)	—	—	—	—	—	—	—	—	—	—	—	—
Cholest-16-ene	—	—	—	—	—	—	—	—	—	—	—	—
Cholesta-3,5-diene	—	—	—	—	—	—	—	—	—	—	—	—
Cholest-4-en-3-one	—	—	—	—	—	—	—	—	—	—	—	—
Cholest-5-en-3-one	—	—	—	—	—	—	—	—	—	—	—	—
Cholesta-1,4-dien-3-one	—	—	—	—	—	—	—	—	—	—	—	—
Cholesta-4,6-dien-3-one	—	—	—	—	—	—	—	—	—	—	—	—
Cholesta-3,5-dien-7-one	—	—	—	—	—	—	72	—	—	—	—	—
Cholest-4-ene-3,6-dione	—	—	—	—	—	100	—	—	—	—	—	—
Cholesterol (9)	100	100	50	25	40	100	50	—	—	—	—	—
Lathosterol (10)	—	95	—	—	39	—	—	—	—	—	—	—
Epicholesterol (11)	145	141	—	29	—	—	—	—	—	—	—	—
Allocholesterol (12)	—	—	53	—	—	—	—	—	—	—	—	—
7-Dehydrocholesterol (13)	92	90	—	—	—	—	48	—	66	—	—	—
Desmosterol (14)	—	—	50	25	—	—	—	—	—	—	—	—

Compound									
Zymosterol (15)	95	91	—	—	—	—	—	—	—
4β-Hydroxycholesterol	—	—	—	—	—	—	—	—	—
7α-Hydroxycholesterol	—	—	—	—	—	—	—	—	—
7β-Hydroxycholesterol	—	—	—	—	—	57	—	—	—
20α-Hydroxycholesterol	—	—	—	—	62	—	—	—	—
25-Hydroxycholesterol	—	—	—	—	—	—	—	—	—
26-Hydroxycholesterol	—	—	—	—	—	—	35	—	—
20,22-Dihydroxycholesterol	—	—	—	—	19	—	—	—	—
7-Ketocholesterol	—	—	—	—	19	—	—	—	—
6-Ketocholestanol	—	—	—	—	—	—	—	—	—
6-Ketolathosterol	—	—	—	—	—	—	—	75	—
3β,5α-Dihydroxycholestan-6-one	—	—	—	—	—	—	—	52	—
6β-Hydroxylathosterol	—	—	—	—	—	—	—	45	—
6α-Hydroxylathosterol	—	—	—	—	—	—	—	—	—
Cholestane-3β,5α,6β-triol	—	—	—	—	—	—	—	—	—
C. 24-Methylcholestanes (ergostanes)									
Campesterol (16)	—	—	50	25	—	—	—	—	—
Campestanol (17)	—	—	51	25	—	—	—	—	—
Ergosterol (18)	94	89	52	—	—	—	—	—	—
Pyrocalciferol (10α-ergosterol)	138	161	—	—	—	—	—	—	—
Isopyrocalciferol (9β-ergosterol)	93	96	—	—	—	—	—	—	—
Ergocalciferol (vitamin D_2) (19)	100	100	—	—	—	—	—	—	—
D. 24b-Ethylcholestanes (stigmastanes)									
Stigmastanol (β-sitostanol) (20)	97	—	49	25	—	—	—	—	—
β-Sitosterol (21)	—	—	50	25	—	—	—	—	—
Stigmasterol (22)	100	—	52	25	—	—	—	—	—
Stigmasta-5,25-dien-3β-ol	75	59	—	—	—	—	—	—	—
Stigmast-5-ene-3β,25-diol	45	19	—	—	—	—	—	—	—
3β-Hydroxy-27-norstigmast-5-en-25-one	—	—	—	—	—	—	—	—	40

(Continued)

(Footnotes appear on p. 60.)

TABLE II (Continued)

a Systematic names: (1): 3β-hydroxy-27-norcholest-5-en-25-one; (2): 5α-cholestan-3β-ol; (3): 5α-cholestan-3α-ol; (4): 5β-cholestan-3β-ol; (5): 5β-cholestan-3α-ol; (6): 5α-cholestan-3-one; (7): 5β-cholestan-3-one; (8): 5α,6β-dibromocholestan-3β-ol; (9): cholest-5-en-3β-ol; (10): cholest-7-en-3β-ol; (11): cholest-5-en-3α-ol; (12): cholest-4-en-3β-ol; (13): cholesta-5,7-dien-3β-ol; (14): cholesta-5,24-dien-3β-ol; (15): 5α-cholesta-8,24-dien-3β-ol; (16): 24a-methylcholest-5-en-3β-ol; (17): 24a-methyl-5α-cholestan-3β-ol; (18): ergosta-5,7,22-trien-3β-ol; (24b-methylcholesta-5,7,22-trien-3β-ol); (19): 9,10-secoergosta-5,7,10(19),22-tetraen-3β-ol; (20): 5α-stigmastan-3β-ol; (21): stigmast-5-en-3β-ol; (22): stigmasta-5,22-dien-3β-ol.

b Abbreviations used in Tables II–XXXIII: n-TLC = normal TLC, ascending one-dimensional TLC, using 20 × 20 cm plates, 250–275 μ thick layers, 22 × 22 × 7 cm tanks, and 200–250 ml of solvent, unless otherwise stated; I.l. = immersion line; S.l. = starting line, in centimeters, from the lower edge of the plate; Fr. = front of the solvent; S.v. = solvent volume; Dv. = development (time, in minutes or hours; distance, in centimeters); Dv-distance is used when the expression "front" is not employed by the author or cannot be calculated from the figures; Dg. imp. = degree of impregnation, i.e., the ratio stationary phase/adsorbent (g/g).

Solvent systems: Bz = benzene; Tol = toluene; Ch = cyclohexane; Hex = n-hexane; Hpt = n-heptane; TMeP = trimethyl-pentane; [LgPt = light petroleum; iOct = i-octane; Xl = xylene; Chf = chloroform; CCl₄ = carbon tetrachloride; ClBz = mono-chlorobenzene; dClMe = dichloromethane; EtAc = ethyl acetate; iAmAc = i-amyl acetate; BuAc = butyl acetate; DEtOx = diethyl oxalate; Dx = dioxane; ThF = tetrahydrofuran; Atn = acetone; Eth = diethyl ether; PtEth = petroleum ether; iPrEth = diisopropyl ether; nBuEth = di-n-butyl ether; MEK = methyl ethyl ketone; H₂O = water; MeOH = methanol; EtOH = ethanol; BuOH = n-butanol; iPrOH = isopropyl alcohol; nPrOH = n-propyl alcohol; AcOH = acetic acid; Py = pyridine.

References: These are indicated in the tables by italic lower-case superscript letters and given in full in corresponding footnotes to the tables.

Comments on the experimental conditions: **I**:*c* n-TLC, silica gel G; I.l. = 0.5 cm; S.l. = 3 cm; Fr. = 15 cm; R_S × 100 values; bromination: at the S.l., with 0.1% bromine-chloroform (ratio bromine/sample = 3:1, w/w). Systems: Bz/Chf type, **Ia**, 2:1; R_f(cholesterol) = 0.67; **Ib**, 19:1; R_f(cholesterol) = 0.19.

II:*d* n-TLC, silica gel G; I.l. = 0.5 cm; S.l. = 2 cm; Fr. = 16.5 cm; tank 21 × 24 × 10 or 20 cm; conditions of saturation not indicated. R_f × 100 values. Systems: **IIa**, Bz/EtAc 2:1; **IIb**, Tol/EtAc 9:1.

III:*e* n-TLC, silica gel G; S.l. = 2 cm; Fr. = 14 cm; R_f × 100 values. System: EtAc/Ch 7:3.

IV:*f* n-TLC, silica gel G; tank, 30.5 × 9.9 × 27.6 cm; S.v. = 15 ml; Fr. = 10 cm; Dv. = 29 min; R_f × 100 values; iOct/CCl₄ 19:1.

V:*g-i* Normal or overrun TLC, silica gel G; 20 × 20 cm plates, 450 μ thick; Dv. = until the solvent reaches the upper edge of the plate; thereafter, the tank is opened ca. 15 mm and the plate is left another 45 minutes for overrun. Systems: **Va–c**, Bz/EtAc: **a**, 9:1; **b**, 2:1; **c**, 1:2; **Vd**,*i* Tol/EtAc 9:1. R_S × 100 values (S = cholesterol) has R_f × 100 values:38(**a**), 86(**b**), 79(**c**) and 41(**d**); it is not clear how the front has been measured for the calculation of the R_f values.

VI:*i* n-TLC, silica gel G; Fr. = 12 cm. System: Tol/EtAc 9:1.

VII:*k,l* Wedge-TLC after Matthias, kieselguhr G; plates 14 × 24 cm, 400 μ thick; tank, 19 × 7 × 30 cm, lined with paper; Dv. = 20 cm, 2–3 hours (23°); R_S × 100 values (S = cholesterol). System: Ch/EtAc 99.5:0.5.

VIII:[k] As **VII**, but with silica gel G. Systems: **VIIIa**, Hex/EtAc 8:2; **VIIIb**, ClBz/AcOH 9:1; **VIIIc**, Ch/EtAc 8:2.

IX:[k] As **VII**, but with aluminum oxide. Systems: **IXa**, aluminum oxide G; Hex/EtAc 8:2; **IXb**, aluminum oxide, iOct/EtAc 8:2.

X:[m] n-TLC, alumina G; activation = 30 min/125°; S.l. = 2 cm; Dv. = 12 cm, Bz/EtOH 19:0.4; $R_f \times 100$ values.

XI:[n] n-TLC, silica gel G; activation = 125–130°/45 min; Dv. = 15 cm (25–27°). System: Chf.

XII:[o] n-TLC, silica gel G; Fr. = 15 cm. System: Chf/Atn 98:2; $R_f \times 100$ values (mean of 4 replicates, each two plates run simultaneously in the same tank).

XIII:[p] One-dimensional multiple TLC, silica gel H, run twice: (1), 75% of plate length, from the origin, with PtEth/Eth/AcOH 75:25:2; (2) run until the top of the plate, with the same solvents but in the proportions 65:35:2; $R_f \times 100$ values (S = cholesterol).

XIV:[q] n-TLC, silica gel G. System: EtAc/Ch 1:1; $R_f \times 100$ values (lanosterol = 61). Experimental conditions after Lisboa.[r]

XV:[r] n-TLC, silica gel G. System: EtAc.

XVI:[t] n-TLC, silica gel H, 8 × 2 in. plates. System: EtAc/LgPt (b.p. = 60–80°) 1:1.

XVII:[u] One-dimensional multiple TLC, silica gel G, $R_f \times 100$ values, Systems: **XVIIa**, Ch/iPrEth 6:4 (2 runs); **XVIIb**, Ch/iPrEth 5:5 (3 runs); **XVIIc**, Chf (2 runs).

[c] D. I. Cargill, *Analyst* **87**, 865 (1962).

[d] M. J. D. Van Dam, G. J. de Kleuver, and J. G. de Heus, *J. Chromatog.* **4**, 26 (1960).

[e] E. Iseli, M. Kotake, E. Weiss, and T. Reichstein, *Helv. Chim. Acta* **48**, 1093 (1965).

[f] R. D. Bennett and E. Heftmann, *J. Chromatog.* **9**, 359 (1962).

[g] C. Tamm, A. Gutler, G. Juhasz, E. Weiss-Berg, and W. Zürcher, *Helv. Chim. Acta* **46**, 889 (1963).

[h] J. R. Claude, *J. Chromatog.* **23**, 267 (1966).

[i] J. R. Claude, *J. Chromatog.* **21**, 189 (1966).

[j] P. Samuel, M. Urivetzky, and G. Kaley, *J. Chromatog.* **14**, 508 (1964).

[k] J. W. Copius-Peereboom, "Chromatographic Sterol Analysis as Applied to the Investigation of Milk Fat and Other Oils and Fats." PUDOC (Centrum voor Landbouwpublikaties en Landbouwdocumentatie), Wageningen, 1963.

[l] J. W. Copius-Peereboom and H. W. Beekes, *J. Chromatog.* **9**, 316 (1962).

[m] R. Ikan, S. Harel, J. Kashman, and E. D. Bergmann, *J. Chromatog.* **14**, 504 (1964).

[n] R. Ikan and M. Cudzinovski, *J. Chromatog.* **18**, 422 (1965).

[o] A. Truswell and W. C. Mitchell, *J. Lipid Res.* **6**, 438 (1965).

[p] E. R. Simpson and G. S. Boyd, *Biochem. Biophys. Res. Commun.* **24**, 10 (1966).

[q] B. P. Lisboa, unpublished results, 1966.

[r] B. P. Lisboa, *Steroids* **7**, 41 (1966).

[s] M. Slaytor and K. Bloch, *J. Biol. Chem.* **240**, 4598 (1965).

[t] P. D. G. Dean and M. W. Whitehouse, *Biochem. J.* **98**, 410 (1966).

[u] W. Sucrow, *Chem. Ber.* **99**, 2765 (1966).

TABLE III

Chromatographic Mobilities (R_f or R_s × 100 Values) Found on TLC for 4α-Methyl Sterols and Sterols of the Lanostane (4,4',14α-Trimethylcholestane), Dammarane (8β-Methyl-18-norlanostane), Euphane (13α,14β-retro-17β-H-lanostane) and Tirucane (20α-Epieuphane) Series

C_{28}-, C_{30}- and C_{31}-sterols	Chromatographic systems[b]												
	I	IIa	IIb	IIc	IIIa	IIIb	IVa	IVb	V	VI	VII	VIIIa	VIIIb
A. 4α-Methyl sterols													
Methostenol (lophenol) (1)[a]	—	—	—	—	—	—	—	—	—	—	60	65	—
4α-Methyl-5α-stigmasta-7,24(28)–dien-3β-ol (citrostadienol)	—	—	—	—	—	—	—	—	—	—	—	65	—
B. Lanostane type													
Lanosterol (2)	137	136	122	141	138	146	33	78	42	67	—	—	—
24-Dihydrolanosterol (lanostenol)	138	136	122	141	138	148	—	77	42	67	—	—	—
Agnosterol (3)	135	137	113	132	142	145	88	95	92	—	—	—	—
24-Dihydroagnosterol	134	—	—	—	—	—	—	—	92	—	—	—	—
Parkeol (4)	—	—	—	—	—	—	20	—	—	—	70	—	—
Cycloartenol (5)	—	—	—	—	—	—	—	—	—	—	70	—	—
Cycloartenone (6)	—	—	—	—	—	—	67	—	—	—	—	—	—
Cyclolaudenol (7)	—	—	—	—	—	—	21	—	—	—	—	—	—
Polyporenic acid A (ungulinic acid) (8)	—	—	—	—	—	—	—	—	—	—	—	—	43t
Polyporenic acid A, methyl ester	—	—	—	—	—	—	—	—	—	—	—	24	—
Polyporenic acid C, methyl ester (9)	—	—	—	—	—	—	—	—	—	—	—	68	—
Tumulosic acid (10)	—	—	—	—	—	—	—	—	—	—	—	—	2
C. Dammarane type													
Dipterocarpol (11)	—	—	—	—	—	—	—	—	—	—	—	61	—
D. Euphane and tirucane types													
Euphol (euphadienol) (12)	—	—	—	—	—	—	20	—	—	—	—	—	—
Butyrospermol (13)	—	—	—	—	—	—	78	—	—	—	—	—	—
Euphone (14)	—	—	—	—	—	—	73	—	—	—	—	—	—

Butyrospermone (15)	78	—	—	—	—	—	—	—	—	—	—
α-Eupherbol (16)	24	—	—	—	—	—	—	—	—	—	—
Euphene (eupha-7-ene)	96	—	—	—	—	—	—	—	—	—	—
Masticadienonic acid (17)	—	—	—	—	—	—	—	—	—	—	47
Isomasticadienonic acid (18)	—	—	—	—	—	—	—	—	—	—	47

^{a} *Systematic names:* (1): 4α-methyl-cholest-7-en-3β-ol; (2): lanosta-8,24-dien-3β-ol; (3): lanosta-7,9(11),24-trien-3β-ol; (4): lanosta-9(11),24-dien-3β-ol; (5): 9,10-cyclo-lanost-24-en-3β-ol; (6): 9,10-cyclo-lanost-24-en-3-one; (7): 9,10-cyclo-24b-methyl-lanost-25-en-3β-ol; (8): 3α,12α-dihydroxy-24-methylene-lanost-8-en-26-oic acid; (9): 16α-hydroxy-3-keto-24-methylene-lanosta-7,9(11)-dien-21-oic acid, methyl ester; (10): 3β,16α-dihydroxy-24-methylene-lanost-8-en-21-oic acid; (11): 20ξ-hydroxy-dammar-24-en-3-one; (12): eupha-8,24-dien-3β-ol; (13): eupha-7,24-dien-3β-ol; (14): euph-8-en-3-one; (15): eupha-7,24-dien-3-one; (16): 24b-methyl-tiruca-8,24(28)-dien-3β-ol; (17): 3-ketotiruca-7,24-dien-26-oic acid; (18): 3-ketotiruca-8,24-dien-26-oic acid.

^{b} *Comments on the experimental conditions:*

I: n-TLC, kieselguhr G, as **VII** in Table II.

II: n-TLC, silica gel G; **IIa-c** as **VIIIa-c** in Table II.

III: n-TLC, aluminum oxide; **IIIa** and **b** as **IXa** and **b** in Table II.

IV: n-TLC, aluminum oxide G; activation = 125°, 30 min; Dv. = 60 min; Fr. = about 10 cm. **IVa**,^{c} Hpt/Bz/EtOH 50:50:5; **IVb**,^{d} Bz/EtOH 19:0.4.

V: n-TLC, silica gel G; as **XI** in Table II.

VI:^{e} n-TLC, silica gel G; Dv. = 1 hour; Bz/EtAc 5:1; R_f × 100 values.

VII:^{f} n-TLC, silica gel G, 13 × 25 cm plates, tank lined with paper; S.l. = 1.5 cm; Fr. = 15 cm; Ch/EtAc 1:1.

VIII:^{g,h} n-TLC, silica gel; 7.5 × 11.5 cm plates, experimental conditions as in Tschesche et al.^{i} R_f × 100 values: t = tailing.

Systems: **VIIIa**, Pr.Eth; **VIIIb**, Pr.Eth/Atn 5:2.

^{c} R. Ikan, J. Kashman, and E. D. Bergmann, *J. Chromatog.* **14**, 275 (1964).

^{d} R. Ikan, S. Harel, J. Kashman, and E. D. Bergmann, *J. Chromatog.* **14**, 504 (1964).

^{e} J. Avigan, D. S. Goodman, and D. Steinberg, *J. Lipid Res.* **4**, 100 (1963).

^{f} K. Schreiber, O. Aurich, and G. Osske, *J. Chromatog.* **12**, 63 (1963).

^{g} R. Tschesche, I. Duphorn, and G. Snatzke, *in* "New Biochemical Separations" (A. T. James and L. J. Morris, eds.), p. 248. Van Nostrand, Princeton, New Jersey, 1964.

^{h} R. Tschesche, F. Lampert, and G. Snatzke, *J. Chromatog.* **5**, 217 (1961).

^{i} R. Tschesche, W. Freytag, and G. Snatzke, *Chem. Ber.* **92**, 3053 (1959).

TABLE IV

CHROMATOGRAPHIC MOBILITIES ON REVERSED-PHASE TLC FOR STEROIDS, STEROLS AND STEROL ACETATES OF THE CHOLESTANE, 24-METHYLCHOLESTANE, 24-ETHYLCHOLESTANE, AND LANOSTANE SERIES

Sterols	Solvent systems used on reversed-phase TLC[b]								Steroid acetates	
	Free steroids									
	Ia	Ib	Ic	Id	Ie	IIa	IIb	III	IIc	IId
A. C27-sterols										
Cholestane	108	110	—	—	—	—	85	—	—	—
Cholestanol	192	190	92	81	91	90	86	39	89	89
Epicholestanol	192	160	82	62	68	—	—	—	—	—
Coprostanol	167	140	92	75	77	—	33	—	—	—
Epicoprostanol	—	—	94	93	96	—	—	—	—	—
Cholestanone	—	630	71	28	34	—	—	—	—	—
Coprostanone	—	810	71	31	34	—	61	—	—	—
Cholesta-4,6,8(14)-triene	—	—	—	—	—	—	128	—	—	—
Cholesterol	100	100	100	100	100	100	100	42	100	100
7-Dehydrocholesterol	—	—	—	—	—	112	120	49	116	116
Epicholesterol	—	—	—	—	—	90	80	—	116	119
3-Dehydrocholesterol (1)[a]	—	—	—	—	—	—	99	—	—	122
Cholest-3-en-3β-ol	—	—	—	—	—	—	63	—	—	100
Lathosterol	—	—	—	—	—	—	102	—	—	99
Desmosterol	—	—	—	—	—	—	124	52	—	128
Zymosterol	—	—	—	—	—	—	125	—	—	128
7-Hydroxycholesterol	—	—	—	—	—	—	203	—	—	—

	1	2	3	4	5	6	7	8	9	10	11
B. 24-Methylcholestanes											
Campesterol	92	—	—	—	—	—	—	—	—	—	—
Ergosterol	119	120	—	—	124	116	—	—	—	—	—
Ergosterol D (2)	111	—	—	—	105	—	—	—	—	—	—
Brassicasterol (3)	107	100	—	—	109	102	—	—	—	—	—
5-Dihydroergosterol	107	—	—	—	111	—	—	—	—	—	—
22-Dihydroergosterol	107	—	—	—	—	—	—	—	—	—	—
Ergost-7-en-3β-ol	92	—	—	—	—	—	—	—	—	—	—
9(11)-Dehydroergosterol	145	—	—	—	107	—	—	—	—	—	—
Pyrocalciferol	—	—	—	—	128	—	—	—	—	—	—
Isopyrocalciferol	—	—	—	—	117	—	—	—	—	—	—
Lumisterol (4)	—	—	—	—	125	—	—	—	—	—	—
Epilumisterol (5)	—	—	—	—	—	—	—	—	—	—	—
C. 24b-Ethylcholestanes											
Stigmasterol	91	91	—	—	—	—	—	—	—	—	—
α-Spinasterol (6)	91	—	—	—	95	—	—	—	—	—	—
β-Sitosterol	83	83	—	—	81	86	—	—	—	—	—
Stigmastanol	73	—	—	—	91	93	—	—	—	—	—
D. 9,10-Secosterols											
Cholecalciferol (vitamin D_3) (7)	141	—	—	—	118	—	—	—	—	—	—
Ergocalciferol (vitamin D_2)	126	—	—	—	122	—	—	—	—	—	—
22-Dihydro vitamin D_2 (vitamin D_4)	134	—	—	—	101	—	—	—	—	—	—
E. Trimethyl sterols											
Lanosterol	100	97	—	38	80	84	—	—	—	—	—
Agnosterol	86	86	—	—	68	76	—	—	—	—	—
24-Dihydrolanosterol	—	—	—	—	57	70	—	—	—	—	—
24-Dihydroagnosterol	—	—	—	—	67	75	—	—	—	—	—

(Footnotes appear on p. 66.)

(Continued)

TABLE IV (*Continued*)

[a] *Systematic names:* (1) cholesta-3,5-dien-3β-ol; (2) ergosta-7,9(11),22-trien-3β-ol; (3) ergosta-5,22-dien-3β-ol; (4) 9β,10α-ergosterol; (5) 9β,10α-ergosta-5,7,22-trien-3α-ol; (6) stigmasta-7,22-dien-3β-ol; (7) 9,10-secocholesta-5,7,10(19)-trien-3β-ol. For the other systematic names, see Tables II and III.

[b] *Comments on the experimental conditions:*   **I:**[c] Silica gel G; impregnation by dipping the plate into a solution of the stationary phase in the volatile organic solvent. $R_S \times 100$ values (S = cholesterol); R_f-values: **Ia**, 0.23; **Ib**, 0.06; **Ic**, 0.61; **Id**, 0.26; **Ie**, 0.31. Conditions as **I** (of Table II). Systems: **Ia**, Hpt/15% 2-phenoxy-EtOH in Atn; **Ib**, Hpt/15% 2-methoxy-EtOH in Atn; **Ic**, MeOH/0.5% liquid paraffin in Eth; **Id**, MeOH/15% undecane in LgPt (l.p. range, 40–60°); **Ie**, MeOH/Eth (94:1)/15% undecane in LgPt (40–60°).

II: Reversed-phase TLC[d] on wedge layers[e] of kieselguhr G. **IIa**,[d,f] undecane/AcOH–H₂O 9:1;Dg.imp. = 0.13 g/g; **IIb**,[d] undecane–tetradecane (8:2)/AcOH–H₂O–acetonitrile (22.5:2.5:5.75); Dg.imp. = 0.25 g/g; **IIc**,[d,f] undecane/AcOH–H₂O 92:8; Dg.imp. = 0.09 g/g; **IId**,[g,h] undecane/AcOH–acetonitrile 25:75; Dg.imp. = 0.04 g/g; Dv. = 1.5 hour. $R_f \times 100$ value of cholesterol acetate = 28; $R_S \times 100$ values of 5α-androstan-3β-ol acetate = 225; $R_S \times 100$ values given by Copius-Peereboom[d] for the acetates of 7-dehydrocholesterol and ergosterol = 128 and 135.

III: Undecane reversed-phase TLC;[i] 275μ-thick silica gel layer (Adsorbil 1, Applied Sci. Lab. Inc., State College, Pennsylvania); impregnation by developing the plate in a 15% solution of undecane in Pt.Eth. $R_f \times 100$ values. Solvent systems: AcOH–acetonitrile 1:1 saturated with undecane/AcOH–acetonitrile 1:1 (70:30).

[c] D. I. Cargill, *Analyst* **87**, 865 (1962).

[d] J. W. Copius-Peereboom, "Chromatographic Sterol Analysis as Applied to the Investigation of Milk Fat and Other Oils and Fats." PUDOC (Centrum voor Landbouwpublikaties en Landbouwdocumentatie), Wageningen, 1963.

[e] V. Prey, H. Berbalk, and M. Kausz, *Mikrochim. Acta*, p. 968 (1961).

[f] J. W. Copius-Peereboom and H. W. Beekes, *J. Chromatog.* **9**, 316 (1962).

[g] J. W. Copius-Peereboom and H. W. Beekes, *J. Chromatog.* **17**, 99 (1965).

[h] J. W. Copius-Peereboom, *in* "Thin-Layer Chromatography" (G. B. Marini-Bettolo, ed.), p. 194. Elsevier, Amsterdam, 1964.

[i] L. Wolfman and B. A. Sachs, *J. Lipid Res.* **5**, 127 (1964).

(separation factor: 2.05). Also Berséus *et al.*[212] separated some of these compounds in silica gel G layers with benzene–ethyl acetate 1:3 as solvent; $R_f \times 100$ values reported are for 5β-cholestane-3α,7α,12α-triol (24), cholest-4-ene-7α,12α-diol-3-one (44), and 5β-cholestane-7α,12α-diol-3-one (83). For the chromatographic mobilities of other 3,7,12-oxygenated sterols, see under bile acids.

There are a number of critical sterol pairs—e.g., cholesterol/desmosterol, cholesterol/lathosterol, zymosterol/desmosterol, Δ^7-cholesterol/cholestanol—which cannot be resolved under normal adsorption TLC conditions. The separation of cholesterol and desmosterol is possible only on partition TLC on silica gel G with undecane as stationary phase,[293] or on silver nitrate-impregnated layers (Table V). The trifluoroacetates of these two important biological substances can be resolved on silica gel G–kieselguhr G (1:1)[7] by chromatography with cyclohexane–heptane 1:1, and their acetates have been separated in 1 hour continuous flow horizontal TLC on silica gel G,[76] using the system benzene–hexane 1:3 ($R_\mathrm{S} \times 100$ value of desmosterol acetate $= 57$; S $=$ cholesterol acetate). Mixtures of cholesterol, lathosterol, and 7-dehydrocholesterol can be resolved by chromatography for 2–3 days on plates 40 cm long[294] with benzene–ethyl acetate 20:1. The solvent evaporates continuously and reaches only three-fourths of the length of the plate. However, in this system cholesterol and cholestanol run together. Better separation of cholesterol and lathosterol can be obtained using silver nitrate-impregnated layers, by single developments with chloroform–light petroleum–acetic acid 25:75:0.5,[51] or with repeated TLC on chloroform (two developments).[295]

Sometimes two different TLC procedures must be used in order to assure the separation of several sterols from each other. Truswell and Mitchell[24] have used two chromatographic steps to obtain the separation of cholesterol, Δ^7-cholestenol and cholestanol isomers. In the first TLC, employing silver nitrate-impregnated layers and the solvent system chloroform–acetone 98:2, they achieved preliminary group separation ($R_f \times 100$ values: cholesterol $= 48$; Δ^7-cholestenol and cholestan-3β-ol $= 57$–59; coprostan-3β- and 3α-ol $= 72$–74; and cholestan-3α-ol $= 80$). In the second, by means of reversed-phase TLC at low temperature (undecane–acetic acid acetonitrile 1:1; front, 15 cm; temperature, $-15°$ to $-20°$), they obtained a complete resolution of Δ^7-cholestenol and cholestanol.

[293] L. Wolfman and B. A. Sachs, *J. Lipid Res.* **5**, 127 (1964).
[294] J. Avigan, D. S. Goodman, and D. Steinberg, *J. Lipid Res.* **4**, 100 (1963).
[295] N. W. Ditullio, C. S. Jacobs, Jr., and W. L. Holmes, *J. Chromatog.* **20**, 354 (1965).

TABLE V

CHROMATOGRAPHIC MOBILITIES OF STEROLS, STEROL ACETATES AND STEROL TRIMETHYLSILYL ETHERS (TMeSi-ETHERS) FOUND ON SILVER NITRATE-IMPREGNATED LAYERS

Sterols[a]	Solvent systems used on silver nitrate-impregnated layers for													
	Free steroids							Steroid acetates					TMeSi-ethers	
	Ia	IIa	III	IV	V	VIa	VIb	Ib	IIb	IIc	VII	VIII	IXa	IXb
A. C$_{27}$-sterols														
Cholestanol	29	114	59	26	—	240	—	—	125	—	133	—	44	42
Coprostanol	—	—	72	42	—	—	—	—	—	—	—	—	—	—
Epicholestanol	—	—	74	—	—	—	—	—	—	—	—	—	—	—
Epicoprostanol	—	—	80	—	—	—	—	—	—	—	—	—	—	—
Cholesterol	19	100	48	23	100	100	100	—	100	—	100	—	41	35
Allocholesterol	—	—	—	28	—	—	—	—	—	—	—	—	—	—
Cholest-6-en-3β-ol	—	—	—	—	—	30	34	—	—	—	—	—	—	—
Lathosterol	—	117	—	—	—	210	—	—	114	—	105	—	—	—
Cholest-8-en-3β-ol	—	—	—	—	—	240	—	—	—	—	—	—	—	—
Cholest-8(14)-en-3β-ol	—	—	—	—	—	250	—	—	—	—	—	—	—	—
7-Dehydrocholesterol	—	44	—	—	—	20	17	—	—	43	—	—	—	—
22-Dehydrocholesterol	—	—	—	—	—	—	—	—	—	—	70	—	—	—
Desmosterol	—	88	—	14	74	90	68	—	—	—	49	79	—	—
Cholesta-8,14-dien-3β-ol	—	—	—	—	—	90	76	—	—	—	—	—	—	—
25-Dehydrocholesterol	—	—	—	—	—	—	—	—	—	—	—	56	—	—
B. 24-Methylcholestanes														
Campesterol	—	—	—	23	—	—	—	—	—	—	102	—	—	—
Campestanol	—	—	—	26	—	—	—	—	—	—	—	—	—	—
Ergosterol	—	44	—	—	—	10	—	—	—	35	—	—	—	—
Frgosterol D	—	83	—	—	—	—	—	—	33	—	—	—	—	—
Brassicasterol	—	87	—	—	—	—	—	—	33	—	84	—	—	—

Compound										
5-Dihydroergosterol	113	—	—	130	128	88	—	—	—	—
22-Dihydrobrassicasterol	—	—	—	—	—	—	—	—	—	—
Ergost-7-en-3β-ol	122	—	—	200	—	121	103	—	—	—
9(11)-Dehydroergosterol	69	—	—	—	—	—	—	—	—	—
C. 24b-Ethylcholesterol										
Stigmasterol	98	23	—	—	—	87	95	—	—	—
β-Sitosterol	100	23	—	—	—	100	102	—	—	35
β-Sitostanol	114	32	—	—	—	130	127	—	—	—
Fucosterol	—	—	—	—	—	—	61	—	—	—
D. 9,10-Secosterols										
Vitamin D₂ (ergocalciferol)	64	—	—	—	—	—	—	—	—	—
Vitamin D₄ (22-dihydro vitamin D₂)	47	—	—	—	—	—	—	—	—	—
E. 4α-Methyl sterols										
4α-Methylergosta-7,24(28)-dien-3β-ol	—	—	—	—	—	—	—	—	—	22
4α-Methylstigmasta-7,24(28)-dien-3β-ol	—	—	—	—	—	—	—	—	—	14
F. 4,4-Dimethyl sterols										
4,4-Dimethylcholest-7-en-3β-ol	—	—	—	360	—	—	—	—	—	—
4,4-Dimethylcholest-8-en-3β-ol	—	—	—	380	—	—	—	—	—	—
G. Trimethyl sterols										
Lanosterol	170	—	—	220	61	78	89	40	—	47
Dihydrolanosterol	45	41	—	370	76	—	153	50	—	70
Agnosterol	168	89	—	—	47	40	—	—	—	—
Dihydroagnosterol	161	95	—	—	71	—	—	—	—	—
Cycloartenol	—	—	—	—	—	—	—	38	—	47
Butyrospermol	—	—	—	—	—	—	—	35	—	47
Lupeol	—	—	—	—	—	—	—	30	—	35
24-Methylencycloartenol	—	—	—	—	—	—	—	29	—	27

(Footnotes appear on p. 70.)

(Continued)

TABLE V (Continued)

[a] *Systematic names:* 22-dihydrobrassicasterol = ergost-5-en-3β-ol; fucosterol = stigmasta-5,24(28)-dien-3β-ol. For the other systematic names, see Tables II–IV.

[b] *Comments on the experimental conditions:* **I:** Silver nitrate–silica gel G–TLC.[c] **Ia,** Chf; **Ib,** Chf/LgPt 7:3.

II: Silver nitrate–silica gel G (13 g:30 g)–TLC,[d] 20 × 20 cm plates, S.l. = 1.5 cm; Dv. = 1.5–2 hours; Fr. = 17 cm, tank equilibrium, 16 hours; R_f × 100 values. Systems: **IIa,** n-TLC, Chf/Eth/AcOH 97:2.3:0.5; **IIb,** n-TLC, Chf/LgPt/AcOH 25:75:0.5; **IIc,** as **IIb,** but with twice development.

III:[e] n-TLC, silica gel G, sprayed with a 25% silver nitrate solution until it is evenly moistened; dried 30 min/110°; Fr. = 15 cm; R_f × 100 values; Chf/Atn 98:2.

IV:[f] n-TLC, silica gel sprayed with a 5% silver nitrate solution; S.l. = 2 cm; Dv. = 15 cm; Chf.

V:[g] Continuous flow horizontal TLC, 20 × 20 cm plates, silica gel G impregnated by dipping in a 5% solution of silver nitrate in acetonitrile, followed by air-drying. System: Bz/EtAc 95:5; Dv. = 6 hours.

VI:[h] n-TLC, neutral alumina AG-7 (Bio-Rad Laboratories, Richmond, California), containing 5% by weight of calcium sulfate; argentation: 10.5 g silver nitrate (10 ml water) was added to 35 g alumina (25 ml water); plates dried in air (4–5 hours, room temp.) and stored in desiccator overnight; no activation. R_s × 100 values, S = cholesterol. Systems: **VIa,** Chf/Atn 95:5; Dv. = 16 min (22°), 12 cm front; **VIb,** Chf/Atn 65:35; Dv. = 26 min (22°), 15 cm front. Amount of steroid: 20 μg/spot.

VII: n-TLC, silica gel H–silver nitrate (impregnation = 12.5%), activation 110°/30 min; S.l. = 1.5 cm; Fr. = 15 cm; unlined tanks, containing 200 ml of the system. Bz/Hx 3:5; R_s × 100 values, S = cholestanyl acetate.

VIII:[i] n-TLC, silica gel G–silver nitrate (impregnation = 20%); Dv. = 19 cm (from the lower edge of the plate). R_f × 100 values. Solvent: Bz.

IX:[k] n-TLC; 1 mm thick; silica gel HF$_{254}$ (15 g)/silver nitrate (10%, 37 ml); layer dried at 110°. Systems: **IXa,** Hx/Bz 2:3; **IXb,** Hx/Bz 5:1.

[c] J. H. Recourt, *3rd Intern. Symp. Chromatog. Brussels, 1964* p. 63. Soc. Belge des Sciences Pharmaceutiques, Brussels, 1964.

[d] J. W. Copius-Peereboom and H. W. Beekes, *J. Chromatog.* **17,** 99 (1965).

[e] A. Truswell and W. C. Mitchell, *J. Lipid Res.* **6,** 438 (1965).

[f] R. Ikan and M. Cudzinovski, *J. Chromatog.* **18,** 422 (1965).

[g] T. M. Lees, M. J. Lynch, and F. R. Mosher, *J. Chromatog.* **18,** 595 (1965).

[h] R. Kammereck, W.-H. Lee, A. Paliokas, and G. J. Schroepfer, Jr., *J. Lipid Res.* **8,** 282 (1967).

[i] H. E. Vroman and C. F. Cohen, *J. Lipid Res.* **8,** 150 (1967).

[j] J. A. Svoboda and M. J. Thompson, *J. Lipid Res.* **8,** 152 (1967).

[k] B. O. Lindgren and C. M. Svahn, *Acta Chem. Scand.* **20,** 1763 (1966).

The separation of the closely related zymosterol (5α-cholesta-8,24-dien-3β-ol) and desmosterol (cholesta-5,24-dien-3β-ol) is possible by chromatography[296] on 40 cm-long silica gel plates for 24 hours with benzene–ethyl acetate 20:1 as solvent system. On 40 cm plates it is also possible to separate the critical pair zymosterol/Δ⁷-cholestenol as acetate derivatives (system: hexane–benzene 6:1; the less polar compound is zymosterol acetate[296]).

The separation on adsorption TLC of steroids differing by the presence of a methyl or ethyl group at C-24 is more difficult. Their resolution can be accomplished satisfactorily by reversed partition TLC using the systems developed by Copius-Peereboom[23] (see Table IV) or on silver nitrate-impregnated supports (Table V) as both free and acetylated sterols.

The steroids of the "critical pairs" lanosterol/dihydrolanosterol and agnosterol/dihydroagnosterol can be completely resolved from each other on silver nitrate-treated layers (Table V); the resolution of the first pair has been reported also on partition TLC (see Table IV).

The use of special techniques of development is also indicated to separate steroids with similar mobilities. The pairs cholesterol/cholestanol and stigmasterol/Δ²²-stigmastenol are resolved by continuous development on silica gel G in the system benzene–methanol 99:1.[77]

Tschesche et al.[297] have used TLC for the separation of tetracyclic triterpenes of the cucurbitane type (10-nor-9-methyllanostanes) related to gratiogenin (3,25-dihydroxy-(20,24)-oxido-10-nor-9-methyllanost-5-en-11-one), an aglycone isolated from Gratiola officinalis. The following $R_f \times 100$ values for gratiogenin (44) and derivatives have been found in the system chloroform–acetone 6:1 on silica gel layers: 11-dihydrogratiogenins (two isomers: 29 and 22); gratiogenin monoacetate (75); diacetate (86); 11-dihydrogratiogenin diacetate (76) and triacetate (84); 16α-hydroxygratiogenin (15) and triacetate (88).

The resolution of the naturally occurring esters of cholesterol was systematically investigated by Van Dam et al.[198] on adsorption TLC and by Kaufmann et al.[82] and Michalec et al.[25] on impregnated layers. Using a two-dimensional procedure, Kaufmann[82] has successfully accomplished the separation of a mixture of cholesterol and the following saturated and unsaturated esters: formate (1), acetate (2), propionate (3), butyrate (4), caproate (5), caprylate (6), caprate (7), laurate (8), myristate (9), palmitate (10), stearate (11), linolenate (12), linoleate (13), and oleate (14). The first chromatogram was run on untreated silica gel

[296] L. Horlick and J. Avigan, J. Lipid Res. **4**, 160 (1963).
[297] R. Tschesche, G. Biernoth, and G. Snatzke, Ann. Chem. **674**, 196 (1964).

TABLE VI

CHROMATOGRAPHIC MOBILITIES FOUND ON THIN-LAYER CHROMATOGRAPHY FOR 5β-CHOLANOIC ACIDS[a]

Structure	Trivial names	Chromatographic systems[b]													
		Ia	Ib	Ic	Id	Ie	If	Ig	Ih	Ii	Ij	Ik	Il	Im	In
No substitution	Cholanoic acid	—	—	—	—	—	—	—	—	—	—	—	—	—	—
Δ³	3-Dehydrocholanoic acid	—	—	—	—	173	158	189	—	—	—	—	—	—	—
3α-ol	Lithocholic acid	128	100	160	227	100	100	100	156	100	—	—	—	—	—
3β-ol		—	104	—	—	113	108	115	—	—	254	—	—	—	—
7α-ol		—	108	—	—	135	124	140	—	112	320	—	—	—	—
7β-ol		—	101	117	—	119	110	121	—	—	—	—	—	—	—
12α-ol		—	112	—	—	137	—	140	—	—	—	—	—	—	—
12β-ol		—	112	—	—	128	—	132	—	—	—	—	—	—	—
3α,7α-ol	Chenodeoxycholic acid	91	82	88	100	—	—	—	100	—	100	—	—	—	—
3β,7α-ol		85	73	95	133	—	—	—	—	—	126	—	—	—	—
3α,7β-ol	Ursodeoxycholic acid	77	77	78	100	—	—	—	—	—	114	—	—	—	—
3α,12α-ol	Deoxycholic acid	100	83	100	100	21	—	—	100	86	100	262	128	259	348
3β,12α-ol		—	81	106	140	—	—	—	—	—	136	—	—	—	—
3α,12β-ol		100	87	109	153	34	—	—	—	—	158	—	—	—	—
3α,6α-ol	Hyodeoxycholic acid	67	71	50	56	10	—	—	—	46	63	186	112	180	202
3α,6β-ol		—	—	—	—	—	—	—	—	—	—	—	—	—	—
3α,11α-ol		—	—	—	—	—	—	—	—	—	—	—	—	—	—
3α,11β-ol		—	—	—	—	—	—	—	—	—	251	—	—	—	—
7α,12α-ol		—	98	162	223	80	84	—	—	—	—	—	—	—	—
3β,6β-ol		—	—	—	—	—	—	—	—	—	—	—	—	—	—
3β,6α-ol		—	—	—	—	—	—	—	—	—	—	—	—	—	—
3α,6α,7α-ol		—	—	—	—	—	—	—	—	—	30	—	—	—	—
3α,6β,7α-ol		—	—	—	—	—	—	—	—	—	—	—	—	—	—
3α,6β,7β-ol		—	—	—	—	—	—	—	—	—	—	—	—	—	—

Substituents	Name														
3α,7α,12α-ol	Cholic acid	76	—	—	—	—	—	—	34	34	17	100	100	100	100
3β,7α,12α-ol		—	—	—	—	—	—	—	—	—	—	127	114	144	100
3α,7β,12α-ol		88	—	72	—	—	—	—	—	—	30	146	117	147	159
3α,7α,16α-ol		—	—	—	—	—	—	—	—	—	67	208	141	250	214
3α,11β,12α-ol		—	—	—	—	—	—	—	—	—	—	—	—	—	—
3α,7α,23ξ-ol		—	—	—	—	—	—	—	—	—	15	62	43	25	85
3α,7α,12α,23ξ-ol		—	—	—	—	—	—	—	—	13	4	17	12	7	13
3-oxo	Dehydrolithocholic acid	—	112	—	—	150	121	135	162	113	310	—	—	—	—
7-oxo		—	116	—	—	162	127	150	—	—	—	—	—	—	—
12-oxo		—	116	—	—	162	127	150	—	—	—	—	—	—	—
3,7-oxo	Dehydrochenodeoxycholic acid	110	85	—	—	93	191	93	156	108	265	—	—	—	—
3,12-oxo	Dehydrodeoxycholic acid	117	91	—	—	93	191	95	156	108	265	—	—	—	—
3,6-oxo	Dehydrohyodeoxycholic acid	—	—	—	—	—	—	—	—	—	—	—	—	—	—
3,7,12-oxo	Dehydrocholic acid	55	46	91	178	46	41	41	158	100	213	—	—	—	—
3α-ol; 7-oxo		88	70	95	144	45	42	—	138	90	163	—	—	—	—
3α-ol; 12-oxo		103	80	117	—	55	54	—	138	—	178	—	—	—	—
7α-ol; 3-oxo		—	93	135	211	74	—	—	—	99	213	—	—	—	—
12α-ol; 3-oxo		122	87	145	—	54	—	—	—	—	185	—	—	—	—
3α,12β-ol; 11-oxo		—	—	—	—	—	—	—	—	—	—	—	—	—	—
7α,12α-ol; 3-oxo		87	81	63	126	—	—	—	—	—	—	282	155	294	228
3α,12α-ol; 7-oxo		66	62	44	60	—	—	—	—	—	131	195	115	192	183
3α,7α-ol; 12-oxo		66	62	48	—	—	—	—	—	—	45	195	115	194	185
3β-ol; 7,12-oxo	Isoreductodehydrocholic acid	—	—	—	—	—	—	—	—	66	45	—	—	—	—
3α-ol; 7,12-oxo	Reductodehydrocholic acid	54	53	52	—	—	—	—	116	85	107	256	130	258	194
7α-ol; 7,12-oxo		—	—	—	—	—	—	—	—	—	—	—	—	—	—

(Footnotes appear on p. 76.)

(Continued)

TABLE VI (*Continued*)

Structure	Trivial names	Io	Ip	Iq	IIa	IIb	IIIa	IIIb	IIIc	IVa	IVb	IVc	IVd	V
							Chromatographic systems[b]							
No substitution	Cholanoic acid	—	—	—	—	—	—	—	—	—	—	—	—	—
Δ^3	3-Dehydrocholanoic acid	—	—	—	100	23	—	—	—	—	—	—	—	—
3α-ol	Lithocholic acid	—	167	233	84	—	206	—	—	89	96	—	—	90
3β-ol	—	—	—	—	89	—	—	—	—	—	—	—	—	—
7α-ol	—	—	—	—	—	—	—	—	—	—	—	—	—	—
7β-ol	—	—	—	—	—	—	—	—	—	—	—	—	—	—
12α-ol	—	—	—	—	—	—	—	—	—	—	—	—	—	—
12β-ol	—	—	—	—	—	—	—	—	—	—	—	—	—	—
3α,7α-ol	Chenodeoxycholic acid	—	—	—	44	—	79	98	—	37	89	—	—	67
3β,7α-ol	—	—	—	—	—	—	—	—	—	—	—	—	—	—
3α,7β-ol	—	—	—	—	—	—	68	106	—	—	—	—	—	61
3α,12α-ol	Ursodeoxycholic acid / Deoxycholic acid	368	124	170	44	—	100	100	—	43	89	—	—	74
3β,12α-ol	—	—	—	—	—	—	—	—	—	—	—	—	—	—
3α,12β-ol	—	—	—	—	—	—	—	—	—	—	—	—	—	—
3α,6α-ol	Hyodeoxycholic acid	224	—	—	24	—	50	61	—	27	83	—	—	—
3α,6β-ol	—	—	—	—	44	—	63	103	—	—	—	—	—	—
3α,11α-ol	—	—	—	—	—	—	85	120	—	—	—	—	—	—
3α,11β-ol	—	—	—	—	—	—	132	232	—	—	—	—	—	—
7α,12α-ol	—	—	—	—	—	—	—	—	—	—	—	—	—	—
3β,6β-ol	—	—	—	—	44	—	—	—	—	—	—	—	—	—
3β,6α-ol	—	—	—	—	24	—	—	—	—	—	—	—	—	—
3α,6α,7α-ol	—	—	—	—	—	—	—	21	100	—	—	—	—	—
3α,6β,7α-ol	—	—	—	—	—	—	—	24	95	—	—	—	—	—
3α,6β,7β-ol	—	—	—	—	—	—	—	31	101	—	—	—	—	—

3α,7α,12α-ol	Cholic acid	100	100	100	6	—	21	100	8	76	—	—	42
3β,7α,12α-ol	—	100	87	100	—	—	—	—	—	—	—	—	—
3α,7β,12α-ol	—	175	122	140	—	—	—	—	—	—	—	—	—
3α,7α,16α-ol	—	254	162	179	—	—	—	—	—	—	—	—	—
3α,11β,12α-ol	—	—	—	—	—	—	—	117	—	—	—	—	—
3α,7α,23ξ-ol	—	43	26	28	—	—	—	—	—	—	—	—	—
3α,7α,12α,23ξ-ol	—	28	6	8	—	—	—	—	—	—	—	—	—
3-oxo	Dehydrolithocholic acid	—	—	—	95	—	—	—	—	—	85	—	—
7-oxo	—	—	—	—	—	—	—	—	—	—	—	—	—
12-oxo	—	—	—	—	—	—	—	—	—	—	—	—	—
3,7-oxo	Dehydrochenodeoxycholic acid	—	—	—	—	—	—	—	—	—	47	—	—
3,12-oxo	Dehydrodeoxycholic acid	—	—	—	—	—	—	—	—	—	46	—	—
3,6-oxo	Dehydrohyodeoxycholic acid	—	—	—	14	—	148	—	—	—	51	—	—
3,7,12-oxo	Dehydrocholic acid	—	—	—	48	—	—	—	—	—	21	70	—
3α-ol; 7-oxo	—	—	—	—	—	77	—	—	—	—	—	—	—
3α-ol; 12-oxo	—	—	—	—	65	98	—	—	—	—	—	—	—
7α-ol; 3-oxo	—	—	—	—	—	111	—	—	—	—	—	—	—
12α-ol; 3-oxo	—	—	—	—	—	103	—	—	—	—	—	—	—
3α,12β-ol; 11-oxo	—	—	—	—	—	—	—	—	—	—	—	—	—
7α,12α-ol; 3-oxo	—	302	166	204	24	24	55	—	—	—	—	—	—
3α,12α-ol; 7-oxo	—	232	137	163	—	23	—	—	—	—	—	—	—
3α,7α-ol; 12-oxo	—	232	137	163	—	23	—	—	—	—	—	—	—
3β-ol; 7,12-oxo	Isoreductodehydrocholic acid	—	—	—	—	—	85	—	—	—	—	52	—
3α-ol; 7,12-oxo	Reductodehydrocholic acid	265	152	139	37	—	—	—	—	—	—	47	—
7α-ol; 7,12-oxo	—	—	—	—	—	—	139	—	—	—	—	—	—

(Footnotes appear on p. 76.)

(Continued)

TABLE VI (Continued)

[a] 5β-Cholanoic acid = 5β-cholan-24-oic acid (isopentyl side chain; C_{24}-acids).

[b] *Comments on the experimental conditions:* **I:**[c] n-TLC, silica gel G, Dv. (time) varied with the system, until 3 hours; Fr. = 17–18 cm; no previous saturation of the tanks was used. $R_S \times 100$ values [S = deoxycholic (D), lithocholic (L), or cholic (C) acids; for each system their mobility is given in cm]. **Ia,** TMeP/iPrOH/AcOH 30:10:1; D = 9.1 cm; **Ib,** TMeP/iPrOH/AcOH 60:20:0.5; L = 12.0 cm; **Ic,** TMeP/EtAc/AcOH 10:10:2; D = 9.1 cm; **Id,** TMeP/EtAc/AcOH 5:25:0.2; D = 5.7 cm; **Ie,** TMeP/EtAc/AcOH 50:50:0.7; L = 9.1 cm; **If,** TMeP/EtAc/AcOH 10:10:0.25; L = 8.9 cm; **Ig,** TMeP/EtAc/AcOH 10:10:0.1; L = 7.5 cm; **Ih,** DEtOx/Dx 40:10; D = 5.0 cm; **Ii,** DEtOx/Dx 48:8; L = 9.7 cm; **Ij,** Bz/Dx/AcOH 75:20:2; D = 5.4 cm; **Ik,** Bz/Dx/AcOH 20:10:2; C = 4.7 cm; **Il,** Bz/Dx/AcOH 15:5:2; C = 8.5 cm; **Im,** Bz/Dx/AcOH 55:40:2; C = 4.0 cm; **In,** Ch/EtAc/AcOH 10:15:4; C = 5.2 cm; **Io,** Ch/EtAc/AcOH 7:23:3; C = 5.1 cm; **Ip,** Bz/iPrOH/AcOH 30:10:1; C = 9.0 cm; **Iq,** Ch/iPrOH/AcOH 30:10:1; C = 4.7 cm.

II:[d] n-TLC, Wakogel (Wako Chemicals Co. Ltd. Tokyo); Dv. = 15 cm (20°); $R_f \times 100$ values. Systems: **IIa,** Eth/AcOH 96.6:0.4; **IIb,** Bz/AcOH 95:5.

III:[e] n-TLC, silica gel G; Fr. = 10 cm; S.l. = 1.5 cm; tanks lined with filter paper; equilibrium time, at least 1 hour; S.v. = 100 ml; $R_S \times 100$ values. Systems: **IIIa,** AcOH/iPrEth/iOct 25:25:50; S = deoxycholic acid $(R_f = 0.27)$; **IIIb,** AcOH/nPrOH/Bz/CCl$_4$/iPrEth/isoamyl acetate 5:10:10:20:30:40; S = deoxycholic acid $(R_f = 0.33)$; **IIIc,** H$_2$O/n-PrOH/propionic acid/isoamyl acetate 5:10:15:20; S = cholic acid $(R_f = 0.73)$.

IV:[f] n-TLC, silica gel G; $R_f \times 100$ values. Systems: **IVa,** Bz/AcOH 8:2; **IVb,** EtAc/AcOH/MeOH 85:5:10; **IVc,** Bz/AcOH 9:1; **IVd,** Bz/AcOH 7:3.

V:[g] n-TLC on wedge-silica gel layers; $R_f \times 100$ values. iOct/iPrEth/AcOH/iPrOH 2:1:1:1.

[c] P. Eneroth, *J. Lipid Res.* **4,** 11 (1963).

[d] S. Hara and M. Takeuchi, *J. Chromatog.* **11,** 565 (1963).

[e] A. F. Hofmann, *in* "New Biochemical Separations" (A. T. James and L. J. Morris, eds.), p. 262. Van Nostrand, Princeton, New Jersey, 1964.

[f] T. Usui, *J. Biochem. (Tokyo)* **54,** 283 (1963).

[g] J. A. Gregg, *J. Lipid Res.* **7,** 579 (1966).

layers (system tetralin–hexane 2.5:7.5), for the separation of the lower fatty acid esters (1–5), and the fatty acid esters (11–14). After partial hydrophobization of the layer by immersion in a 5% solution of paraffin in petroleum ether, the plate was run in methyl ethyl ketone–acetonitrile 7:3 (reversed-phase TLC) for the separation of the higher fatty acid esters of cholesterol. A mixture of the cholesterol and some of its esters (2-5-6-8-9-11) has been resolved on paraffin-impregnated silica gel layers by development in acetic acid.[25] The esters of cholesterol with the un-saturated fatty acid oleate, linolate, linolenate, and erucate can be separated on silica gel G with tetralin–hexane (1:1) or (2.5:7.5) as solvent systems; no separation was obtained for the critical pair erucate/stearate esters.

2. Bile Acids and Related Steroid Acids

The mobilities on TLC for a great number of the most frequently encountered 5β-cholanoic acids and their methyl esters are summarized in the Tables VI and VII.

For monosubstituted 5β-cholanoic acids, the migrations reported in the systems employed by Eneroth[103] indicate the following order of polarity for the substituents:

3α-ol(equatorial) $> 3\beta$-ol(axial) $> 7\beta$-ol $> 12\beta$-ol $> 7\alpha$-ol $\geqslant 12\alpha$-ol

$$> 3\text{-oxo} > 7\text{-oxo} \geqslant 12\text{-oxo}$$

The results found by Hofmann[8] for 5β-cholanoic acids and methyl cholanoates have shown that a 6β-hydroxyl is less polar than the 6α-, even when a vicinal 7-hydroxyl group is present. When hydroxyl groups are in the positions C-7 and C-11, the α-configuration is also the more polar. A 12α-hydroxyl group in a trihydroxy-5β-cholanoic acid confers much greater polarity than a 16α-group, but a 23-hydroxyl group renders the compound even more polar than its 12α-hydroxy analog.

It is noteworthy that generally the bile acids with a ring A/B-*cis* configuration (5β) travel faster than those with a *trans*-configuration (5α). This difference of polarity allows the separation of bile acids or their methyl esters differing in their A/B-ring junction, such as cholic acid and allocholic acid (R_f values, 0.68 and 0.59; silica gel G, system: ethyl acetate–acetic acid–water 85:10:5);[298] methylcholanoate and methylallocholanoate ($R_s \times$ values: 100 and 70; silica gel G, cyclohexane–ethyl acetate–acetic acid 7:23:20);[299] and the pairs $3\beta,6\beta$-dihydroxy-$5\alpha/5\beta$-cholanoic acids ($R_f = 0.41$ and 0.44) and $3\beta,6\alpha$-dihydroxy-$5\alpha/5\beta$-

[298] T. Sasaki, *Hiroshima J. Med. Sci.* **14**, 85 (1965).
[299] A. Kallner, *Acta Chem. Scand.* **18**, 1502 (1964).

TABLE VII

CHROMATOGRAPHIC MOBILITIES (R_f or R_S × 100 VALUES) FOUND ON TLC FOR METHYL ESTER DERIVATIVES OF 5β-CHOLAN-24-OIC ACIDS (5β-METHYLCHOLANOATES)

Structure	Chromatographic systems[a]												
	Ia	Ib	Ic	IIa	IIb	IIc	IId	IIe	IIIa	IIIb	IIIc	IVa	IVb
No substitution	—	—	—	—	—	—	—	—	96	—	—	—	—
3α-ol	32	41	—	8	100	—	—	—	—	100	—	93	97
3β-ol	36	54	—	13	104	—	—	—	—	117	—	—	—
7α-ol	82	83	—	32	155	—	—	—	—	166	—	—	—
12α-ol	82	83	—	40	—	—	—	—	—	—	—	—	—
3α,7α-ol	—	—	—	—	—	94	—	—	—	—	89	46	91
3α,7β-ol	—	—	—	—	—	114	—	—	—	—	112	—	—
3α,12α-ol	—	—	—	—	18	100	—	—	—	—	100	—	—
7α,12α-ol	—	—	—	—	69	200	—	—	—	78	136	—	86
3α,6α-ol	—	—	—	—	—	51	—	—	—	—	55	34	—
3α,6β-ol	—	—	—	—	—	103	—	—	—	—	115	—	—
3α,11α-ol	—	—	—	—	—	122	—	—	—	—	107	—	—
3α,11β-ol	—	—	—	—	64	221	—	—	—	71	162	—	—
3β,12α-ol	—	—	—	—	—	138	—	—	—	—	139	—	—
3α,7α,12α-ol	0	0	0	—	—	13	—	100	—	—	17	8	80
3α,6α,7α-ol	—	—	—	—	—	24	—	149	—	—	22	—	—
3α,6β,7α-ol	—	—	—	—	—	26	—	118	—	—	32	—	—
3α,6β,7β-ol	—	—	—	—	—	31	—	178	—	—	43	—	—
3α,11β,12α-ol	—	—	—	—	—	67	—	—	—	—	76	—	—
3-oxo	82	83	—	32	173	—	—	—	—	141	—	—	—
7-oxo	89	90	—	54	192	—	—	—	—	169	—	—	—

Compound	Ia	Ib	Ic	IIa	IIb	IIc	IId	IIe	IIIa	IIIb	IIIc	IVa	IVb
12-oxo	41	—	66	—	—	—	—	—	—	—	—	—	—
3,7-oxo	—	32	—	119	—	—	—	—	70	—	—	—	—
3,12-oxo	—	—	—	115	—	—	—	—	71	—	—	—	—
7,12-oxo	—	—	—	148	—	—	—	—	104	—	—	—	—
3,6-oxo	—	—	—	139	—	—	—	—	66	—	—	—	—
3,7,12-oxo	14	9	—	66	—	—	—	39	25	—	120	—	—
3α-ol, 12-oxo	—	—	—	49	—	—	—	—	43	—	—	—	—
12α-ol, 3-oxo	—	—	—	42	—	—	—	—	53	—	—	—	—
3α-ol, 7-oxo	—	—	—	38	—	—	—	—	35	—	—	—	—
7α-ol, 3-oxo	—	—	—	36	—	—	—	—	66	—	—	—	—
7α,12α-ol, 3-oxo	0	3	—	—	—	—	—	88	—	—	85	—	—
3α,12α-ol, 7-oxo	0	1	—	—	—	—	—	60	—	—	56	—	—
3α,7α-ol, 12-oxo	—	—	—	—	—	—	—	62	—	—	62	—	—
3α-ol, 7,12-oxo	—	—	—	—	—	—	—	126	—	—	71	—	—
7α-ol, 3,12-oxo	—	—	—	—	—	—	—	103	—	—	123	—	—

a Comments on the experimental conditions: **I:**[b] Conditions as **II** of Table VI. Systems: **Ia,** Bz/EtOH 80:20; **Ib,** Hex/EtOH 80:20; **Ic,** Hex/EtAc 70:30.

II:[c] Conditions as for **III** of Table VI. Systems: **IIa,** Eth/Hpt 30:70; S = methyl oleate (R_f = 0.63); **IIb,** Atn/Bz 15:85; S = methyl lithocholate (R_f = 0.40); **IIc,** Atn/Bz 30:70; S = methyl deoxycholate (R_f = 0.38); **IId,** AcOH/iPrEth/iOct 25:25:50; S = deoxycholic acid (R_f = 0.27); **IIe,** MeOH/Atn/Chf 5:25:70; S = methyl cholate (R_f = 0.16).

III:[c] Conditions as for **III** of Table VI, but Anasyl B (Anal. Eng. Labs. Inc., Hamden, Connecticut) is employed as adsorbent. Systems: **IIIa,** Eth/Hpt 8:92; S = methyl oleate (R_f = 0.35); **IIIb,** Atn/nBuEth 15:85; S = methyl lithocholate; **IIIc,** Atn/nBuEth 30:70; S = methyl deoxycholate (R_f = 0.46).

IV:[d] Conditions and systems as **IVa** and **IVb** of Table VI.

b S. Hara and M. Takeuchi, *J. Chromatog.* **11,** 565 (1963).

c A. F. Hofmann, *in* "New Biochemical Separations" (A. T. James and L. J. Morris, eds.), p. 262. Van Nostrand, Princeton, New Jersey, 1964.

d T. Usui, *J. Biochem. (Tokyo)* **54,** 283 (1963).

cholanoic acids ($R_f = 0.22$ and 0.44), both on silica gel layers, system: ether acetic acid (99.6:0.4).[300]

However, the results obtained by Kallner[301] on silica gel layers show that in the system[103] trimethylpentane–ethyl acetate–acetic acid 5:25:2 the allo-acids are less polar than their 5β-isomers, with the following sequence of polarity: 3α,12α-dihydroxy-5β-cholanoic acid ($R_s \times 100$ values = 100) > 3β,12α-dihydroxy-5β-cholanoic acid (125) > 3α,12α-di-hydroxy-5α-cholanoic acid (133) > 3β,12α-dihydroxy-5α-cholanoic acid (146) > 12α-hydroxy-3-oxochol-4-enoic acid (200) > 12α-hydroxy-3-oxo-5β-cholanoic acid (231) > 12α-hydroxy-3-oxo-5α-cholanoic acid (250). Under the same conditions no inversion of polarity was found for the methyl ester derivatives:[302] methylchenodeoxycholate ($R_f \times 100 =$ 46) is less polar than its 5α-isomer methylallochenodeoxycholate ($R_f \times 100 = 36$).

For the separation of some pairs of bile acids of biochemical interest, Eneroth[103] has found the methyl esters to offer a distinct advantage. For instance, the methyl esters of chenodeoxycholic and deoxycholic acids have very different migrations in systems of cyclopentane–tetrahydro-furane–acetic acid-type, in the proportions 20:8.5:0.25 (R_s = values are, respectively, 1.46 and 1.77; S = deoxycholic acid). The formation of esters can be useful for the characterization of bile acids. In the system trimethylpentane–isopropyl alcohol–acetic acid 30:30:10, the methyl and ethyl ester derivatives of 7-oxodeoxycholic acid show R_s values, respectively, of 0.77 and 0.84 (S = deoxycholic acid).

The chromatographic migration characteristics of a number of steroid acids with short side chains, C_{20}-, C_{22}-, and C_{23}-steroids related to the bile acids and some steroid acids related to the C_{29}-sterols have been summarized in Table VIII. The mobilities of their methyl ester derivatives are given in Table IX. They are important as degradation products of steroids. As observed also for the 5β-cholanoic acids, stero-acids with an 11α-hydroxyl group are less strongly adsorbed than the 12α-isomer.

Morimoto[303] has investigated the behavior of eight homologous 3α,-7α,12α-trihydroxy-sterobile acids on silica gel G; the R_f values which have been found for these acids and their methyl esters (Table X) show that the distance traveled by a steroid acid increases with the number of carbons of the side chain.

The following are the chromatographic mobilities on silica gel G layers

[300] S. Hara and M. Takeuchi, *J. Chromatog.* **11**, 565 (1963).
[301] A. Kallner, *Arch. Kemi* **26**, 555 (1967).
[302] A. Kallner, *Arch. Kemi* **26**, 567 (1967).
[303] K. Morimoto, *J. Biochem.* (*Tokyo*) **55**, 410 (1964).

found for haemulcholic acid ($3\alpha,7\alpha,22\beta$-trihydroxy-5β-cholanic acid), recently isolated from the bile of a teleostian fish (*Parapristipoma trilineatum*) :[304] benzene-dioxane-acetic acid 55:40:2, 0.19 (cholic acid: 0.23); benzene-2-propanol-acetic acid 30:10:1, 0.23 (cholic acid: 0.51).

Conjugated di- and trihydroxy bile acids with glycine and taurine can be easily separated by TLC.[133, 144, 298, 304, 305] For taurocholic, taurodeoxycholic, glycocholic and glycodeoxycholic acids on silica gel G layers, the following $R_f \times 100$ values have been found: butanol–acetic acid–water (70:10:20) :[144] 12, 25, 56, and 73; butanol–acetic acid–water (10:1:1) :[133] 22, 34, 51, and 69; and chloroform–n-butanol–acetic acid–water (40:30: 20:10) :[298] 20, 32, 60, and 82. For conjugated bile acids, with identical numbers of hydroxyl groups, separation has not so far been achieved (taurocholic/tauroallocholic acids; taurodeoxycholic/taurochenodeoxycholic acids) or has been only partially realized ($R_f \times 100$ values for chenodeoxycholic and ursodeoxycholic in the system ethyl acetate–acetic acid–water, 85:10:5 on silica gel G: respectively, 46 and 49). Taurocholic (I), taurochenodeoxycholic (III), and taurohaemulcholic (II) acids present on silica gel layers[304] using solvent system n-butanol–acetic acid– water 85:10:5, gave the following order of mobility (R_f values, front = 15.5): I (0.21) < II (0.26) < III (0.41).

Gregg[305] has shown the advantages of wedge strip TLC over the normal techniques for the separation of conjugated bile acids. Using isooctane–isopropyl ether–acetic acid–isopropyl alcohol (2:1:1:1), a complete separation could be obtained for glycochenodeoxycholic ($R_f \times 100 = 36$) and glycodeoxycholic (39) acids, both with two hydroxyl groups.

3. Bile Alcohols

Kazuno and Hoshita[102] have studied the behavior of 40 bile alcohols on silica gel layers, and compared the resolution obtained with other methods of fractionation, such as partition paper and reversed-phase column chromatography. Sasaki[298] has examined a series of systems for the separation of polyhydroxy bile alcohols and their sulfates.

Most of the penta and tetrahydroxy isomers differing only by the position of the primary or secondary hydroxyl groups on the side chain are separated on TLC (Table X), and only the two pairs $3\alpha,7\alpha,12\alpha$, 24,26-/$3\alpha,7\alpha,12\alpha$, 25,26-pentahydroxycoprostanes and 26-deoxy-/25-deoxy-5β-bufols remained unresolved in the investigated systems. Unsaturated isomers differing by the position of the double bond on the

---

[304] T. Hoshita, S. Hirofuji, T. Sasaki, and T. Kazuno, *J. Biochem.* (*Tokyo*) **61**, 136 (1967).

[305] J. A. Gregg, *J. Lipid Res.* **7**, 579 (1966).

TABLE VIII

Chromatographic Mobilities Found on TLC for Stero-Bile Acids of the Etiocholane, Bisnorcholane, Norcholane, Cholane, Homocholane, Bishomocholane, Coprostane, 24-Methylcoprostane, and 24-Ethylcoprostane Series

Stero-bile acids	Mobility in the solvent systems[c]					
	Ia	Ib	II	IIIa	IIIb	IV
C-20						
3β-Hydroxyetiochol-5-enoic	21	46	—	—	—	—
3β-Acetoxyetiochol-5-enoic	32	56	44	—	—	—
3β-Hydroxyetiochola-5,16-dienoic	13	31	—	—	—	—
C-22						
3α,12α-Dihydroxy-5β-etiocholanoic	—	—	—	72	67	—
3β-Hydroxybisnorchol-5-enoic	14	38	—	—	—	—
3β-Acetoxybisnorchol-5-enoic	27	—	—	—	—	—
3α,11α-Dihydroxybisnorcholanoic	—	—	—	75	87	—
3α,12α-Dihydroxybisnorcholanoic	—	—	—	92	84	—
3α,7α,12α-Trihydroxybisnorcholanoic	—	—	—	—	—	15
C-23						
3α,11α-Dihydroxynorcholanoic	—	—	—	83	113	—
3α,12α-Dihydroxynorcholanoic	—	—	—	97	110	—
3α,7α,12α-Trihydroxynorcholanoic	—	—	—	—	—	18
C-24						
3β-Hydroxy-5β-cholanoic[a]	9	23	—	—	—	—
3β-Hydroxychol-5-enoic	10	23	—	—	—	—
3β-Acetoxy-5β-cholanoic	23	48	—	—	—	—
3β-Acetoxychol-5-enoic	20	46	23	—	—	—
3α,11α-Dihydroxycholanoic	—	—	—	85	120	—
3α,12α-Dihydroxycholanoic	—	—	—	100	100	—
3α,7α,12α-Trihydroxycholanoic	—	—	—	—	—	21
C-25						
3α,7α,12α-Trihydroxyhomocholanoic	—	—	—	—	—	27
C-26						
3α,7α,12α-Trihydroxybishomocholanoic	—	—	—	—	—	32
C-27						
3α,7α,12α-Trihydroxycoprostanoic	—	—	—	—	—	40
C-28						
3α,7α,12α-Trihydroxy-24-methylcoprostanoic[b]	—	—	—	—	—	45
C-29						
3α,7α,12α-Trihydroxy-24-ethylcoprostanoic	—	—	—	—	—	43

[a] For other chromatographic mobilities, see Table VII.

[b] Two isomers: trihydroxyisosterocholanoic and trihydroxybufosterocholanoic acids.

[c] Comments on the experimental conditions: **I:** Spread TLC, aluminum oxide (with 2.5% AcOH).[d] Systems: **Ia,** Chf; **Ib,** Chf/EtOH 98:2. $R_f \times 100$ values.

II:[e] Spread TLC, aluminum silicate (Florisil Lachema, 100/200 mesh), with 1.5% AcOH; 600 μ thick. $R_f \times 100$ values. Chf/EtOH 96:4.

III:[f] n-TLC, silica gel G; S.l. = 1.5 cm; Fr. = 10 cm; $R_S \times 100$ values. Systems: **IIIa,** AcOH/iPrEth/iOct 25:25:50 (S = deoxycholic acid, R_f = 0.27); **IIIb,** AcOH/n-PrOH/Bz/CCl$_4$/iPrEth/iAmAc 5:10:10:20:30:40 (S = deoxycholic acid, R_f = 0.33).

IV:[g] n-TLC, silica gel; Chf/EtAc/AcOH 45:45:10; $R_f \times 100$ values; Dv. = 12 cm.

[d] S. Heřmánek, V. Schwarz, and Z. Čekan, Collection Czech. Chem. Commun. **26,** 1669 (1961).

[e] V. Schwarz, Pharmazie **18,** 122 (1963).

[f] A. F. Hofmann, in "New Biochemical Separations" (A. T. James and L. J. Morris, eds.), p. 262. Van Nostrand, Princeton, New Jersey, 1964.

[g] K. Morimoto, J. Biochem. (Tokyo) **55,** 410 (1964).

TABLE IX

CHROMATOGRAPHIC MOBILITIES FOUND ON TLC FOR THE METHYL ESTER DERIVATIVES OF STEROID ACIDS OF THE ETIOCHOLANE, BISNORCHOLANE, NORCHOLANE, AND CHOLANE SERIES

Steroid acids, methyl ester	Mobility in the solvent systems[a]										
	Ia	Ib	Ic	Id	IIa	IIb	III	IVa	IVb	Va	Vb
C-20											
3β-Hydroxyetiochol-5-enoic acid	—	—	—	—	54	50	—	—	—	—	—
3β-Acetoxy-5β-etiochol-14-enoic acid	47	—	—	—	—	—	—	—	—	—	—
3β-Acetoxy-5β-etiochol-8(14)-enoic acid	47	—	—	—	—	—	—	—	—	—	—
3β-Acetoxy-5β-etiocholanoic acid	47	—	—	—	—	—	—	—	—	—	—
3β-Acetoxy-14β-hydroxy-5β-etiocholanoic acid	22	—	—	—	—	—	76	—	—	—	—
3β-Acetoxy-14α-hydroxy-5β-etiocholanoic acid	—	—	—	—	—	—	65	—	—	—	—
3β-Acetoxy-14α,15α-oxido-5β-etiocholanoic acid	23	—	—	—	—	—	—	—	—	—	—
3β-Acetoxy-8α,14α-oxido-5β-etiocholanoic acid	18	—	—	—	—	—	87	—	—	—	—
3β-Acetoxy-15α-hydroxy-5β-etiocholanoic acid	16	—	—	—	—	—	—	—	—	—	—
3β,12β-Dihydroxy-5β-etiocholanoic acid	—	54	8	—	—	—	—	—	—	—	—
3β-Acetoxy-12β-hydroxy-5β-etiocholanoic acid	—	—	40	—	—	—	—	—	—	—	—
3β,12β-Diacetoxy-5β-etiocholanoic acid	—	—	52	—	—	—	—	—	—	—	—
3β,12β-Diacetoxy-14β-hydroxy-5β-etiocholanoic acid	—	—	42	41	—	—	—	—	—	—	—
3α-Acetoxy-14β-hydroxy-12-oxo-5β-etiocholanoic acid	—	—	—	—	—	—	—	—	—	—	—
3β-Acetoxy-12-oxo-5β-etiochol-14-enoic acid	—	—	48	—	—	—	—	—	—	—	—
3β,12β-Diacetoxy-11-oxo-5β-etiochol-14-enoic acid	—	—	32	—	—	—	—	—	—	—	—
3α-Acetoxy-12-oxo-5β-etiocholanoic acid	—	—	58	—	—	—	—	—	—	—	—
3α-Acetoxy-12α-hydroxy-5β-etiocholanoic acid	—	—	52	—	—	—	—	—	—	—	—
3β,12β,14β-Trihydroxy-5β-etiocholanoic acid	—	—	—	—	—	—	—	—	—	35	—
12β,14β-Dihydroxy-3-oxo-5β-etiocholanoic acid	—	—	—	—	—	—	—	—	—	42	—
14β-Hydroxy-3,12-dioxo-5β-etiocholanoic acid	—	—	—	—	—	—	—	—	—	54	—
12β-Acetoxy-14β-hydroxy-3-oxo-5β-etiocholanoic acid	—	—	—	—	—	—	—	—	—	—	67
12β-Acetoxy-14β-hydroxy-3-oxo-5β-etiochol-4-enoic acid	—	—	—	—	—	—	—	—	—	—	60

Compound						
12β-Acetoxy-14β-hydroxy-3-oxo-4β-brom-5β-etiocholanoic acid	—	—	—	—	—	79
C-22 3α,12α-Dihydroxyetiocholanoic acid	52	46	—	62	71	—
3β-Hydroxybisnorchol-5-enoic acid	—	—	—	—	—	—
3α,11α-Dihydroxybisnorcholanoic acid	—	—	—	100	101	—
3α,12α-Dihydroxybisnorcholanoic acid	—	—	—	84	87	—
C-23 3α,11α-Dihydroxynorcholanoic acid	—	—	—	119	116	—
3α,12α-Dihydroxynorcholanoic acid	—	—	—	89	96	—
C-24 3α,11α-Dihydroxycholanoic acid	—	—	—	122	115	—
3α,12α-Dihydroxycholanoic acid	—	—	—	100	100	—
3β-Hydroxychol-5-enoic acid	50	45	—	—	—	—

[a] *Comments on the experimental conditions:* **I:**[b] n-TLC, silica gel G; S.l. = 1.5 cm; I.l. = 0.5 cm; Fr. = about 18 cm; $R_f \times 100$ values. Systems: **Ia,** EtAc/Ch 15:85; **Ib,** EtAc/Hex 75:25; **Ic,** EtAc/Ch 30:70; **Id,** EtAc/Ch 25:75. **II:**[c] Spread TLC, as for **I** in Table VIII. Systems: **IIa,** Bz/EtOH 97:3; **IIb,** PtEth/Chf 1:1. **III:**[d] n-TLC as for **I.** EtAc/Ch 2:1. **IV:**[e] n-TLC; Fr. = 10 cm; S.l. = 1.5 cm; $R_S \times 100$ values. Systems: **IVa,** silica gel G; Atn/Bz 30:70 (S = methyl deoxycholate, $R_f = 0.38$); **IVb,** anasil B; Atn/nBuEth 30:70 (S = methyl deoxycholate, $R_f = 0.46$). **V:**[f] n-TLC, ribbed-glass/silica gel G; $R_f \times 100$ values. **Va,** EtAc; **Vb,** EtAc/Ch 2:1.

[b] M. Barbier, H. Jäger, H. Tobias, and E. Wyss, *Helv. Chim. Acta* **42,** 2440 (1959).

[c] S. Heřmánek, V. Schwarz, and Z. Čekan, *Collection Czech. Chem. Commun.* **26,** 1669 (1961).

[d] A. Lardon and T. Reichstein, *Helv. Chim. Acta* **46,** 392 (1963).

[e] A. F. Hofmann, *in* "New Biochemical Separations" (A. T. James and L. J. Morris, eds.), p. 262. Van Nostrand, Princeton, New Jersey, 1964.

[f] J. Von Euw and T. Reichstein, *Helv. Chim. Acta* **46,** 142 (1963).

TABLE X

CHROMATOGRAPHIC MOBILITIES FOUND ON SILICA GEL LAYERS FOR BILE POLYHYDROXYSTEROLS AND BILE KETOSTEROLS

Bile sterols	Ring and carbons[a]	$R_f \times 100$ values in solvent systems[c]								
		Ia	Ib	Ic	Id	Ie	If	Ig	II	III
A. Bile hydroxysterols										
3α,7α,12α,24,26,27-Hexahydroxycoprostane (1)[b]	5β C$_{27}$	5	16	13	—	—	—	—	21	—
3α,7α,12α,24,25-Pentahydroxyhomocholane	5β C$_{25}$	8	28	22	—	—	—	—	—	—
3α,7α,12α,24,26-Pentahydroxybishomocholane (2)	5β C$_{26}$	8	34	25	—	—	—	—	32	—
3α,7α,12α,24,25-Pentahydroxycoprostane	5β C$_{27}$	16	53	31	—	—	—	—	—	—
3α,7α,12α,24,26-Pentahydroxycoprostane (3)	5β C$_{27}$	15	45	26	—	—	—	—	—	—
3α,7α,12α,25,26-Pentahydroxycoprostane (4)	5β C$_{27}$	15	52	31	—	—	—	—	43	—
3α,7α,12α,26,27-Pentahydroxycoprostane (5)	5β C$_{27}$	15	49	31	—	—	—	—	43	—
3α,7α,12α,26,27-Pentahydroxycholestane (6)	5α C$_{27}$	11	46	29	—	—	—	—	40	—
3α,7α,12α,26,27-Tetrahydroxy-24,27-oxidocoprostane (7)	5β C$_{27}$	10	56	34	—	—	—	—	—	—
3α,7α,12α,24-Tetrahydroxycholane	5β C$_{24}$	21	61	40	—	—	—	—	—	10
3α,7α,12α,25-Tetrahydroxyhomocholane	5β C$_{25}$	28	70	47	—	—	—	—	—	—
3α,7α,12α,24-Tetrahydroxybishomocholane (8)	5β C$_{26}$	35	76	56	—	—	—	—	69	—
3α,7α,12α,26-Tetrahydroxybishomocholane (9)	5β C$_{26}$	34	76	55	—	—	—	—	—	—
3α,7α,12α,24-Tetrahydroxycoprostane	5β C$_{27}$	45	85	64	—	—	—	—	—	—
3α,7α,12α,25-Tetrahydroxycoprostane (10)	5β C$_{27}$	37	79	57	—	—	—	—	—	—
3α,7α,12α,26-Tetrahydroxycoprostane (11)	5β C$_{27}$	38	80	59	—	—	—	—	82	—
3α,7α,12α,26-Tetrahydroxycholestane (12)	5α C$_{27}$	33	77	54	—	—	—	—	75	—
3α,7α,12α,24-Tetrahydroxy-24-ethylbishomocholane	5β C$_{28}$	42	82	60	—	—	—	—	—	—
3α,7α,12α,24-Tetrahydroxy-24-ethylcoprostane	5β C$_{29}$	42	82	60	—	—	—	—	—	—
3α,12α,24-Trihydroxycholane	5β C$_{24}$	—	—	—	53	22	43	—	—	34
3α,7α,12α-Trihydroxycholane	5β C$_{24}$	—	—	—	54	23	44	—	—	—
3α,7α,12α-Trihydroxyhomocholane	5β C$_{25}$	—	—	—	53	22	43	—	—	—
3α,7α,12α-Trihydroxyhomochol-24-ene	5β C$_{25}$	—	—	—	56	24	44	—	—	—
3α,7α,12α-Trihydroxybishomocholane	5β C$_{26}$	—	—	—	—	—	—	—	—	—

Compound	Config.									
3α,7α,12α-Trihydroxybishomochol-24-ene	5β C_{26}	—	54	22	44	—	—	—	—	—
3α,7α,12α-Trihydroxycoprostane	5β C_{27}	—	58	25	47	—	—	—	—	—
3α,7α,12α-Trihydroxycoprost-24-ene	5β C_{27}	—	56	23	47	—	—	—	—	—
3α,7α,12α-Trihydroxycoprost-25-ene	5β C_{27}	—	56	23	47	—	—	—	—	—
3α,7α,12α-Trihydroxy-24-methylbishomocholane	5β C_{27}	—	56	25	45	—	—	—	—	—
3α,7α,12α-Trihydroxy-24-methylcoprostane	5β C_{28}	—	59	27	49	—	—	—	—	—
3α,7α,12α-Trihydroxy-24-ethylbishomocholane	5β C_{28}	—	60	28	51	—	—	—	—	—
3α,7α,12α-Trihydroxy-24-n-butylcholane	5β C_{28}	—	58	26	49	—	—	—	—	—
3α,7α,12α-Trihydroxy-24-n-pentylcholane	5β C_{29}	—	63	30	52	—	82	—	—	—
3α,24-Dihydroxycholane	5β C_{24}	—	—	—	—	—	—	—	—	—
B. Bile ketosterols										
3α,7α,12α-Trihydroxycholan-24-al	5β C_{24}	40	—	—	17	—	—	—	—	—
3α,7α,12α-Trihydroxyhomocholan-24-one	5β C_{25}	37	—	—	25	—	—	—	—	—
3α,7α,12α-Trihydroxybishomocholan-24-one	5β C_{26}	48	—	—	36	—	—	—	—	—
3α,7α,12α-Trihydroxybishomocholan-25-one	5β C_{26}	42	—	—	30	—	—	—	—	—
3α,7α,12α-Trihydroxycoprostan-24-one	5β C_{27}	53	—	—	38	—	—	—	—	—
Homocholane-3,7,12,24-tetraone	5β C_{25}	—	—	—	—	37	—	—	—	—
Bishomocholane-3,7,12,24-tetraone	5β C_{26}	—	—	—	—	48	—	—	—	—
Coprostane-3,7,12,24-tetraone	5β C_{27}	—	—	—	—	56	—	—	—	—

[a] Configuration of ring A/B and number of carbons.

[b] Trivial names: (1) 5β-scymnol; (2) 5β-ranol; (3) 27-deoxyscymnol; (4) 5β-bufol; (5) 24-deoxyscymnol; (6) 5α-cyprinol; (7) anhydroscymnol; (8) 26-deoxy-5β-ranol; (9) 24-deoxy-5β-ranol; (10) 26-deoxy-5β-bufol; (11) 25-deoxy-5β-bufol; (12) 27-deoxy-5α-cyprinol.

[c] Comments on the experimental conditions: I:[d] n-TLC, silica gel (with 5% gypsum); Dv. = 10–12 cm. Systems: Ia, EtAc/Atn 70:30; Ib, Chf/EtOH 80:20; Ic, Chf/Atn/EtOH 70:15:15; Id, EtAc/Atn 80:20; Ie, Bz/EtAc 40:60; If, Chf/EtOH 90:10; Ig, Bz/EtAc 60:40.

II:[e] n-TLC, silica gel G; Sl. = 2 cm; Fr. = 13 cm. EtAc/AcOH/H_2O 85:10:5.

III:[f] n-TLC, Wakogel (Wako Chemical Co., Ltd., Tokyo); Dv. = 15 cm (20°). Chf/MeOH 9:1.

[d] T. Kazuno and T. Hoshita, Steroids **3**, 55 (1964).

[e] T. Sasaki, Hiroshima J. Med. Sci. **14**, 85 (1965).

[f] S. Hara and M. Takeuchi, J. Chromatog. **11**, 565 (1963).

side chain (C-24/C-25) and the pairs Δ^{24} or Δ^{25}-unsaturated/saturated bile alcohols have similar chromatographic mobilities. The two 24-epimeric 5β-cholestane-3α,7α,12α,24-tetrols have been synthesized and separated on silica gel TLC recently by Masui and Staple;[306] following are the $R_f \times 100$ values found, respectively, for the 24α- and 24β-isomers in the systems (a) benzene–isopropanol–acetic acid 30:10:1, 39 and 44; (b) ethyl acetate–acetone 70:30, 20 and 28; (c) chloroform–acetone–ethanol 70:15:15, 37 and 48; and (d) chloroform–ethanol 80:20, 70 and 85.

The length of the side chain influences the polarity of the steroid: the mobility of the bile alcohols increases with the number of carbons. Sometimes the difference of migration of two homologous alcohols is too small to permit their separation, such as for the steroid pairs trihydroxycholane/trihydroxyhomocholane and tetrahydroxy-24-ethylcholane/tetrahydroxyhomocholane. More polar homologous pairs, such as 5β-ranol/26-deoxyscymnol and 24-deoxy-5β-ranol/25-deoxy-5β-bufol, are completely separated.

From the TLC data of Kazuno and Hoshita[102] for tetrahydroxycoprostanes (Table X) the following order of polarity can be established for hydroxyl groups in the 5β-C_{27}-bile alcohol series: 25-ol (tertiary) $\geqslant$ 26-ol (primary) > 24-ol (secondary). The presence of vicinal hydroxyl groups, as in the pentahydroxy alcohols, alters this rule: a 26-hydroxyl can be much more polar than a 25-hydroxyl vicinal to a 24-group, and the polarities of 24- and 26-hydroxyls are similar in the presence of a vicinal 25-hydroxyl group. In the bishomocholane series, secondary 24- and primary 26-hydroxyl groups lead to the same polarity.

The use of the systems of Sasaki[298] for bidimensional-TLC on silica gel layers gives satisfactory separations of some sulfonic acid derivatives of bile alcohols. In the system ethyl acetate–methanol–acetic acid–water (65:20:10:5), it is possible to resolve the pairs 5β-ranol-24-sulfonic/5β-cyprinol-26-sulfonic acids and 5α-cyprinol-26-sulfonic/5β-bufol-26-sulfonic acids, whereas only a partial separation between the epimeric 5α- and 5β-cyprinol derivatives was obtained. Good separations of 5β-scymnol-26-sulfonic and 5β-ranol-24-sulfonic acids are accomplished in the systems n-butanol–acetic acid–water (85:10:5) (R_f values, 0.39 and 0.45) and chloroform–methanol–acetic acid–water (65:20:10:5) (R_f values, 0.19 and 0.26). This latter system permits a good resolution of 5β-bufol-26-sulfonic acid and 26-deoxy-5β-ranol-24-sulfonic acid (R_f values: 0.28 and 0.35).

[306] T. Masui and E. Staple, *Steroids* **9**, 443 (1967).

4. Cardenolides and Bufogenins

Adsorption TLC[122, 307] and partition TLC[109] applied to the cardenolides allow good resolution of the steroid pairs differing in their A/B-ring junction as well as in their side-chain configuration. Uzarigenin, 3-epi-uzarigenin, and allouzarigenin can be separated[122] either as the free compounds ($R_f \times 100$ values in ethyl acetate: 100, 89, and 81) or as the acetates (R_s = allouzarigenin) $\times 100$ in butyl acetate: 400, 130, and 270). In some cases, overrun TLC has been successfully employed to separate steroids of similar mobility. Janiak et al.[87] using the Brenner-Niederwieser device have obtained in 7 hours overrun TLC a complete separation on silica gel of the two following pairs of acetates: dihydro-digitoxigenin/$\Delta^{8(14)}$-anhydrodihydrodigitoxigenin (13.9 and 9.3 cm migration in ethyl acetate/cyclohexane 2:1) and adynerigenin/14α-adynerigenin (9.7 and 10.6 cm migration in ethyl acetate–cyclohexane 1:2).

The chromatographic mobilities reported for 36 cardenolides are summarized in Table XI. 11β-Hydroxycardenolides (e.g., 11-episarmento-genin) move faster than the 11α-compound (sarmentogenin). Cardeno-lides with a 17α side chain (17β-H, allo series) are more polar than those with a 17β-butenolide group (natural or normal series).

For the evaluation of the stability of helveticoside (digitoxose-strophanthidin) in pharmaceutical preparations, Lutomski et al.[307a] separated helveticoside from the cardenolides helveticosol (I), strophan-thidin (II), and strophanthidol (III) on 500 μ-thick layers of silica gel–gypsum (85:15). The following are the mobilities ($R_f \times 100$ values) found for these substances in the solvent systems (a) chloroform–ethanol 3:1 and (b) acetone–benzene 5:3: helveticoside: system a = 63, system b = 60; cardenolides: (I) system a = 47, system b = 52, (II) system a = 72, system b = 75, and (III) system a = 66, system b = 70.

Tschesche et al.[308] have submitted cardenolide–methyl ether deriva-tives to TLC for the characterization of canarigenin and 3-epicanarigenin. These methyl ethers show on silica gel G $R_s \times 100$ values, respectively, of 84 and 90 (S = anhydrocanarigenin) on development in acetic acid. They could be resolved also after two successive developments in benzene–ethyl acetate (first run, 70:30; second run, 60:40), with $R_s \times 100$ values of 62 (canarigenin methyl ether) and 68 (3-epi compound). Zelnik and Ziti[309] have investigated the resolution of bufogenins on silica gel G

[307] I. Herrmann and K. Repke, Arch. Exptl. Pathol. Pharmakol. 248, 351 (1964).
[307a] J. Lutomski, Z. Kowalewski, and M. Kortus, Dissertationes Pharm. 18, 409 (1966).
[308] R. Tschesche, G. Snatzke, J. Delgado, and A. G. Gonçales, Ann. Chem. 663, 157 (1963).
[309] R. Zelnik and L. M. Ziti, J. Chromatog. 9, 371 (1962).

TABLE XI

Cardenolides	Structure		Chromatographic systems[a]		
			I	IIa	IIb
14α,15α-Oxido-14-anhydrodigitoxigenone	5β	3-oxo, 14α,15α-oxido	—	—	—
Adynerigenin	5β	3β-ol, 8,14β-oxido	—	—	—
Uzarigenone	5α	14β-ol, 3-oxo	—	—	—
14α,15α-Oxido-14-anhydrodigitoxigenin	5β	3β-ol, 14α,15α-oxido	—	—	—
14β,15β-Oxido-14-anhydrodigitoxigenin	5β	3β-ol, 14β,15β-oxido	—	—	—
β-Anhydrodigitoxigenin	5β	$\Delta^{14(15)}$-3β-ol	—	—	—
Dianhydrogitoxigenin		$\Delta^{14(15):16}$-3β-ol	79	—	—
Anhydroallouzarigenin	5α	$\Delta^{14(15)}$-3β-ol, 17β(H)	—	—	—
Anhydroadynerigenin	5β	$\Delta^{8(9):14(15)}$-3β-ol	—	—	—
Anhydrocanarigenin	—	$\Delta^{4:14(15)}$-3β-ol	—	—	—
Anhydroallocanarigenin	—	$\Delta^{4:14(15)}$-3β-ol, 17β(H)	—	—	—
4-Dehydrodigitoxigenin	—	Δ^{4}-14β-ol, 3-oxo	—	—	—
15-Oxodigitoxigenin	5β	14β-ol, 3,15-oxo	—	—	—
Anhydrodigoxigenin	5β	$\Delta^{14(15)}$-3β,12β-ol	—	—	—
Digitoxigenin	5β	3β,14β-ol	60	60	82
Canarigenin	—	Δ^{4}-3β,14β-ol	—	—	—
Uzarigenin	5α	3β,14β-ol	—	—	—
Xysmalogenin	—	Δ^{5}-3β,14β-ol	—	—	—
3-Epiuzarigenin	5α	3α,14β-ol	—	—	—
Allouzarigenin	5α	3β,14β-ol, 17β(H)	—	—	—
11-Oxouzarigenin	5α	3β,14β-ol, 11-oxo	—	—	—
7β-Hydroxy-4-dehydrodigitoxigenone	—	Δ^{4}-7β,14β-ol, 3-oxo	—	—	—
7β-Hydroxydigitoxigenin	5β	3β,7β,14β-ol	—	—	—
Sarmentogenin	5β	3β,11α,14β-ol	—	—	—
3-Episarmentogenin	5β	3α,11α,14β-ol	—	—	—
11-Episarmentogenin	5β	3β,11β,14β-ol	—	—	—
Mallogenin	5α	3β,11β,14β-ol	—	—	—
Digoxigenin	5β	3β,12β,14β-ol	45	39	66
3-Epidigoxigenin	5β	3β,12β,14α-ol	—	—	—
15α-Hydroxydigitoxigenin	5β	3β,14β,15α-ol	—	—	—
Gitoxigenin (deacetyloleandrigenin)	5β	3β,14β,16β-ol	50	45	72
3-Epigitoxigenin	5β	3α,14β,16β-ol	—	—	—
Periplogenin	5β	3β,5β,14β-ol	—	—	—
7β,12β-Dihydroxy-4-dehydrodigitoxigenone	—	Δ^{4}-7β,12β,14β-ol, 3-oxo	—	—	—
Diginatigenin	5β	3β,12β,14β,16β-ol	32	—	—
Oleandrigenin	5β	3β,14β-ol, 16β-acetyl	—	—	—

[a] *Comments on the experimental conditions:* **I:**[b] Spread TLC, silica gel (water deactivated, 25%); $R_f \times 100$ values. Bz/EtOH 3:1.

    II:[c] Uniform saturation TLC; silica gel G; I.l. = 1 cm; S.l. = 1.7 cm. Systems: dClMe/MeOH/H₂O; **IIa,** 84:15:1; **IIb,** 75:23:2.

    III:[d] n-TLC, silica gel G, activation 30 min/130°; I.l. = 0.5 cm; S.l. = 1.0 cm;

CHROMATOGRAPHIC MOBILITIES FOUND ON ADSORPTION TLC AND PARTITION TLC FOR CARDENOLIDES OF THE NORMAL (17α-H)- AND ALLO (17β-H)-SERIES

Chromatographic systems[a]												
III	IVa	IVb	V	VIa	VIb	VII	VIII	IX	X	XI	XII	XIII
—	52	—	—	—	—	—	—	—	—	—	—	—
100	—	—	—	—	—	—	—	—	—	—	—	—
108	—	—	—	—	—	—	—	—	—	—	—	—
—	47	—	—	—	—	—	—	—	—	—	—	—
—	—	—	42	—	—	—	—	—	—	—	—	—
—	60	—	—	—	—	—	—	—	90	—	—	—
—	—	—	—	—	—	—	—	—	—	—	—	—
133	—	—	—	—	—	—	—	—	—	—	—	—
130	—	—	—	—	—	—	—	—	—	—	—	—
—	—	—	—	100	100	—	—	—	—	—	—	—
—	—	—	—	—	75	—	—	—	—	—	—	—
—	—	—	—	—	—	—	—	100	—	—	—	—
—	26	—	—	—	—	—	—	—	—	—	—	—
—	—	—	—	—	—	—	—	—	75	—	—	—
106	27	56	—	—	—	—	100	—	72	—	—	—
—	—	—	—	69	31	—	—	—	—	—	—	—
100	—	—	—	63	28	—	—	—	—	—	—	—
104	—	—	—	69	32	—	—	—	—	—	—	—
89	—	—	—	—	—	—	—	—	—	—	—	—
81	—	—	—	—	—	—	—	—	—	—	—	—
—	—	—	—	—	—	—	—	—	—	—	—	49
—	—	—	15	—	—	—	—	81	—	—	—	—
—	—	—	—	—	—	—	87	—	—	—	—	—
—	—	—	—	—	—	—	67	—	—	100	40	—
—	—	—	—	—	—	—	53	—	—	65	—	—
—	—	—	—	—	—	—	80	—	—	—	—	—
—	—	—	—	—	—	—	—	—	—	—	—	44
40	—	—	—	—	—	—	72	—	35	—	—	—
—	—	—	—	—	—	—	58	—	—	—	—	—
—	1	35	—	—	—	—	79	—	—	—	—	—
47	—	—	—	—	—	100	—	—	45	—	—	—
—	—	—	—	—	—	—	61	—	—	—	—	—
—	—	—	—	—	—	—	—	—	—	—	48	—
—	—	—	—	—	—	—	—	39	—	—	—	—
—	—	—	—	—	—	—	38	—	—	—	—	—
—	—	—	—	—	—	185	—	—	—	—	—	—

Dv. = 12–13 cm; $R_S \times 100$ values (S = uzarigenin). EtAc. Similar chromatographic behavior on iPrEth/Atn 3:1.

IV:[e] n-TLC, aluminum oxide. Systems: Chf/MeOH; **IVa,** 96:4; **IVb,** 90:10.

V:[f] n-TLC, silica gel G. MeCl/MeOH 93:7.

VI:[g] Normal or multiple TLC; silica gel G, conditions as for **III**; $R_S \times 100$ values

TABLE XI *(Continued)*

(S = anhydrocanarigenin). Systems: **VIa,** AcOH; **VIb,** Bz/EtAc 7:3 and 6:4, repeated development.

VII:[h] n-TLC, silica gel G; Dv. = 80–90 min (24°); $R_S \times 100$ values (S = gitoxigenin, R_f = 0.30). EtAc/Chf 9:1.

VIII:[i] n-TLC, silica gel G; $R_S \times 100$ values (S = digitoxigenin); Fr. = 12 cm. System: Atn/Ch 9:6. Variation of R_S values, about 12%; see the values in footnote j reference.

IX:[k] n-TLC, silica gel G, $R_S \times 100$ values (S = 4-anhydrodigitoxigenon). EtAc.

X:[l] n-TLC, formamide-impregnated kieselguhr G; Xl/MEK/formamide 50:50:4. Impregnation: 10% formamide/Atn.

XI:[m] Overrun TLC with "reabsorbed front"; ribbed glass/silica gel G; Dv. = 4 hours; $R_S \times 100$ values (S = sarmentogenin: 6.9 cm). EtAc.

XII:[n] n-TLC on ribbed glass/silica gel G layers, 20 × 7.5 cm; Dv. = 20–30 min; S.l. = 1.6 cm; Fr. = 11 cm. EtAc.

XIII:[o] n-TLC, silica gel G; S.l. = 2 cm; Fr. = 14 cm; $R_f \times 100$ values. Chf/MeOH 1:1.

[b] J. Reichelt and J. Pitra, *Collection Czech. Chem. Commun.* **27,** 1709 (1962).

[c] L. Fauconnet and M. Waldesbühl, *Pharm. Acta Helv.* **38,** 423 (1963).

[d] R. Tschesche, W. Freytag, and G. Snatzke, *Chem. Ber.* **92,** 3053 (1959).

[e] M. Okada and M. Hasunuma, *Yakugaku Zasshi* **85,** 822 (1965).

[f] M. Schüpbach and C. Tamm, *Helv. Chim. Acta* **47,** 2217 (1964).

[g] R. Tschesche, G. Snatzke, J. Delgado, and A. G. Gonçales, *Ann. Chem.* **663,** 157 (1963).

[h] B. Görlich, *Planta Med.* **9,** 237 (1961).

[i] I. Herrmann and K. Repke, *Arch. Exptl. Pathol. Pharmakol.* **248,** 351 (1964).

[j] I. Herrmann and K. Repke, *Arch. Exptl. Pathol. Pharmakol.* **248,** 370 (1964).

[k] T. Okumura, Y. Nozaki, and D. Satoh, *Chem. Pharm. Bull. (Tokyo)* **12,** 1143 (1964).

[l] D. Sonanini, *Pharm. Acta Helv.* **39,** 673 (1964).

[m] M. L. Lewbart, W. Wehrli, and T. Reichstein, *Helv. Chim. Acta* **46,** 505 (1963).

[n] M. L. Lewbart, W. Wehrli, H. Kaufmann, and T. Reichstein, *Helv. Chim. Acta* **46,** 517 (1963).

[o] K. D. Roberts, E. Weiss, and T. Reichstein, *Helv. Chim. Acta* **49,** 316 (1966).

layers with systems containing ethyl acetate. Table XII gives the $R_f \times 100$ values found on TLC for several bufogenins.

5. Cardiac Glycosides

The best and quickest separations of complex mixtures of cardenolide glycosides are obtained on silica gel layers by use of partition TLC. Herbal extracts containing up to 15 steroid glycosides[200] obtained from *Convallaria majalis* L., *Scilla maritima* L., *Nerium oleander* L., and *Adonis vernalis* L. have been fractionated into their steroid constituents by ascending one-dimensional TLC in the system methyl ethyl ketone–toluene–water–methanol–acetic acid (40:5:3:2.5:1). The sequential combination of partition TLC, using this last system, with adsorption TLC (ethyl acetate–chloroform 9:1) in a bidimensional procedure gives excel-

TABLE XII
CHROMATOGRAPHIC MOBILITIES OF BUFOGENINS ON TLC

5β-Bufa-20,22-dienolides	Structure	$R_f \times 100$ values in the systems[a]							
		Ia	Ib	Ic	Id	IIa	IIb	III	IV
Resibufogenin	3β-ol, 14,15β-oxido	61	61	60	66	—	—	45	—
Bufalin	3β,14β-ol	62	61	64	62	—	—	—	—
Marinobufogenin	3β,5β-ol, 14,15β-oxido	42	33	47	50	—	—	—	36
12β-Hydroxymarinobufogenin	—	—	—	—	—	—	—	—	15
12β-Hydroxyresibufogenin	3β,12β-ol, 14,15β-oxido	—	—	—	—	—	—	12	—
Bufotalinin	3β,5β-ol, 14,15β-oxido, 19-oxo	31	22	39	34	—	—	—	—
Gamabufotalin	3β,11α,14β-ol	31	26	37	47	—	—	—	—
12β-Hydroxybufalin	3β,12β,14β-ol	—	—	—	—	33	—	—	—
Telocinobufogenin	3β,5β,14β-ol	34	23	28	37	—	—	—	—
Hellebrigenin (bufotalidin)	3β,5β,14β-ol, 19-oxo	28	18	25	30	—	—	—	—
Hellebrigenol	3β,5β,14β,19-ol	9	7	17	23	—	—	—	—
12β-Hydroxyresibufogenin diacetate	3β,12β-acetate, 14,15β-oxido	—	—	—	—	—	79	—	—

[a] *Comments on the experimental conditions:* **I:**[b] n-TLC, silica gel G, 10 × 18 cm plates; S.l. = 3 cm. Systems: **Ia,** EtAc; **Ib,** EtAc/Ch 8:2; **Ic,** EtAc/Atn 9:1; **Id,** EtAc/water saturated. **II:**[c] n-TLC, aluminum oxide G. Systems: Chf/MeOH; **IIa,** 96:4 **IIb,** 99:1. **III:**[d] n-TLC, silica gel. MeCl/MeOH 93:7. **IV:**[e] Same as **III.**

[b] R. Zelnik and L. M. Ziti, *J. Chromatog.* **9,** 371 (1962).

[c] M. Okada, M. Hasunuma, and Y. Saito, *Yakugaku Zasshi* **85,** 1092 (1965).

[d] M. Schüpbach and C. Tamm, *Helv. Chim. Acta* **47,** 2217 (1964).

[e] M. Schüpbach and C. Tamm, *Helv. Chim. Acta* **47,** 2226 (1964).

lent resolution of mixtures containing steroid monoglycosides, steroid diglycosides and their aglycone, as Görlich[85] has shown for *Oleander purpurates*.

For the separation in two-dimensional TLC of cardiac glycosides of *Digitalis* on silica gel layers, Sjöholm[174] has chosen the systems ethyl acetate–methanol–water (80:5:5) and chloroform–pyridine (6:1). In these two systems the monodigitoxoses of digitoxigenin, gitoxigenin, and gitaloxigenin (lanadoxin) are completely separated and present, respectively, the following $R_S \times 100$ values (S = digitoxin): 127, 102, and 103 in the system ethyl acetate–methanol–water 16:1:1; and 141, 86, and 136 in the system chloroform–pyridine 6:1. For the separation of the deacetyllanatosides A (purpurea glycoside A), B (purpurea glycoside B), and D (glucogitaloxin), Sjöholm has employed methyl ethyl ketone–chloroform–formamide 5:2:1 as solvent system; $R_f \times 100$ values found: A = 100, B = 67, and D = 89.

Sonanini[109] has employed formamide-impregnated layers of silica gel G to separate secondary *Digitalis* glycosides, and has obtained excellent resolution in the system methyl ethyl ketone–xylol–formamide 50:50:4. Heusser[80] prefers the use of ascending overrun TLC, according to Bennett and Heftmann,[79] to fractionate glycosides in his assay method for digitoxin and gitoxin in *Digitalis* leaves. Overrun development on silica gel plates with chloroform–methanol 92:8 gives a complete separation of several glycosides, e.g., gitoxin, strospeside, digotoxin, and gitaloxin.

The lanatosides A, B, C, and D have been separated by means of several different TLC procedures: on silica gel layers developed according to the "uniform-saturation" device of Fauconnet and Waldesbühl,[65] on spread TLC using silica gel deactivated layers (25% water),[205] or by partition TLC on talc layers.[21] Fractionation of the desacetyllanatosides can be carried out also on loose silica gel layers deactivated with aqueous acetic acid.[205]

The mobility values for cardenolide glycosides under several TLC conditions are summarized in Table XIII.

Mixtures of a cardenolide glycoside and the corresponding 20,22-cardanolide glycoside have been separated on analytical and preparative cellulose layers[309a] (MN-cellulose powder 300, Macherey, Nagel and Co., Düren, Germany) by using multiple one-dimensional TLC. The following $R_f \times 100$ values for the investigated glycoside pairs have been found on analytical cellulose layers after three developments: digitoxin/20,22-dihydrodigitoxin 25 and 45 (propyleneglycol–acetone 1:4/benzene–ethyl acetate 9:1, saturated with propyleneglycol); digoxin/20,22-dihydrodi-

---

[309a] G. Rabitzsch, *J. Chromatog.* **35**, 122 (1968).

goxin 55 and 71 (formamide–acetone 1:4/chloroform–benzene 95:5, saturated with formamide).

The resolution of steroid glycosides of *Gongronema gazeuse* B. on silica gel G using ethyl acetate has been accomplished on ribbed glass plates with different TLC techniques: after a single run (30 minutes)[310] or with 4–6 hours overrun.[78] The separation of bufadienolide glycosides and C_{21}-steroid glycosides on TLC has also been investigated. Some chromatographic values are given in the Tables XIV and XV.

Reichstein[310a] has separated the cardenolide glycosides isolated from the latex of *Calotropis procera* L. by employing formamide-impregnated paper chromatography combined with thin-layer chromatography developed on ribbed glass plates. The following are the mobilities (in cm) found on silica gel HF_{254} (solvent: ethyl acetate; front = 18 cm, starting line = 2 cm) for the *Calotropis* glycosides and their common aglycone calotropagenin (19-aldo-5α-cardenolide-3β,12ξ, 14β-triol): uscharidin ($C_{29}H_{38}O_9$), 13.6; uscharin ($C_{31}H_{41}O_8NS$), 10.8; calactin ($C_{29}H_{40}O_9$), 10.2; calotropin ($C_{29}H_{40}O_9$), 8.6; calotoxin ($C_{29}H_{40}O_{10}$), 8.1; voruscharin ($C_{31}H_{43}O_8NS$), 4.8, and calotropagenin, 5.0. Uscharidin gives on TLC, under the conditions described above, one compact spot, as opposed to the considerable tailing observed on paper chromatography.

6. *Sapogenins*

Systematic separation of steroid sapogenins by means of TLC techniques has been described by Tschesche *et al.*[53, 123, 167] and by Matsumoto[105] and Takeda and co-workers.[311] The mobilities of a number of these compounds in several systems are shown in Table XVI.

Sapogenins differing only in their structures at C-25 (25β-methyl-neo series and the 25α-methyl-iso series) are not separable using normal TLC techniques. However, a complete separation of tigogenin/neotigogenin and sarsasapogenin/smilagenin was obtained by Bennett and Heftmann on silica gel G (dichloromethane–ether 199:1) by ascending continuous-development TLC in 6 hours.[77] These important epimers can also be separated after the formation of trifluoroacetates or acetates. The acetate of tigogenin moves faster than that of neotigogenin on kieselguhr G (hexane–ethanol–water 40:3:7) and the pairs tigogenin/neotigogenin and smilagenin/sarsasapogenin can be resolved as trifluoroacetate derivatives (the 25α-epimer is less polar) on silica gel layers in the solvent system chloroform–toluene 9:1.

[310] M. L. Lewbart, W. Wehrli, H. Kaufmann, and T. Reichstein, *Helv. Chim. Acta* **46**, 517 (1963).

[310a] T. Reichstein, *Naturwissenschaften* **20**, 499 (1967).

[311] K. Takeda, S. Hara, A. Wada, and N. Matsumoto, *J. Chromatog.* **11**, 562 (1963).

TABLE XIII

CHROMATOGRAPHIC MOBILITIES FOUND ON ADSORPTION TLC AND PARTITION TLC FOR 5β- AND 5α-CARDENOLIDE GLYCOSIDES

Cardenolide glycosides			Chromatographic systems[e]										
Glycoside	Aglycone[a]	Sugars[b]	I	II	III	IV	Va	Vb	VI	VIIa	VIIb	VIIc	VIId
A. Digitoxigenin (5β-cardenolide-3β,14β-diol) derivatives													
Digitoxin	DT	dgx; dgx; dgx	—	63	—	54	45	77	72	—	—	—	—
Lanatoside A (1)[c]	DT	adgb; dgx; dgx	—	42	38	—	30	61	52	46	51	79	70
Deacetyllanatoside A (2)	DT	dgb; dgx; dgx	—	32	—	24	18	48	—	—	—	—	—
Acetyl-β-digitoxin	DT	adgb; dgx	—	—	—	—	67	87	—	—	—	—	—
Acetyl-α-digitoxin	DT	adgb; dgx	—	—	—	—	64	87	—	—	—	—	—
Acetyldigitoxin (3)	DT	adgb; dgx	—	85	—	58	—	—	82	—	—	—	—
Gitoxin	GT	dgx; dgx; dgx	—	54	—	43	40	70	—	—	—	—	—
Lanatoside B	GT	dgx; dgx; adgb	—	41	24	—	25	55	41	35	42	46	45
Deacetyllanatoside B	GT	dgx; dgx; dgb	—	23	—	18	14	40	—	—	—	—	—
Acetyl-β-gitoxin	GT	dgx; dgx; adgb	—	—	—	—	54	81	—	—	—	—	—
Acetyl-α-gitoxin	GT	dgx; dgx; adgb	—	—	—	—	51	81	—	—	—	—	—
Digitalinum verum	GT	gl; dgt	—	—	—	—	11	37	—	—	—	—	—
Digoxin	DG	dgx; dgx; dgx	—	53	—	40	35	66	62	—	—	—	—
Lanatoside C	DG	dgx; dgx; adgb	—	36	19	—	19	49	36	25	30	33	21
Deacetyllanatoside C	DG	dgx; dgx; dgb	—	20	—	11	19	36	27	—	—	—	—
Acetyl-β-digoxin	DG	ac.dgx; dgx; dgx	—	—	—	—	51	81	—	—	—	—	—
Acetyl-α-digoxin	DG	ac.dgx; dgx; dgx	—	—	—	—	45	77	—	—	—	—	—
Diginatin	DN	dgx; dgx; dgx	—	—	—	30	—	—	—	—	—	—	—
Lanatoside D	DN	adgb; dgx; dgx	—	—	10	—	14	39	—	7	10	14	10
Deacetyllanatoside D	DN	dgx; dgx; dgb	—	—	—	5	—	—	—	—	—	—	—
Gitaloxin	GL	dgx; dgx; dgx	—	—	—	—	—	—	—	—	—	—	—
Lanatoside E	GL	adgb; dgx; dgx	—	—	—	—	30	63	—	—	—	—	—
Oleandrin	OL	old	100	—	—	—	—	—	—	—	—	—	—
Oleandrin diglycoside	OL	old; gl; gl	—	—	—	—	—	—	—	—	—	—	—
Urechitoxin	OL	old; gl	—	—	—	—	—	—	—	—	—	—	—
Deacetyloleandrin	GT	old	57	—	—	—	—	—	—	—	—	—	—
Adynerin	AG	dgm	70	—	—	—	—	—	—	—	—	—	—
Adynerin diglycoside	AG	dgm; gl; gl	—	—	—	—	—	—	—	—	—	—	—
Stropeside	GT	dgt	—	—	—	—	—	—	—	—	—	—	—
Verodoxin (4)	OL	dgt	—	—	—	—	—	—	—	—	—	—	—

Convalloside	SN	scb			10	
Convallotoxin	SN	rhm			25	
Glucoconvalloside	SN	gl; scb				
Helveticoside	SN	dgx			42	
Deglucocheirotoxin	SN	gmt				
Convallotoxol	SL	rhm				
Glucoconvallatoxoloside	SL	scb; gl				
Vallarotoxin (C$_{29}$H$_{42}$O$_9$)	—	rhm				
Majaloside (C$_{35}$H$_{52}$O$_{14}$)	—	rhm-gl			7	
Ouabain	OU	rhm	9		54	
Periplocymarin	PR	cym				
Periplogenin digitoxoside	PR	dgx^d				
Gongronema glycoside L	PR	cym-dgx^d				
Gongronema glycoside C	SR	cym-cym-cym^d				
Gongronema glycoside E	SR	cym-cym^d				
Gongronema glycoside G	SR	cym-cym-dgx^d				
Gongronema glycoside R	SR	old-dgx-dgx^d				
Gongronema glycoside M	SR	cym-dgx^d				
Sarmentogenin digitoxoside	SR	dgx^d				
Sarmentocymarin	SR	cym^d			25	
Adonitoxin	AD	rhm				
Acetyladonitoxin	AD	ac.rhm				74
Cymarin	SN	cym			50	
Cymarol	SL	cym			47	
K-Strophanthidin-β	SN	ppb				
K-Strophanthoside	SN	gl; ppb			52	
Neriifolin	DT	tvl				
Erysimoside	SN	dgx-gl				
B. Uzarigenin (5α-cardenolide-3β,14β-diol) derivatives						
Coroglaucigenin rhamnoside	CO	rhm			29	
Frugoside	CO	dal				
Corotoxigenin rhamnoside	CX	rhm				
Panoside	PN	rhm				
Glucopanoside	PN	scb				
Malloside	ML	rhm				

(*Footnotes appear on p. 100.*)

(*Continued*)

TABLE XIII (*Continued*)

| Cardenolide glycosides | | | Chromatographic systems[e] | | | | | | | | | | | |
Glycoside	Aglycone[a]	Sugars[b]	VIIe	VIII	IXa	IXb	X	XI	XII	XIII	XIV	XVa	XVb	XVc
A. Digitoxigenin (5β-cardenolide-3β,14β-diol) derivatives														
Digitoxin	DT	dgx; dgx; dgx	—	—	—	58	—	—	—	—	—	100	100	151
Lanatoside A (1)[c]	DT	adgb; dgx; dgx	79	—	65	—	—	—	—	—	—	26	18	100
Deacetyllanatoside A (2)	DT	dgb; dgx; dgx	—	—	29	—	—	—	—	—	—	14	5	—
Acetyl-β-digitoxin	DT	adgb; dgx	—	—	—	—	—	—	—	—	—	142	203	—
Acetyl-α-digitoxin	DT	adgb; dgx	—	—	—	—	—	—	—	—	—	132	195	—
Acetyldigitoxin (3)	DT	adgb; dgx	—	—	73	—	—	—	—	—	—	—	—	—
Gitoxin	GT	dgx; dgx; dgx	—	—	—	25	—	—	—	—	—	75	59	—
Lanatoside B	GT	dgx; dgx; adgb	59	—	30	—	—	—	—	—	—	16	12	114
Deacetyllanatoside B	GT	dgx; dgx; dgb	—	—	11	—	—	—	—	—	—	9	2	67
Acetyl-β-gitoxin	GT	dgx; dgx; adgb	—	—	—	—	—	—	—	—	—	117	145	—
Acetyl-α-gitoxin	GT	dgx; dgx; adgb	—	—	—	—	—	—	—	—	—	104	139	—
Digitalinum verum	GT	gl; dgt	—	—	—	—	—	—	—	—	—	6	2	50
Digoxin	DG	dgx; dgx; dgx	—	—	—	18	—	—	—	—	—	68	62	—
Lanatoside C	DG	dgx; dgx; adgb	40	—	20	—	—	—	—	—	—	15	8	105
Deacetyllanatoside C	DG	dgx; dgx; dgb	—	—	9	—	—	—	—	—	—	—	—	—
Acetyl-β-digoxin	DG	ac.dgx; dgx; dgx	—	—	—	—	—	—	—	—	—	113	157	—
Acetyl-α-digoxin	DG	ac.dgx; dgx; dgx	—	—	—	—	—	—	—	—	—	100	138	—
Diginatin	DN	dgx; dgx; dgx	—	—	—	—	—	—	—	—	—	—	—	—
Lanatoside D	DN	adgb; dgx; dgx	19	—	—	—	—	—	—	—	—	—	—	—
Deacetyllanatoside D	DN	dgx; dgx; dgb	—	—	—	—	—	—	—	—	—	—	—	—
Gitaloxin	GL	dgx; dgx; dgx	—	—	—	45	—	—	—	—	—	101	92	—
Lanatoside E	GL	adgb; dgx; dgx	—	—	—	—	—	—	—	—	—	—	—	—
Oleandrin	OL	old	—	—	—	—	—	—	—	76	—	—	—	—
Oleandrin diglycoside	OL	old; gl; gl	—	—	—	—	—	—	—	10	—	—	—	—
Urechitoxin	OL	old; gl	—	—	—	—	—	—	—	23	—	—	—	—
Deacetyloleandrin	GT	old	—	—	—	—	—	—	—	76	—	—	—	—
Adynerin	AG	dgn	—	—	—	—	—	—	—	76	—	—	—	—
Adynerin diglycoside	AG	dgn; gl; gl	—	—	—	—	—	—	—	3	—	—	—	—
Stropeside	GT	dgt	—	—	40	—	—	—	—	—	—	50	91	—
Verodoxin (4)	OL	dgt	—	—	65	—	—	—	—	—	—	77	141	—

Compound								
Convalloside	SN	scb						28
Convallotoxin	SN	rhm						39
Glucoconvalloside	SN	gl; scb						22
Helveticoside	SN	dgx						—
Deglucocheirotoxin	SN	gmt						46
Convallotoxol	SL	rhm						31
Glucoconvallatoxoloside	SL	scb; gl						25
Vallarotoxin ($C_{29}H_{42}O_9$)	—	rhm						9
Majaloside ($C_{35}H_{52}O_{14}$)	—	rhm-gl						59
Ouabain	OU	rhm						52
Periplocymarin	PR	cym			17			
Periplogenin digitoxoside	PR	dgx^d		50	26			
Gongronema glycoside L	PR	cym-dgx^d		130	63			
Gongronema glycoside C	SR	cym-cym-cym^d		130	63			
Gongronema glycoside E	SR	cym-cym^d		107	52			
Gongronema glycoside G	SR	cym-cym-dgx^d		72	35			
Gongronema glycoside R	SR	old-dgx-dgx^d		108	52			
Gongronema glycoside M	SR	cym-dgx^d		80	49			
Sarmentogenin digitoxoside	SR	dgx^d			72			
Sarmentocymarin	SR	cym^d						
Adonitoxin	AD	rhm						44
Acetyladonitoxin	AD	ac.rhm						64
Cymarin	SN	cym				11.6	91	70
Cymarol	SL	cym						
K-Strophanthidin-β	SN	ppb				4.1	58	22
K-Strophanthoside	SN	gl; ppb				0.7	17	1
Neriifolin	DT	tvl						
Erysimoside	SN	dgx-gl				2.5	51	
B. Uzarigenin (5α-cardenolide-3β,14β-diol) derivatives								
Coroglaucigenin rhamnoside	CO	rhm						20
Frugoside	CO	dal						
Corotoxigenin rhamnoside	CX	rhm						34
Panoside	PN	rhm						20
Glucopanoside	PN	scb						9
Malloside	ML	rhm						30

(Footnotes appear on p. 100.)   *(Continued)*

TABLE XIII (Continued)

[a] DT = digitoxigenin; GT = gitoxigenin; DG = digoxigenin; DN = diginatigenin; GL = gitaloxigenin; OL = oleandrigenin; AG = adynerigenin; SN = strophanthidin; SL = strophanthidol; OU = ouabagenin; PR = periplogenin; SR = sarmentogenin; AD = adonitoxigenin; CO = coroglaucigenin; CX = corotoxigenin; PN = panogenin; ML = mallogenin.

[b] dgx = D-digitoxose (2,6-dideoxy-D-ribohexose); ac.dgx = D-acetyldigitoxose; ac.dgb = acetyldigilanidobiose (ac.dgx; gl); dgb = digilanidobiose (dgx; gl); dgt = D-digitalose 3-O-methyl-6-deoxy-D-galactose; gl = D-glucose; adgb = acetyldigilanidobiose (ac.dgx; gl); nose); dgn = D-diginose (2-deoxy-digitalose); scb = scillabiose (L-rhm; D-gl); cym = D-cymarose (3-O-methyl-digitoxose); gmt = D-gulomethylose, antiarose (6-deoxy-D-gulose); rhm = L-rhamnose; ac.rhm = 3-O-acetyl-L-rhamnose; ppb = periplobiose, strophanthobiose (D-gl; D-cym); tvl = L-thevelose 3-O-methyl-6-deoxy-L-glucose; dal = 6-deoxy-D-allose.

[c] Remarks: (1) glucosidodigitoxin; (2) purpuraglycoside A, digilanide A; (3) α- or β-structure was not given; (4) 16-formyl strospeside.

[d] D- or L-form is not determined.

[e] Comments on the experimental conditions: I:[f] n-TLC, silica gel G; Dv. = 80–90 min (24°); $R_B \times 100$ values (S = oleandrin; R_f = 0.62). EtAc/Chf 90:10.

II:[g] n-TLC, silica gel G; Dv. = 10 cm. MeCl/MeOH/formamide 80:19:1.

III:[h] Spread TLC; silica gel deactivated (25% water). Bz (water saturated)/EtOH 3:1.

IV:[h] Spread TLC; silica gel deactivated (43% aqueous AcOH, 50%). Bz (water saturated)/EtOH 3:1.

V:[j] "Uniform saturation" TLC, Dv. = 13 cm. System: dClMe/MeOH/H₂O, 84:15:1; **Vb**, 75:23:2.

VI:[j] n-TLC, $R_f \times 100$ values; Dv. not referred. Bz/95% EtOH 7:3.

VII:[k] n-TLC, formamide (30% in Atn) impregnated talc; S.l. = 1.5 cm; I.l. = 0.5 cm; Fr. = 14–15 cm; $R_f \times 100$ values. Systems: **VIIa**, EtAc/Chf 2:8; **VIIb**, Dx/Chf/BuOH 20:70:5 (formamide saturated); **VIIc**, ThF/Chf 5:5 (formamide saturated); **VIId**, Chf/Bz/EtOH/formamide 69:20:10:1; **VIIe**, Chf/Bz/EtOH/formamide 79:10:10:1.

VIII:[l] Overrun TLC with "reabsorbed front"; ribbed glass silica gel plate; Dv. = 4 hours; $R_B \times 100$ values (S = sarmentogenin; 6.9 cm). EtAc.

IX:[m] n-TLC, formamide (10% in Atn) impregnated kieselguhr G; Fr. = 10–15 cm; $R_f \times 100$ values. Systems: **IXa**, Chf/ThF/formamide 50:50:6; **IXb**, MEK/XI/formamide 50:50:4.

X:[n] As **XII** of Table XI.

XI:[o] Multiple TLC, twice development ($\uparrow$ = 13 cm; $\uparrow\uparrow$ = 8 cm); silica gel G, 10 × 20 cm, 200-μ plates. Values in cm (strophanthidin = 7.54 cm). Chf/AcOH/MeOH 85:2:13.

XII:[p] n-TLC, silica gel G, 10 × 20 cm, 200-μ plates; tank: 18 × 12 × 12 cm. R_f × 100 values. EtAc/Py/H$_2$O 5:1:4 (superior phase).

XIII:[q] n-TLC, kieselgel G; Dv. = 50 min (22°). R_f × 100 values. MEK/Tol/H$_2$O/MeOH/EtOH 40:5:3:2.5:1.

XIV:[r] As **XIII** of Table XI.

XV:[s] n-TLC, silica gel G; Fr. = 12–14 cm; S.l. = 1.5 cm. Systems: **XVa**, EtAc/MeOH/H$_2$O 16:1:1; **XVb**, Chf/Py 6:1; **XVc**, MEK/Chf/formamide 5:2:1.

[f] B. Görlich, *Planta Med.* **9**, 237 (1961).

[g] E. Stahl and U. Kaltenbach, *J. Chromatog.* **5**, 458 (1961).

[h] J. Reichelt and J. Pitra, *Collection Czech. Chem. Commun.* **27**, 1709 (1962).

[i] L. Fauconnet and M. Waldesbühl, *Pharm. Acta Helv.* **38**, 423 (1963).

[j] E. Johnston and A. L. Jacobs, *J. Pharm. Sci.* **55**, 531 (1966).

[k] J. Zurkowska and A. Ozarowski, *Planta Med.* **12**, 222 (1964).

[l] M. L. Lewbart, W. Wehrli, and T. Reichstein, *Helv. Chim. Acta* **46**, 505 (1963).

[m] D. Sonanini, *Pharm. Helv. Acta* **39**, 673 (1964).

[n] M. L. Lewbart, W. Wehrli, H. Kaufmann, and T. Reichstein, *Helv. Chim. Acta* **46**, 517 (1963).

[o] G. L. Corona and M. Raiteri, *J. Chromatog.* **19**, 435 (1965).

[p] G. L. Corona, M. Raiteri, and G. Tieghi, *Farmaco (Pavia), Ed. Sci.* **19**, 574 (1964).

[q] B. Görlich, *Arzneimittel-Forsch.* **15**, 493 (1965).

[r] K. D. Roberts, E. Weiss, and T. Reichstein, *Helv. Chim. Acta* **49**, 316 (1962).

[s] I. Sjöholm, *Svensk Farm. Tidskr.* **66**, 321 (1962).

TABLE XIV

Chromatographic Mobilities of Bufadienolide Glycosides on TLC

Bufadienolide glycosides	Aglycone	Structure[a]	Sugar[b]	$R_f \times 100$ values in the systems[e] I	II	IIIa	IIIb
Scillaren A	Scillarenin	Δ^4	rhm,gl	20	28	—	—
Proscillaren A	Scillarenin	Δ^4	rhm	45	55	—	—
Hellebrin[d]	Hellebrigenin[c]	5β-ol, 19-oxo	rhm,gl	—	—	37	62
Glucoscillaren A	Scillarenin	Δ^4	rhm,gl,gl	—	14	—	—
Scilliglaucosid (scillaren F)	Scilliglaucosidin	19-al	gl	—	34	—	—
Scillicyanside	—	—	gl	—	31	—	—

[a] Substitutions on the bufadienolide-$3\beta,14\beta$-diol molecule.

[b] rhm = L-rhamnose; gl = D-glucose.

[c] Also called bufatalidin.

[d] Other $R_f \times 100$ values:[f] 50 (Chf/MeOH/Py 21:9:5); 51 (Bz/MeOH 3:2); 45 (EtAc/MeOH/H$_2$O 7:3:2).

[e] *Chromatographic conditions:* Ascending TLC on silica gel G.   **I:**[g] MeCl/MeOH/formamide 80:19:1; Dv. = 10 cm. **II:**[h] MEK/Tol/H$_2$O/MeOH/AcOH 40:5:3:2.5:1; Dv. = 50 min/22°. **III:**[f] **IIIa,** Chf/MeOH/H$_2$O 65:33:10; **IIIb,** iPrAc/MeOH/H$_2$O 9:9:1.

[f] R. Tschesche and B. Brassat, *Z. Naturforsch.* **20b,** 707 (1965).

[g] E. Stahl and U. Kaltenbach, *J. Chromatog.* **5,** 458 (1961).

[h] B. Görlich, *Arzneimittel-Forsch.* **15,** 493 (1965).

TABLE XV
CHROMATOGRAPHIC MOBILITIES OF C_{21}-STEROID GLYCOSIDES
ON SILICA GEL G LAYERS

C_{21}-steroid glycosides	Aglycone	Sugar	Solvent systems[d] Ia	Ib
Digifolein	Digifologenin[a]	Diginose	13	42
Digitalonin	Diginigenin[b]	Digitalose	18	37
Digipronin	Digiprogenin[c]	Digitalose	35	39
Diginin	Diginigenin	Diginose	35	62

[a] $2\beta,3\beta$-Dihydroxy-$12\alpha,20\alpha$-oxido-$17\beta(H),14\beta(H)$-pregn-5-ene-11,15-dione.
[b] 3β-Hydroxy-$12\alpha,20\alpha$-oxido-$17\beta(H),14\beta(H)$-pregn-5-ene-11,15-dione.
[c] $3\beta,17\beta$-Dihydroxy-$14\beta(H)$-pregn-5-ene-11,15,20-trione.
[d] Silica gel G; $R_f \times 100$ values. Systems:[e] **Ia,** EtAc/MeOH 96:4; **Ib,** Chf/MeOH 9:1.
[e] R. Tschesche and G. Brügmann, unpublished results (1962) quoted by R. Tschesche, G. Wulff, and K. H. Richert, in "New Biochemical Separations" (A. T. James and L. J. Morris, eds.), p. 198. Van Nostrand, Princeton, New Jersey, 1964.

A 5α-sapogenin adsorbs more strongly than the 5β-epimer, and the separation of the 5-epimeric pairs is easy to accomplish as free steroids or as trifluoroacetates (with chloroform–toluene 9:1 on silica gel),[312] but impossible as acetate derivatives using conventional TLC. However, after 5 hours of continuous development with hexane–dichloromethane 3:7 on silica gel,[77] the acetates of 5-epimeric spirostan steroids show different migrations.

The separation of the pairs with Δ^5-/5α-configurations is much more difficult and sometimes only partial, either as the free steroids or as the esters. Gentiogenin and hecogenin have different mobilities on silica gel layers in several systems; their $R_f \times 100$ values are in cyclohexane–acetone (1:1) 58 and 49; cyclohexane–ethyl acetate–water (600:400:1): 30 and 25; and chloroform–methanol–water (485:15:1): 58 and 49, respectively.[312] The resolution of a mixture of diosgenin (Δ^5-25α), tigogenin ($5\alpha,25\alpha$), and smilagenin ($5\beta,25\alpha$) is complete with solvent system hexane–toluene–ethanol–water 100:50:5:45, with the following sequence of polarity: Δ^5-$25\alpha > 5\alpha,25\alpha > 5\beta,25\alpha$.

By means of continuous development on silica gel G, diosgenin and tigogenin were well separated as free compounds (benzene–methanol 399:1, 4 hours) or as acetates (hexane–dichloromethane 3:7, 5 hours).

The separation of isomeric sapogenins differing by the configuration at C-3 is possible on TLC with layers of kieselguhr G–silica gel G (1:1). Under the conditions used by Bennett and Heftmann, sarsasapogenin

[312] R. D. Bennett and E. Heftmann, *J. Chromatog.* **11,** 562 (1963).

TABLE XVI
CHROMATOGRAPHIC MOBILITIES ($R_f \times 100$ VALUES) FOUND ON TLC FOR STEROID SAPOGENINS

Sapogenins	Spirostane structure			Chromatographic systems[a]									
	C_5	C_{25}	Substituents	I	IIa	IIb	IIc	III	IVa	IVb	IVc	IVd	IVe
Δ^2-Spirosten	5α	25α	Δ^2	—	68	—	—	—	—	—	—	—	—
Luvigenin	—	25α	$\Delta^{1,3,5}$ 4-Me	—	—	—	—	—	72	87	76	78	93
Neometeogenin	—	25β	$\Delta^{1,3,5}$ 1-Me, 11α-ol	—	—	—	—	—	48	81	52	48	85
Meteogenin	—	25α	$\Delta^{1,3,5}$ 1-Me, 11α-ol	—	—	—	—	—	48	81	52	48	85
Δ^2-5α-Spirosten-1-one	5α	—	1-one	—	—	95	—	—	—	—	—	—	—
Δ^2-5β-Spirosten-1-one	5β	—	1-one	—	—	95	—	—	—	—	—	—	—
Tigogenone	5α	25α	3-one	50	—	—	—	—	—	—	—	—	—
Sarsasapogenin	5β	25β	3β-ol	43	—	—	62	61	31	65	39	42	67
3-Episarsasapogenin	5β	25β	3α-ol	44	—	—	—	—	—	—	—	—	—
Diosgenin	Δ^5	25α	3β-ol	44	—	—	55	55	25	59	35	34	57
Tigogenin	5α	25α	3β-ol	—	—	55	56	55	23	59	35	29	53
Smilagenin	5β	25α	3β-ol	—	—	—	62	61	—	—	—	—	—
Neotigogenin	5α	25β	3β-ol	—	—	—	56	55	—	—	—	—	—
Yamogenin	Δ^5	25β	3β-ol	—	—	—	55	55	—	—	—	—	—
5α,25α-Spirosten-2-en-15β-ol	5α	25α	Δ^2, 15β-ol	—	58	—	—	—	—	—	—	—	—
3-Dehydrodiosgenin	Δ^5	25α	Δ^3	—	—	—	—	—	—	—	—	—	—
5α-Spirosten-3β-ol-1-one	5α	—	3β-ol-1-one	—	—	51	—	—	—	—	—	—	—
5β-Spirosten-3β-ol-1-one	5β	—	3β-ol-1-one	—	—	48	—	—	—	—	—	—	—
Gentiogenin	Δ^5	25α	3β-ol-12-one	32	—	47	41	32	24	49	24	16	42
Hecogenin	5α	25α	3β-ol-12-one	28	—	47	41	32	25	46	32	16	50

(Continued)

Compound	5	25	Substituent										
9-Dehydrohecogenin	5α	25α	$\Delta^{9(11)}$ 3β-ol-12-one	—	—	—	—	—	55	—	—	—	—	—
Sisalagenin	5α	25β	3β-ol-12-one	—	—	—	—	—	32	—	—	—	—	—
Pennogenin	Δ^5	25α	3β-17α(H)-ol	—	—	—	50	—	—	27	53	26	22	50
Convallamarogenin	5β	Δ^{25}	1β,3β-ol	—	—	—	—	—	—	22	39	21	22	42
Isorhodeasapogenin	5β	25α	1β,3β-ol	—	—	—	—	—	—	24	39	18	22	47
Rhodeasapogenin	5β	25β	1β,3β-ol	—	—	—	—	—	—	22	37	18	22	47
Neoruscogenin	Δ^5	25β	1β,3β-ol	—	—	—	50	—	—	—	—	—	—	—
Yonogenin	5β	25α	2β,3α-ol	—	—	—	30	16	—	18	21	11	9	30
Gitogenin	5α	25α	2α,3β-ol	—	—	—	—	—	21	15	19	7	11	35
Chlorogenin	5α	25α	3β,6α-ol	6	—	—	—	—	—	—	—	—	—	—
Nogiragenin	5β	25α	3β,11α-ol	—	—	—	—	—	—	18	27	11	15	37
Heloniogenin	Δ^5	25α	3β,12α-ol	—	—	—	—	—	—	18	26	7	11	30
Rockogenin	5α	25α	3β,12β-ol	—	—	—	—	28	—	—	—	—	—	—
Isochiapugenin	Δ^5	25α	3β,12β-ol	17	—	—	—	—	—	—	—	—	—	—
12-Epirockogenin	5α	25α	3β,12α-ol	—	—	—	—	22	—	—	—	—	—	—
Yuccagenin	Δ^5	25α	2α,3α-ol	—	—	—	30	16	49	—	—	—	—	—
Digalogenin	5α	25α	3β,15β-ol	—	—	—	—	45	—	—	—	—	—	—
25β-Digalogenin	5α	25β	3β,15β-ol	—	—	—	—	44	—	—	—	—	—	—
3α-Digalogenin	5α	25α	3α,15β-ol	—	—	—	—	58	—	—	—	—	—	—
Tokorogenin	5β	25α	1β,2β,3α-ol	—	—	—	—	—	—	10	9	2	4	27
Metagenin	5β	25α	2β,3β,11α-ol	—	—	—	—	—	—	7	6	1	1	23
Digitogenin	5α	25α	2α,3β,15β-ol	—	—	—	—	11	—	—	—	—	—	—
25β-Digitogenin	5α	25β	2α,3β,15β-ol	—	—	—	—	11	—	—	—	—	—	—
Kitigenin	5β	25α	1β,3β,4β,5β-ol	—	—	—	—	—	—	—	4	1	1	20
Kogagenin	5β	25α	1β,2β,3β,5β-ol	—	—	—	—	—	—	5	3	1	—	23
Tetrahydroneoruscogenin	Δ^5	—	1β,3β,16β,27-ol	—	—	—	30	—	—	—	—	—	—	—
Kryptogenin	Δ^5	25α	3β,26-ol, 16,22-one	—	—	—	—	—	—	—	—	—	—	—

(Footnotes appear on p. 108.)

TABLE XVI (*Continued*)

Sapogenins	Chromatographic systems[a]													
	IVf	IVg	IVh	V	VI	VII	VIII	IX	Xa	Xb	Xc	XIa	XIb	XII
Δ²-Spirosten	—	—	—	—	—	—	—	—	—	—	—	—	—	—
Luvigenin	80	85	80	—	—	—	—	—	—	—	—	—	—	—
Neometeogenin	64	66	63	—	—	—	—	—	—	—	—	—	—	—
Meteogenin	64	66	63	—	—	—	—	—	—	—	—	—	—	—
Δ²-5α-Spirosten-1-one	—	—	—	—	—	—	—	—	—	—	—	—	—	—
Δ²-5β-Spirosten-1-one	—	—	—	—	—	—	66	—	—	—	—	—	—	—
Tigogenone	—	—	—	—	—	19	—	—	—	—	—	—	—	—
Sarsasapogenin	51	51	46	—	81	—	—	—	—	—	—	—	—	—
3-Episarsasapogenin	—	—	—	—	—	—	—	28	—	—	—	—	—	—
Diosgenin	46	42	41	33	71	—	—	—	18	67	—	63	60	—
Tigogenin	46	39	39	33	75	8	40	—	18	67	—	68	65	79
Smilagenin	—	—	—	—	82	22	—	—	26	72	—	—	—	—
Neotigogenin	—	—	—	—	72	12	—	—	—	—	—	—	—	—
Yamogenin	—	—	—	—	—	—	—	—	—	—	—	—	—	—
5α,25α-Spirosten-2-en-15β-ol	—	—	—	—	—	—	—	—	—	—	—	—	—	—
3-Dehydrodiosgenin	—	—	—	—	—	—	—	85	—	—	—	—	—	—
5α-Spirosten-3β-ol-1-one	—	—	—	—	—	—	—	—	—	—	—	—	—	—
5β-Spirosten-3β-ol-1-one	—	—	—	—	—	—	—	—	—	—	—	—	—	—
Gentiogenin	26	32	25	—	—	—	—	—	2	31	—	—	—	—
Hecogenin	21	37	22	29	—	—	—	—	—	—	58	—	—	—

9-Dehydrohecogenin	—	—	—	—	—	—	—	—	—	—	—	—	—	—
Sisalagenin	—	—	—	—	—	—	—	—	—	—	—	—	—	—
Pennogenin	35	42	30	—	—	—	—	—	4	48	68	—	—	—
Convallamarogenin	28	30	24	—	—	—	—	—	—	—	—	—	—	—
Isorhodeasapogenin	28	28	25	—	—	—	—	—	—	—	—	—	—	—
Rhodeasapogenin	28	32	25	—	—	—	—	—	—	—	—	—	—	—
Neoruscogenin	—	—	—	—	—	—	—	—	—	—	—	—	—	—
Yonogenin	13	17	11	—	—	—	—	—	—	—	—	35	—	—
Gitogenin	9	16	11	—	—	—	—	—	—	—	—	—	25	39
Chlorogenin	—	—	—	14	—	—	—	—	—	4	18	—	—	—
Nogiragenin	18	20	15	—	—	—	—	—	—	—	—	—	—	—
Heloniogenin	16	16	13	—	—	—	—	—	—	—	—	—	—	—
Rockogenin	—	—	—	—	—	—	—	—	—	—	—	—	—	—
Isochiapugenin	—	—	—	—	—	—	—	—	—	—	—	—	—	—
12-Epirockogenin	—	—	—	—	—	—	—	—	—	—	—	—	—	—
Yuccagenin	—	—	—	16	—	—	—	—	—	—	—	—	—	—
Digalogenin	—	—	—	—	—	—	—	—	—	—	—	—	—	66
25β-Digalogenin	—	—	—	—	—	—	—	—	—	—	—	—	—	—
3α-Digalogenin	—	—	—	—	—	—	—	—	—	—	—	—	—	—
Tokorogenin	3	12	2	—	—	—	—	—	—	—	—	—	—	—
Metagenin	2	9	1	—	—	—	—	—	—	—	—	—	—	—
Digitogenin	—	—	—	—	—	—	—	—	—	—	—	—	—	33
25β-Digitogenin	—	—	—	—	—	—	—	—	—	—	—	—	—	—
Kitigenin	—	10	1	—	—	—	—	—	—	—	—	—	—	—
Kogagenin	—	8	1	—	—	—	—	—	—	—	—	—	—	—
Tetrahydroneoruscogenin	—	—	—	—	—	—	—	—	—	—	—	—	—	—
Kryptogenin	—	—	—	—	—	—	—	—	—	22	25	—	—	—

(Footnotes appear on p. 108.)

(Continued)

TABLE XVI (Continued)

[a] Comments on the experimental conditions: **I:**[b] As **IV** in Table II. System: dClMe/MeOH/formamide 93/6/1.
II: n-TLC, silica gel G, plates 8×14 cm; S.l. = 1 cm; $R_f \times 100$ values; experimental conditions, as given by Tschesche et al.[c]
Systems: Chf/Atn/type: **IIa,**[d] 99:1; **IIb,**[e] 5:1; **IIc,**[f] 8:2.

III:[g] As **VII** in Table III. Ch/EtAc 1:1.

IV:[g,i] n-TLC, Wakogel B-5 (silica gel with 5% gypsum); activation 10 min/130°; $R_f \times 100$ values in benzene for butter yellow
and indophenol: 65 and 11, respectively; S.l. = 1.5 cm; I.l. = 1 cm; Dv. = 15 cm; tank lined with paper. Systems: **IVa,** Bz/EtOH
92:8; **IVb,** Chf/EtOH 95:5; **IVc,** Chf/Atn 9:1; **IVd,** Hex/Atn 8:2; **IVe,** Bz/EtOH 85:15; **IVf,** Hex/EtAc 50:50; **IVg,** Bz/MeOH
92:8; **IVh,** Bz/Atn 85:15.

V:[j] Unbound TLC, magnesium silicate (Florisil Lachema, 100/200 mesh) with 1.5% AcOH, 600 μ thick; experimental conditions
as given by Heřmánek et al.[k]; $R_f \times 100$ values (average of 3–4 experiments). Bz/EtOH 98:2.

VI:[b] n-TLC, kieselguhr G, Dv. = 42 min; conditions as **IV** of Table II. System: Hex/Tol/EtOH/H₂O 100:50:5:45.

VII:[b] n-TLC, kieselguhr G/silica gel G 1:1; Dv. = 34 min; conditions as **IV** of Table II. System: Chf/Tol 9:1.

VIII:[g] n-TLC on silica gel; conditions as **VII** of Table III. System: PtEt (b.p. = 80–90°)/Bz/EtAc 85:5:10.

IX:[l] n-TLC, silica gel/gypsum (26:4) Hpt/Chf 2:1.

X:[m] n-TLC, silica gel/rice starch (acid). Systems: **Xa,** Hex/EtAc 4:1; **Xb,** Hex/EtAc 1:1; **Xc,** EtAc.

XI:[n] n-TLC, silica gel G. Systems: **XIa,** Bz/EtAc 1:1; **XIb,** Chf/MeOH 19:1.

XII:[o] n-TLC on ribbed glass, silica gel H. System: iPrEth/Ch 2:1.

[b] R. D. Bennett and E. Heftmann, J. Chromatog. **11,** 562 (1963).

[c] R. Tschesche, W. Freytag, and G. Snatzke, Chem. Ber. **92,** 3053 (1959).

[d] R. Tschesche and G. Wulff, Chem. Ber. **94,** 2019 (1961).

[e] R. Tschesche, H. Schwarz, and G. Snatzke, Chem. Ber. **94,** 1699 (1961).

[f] R. Tschesche, G. Wulff, and K. H. Richert, in "New Biochemical Separations" (A. T. James and L. J. Morris, eds.), p. 198. Van
Nostrand, Princeton, New Jersey, 1964.

[g] K. Schreiber, O. Aurich, and G. Osske, J. Chromatog. **12,** 63 (1963).

[h] N. Matsumoto, Chem. Pharm. Bull. (Tokyo) **11,** 1189 (1963).

[i] K. Takeda, S. Hara, A. Wada, and N. Matsumoto, J. Chromatog. **11,** 562 (1963).

[j] V. Schwarz, Pharmazie **18,** 122 (1963).

[k] S. Heřmánek, V. Schwarz, and Z. Čekan, Collection Czech. Chem. Commun. **26,** 1669 (1961).

[l] P. Bite and G. Magyar, Acta Chim. Acad. Sci. Hung. **48,** 255 (1966).

[m] L. L. Smith and T. Foell, J. Chromatog. **9,** 339 (1962).

[n] R. Tschesche and H. Hulpe, Z. Naturforsch. **21b,** 494 (1966).

[o] J. Von Euw and T. Reichstein, Helv. Chim. Acta **49,** 1468 (1966).

$(5\beta/25\beta; 3\beta$-ol) moves ahead of the 3α-epimer (3-episarsasapogenin) as both free steroid and acetate derivatives.[312] By means of two developments with cyclohexane–ethyl acetate 4:1 on silica gel G, Ripperger et al.[313] have separated neotigogenin $(R_f \times 100 = 47)$ from its less polar 3α-epimer 3-epineotigogenin (56),[313] which was isolated for the first time in this study.

7. Saponins

The R_f or R_S values of several steroid saponins are given in Table XVII.

A systematic study of spirostane saponins, tomatine, and their peracetates and permethylates on silica gel gypsum layers has been undertaken by Kawasaki and Miyahara.[314] The use of butanolic or aqueous systems for the TLC of highly polar saponins and of a system of the chloroform–methanol or ethanol type for the less polar compounds and their acetates was found to give satisfactory separations. Saponins with the same kind and number of sugars, such as gitonin and deglucodigitonin or F-gitonin and degalactotigonin have been resolved by TLC. For the resolution of sapogenin permethylates, a system has been developed for silica gel G (benzene–acetone 80:20). Some examples of separations with the system are: $(R_f \times 100$ values): timosaponin A heptamethylate, 34; dioscin octamethylate, 36; gracillin nonamethylate, 46; and gitonin tridecamethylate, 14.

The systems of Kawasaki and Miyahara give particularly good separations of neutral saponins (e.g., of *Digitalis*, *Discorea*, and *Anemarrhena* species), and could be satisfactorily employed for preparative fractionations of crude sapogenins of *Hedera helix;*[315] however, their application for acid saponins is less effective.[315]

Khorlin et al.[316, 317] have employed partition TLC on unactivated silica gel for the separation of acid saponins with a high number of sugars. They have fractionated, for instance, saponins of *Gypsophila* and *Aralia* species, using a solvent system of the *n*-butanol–ethanol–ammonium hydroxide type.[315] From *Solanum paniculatum* L., Ripperger et al. have isolated two new saponins,[318] β-D-xylosyl-α-L-rhamnosylpani-

[313] H. Ripperger, H. Budzikiewicz, and K. Schreiber, *Chem. Ber.* **100**, 1725 (1967).
[314] T. Kawasaki and K. Miyahara, *Chem. Pharm. Bull. (Tokyo)* **11**, 1546 (1963).
[315] R. Tschesche and G. Wulff, *Planta Med.* **12**, 272 (1964).
[316] A. J. Khorlin, Y. S. Ovodov, and N. K. Kochetkov, *Zh. Obshch. Khim.* **32**, 782 (1962).
[317] A. J. Khorlin, L. V. Bakinovskii, V. E. Vaskovskii, A. G. Venjaminova, and Y. S. Ovodov, *Isv. Akad. Nauk SSSR Otd. Khim. Nauk* p. 2008 (1963).
[318] H. Ripperger, K. Schreiber, and H. Budzikiewicz, *Chem. Ber.* **100**, 1741 (1967).

TABLE XVII

CHROMATOGRAPHIC MOBILITIES FOUND ON ADSORPTION TLC AND PARTITION TLC FOR SAPONINS

Saponin	Aglycone[b]	Sugar[c]	Chromatographic systems[e]											
			I	IIa	IIb	III	IV	Va	Vb	Vc	Vd	VI	VII	VIII
Solasonine (1)[a]	SA Solasodine	sol	22	—	—	—	—	—	—	—	—	—	—	—
Tomatine	SA Tomatidine	lyt	—	—	—	—	—	25	44	—	—	—	—	—
Solanine (2)	SA Solanidine	sol	—	—	—	—	—	—	—	—	60	—	—	—
Choconine	SA Solanidine	gl-rhm[d]	—	—	—	—	21t	—	—	—	—	—	—	—
Hederasaponin B	T Oleanolic acid	rhm-ar-scl-gl	—	—	30	—	26t	—	—	—	—	—	—	—
Hederasaponin C (3)	T Hederagenine	rhm-ar-scl-gl	—	—	21	—	—	—	—	—	—	—	—	—
α-Hederin	T Hederagenine	rhm-ar	—	—	47	—	—	—	—	—	—	—	—	—
Hederacoside A	T Hederagenine	gl-ar	—	—	42	—	—	—	—	—	—	—	—	—
Chinovin glycoside A	T Chinovanic acid	chin	—	—	—	48	—	—	—	—	—	—	—	—
Chinovin glycoside B	T Cincolic acid	chin	—	—	—	36	—	—	—	—	—	—	—	—
Chinovin glycoside C	T Chinovanic acid	gl	—	—	—	33	—	—	—	—	—	—	—	—
Trillin	S Diosgenin	gl	—	—	—	—	—	100	100	100	100	—	—	—
Yononin	S Yonogenin	ar	—	—	—	—	—	97	108	107	—	—	—	—
Timosaponin A-I	S Sarsagenin	gal	—	—	—	—	—	85	86	103	—	—	—	—
Timosaponin A-III	S Sarsagenin	gal-gl	—	—	—	—	—	52	67	46	84	—	—	—
Tokoronin	S Tokorogenin	ar	—	—	—	—	—	77	89	87	98	—	—	—
Dioscinprosapogenin A	S Diosgenin	scl	—	—	—	—	—	68	72	55	—	—	—	—
Dioscin	S Diosgenin	scl-rhm	—	—	—	—	—	46	61	—	77	50	—	120
Gracillin	S Diosgenin	scl-gl	—	—	—	—	—	45	58	—	76	45	—	100
Degalactotigonin	S Tigogenin	gal-gl-gl-xl	—	—	—	—	—	40	55	—	71	—	—	—
F-gitonin	S Gitogenin	gal-gl-gl-xl	—	—	—	—	—	33	50	—	67	—	—	—
Gitonin	S Gitogenin	gal-gal-gl-xl	—	—	—	—	—	34	54	—	—	—	—	—
Deglucodigitonin	S Digitogenin	gal-gal-gal-xl	—	—	—	—	—	30	51	—	—	—	—	—
Kikubasaponin	S Diosgenin	gl-gl-gl-rhm	—	—	—	—	—	19	33	—	—	—	—	—
Avenacoside	S Nuatigenin (4)	gl-gl-gl-gl-rhm	—	—	—	—	—	—	—	—	—	—	12	—
Deglucoavenacoside	S Nuatigenin (4)	gl-gl-gl-rhm	—	—	—	—	—	—	—	—	—	—	18	—

^{a} (1) Solancarpine; (2) solatunine; (3) hederacoside C; (4) isonuatigenin, perhaps as isomerization product formed during the acid hydrolysis.

^{b} SA = steroid alkaloids; T = triterpenes; S = spirostanols.

^{c} sol = solanose (solatriose: L-rhm; D-gal; D-gl); lyt = β-lycotetraose (L-gl; 2 D-gl; D-gal; D-gl); scl = scillabiose (L-rhm; (1α,2)-D-gl); rhm = rhamnose; ar = arabinose; gl = glucose; gal = galactose; xl = xylose; chin = D-chinovose (D-glucomethylose).

^{d} L-rhm (1α,4)-D-gl.

^{e} Comments on the experimental conditions: **I:**^{f} n-TLC, silica gel G; $R_f \times 100$ values. System: AcOH/EtOH (95%) H₂O 75:15:10.

II:^{g} n-TLC, silica gel G; $R_f \times 100$ values. Systems: Chf/MeOH/H₂O; **IIa,** 65:30:10; **IIb,** 65:35:10.

III:^{h} partition TLC; silica gel G, experimental conditions as given by Tschesche et al.,^{i} the layer is sprayed with the stationary phase (ammonium hydroxide); $R_f \times 100$ values. System: BuOH/2 N aqueous ammonium hydroxide.

IV:^{j} n-TLC, silica gel G; Dv. = 12–15 cm; conditions of saturation are not stated; $R_f \times 100$ values. System: AcOH/EtOH (95%) 1:1; t = oval spots, moderate tailing.

V:^{k} n-TLC, silica gel G; S.l. = 3–4 cm; Dv. = 12–14 cm; I.l. = 1–2 cm; tank lined with filter paper. Systems: **Va,** BuOH, water saturated; **Vb,** Chf/MeOH/H₂O 65:35:10 (lower phase); **Vc,** Chf/MeOH 80:20; **Vd,** BuOH/AcOH/H₂O 4:1:5 (upper phase); $R_S \times 100$ values (systems **Va–d:** S = trillon; $R_f \times 100$ values: 45 (**a**), 82 (**b**), 63 (**c**), and 78 (**d**).

VI: n-TLC^l silica gel G, $R_f \times 100$ values. Bz/EtAc 1:1. $R_f \times 100$ values for some spirostanes in this system: 25-D-spirosta-3,5-diene = 81; diosgenin = 58; gitogenin = 20; digitogenin = 10.

VII: n-TLC,^{m} silica gel G, $R_f \times 100$ values; experimental conditions as given by Tschesche et al.^{i} System: Chl/MeOH/H₂O 65:70:35.

VIII:^{n} n-TLC, silica gel G, water-saturated n-BuOH/MeOH 4:1. $R_S \times 100$ values.

^{f} K. Schreiber and H. Rönsch, Tetrahedron Letters p. 329 (1963).

^{g} R. Tschesche, W. Schmidt, and G. Wulff, Z. Naturforsch. **20b,** 708 (1965).

^{h} R. Tschesche, I. Duphorn, and G. Snatzke, Ann. Chem. **667,** 151 (1963).

^{i} R. Tschesche, W. Freytag, and G. Snatzke, Chem. Ber. **92,** 3053 (1959).

^{j} R. Paquin and M. Lepage, J. Chromatog. **12,** 57 (1963).

^{k} T. Kawasaki and K. Miyahara, Chem. Pharm. Bull. (Tokyo). **11,** 1546 (1963).

^{l} T. Kawasaki, I. Nishioka, T. Tsukamoto, and K. Mihashi, Yakugaku Zasshi **86,** 673 (1966).

^{m} R. Tschesche and W. Schmidt, Z. Naturforsch. **21b,** 896 (1966).

^{n} G. Held and D. Vágujfalvi, Botan. Kozlem **52,** 201 (1965).

culogenin and bis-α-L-rhamnosylpaniculogenin, with the $R_f \times 100$ values of 36 and 28, respectively, in the system chloroform–methanol 4:1 (silica gel G). These saponins were not completely pure and after hydrolysis give, together with the principal aglycone, paniculogenin (25β,5α-spirostane-3β,6α,23-triol), some amounts of neochlorogenin (25β,5α-spirostane-3β,6α-diol) ; these sapogenins show $R_f \times 100$ values of 58 and 54, respectively, under the same chromatographic conditions.

Triterpenic saponins have been analyzed on silica gel layers by Tschesche et al.[319, 320] A typical separation is that of the hederasaponins B and C,[319] both with two identical sugar moieties. The separation of the chinovin glycosides A, B, and C[320] on the plates could not be realized by paper chromatography, even if the same solvent system was used, showing that partition TLC on some occasions has a greater resolving power for closely related compounds than other fractionation procedures. The resolution of saponins with similar mobility is improved further by using special TLC techniques, and the V-shaped wedge-strip technique. Partition TLC on wedged silica gel layers[39] has permitted the fractionation of a great number of saponins from *Brademeyera floribunda* Willd., *Gypsophila paniculata,* and *Quillaja saponaria* (stationary phase: 2 N ammonium hydroxide; mobile phase: butanol–methanol 10:1, saturated with the stationary phase).

Silica gel containing boric acid has been used for the fractionation of saponins during isolation studies of the alkaloid glycosides from *Solanum dulcamara* L. by Schreiber and Rönsch.[190] Four substances have been separated with R_f values of 0.22, 0.32, 0.44, and 0.82 (acetone–methanol–water 75:15:10), one of them identical to solasonin.

Also from the roots of *Solanum paniculatum* L., Schreiber and Ripperger[321] have isolated a new steroid saponin, jurubin, which on silica gel layers in the system chloroform–ethanol–conc. ammonium hydroxide 5:5:1.3[313] gives a R_f value of 0.63; this saponin contains glucose and jurubidine in the proportion 1:1 (3β-amino-O(26)-β-D-glucopyranosyl-25β,5α-furostane-22α,26-diol).

Chromatography on silica gel–gypsum (26:4) layers was employed by Bite and Rettegi[321a] for the separation of synthetic steroid glycosides of solasodine and diosgenin. $R_f \times 100$ values found, respectively, in systems chloroform–ethanol (20:1) and n-butanol saturated with water are: solasodine (45; 95) ; diosgenin (80; 100) ; solasodine-3β-D-glucoside(I) (0; 60) ; N-nitrososolasodine-3β-D-glucoside(II) (0; 75) ; solasodine-3β-

[319] R. Tschesche, W. Schmidt, and G. Wulff, *Z. Naturforsch.* **20b,** 708 (1965).
[320] R. Tschesche, I. Duphorn, and G. Snatzke, *Ann. Chem.* **667,** 151 (1963).
[321] K. Schreiber and H. Ripperger, *Tetrahedron Letters* p. 5997 (1966).
[321a] P. Bite and T. Rettegi, *Acta Chim. Acad. Sci. Hung.* **52,** 79 (1967).

D-glucoside tetraacetate(III) (65; 95); solasodine-3β-D-glucoside(IV) (0; 75); solasodine-3β-D-galactoside tetraacetate (V) (45; 95); diosgenin-3β-D-glucoside tetraacetate(VI) (90; 100); diosgenin-3β-D-galactoside tetraacetate(VII) (90; 100); solasodine-3β-D-galactoside(VIII) (0; 65); diosgenin-3β-D-glucoside(IX) (0; 85); diosgenin-3β-D-galactoside(X) (0; 80); and N-nitrososolasodine-3β-D-galactoside(XI) (0; 80). No separation was obtained for the glycoside pairs II/IV, VI/VII, and X/XI.

8. Steroid Alkaloids

The chromatographic mobilities of steroid alkaloids on TLC, using various systems, are given in Table XVIII.

A systematic investigation of the steroid bases of the Indian shrub *Holarrhena antidysenterica*, a group of saturated and Δ^5-C$_{21}$-steroids with amino substituents in the positions 3, 18, and 20, has been undertaken by Lábler and Černý[322, 323] on silica gel G layers. In addition, some cholestane bases have been investigated by these authors.[322] From their findings it can be concluded that a 3α-dimethylamino-Δ^5-structure is less polar than the corresponding 3β-epimer. For instance, conessine is more polar than concuressine. In both the cholestane and pregnane series, the axial 3α/5α- and 3β/5β-structures are less polar than the corresponding equatorial 3β/5α and 3α/5β. The following is the order of polarity found for 3-dimethylamino-substituted steroids:

$$3\alpha\text{-series}: 5\alpha < \Delta^5 < 5\beta; \text{and } 3\beta\text{-series}: 5\beta < \Delta^5 < 5\alpha$$

For the 20-substituted bases, it was found that a 20β-dimethylamino alkaloid was less polar than the 20α-epimer.

The separation of steroid alkaloids isolated from Solanaceae, which are related to spirosolanol, solanidine, and solanocapsine has been investigated on silica gel, aluminum oxide, and silicic acid TLC by Schreiber *et al.*[324] For the separation of the closely related steroids solanidine and demissidine, silver nitrate-impregnated layers also have been employed.

The recently isolated steroid alkaloid jurubidine (3β-amino-25β,5α-spirostane) with a new structure type related to neotigogenin gives, on silica gel G layers, an R_f value of 0.50 when developed with the mixture concentrated ammonium hydroxide–chloroform–methanol 5:4:1;[313] under the same conditions paniculidine (9α-hydroxyjurubidine)[313] shows a migration of 0.28.

[322] L. Lábler and V. Černý, *in* "Thin-Layer Chromatography" (G. B. Martini-Bettòlo, ed.), p. 144. Elsevier, Amsterdam, 1964.
[323] L. Lábler and V. Černý, *Collection Czech. Chem. Commun.* **28**, 2932 (1963).
[324] K. Schreiber, O. Aurich, and G. Osske, *J. Chromatog.* **12**, 63 (1963).

TABLE XVIII

CHROMATOGRAPHIC MOBILITIES FOUND ON TLC FOR STEROID ALKALOIDS

Steroid alkaloids	Chromatographic systems[a]								
	Ia	Ib	II	IIIa	IIIb	IVa	IVb	V	VIa
3α-Dimethylamino-5α-cholestane	—	—	—	—	—	—	—	—	82
3β-Dimethylamino-5α-cholestane	—	—	—	—	—	—	—	—	14
3α-Dimethylamino-5β-cholestane	—	—	—	—	—	—	—	—	22
3β-Dimethylamino-5β-cholestane	—	—	—	—	—	—	—	—	71
3α-Dimethylamino-cholest-5-ene	—	—	—	—	—	—	—	—	75
3β-Dimethylamino-cholest-5-ene	—	—	—	—	—	—	—	—	20
5α-Solanidan-3β-ol (demissidine)	30	72	295	85	57	—	—	—	—
Solanid-5-en-3β-ol (solanidine)	30	72	210	85	57	51	64	47	—
5α-Solanidan-3-one	—	—	—	—	89	—	—	—	—
5α-Tomatidan-3β-ol (tomatidine)	17	85	—	66	29	—	—	—	—
Tomatid-5-en-3β-ol (tomatidenol)	17	85	—	66	29	—	—	—	—
Solasodan-3β-ol (22,25-iso-tomatidine, soladulcidine)	6	80	—	53	28	—	—	—	—
Solasod-5-en-3β-ol (solasodine)	6	80	—	53	28	—	—	—	—
Solaso-3,5-diene	24	82	—	87	—	—	—	—	—
5α-Tomatidan-3-one	—	—	—	—	74	—	—	—	—
5α-Solasodan-3-one	—	—	—	—	53	—	—	—	—
3β-Amino-22,26-imino-16β,23-oxido-5α-cholestan-23-ol (solanocapsine)	0	9	—	—	0	—	—	—	—
Solasodine tosylate	0	66	—	—	—	—	—	—	—
3β-Hydroxy-22,26-imino-17α,23-oxido-jerva-5,12-dien-11-one (jervine)	—	—	—	—	—	—	—	—	—
Tomatidan-3β,15α-diol (15α-hydroxytomatidine)	—	—	—	—	—	—	—	56	—
Solasodan-3β,15α-diol (15α-hydroxysoladulcidine)	—	—	—	—	—	—	—	32	—
18-Dimethylamino-5α-pregnan-3-one	—	—	—	—	—	—	—	—	69
3α-Amino-20α-hydroxy-5α-pregnane (funtumidine)	—	—	—	—	—	—	—	—	—
20α-Dimethylamino-5α-pregnan-3-one	—	—	—	—	—	—	—	—	54

Compound								
20β-Dimethylamino-5α-pregnan-3-one	—	—	—	—	—	—	—	64
3β-Acetoxy-20α-dimethylamino-5α-pregnane	—	—	—	—	—	—	—	63
3β-Acetoxy-20β-dimethylamino-5α-pregnane	—	—	—	—	—	—	—	75
3β,20α-Diaminopregn-5-en-18-ol (holarrhimine)	—	—	—	—	—	—	—	—
3β,20α-Bisdimethylamino-5α-pregnane	—	—	—	—	—	—	—	8t
3β,20β-Bisdimethylamino-5α-pregnane	—	—	—	—	—	—	—	8t
3α,20α-Bisdimethylaminopregn-5-ene	—	—	—	—	—	—	—	48t
3β,20α-Bisdimethylaminopregn-5-ene	—	—	—	—	—	—	—	22t
3α,20α-Bisdimethylaminopregn-5-en-18-ol (tetramethylholarrhidine)	—	—	—	—	—	—	—	17t
3β,20α-Bisdimethylaminopregn-5-en-18-ol (tetramethylholarrhimine)	—	—	—	—	—	—	—	7
5α-Conanine (18,20-methylimino-5α-pregnane)	—	—	—	—	—	—	—	71
$\Delta^{N(20)}$-5α-Conaninene	—	—	—	—	—	—	—	43
N-Demethyl-5α-conanine (18,20-imino-5α-pregnane)	—	—	—	—	—	—	—	10
3β-Hydroxy-$\Delta^{N(20)}$-5α-conaninene	—	—	—	—	—	—	—	—
5α-Conanin-3-one (18,20-methylimino-5α-pregnan-3-one)	—	—	—	—	—	—	—	47
3β-Hydroxy-N-demethyl-5α-conanine	—	—	—	—	—	—	—	—
3β-Hydroxy-5α-conanine	—	—	—	—	—	—	—	—
3α-Acetoxy-5α-conanine	—	—	—	—	—	—	—	—
3β-Methylamino-5α-conanine (dihydroisoconessimine)	—	—	—	—	—	—	—	—
3β-Methylaminoconanine-5-ene (isoconessimine)	—	—	—	—	—	—	—	—
3α-Dimethylamino-5α-conanine (dihydroconcurassine)	—	—	—	—	—	—	—	64
3β-Dimethylamino-5α-conanine (dihydroconessine)	—	—	—	—	—	—	—	8t
3α-Dimethylaminoconanin-5-ene (concuressine)	—	—	—	—	—	—	86	54
3β-Dimethylaminoconanin-5-ene (conessine)	—	—	—	—	—	—	0	13t
3β-Aminoconanin-5-ene (conamine)	—	—	—	—	—	—	—	—
3β-Dimethylaminoconanin-5-en-12β-ol (12β-hydroxyconessine, holarrhenine)	—	—	—	—	—	—	—	—
3β-Dimethylamino-N-demethylconanin-5-ene (conessimine)	—	—	—	—	—	—	—	—
3β-Methylamino-N-demethylconanin-5-ene (conimine)	—	—	—	—	—	—	—	—

(Footnotes appear on p. 118.)

(Continued)

TABLE XVIII (Continued)

Steroid alkaloids	Chromatographic systems[a]							
	VIb	VIc	VII	VIIIa	VIIIb	VIIIc	IXa	IXb
3α-Dimethylamino-5α-cholestane	—	—	—	—	—	—	70–86	—
3β-Dimethylamino-5α-cholestane	—	—	—	—	—	—	10–25	—
3α-Dimethylamino-5β-cholestane	—	—	—	—	—	—	17–36	—
3β-Dimethylamino-5β-cholestane	—	—	—	—	—	—	59	—
3α-Dimethylamino-cholest-5-ene	—	—	—	—	—	—	67	—
3β-Dimethylamino-cholest-5-ene	—	—	—	—	—	—	32	—
5α-Solanidan-3β-ol (demissidine)	—	—	—	—	—	—	—	—
Solanid-5-en-3β-ol (solanidine)	—	—	—	—	—	—	—	—
5α-Solanidan-3-one	—	—	—	—	—	—	—	—
5α-Tomatidan-3β-ol (tomatidine)	—	—	—	—	—	—	—	—
Tomatid-5-en-3β-ol (tomatidenol)	—	—	—	—	—	—	—	—
Solasodan-3β-ol (22,25-iso-tomatidine, soladulcidine)	—	—	35	—	—	—	—	—
Solasod-5-en-3β-ol (solasodine)	—	—	50	—	—	—	—	—
Solaso-3,5-diene	—	—	—	—	—	—	—	—
5α-Tomatidan-3-one	—	—	—	—	—	—	—	—
5α-Solasodan-3-one	—	—	—	—	—	—	—	—
3β-Amino-22,26-imino-16β,23-oxido-5α-cholestan-23-ol (solanocapsine)	—	—	—	—	—	—	—	—
Solasodine tosylate	—	—	43	—	—	—	—	—
3β-Hydroxy-22,26-imino-17α,23-oxido-jerva-5,12-dien-11-one (jervine)	—	—	—	—	—	—	—	—
Tomatidan-3β,15α-diol (15α-hydroxytomatidine)	—	—	—	—	—	—	—	—
Solasodan-3β,15α-diol (15α-hydroxysoladulcidine)	—	—	—	—	—	—	—	—
18-Dimethylamino-5α-pregnan-3-one	—	—	—	66	—	—	—	—
3α-Amino-20α-hydroxy-5α-pregnane (funtumidine)	—	—	—	—	80	—	—	—
20α-Dimethylamino-5α-pregnan-3-one	—	—	—	—	—	69	—	—

Compound						
20β-Dimethylamino-5α-pregnan-3-one	—	—	—	—	—	—
3β-Acetoxy-20α-dimethylamino-5α-pregnane	—	—	—	—	—	—
3β-Acetoxy-20β-dimethylamino-5α-pregnane	—	—	—	—	—	—
3β,20α-Diaminopregn-5-en-18-ol (holarrhimine)	76	30-70	38	55	—	—
3β,20α-Bisdimethylamino-5α-pregnane	80	—	—	—	—	—
3β,20β-Bisdimethylamino-5α-pregnane	—	—	—	—	—	—
3α,20α-Bisdimethylaminopregn-5-ene	—	—	—	—	—	—
3β,20β-Bisdimethylaminopregn-5-ene	—	—	—	—	—	—
3α,20α-Bisdimethylaminopregn-5-en-18-ol (tetramethylholarrhidine)	83	86	78	—	—	—
3β,20α-Bisdimethylaminopregn-5-en-18-ol (tetramethylholarrhimine)	36	—	—	17	—	—
5α-Conanine (18,20-methylimino-5α-pregnane)	—	—	—	—	—	—
$\Delta^{N(20)}$-5α-Conaninene	—	—	—	—	—	—
N-Demethyl-5α-conanine (18,20-imino-5α-pregnane)	57	—	—	—	—	—
3β-Hydroxy-$\Delta^{N(20)}$-5α-conaninene	—	—	—	—	—	—
5α-Conanin-3-one (18,20-methylimino-5α-pregnan-3-one)	12	—	—	—	—	82
3β-Hydroxy-N-demethyl-5α-conanine	92	—	—	—	—	—
3β-Hydroxy-5α-conanine	—	—	—	—	—	—
3α-Acetoxy-5α-conanine	—	—	—	—	—	—
3β-Methylamino-5α-conanine (dihydroisoconessimine)	28	—	—	—	—	—
3β-Methylaminoconanine-5-ene (isoconessimine)	62	—	—	—	—	1.00
3α-Dimethylamino-5α-conanine (dihydroconcurassine)	24	—	—	—	—	11
3β-Dimethylamino-5α-conanine (dihydroconessine)	64	—	—	—	—	9-22
3α-Dimethylaminoconanin-5-ene (concuressine)	—	—	—	—	—	—
3β-Dimethylaminoconanin-5-ene (conessine)	78	1.00	1.00	17	—	30-50
3β-Aminoconanin-5-ene (conamine)	—	67	57	17	—	—
3β-Dimethylaminoconanin-5-en-12β-ol (12β-hydroxyconessine, holarrhenine)	—	—	—	—	—	—
3β-Dimethylamino-N-demethylconanin-5-ene (conessimine)	7	—	—	—	—	—
3β-Methylamino-N-demethylconanin-5-ene (conimine)	3	—	—	—	—	—

(Footnotes appear on p. 118.)

(Continued)

TABLE XVIII (Continued)

a Comments on the experimental conditions. I:b n-TLC as VII in Table III. Systems: Ia, Ch/EtAc 1:1; Ib, Chf/MeOH 6:4.

II:b Overrun TLC, according to Brenner and Niederwieser.c Dv. = 16 hours, values in cm; Silver nitrate impregnated silica gel. EtAc/Ch/EtOH 96%, 50:40:50.

III:b n-TLC, aluminum oxide/gypsum 10%; activation: 120°/2 hours; experimental conditions as for VII (in Table III); $R_f \times 100$ values. Systems: IIIa, Ch/EtAc 1:1; IIIb, Hex/triethylamine 15:1.

IV:d As IV in Table XVII. Systems: IVa, Chf/AcOH/MeOH 85:2:13; IVb, AcOH/EtOH 95%, 1:1.

V:e n-TLC, silica gel G; conditions as VII in Table III. Eth/iPrOH 9:1.

VI:f,g n-TLC, silica gel/gypsum 15%; Fr. = 10 cm; $R_f \times 100$ values; tank atmosphere saturated with ammonia; the R_f values depend on the degree of ammonia saturation; t = moderate tailing ($R_f \times 100 \pm 3$). Systems: VIa, Bz; VIb, Eth; VIc, Eth (Fr. = 16–17 cm).

VII:h n-TLC, silica gel/gypsum (26:4); $R_f \times 100$ values. Chf/MeOH 20:1.

VIII:i n-TLC; experimental conditions according to Tschesche et al.i $R_B \times 100$ values (S = conessine) for VIIIa and VIIIb; $R_f \times 100$ values for VIIIc, in cases of streaking two values are quoted that represent the lower and upper limits of the spot. Systems: VIIIa, Chf/Ch/diethylamine 5:4:1 (Dv. = 15 cm, 105 min); VIIIb, Chf/MeOH/conc. ammonium hydroxide 60:40:1.5 (Dv. = 15 cm, 60 min); VIIIc, upper phase of BuOH/AcOH/H2O 4:1:5.

IX: Spread TLC,k plates 20 × 10 cm, 1 mm thick; Aluminum oxide, grade V; S.l. = 2 cm, angle of the plate with the bottom = 20°; Dv. = 10–20 min; in cases of streaking two $R_f \times 100$ values are given (see VIII). Systems: IXa, Bz/Eth 7:3; IXb, Eth/EtOH 98:2.

b K. Schreiber, O. Aurich, and G. Osske, J. Chromatog. 12, 63 (1963).

c M. Brenner and A. Niederwieser, Experientia 17, 237 (1961).

d R. Paquin and M. Lepage, J. Chromatog. 12, 57 (1963).

e H. Rönsch and K. Schreiber, Tetrahedron Letters p. 1947 (1965).

f L. Lábler and V. Černý, in "Thin-Layer Chromatography" (G. B. Martini-Bettòlo, ed.), p. 144. Elsevier, Amsterdam, 1964.

g L. Lábler and V. Černý, Collection Czech. Chem. Commun. 28, 2932 (1963).

h P. Bite and G. Magyar, Acta Chim. Acad. Sci. Hung. 48, 255 (1966).

i R. Tschesche and H. Ockenfels, unpublished results (1962), quoted by R. Tschesche, G. Wulff, and K. H. Richert, in "New Biochemical Separations" (A. T. James and L. J. Morris, eds.), p. 198. Van Nostrand, Princeton, New Jersey, 1964.

j R. Tschesche, W. Freytag, and G. Snatzke, Chem. Ber. 92, 3053 (1959).

k V. Černý, J. Joska, and L. Lábler, Collection Czech. Chem. Commun. 26, 1658 (1961).

9. Triterpenes

Besides the tetracyclic triterpenes, which have been considered together with the sterols, TLC techniques have been successfully employed to separate the pentacyclic terpenes. The $R_f \times 100$ values found for the neutral terpenes related to the ursane, oleane, α-lupene, friedelane, taraxastane, and zeorane structures are given in Table XIX. The mobilities of triterpenic acids and some of their derivatives are summarized in Tables XX and XXI.

Neutral or acidic tetracyclic terpenes differing only by the isomerism of the C-30 and C-31 methyl groups in ring E (19,20-dimethyl: ursane series; 20,20-dimethyl: oleane series), such as the pairs α-amyrin/β-amyrin and ursolic acid/oleanolic acid, cannot be separated by adsorption TLC. In both the ursane and oleane series, a very good separation is possible between mono- and dicarboxylic acids: quinovic acid (3β-hydroxyurs-12-ene-27,28-dioic) and pyroquinovic acid (27-nor-3β-hydroxyurs-13-en-28-oic), cincholic acid (3β-hydroxyolean-12-ene-27,28-dioic) and pyrocincholic acid (27-nor-3β-hydroxyolean-13-en-28-oic) or their acetates. There are also differences in the mobilities of two triterpenic acids differing by the presence of an angular methyl group at C-14, such as ursolic acid (3β-hydroxyurs-12-en-28-oic) and pyrocincholic acid.

Closely related substances with isomeric groups, such as β-boswellic acid (3α-hydroxyurs-12-en-24-oic) and ursolic acid are separable by partition TLC on silica gel G using n-butanol–$2\,N$ ammonium hydroxide;[40] thus the plate is sprayed with the $2\,N$ ammonium hydroxide, the R_f values depend on the degree of impregnation. In this system, even the isomeric α-amyrin/β-amyrin derivatives such as quinovic acids, can be resolved.[320]

The monohydroxy derivatives of oleanic acid isomers as bredenolic ($3\beta,24$- or $3\alpha,23$-dihydroxyolean-12-en-28-oic), siaresinolic ($3\beta,19$-hydroxyolean-12-en-28-oic), and cochalic ($3\beta,16\beta$-dihydroxyolean-12-en-28-oic) acids have different mobilities in diisopropyl ether–acetone (5:2).[54] Also separable on TLC are the 2-hydroxyhederagenins arjunolic acid ($2\alpha,3\beta,23$-trihydroxyolean-12-en-28-oic) and bayogenin (2β-epimer). They can also be separated as lactones.

Stöcklin[324a] has employed silica gel layers (front = 17.2 cm; system chloroform–methanol 9:1; mobilities in centimeters) for the separation of the following hexahydroxy-Δ^{12}-oleane derivatives ($C_{30}H_{50}O_6$): dihydrotheasapogenol (I), 2.1; gymnemagenin (II), 2.3; protoäscigenin (IV),

[324a] W. Stöcklin, *Helv. Chim. Acta* **50**, 491 (1967).

TABLE XIX
CHROMATOGRAPHIC MOBILITIES FOUND ON TLC FOR NEUTRAL PENTACYCLIC TERPENES OF THE URSANE, OLEANE, α-LUPENE, FRIEDELANE, TARAXASTANE, AND ZEORANE SERIES

Neutral terpenes[a]	Structure[b]	Chromatographic systems[c]						
		Ia	Ib	Ic	Id	Ie	IIa	IIh
α-Amyrin (α-amyrenol, α-alban)	U Δ12; 3β-ol	—	—	—	—	—	—	—
Uvaol (28-hydroxy-α-amyrin)	U Δ12; 3β-28-ol	—	—	—	—	—	—	—
Brein	U Δ12; 3β,16β-ol	—	—	—	—	—	—	—
Breindione	U Δ12; 3,16-oxo	—	—	—	—	—	—	—
α-Amyrone (α-amyrenone)	U Δ12; 3-oxo	—	—	—	—	—	—	—
24-Nor-β-boswellenone	U Δ12; 3-oxo; 24-nor	—	—	—	85	—	—	—
Aescigenin (C₃₅H₅₈O₇)	Five hydroxyl groups	—	—	—	—	—	—	—
β-Amyrene	O Δ12	—	—	—	—	—	—	—
β-Amyrin (β-amyrenol)	O Δ12; 3β-ol	—	—	—	—	—	—	—
β-Amyrin, ac	—	—	—	—	—	—	—	—
β-Amyrin, bz	—	—	—	—	—	—	—	—
Epi-β-amyrin	O Δ12; 3α-ol	—	—	—	—	—	—	—
Germanicol	O Δ18; 3β-ol	—	—	—	—	—	—	—
Taraxenol	O Δ14; 3β-ol	—	—	—	—	—	—	—
Taraxenol, ac		—	—	—	—	—	63	—
Dihydrotaraxene	O 3β-ol	—	—	—	—	76	38	—
Erythrodiol	O Δ12; 3β,28-ol	—	—	—	—	84	84	55
28-Hydroxyerythrodiol, 3,28-dime	—	—	—	—	—	—	—	88
28-Hydroxyerythrodiol, 3,28-dime, 27-ts	—	—	—	—	—	—	—	90
Erythrodiol, dime	O Δ13; 3,28-ol; 27-nor	—	—	—	—	—	60	—
Pyrocincholdiol	O Δ12,14; 3,28-ol; 27-nor	—	—	—	—	—	90	—
27-Nor-14-dehydroerythrodiol, diac		—	—	—	—	—	—	—
Primulagenin A	O Δ12; 3β,16α,28-ol	—	—	—	—	—	30	—

Compound		Structure								
Primulagenin D (28-dehydroprimulagenin A)	O	Δ^{12}; 3β,16α-ol; 28-al	—	—	—	—	—	—	—	—
Priverogenin A	O	Δ^{12}; 3β,16α,22β-ol; 28-al	—	—	—	—	—	—	—	—
Priverogenin B	O	3β,16α,22β-ol; (13β → 28)oxido	—	—	—	—	—	—	—	—
Priverogenin A, 16-ac		—	—	—	—	—	—	—	—	—
Priverogenin B, monoac		—	—	—	—	—	—	—	—	—
Stimulagenin A, triac			—	—	—	—	—	—	—	—
Δ^{21}-Primulagenin A, triac			—	—	—	—	—	—	—	—
Dihydropriverogenin A			—	—	—	—	—	—	—	—
Betulin	L	3β,28-ol	—	—	—	—	75	—	—	—
Allobetulin [(19 → 28)-oxido-13β(H)-β-amyrin]	O	3β-ol; (19 → 28)oxido	—	—	—	—	—	—	—	—
Allobetulin, 3-ac		—	—	—	—	—	—	—	—	—
Betulin, diac		—	42	—	—	—	—	—	—	—
Allobetulone (3-dehydroallobetulin)	O	3-oxo; (19 → 28) oxido	—	—	—	—	—	—	—	—
Lupeol (lupenol)	L	3β-ol	80	67	—	—	—	—	—	—
Lupeol, ac			85	—	—	—	—	—	—	—
Lupeol, bz	L		—	—	—	—	—	—	—	—
Lupenal, diac	L	3β,28-ol (ac); 30-al	—	—	—	68	—	—	—	—
Friedelin (3-dehydrofriedelanol)	F	3-oxo	—	—	62	—	—	83	—	—
Friedelanol	F	3β-ol	—	—	—	—	—	—	—	—
Epifriedelanol, ac	F	3α-ol (ac)	—	—	—	—	—	—	—	—
Taraxasterol	T	$\Delta^{20(30)}$; 3β-ol	—	—	—	—	—	—	—	—
Zeorin	Z	6α,22-ol	77	—	—	—	95	34	—	—
22-Deoxyzeorin	Z	6α-ol	14	—	—	—	—	—	—	—
Zeorin, ac	Z	—	—	—	—	—	—	—	—	—
Zeorininon	Z	$\Delta^{17(21)}$; 6-oxo	89	—	—	—	—	—	—	—
Leucotylin (3β-hydroxyzeorin)	Z	3β,6α,22-ol	—	—	—	—	32	2	—	—
Leucotylin, diac	Z	—	9	—	—	—	—	—	—	—
Epizeorin	Z	3β,6β,22-ol	—	—	—	—	97	—	—	—

(Footnotes appear on p. 124.)

(Continued)

TABLE XIX *(Continued)*

Neutral terpenes[a]	Chromatographic systems[c]									
	IIi	IIj	IIk	IIl	IIm	IIn	IIo	III	Va	Vb
α-Amyrin (α-amyrenol, α-alban)	—	—	—	—	—	—	—	26	—	—
Uvaol (28-hydroxy-α-amyrin)	—	—	—	—	—	—	—	3	—	—
Brein	—	—	—	—	—	—	—	4	—	—
Breindione	—	—	—	—	—	—	—	40	—	—
α-Amyrone (α-amyrenone)	—	—	—	—	—	—	31	—	—	—
24-Nor-β-boswellenone	—	30	—	—	—	—	—	—	—	—
Aescigenin (C$_{35}$H$_{58}$O$_7$)	—	—	—	—	—	—	—	—	—	—
β-Amyrene	—	—	—	—	—	82	—	—	—	—
β-Amyrin (β-amyrenol)	—	—	—	—	—	—	12	24	—	—
β-Amyrin, ac	—	—	—	—	38	—	45	87	—	—
β-Amyrin, bz	—	—	—	—	—	—	—	92	—	—
Epi-β-amyrin	—	—	—	—	—	—	—	53	—	—
Germanicol	—	—	—	—	—	—	—	17	—	—
Taraxenol	—	—	—	—	—	—	—	30	—	—
Taraxenol, ac	—	—	—	—	—	—	—	91	—	—
Dihydrotaraxene	—	—	—	—	—	—	—	35	—	—
Erythrodiol	—	—	—	—	—	—	—	—	—	—
28-Hydroxyerythrodiol, 3,28-dime	—	—	—	85	—	—	—	—	—	—
28-Hydroxyerythrodiol, 3,28-dime, 27-ts	—	—	—	—	—	—	—	—	—	—
Erythrodiol, dime	37	—	—	—	—	—	—	—	—	—
Pyrocincholdiol	—	—	—	—	—	—	—	—	—	—
27-Nor-14-dehydroerythrodiol, diac	—	—	—	—	—	—	—	—	—	—
Primulagenin A	—	—	—	—	—	—	—	—	50	89

Compound									
Primulagenin D (28-dehydroprimulagenin A)	216	114	—	—	—	—	—	—	—
Priverogenin A	89	50	—	—	—	—	—	—	—
Priverogenin B	55	39	—	—	—	—	—	—	—
Priverogenin A, 16-ac	100	100	—	—	—	—	—	—	—
Priverogenin B, monoac	150	100	—	—	—	—	—	—	—
Stimulagenin A, triac	—	—	—	—	—	—	73	—	—
Δ²¹-Primulagenin A, triac	—	—	—	—	—	—	74	—	—
Dihydropriverogenin A	33	29	—	—	—	—	—	—	—
Betulin	—	—	14	—	—	—	—	—	—
Allobetulin [(19 → 28)-oxido-13β(H)-β-amyrin]	—	—	38	—	—	—	—	—	—
Allobetulin, 3-ac	—	—	91	—	—	—	—	—	—
Betulin, diac	—	—	—	—	—	—	—	—	—
Allobetulone (3-dehydroallobetulin)	—	—	78	—	—	—	—	—	—
Lupeol (lupenol)	—	—	—	—	—	—	—	—	—
Lupeol, ac	—	—	—	—	—	—	—	—	—
Lupeol, bz	—	—	—	—	—	—	—	—	—
Lupenal, diac	—	—	—	—	—	—	—	—	—
Friedelin (3-dehydrofriedelanol)	—	—	88	—	—	—	—	—	—
Friedelanol	—	—	50	—	—	—	—	—	—
Epifriedelanol, ac	—	—	89	—	—	—	—	—	—
Taraxasterol	—	—	15	—	—	—	—	—	—
Zeorin	—	—	—	—	—	—	—	—	—
22-Deoxyzeorin	—	—	—	—	—	—	—	—	—
Zeorin, ac	—	—	—	—	—	—	—	—	—
Zeorininon	—	—	—	—	—	—	—	—	—
Leucotylin (3β-hydroxyzeorin)	—	—	—	—	—	—	—	—	—
Leucotylin, diac	—	—	—	—	—	—	—	—	—
Epizeorin	—	—	—	—	—	—	—	—	—

(Continued)

(Footnotes appear on p. 124.)

TABLE XIX (Continued)

[a] ac = acetate; me = methyl ether; ts = tosylate; bz = benzoate.

[b] U = ursane; O = oleane; L = α-lupene; F = friedelane; T = taraxastane (heterolupane); Z = zeorane.

[c] Comments on the experimental conditions for the Tables XIX–XXI: **I:** [a] n-TLC, aluminum oxide (activation 110°/3 min); plate 28.3 × 5 cm; Sl. = 3.5–4 cm; R_f × 100 values. Systems: **Ia,** Bz; **Ib,** Eth; **Ic,** Hex/Bz 8:1; **Id,** Ch/EtAc 8.5:1.5; **Ie,** Eth/EtOH 98:2.
II: n-TLC, silica gel; experimental conditions as given by Tschesche et al.;[e] R_f × 100 values. Systems: **IIa,**[f–h] iPrEth; **IIb,**[i] iPrEth/Atn 19:1; **IIc,**[f,a,i,i] iPrEth/Atn 5:2 (R_f values referred to that of oleanonic acid = 0.68); **IId,**[f] as **IIc,** with 5% Py; **IIe**[f] ClBz/AcOH 9:1; **IIf,**[f] AcOH/MeOH/diethylamine 14:4:3; **IIg,**[g] dClMe/Py 7:2; **IIh,**[h] Bz/MeOH 99:1; **IIi,**[h] Chf/Ch 2:1; **IIj,**[k] Chf/MeOH 10:1; **IIk,**[l] Bz/Atn 92:5:7.5; **IIl**[k] Bz/iPrEth 9:1; **IIm,** dClMe; **IIn,** Ch; **IIo,** Bz.

III: n-TLC, aluminum oxide G;[l] Hep/Bz/EtOH 50:50:0.5.

IV: n-TLC, experimental conditions as **II** and **III.** Systems:[m] **IVa,** iPrEth/Atn 5:2; **IVb,** iPrEth/Atn 2:1; **IVc,** Bz/Chf (1:1) + 10% MeOH; **IVd,** Bz/Chf (1:1) + 15% MeOH; **IVe,** Bz/Chf (1:1) + 20% MeOH; **IVf,** Bz/Chf (1:1) + 25% MeOH; **IVg,** iPrEth/Atn (5:2) + 7% AcOH.

V: n-TLC, R_S × 100 values (S = primulagenin A, 16 acetate);[n] experimental conditions as **II.** Systems: **Va,** Chf/Atn 5.5:1; **Vb,** Bz/Ac 12:1.

VI:[o] System Eth/Bz 4:1.

VII: Partition TLC.[o] Systems: n-BuOH/2 N-ammonium hydroxide; R_f values vary with the kind of impregnation and have been related to that of quinovic acid (0.76).

[d] S. Huneck, J. Chromatog. **7,** 56 (1962).

[e] R. Tschesche, W. Freytag, and G. Snatzke, Chem. Ber. **92,** 3053 (1959).

[f] R. Tschesche, F. Lampert, and G. Snatzke, J. Chromatog. **5,** 217 (1961).

[g] R. Tschesche, I. Duphorn, and G. Snatzke, in "New Biochemical Separations" (A. T. James and L. J. Morris, eds.), p. 248. Van Nostrand, Princeton, New Jersey, 1964.

[h] R. Tschesche, I. Duphorn, and G. Snatzke, Ann. Chem. **667,** 151 (1963).

[i] R. Tschesche and A. K. Sen Gupta, Chem. Ber. **93,** 1903 (1960).

[j] R. Tschesche and W. Schmidt, Z. Naturforsch. **21b,** 896 (1966).

[k] R. Tschesche, V. Axen, and G. Snatzke, Ann. Chem. **669,** 171 (1963).

[l] R. Ikan, J. Kashman, and E. D. Bergmann, J. Chromatog. **14,** 275 (1964).

[m] J. Jacob, Über die Aglykone der Saponine der levantinischen Seifenwurzel. Inaugural Dissertation, Univ. of Bonn, Germany, 1963.

[n] R. Tschesche, B. T. Tjoa, and G. Wulff, Ann. Chem. **696,** 160 (1966).

[o] A. J. Khorlin, Y. S. Ovodov, and N. K. Kochetkov, Zh. Obshch. Khim. **32,** 782 (1962).

3.1; 7α-hydroxy-A_1-barrigenol (III), 3.0; and tanginol (V), 4.6. In this system the genin IV can be separated from barringtogenol C (äscinidin, theasapogenol B). After acetylation a complete resolution was achieved for the genins I (mobility: 8.1) and II (10.4) in the system chloroform–ethyl acetate–methanol 97:2:1; furthermore, these two acetates can be separated in the system benzene–ethyl acetate 1:1 from that of genin IV (mobilities: I-ac, 3.4; II-ac, 3.4; IV-ac, 8.1).

The chromatographic mobilities for several lactones, bromolactones, acetonides, and methyl ester derivatives of triterpenoid acids are summarized in Table XXI.

10. C_{18}-Steroids

Chromatographic data for more than 50 steroid estrogens in eight principal solvent systems have been reported by Lisboa[325] and are summarized in Table XXII. Lisboa and Diczfalusy[83, 84] have developed a systematic analytical procedure for the separation and characterization of a large number of estrogens, based on their chromatographic mobilities, derivative formation, and several color reactions.

The best resolution of the epimeric estradiols was obtained on silica gel layers by the use of benzene–ethyl acetate 50:50.[325] Mixtures of both isomers have been also resolved on silica gel G with overrun horizontal TLC[326] according to the technique of Brenner and Niederwieser.[86] After a 2-hour and 35-minute run with chloroform–ethanol 95.5:0.5, 17α- and 17β-estradiols have shown mobilities of 4.2 and 3.6 cm, respectively.

Epimeric estriols and 15α-hydroxyestradiol are completely resolved on multiple one-dimensional TLC after three developments in the system ethyl acetate–cyclohexane 50:50.[325] After two developments with ethyl acetate–n-hexane–acetic acid–ethanol 78:13.5:10:4.5, Lisboa et al.[159] have obtained optimal separations of estriol and 15α-hydroxyestradiol, even on preparative scale (1 mg mixture on 20 × 20 cm plate). A separation of trans- and cis-glycols, such as estriol (estra-1,3,5(10)-triene-$3,16\alpha,17\beta$-triol) and 16-epiestriol, can be accomplished also on boric acid-treated silica gel layers.[31, 286]

Golab and Layne[199] have investigated the behavior on silica gel layers in systems on the ethyl acetate–cyclohexane type of a great number of synthetic 19-nor steroids. The R_f values are summarized, together with those found for norsteroids by other investigators,[154] in Table XXIII.

Working with 19-nor steroids of unnatural configuration, Smith et

[325] B. P. Lisboa, Clin. Chim. Acta 13, 179 (1966).
[326] M. Maugras, C. Robin, and R. Gay, Bull. Soc. Chim. Biol. 44, 61 (1962).

TABLE XX

CHROMATOGRAPHIC MOBILITIES FOUND ON TLC FOR TRITERPENIC ACIDS

Triterpenic acids	Chromatographic systems[a]															
	IIa	IIc	IId	IIe	IIf	IIg	III	IVa	IVb	IVc	IVd	IVe	IVf	IVg	VI	VII
Ursolic acid	—	68	64	—	50	—	—	—	—	—	—	—	—	—	—	84
Acetylursolic acid	—	76	—	—	—	—	—	—	—	—	—	—	—	—	—	—
β-Boswellic acid	85	—	—	—	—	—	—	—	—	—	—	—	—	—	—	80
Acetyl-β-boswellic acid	—	—	—	—	—	—	—	—	—	—	—	—	—	—	—	85
Quinovic acid	—	55	—	—	—	—	—	—	—	—	—	—	—	86	—	76
Pyroquinovic acid	—	71t	—	—	—	—	—	—	—	—	—	—	—	—	—	—
Oleanolic acid	—	68	64	48	50	96	—	—	—	35	61	—	—	—	—	81
Oleanonic acid	—	68	—	—	—	—	—	—	—	—	—	—	—	—	—	—
Bredenolic acid	—	59t	62	—	—	82	—	—	—	—	—	—	—	—	—	—
Hederagenin	10	—	—	—	—	—	—	—	—	31	—	—	—	—	—	—
Erythrodiol-27-oic acid	43	—	—	—	—	—	—	—	—	—	—	—	—	—	—	—
Diacetylerythrodiol-27-oic acid	50	—	—	—	—	—	—	—	—	—	—	—	—	—	—	—
Erythrodiol-27-oic acid, dimethyl ether	—	—	—	—	—	—	—	—	—	—	—	—	—	—	—	—
Cincholic acid	—	55	—	—	—	—	—	—	—	—	—	—	—	—	—	67
Acetylcincholic acid	—	65	—	—	—	—	—	—	—	—	—	—	—	—	—	—
Pyrocincholic acid	—	71	—	—	—	—	—	—	—	—	—	—	—	—	—	86
Acetylpyrocincholic acid	—	78	—	—	—	—	—	—	—	—	—	—	—	—	—	—
Morolic acid	—	59	—	—	—	—	—	—	—	—	—	—	—	—	—	—
Arjunolic acid	—	—	—	—	—	—	—	20	—	—	38	—	—	—	—	—
Bayogenin	—	—	—	—	—	—	—	—	14	—	57	—	—	—	—	—
Mirtillogenic acid	—	—	—	—	—	—	—	—	—	—	—	—	—	68	—	—

Compound	1	2	3	4	5	6	7	8	9	10	11	12	13	14
Asiatic acid	—	—	—	—	—	—	—	—	—	—	—	—	—	—
Quillajic acid	—	—	—	—	—	—	58	36	38	37	—	—	—	—
Quillolic acid	—	—	—	—	—	—	14	72	—	—	51	74	—	25
23-Dihydroquillajic acid	—	—	—	—	—	—	—	24	—	—	45	—	—	15
23-Dihydroquillolic acid	—	—	—	—	—	—	—	—	18	17	—	—	—	—
Siaresinolic acid	66	15t	—	—	—	—	—	—	—	—	—	—	—	—
Cochalic acid	18t	—	29	45	55t	—	—	—	—	—	—	—	—	—
Gypsogenin	—	—	—	—	—	—	—	—	—	—	—	79	—	—
12-Dihydro-13α-hydroxygypsogenin	—	—	—	—	—	—	—	—	—	—	—	—	—	—
Machaeric acid	23t	27	23	25	54	—	—	—	—	—	—	—	—	—
Castanogenin	59	—	—	—	—	—	—	—	54	—	—	—	—	—
Emmolic acid	—	—	—	—	—	—	—	—	—	—	—	89	—	—
Bassiac acid	—	—	—	—	—	—	39	51	—	—	—	—	—	—
Acantolic acid	29	—	—	—	—	—	37	—	—	—	—	—	—	—
Guajavolic acid	35	—	—	—	—	—	—	—	—	—	—	—	—	—
Ceanolic acid	59	—	—	—	—	—	—	—	—	—	—	—	—	—
Acacic acid	63	—	—	—	—	—	—	—	—	—	—	—	—	—
Acetylacacic acid	83	—	—	—	—	—	—	—	—	—	—	—	—	—
Rehmannic acid	82	—	—	—	—	—	—	—	—	—	—	—	—	—
Proceric acid	4	—	—	—	—	—	—	—	—	—	—	—	—	—
Medicagenic acid	29t	38t	10t	35	71	—	—	—	—	—	—	—	—	—
Betulinic acid[b]	68	—	—	—	—	—	—	—	—	—	—	—	—	—
Oxyallobetulinic acid[c]	—	—	—	—	—	55	—	—	—	—	—	—	—	—

[a] For conditions see footnote c to Table XIX.

[b] Or betulic acid (3β-hydroxy-α-lupen-28-oic acid).

[c] 3β-Hydroxy-(19 → 28) lactone-13β(H)-olean-28-oic acid.

TABLE XXI

CHROMATOGRAPHIC MOBILITIES FOUND ON TLC FOR METHYL ESTER, LACTONE, BROMOLACTONE, AND ACETONIDE DERIVATIVES OF TRITERPENIC ACIDS

Triterpenoids acids, derivatives	Chromatographic systems[a]											
	Ia	Ib	IIa	IIb	IIc	IIm	IIo	IVa	IVc	IVd	IVe	VI
Ursolic acid, Me[b]	—	—	85	—	—	51	—	—	—	—	—	—
Acetylursolic acid, Me	—	—	—	—	—	77	26	—	—	—	—	—
Uvaol-27-oic acid, Me	—	—	28	—	—	—	—	—	—	—	—	—
Quinovic acid, diMe	—	—	40	—	—	—	—	—	—	—	—	—
Oleanolic acid, Me	73	—	60	—	—	77	24	—	—	—	—	—
Acetyloleanolic acid, Me	—	76	—	—	—	—	—	—	—	—	—	—
Oleanolic acid, lactone	—	—	—	—	—	—	—	—	62	—	—	—
Bredenolic acid, Me	—	—	—	28	—	—	—	—	—	77	—	—
Bredenolic acid, Me, acetonide	—	—	—	88	—	—	—	—	—	—	—	—
Bredenolic acid, bromolactone	—	—	—	—	51	—	—	—	—	—	—	—
Hederagenin, lactone	—	—	—	—	—	—	—	—	45	51	—	—
Erythrodiol-27-oic-acid, Me	—	—	30	—	—	—	—	—	—	—	—	—
Erythrodiol-dimethylether-27-oic acid, Me	—	—	92	—	—	—	—	—	—	—	—	—
Erythrodiol-monomethylether-27-oic acid, Me												
(isomer I)	—	—	48	—	—	—	—	—	—	—	—	—
(isomer II)	—	—	43	—	—	—	—	—	—	—	—	—
Cincholic acid, diMe	—	—	40	—	76	—	—	—	—	—	—	—
Acetylcincholic acid, diMe	—	—	80	—	—	—	—	—	—	—	—	—
Pyrocincholic acid, Me	—	—	—	—	78	—	—	—	—	—	—	—
Acetylpyrocincholic acid, Me	—	—	—	—	95	—	—	—	—	—	—	—
3β-Hydroxy-14-brom(28 → 23)lactone,14-nor-olean-28-oic acid	—	—	—	—	82	—	—	—	—	—	—	—

Compound										
Quillajic acid, Me	—	—	—	—	77	—	—	—	—	—
Quillajic acid, lactone	—	73	—	—	—	—	—	—	—	—
23-Dihydroquillajic acid, Me	—	—	68	—	—	—	—	—	—	—
23-Dihydroquillajic acid, Me, ac	—	—	89	—	—	—	—	—	—	—
23-Dihydroquillajic acid, lactone	—	—	47	31	—	—	—	—	—	—
23-Dihydroquillajic acid, lactone, ac	—	—	76	—	—	—	—	—	—	—
Quillolic acid, Me[c]	—	70	—	—	63	—	—	—	—	—
23-Dihydroquillolic acid, Me	—	—	56	—	—	—	—	—	—	—
23-Dihydroquillolic acid, Me, ac	—	—	93	—	—	—	—	—	—	—
Albigen, Me	—	—	66	—	—	—	—	—	—	—
Acantholic acid, Me	—	—	—	—	53	—	—	73	—	—
Acetylacantholic acid, Me	—	—	—	—	—	—	—	—	—	—
Arjanolic acid, lactone	—	—	47	—	—	—	—	—	—	—
Asiatic acid, lactone	—	—	52	—	—	—	—	—	—	—
Bayogenin, lactone	—	—	70	—	—	—	—	—	—	—
Echinocystic acid, Me	—	—	—	—	—	—	15	—	—	—
Gypsogenin, lactone	85	—	—	—	—	—	—	—	—	—
Emmolic acid, diMe	—	—	—	—	—	—	—	—	73	—
Diacetyltennifolic acid diMe	—	—	—	—	—	—	—	75	—	—
Monoacetylcratalgolic acid, Me	—	—	—	—	—	—	13	80	80	40
Diacetylcratalgolic acid, Me	—	—	—	—	—	—	37	92	92	72
Diacetylcratalgolic acid, Me	—	—	—	—	—	—	39	—	—	69
11-Oxomonoacetylcratalgolic acid, Me	—	—	—	—	—	—	—	77	—	—

[a] For conditions, see footnote c to Table XIX.

[b] Me = methyl ester; ac = acetate.

[c] Quillolic acid (Quillolsäure), isolated by Tschesche group (Jacob)[d] from *Quillaja saponaria*, is a 3β-hydroxy-23-oxoolean-13(18)-en-28-oic acid derivative ($C_{30}H_{46}O_5$) with a hydroxyl group probably at C-16.

[d] J. Jacob, Über die Aglykone der Saponine der levantinischen Seifenwurzel Inaugural Dissertation, Univ. of Bonn, Germany, 1963.

TABLE XXII

$R_f \times 100$ Values Found for Steroid Estrogens on Silica Gel G Layers by Use of Ascending One-Dimensional TLC

Steroids	Solvent systems[b]														
	I	II	III	IV	V	VI	VIII	IX	X	XI	XIV	XV	XVII	XVIII	XIX
E[a]	72	—	63	44	67	53	84	76	61	56	—	—	—	—	—
E^{16}	65	—	48	18	63	44	78	67	56	51	—	—	—	—	—
16β,17β-oxido-E	69	—	56	27	66	47	80	70	59	54	—	—	—	—	—
16α,17α-oxido-E	67	72	54	26	66	47	79	69	60	51	—	—	—	—	—
E-17-one	69	—	53	20	65	46	80	69	59	51	60	44	47	56	50
E^6-17-one	62	—	—	—	—	—	75	—	—	—	—	—	—	—	—
E^7-17-one	68	—	—	—	66	—	78	—	—	—	60	—	—	55	—
E$^{9(11)}$-17-one	65	—	50	17	64	45	79	68	56	52	—	—	—	—	—
E^{6,8}-17-one	63	—	47	16	63	42	77	65	58	50	—	—	—	—	—
E-6,17-one	60	60	33	—	—	—	—	—	—	—	44	22	—	38	30
E-7,17-one	61	61	36	5	55	44	72	56	55	48	—	—	—	—	—
E-11,17-one	59	—	34	—	61	38	74	58	—	45	—	—	—	—	—
E-16,17-one	66	68	43	—	64	45	—	—	—	41	58	39	40	44	36
4-ol-E-17-one	54	53	33	—	39	34	76	61	—	42	—	—	—	—	—
6α-ol-E-17-one	47	51	13	—	35	—	65	—	—	—	—	—	—	—	—
6β-ol-E-17-one	46	—	12	—	34	17	62	—	—	—	—	—	—	—	—
7α-ol-E-17-one	42	—	12	—	40	—	60	41	47	30	—	—	—	—	—
7β-ol-E-17-one	43	—	—	—	—	—	59	—	—	—	—	—	—	—	—
11α-ol-E-17-one	51	—	—	—	—	—	—	—	—	—	—	—	—	—	—
11β-ol-E-17-one	56	—	27	4	50	26	72	54	51	42	—	—	—	—	—
15α-ol-E-17-one	48	—	18	—	46	21	66	46	45	30	—	—	—	—	—
16α-ol-E-17-one	57	60	32	6	58	33	68	—	—	—	—	21	21	37	21
16β-ol-E-17-one	55	63	28	—	—	—	—	—	—	—	—	—	—	—	—

2-MeO-E-17-one	68	70	50	18	73	66	—	69	—	—	—	—	—	—	—
4-ol, 3-MeO-E-17-one	66	—	53	24	76	69	79	68	67	63	—	—	—	—	—
17β-ol-E	61	64	40	10	52	30	74	61	53	42	44	24	25	38	30
17α-ol-E	61	—	43	11	55	34	75	62	52	45	51	27	30	43	35
17β-ol-E[6]	57	—	36	8	52	29	73	59	49	41	—	—	—	—	—
17β-ol-E[7]	57	50	36	7	53	30	74	59	50	41	—	—	—	—	—
17β-ol-E-6-one	52	—	23	—	—	—	—	—	—	—	—	—	—	—	—
17β-ol-E-7-one	44	—	—	—	41	22	—	50	—	—	—	—	—	—	—
17β-ol-E-16-one	59	60	32	—	55	—	69	—	—	40	40	19	—	35	23
E-4,17β-ol	—	—	—	—	—	—	—	—	—	—	—	—	—	—	—
E-2,17β-ol	47	48	21	2	24	—	70	53	—	—	—	—	—	—	—
E-6α,17β-ol	45	47	12	—	23	12	61	47	—	—	—	—	—	—	—
E-6β,17β-ol	44	45	12	—	23	10	61	44	—	—	—	—	—	—	—
E-7α,17β-ol	40	42	9	—	22	8	59	43	—	—	—	—	—	—	—
E-11β,17β-ol	34	—	8	—	29	9	54	40	37	24	7	2	—	4	1
E-16α,17β-ol	29	30	6	—	21	7	48	34	39	20	—	—	—	—	—
E-16β,17β-ol	43	40	18	—	26	17	54	29	38	17	16	5	—	15	8
E-16α,17α-ol	45	42	20	—	30	21	58	38	44	—	—	—	—	—	—
E-16β,17α-ol	34	34	8	—	18	—	50	40	42	—	5	2	—	2	—
E-15α,17β-ol	26	25	—	—	16	—	38	20	33	—	37	21	—	34	32
2-MeO-E-17β-ol	58	59	36	8	58	41	72	57	53	45	—	—	—	—	—
16β,17β-ol-E-6-one	21	—	—	—	—	—	34	18	—	—	—	—	—	—	—
E-2,16α,17β-ol	16	17	5	—	14	—	36	20	28	—	—	—	—	—	—
E-6,7,17β-ol	23	26	—	—	12	—	40	24	33	—	—	—	—	—	—
E-6α,16α,17β-ol	12	11	1	—	5	1	24	10	22	—	—	—	—	—	—
2-MeO-E-16α,17β-ol	29	29	6	—	—	13	43	26	—	—	—	—	—	—	—
Doysinolic acid	31	—	—	—	30	—	78	—	—	—	—	—	—	—	—
E-15α,16α,17β-ol	17	15	—	—	13	—	33	17	26	—	—	—	—	—	—

(Footnotes appear on p. 132.)

(Continued)

TABLE XXII (Continued)

[a] E = estra-1,3,5(10)-trien-3-ol; MeO = methoxy.

[b] Experimental conditions for Tables XXII, XXIV–XXVIII and XXX–XXXIII: n-TLC, silica gel G; tank lined with paper and equilibrated before use for at least 3 hours. I.l. = 0.8–1.0 cm; S.l. = 2.5 cm; Fr. = 15–16 cm. Solvent systems: **I,**[a] Ch/EtAc/EtOH 45:45:10; **II,**[a] EtAc/Hex/EtOH 80:15:5; **III,**[a] Ch/EtAc 50:50; **IV,**[e] Hex/EtAc 75:25; **V,**[a] Chf/EtOH 90:10; **VI,**[f] Chf/EtOH 95:5; **VII,**[g] Chf/EtOH 98:2; **VIII,**[a] EtAc/Hex/AcOH/EtOH 72:13.5:10:4.5; **IX,**[e] EtAc/Hex/AcOH 75:20:5; **X,**[f] Bz/EtOH 80:20; **XI,**[f] Bz/EtOH 90:10; **XII,**[h] Bz/EtOH 95:5; **XIII,**[h] Bz/EtOH 98:2; **XIV,**[h] Bz/EtAc 50:50; **XV,**[h] Bz/EtAc 75:25; **XVI,**[i] Bz/EtAc 80:20; **XVII,**[h] Chf/EtAc 75:25; **XVIII,**[h] Bz/Eth 50:50; **XIX,**[h] Chf/Eth 75:25.

[c] B. P. Lisboa, Steroids **7,** 41 (1966).

[d] B. P. Lisboa and E. Diczfalusy, Acta Endocrinol. **40,** 60 (1962).

[e] B. P. Lisboa, J. Chromatog. **13,** 391 (1964).

[f] B. P. Lisboa, Acta Endocrinol. **43,** 47 (1963).

[g] B. P. Lisboa and R. Palmer, Anal. Biochem. **20,** 77 (1967).

[h] B. P. Lisboa, Clin. Chim. Acta **13,** 179 (1966).

[i] L. Cédard, B. Fillmann, R. Knuppen, B. P. Lisboa, and H. Breuer, Z. Physiol. Chem. **338,** 89 (1964).

$al.$[327] obtained on rice starch–silica gel layers with ethyl acetate as the solvent, the following sequence of polarity for hydroxyl groups:

$$11\alpha > 1\beta > 6\beta > 10\beta$$

11. C_{19}-Steroids

Eight principal solvent systems have been developed by Lisboa for the separation on silica gel G of Δ^4-3-oxo-, Δ^5-3β-hydroxy-, and saturated C_{19}-steroids (Tables XXIV–XXVIII). The resolution of isomeric steroids differing in the C-3 and C-5 configuration is achieved by using the techniques of multiple development in one or several chromatographic systems. This allows the separation of the four 3-hydroxyandrostan-17-ones and the androstane-3,17β-diols after three runs in the system n-hexane–ethyl acetate 75:25. More difficult is the separation of isomeric androstanolones differing in the position of the ketone group at C-3 or C-17: etiocholanolone and 5β-dihydrotestosterone (17β-hydroxy-5β-androstan-3-one) can be resolved after 5 runs in n-hexane–ethyl acetate 75:25. Nienstedt[328] has worked out systems which give a complete separation of the 5α-androstanolone isomers on magnesium silicate layers; the following $R_f \times 100$ values have been found for androsterone (A), epiandrosterone (epi-A) and 5α-dihydrotestosterone (5α-T) in the systems: (a) 20% $tert$-butanol–n-hexane: A = 59, epi-A = 35, 5α-T = 41; (b) n-butanol–hexane, 5:95: A = 29, epi-A = 17, 5α-T = 24; (c) ether: A = 91, epi-A = 71, 5α-T = 82.

Lisboa and Palmer[29] have investigated the resolution of weakly polar C_{19}-steroids by TLC techniques; they have observed a complete separation of Δ^{16}-androstenols from the corresponding saturated compounds on silver nitrate-impregnated silica gel layers.

Multiple chromatography on silica gel layers with two developments in system ethyl acetate–cyclohexane 60:40 allows an optimal separation of androsterone and etiocholanolone. This resolution was utilized by Böttiger and Lisboa[329] in their method for the quantitative determination of urinary 17-ketosteroids after fractionation by means of gradient-column and TL chromatography.

$R_f \times 100$ values of 37 and 47 have been found, respectively, for testosterone (T) and epitestosterone (epi-T, 17α-hydroxyandrost-4-en-3-one) on aluminum oxide layers in benzene–ether 1:1;[214] the complete separation of these isomers on alumina was reported by Galletti[330] after

[327] L. L. Smith, G. Greenspan, R. Rees, and T. Foell, *J. Am. Chem. Soc.* **88**, 3120 (1966).

[328] W. Nienstedt, *Acta Endocrinol. Suppl.* **114**, 53 pp. (1967).

[329] L. E. Böttiger and B. P. Lisboa, *Z. Klin. Chem. Klin. Biochem.* **5**, 176 (1967).

[330] F. Galletti, *Res. Steroids (Rome)* **2**, 189 (1966).

TABLE XXIII

CHROMATOGRAPHIC MOBILITIES ($R_f \times 100$ VALUES) FOUND ON SILICA GEL LAYERS FOR 19-NORSTEROIDS

19-Norsteroids	Chromatographic systems[a]											
	Ia	Ib	IIa	IIb	IIc	IId	IIe	IIf	IIg	IIh	IIi	III
Estr-4-ene-3,17-dione	25	13	51	26	38	71	66	61	52	61	53	—
Estr-5(10)-ene-3,17-dione	71	50	—	—	—	—	—	—	—	—	—	—
17β-Hydroxyestr-4-en-3-one	36	14	42	19	21	60	63	52	37	39	50	60
17β-Hydroxyestr-5(10)-en-3-one	51	29	—	—	—	—	—	—	—	—	—	—
17β-Hydroxy-17α-methylestr-4-en-3-one	40	21	—	—	—	—	—	—	—	—	—	—
17β-Hydroxy-17α-ethylestr-4-en-3-one	47	27	—	—	—	—	—	—	—	—	—	—
17β-Hydroxy-17α-ethenylestr-4-en-3-one	56	31	—	—	—	—	—	—	—	—	—	—
17β-Hydroxy-17α-ethynylestr-4-en-3-one	55	30	—	—	—	—	—	—	—	—	—	—
17β-Hydroxy-17α-propylestr-4-en-3-one	41	19	—	—	—	—	—	—	—	—	—	—
17β-Hydroxy-17α-ethylestr-5(10)-en-3-one	68	46	—	—	—	—	—	—	—	—	—	—
17β-Hydroxy-17α-ethynylestr-5(10)-en-3-one	74	51	—	—	—	—	—	—	—	—	—	—
11β-Hydroxyestr-4-ene-3,17-dione	25	11	—	—	—	—	—	—	—	—	—	—
10β-Hydroxyestra-1,4-diene-3,17-dione	36	21	—	—	—	46	—	—	23	18	—	29
1β,17β-Dihydroxyestr-4-en-3-one	—	—	—	—	—	—	—	—	—	—	—	42
6β,17β-Dihydroxyestr-4-en-3-one	—	—	—	—	—	19	—	—	14	5	—	51
10β,17β-Dihydroxyestr-4-en-3-one	—	—	—	—	—	28	35	—	17	10	16	21
11α,17β-Dihydroxyestr-4-en-3-one	—	—	—	—	—	37	52	41	23	12	41	—
11β,17β-Dihydroxyestr-4-en-3-one	28	10	31	—	—	—	—	—	—	—	—	54
10β,17β-Dihydroxy-17α-ethynylestr-4-en-3-one	—	—	—	—	—	—	—	—	—	—	—	42
17β-Hydroxy-13β-ethylgon-4-en-3-one	—	—	—	—	—	—	—	—	—	—	—	21
17β-Hydroxy-13β-n-propylgon-4-en-3-one	—	—	—	—	—	—	—	—	—	—	—	12
1β,17β-Dihydroxy-13β-ethylgon-4-en-3-one	—	—	—	—	—	—	—	—	—	—	—	29
1β,17β-Dihydroxy-13β-n-propylgon-4-en-3-one	—	—	—	—	—	—	—	—	—	—	—	18
6β,17β-Dihydroxy-13β-ethylgon-4-en-3-one	—	—	—	—	—	—	—	—	—	—	—	41
6β,17β-Dihydroxy-13β-n-propylgon-4-en-3-one	—	—	—	—	—	—	—	—	—	—	—	24
10β,17β-Dihydroxy-13β-ethylgon-4-en-3-one	—	—	—	—	—	—	—	—	—	—	—	14
10β,17β-Dihydroxy-13β-n-propylgon-4-en-3-one	—	—	—	—	—	—	—	—	—	—	—	—
11α-Dihydroxy-13β-ethylgon-4-en-3-one	—	—	—	—	—	—	—	—	—	—	—	—

Compound		
11α,17β-Dihydroxy-13β-n-propylgon-4-en-3-one	—	6
17β-Hydroxy-5α-estran-3-one	59	36
17β-Hydroxy-5α,10α-estran-3-one	63	40
17β-Hydroxy-17α-methyl-5α-estran-3-one	59	36
17β-Hydroxy-17α-ethyl-5α-estran-3-one	64	39
17β-Hydroxy-17α-ethenyl-5α-estran-3-one	70	41
17β-Hydroxy-17α-ethynyl-5α-estran-3-one	72	42
17β-Hydroxy-17α-ethyl-5α,10α-estran-3-one	63	38
17β-Hydroxy-17α-ethyl-5β-estran-3-one	65	39
17β-Hydroxy-17α-propyl-5ξ-estran-3-one	76	53
Estr-4-ene-3β,17β-diol	33	16
Estr-5-ene-3β,17β-diol	38	23
Estr-5(10)-ene-3β,17β-diol	36	21
5α-Estrane-3β,17β-diol	39	25
5β-Estrane-3β,17β-diol	54	28
5α,10α-Estrane-3β,17β-diol	48	25
5α,10α-Estrane-3α,17β-diol	42	24
17α-Methylestr-5-ene-3β,17β-diol	43	26
17α-Ethynylestr-5-ene-3β,17β-diol	54	29
17α-Ethenyl-5α-estrane-3β,17β-diol	53	30
17α-Ethynyl-5α-estrane-3β,17β-diol	53	31
17α-Ethyl-5α,10α-estrane-3β,17β-diol	65	41
17α-Ethyl-5α,10α-estrane-3α,17β-diol	54	32

a Comments on the experimental conditions: I: b n-TLC, silica gel G, 300 μ thick; I.l. = 1 cm; S.l. = 2 cm; Fr. = 17 cm; Dv. = 90–105 min. Solvent: EtAc/Ch; Ia, 1:1; Ib, 3:7.

II: c n-TLC, silica gel G; I.l. = 0.8–1.0 cm; S.l. = 2.5 cm; Fr. = 15–16 cm; tank lined with paper and equilibrated for 3 hours prior to use. Systems: IIa, EtAc/Ch/EtOH 45:45:10; IIb, EtAc/Ch 50:50; IIc, Bz/EtOH 95:5; IId, Chf/EtOH 90:10; IIe, EtAc/Hex/AcOH/EtOH;72:13.5:10:4.5; IIf, Bz/EtOH 80:20; IIg, Bz/EtOH 90:10; IIh, Chf/EtOH 95:5; IIi, EtAc/Hex/AcOH 75:20:5.

III: d n-TLC, silica gel-rice starch. System: EtAc. Experimental conditions as given by Smith and Foell.e

b T. Golab and D. S. Layne, J. Chromatog. 9, 321 (1962).
c B. P. Lisboa, J. Chromatog. 19, 81 (1965).
d L. L. Smith, G. Greenspan, R. Rees, and T. Foell, J. Am. Chem. Soc. 88, 3120 (1966).
e L. L. Smith and T. Foell, J. Chromatog. 9, 339 (1962).

TABLE XXIV

$R_f \times 100$ Values Found for Ketonic Steroids of the Androstane Series on Silica Gel G Layers by Use of Ascending One-Dimensional TLC[a]

Steroids	Solvent systems													
	I	II	III	IV	V	VI	VII	VIII	IX	X	XI	XII	XIII	XIV
5α-A-11-one	—	—	69	64	—	72	—	—	—	—	72	66	—	—
5β-A-11-one	—	—	69	64	—	71	—	—	—	—	73	66	—	—
5α-A-3-one	74	—	69	53	78	—	69	87	80	—	—	62	53	67
5α-A-17-one	76	—	72	57	77	76	69	—	—	77	77	65	56	71
Δ4,16-3-one	—	—	—	—	74	72	—	—	—	74	74	—	—	—
5α-A-3,17-one	66	—	46	17	74	69	—	78	69	66	64	53	—	48
5β-A-3,17-one	64	—	43	14	74	67	—	78	66	67	63	49	—	46
5α-Δ1-3,17-one	61	—	38	13	76	67	—	75	67	66	58	45	—	—
Δ4-3,17-one	53	58	30	8	73	65	52	68	60	64	55	42	17	39
5α-Δ2-7,17-one	67	—	52	27	77	70	—	81	74	70	61	55	—	—
Δ1,4-3,17-one	49	—	20	—	71	58	—	62	51	58	47	32	—	—
Δ1,4,6-3,17-one	49	—	22	—	71	61	—	—	50	60	48	35	—	—
5β-A-3,11,17-one	52	—	25	—	69	58	—	70	56	66	50	36	—	—
19-al-Δ4-3,17-one	48	—	19	—	70	61	—	63	49	64	54	41	—	—
Δ4-3,11,17-one	39	—	14	—	67	53	—	60	43	56	48	30	—	—

[a] In this table, one or more ketonic groups are indicated by "-one" and an aldo group by "-al". For the experimental conditions and systems, see footnote b of Table XXII.

TABLE XXV

$R_f \times 100$ Values Found for Alcoholic Steroids of the Androstane Series on Silica Gel G Layers by Use of Ascending One-Dimensional TLC[a]

Steroids	Solvent systems														
	I	II	III	IV	V	VI	VII	VIII	IX	X	XI	XII	XIII	XIV	XV
5α-A-17β-ol	64	—	56	31	64	57	—	78	73	62	57	46	—	—	—
5α-A-3α-ol	69	—	59	31	—	—	48	—	—	—	—	37	30	50	—
5β-A-3β-ol	69	—	57	30	—	—	46	—	—	—	—	35	27	50	—
5α-A-3β-ol	65	—	50	21	—	—	38	—	—	—	—	25	20	44	—
5β-A-3α-ol	68	—	55	25	—	—	43	—	—	—	—	33	25	50	—
Δ5-3β-ol	65	—	49	23	62	52	39	75	71	60	44	28	20	45	—
5α-A16-3α-ol	68	—	59	29	—	—	48	—	—	—	—	36	28	51	—
5α-A16-3β-ol	66	—	51	20	—	—	36	—	—	—	—	25	18	41	—
5β-A16-3α-ol	66	—	56	24	—	—	42	—	—	—	—	30	23	48	—
Δ5,16-3β-ol	66	—	53	21	—	—	38	—	—	—	—	27	15	45	—
5α-A-3α,17β-ol	52	53	26	—	50	32	—	70	58	48	33	18	28	28	—
5α-A-3β,17β-ol	51	50	25	—	47	29	—	68	56	49	29	16	—	26	—
5β-A-3α,17β-ol	46	47	17	—	45	23	—	63	50	48	25	13	—	19	—
5β-A-3β,17β-ol	53	56	30	—	50	32	—	70	57	50	34	—	—	30	—
Δ4-3β,17β-ol	50	—	26	—	47	32	—	70	58	50	30	17	—	29	—
Δ5-3β,17β-ol	49	—	26	—	48	31	—	69	56	50	29	17	—	—	—
Δ5-3β,17α-ol	50	—	26	—	49	32	—	71	57	50	30	18	—	—	—
Δ5,7-3β,17β-ol	—	—	—	—	48	32	—	—	—	—	29	—	—	—	—
5α-A-3α,11β,17β-ol	—	—	—	—	21	—	—	43	32	—	15	—	—	—	—
5β-A-3α,11β,17β-ol	—	—	—	—	17	7	—	37	26	—	12	—	—	—	—
Δ5-3β,7β,17β-ol	28	—	4	—	23	7	—	43	28	29	9	3	—	—	13
Δ5-3β,7α,17β-ol	19	—	—	—	17	5	—	33	15	31	11	—	—	—	—
Δ5-3β,11β,17β-ol	31	—	6	—	22	7	—	47	31	33	13	4	—	—	—

[a] In this table "-ol" signifies one or more hydroxyl groups. For the experimental conditions and systems, see footnote b of Table XXII.

TABLE XXVI

$R_f \times 100$ Values Found for Monohydroxymonoketonic Steroids of the Androstane Series on Silica Gel G Layers by Use of Ascending One-Dimensional TLC[a]

Steroids	Solvent systems													
	I	II	III	IV	V	VI	VII	VIII	IX	X	XI	XII	XIII	XIV
17β-ol-5αA-3-one	56	—	35	—	61	49	—	73	63	57	42	31	—	40
17β-ol-5βA-3-one	55	59	30	—	61	46	—	71	63	59	42	28	—	34
17β-ol-5αA¹-3-one	55	—	—	—	60	45	—	70	56	—	38	—	—	—
17β-ol-A⁴-3-one	50	—	22	4	88	43	—	64	53	52	37	23	—	26
17α-ol-A⁴-3-one	51	52	23	—	60	43	—	64	55	52	39	23	—	—
17β-ol-A¹,⁴-3-one	41	—	14	—	55	34	—	58	46	49	31	16	—	—
17β-ol-A⁴,⁶-3-one	46	—	—	—	55	38	—	—	—	—	36	—	—	—
17β-ol-A¹,⁴,⁶-3-one	—	—	—	—	49	—	—	65	—	—	30	—	—	—
3α-ol-5αA-17-one	56	59	33	—	63	50	—	74	62	58	44	29	—	37
3β-ol-5αA-17-one	52	56	30	—	58	44	—	70	60	56	40	23	—	34
3α-ol-5βA-17-one	53	54	24	—	58	43	—	71	57	55	39	27	—	30
3β-ol-5βA-17-one	57	59	36	—	62	50	—	75	64	59	44	33	—	40
3β-ol-A⁴-17-one	53	—	34	9	—	—	36	—	—	—	—	18	14	38
3α-ol-A⁴-17-one	53	—	30	5	—	—	34	—	—	—	15	15	19	31
3β-ol-A⁵-17-one	53	—	36	8	57	43	33	73	62	55	39	20	14	35
3α-ol-A⁵-17-one	54	—	26	—	67	52	—	68	55	64	49	29	—	—

[a] In this table and in Tables XXVII–XXXI, "-ol" and "-one" signify one or more hydroxyl and ketonic groups, respectively; "al" signifies an aldo group. For the experimental conditions and systems see footnote b of Table XXII.

TABLE XXVII

$R_f \times 100$ Values Found for Monohydroxydiketonic Steroids of the Androstane Series on Silica Gel G Layers by Use of Ascending One-Dimensional TLC[a]

Steroids	Solvent systems									
	I	III	V	VI	VIII	IX	X	XI	XII	XVI
3α-ol-5αA-7,17-one	40	12	53	32	54	40	47	27	16	—
3β-ol-5αA-7,17-one	33	7	50	27	49	34	45	23	13	—
3α-ol-5αA-11,17-one	44	14	59	40	60	48	59	37	—	—
3α-ol-5βA-11,17-one	38	9	52	31	61	43	55	33	18	—
11β-ol-5αA-3,17-one	57	27	63	46	75	59	64	44	27	—
17β-ol-5αA-3,7-one	40	12	56	36	57	42	49	—	18	—
4-ol-Δ⁴-3,17-one	59	41	72	67	76	68	70	62	52	—
6β-ol-Δ⁴-3,17-one	36	8	52	30	—	31	54	34	16	—
6α-ol-Δ⁴-3,17-one	32	9	49	25	—	27	54	32	14	—
11β-ol-Δ⁴-3,17-one	47	17	58	42	64	48	55	41	21	—
11α-ol-Δ⁴-3,17-one	37	8	49	26	49	33	50	29	13	—
12β-ol-Δ⁴-3,17-one	—	—	66	50	—	—	—	41	—	—
14α-ol-Δ⁴-3,17-one	39	10	54	33	54	38	52	33	15	—
15α-ol-Δ⁴-3,17-one	32	6	47	24	45	28	46	29	11	—
16α-ol-Δ⁴-3,17-one	38	11	59	42	58	39	53	36	19	—
19-ol-Δ⁴-3,17-one	31	6	44	23	13	26	51	26	11	—
17β-ol-19al-Δ⁴-3-one	39	10	53	35	56	40	54	38	20	—
3β-ol-Δ⁵-7,17-one	35	8	49	28	49	36	45	30	14	26
3β-ol-Δ⁵-11,17-one	47	19	59	41	62	48	51	37	21	—

[a] For particulars, see the footnotes to Tables XXII and XXVI.

TABLE XXVIII

$R_f \times 100$ Values Found for Dihydroxymonoketonic Steroids of the Androstane Series on Silica Gel G Layers by Use of Ascending One-Dimensional TLC[a]

Steroids	Solvent systems									
	I	III	V	VI	VIII	IX	X	XI	XII	XVI
3α,11β-ol-5αA-17-one	49	17	51	29	66	53	57	34	13	—
3β,11β-ol-5αA-17-one	45	13	43	22	68	51	53	28	13	—
3α,11β-ol-5βA-17-one	44	12	45	24	64	48	52	30	—	—
2α,17β-ol-A⁴-3-one	34	11	46	27	53	39	44	25	13	—
4,17β-ol-A⁴-3-one	52	28	58	46	70	59	61	47	28	—
6β,17β-ol-A⁴-3-one	—	5	41	19	—	—	47	26	9	—
11,17β-ol-A⁴-3-one	31	5	34	11	46	26	45	22	7	—
11β,17β-ol-A⁴-3-one	18	2	26	5	27	12	33	13	2	—
14α,17β-ol-A⁴-3-one	27	4	31	11	39	21	39	19	5	—
16α,17β-ol-A⁴-3-one	19	2	26	9	33	15	38	18	5	—
17β,19-ol-A⁴-3-one	28	6	31	12	39	22	37	16	4	—
3β,17α-ol-A⁵-17-one	22	2	27	9	35	17	36	14	4	—
3β,11β-ol-A⁵-17-one	47	17	48	25	62	49	50	29	4	—
3β,16α-ol-A⁵-17-one	43	16	46	30	67	51	51	19	14	—
3β,17β-ol-A⁵-7-one	—	—	39	18	—	—	—	21	—	20
3β,17β-ol-A⁵-11-one	37	10	38	17	50	39	40	20	8	—

[a] For particulars, see the footnotes to Tables XXII and XXVI.

three runs in this system (migration, in cm: T, 5.7, epi-T, 2.6) or two runs in carbon tetrachloride–ethyl acetate 1:1 (T, 9.1; epi-T, 7.0).

On kieselguhr G-layers impregnated with propyleneglycol (10% in ethanol), Sonanini and Anker,[331] by using cyclohexane–toluene 8:2, have separated testosterone ($R_f \times 100$, 27), nortestosterone (21), epitestosterone (36), and dehydroepiandrosterone (46). Galletti[330] has investigated the migration of nine androstane-3,16,17-triols and the two Δ^5-androstene-3β,16,17β-triols on aluminum oxide and silica gel layers. Pairs of steroids differing only by the structures Δ^5-3β/5α(H)-3β-ol have resisted separation. Isomers differing by one epimeric hydroxy group at C-16 or C-17 were separated on silica gel G in the system chloroform–acetone 7:3; examples of $R_f \times 100$ values: 5α-androstane-3α,16α,17α-triols = 13; 17β-epimers = 7; 5α-androstane-3β,16α,17α-triol = 17; 17β-epimer, 8; 5α-androstane-3β,16α,17β-triol = 8; 16β-epimer = 19. Steroids differing epimerically at C-5 have different mobility in the system benzene–ethanol 9:1, in both alumina and silica gel layers; the $R_f \times 100$ values on aluminum oxide for 5α-androstane-3α,16α,17β-triol and for 5β-androstane-3α,16α,17β-triol are, respectively, 15 and 9 in this system. It is noteworthy that the mobility order between androstane-3α,16α,17α-triols and their 17β-epimers is reversed in one change from silica gel to aluminum oxide layers: the $R_f \times 100$ values on the system benzene–ethanol 9:1 for 5α-androstane-3α,16α,17α-triol, and its 17β-isomer, are, respectively, 7 and 15 on alumina layers and 17 and 13 on silica gel layers.

The formation of dinitrophenylhydrazones of C_{19}-steroids has been utilized for their characterization[332–335] or quantitation.[276] The mobilities found for these derivatives are summarized in the Table XXIX. These data indicate that a good group separation can be obtained between the diketonic C_{19}-steroids and the monohydroxylated monoketonic C_{19}-steroids; moreover, several of the isomeric 3(or 17)-hydroxy-17(or 3)-ketonic C_{19}-steroids can be resolved as DNP-hydrazones.

12. C_{21}-Steroids

A systematic study of the behavior of a great number of 21-deoxypregnanesteroids, Δ^4-3-oxopregnane, and Δ^5-3β-hydroxypregnane steroids

[331] D. Sonanini and L. Anker, *Pharm. Acta Helv.* **42**, 54 (1967).

[332] T. Feher, *Mikrochim. Acta*, p. 105 (1965).

[333] P. Knapstein, W. Rindt, F. Wendlberger, and G. W. Oertel, *Z. Physiol. Chem.* **348**, 93 (1967).

[334] J. R. Kent and A. B. Rawitch, *J. Chromatog.* **20**, 614 (1965).

[335] W. R. Starnes, A. H. Rhodes, and R. H. Lindsay, *J. Clin. Endocrinol. Metab.* **26**, 1245 (1966).

TABLE XXIX

CHROMATOGRAPHIC MOBILITIES FOUND FOR C_{19}-STEROID DNP-HYDRAZONES ON SILICA GEL G AND ALUMINA G LAYERS BY USE OF ASCENDING TLC[a]

Steroid DNP-hydrazones	I	IIa	IIb	IIc	III	IV	Va	Vb	Vc	Vd	Ve	Vf
5βA-3,17-one	76 (92)	85	81	73	16.5	—	—	—	—	—	—	—
5αA-3,17-one	71 (86)	87	82	75	16.5	—	80	67	66	—	—	93
A^4-3,17-one	73 (85)	87	79	74	17.3	60	78	66	77	—	—	94
$A^{1,4}$-3,17-one	64 (80)	—	—	—	—	—	—	—	—	—	—	—
A^4-3,11,17-one	49 (86)	86	75	70	11.7	—	52	33	66	—	—	89
3α-ol-5αA-17-one	33	68	50	54	—	39	36	21	44	25	72	78
3α-ol-5βA-17-one	30	64	42	50	3.8	29	28	14	37	17	63	72
3β-ol-5αA-17-one	21	—	—	—	—	—	29	19	31	21	55	61
3β-ol-5βA-17-one	—	59	37	45	5.6	—	43	30	47	—	—	73
3β-ol-A^5-17-one	22	—	—	—	5.0	—	29	19	31	18	54	65
17β-ol-5βA-3-one	18	65	41	51	6.4	—	—	—	—	—	—	—
17β-ol-A^4-3-one	—	—	—	—	—	37	35	21	38	24	67	68
17α-ol-A^4-3-one	11	55	36	43	—	45	—	—	—	—	—	—
3α,11β-ol-5αA-17-one	14	52	33	40	—	—	18	7	28	10	52	61
3α,11β-ol-5βA-17-one	—	—	—	—	—	—	14	4	23	7	38	56
3β,11β-ol-5αA-17-one	—	—	—	—	1.0	—	6	2	12	—	—	32
3β,16α-ol-A^5-17-one	—	79	73	66	—	—	—	—	—	—	—	—
11β-ol-A^4-3,17-one	27 (70)	63	38	44	—	—	—	—	—	—	—	—
3α-ol-5αA-11,17-one	14	61	34	41	—	—	10	15	21	29	34	47
3α-ol-5βA-11,17-one	—	35	12	27	—	—	9	3	18	7	33	47
(Reagents)	—	—	—	—	—	—	—	—	—	—	—	—

[a] *Experimental conditions:* **I:**[b] n-TLC, tanks not lined; silica gel G. System Bz/EtOH 98:2; $R_f \times 100$ values.
II:[c] n-TLC, 300 μ. Systems: **IIa,** silica gel G, Chf/EtOH 19:1; **IIb,** alumina G, Ch/BuAc/BuOH 50:45:5; **IIc,** silica gel G, Bz/Dx8:2; $R_f \times 100$ values.
III:[d] Multiple TLC with 4 runs in chloroform; silica gel G; mobility in centimeters.
IV:[e] n-TLC; 18 × 18 cm plates, 200 μ; S.l. = 3 cm; Fr. = 15 cm; silica gel; EtAc/Hex 15:85.
V:[f] n-TLC, 500 μ, S.l. = 2 cm; Fr. = 18 cm (Dv. = 40–60 minutes); aluminum oxide G, activation: 110°/4 hours. Systems: **Va,** Bz/Hex/EtAc 300:100:100; **Vb,** Bz/Hex/EtAc 300:25:25; **Vc,** Tol/EtAc 300:100; **Vd,** Tol/Chf 50:50; **Ve,** Bz/Hex/EtAc 75:75:75; **Vf,** Bz/Chf/EtAc 100:25:50.

[b] T. Feher, *Mikrochim. Acta*, p. 105 (1965).
[c] L. Treiber and G. W. Oertel, *Z. Klin. Chem. Klin. Biochem.* **5**, 83 (1967).
[d] P. Knapstein, W. Rindt, F. Wendlberger, and G. W. Oertel, *Z. Physiol. Chem.* **348**, 93 (1967).
[e] J. R. Kent and A. B. Rawitch, *J. Chromatog.* **20**, 614 (1965).
[f] W. R. Starnes, A. H. Rhodes, and R. H. Lindsay, *J. Clin Endocrinol. Metab.* **26**, 1245 (1966).

TABLE XXX

$R_f \times 100$ Values Found for Saturated Alcoholic Steroids of the Pregnane Series on Silica Gel G Layers by Use of Ascending One-Dimensional TLC[a]

Steroids	Solvent systems												
	I	III	IV	V	VI	VII	VIII	IX	X	XI	XII	XIII	XIV
5α-P	82	78	—	82	82	—	89	84	79	79	—	—	—
5α-P-3β-ol	66	48	23	62	53	38	76	66	64	48	27	18	45
5β-P-3α-ol	67	55	27	65	59	45	77	71	67	52	33	25	49
5β-P-3α,6α-ol	31	6	—	29	12	—	48	33	58	23	—	—	—
5α-P-3α,20α-ol	59	32	—	55	37	—	68	—	58	35	—	—	—
5α-P-3α,20β-ol	52	34	—	55	34	—	70	59	56	34	—	—	—
5α-P-3β,20α-ol	57	27	—	50	33	—	65	52	53	30	—	—	—
5α-P-3β,20β-ol	49	31	—	52	35	—	70	56	56	33	—	—	—
5β-P-3α,20α-ol	54	22	—	50	28	—	65	50	52	33	—	—	—
5β-P-3α,20β-ol	56	25	—	52	31	—	68	56	54	36	—	—	—
5β-P-3β,20α-ol	—	34	—	—	37	—	70	59	54	40	—	—	—
5β-P-3β,20β-ol	59	37	—	57	40	—	70	61	57	41	—	—	—
5α-P-3β,6β,20β-ol	30	—	—	30	11	—	47	34	40	14	—	—	—
5β-P-3α,6β,20β-ol	27	—	—	30	13	—	47	34	39	13	—	—	—
5α-P-3α,6α,20β-ol	24	—	—	26	9	—	39	26	41	14	—	—	—
5β-P-3α,6α,20β-ol	21	—	—	24	7	—	38	25	37	11	—	—	—
5β-P-3α,6α,20α-ol	18	—	—	20	5	—	32	18	36	10	—	—	—
5β-P-3α,17α,20α-ol	15	5	—	16	4	—	30	16	32	8	—	—	—
5β-P-3β,17α,20α-ol	26	11	—	23	7	—	45	30	39	20	—	—	—
5β-P-3α,17α,20β-ol	34	8	—	35	15	—	55	39	47	26	—	—	—
5β-P-3β,17α,20β-ol	32	16	—	32	12	—	51	37	44	23	—	—	—
5α-P-3β,17α,20α-ol	39	10	—	41	19	—	61	45	49	29	—	—	—
5α-P-3β,17α,20β-ol	34	13	—	33	14	—	58	41	45	25	—	—	—
5α-P-3β,6α,16β,20α-ol	9	—	—	4	18	—	13	3	17	27	—	—	—
5α-P-3β,6β,16β,20α-ol	7	—	—	5	—	—	10	4	19	—	—	—	—
5α-P-3β,5,6β,16β,20α-ol	3	—	—	1	—	—	6	2	8	—	—	—	—

[a] For particulars, see the footnotes to Tables XXII and XXVI.

TABLE XXXI

$R_f \times 100$ Values Found for Saturated Ketonic Steroids of the Pregnane Series on Silica Gel G Layers by Use of Ascending One-Dimensional TLC[a]

Steroids	I	III	IV	V	VI	VII	VIII	IX	X	XI	XII	XIII	XIV
5β-P-3-one	73	70	57	79	80	69	84	80	76	75	65	58	66
20α-ol-5α-P-3-one	57	38	—	—	50	—	—	—	62	47	—	—	—
20β-ol-5α-P-3-one	64	42	—	71	60	—	76	65	65	51	—	—	—
3β-ol-5α-P-20-one	55	33	—	62	45	—	71	63	58	42	—	—	—
3α-ol-5α-P-20-one	59	36	—	67	52	—	74	—	59	47	—	—	—
3β-ol-5β-P-20-one	56	39	—	—	49	—	—	—	61	47	—	—	—
3α-ol-5β-P-20-one	52	29	—	—	45	—	—	—	61	44	—	—	—
3α,6α-ol-5α-P-20-one	26	—	—	33	12	—	38	25	46	18	—	—	—
3α,6α-ol-5β-P-20-one	23	4	—	29	9	—	37	22	43	15	—	—	—
3α,12α-ol-5β-P-20-one	31	4	—	32	10	—	44	29	49	22	—	—	—
3α,17α-ol-5β-P-20-one	46	15	—	42	20	—	64	52	57	30	—	—	—
3β,17α-ol-5β-P-20-one	50	25	—	52	33	—	68	58	61	36	—	—	—
3β,12α-ol-5β-P-20-one	47	22	—	47	30	—	65	54	57	33	—	—	—
3α,17α,20α-ol-5β-P-11-one	24	—	—	25	—	—	—	—	—	—	—	—	—
P4-3,20-one	62	38	12	75	67	—	72	63	68	56	42	20	—
5α-P-3,20-one	66	53	25	75	70	63	78	68	70	63	48	33	55
5β-P-3,20-one	64	50	21	73	69	—	76	70	70	63	—	32	—
12α-ol-5β-P-3,20-one	51	13	—	56	38	—	58	43	56	31	—	—	—
17α-ol-5β-P-3,20-one	66	38	—	71	57	—	75	66	—	47	—	—	—
3α-ol-5β-P-11,20-one	42	10	—	54	38	—	54	40	54	31	—	—	—
3β-ol-5α-P-11,20-one	48	14	—	57	40	—	59	47	—	33	—	—	—
3α,17α-ol-5β-P-11,20-one	49	8	—	44	19	—	62	49	49	19	—	—	—
3β,17α-ol-5α-P-11,20-one	45	12	—	37	26	—	57	41	47	24	—	—	—
5α-P-3,11,20-one	56	26	—	76	69	—	68	56	67	53	—	—	—
5β-P-3,11,20-one	53	22	—	73	66	—	65	54	68	49	—	—	—
5β-P-3,6,20-one	53	27	—	74	66	—	65	55	67	51	—	—	—
5β-P-3,12,20-one	61	39	—	77	69	—	72	63	71	57	—	—	—

(Solvent systems)

[a] For particulars, see the footnotes to Tables XXII and XXVI. The mobilities of progesterone are given for comparison.

TABLE XXXII

$R_f \times 100$ Values Found for Δ^4-3-Oxopregnanesteroids on Silica Gel G Layers by Use of Ascending One-Dimensional TLC[a]

Steroids	\multicolumn{11}{c}{Solvent systems}										
	I	III	IV	V	VI	VIII	IX	X	XI	XII	XIII
P4-3-one	70	59	40	75	72	83	—	72	64	52	—
P4,17(20)-3-one	69	56	35	75	72	83	—	70	62	53	35
20β-ol-P4-3-one	56	27	—	63	55	70	58	56	38	20	—
20β-ol-P4,6-3-one	56	—	—	61	55	69	—	57	—	—	—
20α-ol-P4-3-one	50	24	—	60	52	67	—	54	36	18	—
6β,20β-ol-P4-3-one	44	16	—	51	27	61	50	55	32	—	—
6α,20β-ol-P4-3-one	39	10	—	48	19	65	44	51	27	—	—
11β,20β-ol-P4-3-one	34	5	—	38	11	—	—	—	22	—	—
17α,20β-ol-P4-3-one	41	12	—	51	27	—	42	—	32	—	—
17α,20α-ol-P4-3-one	36	9	—	48	23	—	37	—	28	—	—
17α,20β,21-ol-P4-3-one	15	—	12	21	—	28	—	30	13	41	20
11β,17α,10β,21-ol-P4-3-one	7	—	10	8	—	17	—	27	—	40	18
11α,17α,20β,21-ol-P4-3-one	3	—	15	4	—	9	—	12	—	40	19
P4-3,20-one	60	38	14	74	67	74	64	68	56	40	19
P4,6-3,20-one	57	34	—	71	67	73	—	66	55	46	22
P4,11-3,20-one	60	39	—	74	66	—	—	—	57	—	—
P4,16-3,20-one	59	38	—	72	68	74	—	65	55	50	28
16α,17α-oxido-P4-3,20-one	58	40	—	76	67	—	66	68	59	—	—
2β-ol-P4-3,20-one	47	18	—	62	53	63	50	56	40	—	—
2α-ol-P4-3,20-one	47	18	—	62	53	62	51	56	40	—	—
4-ol-P4-3,20-one	60	41	20	68	68	61	—	—	60	—	—
6β-ol-P4-3,20-one	49	15	—	55	37	59	46	55	30	14	—
7β-ol-P4-3,20-one	41	—	—	55	32	60	—	—	33	14	—
11β-ol-P4-3,20-one	46	14	—	57	40	—	—	54	32	—	—

Compound	1	2	3	4	5	6	7	8	9	10	11
11α-ol-P⁴-3,20-one	31	6	—	43	24	43	—	45	20	7	—
12α-ol-P⁴-3,20-one	38	—	—	55	30	56	—	—	30	—	—
14α-ol-P⁴-3,20-one	47	14	—	55	40	58	41	52	30	—	—
15β-ol-P⁴-3,20-one	41	—	—	59	35	59	32	—	33	—	—
15α-ol-P⁴-3,20-one	31	5	—	50	21	43	34	—	26	8	—
16α-ol-P⁴-3,20-one	29	6	—	47	35	44	—	52	25	22	—
17α-ol-P⁴-3,20-one	56	27	—	64	52	72	51	58	40	—	—
18-ol-P⁴-3,20-one	48	17	—	57	42	—	38	55	38	—	—
19-ol-P⁴-3,20-one	39	9	—	53	27	51	46	54	26	—	—
21-ol-P⁴-3,20-one	40	13	—	62	51	55	11	55	38	20	—
7α,15β-ol-P⁴-3,20-one	14	—	—	24	6	21	8	—	10	—	—
9α,15α-ol-P⁴-3,20-one	16	1	—	21	5	20	14	29	8	—	—
12β,14α-ol-P⁴-3,20-one	23	2	—	33	10	28	11	35	8	—	—
12β,15α-ol-P⁴-3,20-one	12	—	—	32	7	20	—	—	11	—	—
11β,17α-ol-P⁴-3,20-one	43	11	—	40	20	62	26	50	22	7	—
14α,15β-ol-P⁴-3,20-one	—	—	—	30	7	35	—	—	17	—	—
11β,21-ol-P⁴-3,20-one	22	—	—	31	15	37	9	38	16	5	—
11α,21-ol-P⁴-3,20-one	13	—	—	20	7	22	—	26	9	—	—
16α,17α-ol-P⁴-3,20-one	26	6	—	56	35	—	20	—	29	—	—
14α,21-ol-P⁴-3,20-one	28	—	—	37	17	36	—	38	15	—	—
17α,19-ol-P⁴-3,20-one	34	—	—	42	14	—	—	—	22	—	—
16α,21-ol-P⁴-3,20-one	14	—	—	22	9	26	39	31	11	—	—
17α,21-ol-P⁴-3,20-one	38	9	—	41	22	57	24	49	23	8	—
18,21-ol-P⁴-3,20-one	31	—	—	38	—	42	15	42	22	—	—
19,21-ol-P⁴-3,20-one	16	—	—	30	7	30	13	31	16	—	—
18-al-11β,21-ol-P⁴-3,20-one	16	—	—	26	—	26	—	32	16	3	—
17α,20β,21-ol-P⁴-3,11-one	9	—	—	16	—	20	8	25	—	—	—
17α,20α,21-ol-P⁴-3,11-one	9	—	—	17	—	21	—	25	—	—	—
6β,11β,21-ol-P⁴-3,20-one	12	—	—	14	—	26	—	24	—	—	—
7α,14α,21-ol-P⁴-3,20-one	24	—	—	30	10	28	14	38	14	—	—

(Continued)

TABLE XXXII (Continued)

Steroids	Solvent systems										
	I	III	IV	V	VI	VIII	IX	X	XI	XII	XIII
11β,17α,21-ol-P^4-3,20-one	28	—	—	18	7	44	26	37	12	—	—
11α,17α,21-ol-P^4-3,20-one	16	—	—	11	3	28	14	25	6	—	—
17α,19,21-ol-P^4-3,20-one	18	—	—	16	—	23	—	26	10	—	—
16α,17α,21-ol-P^4-3,20-one	—	—	—	29	—	44	28	—	—	—	—
2β,11β,17α,21-ol-P^4-3,20-one	17	—	—	7	2	32	15	23	7	—	—
2α,11β,17α,21-ol-P^4-3,20-one	20	—	—	11	3	36	19	26	8	—	—
16α,11β,17α,21-ol-P^4-3,20-one	—	—	—	14	—	39	21	—	—	—	—
P^4-3,11,20-one	44	16	2	70	60	62	23	60	44	27	—
21-ol-P^4-3,11,20-one	21	4	—	49	27	37	29	42	23	8	—
17α,21-ol-P^4-3,11,20-one	31	5	—	34	13	47	—	43	17	5	—
6β,17α,21-ol-P^4-3,11,20-one	17	—	—	16	—	31	—	28	—	—	—
16α,17α,21-ol-P^4-3,11,20-one	—	—	—	24	5	35	20	—	—	—	—

ᵃ For particulars, see the footnotes to Tables XXII and XXVI.

TABLE XXXIII

$R_f \times 100$ Values Found for Δ^5-3-Hydroxypregnanesteroids on Silica Gel G Layers by Use of Ascending One-Dimensional TLC[a]

Steroids	Solvent systems							
	I	III	V	VI	VIII	IX	X	XI
3β-ol-P^5-20-one	54	37	58	48	72	—	56	42
3β-ol-P^{5,16}-20-one	53	37	57	49	71	—	55	42
3β,7α-ol-P^5-20-one	26	—	31	12	44	24	41	22
3β,15α-ol-P^5-20-one	37	8	38	17	51	36	46	23
3β,16α-ol-P^5-20-one	35	9	41	21	52	38	47	26
3β,17α-ol-P^5-20-one	51	28	50	33	70	58	53	35
3β,21-ol-P^5-20-one	48	19	52	36	63	48	52	36
3β,17α,20α-ol-P^5-11-one	44	13	37	13	66	49	46	31
3β,11β,17α-ol-P^5-20-one	34	7	31	11	51	33	40	23
3β,16α,17α-ol-P^5-20-one	35	12	47	26	52	44	—	25
3β,17α,21-ol-P^5-20-one	35	14	34	17	61	—	43	25
3β-ol-P^5-15,20-one	44	16	51	37	60	46	54	40
3β,11α-ol-P^5-7,20-one	20	—	19	5	30	—	—	14
3β-ol-P^5	64	51	61	50	79	—	65	50
3β,20α-ol-P^5	50	27	48	33	66	—	49	33
3β,20β-ol-P^5	53	31	50	36	69	—	50	34
3β,15α,20α-ol-P^5	19	2	15	6	35	15	32	10
3β,15α,20β-ol-P^5	30	6	28	12	48	29	38	17
3α,16α,20α-ol-P^5	30	2	32	11	42	26	42	19
3α,16α,20β-ol-P^5	27	2	22	8	41	21	48	16
3β,16α,20α-ol-P^5	30	14	38	15	46	27	43	22
3β,16α,20β-ol-P^5	25	8	26	10	44	26	36	14
3β,17α,20α-ol-P^5	38	11	36	14	57	40	44	23
3β,17α,20β-ol-P^5	41	16	40	22	61	43	47	30
3β,18,20β-ol-P^5	41	11	36	18	58	—	44	22
3β,20α,21-ol-P^5	26	5	24	10	46	28	36	15
3β,20β,21-ol-P^5	29	7	26	10	48	31	40	18
3β,11β,17α,20α-ol-P^5	29	5	21	6	51	31	36	19

[a] For particulars, see the footnotes to Tables XXII and XXVI.

on silica gel G in eight principal solvent systems has been undertaken by Lisboa[63, 69, 155, 156, 336] (Tables XXX–XXXIII) by the use of one-dimensional and two-dimensional TLC, with single and multiple developments. For the separation of weakly polar C_{21}-steroids some other systems have been developed,[156] and silver nitrate-impregnated layers have been employed to resolve progesterone, 11-dehydro- and 6-dehydroprogesterones.[156]

The separation of the eight pregnanediol and allopregnanediol isomers

[336] B. P. Lisboa, J. Chromatog., in press (1969).

was investigated by using multiple developments on silica gel layers in the system ethyl acetate–cyclohexane $50:50$,[155] and only the pair 5α-pregnane-3α,20α/3β,20β-diols resisted chromatographic fractionation. However, Galletti[330] has separated these two steroids on alumina layers after three runs on chloroform–ether $4:6$ (migration, in cm: 4.2 and 8.2, respectively, for 5α-pregnane-3α,20α-diol and 5α-pregnane-3β,20β-diol). The sequence of polarity for stereoisomers varies with the adsorbent employed; regarding the configuration at C-3 and C-5, the following order of polarity was found on silica gel layers:

$$5\beta(3\alpha) > 5\alpha(3\beta) > 5\alpha(3\alpha) > 5\beta(3\beta)$$

on aluminum oxide layers an inversion in this sequence was observed:

$$5\beta(3\alpha) > 5\alpha(3\alpha) > 5\alpha(3\beta) > 5\beta(3\beta)$$

The rearrangement of milligram amounts of 17α,20-ketolic steroids (I) to $17a\alpha$-hydroxy-$17a\beta$-methyl-17-oxo-D-homoandrostane (II)-steroids after reflux with 5% methanolic potassium hydroxide has been employed for the characterization of this class of C_{21}-steroids.[337] $R_f \times 100$ values (starting line $= 1.5$ cm; front $= 10$ cm) obtained for the starting material (I) and the major rearrangement product (II) in the systems: (a) ethyl acetate–chloroform $1:9$, silica gel G, for 17α-hydroxypregn-5-en-20-one. $I = 32$, $II = 39$; and 3α,17α-dihydroxy-5β-pregnan-20-one. $I = 14$, $II = 25$; and (b) benzene–chloroform $2:3$, alumina G, for 17α-hydroxyprogesterone: $I = 43$, $II = 62$; 3β,17α-dihydroxy-5α-pregnan-20-one: $I = 28$, $II = 42$; 3β,17α-dihydroxypregn-5-en-20-one: $I = 27$, $II = 42$; and 3α,17α-dihydroxy-5β-pregn-20-one: $I = 6$, $II = 32$. Also minor amounts of the $17a\alpha$-methyl-D-homo-isomer is formed during the alkali treatment; the order of mobility of the homoannulation steroids on both supports is: $17a\alpha$-methyl-D-homo-steroid $> 17\alpha$-hydroxy-20-oxo-21-deoxy-steroid $> 17a\beta$-methyl-D-homo-steroid.

Bennett and Heftmann,[261] Adamec et al.,[229] and Quesenberry and Ungar[338] have developed systems for the separation of saturated dihydrocorticosteroids on silica gel G. In the system[261] ethyl acetate–chloroform–water $90:10:1$ the $R_f \times 100$ values found for cortisone, 11-dehydrocorticosterone, and their dihydro derivatives give the following sequence of mobility:

$$\Delta^4\text{-3-oxo-} < 5\beta\text{-dihydro} < 5\alpha\text{-dihydro(allo)}$$

On spread TLC with silica gel, Adamec et al.[229] have obtained a good separation for cortisone (compound E), cortisol (compound F) and their

[337] J. Chamberlain and G. H. Thomas, Anal. Biochem. 8, 104 (1964).
[338] R. O. Quesenberry and F. Ungar, Anal. Biochem. 8, 192 (1964).

tetrahydro derivatives, with chloroform–ethanol 95:5; the following relative mobilities ($R_S \times 100$ values) have been found: E, 161; F, 100; THE, 74; and THF, 43. Recently Quesenberry and Ungar[338] obtained the resolution of several tetrahydro compounds (A, B, E, and F of Kendall) on silica gel layers (also on polyester sheets[18]); $R_f \times 100$ values found on system dichloromethane–methanol–water–glycerol 150:10:1:0.4: THA, 58; THB, 51; THE, 29; and THF, 20.

TLC of synthetic steroids has been investigated on magnesium silicate[339] and aluminum oxide layers.[340] The following $R_f \times 100$ values are found on partition TLC with aluminum oxide layers in the system benzene–dimethyl formamide 9:1 for some clinically important corticosteroids: cortisol (53), cortisone (67), prednisone (62), prednisolone (50), 6α-methylprednisolone (46), triamcinolone (= 9α-fluoro-16α-hydroxyprednisolone) (16), and dexamethasone (= 9α-fluoro-16α-methylprednisolone) (48).

TLC techniques have been shown to be useful in the isolation and identification of C_{21}-steroids in plants.[50, 341] Tschesche and Brügmann[342] have obtained separation on silica gel G for the 12-epimeric digitopurpurigenins (I, isoramanon = 3β,12β,14β-trihydroxypregn-5-en-20-one; II = the 12α-epimer) in the system ethyl acetate–methanol 96:4 ($R_f \times$ 100 values: I = 33, II = 39).

13. Steroid Conjugates

Very useful separation of free steroids, their sulfate esters and glucosiduronate conjugates can be achieved on TLC; in contrast, only in few cases has a satisfactory resolution of different steroid sulfates or different steroid glucosiduronates been obtained with this fractionation procedure.

During the investigation of glucosiduronation of estrogens by the human intestinal tract, Diczfalusy et al.[220] employed TLC on silica gel G layers for the characterization of estrogen conjugates and derivatives. In water-saturated n-butanol, both estrone and estradiol glucosiduronates showed an R_f value of 0.14, whereas R_f values of 0.33–0.35 have been found for estrone sulfate, estradiol 17β-sulfate and estradiol 3-sulfate, and 0.69 for the much less polar free steroids. In water-saturated benzene–methanol 50:50, the estrogen glucosiduronates remain close to the start-

[339] V. Schwarz, Pharmazie 18, 122 (1963).

[340] V. Schwarz and K. Syhora, Collection Czech Chem. Commun. 28, 101 (1936).

[341] V. Eppenberger, W. Vetter, and T. Reichstein, Helv. Chim. Acta 49, 1505 (1966).

[342] R. Tschesche and G. Brügmann, unpublished results, 1962. Quoted by R. Tschesche, G. Wulff, and K. H. Richert, in "New Biochemical Separations" (A. T. James and L. J. Morris, eds.), p. 198. Van Nostrand, Princeton, New Jersey, 1964.

ing line, whereas the free steroids show R_f values greater than 0.70 (estrone; 0.73), close to the triacetyl methyl ester of glucosiduronates (estrone glucosiduronate triacetyl methyl ester, 0.75). In this system the sulfate esters are well separated from both steroid glucosiduronates and the parent steroids,[343] as can be seen from their mobilities: estrone sulfate, 0.59; estradiol 3-(or 17β-) sulfate, 0.54; and estriol 3-sulfate, 0.43.

Oertel et al.[343a] have investigated the separation of steroid conjugates on anion exchange cellulose layers (MN-300 G/DEAE, ion-exchange capacity = 0.7 meq/g; MN-300 G/Ecteola, ion-exchange capacity = 0.35 meq/g; Macherey, Nagel and Co., Duren, Germany), using acetate buffers, isopropanol–water–formic acid 65:33:2, ethanol–water–acetic acid 80:15:3, and methanol–water–acetic acid 75:15:10 as solvent systems. On both cellulose layers a complete group separation was achieved, in all the systems, between sulfates and glucosiduronates, especially in the two acetic acid systems. Valuable separations of glucosiduronates as well as of sulfates were described; the mobilities ($R_f \times 100$ values) found on DEAE-cellulose after development with 1.0 M acetate buffer (pH 4.75) were, for the glucosiduronates of dehydroepiandrosterone (41), etiocholanolone (45), and androsterone (48) and for the sulfates of estrone (7), 17α-hydroxypregnenolone (9), dehydroepiandrosterone (14), androsterone (22), and pregnenolone (27). The most satisfactory resolution of the glucosiduronates of C_{19}-steroids was achieved on Ecteola layers in the systems ethanol–water–acetic acid 80:11:3 (glucosiduronates of dehydroepiandrosterone, 64; etiocholanolone, 75; and androsterone, 77) and methanol–water–acetic acid 75:15:10 (glucosiduronates of dehydroepiandrosterone, 64; etiocholanolone, 75; and androsterone, 77).

Group separation of 17-oxo steroids, their sulfates and glucosiduronates, was achieved on silica gel H layers after development with ethyl acetate–ethanol–15 N ammonium hydroxide 5:5:1 by Sarfaty and Lipsett.[344] In this system androsterone (A), etiocholanolone (E), and dehydroepiandrosterone (DHA) have R_f values of 0.79 to 0.81, their sulfates of 0.60 (E) and 0.64 (A, DHA), and their glucosiduronates of only 0.08; though the sulfates show tailing, they can be well separated from the much more polar glucosiduronates. The separation of all C_{19}-steroid glucosiduronates from the parent steroids was also obtained on silica gel G layers with the system 16 N ammonium hydroxide–methanol–n-butanol 15:15:70 by Schriefers et al.;[345] the free steroids run with the front, well separated from the very polar glucosiduronates.

[343] B. P. Lisboa, unpublished results, 1964.
[343a] G. W. Oertel, M. C. Tornero, and K. Groot, J. Chromatog. 14, 509 (1964).
[344] G. A. Sarfaty and M. B. Lipsett, Anal. Biochem. 15, 184 (1966).
[345] H. Schriefers, H. K. Kley, and M. Otto, Z. Physiol. Chem. 341, 215 (1965).

For TLC of steroid sulfate esters of several classes on silica gel G layers, Wusteman et al.[346] prefer the systems (a) benzene–ethyl methyl ketone–ethanol–water 3:3:3:1 and (b) 2-propanol–chloroform–methanol–10 N-ammonium hydroxide 10:10:5:2. The mobilities found for the sterol and steroid sulfates investigated in these systems are as follows: ($R_f \times$ 100 values): cholesterol sulfate (a = 58; b = 67), cortisone, 21-sulfate (a = 44, b = 58), dehydroepiandrosterone sulfate (a = 49, b = 66), estradiol, 3-sulfate (a = 48, b = 60), estradiol, 17β-sulfate (a = 46, b = 60), estradiol disulfate (a = 16, b = 27), estriol 3-sulfate (a = 35, b = 48), estrone sulfate (a = 54, b = 71), and scymnol sulfate (a = 29, b = 35); the free parent steroids show in these systems values of 85–95 (a) and 100 (b).

Partition TLC on kieselguhr G–silica gel G 9:1 has been used for the separation of C_{19}- and C_{21}-steroid conjugates by Crépy et al.[347] With the system toluene–tert. butanol–acetic acid–water 82:18:30:70 (starting line, 1.5 cm; front, 15 cm; tank saturated with both phases; lining paper impregnated with the aqueous phase; $R_f \times$ 100 values), the resolution of the paired C-5-isomers of the 3-glucosiduronates of pregnanolone and pregnanediol was achieved: 5α-pregnan-3α-ol-20-one, 47; 5β-pregnan-3α-ol-20-one, 38; 5α-pregnan-3α,20α-diol, 32; and 5β-pregnan-3α,20α-diol, 24. For the sulfate esters, Crépy et al.[347] have used systems of butyl acetate–toluene–4 N-ammonium hydroxide–methanol (a =85:35:50:70; b = 110:90:120:160), in tanks equilibrated with the organic phase and plates preequilibrated for 2 hours in an atmosphere saturated with the aqueous phase. The mobilities found by Crépy et al. for the C_{19}-3-sulfates are: dehydroepiandrosterone (a = 85, b = 50), androsterone (a = 82, b = 51), epiandrosterone (a = 87, b = 55), androst-5-ene-3α,17β-diol (a = 61, b = 38), and androstane-3α,17β-diol (a = 76, b = 40).

Pierrepoint[348] first separates sulfate esters and free steroids on silica gel H with the system tert-butanol–ethyl acetate–5 N ammonium hydroxide 41:50:20. The sulfates move with an R_f value of 0.32, while the free parent steroids move with the front. In the above system, but with the use of anion-exchange cellulose layer (MN-300 G/Ecteola), pregnenolone sulfate ran with a mobility of 0.63 and could be isolated from the other sulfates (R_f values between 0.43 and 0.58). The further separation of these sulfates is achieved by running them on the cellulose layers with a solution of 4 M urea in 3 N ammonium hydroxide: the sulfates of testosterone, dehydroepiandrosterone, and androst-5-enediol shown

[346] F. S. Wusteman, K. S. Dodgson, A. G. Lloyd, F. A. Rose, and N. Tudball, J. Chromatog. 16, 334 (1964).
[347] O. Crépy, O. Judas, and B. Lachese, J. Chromatog. 16, 340 (1964).
[348] C. G. Pierrepoint, Anal. Biochem. 18, 181 (1967).

R_f values of 68, 60, and 45, but 17α-hydroxypregnenolone presents very strong tailing (00–0.45). The authors could separate the sulfates of androstenediol and 17α-hydroxypregnenolone only after the conversion of the former to DHA-sulfate.

To establish the purity of steroid sulfates (as potassium or ammonium salts), Joseph et al.[349] have employed TLC on silica gel layers, with the system (benzene–acetone–water 2:1:2)–methanol (70:30). Neither in this system nor in other very polar systems, such as n-butyl acetate–n-butyl alcohol–4 N-ammonium hydroxide 25:75:100, was the double conjugate 3-sulfooxyestra-1,3,5(10)-trien-17β-yl-β-D-glucosiduronic acid, dipotassium salt, mobile on silica gel layers,[350] in contrast to paper electrophoresis.

Urinary estriol and estradiol glucosiduronic acids have been separated by Fishman et al.[351] by two-dimensional TLC on layers of silica gel H by means of four developments in three consecutively applied systems: (a) chloroform–isopropyl alcohol 3:1, (b) chloroform–isopropyl alcohol–formic acid 15:5:3, and (c) ethyl acetate–pyridine–glacial acetic acid–water 6:5:1:1. After purification of the extract material in system (a), in which the conjugates remain on the starting point, the layer is submitted to two developments in system (b) in one direction followed by a single development in system (c) in the other direction. Estradiol and estriol glucosiduronic acids show $R_f \times 100$ values, respectively, of 60 and 45 (migration 9.4 and 7.7 cm) in system (b) and 60 and 40 in system (c), after 10 cm development. After development with system (b) of Fishman et al., Ferrara et al.[352] found R_f values of 0.60 and 0.42, respectively, for the sodium salts of estradiol-17- and 6-dehydroestradiol-17-β-D-glucosiduronates on silica gel G layers. Lisboa[286] prefers systems of ethyl acetate–methanol–acetic acid type (a = 70:40:10, b = 70:15:15) for the separation of the estradiol glucosiduronates; in these systems the 17β-monoglucosiduronate (a = 57, b = 42) is less polar than the 3-monoglucosiduronate (a = 52, b = 39); estradiol diglucosiduronate shows in these systems a migration of 0.14 (a) and 0.11 (b).[286,353]

VI. Quantitation with TLC

Several methods for the quantitation of steroids and steroidal compounds have been described, in which TLC has been used alone or

[349] J. P. Joseph, J. P. Dusza, and S. Bernstein, Steroids 7, 577 (1966).
[350] E. W. Cantrall, M. G. McGrath, and S. Bernstein, Steroids 8, 967 (1966).
[351] W. H. Fishman, F. Harris, and S. Green, Steroids 5, 375 (1965).
[352] G. Ferrara, C. Boffi, G. Torti, and A. Corbellini, Steroids 8, 111 (1966).
[353] B. P. Lisboa, unpublished results, 1964.

together with other fractionation procedures for the isolation and purification of the substances to be measured. Recently Lisboa[354] has reviewed the quantitative aspects of the application of TLC for the analysis of steroids and sterols, and Eneroth[355] has discussed the quantitation of bile acids with this technique.

Unless radioactive carrier is added to correct the loss of material during chromatography and extraction, or direct densitometry of the fractionated substances is carried out, the recovery of the steroid after TLC remains the principal problem in the application of the method to quantitative analysis.

High recovery values have been reported for steroids, sterols, and steroidal compounds after TLC with direct extraction of the material followed by centrifugation, or with elution from the adsorbent through a tube containing a cotton or fritted-glass disk. Both polar solvents and less polar mixtures of solvents have been used.

Lisboa[280] has found recovery values of 90–95% for several C_{19}- and C_{21}-steroids, such as cortisol, cortisone, cortexone, 17α-hydroxyprogesterone, 17α-hydroxycortexone, androstenedione, and adrenosterone, eluted from silica gel G and measured by the use of specific color reactions, after extraction twice with absolute ethanol. Matthews et al.[214] have recovery values of 97% for progesterone from alumina and from silica gel G, after elution through a sintered disk with the same solvent and quantitated by UV absorption at 240 mμ. Scavino and Chiaramonti[356] recovered 98% of β-methasone (9α-fluoro-16β-methylprednisolone) and dexamethasone (9α-fluoro-16α-methylprednisolone), also measured by spectrometry at 239 mμ, after mechanical shaking of the silica gel with methanol, followed by filtration through a fritted-glass disk. Recovery values of 97% were obtained by Hutton and Boyd[357] for amounts of 7α-hydroxycholesterol and 7α-hydroxycholest-4-en-3-one varying from 5 to 190 μg extracted with ether–alcohol 1:1 from developed silica gel plates and measured by the Lifschütz reaction (7α-hydroxycholesterol) or by absorption in ethanol at 242 mμ.

Working with labeled steroids chromatographed on silica gel G layers, Ertel and Ungar[358] have recovered 90–95% of the spotted progesterone

[354] B. P. Lisboa, in "Chromatographic Analysis of Lipids" (G. V. Marinetti, ed.), p. 57. Bd. II. Dekker, New York, 1968.

[355] P. Eneroth, in "Chromatographic Analysis of Lipids" (G. V. Marinetti, ed.), p. 149. Bd. II. Dekker, New York, 1968.

[356] C. Scavino and D. Chiaramonti, 3rd Intern. Symp. Chromatog. Brussels, 1964, p. 169. Soc. Belge des Sciences Pharmaceutiques, Brussels, 1964.

[357] H. R. B. Hutton and G. S. Boyd, Biochim. Biophys. Acta 116, 336 (1966).

[358] R. J. Ertel and F. Ungar, Endocrinology 75, 949 (1964).

by shaking with methanol, and Idler *et al.*[359] have obtained average recoveries between 86.5 and 96.6% for testosterone, corticosterone, cortisol, and the acetates of testosterone and cortisol with the use of methanol and elution through a sintered-glass disk. Slightly inferior were the recoveries found for tritium-labeled deoxycorticosterone (85%), aldosterone (88%), and corticosterone (91%) after chromatography on TLC sheets.[338]

The recovery of 0.5–1.0 μg amounts of several sterols and steroids from silica gel–gypsum (10%) layers (Adsorbosil 4, Applied Science Laboratories, State College, Pennsylvania), measured by gas chromatography, was investigated by Vandenheuvel.[360] The extraction from the adsorbent was carried out in a specially developed device for microelution, in which the substance was eluted with chloroform–methanol 2:1 and filtered through a medium-porosity fritted-glass disk. Recovery values between 95.8 and 99.5% (average: 97.4%) have been found for the fifteen investigated substances (cholestane, coprostane, cholestanone, cholesterol, cholestanol, pregnanolone, pregnanediol, pregnanetriol, androsterone, androstane-3α,17β-diol, dehydroepiandrosterone, testosterone, estrone, estradiol and estriol). The low recoveries found under the same conditions when methanol was employed as eluent, is explained by the elution of fine particles of the adsorbent, which interferes with the formation of trimethyl silyl ethers used in the further gas-chromatography quantitation step.

The choice of the solvent to be used for elution depends upon the polarity of the steroids concerned. For the recovery of estradiol and estriol glucosiduronic acids (recovery values: 83.4 ± 11.6 and $85.5 \pm 15.0\%$, respectively), Fishman[351] shakes the silica gel with 2.5 ml of 1% sodium bicarbonate and after centrifugation transfers the supernatant to a second tube, adjusts its pH to 4.0, adds 0.5 ml of 1 M acetate buffer (pH 4.0), and finally extracts with ethyl acetate. Sarfaty and Lipsett[343] eluted free dehydroepiandrosterone with methanol, its sulfate with water, and its glucosiduronate with 0.1 M acetate buffer (pH 5) and obtained recovery values of 90, 72, and 63%, respectively, measured as free dehydroepiandrosterone. For the extraction of steroid sulfates from silica gel and cellulose layers, Pierrepoint[348] prefers to add to the adsorbent 2 ml of a saturated sodium chloride solution, and then to shake twice with 2 ml of ethyl acetate.

The use of highly polar solvents for the extraction of free steroids from the layer is limited by the method used for their further quantita-

---

[359] D. R. Idler, N. R. Kimball, and B. Truscott, *Steroids* **8**, 865 (1966).
[360] F. A. Vandenheuvel, *J. Lab. Clin. Med.* **69**, 343 (1967).

tion, since these solvents extract fine particles of the adsorbent which can interfere with these methods. Besides, if successive chromatograms are necessary, these fine soluble materials of the adsorbents sometimes cause an important destruction of the steroid during the concentration of the eluate for the second chromatography,[359] even if fine synthetic membrane filters are used. Therefore several investigators[359-361] prefer the use of less polar solvent mixtures for the elution of the steroids, such as dichloromethane–methanol (9:1),[359] chloroform–methanol (2:1),[360] or mixtures of toluene or benzene–methanol or ethyl acetate.[360] The instability of the steroids on the adsorbent is an important factor which must be considered when TLC is used for the purification of materials prior to biological experiments, especially if they are labeled. It is extremely important to reduce as much as possible the time of contact between the steroid and the layer when ketolic steroids, 11-substituted pregnane steroids, and 3-hydroxyestrogens with additional 6-keto, 2- or 4-hydroxy, or 2-methoxy groups are submitted to TLC.

The destruction of the steroid in the presence of the adsorbent is also indicated in the data of Vandenheuvel;[360] whereas high recovery values are found for steroids extracted within 12 hours of development from plates conserved in the darkness, with increasing storage time these values decreases markedly to only 30% after 5 days.

Sometimes the solvents or reagents in which the steroid is quantitated also can be used for their extraction, but in these cases a special pre-purification of the adsorbent is necessary. Using silica gel G extensively treated with diluted sulfuric acid (1:1), diluted hydrochloric acid (1:1), and water until the washings were neutral, Kottke et al.[362] eluted bile acids with 65% sulfuric acid and after centrifugation read the colorless supernatant solutions at 320 mμ against a blank; recovery values of 90.4, 83.4, and 89.0% were obtained for taurocholic, glycocholic, and cholic acids, respectively, and the silica gel blank gave an optical density of only 0.010–0.035. It must be emphasized that the higher recovery found by other workers under similar conditions[47, 363] resulted from the use of unpurified silica gel.

Finally, the recovery of steroids after in situ detection with the so-called "nondestructive methods" has been investigated by Varon et al.[364] for estrone, estradiol, and estriol. For 5-μg amounts of estrogens, recovery values greater than 93% have been found after detection of the steroids

[361] J. Attal, S. M. Hendeles, J. A. Engels, and K. B. Eik-Nes, J. Chromatog. 27, 167 (1967).
[362] B. A. Kottke, J. Wollenweber, and C. A. Owen, Jr., J. Chromatog. 21, 439 (1966).
[363] B. Frosch and H. Wagener, Z. Klin. Chem. 2, 7 (1964).
[364] H. H. Varon, H. A. Darnold, M. Murphy, and J. Forsythe, Steroids 9, 507 (1967).

by spraying them with either water or 2′,7′-dichlorofluorescein. In contrast, the values obtained after exposure to iodine vapors were only 44–51%. The infrared spectra of calciferol (vitamin D_2) and cholecalciferol (vitamin D_3) eluted from aluminum oxide layers after exposure to iodine vapors have shown an alteration on the structure of these vitamins.[365] Losses after iodine treatment have been indicated also by Matthews[214] in the recovery of steroids with different functional groups from silica gel G or alumina layers, but they are much less intensive than those observed by Varon et al.[364] for the steroid estrogens.

Acknowledgments

Many thanks are due to Dr. J. W. Reynolds (University of Minnesota) for many helpful discussions during the preparation of this review. Also the competent assistance of Mrs. Ingrid Lisboa is gratefully acknowledged.

[365] B. Kakáč, M. Šaršúnová, Tran Thi Hoang Ba, and J. Vachek, J. Pharmazie 22, 202 (1967).

[2] Gas-Liquid Chromatographic Methods for the Analysis of Steroids and Sterols

By HERBERT H. WOTIZ and STANLEY J. CLARK

The development of gas chromatographic methods for the estimation of steroid hormones has been quite rapid, and a considerable number of reliable methods are now available. Two major reviews have appeared recently, one of them[1] covers the field comprehensively through 1965 and the other[2] describes in detail techniques for the determination of steroid hormones and metabolites in urine and plasma.

As with most analytical techniques, attempts to attain greater sensitivity are constantly being made. Such attempts are certainly justifiable since steroid hormones are often present at extremely low concentrations and sample size is often very limited. This short review is concerned primarily with recent high sensitivity methods, particularly those employing electron capture techniques, as applied to steroids of biological fluids. While this approach is admittedly specialized, it will illustrate in detail the technical problems that are likely to be encountered in a variety of situations.

[1] A. Kuksis, Methods Biochem. Anal. 14, 325 (1966).
[2] H. H. Wotiz and S. J. Clark, "Gas Chromatography in the Analysis of Steroid Hormones." Plenum Press, New York, 1966.

Other special aspects of gas-liquid chromatography (GLC) methodology are discussed elsewhere in this volume. The application of GLC to the analysis of polycyclic triterpenes is discussed in the following chapter by Ikekawa [3], a detailed account of the use of GLC in the analysis of bile acids is given by Eneroth and Sjövall [5], and the application of GLC to the analysis of retinol and derivatives is described by Dunagin and Olson [7].

Factors Affecting Sensitivity

The chromatographic sensitivity attainable is a function of the sensitivity of the detecting device, noise contributed by the rest of the apparatus (which is usually greater than that of the detector), and the quality of the column. The overall sensitivity of the method is also influenced by the preliminary separation and purification procedure and the losses incurred therein.

Although the gas chromatographic column is capable of high resolution, it cannot deal with crude biological samples when the compound sought is present only in trace amounts. In this situation, the compound of interest is usually lost in the background and, in addition, the presence of slow-moving and involatile components leads to a marked increase in noise contributed by the column. Thus it is essential that adequate preliminary separation of the compound from the matrix be performed before gas chromatography is undertaken. This applies equally to the products of *in vitro* incubations with enzymes and tissue preparations as to the constituents of biological fluids, though in the latter case the problem is usually much more difficult because of the low endogenous concentration of most steroids and the complexity of the background materials.

Detectors

The flame ionization detector is the most reliable of the high-sensitivity devices and is to be preferred for this reason. However, even under the best conditions, the limit of detection is of the order of 1 nanogram (ng) and in practice a value of 10 ng is more realistic. If higher sensitivity is needed, it is necessary to use the electron capture detector, which is potentially capable of detecting about 1 picogram (pg) and, with suitable steroid derivatives, has a practical limit of 10–100 pg.

Flame Ionization. The general characteristics of the flame ionization detector have been described repeatedly[3-5] and will not be discussed here.

---

[3] I. G. McWilliam and R. A. Dewar, *in* "Gas Chromatography 1958" (D. H. Desty, ed.), p. 142, Butterworth, London, 1959.

However, it must be emphasized that optimization of conditions, particularly gas flow rates, is necessary for best results and this must be done experimentally. Once conditions have been established to produce maximum detector response, sensitivity can be increased only by decreasing noise. The detector and its associated electronics contribute comparatively little; most of the noise arises from variations in impurity levels and flow rates of the gases reaching the detector. Quality of the flow control system is dependent upon the manufacturer, and little can be done about it except to ensure that the system is completely free from leaks. Removal of impurities from gases is best achieved by inserting traps containing 5 Å molecular sieve in each of the gas lines. The traps must be replaced or regenerated regularly to ensure their continued efficiency.

Electron Capture. Although it is capable of much greater sensitivity than the flame detector, the electron capture detector is considerably more difficult to operate reliably in a routine manner. Thus some discussion of the practical problems involved is in order.

Several varieties of detector are available, all, with one exception, employing a radioisotope as the source of electrons. The exception employs a rare gas discharge to produce electrons, which are then transferred to the ionization chamber where interaction with the sample occurs. The potential advantages of the system are that it can be operated at higher temperatures than the conventional detector and that it should be largely immune to contamination. The system is considerably more complicated than any other, and its long-term reliability and precision remain to be demonstrated.

Other detectors employing radioactive sources differ in geometry and in the nature of the radioisotope. Early detectors were, in fact, argon detectors operated at low voltages and had a concentric configuration with the anode located along the axis of the cell and surrounded by a cylindrical cathode to which was attached the radioactive source. Detectors of this type are still supplied by some manufacturers, but the majority are now of the plane-parallel design in which anode and cathode are disks disposed parallel to each other at the ends of a Teflon ionization chamber.

There appears to be little difference in performance between the two types, provided due attention is paid to the peculiarities of each and care is taken to optimize operating conditions. The literature should be con-

[4] L. Onkiehong, *in* "Gas Chromatography 1960" (R. P. W. Scott, ed.), p. 7. Butterworth, London, 1960.
[5] R. D. Condon, P. R. Scholly, and W. Averill, *in* "Gas Chromatography 1960" (R. P. W. Scott, ed.), p. 30. Butterworth, London, 1960.

sulted for details of detector characteristics.[6,7] The overriding practical consideration is that *all* electron capture detectors are subject to troubles arising from impurities carried in the gas stream, and all detectors employing radioactive sources are subject to contamination of the source.

Trouble from contamination of the gas stream manifests itself in loss of response, increase in noise, and, often, baseline drift. It may be minimized by ensuring that all gases are thoroughly dried, that only stable columns with a low bleed rate are employed, and that only relatively clean samples are injected into the chromatograph.

Contamination of the radioactive source arises mainly from deposition on the surface of the source of material bleeding from the column. It gives rise to peak distortion which usually consists of a negative signal on the trailing edge of the peak. The distortion presumably is due to sorption and desorption of the sample vapor passing through the detector causing changes in the electrical double layer set up on the contaminated source. Tritium is the isotope most commonly used as source material and, although its energy characteristics are suitable and it can be obtained cheaply at high specific activity, tritium sources are particularly prone to contamination since they cannot be safely heated above about 210°. An improvement in this respect can be obtained by using nickel sources. The temperature limit is then set by the material of construction of the detector, but in any case the detector can be operated at 250°. The main disadvantages of nickel are its high price and its comparatively low specific activity. In this latter respect, the concentric geometry is superior since it is possible to use sources of larger area than in the plane-parallel design and thus a greater total activity can be used.

The rate at which symptoms of source contamination appear is governed to some extent by the type of detector and its mode of operation. Plane-parallel detectors operated in the pulsed mode or concentric detectors operated in pulsed or direct-current (d.c.) modes are superior to plane-parallel detectors operated in the d.c. mode. The contact potential set up at the surface of the contaminated source acts to oppose the operating potential of the detector, and since plane-parallel d.c. detectors operate at lower potentials, the effect is observed sooner. However, contamination will occur sooner or later, and it cannot be too often reiterated that care in using stable well-conditioned columns is the most important single factor in maintaining the detector in proper operating condition.

[6] J. E. Lovelock, *Anal. Chem.* **35**, 474 (1963).
[7] S. J. Clark, *in* "Residue Reviews" (F. A. Gunter, ed.), Vol. 5, p. 32. Springer, Berlin, 1964.

Columns

The quality of the column is still the major factor in determining the success or failure of the analysis. Much empirical information has been acquired concerning the characteristics of columns suitable for steroid gas chromatography, but the preparation of good columns is still largely an art (cf. Eneroth and Sjövall, this volume [5], p. 256).

Support. The influence of the support is particularly noticeable in columns with low liquid loadings such as are used in steroid gas chromatography. Losses of sample on the column can be attributed mainly to adsorption by the support material, and much effort has been expended in producing a suitably deactivated support.[8]

Silanization of the support is the preferred method of deactivation; considerable care is required to produce acceptable material, and the process is laborious. Commercial supports of acceptable quality are now available[9] and, although expensive, are worth the price in terms of time and effort saved.

Stationary Phase. Even the most inert support is not entirely free of adsorptive properties, and it is important that sufficient stationary phase be used to cover the support adequately. The extent to which losses occur on very lean columns depends to a large extent on the nature of the stationary phase and of the compound being chromatographed. For example, it is perfectly possible to chromatograph trimethylsilyl ethers of 17-ketosteroids on a 1% QF-1 column, whereas losses of cholesterol on SE-30 columns increase rapidly when less than about 2% of stationary phase is used. Moreover, column efficiencies decrease quite markedly at concentrations of less than 2% of stationary phase. Thus, unless a pressing reason exists, columns containing less than about 2% of stationary phase are not generally to be recommended.

Coating. Two requirements must be satisfied by the method of coating the support: (1) that the stationary phase be distributed as uniformly as possible, and (2) that mechanical handling be kept to a minimum. The support material is fragile, and attrition leads to production of fines and the exposure of fresh adsorptive surfaces. The most satisfactory method in our hands is the solution coating technique[10] or its later modification.[11]

Conditioning. Processes occurring during column conditioning are not

---

[8] W. R. Supina, R. S. Henly, and R. F. Kruppa, *J. Am. Oil Chemist's Soc.* **43**, 2042, 228 (1966).

[9] Gas Chrom Q (Applied Science Corp., State College, Pennsylvania) and Diatoport S (F & M Scientific Corp., Avondale, Pennsylvania).

[10] E. C. Horning, E. A. Moscatelli, and C. C. Sweeley, *Chem. & Ind.* p. 751, (1959)

[11] J. F. Parcher and P. Urone, *J. Gas Chromatog.* **2**, 184 (1964).

well understood, but apparently more is involved than the stripping of volatile components from the stationary phase. With some phases, namely QF-1, XE-60, and SE-52, conformational changes occur that appear to render the phases less polar at higher temperatures.[12] Moreover, the particular conformation of the polymer is dependent upon the thermal history of the column. Thus, care must be exercised, both in conditioning such columns and in their use in temperature-programmed operation, if reproducible retention times are to be achieved. Increased efficiency can be obtained from SE-30 columns by conditioning for 2–3 hours at 300° with no carrier gas flowing.[13] However, this treatment appears to be harmful to columns coated with other stationary phases.

For most columns, conditioning is a matter of heating at slightly above the maximum operating temperatures, with carrier gas flowing until the column is stable. However, as the above examples show, the behavior of the high polymers used as stationary phases in steroid work is sometimes unpredictable.

Choice of Derivative

Although many of the steroids of interest can be chromatographed intact in the free state, it is usually better to form a derivative. Several advantages accrue from the use of derivatives: increased thermal stability; decreased interaction with the column support; greater volatility; improved separation of certain compounds; conferment of special properties such as electron-capturing ability. Not all these advantages will be necessary for any particular analysis, and the particular derivative used must depend upon the requirements of the analysis. Short summaries of the characteristics of the most commonly used derivatives follow (cf. Eneroth and Sjövall, this volume [5], pp. 258–260).

Acetates.[14] Primary use has been in the rapid determination of estriol, pregnanetriol, and pregnanediol,[15, 16] although many other steroid acetates have been chromatographed successfully. The derivatives are readily formed and stable. Volatilities are comparable with those of the parent compounds. Some interaction with the column support may occur although to a lesser extent than with the free steroid.

[12] C. Chen and O. Gaeke, *Anal. Chem.* 36, 72 (1964).
[13] M. B. Whittier, L. Mikkelsen, and S. Spencer, *Technical Paper No. 13*, F. & M. Scientific Corporation, Avondale, Pennsylvania.
[14] H. H. Wotiz and H. F. Martin, *Am. Chem. Soc. 138th Natl. Meeting, Abstr.*, p. 58C (1960).
[15] H. H. Wotiz and S. J. Clark, "Gas Chromatography in the Analysis of Steroid Hormones," p. 232. Plenum Press, New York, 1966.
[16] H. H. Wotiz and S. J. Clark, "Gas Chromatography in the Analysis of Steroid Hormones," p. 148. Plenum Press, New York, 1966.

Trifluoroacetates.[17] These compounds are generally easy to prepare but somewhat unstable, being readily hydrolyzed. They are considerably more volatile than the parent compounds. These derivatives do not capture electrons very strongly and are not suitable for use with the electron capture detector.

Heptafluorobutyrates.[18] These are generally similar in properties to the trifluoroacetates, but rather less stable and less volatile. Their most important characteristic is their strong electron-capturing ability. Both the trifluoroacetates and heptafluorobutyrates interact only to a very small extent with the column support so that it is possible to chromatograph less than nanogram amounts.

Monochloroacetates.[19] Considerably less volatile than the acetates and other halogenated esters, monochloroacetates are valuable because they capture electrons reasonably strongly.

The preparation of the halogenated esters is relatively straightforward. However, strong acid is liberated during the reaction, and side reactions may occur, particularly with those compounds capable of undergoing enolization.

Trimethylsilyl Ethers.[20] Perhaps the most generally useful derivatives, these ethers are easily prepared, volatile, and react only to a very small extent with the column support. They are easily hydrolyzed, however, and adequate precautions must be taken to exclude moisture during preparation and subsequently.

Structure and Retention Time

The thermodynamic behavior of a solute in a partition chromatographic system determines the degree of retention of the solute in the system. Obviously, an understanding of the factors governing the rate of movement through the partitioning medium is of great theoretical and practical importance.

Detailed studies of structure-retention relationships have been made by a number of workers for most of the steroids that are currently of analytical importance, and this work has been reviewed by Hartman[21] and Kuksis.[22] Attempts at quantitative description of retention charac-

[17] W. J. A. VandenHeuvel, J. Sjövall, and E. C. Horning, *Biochim. Biophys. Acta* **48**, 596 (1961).

[18] S. J. Clark and H. H. Wotiz, *Steroids* **2**, 540 (1963).

[19] R. A. Landowne and S. R. Lipsky, *Anal. Chem.* **35**, 532 (1963).

[20] T. Luukkainen, W. J. A. VandenHeuvel, E. O. A. Haahti, and E. C. Horning, *Biochim. Biophys. Acta* **52**, 599 (1961).

[21] I. S. Hartman, *in* "Gas Chromatography in the Analysis of Steroid Hormones" (H. H. Wotiz and S. J. Clark), p. 99. Plenum Press, New York, 1966.

[22] A. Kuksis, *Methods Biochem. Anal.* **14**, 325 (1966).

teristics have been generally successful, at least for the simpler steroids. The theory of Martin,[23] that the contribution of substituent groups to the partition coefficient, and hence the retention value, are additive and noninteracting, has been employed. Liquid–liquid partition systems have been described by Bush[24] and gas-liquid systems by Clayton[25, 26] and Knights and Thomas.[27-30] The latter workers have used the method successfully as an aid to qualitative analysis of steroid mixtures.

Besides the application to qualitative analysis, retention data are valuable in selecting the appropriate conditions for the separation of known compounds. In the following short discussion, information is arranged in tabular form to emphasize the regularities of behavior of groups of compounds.

The Hydrocarbon Nucleus

In the absence of groups having specific interactions with the stationary phase, partition is determined by molecular size and shape. The effect of molecular shape is illustrated by the behavior of a series of cholestanes[31] (Table I).

TABLE I
RELATIVE RETENTION TIMES OF CHOLESTANES

Compound	SE-30[a]	EGSS-X[b]
Coprostane (5β; A/B *cis*)	0.9	0.9
Cholestane (5α; A/B *trans*)	1.0	1.0
Cholest-5-ene	1.0	1.2
Cholesta-3,5-diene	1.1	1.8

[a] 1% SE-30; 212°.   [b] 1% EGSS-X; 187°.

As the structure becomes more nearly planar, interaction with the solute (SE-30) increases, as does the retention time. The order of elution of the saturated hydrocarbons also applies to the androstanes, preg-

[23] A. J. P. Martin, *Biochem. Soc. Symp.* (Cambridge, Engl.) 3, 4 (1949).
[24] I. E. Bush, "The Chromatography of Steroids," Chapters I and II. Pergamon, New York, 1961.
[25] R. B. Clayton, *Nature* 190, 1071 (1961) *ibid.*, 192, 524 (1961).
[26] R. B. Clayton, *Biochemistry* 1, 357 (1962).
[27] B. A. Knights and G. H. Thomas, *Anal. Chem.* 34, 1046 (1962).
[28] B. A. Knights and G. H. Thomas, *J. Chem. Soc.* p. 3477 (1963).
[29] B. A. Knights and G. H. Thomas, *Nature* 194, 833 (1962).
[30] B. A. Knights and G. H. Thomas, *Chem. & Ind.* p. 43 (1963).
[31] W. J. A. VandenHeuvel and E. C. Horning, *in* "Biomedical Application of Gas Chromatography" (H. Szymanski, ed.), p. 89. Plenum Press, New York, 1964.

nanes, and cholanes. The 5β isomers have the shorter retention times.[33]

Table II also illustrates the effect of molecular size (or weight). The molecules differ only in the length of side chains at C_{17}, and the elution order is related directly to the length of the side chain. It should be noted that QF-1 is a much poorer solvent for the hydrocarbons than is SE-30, although this is not obvious from the relative retentions. Under the conditions of the experiment, retention values were four to five times greater on the SE-30 column than on QF-1.

TABLE II

RELATIVE RETENTION TIMES OF 5α AND 5β STEROIDS[a,b]

Compound	SE-30	QF-1
C_{19} 5β-Androstane	0.07	—
5α-Androstane	0.08	—
C_{21} 5β-Pregnane	0.13	0.18
5α-Pregnane	0.15	0.21
C_{24} 5β-Cholane	0.37	0.43
5α-Cholane	0.40	0.48
C_{27} 5β-Cholestane	0.90	0.90
5α-Cholestane	1.00	1.00

[a] See footnote 32.

[b] Column temperature 200°. Cholestane ret. times: SE-30, 30–33 min; QF-1, 5.5–6.0 min. Length of column: 1.3 m. Liquid phases: 1% concentration.

Double Bonds

The effect of unsaturation in the hydrocarbon skeleton is small with nonselective phases such as SE-30. However, on phases showing specific interactions with double bonds (polyesters and polyester-silicone polymers) the unsaturated compounds are retained longer than the saturated (Table I). The effect varies depending upon the position of the double bond in the molecule, but seems to be essentially constant for the same position in different series.[34]

Monosubstituted Compounds

For these compounds, the general rule applies that the less sterically hindered equatorial substitutent interacts more strongly with the stationary phase and is eluted later than its corresponding axial epimer.

The effect is exemplified by the 3-hydroxy steroids (Table III). Note that, as with the parent hydrocarbons, the 5β-series is always eluted before the 5α.[35]

[32] Retention times in Tables II–VIII are relative to 5α-cholestane.

[33] C. J. W. Brooks and L. Hanaineh, *Biochem. J.* 87, 151 (1963).

[34] R. B. Clayton, *Biochemistry* 1, 357 (1962).

[35] I. S. Hartman and H. H. Wotiz, *Biochim. Biophys. Acta* 90, 334 (1964).

TABLE III
RELATIVE RETENTION TIMES OF 3α AND 3β HYDROXYSTEROIDS OF
THE 5α AND 5β SERIES[a]

A/B ring fusion	5β				5α			
Hydroxyl:	3β (ax)		3α (eq)		3α (ax)		3β (eq)	
	SE-30	QF-1	SE-30	QF-1	SE-30	QF-1	SE-30	QF-1
Androstanol	—	—	0.15	0.39	—	—	0.16	0.43
Pregnanol	0.28	0.57	0.28	0.65	0.30	0.65	0.31	0.73
Cholestanol	1.89	2.83	1.92	3.24	2.12	3.24	2.14	3.66

[a] See Table II.

QF-1 interacts more strongly than SE-30 with the hydroxyl group so that the alcohols are retarded further on this phase relative to the parent hydrocarbon.

The general pattern is similar for the 3-keto substituents, but these compounds are selectively retarded by QF-1 and XE-60 (Tables IV and V).

TABLE IV
RELATIVE RETENTION TIMES FOR 3-KETO AND 3-HYDROXY CHOLESTANES[a]

Compound	5β		5α	
	SE-30	QF-1	SE-30	QF-1
Cholestan-3α-ol	1.92	3.24	2.12	3.24
Cholestan-3β-ol	1.89	2.83	2.14	3.66
Cholestan-3-one	2.06	6.28	2.29	6.95

[a] See Table II.

TABLE V
RELATIVE RETENTION TIMES FOR 3-KETO AND 3-HYDROXY ANDROSTANES[a]

Compound	5β				5α			
	SE-30[b]	XE-60[b]	NGSeb[c]	HiEff-8B[d]	SE-30[b]	XE-60[b]	NGSeb[c]	HiEff-8B[d]
Androstane	0.20	0.25	0.08	0.12	0.21	0.27	0.09	0.13
Androstan-3β-ol	0.29	0.53	0.38	0.61	0.31	0.65	0.47	0.75
Androstan-3-one	0.32	0.68	0.42	0.67	0.33	0.77	0.49	0.73

[a] All liquid phases 3%; [b] 260°; [c] 232°; [d] 238°.

TABLE VI
RELATIVE RETENTION TIMES OF 17-HYDROXY AND 17-KETO ANDROSTANES[a]

	5β				5α			
Compound	SE-30[b]	XE-60[b]	NGSeb[c]	HiEff-8B[d]	SE-30[b]	XE-60[b]	NGSeb[c]	HiEff-8B[d]
Androstane	0.20	0.25	0.08	0.12	0.21	0.27	0.09	0.13
Androstan-17β-ol	0.41	0.89	0.84	1.48	0.32	0.60	0.47	0.71
Androstan-17-one	0.42	0.95	0.71	1.15	0.32	0.65	0.39	0.61

[a] All liquid phases 3%; [b] 260°; [c] 232°; [d] 238°.

The 17-hydroxy and 17-keto androstanes behave anomalously and reverse the order of elution of the parent hydrocarbons (Table VI.)[36]

Disubstituted Compounds

Table VII gives retention data for a number of 3,17-disubstituted androstanes. The normal order of elution is retained, the 5β isomers being eluted first.

TABLE VII
RELATIVE RETENTION DATA FOR 3,17-DISUBSTITUTED ANDROSTANES[a]

	5β				5α			
Compound	SE-30[b]	XE-60[b]	NGSeb[c]	HiEff-8B[d]	SE-30[b]	XE-60[b]	NGSeb[c]	HiEff-8B[d]
Androstane-								
3α,17β-diol	0.51	1.68	2.08	3.85	0.54	1.67	2.27	3.94
3β,17β-diol	0.49	1.56	1.93	3.50	0.59	1.89	2.44	4.19
3α-ol, 17-one	0.47	1.90	1.80	3.58	0.51	2.00	1.83	3.73
3β-ol, 17-one	—	1.98	1.88	—	0.55	2.10	2.10	3.97
3,17-dione	0.50	2.48	1.98	3.64	0.58	2.64	2.27	4.00

[a] All liquid phases 3%; [b] 260°; [c] 232°; [d] 238°.

Another irregularity should be noted: In the 5β series, the axial 3β-ol, 17-one is retained longer than the equatorial 3α-ol, 17-one.

Effect of Blocking Functional Groups

It is common practice to prepare derivatives to confer improved chromatographic properties on the molecule (see pp. 163, 258). The effect

[36] I. S. Hartman and H. H. Wotiz, *Biochim. Biophys. Acta* **90**, 334 (1964).

of derivative formation on the disubstituted androstanes is shown in Table VIII.

The elution order for some of the isomeric pairs is again reversed (italicized values in Table VIII). The retention times of the trimethylsilyl ethers on the more selective columns are decreased compared with the free compounds, but the keto compounds tend to be retained to a much greater extent relative to the dihydroxy compounds.

Aromatics

The presence in the molecule of an aromatic A ring is characteristic of the estrogens. Little systematic information is available about the behavior of this group of compounds for two main reasons. First, estratrienes having single functional groups in the appropriate position are not generally available, so that it is difficult to assess the effect of any particular group. Second, interaction with the support tends to be strong, particularly with the polyhydroxy compounds, so, as a practical matter, the estrogens are usually chromatographed as derivatives. Indeed, some of the results obtained must be treated with reserve since the extent of adsorption on the column cannot be assessed. Some of the values in Table IX[37] for the estriols illustrate the point. The relative retentions of the trimethylsilyl ethers on 0.75% SE-30 are greater than those on 3% SE-30. Values for the other compounds are in good agreement, considering the quite different experimental conditions, and it must be concluded that absorption of the estriol derivatives is greater on the leaner column.

On the nonselective phase SE-30 order of elution is primarily a function of molecular weight although the reversal of the order of the α-ketols (16α-hydroxyestrone and 16-keto estradiol) run as derivatives indicates some interaction between the 16 and 17 positions. The same reversal occurs on the selective QF-1 phase and, in addition, the expected retardation of the ketones is seen.

The foregoing discussion, admittedly incomplete, shows that enough is known about the chromatographic behavior of steroids to enable conditions for the separation of particular compounds to be achieved without excessive preliminary experimentation. Unfortunately, only qualitative predictions can be made, partly because of the complex nature of the chromatographic process and partly because of imperfections in some of the experimental work.

Reference has already been made to the problem of interaction with the support. Because of this problem it is quite certain that some of the

---

[37] H. H. Wotiz and S. C. Chattoraj, in "Gas Chromatography of Steroids in Biological Fluids" (M. B. Lipsett, ed.) p. 198. Plenum Press, New York, 1965.

TABLE VIII
RETENTION DATA FOR DERIVATIVES OF DISUBSTITUTED ANDROSTANES

Compound	Relative retention[a]							
	5β				5α			
	SE-30[b]	XE-60[b]	NGSeb[c]	HiEff-8B[d]	SE-30[b]	XE-60[b]	NGSeb[c]	HiEff-8B[d]
Acetates of androstane-								
3α,17β-diol	0.87	1.95	1.89	2.82	0.89	2.05	1.69	2.48
3β,17β-diol	0.86	1.90	1.66	2.42	0.98	2.21	2.20	3.18
3α-ol, 17-one	0.56	2.15	1.76	2.86	0.58	2.14	1.61	2.62
3β-ol, 17-one	—	2.98	—	—	0.66	2.44	2.00	3.44
Trimethylsilyl ethers of androstane-								
3α,17β-diol	0.61	0.50	0.35	0.37	0.58	0.49	0.31	0.28
3β,17β-diol	0.56	0.51	0.28	0.28	0.69	0.66	0.44	0.42
3α-ol, 17-one	0.46	0.94	0.72	1.03	0.47	0.87	0.58	0.74
3β-ol, 17-one	—	0.86	—	—	0.55	1.16	0.91	1.20

[a] All liquid phases 3%; [b] 260°; [c] 232°; [d] 238°.

TABLE IX
RELATIVE RETENTION DATA FOR ESTROGENS AND DERIVATIVES

Compound	SE-30				QF-1		
	Free	TMSi	TMSi	Acetate	Free	TMSi	Acetate
Estrone	0.96	0.76	0.78	0.69	1.62	2.08	1.00
Estradiol	1.00	1.00	1.00	1.00	1.00	1.00	1.00
2-Methoxyestrone	1.33	1.15	1.12	1.02	2.32	3.27	1.61
16α-Hydroxyestrone	1.41	1.44	1.35	1.43	2.27	2.30	2.39
16-Ketoestradiol	1.31	1.47	1.47	1.48	1.82	3.35	2.86
Estriol	2.06	2.03	1.82	1.85	2.66	1.78	2.19
16-Epiestriol	2.10	2.21	1.94	2.16	2.74	2.08	3.36
% Phase	0.75	0.75	3	3	1	1	4
Column temperature (°C)	205	205	239	239	195	195	208

reported values are erroneous. Moreover, in much of the work insufficient attention has been paid to the fundamentals of chromatography. Strictly speaking, retention measurements should be reported as specific retention volumes. That is, the measured retention volume should be corrected for the dead volume of the chromatograph, the pressure drop in the column, and the nonlinearity of the absorption isotherm and should be reduced to a standard temperature and unit weight of solvent. If, in addition, the temperature coefficient of the partition is reported, the solute-solvent pair is defined independently of the experimental conditions. Much of the published work reports only gross retention values. Although it is unlikely that the qualitative picture would be much changed, use of the correct procedure would give more meaningful values and would facilitate comparison of results from different sources.

The selection of experimental conditions must also take into account the impurities present in biological samples. Compounds to be measured, whose behavior is presumably known, must be separated from steroidal and nonsteroidal components whose behavior is not known. Selection of experimental conditions, although then largely a matter of trial and error, may be expedited in one of a number of ways. Some practical examples follow:

Thomas[38] reported that, in the estimation of urinary 17-ketosteroids, androsterone was consistently overestimated. He showed that a non-ketonic impurity traveling with androsterone could be separated by sub-

[38] B. S. Thomas, *in* "Gas Chromatography of Steroids in Biological Fluids" (M. B. Lipsett, ed.), p. 1. Plenum Press, New York, 1965.

stituting a mixed stationary phase (JXR + HiEff-8B) for the selective phase (HiEff-8B) originally used.

The retention times of compounds having different heats of solution will change at different rates as column temperature is changed. Thus, separation may be improved, or worsened, by changing column temperature. Lurie has used this principle for the separation of progesterone from impurities derived from sebum.[39]

In studies on pregnant rats, measurement of estrone and estriol were made by the method of Wotiz and Chattoraj.[40] Results are shown in Table X.

TABLE X

MEASUREMENTS OF ESTRONE AND ESTRIOL IN PREGNANT RAT URINE
BY THE METHOD OF WOTIZ AND CHATTORAJ

	Concentration (μg/24 hours)[a]			
	SE-30		QF-1	
Compound	Acetate	TMSi	Acetate	TMSi
Estrone	8.30	1.06	1.20	1.40
Estriol	1.69	1.40	1.58	1.50

[a] Collection from 9 animals.

It is obvious that estrone, run as the acetate on an SE-30 column, is grossly overestimated. This result which could not have been predicted, emphasizes the importance of running samples under several different sets of conditions so that the specificity of the method may be improved.

Qualitative and Quantitative Analysis

Specificity

Generally speaking, the detection systems used in gas chromatography are nonspecific. Thus it is necessary to rely on the separatory power of the chromatographic column to isolate the component of interest before it enters the detector. For methods of very high sensitivity, the resolution of the column is not sufficient and a considerable amount of preliminary separation is necessary so that the compound to be measured may be presented to the gas chromatograph in relatively pure form.

[39] A. O. Lurie and C. A. Villee, *J. Chromatog.* **21**, 113 (1966).
[40] H. H. Wotiz and S. C. Chattoraj, *Anal. Chem.* **36**, 1466 (1964).

However adequate the preliminary purification may be, the introduction of adventitious impurities may nullify the whole procedure. Virtually all the solvents used contain traces of high-boiling impurities that assume importance when comparatively large volumes of solvent are evaporated during the course of an analysis. Thus it is essential that all solvents used are first carefully purified. Normal purification procedures may not suffice. For example, Kirschner[41] has reported the presence of two impurities that obscure the testosterone heptafluorobutyrate peak and which even the most careful purification of the solvent failed to remove. We have experienced only occasional trouble from this source. The reason may be that we use a QF-1 column for the analysis whereas Kirschner employed SE-30. Accidental introduction of impurities from sources other than solvents may cause difficulty. Lurie et al.[42] found it necessary to take precautions against contamination of the sample by sebum from the fingers of the technician performing the analysis.

Unequivocal proof of the identity of the compound being measured is often difficult to obtain because of the extremely small amounts of material involved. In the more sensitive electron capture methods, nanograms or less are measured—quantities that are below the level of detection of even the most sensitive techniques for confirmatory analysis. Generally, the behavior of the compound in several different chromatographic systems is the only criterion of specificity available. Measurement of the compound sought as three different derivatives on three different columns will usually provide sufficient information as to the specificity of the measurement, particularly when considered in conjunction with the behavior of the compound during the preliminary separation.

Accuracy and Precision

Accuracy of the gas chromatographic procedure is difficult to measure since all the high-sensitivity detectors require calibration. Internal evidence suggests, however, that in a good chromatographic system losses are small. Precision of the gas chromatographic procedure is typically ±3%, which is quite adequate for biomedical work. Most of the errors in any method arise from the preliminary separation and are of the same order of magnitude as those encountered in any other method of biochemical analysis.

[41] M. A. Kirschner, private communication, 1967.
[42] A. O. Lurie, C. A. Villee, and D. E. Reid, J. Clin. Endocrinol. Metab. 26, 742 (1966).

Analytical Methods

URINARY STEROIDS

The speed and resolving power of the gas chromatograph have been used to advantage in the determination of steroids present in urine at comparatively high levels. Methods in successful routine use include those for the assay of estriol,[43] pregnanediol,[44] pregnanediol and pregnanetriol,[45] and 17-ketosteroids.[46, 47] These comparatively simple methods are inadequate for the determination of steroids at lower levels, and it becomes necessary to introduce further purification steps before gas chromatography. Sensitivity is gained, though at the expense of speed. The faster, less-sensitive methods have been adequately treated in the references cited and only selected higher sensitivity methods are described here.

Testosterone and Epitestosterone

Methods for the determination of urinary testosterone have been reported by Futterweit et al.,[48] Ibayashi et al.,[49] Brooks and Giuliani,[50] and Sandberg et al.[51] The recent work of DeNicola et al.[52] suggests that measurement of both testosterone and epitestosterone gives a better correlation between virilizing states and hormone excretion than does measurement of testosterone alone.

The method of Sparagana,[53] based on the procedure of Ibayashi, is described here. A recent paper by Vestergaard et al.[54] compares results obtained with several methods. Considerable variation occurs among the mean values obtained by these different procedures. It is particularly interesting that the values reported by Vestergaard are considerably lower than those reported by Ibayashi and by Sparagana. The reasons

[43] H. H. Wotiz and S. J. Clark, "Gas Chromatography in the Analysis of Steroid Hormones," p. 232. Plenum Press, New York, 1966.

[44] H. H. Wotiz and S. J. Clark, "Gas Chromatography in the Analysis of Steroid Hormones," p. 131, Plenum Press, New York, 1966.

[45] M. A. Kirschner and M. B. Lipsett, Steroids 3, 277 (1964).

[46] M. A. Kirschner and M. B. Lipsett, J. Clin. Endocrinol. Metab. 23, 255 (1963).

[47] I. S. Hartman and H. H. Wotiz, Steroids 1, 33 (1963).

[48] W. Futterweit, N. L. McNiven, L. Marcus, C. Lantos, M. Drosdowsky, and R. I. Dorfman, Steroids 1, 628 (1963).

[49] H. Ibayashi, M. Nakamura, S. Murakawa, T. Uchikawa, T. Tanioka, and K. Nakao, Steroids 3, 559 (1964).

[50] R. V. Brooks and G. Giuliani, Steroids 4, 101 (1964).

[51] D. H. Sandberg, N. Ahmad, W. W. Cleveland, and K. Savard, Steroids 4, 557 (1964).

[52] A. DeNicola, R. I. Dorfman, and E. Forchielli, Steroids 7, 351 (1966).

[53] M. Sparagana, Steroids 5, 773 (1965).

[54] P. Vestergaard, E. Raabo, and S. Vedso, Clin. Chim. Acta 14, 540 (1966).

for the discrepancy are not known, but obviously further investigation is required to resolve the question.

Method of Sparagana

Hydrolysis. One-twelfth of a 24-hour urine is extracted with ether and the extract is discarded. The pH of the urine is adjusted to 5.0 with 50% acetic acid. Three milliliters of 1 M acetate buffer (pH 5.0), 1 ml of chloroform, 500 units/ml of urine of β-glucuronidase, 0.05 g of streptomycin, and 20 mg of Δ^1-testosterone (internal standard) are added, and the mixture is incubated for 96 hours at 37°.

Extraction. The hydrolyzate is twice extracted with an equal volume of ether. The ether extract is twice washed with 0.1 volume of cold 1 N NaOH and twice with 0.1 volume of water. The extract is dried over anhydrous Na_2SO_4, and the solvent is removed in a rotary evaporator. The residue is transferred to a 30-ml centrifuge tube, and the solvent is evaporated in a gentle stream of nitrogen.

Thin-Layer Chromatography. The extract is applied to a 20 × 20 cm thin-layer plate coated with 0.5 mm of silica gel GF_{254}; testosterone and Δ^1-testosterone are used as markers. The chromatogram is developed in a saturated chamber for a distance of 10 cm from the origin. Dichloromethane–methanol (10:1) is the mobile phase. The plate is air-dried, and the markers are located under ultraviolet illumination. The sample area is removed and extracted twice with 15-ml portions of acetone.

Acetylation and Oxidation. The extract is dried under nitrogen and taken up in 0.2 ml of pyridine and 0.1 ml of acetic anhydride. The reaction tube is sealed, and the mixture is allowed to stand for 16 hours at room temperature. Two milliliters of 20% aqueous methanol is added, and the steroid acetates are extracted into 10 ml of dichloromethane. The dichloromethane solution is twice washed with 1-ml portions of water, and the solvent is removed under nitrogen. After addition of 0.5 ml of chromic acid solution (20 mg per 10 ml of 36% acetic acid), the oxidation is allowed to proceed for 1 hour at 45°. The steroid acetates are again extracted with 10 ml of dichloromethane; after washing twice with 1-ml portions of water, the solvent is evaporated under nitrogen.

Thin-Layer Chromatography (TLC). Chromatography of the acetates is carried out as described above, using testosterone and Δ^1-testosterone acetates as markers and benzene–acetone (200:25) as the mobile phase.

Gas Chromatography. The acetone extract from the thin-layer plate is transferred to a 3-ml conical centrifuge tube, the solvent is evaporated, and the residue is dissolved in 100 μl of acetone. A 3-μl aliquot of this solution is injected into the chromatograph.

Chromatographic conditions are; column: 12 feet packed with 3% SE-30 on 80–100 mesh acid-washed, silanized Gas Chrom P, temperature 220°; carrier flow rate, 30 ml/min; injector temperature, 290°; flame ionization detector.

Estrogens

The rapid but relatively crude procedure mentioned above is entirely satisfactory for the determination of estriol in pregnancy urine. Difficulties arise when attempts are made to apply this procedure to the determination of other estrogens or to the assay of low-titer urines. The refinements and modification of the original method to improve sensitivity and versatility well exemplify the approach to the general problem of method development.

The measurement of estrone and estradiol is not possible using the simple procedure, since gas chromatographic separation from interfering impurities is incomplete. Alumina chromatography of the acetylated sample extract provides sufficient purification so that all three of the classical estrogens can then be measured in both high- and low-titer urines.[55]

Extension of the method to include some of the more recently discovered metabolites, namely, 2-methoxyestrone, 16α-hydroxyestrone, 16-ketoestradiol, and 16-epiestriol, poses certain problems. First, the ring D α-ketols tend to isomerize during acidic hydrolysis, and it is therefore necessary to substitute enzymatic hydrolysis. Second, 2-methoxyestrone is only slightly soluble in alkali, and the ring D α-ketols are unstable in alkaline solution. Thus it is no longer possible to isolate the phenolic fraction by extraction with alkali. Instead, a relatively complex system of TLC must be employed to separate the desired compounds.

Thus modified, the method is useful for the analysis of second and third trimester pregnancy urines.[56] Difficulties are encountered in its application to nonpregnancy urines. Here it is necessary to take much larger urine samples, with the result that the TLC systems are overloaded and the estrogens are insufficiently separated from interfering substances. In order to confirm quantitative results, we have adopted the practice of analyzing each compound of interest as the free compound and in the form of two different derivatives on three different columns. If, from the nine results obtained, three are identical within the limits of experimental error and if no other single result is lower than these three, we consider

---

[55] H. H. Wotiz and S. C. Chattoraj, in "Gas Chromatography of Steroids in Biological Fluids" (M. B. Lipsett, ed.), p. 195. Plenum Press, New York, 1965.

[56] H. H. Wotiz and S. C. Chattoraj, Anal. Chem. 36, 1466 (1964).

the analysis to be valid.[57] The method described here is that of Wotiz and Chattoraj[58] for estrone, estradiol, and estriol in low-titer urines.

Method of Chattoraj and Wotiz

Hydrolysis and Extraction. One-fifth of a 24-hour urine is diluted with an equal volume of distilled water, 15 vol % of concentrated HCl is added, and the mixture is refluxed for 90 minutes. The cooled urine is transferred to a separatory funnel and extracted three times with ether (100, 50, 50 ml). The combined ether extract is washed once with 50 ml of 8% $NaHCO_3$. The estrogens are extracted from the ether layer by two 50-ml portions of 1 N NaOH. The alkaline solution is neutralized with HCl and its pH is adjusted to 9.5–10 by addition of solid $NaHCO_3$. The mixture is extracted once with an equal volume of ether and twice with half the volume. The combined ether extract is washed with 8% $NaHCO_3$ solution (5 ml/100 ml) and with water until the discard is neutral. Solvent is evaporated in a rotary evaporator and the residue is transferred to a test tube with a little dichloromethane.

Acetylation. The residue is dissolved in 1 ml of acetic anhydride–pyridine (5:1) and is maintained for 1 hour at 68°. Ten milliliter of water is added while the sample is stirred thoroughly with a glass rod. The sample is transferred to a separatory funnel and extracted twice with 10 ml and once with 5 ml of petroleum ether. The combined extracts are washed with 5 ml of 8% $NaHCO_3$ and then with 2-ml portions of water until neutral. The solution is allowed to stand for 10 minutes, and any water collected at the bottom of the funnel is removed.

Alumina Chromatography. Neutral, activity grade 1 alumina (M. Woelm, Eschwege, Germany) is deactivated by the addition of water (5 ml/100 g), shaken thoroughly to break up any lumps, and allowed to equilibrate overnight in a tightly stoppered bottle.

A 1 $\times$ 15 cm chromatographic column is partly filled with petroleum ether, and 3 g of deactivated alumina is added in a thin stream so that it is freed from air as it settles. The surface of the alumina is leveled by tapping and is covered with a few glass beads. Flow rate is adjusted to approximately 50 drops per minute.

In order to test the efficacy of the alumina, 5 μg of each of the three estrogens in 25 ml of petroleum ether are applied to a column and eluted as described below. Estradiol diacetate should be eluted completely in the second and third fractions of solvent No. 3. Estrone and estriol acetates

[57] H. H. Wotiz, S. C. Chattoraj, and J. L. Gabrilove, *J. Clin. Endocrinol. Metab.* **28,** 192 (1968).
[58] H. H. Wotiz and S. C. Chattoraj, *in* "Gas Chromatography of Steroids in Biological Fluids" (M. B. Lipsett, ed.), p. 195. Plenum Press, New York, 1965.

should be eluted completely in the first 15 ml of No. 4. If the alumina is properly deactivated, estradiol diacetate should be completely separated from estrone acetate and estriol triacetate.

The extract from acetylation is applied to another column, and elution is carried out with the following solvent mixtures, while 5-ml fractions are collected:

> (1) 25% benzene in petroleum ether 10 ml
> (2) 50% benzene in petroleum ether 15 ml
> (3) 75% benzene in petroleum ether 20 ml
> (4) Benzene   20 ml

The separate eluates containing the estrogens are evaporated to dryness in a rotary evaporator. Each residue is transferred to a 2-ml centrifuge tube with petroleum ether or hexane, and the solvent is removed in a stream of nitrogen.

Gas Chromatography. Each residue is dissolved in 20 μl of hexane and a 2 μl aliquot is injected into the chromatograph. Chromatographic conditions: column 6 feet $\times$ 4 mm i.d. glass, packed with 3% QF-1 on 80–100 mesh Diataport S, temperature 222°, flow rate 60 ml/min nitrogen; injector temperature: 260°; flame ionization detector.

Pregnanediol and Pregnanetriol

Hydrolysis of steroid conjugates excreted in the urine is a necessary preliminary to gas chromatography of the steroids. Whichever method of hydrolysis is chosen, problems arise. Acid hydrolysis often results in partial or complete destruction of the compound of interest, while enzyme hydrolysis may be too time-consuming for a routine procedure. Moreover, the problem of enzyme inhibition may cause incomplete cleavage from the conjugate moiety.

Chattoraj and Scommegna[59] have recently proposed a method in which some of the disadvantages of enzyme hydrolysis are eliminated. The steroid conjugates as well as free steroids are precipitated from 70% ammonium sulfate solution, thus achieving separation from inhibitors in the urine. The extract obtained after enzyme hydrolysis of the precipitate is sufficiently pure for direct gas chromatography.

Method of Chattoraj and Scommegna

Precipitation and Hyrolysis. A 20-ml aliquot of a 24-hour urine is placed in a 50-ml centrifuge tube and 14 g of ammonium sulfate is added. The tube is stoppered, shaken for 1–2 minutes, and allowed to stand for 3–5 minutes. The stopper is removed, and the tube is centrifuged at 3000 rpm for 45 minutes.

[59] S. C. Chattoraj and A. Scommegna, *Steroids* 9, 3, 327 (1967).

The supernatant is poured off and the precipitate is dissolved in 5 ml of 1 N NaOH. If necessary, additional 1-ml portions of NaOH are added until solution is complete. Eight milliliters of water and concentrated HCl are added to neutralize the solution, and the pH of the solution is adjusted to 5 with glacial acetic acid. Sufficient 1 M sodium acetate buffer is added to make the solution 0.1 M in buffer. Fifty units of "Ketodase" (Warner-Chilcott, Morris Plains, New Jersey) is added per milliliter and the reaction mixture is incubated overnight at 37°.

Extraction. The reaction mixture is extracted three times with equal volumes of ether. The ether extracts are combined and washed twice with 25 vol % of 1 N NaOH and then with water until the extract is neutral. The solution is dried with anhydrous sodium sulfate, and the ether is evaporated under reduced pressure. The residue is transferred to a conical centrifuge tube with acetone, and the solvent is evaporated under a stream of nitrogen.

Acetylation. Pyridine (0.2 ml) and acetic anhydride (0.2 ml) are added, and the residue is acetylated for 1 hour at 68°. The reagents are removed in a stream of nitrogen, and the sample is dissolved in 100 μl of acetone.

Gas Chromatography. Aliquots (2 μl) of the sample are injected into the gas chromatograph. Instrumental conditions are as follows: flame ionization detector, 280°; flash heater, 270°; column: 6 feet $\times$ 4 mm i.d. glass, packed with 3% SE-30 on 80–100 mesh Diatoport S, 250°; flow rate, 50 ml/minute of N_2.

Corticosteroids

Gas chromatography of the corticosteroids has proved to be more difficult than that of any other group of steroids. Consequently, although much effort has gone into the development of methodology, only a few procedures are applicable.

Generally, the 17α-hydroxylated compounds undergo side-chain cleavage in the gas chromatograph and emerge as the corresponding 17-ketosteroids. Attempts to develop methods based on degradation in the chromatograph have not led to satisfactory results, for two main reasons: conversion to the cleavage product is far from complete and more than one metabolite can lead to the same product. Far more satisfactory are those methods where the compounds are oxidized after preliminary separation, and the reaction products are then chromatographed. This approach is best exemplified by the method of Bailey.[60]

---

[60] E. Bailey, *in* "Gas Chromatography of Steroids in Biological Fluids" (M. B. Lipsett, ed.), p. 57. Plenum Press, New York, 1965.

Although it was early demonstrated[61] that the acetates of at least some 17α-hydroxy compounds could be chromatographed intact, sorption on the column severely limited sensitivity. More recently, Rosenfeld[62] has developed a successful method for the isomeric cortols and cortolones[62a] in which the compounds are chromatographed as their trimethylsilyl ethers.

Method of Rosenfeld

Hydrolysis and Extraction. Twenty to fifty percent of a 24-hour urine is taken, and the pH is adjusted to 5 with sulfuric acid. Ten volumes percent of $1 M$ acetate buffer (pH 5) and 300 units/ml of beef liver glucuronidase are added. The mixture is incubated for 120 hours at $37°$. The hydrolyzate is continuously extracted with diethyl ether for 48 hours, the ether solution is washed once with $2 N$ NaOH and three times with 5% NaCl solution, and finally evaporated to dryness in a rotary evaporator.

The ketonic fraction is removed from the residue as follows: For each 100 mg of neutral extract, 2 ml of ethanol, 100 mg of Girard's Reagent T, and 0.1 ml of glacial acetic acid are added and the mixture is refluxed for 2 hours. Ice is added, followed by 0.75 ml of 10% NaOH solution. The cold suspension is twice extracted with 150-ml portions of ether, and the extract is washed twice with 25-ml portions of water. After drying the extract over anhydrous Na_2SO_4, the ether is evaporated.

Thin-Layer Chromatography. The extract is applied to a 20×20 cm plate coated with 0.5 mm of silica gel G. Cortol and cortolone are used as markers. The chromatogram is developed for a distance of 10 cm using ethyl acetate–cyclohexane (7:3) as the mobile phase. The plate is air-dried, and the markers are located by spraying with phosphomolybdic acid. The sample area is removed and extracted overnight with methanol in a Soxhlet apparatus. The solvent is removed *in vacuo*.

Derivative Formation. The residue is transferred with a little methanol to a 3-ml centrifuge tube, and the methanol is removed in a stream of nitrogen. Eight-tenths milliliter of pyridine, 70 μl of hexamethyldisilazane, and 16 μl of trimethylchlorosilane are added and mixed. The tube is sealed with a Teflon cap and allowed to stand overnight at room temperature. The reagents are removed in a stream of nitrogen at $60°$ and 0.5–2.0 ml of internal standard solution (0.25 mg androstane-3,11,17-trione per milliliter of chloroform) is added, the amount depend-

---

[61] H. H. Wotiz, I. Naukkarinen, and H. E. Carr, *Biochim. Biophys. Acta* **53**, 449 (1961).

[62] R. Rosenfeld, *Steroids* **4**, 147 (1964).

[62a] Cortols: 5β-pregnane-3α,11β,17α,20α(or β), 21-pentol; Cortolones: 3α,17α,20α(or β), 21-tetrahydroxy-5β-pregnan-11-one.

ing upon the expected concentration of $C_{21}O_5$ steroids. The contents are mixed thoroughly, and the tube is centrifuged for 5 minutes to precipitate ammonium chloride formed during derivative formation.

Gas Chromatography. A 2–8-μl aliquot of the chloroform solution is injected into the chromatograph. Chromatographic conditions: column, 6 feet $\times$ 4 mm i.d. packed with 3% QF-1 on 100–140 mesh Gas Chrom P; temperature 231°; carrier gas pressure, 30 psi.

Tetrahydroaldosterone

Urinary aldosterone is presently estimated almost exclusively by double isotope derivative methods. These are the only procedures offering sufficient sensitivity and specificity to measure accurately the small concentrations present in normal and low-titer urines. All these methods, however, require extensive and tedious preliminary purification by paper and column chromatography and are, as a result, very time consuming.

The higher separation efficiency of gas chromatography permits simplification of the isolation procedure. Kliman[63] has shown that application of the technique to the assay of urinary aldosterone results in a considerable saving of time.

The concentration in urine of the metabolite $3\alpha,5\beta$-tetrahydroaldosterone is higher than, but directly proportional to, that of aldosterone. The method[64] for the measurement of tetrahydroaldosterone described below gives results comparable with those obtained by double isotope derivative methods but requires considerably less time. The method has been applied to the measurement of the urinary excretion of tetrahydroaldosterone.

Materials and Methods. The techniques for paper and column chromatography have been described previously.[65]

Plates for thin-layer chromatography are 5 cm wide and are coated with 0.2–0.7 mm of silica gel GF (E. Merck AG). Plates for chromatography immediately preceding gas chromatography are coated with MN-silica gel G-HR (Macherei-Nagel & Co). To remove impurities affecting the response of the electron-capture detector, they are washed by allowing redistilled ethanol to ascend to their upper edges.

Radioactive steroids are located with a model RSC-363 Scanner (Atomic Accessories). The silica gel to be extracted is transferred to disposable Pasteur pipettes containing a plug of cotton wool. Elution is accomplished with three aliquots of acetone.

[63] B. Kliman, *in* "Gas Chromatography of Steroids in Biological Fluids" (M. B. Lipsett, ed.), p. 101. Plenum Press, New York, 1965.
[64] G. L. Nicolis, H. H. Wotiz, and J. L. Gabrilove, *J. Clin. Endocrinol. Metab.,* in press.
[65] G. L. Nicolis and S. Ulick, *Endocrinology* **76,** 514 (1965).

Analytical gas chromatography is done with a Beckman GC5 instrument. A 12-foot glass column is packed with 3% JXR on Gas-Chrom Q, 100–120 mesh (Applied Science Laboratories) or with a 3.8% SE-30 on Diatoport S, 80–100 mesh (F & M Scientific). Column temperatures are 230°–240° for the JXR phase and 265°–285° for SE-30.

The electron capture detector (Beckman No. 134061) is used with CO_2 flow and polarizing voltage adjusted to peak standing current reading. Detector temperature is 300°.

Carrier gas is ultra high purity helium (The Matheson Co.) at 20 p.s.i. Lower grade brands from the same manufacturer are less satisfactory because of considerable decreases in standing current and higher noise levels.

Acetic anhydride-1-^{14}C (5 mC/mmole; New England Nuclear Corp.) is diluted 1:10 with carrier and stored as a 20% solution in benzene. The specific activity is measured by reaction with 0.5 mg of 11-deoxycorticosterone. The steroid acetate is purified chromatographically to constant specific activity and measured by the blue tetrazolium reaction and by the absorption of an ethanolic solution at 240 mμ.

d-Aldosterone-1,2-^{3}H (32.2 C/mmole; New England Nuclear Corp.) is dissolved in ethanol and injected into multidose vials containing 30 ml of water. Aliquots for intravenous use and for analysis are withdrawn with a syringe.

Tritium and carbon-14 are counted in a Nuclear Chicago Mark 1 liquid scintillation counter. To prevent adsorption of trace amounts of steroids to the walls of the counting vials, 1% ethanol is added to the toluene-based scintillator solution containing 0.4% 2,5-diphenyloxazole (PPO) and 0.01% 1,4-bis-2-(phenyloxazolyl)benzene (POPOP).

Solvents are reagent grade and are redistilled. The monochlorodifluoroacetic anhydride (Pierce Chemical Co.) is vacuum distilled and stored in a desiccator.

Preparation of 3α,5β-Tetrahydroaldosterone. Livers are removed from Sprague-Dawley rats (150–250 g), chilled, and homogenized in twice their weight of 0.1 M sodium phosphate buffer (pH 7.4). The homogenate is centrifuged for 15 minutes at 10,000 g, and the supernatant layer is centrifuged for an additional 45 minutes at 100,000 g. The supernatant fraction containing the 5β-reductases is made into an acetone powder and stored in a desiccator.

Trace amounts of d-aldosterone-1,2-^{3}H as well as d-aldosterone carrier (10–100 mg) are incubated for 3 hours at 37° as described by Tomkins[66] using approximately 1 g of the acetone powder for 10 mg of substrate.

---

[66] G. M. Tomkins, see Vol. V, p. 499.

The incubation media are extracted with dichloromethane, and the extracts are chromatographed on a thin-layer plate with acetone–benzene 6:4. Scanning of the plates shows the presence of a single radioactive, blue tetrazolium-reducing zone moving at 0.7 the rate of the unreacted aldosterone. Yields of the reduced steroid range from 35% to 85%.

The tetrahydroaldosterone zone is rechromatographed on paper for 10 hours (ethyl acetate–toluene 7:3/formamide), and the area running like authentic $3\alpha,5\beta$-tetrahydroaldosterone is eluted.

When the entire blue tetrazolium-reducing zone on this chromatogram was eluted and oxidized with HIO_4, gas chromatographic analysis showed the presence of two minor peaks with retention times about two-thirds that of the γ-lactone of $3\alpha,5\beta$-tetrahydroaldosterone. Both compounds had γ-lactone absorption bands at 5.6 μ. They were not gas chromatographic artifacts since they were not present when pure γ-lactone of $3\alpha,5\beta$-tetrahydroaldosterone was chromatographed. Their acetylation products had retention times different from the acetates of the γ-lactones of $3\beta,5\alpha$- and $3\alpha,5\beta$-tetrahydroaldosterone. The two impurities were absent when the tail portion of the tetrahydroaldosterone band on the paper chromatogram was not included in the elution.

The enzymatically prepared tetrahydroaldosterone-1,2-^{3}H is mixed with reference $3\alpha,5\beta$-tetrahydroaldosterone, and the mixture is chromatographed twice, acetylated with acetic anhydride-^{14}C, and again chromatographed twice. Constancy of the specific activity indicates chromatographic homogeneity. After hydrolysis with dilute HCl the ^{3}H/^{14}C ratio changes, as expected, from the removal of the ^{14}C-acetate radical at C-18. The enzymatically prepared $3\alpha,5\beta$-tetrahydroaldosterone-1,2-^{3}H and the enzymatically prepared nonradioactive $3\alpha,5\beta$-tetrahydroaldosterone have been shown to be homogeneous by the same criteria.

Periodic Acid Oxidation. To obtain a derivative with better gas chromatographic properties, $3\alpha,5\beta$-tetrahydroaldosterone is oxidized to its γ-lactone with 0.1 M HIO_4.[67] The oxidation mixture is extracted with dichloromethane and purified by thin-layer (ethyl acetate–benzene 1:1) and paper chromatography (methylcyclohexane–toluene 1:1/formamide).

The γ-lactone is measured by reaction with acetic anhydride-^{14}C and by chromatography of the acetate on thin layer (benzene–ethyl acetate 9:1) and paper (methylcyclohexane–formamide). Specific activities agree within 5% with those of the aldosterone and tetrahydroaldosterone precursors.

Monochlorodifluoroacetate of the γ-Lactone. To increase the response of the electron capture detector the γ-lactone is converted to its monochlorodifluoroacetate. The steroid as a 0.1% solution in tetrahydrofuran

[67] S. Ulick and K. K. Vetter, *J. Biol. Chem.* **237**, 3364 (1962).

is reacted with excess (250:1, w:w) monochlorodifluoroacetic anhydride at 60° for 45 minutes. The excess reagent is removed under a stream of nitrogen, and the mixture is chromatographed on a thin-layer plate (benzene–ethyl acetate 9:1).

The monochlorodifluoroacetate runs as a single sharp peak on the JXR and SE-30 columns with retention times relative to the unreacted γ-lactone of 1.25 and 1.37, respectively. The smallest amount of mono-chlorodifluoroacetate which can be measured is about 0.2 nanogram.

When stored as a 0.001% solution in benzene at 4° the compound is stable for several months.

Determination of Urinary Tetrahydroaldosterone. To measure urinary excretion, $3\alpha,5\beta$-tetrahydroaldosterone-1,2-³H, 1×10^5 cpm (less than 0.5 ng) is injected intravenously and the urine is collected for 24 hours.

Urine (1/10–1/20 of a 24-hour collection) is incubated with 500 units/ml of Ketodase (Warner-Chilcott Co.) for 48 hours at 37° and pH 5 and then is extracted with ethyl acetate. The ethyl acetate is washed three times with 0.1 volumes of $0.2\,N$ NaOH and of water, and evaporated under reduced pressure.

The extracts are chromatographed on 0.7 mm-thick silica gel plates (acetone–benzene 6:4) as 5-cm lines. The radioactive zones are located for elution by scanning. Eluates are oxidized with periodic acid as described above except that the reaction time is reduced to 3 hours and that the methylene chloride extract is washed three times with 0.2 volumes of $1.0\,N$ NaOH and of water.

The oxidation mixtures are chromatographed as lines 5 cm long on 0.25 mm-thick plates (ethyl acetate–benzene 6:4) and the γ-lactone is located by scanning.

The γ-lactone is eluted into 2-ml clinical stoppered tubes using a total volume of 1.5 ml of acetone. After evaporation of the solvent 0.05 ml of tetrahydrofuran and 0.02 ml of monochlorodifluoroacetic anhydride are added. The stoppers are sealed with 2–3 drops of tetrahydrofuran and covered tightly with parafilm (American Can Co.) After 45 minutes at 60° on a heating block, the excess reagent is evaporated under a stream of nitrogen with the aid of 2–3 small additions of benzene.

Because of the excess of halogenated reagent remaining after evaporation, the acetylation mixtures are unsuitable for direct gas chromatography. The residues are chromatographed as small spots on 0.2 mm-thick plates previously washed with ethanol (benzene–ethyl acetate 9:1).

These plates are stored in a desiccator and eluted within 24 hours. Gas chromatography of γ-lactone monochlorodifluoroacetate samples eluted from thin-layer plates that have stood in the open for 3 weeks show partial (25–40%) hydrolysis to the unacetylated compound. This is not observed when the thin-layer plates are stored in a desiccator for a week or less. Eluted samples show no evidence of hydrolysis over periods of up to 10 days when stored in a desiccator containing P_2O_5.

The zone corresponding to the γ-lactone derivative is eluted into a 2-ml conical stoppered tube using about 0.5 ml acetone. Accurate location of the radioactive compounds to be eluted is aided by spotting a radioactive marker close to them and by using a 2 mm-wide collimator.

After evaporation of the acetone, the samples are stored for 1 hour or more in an evacuated desiccator. This is found to reduce considerably the size of the solvent front recorded by the electron capture detector.

The samples are dissolved in 0.1–2.0 ml of ethyl acetate. Identical volumes (3–6 μl) are then injected into the counting vials and into the gas chromatograph. The reproducibility of both injections, as measured by peak height and radioactivity, is better than 5%. About 4 minutes after the injection of each sample, 0.002 μg of γ-lactone monochlorodifluoroacetate standard is injected; the amount of the unknown is calculated from the relative peak heights. Standard curves are obtained at the beginning and at the end of each batch of 6–18 samples and are found to agree within 10%. When the peak height of the unknown falls in the nonlinear region of the standard curve, the analysis is repeated following dilution of the sample.

PLASMA STEROIDS

Although plasma samples generally have fewer inherent impurities than urine, steroid levels are lower and much less sample is available. Thus only the most sensitive and elaborate procedures are of any value for the assay of plasma steroids.

Testosterone

One of the earlier methods for the gas chromatographic determination of testosterone in plasma was that of Guerra-Garcia et al.[68] However, this method, although rapid, is insufficiently sensitive for the analysis of plasma from females. The need for more sensitivity led to the use of the electron-capture detector. In the method of Brownie et al.[69] testosterone is chromatographed and measured as the chloroacetate, and in that of Sarda et al.[70] as the heptafluorobutyrate. In view of the current interest in epitestosterone levels in urine, it is noteworthy that Sarda has failed

[68] R. Guerra-Garcia, S. C. Chattoraj, J. L. Gabrilove, and H. H. Wotiz, Steroids 2, 605 (1963).
[69] A. C. Brownie, H. J. vanderMolen, E. E. Nishizawa, and K. B. Eik-Nes, J. Clin. Endocrinol. Metab. 24, 1091 (1964).
[70] I. R. Sarda, unpublished results, 1967.

to detect this compound in plasma even under conditions of the highest sensitivity.

Method of Brownie et al.[69]

Materials: Solvents and Reagents

Ether, reagent grade, distilled through a 60-cm fractionating column

Methanol, anhydrous, analytical reagent, distilled from 2,4-dinitrophenylhydrazine

Benzene, thiophene-free, analytical reagent, distilled through a 60-cm fractionating column

Toluene, analytical reagent grade

Chloroform and ethyl acetate, analytical reagent grade, distilled through a 60-cm fractionating column

Cyclohexane, reagent grade

Tetrahydrofuran, refluxed for 3 hours with potassium hydroxide pellets and distilled. It is then distilled off sodium and stored over sodium in a dark brown bottle.

Pyridine, refluxed over barium oxide for 4–6 hours and distilled through a fractionating column. The fraction boiling at 115° is collected and stored over calcium chloride in a desiccator.

Monochloroacetic anhydride, dried and stored in a desiccator

2,5-Diphenyloxazole (PPO) and 1,4-bis-[2-(5-phenyloxazolyl)] benzene (POPOP), scintillation grade, used without further purification

Silica gel G (Merck, according to Stahl, for thin-layer chromatography), washed with boiling methanol after the addition of a phosphor (Dupont luminescent chemical, index 609), 30 mg/100 g. The washed silica gel is dried at 100° for 24 hours before use.

Testosterone-1,2-³H, specific activity ~150 μC/μg. A solution (ca. 2500 cpm/ml) of testosterone-1,2-³H in methanol is stored at 5°

Testosterone, recrystallized from hexane with a few drops of methanol to constant melting point

Testosterone chloroacetate. One gram of chloroacetic anhydride and 0.2 ml of pyridine are added to 100 mg of testosterone in 5 ml of tetrahydrofuran. After the solution has stood in the dark for 18 hours, 5 ml of water is added and the solution is extracted 3 times with 5 ml of ethyl ether. The ether extract is washed once with 5 ml of 6 N hydrochloric acid and twice with 5 ml of water. The washed ether extract is filtered through anhydrous sodium sulfate and taken to dryness. The residue is recrystallized several times from aqueous acetone to give crystals with melting points of 124–125° (corrected).

Cholesterol chloroacetate, prepared by the method described for testosterone chloroacetate. From the ether extract a white crystalline residue is obtained which is recrystallized several times from aqueous acetone to give crystals melting at 159–160° (corrected).

Thin-Layer Chromatography. This process is carried out on silica gel plates. The plates are prepared with a spreader, giving layers of approximately 0.30 mm thickness; about 3 g of silica gel and 9 ml of water are needed to prepare each 20×20 cm plate. The plates are dried at 100° for at least 90 minutes before use. Extracts are chromatographed on 2-cm lanes separated from each other by a 1-cm lane and from standards by a 1.5-cm lane. This arrangement allows 4 extracts and 2 standards to be run on each plate. Samples are applied to thin-layer plates with chloroform–methanol (1:1 v/v) using successively 4, 3, and 2 drops of this solution. Chromatograms are developed in an ascending manner. Steroid standards on the chromatograms are detected using a Haines fluorescent scanner.

Assay of Radioactivity. A liquid scintillation spectrophotometer set to give approximately 25% counting efficiency is used to assay ³H. Samples for assay are dried and dissolved in 10 ml of scintillation fluid (4 g of PPO and 40 mg of POPOP in 1-liter of toluene) in 20-ml glass vials.

ESTIMATION OF TESTOSTERONE IN HUMAN BLOOD

Preparation of Blood Plasma. Whole blood is withdrawn into a bottle containing heparin and the plasma is obtained by centrifugation. A solution of testosterone-1,2-³H (2500 cpm) is taken to dryness in a 50-ml glass tube, and 10 ml of plasma is added (5 ml plasma plus 5 ml 0.9% sodium chloride in the case of male plasma).

Extraction. Twenty percent sodium hydroxide, 0.25 ml, is added to the plasma, which is extracted 3 times with 20 ml of diethyl ether. The pooled ether extracts are washed twice with 5 ml of distilled water and concentrated under nitrogen at 40°. The extract is then transferred to a 15-ml conical centrifuge tube, taken to dryness, and then concentrated in the tip of the bottle.

Thin-Layer Chromatography of Plasma Extract. Thin-layer chromatography is carried out on silica gel plates as described. An ascending chromatogram is run in the solvent cyclohexane–ethyl acetate (1:1 v/v). Standards of testosterone (0.5 μg) are chromatographed on separate lanes. At this stage of purification, testosterone in plasma extracts runs slower than testosterone standards; consequently, the area of silica gel scraped off from the plasma extract lanes extends from just beyond the testosterone standard to a point 1 cm nearer the origin than the standards.

Three drops of water are added to the silica gel, and this is then extracted with 1 ml of methanol 3 times. The combined extracts are taken to dryness in a 15 by 150-mm test tube.

Monochloroacetylation. After the extract is dried in a vacuum desiccator, 0.5 ml of a solution of monochloroacetic anhydride in tetrahydrofuran (100 mg/10 ml) and 0.1 ml of pyridine is added. The reaction is carried out overnight in a darkened desiccator. One milliliter of water is added to stop the reaction, and the solution is extracted three times with 1 ml of ethyl acetate. The combined ethyl acetate extracts are washed once with 1 ml of 6 N hydrochloric acid and twice with 1 ml of distilled water and taken to dryness in a 15-ml centrifuge tube.

Thin-Layer Chromatography of Testosterone Monochloroacetate. The extract from the chloroacetylation is chromatographed on silica gel plates using the solvent system benzene–ethyl acetate (4:1 v/v). Standards of authentic testosterone chloroacetate are run and the corresponding areas from the extracts are scraped off. After the addition of 4 drops of water, the silica gel is extracted three times with 1 ml of benzene. High-speed centrifugation is used to obtain a benzene layer free from silica gel. The benzene extracts are transferred to 2.0-ml conical microcentrifuge tubes and dried after each extraction.

Addition of Internal Standard and Sampling for ^3H Counting. The dried extracts in the microcentrifuge tubes are dissolved in 1 ml of methanol containing 0.4 μg/ml cholesterol monochloroacetate. One-tenth aliquot of this mixture is taken for ^3H assay and the remainder is evaporated to dryness.

Gas Chromatography. The dried extract is dissolved in 15 μl of toluene, and as much as possible (usually 8–10 μl) of this solution is injected into a 3-foot gas chromatographic column.

The stationary phase for this column is prepared as described by DePaoli *et al.*,[71] using 1 g of XE-60 dissolved in 100 ml of toluene for 25 g of purified deactivated Gas Chrom P.

The column temperature is kept at 210° with the detector at 220° and the flash heater at 250°. High-purity nitrogen is used as carrier gas and led through a tube filled with molecular sieve (type 13X from Linde); the gas pressure is kept at 40 psi. The electron-capture detector is operated under conditions of maximal sensitivity and adequate linear range. Quantitation is carried out by area measurement of the peak according to the following formula, which allows for the various losses incurred, and by comparison to the added internal standard:

$$S(\mu g) = \frac{X_s}{10X_p} \times \frac{C_s}{C_x} \times \frac{T_x}{T_s} \times \frac{288.4}{364.5} \times 0.01$$

[71] J. C. DePaoli, E. E. Nishizawa, and K. B. Eik-Nes, *J. Clin. Endocrinol. Metab.* 23, 81 (1963).

where X_s is cpm of testosterone-1,2-^{3}H initially added to the plasma, X_p is cpm of ^{3}H in the aliquot prior to gas chromatography, C_s is area (cm^2) of 0.4 μg cholesterol monochloroacetate, C_x is area (cm^2) of cholesterol monochloroacetate in sample, T_s is area (cm^2) of 0.01 μg testosterone monochloroacetate, T_x is area (cm^2) of testosterone monochloroacetate in sample, 288.4 is molecular weight of testosterone, and 364.5 is molecular weight of testosterone monochloroacetate.

Further correction for the mass measurement of a water blank containing an equal amount of ^{3}H-testosterone and carried through the full procedure is also necessary. However, it should be noted that this latter correction can be eliminated because of the availability of very high specific activity testosterone (150 μC/μg).

Progesterone

The application of gas chromatographic techniques has considerably improved the sensitivity and speed of progesterone assays. Using the flame ionization detector and injecting the total sample in dry form, Yannone et al.[72] and Lurie et al.[73] report good results for both pregnant and nonpregnant females. vanderMolen and Groen[74] have developed an electron capture method in which progesterone is first reduced enzymatically to 20β-hydroxypregn-4-en-3-one. The electron capturing monochloroacetate is utilized to provide sufficient sensitivity. More recently, the same workers[75] have investigated the heptafluorobutyrates of the same compound.

Exley and Chamberlain[76] have recently reported that the enol esters of Δ^4-3-ones can be formed directly by reaction with heptafluorobutyric anhydride. However, our experience is that reaction conditions are critical and that much more investigation of the reaction is necessary before this approach can yield reliable quantitative results.

Method of vanderMolen and Groen[74]

Materials

Ether, reagent grade, distilled through a 60-cm fractionating column
Methanol, anhydrous analytical reagent, distilled from 2,4-dinitrophenylhydrazine

[72] M. E. Yannone, D. B. McComas, and A. Goldfien, *J. Gas Chromatog.* **2**, 30 (1964).
[73] A. O. Lurie, C. A. Villee, and D. E. Reid, *J. Clin. Endocrinol. Metab.* **26**, 742 (1966).
[74] H. J. vanderMolen and D. Groen, *in* "Gas Chromatography of Steroids in Biological Fluids" (M. B. Lipsett, ed.), p. 153. Plenum Press, New York, 1965.
[75] H. J. vanderMolen, private communication, 1967.
[76] D. Exley and J. Chamberlain, *Gas Chromatog. Congr. Rome*, 1966.

Benzene, thiophene-free, analytical grade reagent. This solvent is repeatedly washed with concentrated sulfuric acid followed by water. It is then distilled twice through a 100-cm Vigreux column.

Ethyl acetate, analytical grade, distilled through a 60-cm fractionating column

Tetrahydrofuran, refluxed for 3 hours with potassium hydroxide pellets and distilled. It is then distilled from sodium and stored over sodium in a dark brown bottle.

Pyridine, refluxed over barium oxide 4–6 hours and distilled through a fractionating column. It is stored over calcium chloride in a desiccator.

Chloroacetic anhydride, dried and stored in a desiccator

Scintillation fluid, 2,5-diphenyloxazole (PPO) and 1,4-bis-[2-(5-phenyloxazolyl)]benzene (POPOP), scintillation grade. Used without further purification

Silica gel TF 254 (Merck, according to Stahl, containing ultraviolet fluorescence indicator), used without additional purification

Phosphate buffer, 0.15 M pH 5.2, prepared by mixing 97.5 ml of 0.15 M potassium dihydrogen phosphate with 2.5 ml of 0.15 M disodium hydrogen phosphate solution. To 100 ml of this buffer 100 mg of EDTA is added

Tris(hydroxymethyl)aminomethane(Tris) buffers

0.1 M pH 8.1 without EDTA. 50 ml of an 0.1 M solution of Tris and 26 ml of 0.1 N HCl are mixed and diluted with distilled water to make 100 ml total volume

0.005 M at pH 8.2 with EDTA. A 0.005 M buffer solution is prepared by dissolving 300 mg of Tris in 500 ml of distilled water. The pH is adjusted to 8.2, using a meter, by addition of 0.01 N HCl. To 100 ml of this buffer solution is added 100 mg of EDTA to make a final concentration of 2.7 $\times$ 10^{-3} M

Cofactor solution. DPNH, 5 mg, is dissolved in 3 ml of 0.1 M Tris buffer at pH 8.1 without EDTA. The solution can be stored at 4° for 1 week

Enzyme solution. The concentrated suspension of 20β-hydroxy steroid dehydrogenase prepared from *Streptomyces hydrogenans* (Boehringer, Mannheim, Germany) is stable for several months if stored at 4°. For use, a volume corresponding to approximately 4 mg of protein per milliliter of the concentrated suspension is diluted with 4 volumes of 0.005 M Tris buffer at pH 8.2 containing EDTA

7-^{3}H-Progesterone. Tritiated progesterone with a specific activity of approximately 10 C/millimole is used. Prior to the use of tritiated progesterone, thin-layer chromatography in the systems benzene–ethyl acetate 1:1, cyclohexane–ethyl acetate 1:1, and benzene–ethyl acetate 4:1 is used to test the purity of the material Testosterone chloroacetate prepared as follows: A solution containing 1 g of chloroacetic anhydride, 0.2 ml of pyridine, and 100 mg of testosterone in 5 ml of tetrahydrofuran is left standing for 18 hours in the dark. Then 5 ml of water is added and the derivative is extracted 3 times with 5 ml of ethyl ether. The combined ether extracts are washed with 5 ml of distilled water. After the ether extract is filtered through anhydrous sodium sulfate, it is evaporated to dryness. After several recrystallizations of the amorphous residue, crystals with a melting point of 124–125° (corrected) may be obtained.

Thin-Layer Chromatography. Silica gel plates (20 × 20 cm) are prepared to a thickness of approximately 0.30 mm. A mixture of 3 g silica gel in 9 ml of water is used. Plates are activated at 100° for 90 minutes prior to use. Samples are applied from chloroform–methanol (1:1) solution in 2-cm streaks, separated from each other by 1 cm. After ascending development, the standards are detected by means of a Haynes fluorescence scanner and the areas corresponding to the appropriate standard are scraped from the plate with a spatula. *Caution:* Since only trace amounts of steroids are being analyzed, clean tanks and fresh solvent mixtures must be used for each chromatographic plate.

Gas-Liquid Chromatography. Columns approximately 3 ft long and 0.4 cm i.d. are filled with 80–100-mesh Gas Chrom P coated with 1% XE-60 polymer. Chromatography is carried out at between 215° and 220° with the flash heater operated at 250°. The flow rate of carrier gas (nitrogen or helium) should be set to approximately 75 ml/min. Argon with 10% methane at a flow rate of 225 ml/min is used as a purge gas for this specific detector. Injections are made using a 10-μl Hamilton syringe. Quantification is carried out by determination of peak areas.

Measurement of Radioactivity. The samples to be analyzed are evaporated in glass vials and the residues are dissolved in 0.1 ml of ethanol to which 10 ml of scintillation fluid is added. The samples are then counted in a liquid scintillation spectrometer for a sufficiently long time to allow a precision of between 1 and 2%.

Glassware. Because of the extremely low concentration of steroids to be analyzed, particular care must be taken in the cleaning and preparation of glassware to prevent elution of unrelated material and adsorption of progesterone on the glass surface.

Glassware is soaked overnight in chromic acid and rinsed with tap water followed by soaking in detergent, rinsing again with tap water, followed by soaking overnight in dilute HCl, rinsing 10 times with tap water and 10 times with deionized water. The tubes are finally dried at room temperature. The 2-ml tubes used for collection of samples for GLC are siliconized with a 5% solution of dimethyldichlorosilane in benzene.

Method of Assay. Ten milliliters of plasma is added to a 50-ml glass tube containing approximately 10,000 cpm of 7-^{3}H-progesterone. After the plasma is treated with 0.25 ml of 20% sodium hydroxide solution, the mixture is extracted 6 times with 15 ml of ether. The pooled ether extracts are washed twice with 5 ml of water and evaporated to dryness; the residue is chromatographed on a silica gel thin-layer plate in benzene-ethyl acetate (2:1). The area corresponding to authentic progesterone is scraped off and extracted with 95% methanol (3 × 1 ml). After evaporation of the methanol and its concentration in the tip of a 15-ml centrifuge tube, the residual material is dissolved in 1 drop of ethanol, and 0.5 ml of the 0.1 M phosphate buffer at pH 5.2, as well as 0.03 ml of the cofactor solution and 0.03 ml of the diluted enzyme solution are added. The contents are thoroughly mixed and incubated for 2 hours at 37°. Then 1 ml of distilled water is added and the solution is extracted with 4 × 1 ml of ethyl acetate. The combined extract is again evaporated under nitrogen, and the tubes are put in a vacuum desiccator for about 3 hours. To this residue is then added 0.5 ml of a solution of monochloroacetic anhydride in tetrahydrofuran (10 mg/10 ml) and 0.1 ml of pyridine, and the tube is replaced in the desiccator overnight. The reaction is stopped with 1 ml of distilled water, and the steroid chloroacetate is extracted three times with 1 ml of ethyl acetate. The combined extracts are washed once with 1 ml of 6 N HCl and twice with 1 ml of distilled water, after which the extract is evaporated to dryness. After application of the residue to a silica gel thin-layer plate and development in a benzene–ethyl acetate system (6:1), the area corresponding to the standard of 20β-hydroxy-pregn-4-en-3-one chloroacetate is scraped off and 1 ml of benzene is added to the silica gel. After the extract is mixed thoroughly, 0.02 ml of distilled water is added, agitation is repeated, and the immiscible layer is cleared by centrifugation. The benzene layer is transferred to a 2-ml siliconized conical centrifuge tube and dried down after each of three extractions. This residue is then dissolved in 1 ml of methanolic testosterone chloroacetate and the contents of the tube thoroughly mixed. (During the luteal phase of the menstrual cycle 0.04 μg of testosterone chloroacetate is added; in instances where less progesterone would be expected, 0.01 μg of testosterone is added.) One-tenth of the diluted material is now removed for radioactive counting, and the remainder is again taken to dryness, care being taken to concentrate the material in the tip of the tube. The residue thus obtained is dissolved in 10 or 20 μl of benzene (depending on the amount of internal standard added), and 5 μl of the extract is injected into the gas chromatograph.

The amount of progesterone in the plasma samples can then be calculated by the equation

$$P = R \times C \times U \times A \times 0.80$$

where

$$R = \frac{\text{cpm of 7-}^3\text{H-progesterone added to plasma}}{10 \times \text{cpm of tritium in aliquot prior to gas chromatography}}$$

$$C = \frac{\text{peak area of 0.01 } \mu\text{g testosterone chloroacetate}}{\text{peak area of 0.01 } \mu\text{g 20}\beta\text{-hydroxypregn-4-en-3-one}}$$

$$U = \frac{\text{peak area of 20}\beta\text{-hydroxypregn-4-en-3-one in sample}}{\text{peak area of testosterone chloroacetate in sample}}$$

$A = \mu$g testosterone chloroacetate added

$$0.80 = \frac{\text{molecular weight of progesterone}}{\text{molecular weight of 20}\beta\text{-hydroxypregn-4-en-3-one chloroacetate}}$$

$$= \frac{314}{393}$$

Further correction must be made for the total mass of the tritiated progesterone added. This mass may be obtained by adding equal amounts of the progesterone to 10 ml of water and processing exactly as the plasma. The mass obtained in this manner is then subtracted from that calculated by the above equation.

Estrogens

Comparatively few methods have been developed for the determination of plasma estrogens because of the extremely high sensitivity required. Methods based on fluorescence measurement[77-79] and on double isotope labeling[80] have been described. The only gas chromatographic detector potentially sensitive enough for this assay is the electron capture detector. Accordingly, a method based on the measurement of the heptafluorobutyrate derivatives of the estrogens has been developed.[81]

[77] G. Ittrich, Z. Physiol. Chem. **320**, 103 (1960).
[78] G. W. Oertel, Klin. Wochschr. **39**, 492 (1961).
[79] S. Ichii, E. Forchielli, W. H. Perloff, and R. I. Dorfman, Anal. Biochem. **5**, 422 (1963).
[80] R. Svendsen, Acta Endocrinol. **35**, 161 (1960).
[81] H. H. Wotiz, G. Charransol, and I. N. Smith, Steroids **10**, 127 (1967).

Method of Wotiz and Charransol

Materials

Alumina. 125 g of aluminum oxide (Woelm) is washed with a liter of 5 M HCl in 95% ethanol. After air-drying for an hour, the alumina is washed with 3 times 350 ml of petroleum ether and again air-dried for 15 minutes. Next, the powder is washed 3 times with 350 ml of absolute methanol and oven dried at 105° for 24 hours.

Benzene, reagent grade, twice distilled through a 60-cm fractionating column. The fraction boiling at 79–80° is collected.

Chloroform, reagent grade, fractionally distilled. The fraction boiling at 60–61° is collected.

Ethyl ether; reagent grade, anhydrous. Each liter of reagent is treated with 50 ml of a 13% $AgNO_3$ solution (w/v), then with 100 ml of N NaOH; and after washing to neutral with water, the ether is distilled through a 60-cm fractionating column.

Ethyl acetate, reagent grade, twice fractionally distilled. The fraction boiling at 76–77° is collected.

Heptafluorobutyric anhydride (Peninsular Chemical Research, Gainesville, Florida), refluxed over P_2O_5 and distilled. The fraction boiling at 109° collected.

Hexane, reagent grade, twice fractionally distilled. The fraction boiling at 68–69° is collected.

Methanol, reagent grade, twice fractionally distilled

Sodium carbonate. To a liter of M $NaHCO_3$ is added 130 ml of 5 N NaOH, both reagent grade

Tetrahydrofuran, reagent grade. The material is refluxed over calcium chloride for 2 hours, distilled off and refluxed over KOH before final distillation through a 60-cm column.

Petroleum ether, reagent grade. Twice fractionally distilled through a 60-cm column, the fractions boiling between 30° and 60° being collected.

β-Glucuronidase. "Glusulase" (Endo Laboratories, Inc., Garden City, New York) 100,000 units/ml β-glucuronidase, 50,000 units/ml sulfatase

Glass Fiber Paper, No. 934 AH (Reeve Angel Co., Clifton, New Jersey). The paper strips are heated to 350° for 1 hour in a muffle furnace. The day before intended use, the strips are impregnated with 0.1 M KH_2PO_4 and dried overnight at room temperature

Acetic acid, reagent grade

Toluene, spectrograde

Scintillation fluid. "Liquiflor" (New England Nuclear Corp., Boston, Massachusetts), 42 ml added to 1 liter of toluene

Isooctane, reagent grade, distilled fractionally. Fraction boiling between 99.0° and 99.5° was collected.

Steroids. High specific activity tritiated steroids (~40 C/millimole for estrone and estradiol and ~15 C/millimole for estriol) were obtained from New England Nuclear Corp., Boston, Massachusetts and utilized within 1–2 weeks after synthesis. Otherwise a preliminary TLC purification was interposed.

Gas Chromatograph. An F & M Model 400 series gas chromatograph equipped with a tritium foil electron capture cell, operated in pulsed mode, was used throughout; pulsing was every 50 μsec with a pulse width of 1 μsec.

Preparation of GLC Column. A 6-foot (4 mm i.d.) glass column is packed with 3% QF-1 or 2% XE-60 on 80–100-mesh Gas Chrom Q (Applied Science Laboratories, State College, Pennsylvania), by gravity and gentle tapping. The column is cured at 240° with a flow of about 10 ml/min of carrier gas for at least 72 hours before connecting it to the detector.

Radioactivity Measurement. The aliquots from each fraction are added to 20 ml scintillation vials containing 10 ml of scintillation fluid. Recoveries are then determined in a scintillation spectrometer.

METHODS

Protein Precipitation. To a 50-ml centrifuge tube are first added 0.050 mμg of estrone-6,7-³H and estradiol-6,7-³H and 0.1 mμg of estriol-6,7-³H (about 6000 cpm) each in 0.1 ml of benzene. This is followed by 10 ml of cold methanol, 10 ml of the plasma, and again 10 ml of methanol. The mixture is kept in the freezer overnight. It is then centrifuged and decanted, the residue is twice washed with 5 ml of methanol. The combined alcoholic extracts are evaporated to dryness at 40° in a rotating still *in vacuo*.

Hydrolysis. The residue is redissolved in 4 ml of water and 4 ml of acetate buffer at pH 4.5. "Glusulase" (400 units of β-glucuronidase activity) is added as well as 3 drops of chloroform, and incubation is carried out at 37° for 18 hours.

Extraction. The free estrogens are extracted with first 2 × 10 ml and then 1 × 5 ml of ethyl acetate. The organic layer is washed with 8 ml of saturated NaHCO₃ and then with two 8-ml portions of water. After filtration through Na₂SO₄, the ethyl acetate is evaporated in a flash evaporator.

Partition of Phenols. The residue is dissolved in 1 ml of ethanol, then 5 ml each of benzene and petroleum ether are added. The solution is extracted twice with 10 ml and once with 5 ml of water to yield the estriol fraction. Next, estrone and estradiol are extracted from the organic layer twice with 10 ml and once with 5 ml of 1 N NaOH. The estriol fraction is brought to pH 8.2 with $NaHCO_3$ and the hydroxide extract is neutralized by adding enough HCl to bring it to approximately pH 7, whereupon it is saturated with $NaHCO_3$.

Extraction. Each of the two aqueous solutions is extracted first with 25 ml, then with 10 ml, and finally with 5 ml of ether. The combined ether layers are washed with 8 ml of 0.2 N sulfuric acid and twice with 8 ml of water. The ether is filtered through Na_2SO_4, then the solvent is removed in the flash evaporator.

Alumina Chromatography. Three grams of fully active alumina is deactivated with 5% water (w/v) and a column (8 mm diameter) is filled by the method of Brown.[82] The extract containing the weak phenols is applied to the column with three 2-ml portions of benzene. Elution is carried out with (1) 10 ml of benzene (discard); (2) 15 ml of 2.5% (v/v) ethyl acetate in benzene (E_1); (3) 15 ml of 5% (v/v) ethyl acetate in benzene (E_1); and (4) 35 ml of 10% (v/v) ethyl acetate in benzene (E_2). The estrone fractions are pooled, then the estrone and the estradiol eluate are individually evaporated to dryness in a rotary still.

Glass Fiber Paper Chromatography (GFPC). A 12 $\times$ 15 inch sheet of glass fiber paper is cut into $\frac{1}{2} \times 15$ inch strips. The estrone, estradiol, and estriol fractions are spotted on separate strips with a microsyringe. Ascending development is carried out as follows:

For estrone and estradiol: 200 ml of toluene–petroleum ether (1:1) and 0.3 ml of acetic acid:

$$R_f E_1 = 0.77$$
$$R_f E_2 = 0.62$$

For estriol: 200 ml of chloroform and 0.3 ml of acetic acid

$$R_f E_3 = 0.68$$

The average length of the spots varies from 2 to 3 cm.

Radioactivity Scanning and Elution. After drying in the air, the strips are quickly run through a Tracerlab paper strip scanner. Caution must be advised since the GF paper is very brittle and steroid may be scraped off by running the paper through this instrument without adequate protection. A thin metal holder with center cutout is being used to prevent damage to the fibers, without losing sensitivity.

---

[82] J. B. Brown, *Biochem. J.* **60**, 185 (1955).

The radioactive areas of the strips are cut out and eluted into a small beaker by suspending the paper strip and eluting through a syringe with a No. 22 needle stuck into the paper. A total of 12 ml of ethyl acetate in 4 portions is used.

Second GFPC. After evaporation the residues are again applied to GFP strips as before and development is carried out as follows:

For estrone and estradiol: pet. ether–*i*-octane–benzene (3:2:1)

$$R_f E_1 = 0.60$$
$$R_f E_2 = 0.48$$

For estriol: benzene–ethyl acetate (1:1)

$$R_f E_3 = 0.30$$

Scanning and Elution. After drying, the strips are scanned for radioactivity and eluted as described before.

Determination of Recovery. Each of the fractions obtained is dissolved in 1 ml of methanol, and 0.1 ml of each is removed for scintillation counting. The remaining portions are taken to dryness under nitrogen after transfer to small test tubes with glass stoppers.

Derivative Formation. To the dried residue of each of the estrogenic fractions is added 0.1 ml of hexane, 1 μl of tetrahydrofuran, and 2 μl of heptafluorobutyric anhydride. The tubes are stoppered and heated to 60° for 30 minutes. The solvent is then evaporated under a gentle stream of nitrogen at about 40–50°C. The dried residue is dissolved in 50 μl of hexane.

Gas Chromatography. The gas chromatograph, containing a properly conditioned 6 foot 3% (w/w) QF-1 or 2% (w/w) XE-60 column, is set to allow optimal sensitivity at a column temperature of 200°. Helium flow is adjusted for maximal efficiency commensurate with a retention time for estriol of about 20–25 minutes. Argon-methane (10%) purge gas is adjusted for maximum sensitivity under those conditions (120 ml/min total flow). The detector temperature is set between 210 and 215° and the vaporizer at approximately 250°.

When the above conditions have been attained, a series of estrogen heptafluorobutyrates of exactly known concentrations, in the range anticipated in the plasmas to be analyzed, are injected and peak areas and retention times are determined. When a proper linear response curve for each steroid has been obtained, 5 μl from the above extracts is injected for analysis.

Quantitative determinations of the proper peaks (identified by comparison of retention times with standards) is made by comparing peak areas of the unknowns with those of the standards. Care must be taken to correct all values to equal attenuation.

Quantification. The amount of each estrogen is determined as follows:

$$E = \frac{A_u \times C_s \times 100 \times T}{A_s \times 9}$$

where E is the amount of each estrogen in micrograms per 10 ml plasma; A_u is the peak area of the unknown sample; A_s is that for the corresponding standard; C_s is the amount of standard injected expressed in micrograms; T is the fraction of radioactive steroid recovered.

Miscellaneous Methods

The foregoing methods are exclusively concerned with the determination of steroid hormones and metabolites in humans. Obviously, gas chromatographic techniques are more broadly useful and in this section some of the more interesting applications are pointed out. It is worth emphasizing that a successful method will almost certainly require some modification before it can be applied to samples much different from those for which it was originally designed. The extent of modifications will depend upon the nature of the sample and obviously cannot be predicted. This short account is of course not complete, but merely illustrative of the variety of applications. For a more extensive survey, reference should be made to the recent review by Kuksis.[83]

Kittinger[84] has determined 17-deoxycorticosteroids produced by the rat adrenal. The steroid mixture is oxidized with periodic acid to convert α-ketols to the corresponding etiocholenic acids. Aldosterone and 18-hydroxydeoxycorticosterone form internal esters. Chromatograms are run before and after esterification with diazomethane. Compounds determined were progesterone, aldosterone, 17-hydroxydeoxycorticosterone, 11β-hydroxyandrostenedione, deoxycorticosterone, dehydrocorticosterone, 11β-hydroxyprogesterone, and corticosterone.

Cholesterol and related compounds were determined in normal and pathological plasma by Kanai.[85]

The free sterols of dermatophytes have been isolated and determined by gas chromatography combined with column and paper chromatography.[86] The "house-fly *sterol*" has been identified as *campesterol*.[87] The sterol esters of house fly eggs have been isolated and identified.[88] The

[83] A. Kuksis, *Methods Biochem. Anal.* 14, 325 (1966).
[84] G. W. Kittinger, *Steroids* 3, 21 (1964).
[85] M. Kanai, *J. Biochem. (Tokyo)* 56, 266 (1964).
[86] F. Blank, F. E. Shortland, and G. Just, *J. Invest. Dermatol.* 39, 91 (1962).
[87] M. J. Thompson, S. J. Louloudes, and W. E. Robbins, *Biochem. Biophys. Res. Commun.* 9, 113 (1962).
[88] R. C. Dutky, W. E. Robbins, J. N. Kaplanis, and T. J. Sharting, *Comp. Biochem. Physiol.* 9, 251 (1963).

bufadienolides of Ch'an Su were chromatographed as the trimethylsilyl ethers.[89]

The identification of plant sterols using the Δ_{R^M} function has been reported by Knights.[90] VandenHeuvel[91] has proposed an extensive scheme for identification of human steroid metabolites.

Particularly interesting is the report[92] of the isolation and identification of cholesterol, campesterol, stigmasterol, and β-sitosterol from *Escherichia coli*, presenting the first concrete evidence for the existence of sterols in bacteria.

A study involving silicic acid column chromatography and GLC resulted in a virtually complete resolution of the complex mixture of sterols present in rat skin.[93] Since the components of this mixture probably represent all the sterol intermediates in cholesterol biosynthesis, the techniques used in this work have special importance for this area of research.

The *in vitro* conversion of cholesterol and 20α-hydroxycholesterol by bovine corpus luteum has been studied by gas chromatography.[94] 5α-Pregnane-3α,6α,20α-triol has been identified as a metabolite of progesterone in the rabbit.[95] 17α and 17β-Pregnane-3,20-diols have been identified as progesterone metabolites in man and in the monkey, rabbit, and guinea pig.[96] The determination of progesterone and 20α-hydroxypregn-4-en-3-one has been used in *in vitro* and *in vivo* studies of animal tissues.[97]

The technique has been used to study estrogen metabolism in normal and pregnant women.[98] The identification and estimation of estradiol in a single rat ovary has recently been achieved.[99] Neither estrone nor estriol could be detected.

The identification of 17-ketosteroids in urine from man and the rhesus monkey has been reported.[100] The biosynthesis of androst-16-en-3α-ol in rabbit and guinea pig testicular slices has been studied.[101] Com-

[89] K. N. Sakurai, E. Yoshii, and K. Kubo, *Yakugaku Zasshi* **84**, 1166 (1964).

[90] B. A. Knights, *J. Gas. Chromatog.* **2**, 160 (1964).

[91] F. A. VandenHeuvel, G. J. Hinderks, J. C. Nixon, and W. G. Layng, *J. Am. Oil Chemist's Soc.* **42**, 283 (1965).

[92] K. Schubert, G. Rose, R. Tummler, and N. Ikekawa, *Z. Physiol. Chem.* **339**, 293 (1965).

[93] R. B. Clayton, A. N. Nelson, and I. D. Frantz, Jr., *J. Lipid Res.* **4**, 166 (1963).

[94] P. F. Hall and S. B. Koritz, *Biochemistry* **3**, 129 (1964).

[95] B. A. Knights, A. W. Rogers, and G. H. Thomas, *Biochem. Biophys. Res. Commun.* **8**, 253 (1962).

[96] J. Chamberlain, B. A. Knights, and G. H. Thomas, *J. Endocrinol.* **28**, 235 (1964).

[97] J. D. Neill, B. N. Day, and G. W. Duncan, *Steroids* **4**, 699 (1964).

[98] J. Fishman, J. B. Brown, L. Hellman, B. Zumoff, and T. F. Gallagher, *J. Biol. Chem.* **237**, 1489 (1962).

[99] H. Jonsson and H. H. Wotiz, unpublished results, 1966.

[100] J. Chamberlain, B. A. Knights, and G. H. Thomas, *J. Endocrinol.* **26**, 367 (1963).

[101] D. B. Grower and G. A. D. Haslewood, *J. Endocrinol.* **23**, 253 (1961).

parative studies have been reported of the amounts of androstenedione and ring A reduced metabolites in the urine of normal and adrenalectomized animals.[102] Androstanediols and androstanediones were identified in the urine of pregnant cows.[103]

[102] G. H. Thomas, E. Forchielli, and K. Brown-Grant, *Nature* **202**, 260 (1964).
[103] R. Heitzman and G. H. Thomas, *Biochem. J.* **96**, 22P (1965).

[3] Gas-Liquid Chromatography of Polycyclic Triterpenes

By Nobuo Ikekawa

The behavior of polycyclic triterpenes under gas-liquid chromatographic conditions has not so far received the attention that has been paid to steroids, though, as was already clear from the early report of Eglinton *et al.*,[1] these compounds may be expected to behave similarly to steroids on GLC.[2,3] This expectation has been borne out in a number of more recent reports. The aim of the present article will be to draw together some data and observations on methodology that should be a useful guide in evaluating the usefulness of GLC for biochemical studies involving the polycyclic triterpenes.

GLC of polycyclic triterpenes may be accomplished by techniques closely similar to those described for steroids in the preceding chapter,[4] and those aspects of the methodology that have been outlined there will not be reiterated here.

In general, the same liquid phases may be used for triterpenes as for steroids. These include: SE-30, JXR, OV-1 (methylsiloxane polymer), SE-52 (methylphenylsiloxane polymer), OV-17 (phenylmethylsiloxane polymer, stable to 350°), F-60 (methyl-*p*-chlorophenylsiloxane polymer), and selective phases QF-1 (fluoroalkylsilicon polymer), XE-60 (methyl-β-cyanoethylsiloxane polymer), and polyester phases of NGS (neopentylglycol succinate), NGA (neopentylglycol adipate), or DEGS (diethyleneglycol succinate), HiEff-8b (cyclohexanedimethanol succinate). Thin-film column packings can be prepared as described by Horning[2] by filtration of a 1 ~ 3% solution of the liquid phase through the solid support which has been previously washed with hydrochloric

[1] G. Eglinton, R. J. Hamilton, R. Hodges, and R. A. Raphael, *Chem. & Ind.* p. 955 (1959).
[2] E. C. Horning, W. J. A. VandenHeuvel, and B. G. Creech, *Methods Biochem. Anal.* **11**, 69 (1963); A. Kuksis, *ibid.* **14**, 325 (1966).
[3] E. C. Horning and W. J. A. VandenHeuvel, *Advan. Chromatog.* **1**, 153 (1965).
[4] H. H. Wotiz and S. J. Clark, this volume [2].

acid and silanized with dimethyldichlorosilane in toluene. The use of glass capillary columns coated with SE-30 has also been described.[5]

Typically, column temperatures in the range 220–240° and carrier gas flow rates of 50–100 ml/min are found satisfactory. For the separation of more highly polar triterpenes, column temperatures of 240–250° are necessary, and SE-30, JXR, OV-1, or OV-17 are preferable liquid phases. Glass columns with "on column" injection systems are to be preferred to stainless steel columns, especially for sensitive compounds at high temperatures.

Tri- and tetrahydroxy compounds frequently show excessive retention times or undergo degradation, as evidenced by the emergence of badly tailing peaks. The trimethylsilyl ether derivatives of such compounds, however, show sharp peaks and short retention times. Thus, the trimethylsilyl ethers of polyhydroxy triterpenes, such as primulagenin (compound No. 36; see tables for compounds), longispinogenin (37), and soyasapogenol A (38), are conveniently separated by GLC. The trimethylsilyl ethers may be prepared by treatment of the hydroxy compound with hexamethyldisilazane and trimethylchlorosilane in pyridine or tetrahydrofuran,[6] or more conveniently, according to a recent report,[7] by treatment with bis(trimethylsilyl)acetamide. Methyl ether derivatives may also be used as described by Clayton[8] for some triterpene alcohols (see Table VI).

Dehydration of tertiary hydroxyl groups can be minimized by using glass columns, and the conversion of such compounds to their trimethylsilyl derivatives has been found to prevent decomposition. Thus in a recent study in our laboratory, retention times (r.r.t.)[9] of 22-hydroxyhopane derivatives [66–69] in Table III[10] differed from those expected on the basis of their molecular structures but corresponded more closely with those of the dehydration products hopene-a and hopene-b. The eluted products showed the same retention times in a second gas chromatogram, and on thin-layer chromatography they had the same R_f value as hopene-a. These observations suggest that the 22-hydroxyl group is lost

[5] I. R. Hills and E. V. Whitehead, *Nature* **209**, 977 (1966); I. R. Hills, E. V. Whitehead, D. A. Anders, J. J. Cummins, and W. E. Robinson, *Chem. Commun.*, p. 752 (1966).

[6] C. C. Sweeley, R. Bentley, M. Makita, and W. W. Wells, *J. Am. Chem. Soc.* **85**, 2497 (1963).

[7] J. F. Klebe, H. Finkbeiner, and D. M. White, *J. Am. Chem. Soc.* **88**, 3390 (1966).

[8] R. B. Clayton, *Biochemistry* **1**, 357 (1962).

[9] All relative retention times (r.r.t.) are given in reference to cholestane (r.r.t. = 1).

[10] N. Ikekawa, S. Natori, H. Ageta, K. Iwata, and M. Matsui, *Chem. Pharm. Bull. (Tokyo)* **13**, 320 (1965).

in the flash heater zone or in the column to form the Δ^{21}-compound. It is not clear whether the occasional appearance of a small peak showing the r.r.t. of 4.3 is due to the presence of the unchanged substance or to some unidentified type of dehydration product. However, the trimethylsilyl ether of hydroxyhopane gave r.r.t. 5.3, indicating an absence of elimination. Similarly, the r.r.t. of peaks obtained on injection of hydroxyhopanone, zeorin, and zeorinone (Table III) are actually those of dehydrated products, and evidence has also been discussed by Shibata et al.[11] that under conditions which cause dehydration of free zeorin, its trimethylsilyl ether is stable.

Retention Times and the Structures of Polycyclic Triterpenes

Tables I–VI summarize the relative retention times[12] of a number of polycyclic triterpenes. In Tables I–IV,[10,13,14] an attempt has been made to arrange the compounds in groups of closely related structures. In all cases, behavior on columns of SE-30 is recorded, since in general, each compound shows a single sharp peak on this phase and separation of compounds with closely similar structures is frequently possible. Relative retention times of several triterpenes using SE-30 have also been reported by Shimizu et al.[15] and by Capella et al.[16]

In some cases (Tables III–VI) comparative data for the behavior of compounds on different liquid phases has been recorded and, where available, retention times for free hydroxyl compounds and different derivatives (acetates or trimethylsilyl ethers) are listed. The retention times of some related trimethylsilyl ethers on different liquid phases are shown in Table V,[14] and those of a number of triterpene methyl ethers are given in Table VI.[17]

It should be noted that the reporting of retention times relative to that of a single hydrocarbon standard such as cholestane, as in this paper, is not entirely satisfactory for interlaboratory comparison purposes, since such values are significantly temperature dependent. It is, however, the most frequently used method of reporting at the present time on account

[11] S. Shibata, T. Furuya, and H. Iizuka, *Chem. Pharm. Bull. (Tokyo)* **13**, 1254 (1965).
[12] All the relative retention times in the tables were determined by Shimadzu Seisakusho Model GC-1B and GC-1C gas chromatographs equipped with a hydrogen flame detector.
[13] N. Ikekawa, S. Natori, and O. Tanaka, unpublished data, 1966.
[14] N. Ikekawa, S. Natori, H. Itokawa, S. Tobinaga, and M. Matsui, *Chem. Pharm. Bull. (Tokyo)* **13**, 316 (1965).
[15] M. Shimizu, F. Uchimaru, and G. Ohta, *Chem. Pharm. Bull. (Tokyo)* **12**, 74 (1964).
[16] P. Capella, E. Fedeli, and M. Cirimele, *Chem. & Ind.* p. 1590 (1963).
[17] T. A. Bryce, M. Martin-Smith, G. Osske, K. Schreiber, and G. Subramanian, *Tetrahedron* **23**, 1283 (1967).

of its convenience. For greater accuracy, methylene unit values[18] or retention index values[19] may be used.

It can be seen from the tables that in favorable instances, good separations of isomeric compounds are possible, where older methods would succeed with difficulty, if at all. α-Amyrin and taraxasterol, in which the double bond lies, respectively, in the ring system and in an exomethylene position, and where the configurations at C-18 are different, are a case in point. However, compounds isomeric with respect to the positions of groups often have very similar retention times and a mixture of α-amyrin, β-amyrin, and taraxerol cannot be separated on the SE-30 phase. Trimethylsilyl derivatives of α-amyrin and β-amyrin can be separated on QF-1 or NGS, but β-amyrin and taraxerol show the same retention time on these phases. The nitrile silicone phase XE-60 is most suitable for the separation of these three compounds, as shown in Table V.[14] Lupeol (44) and α-amyrin (23) show similar retentions on SE-30 or QF-1, but these can be separated on XE-60.[20]

Extensive correlations between structure and retention time in the polycyclic triterpenes have not yet been possible because of the limited number of available compounds that lend themselves to such studies. However, the results so far obtained indicate that the effect of substituents on the retention times are similar to those observed in the steroid series, which are discussed in more detail in the preceding article.[4] Since the introduction of substituents into the skeleton causes a relatively large increase in retention time, the behavior of these compounds on GLC can give information concerning the degree and type of substitution. A very approximate, though useful, relationship between number of substituents, such as hydroxyl, ketone, and methoxycarbonyl groups, etc., and retention time on SE-30 is given in the accompanying tabulation.

	r.r.t.
No substituent	1.4–3
One substituent	3.0–6
Two substituents	4.5–9
Three substituents	8.5–10
Diacetoxy compounds	10.0–12
Triacetoxy compounds	12.0–15

[18] W. J. A. VandenHeuvel, W. L. Gardiner, and E. C. Horning, *Anal. Chem.* **36,** 1550 (1964); *J. Chromatog.* **19,** 2631 (1965); M. G. Horning, K. L. Knox, C. E. Dalgliesh, and E. C. Horning, *Anal. Biochem.* **17,** 244 (1966); *Biochem. J.* **101,** 792 (1966).

[19] E. Kovats, *Helv. Chim. Acta* **41,** 1915 (1958); B. A. Knights, *J. Gas Chromatog.* **4,** 329 (1966).

[20] M. Saito-Suzuki and N. Ikekawa, *Chem. Pharm. Bull. (Tokyo)* **14,** 1049 (1966).

TABLE I

RELATIVE RETENTION TIMES OF TETRACYCLIC TRITERPENES[a]

(1, 2)

(3–7)

(8–14)

Compound		1.5% SE-30[b]
(I) Lanostane group		
(1) Lanosterol	Δ^{24}	2.70
(2) Dihydrolanosterol		2.48
(3) Cycloartanol	R = H	2.87
(4) Cycloartenol	R = H, Δ^{24}	3.12
(5) 24-Methlenecycloartanol	R = =CH$_2$	3.58
(6) 24-Methylcycloartanol	R = —CH$_3$	3.66
(7) Cyclolaudenol	R = —CH$_3$, Δ^{25}	3.45
(8) Methyl eburicoate acetate	R$_1$ = H, R$_2$ = =CH$_2$, R$_3$ = H	4.50
(9) Methyl dihydroeburicoate acetate	R$_1$ = H, R$_2$ = —CH$_3$, R$_3$ = H	6.15
		4.94
		6.20
(10) Methyl eburiconate	C-3—O, R$_2$ = =CH$_2$, R$_3$ = H	4.42
(11) Methyl tumulosate	R$_1$ = H, R$_2$ = =CH$_2$, R$_3$ = OH	7.30
(12) Methyl pachymate acetate	R$_1$ = AC, R$_2$ = =CH$_2$, R$_3$ = OH	9.40
		10.0
(13) Methyl dihydropachymate	R$_1$ = Ac, R$_2$ = CH$_3$, R$_3$ = OH	10.2[c]
		11.3
(14) Methyl acetyldehydrotumulosate	R$_1$ = Ac, R$_2$ = CH$_2$, R$_3$ = OAc $\Delta^{7,9(11)}$	9.80

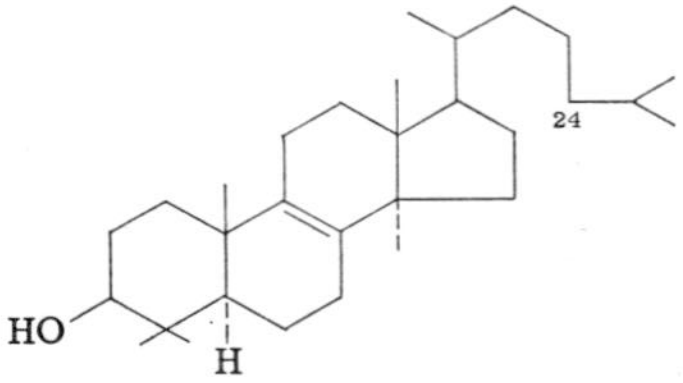

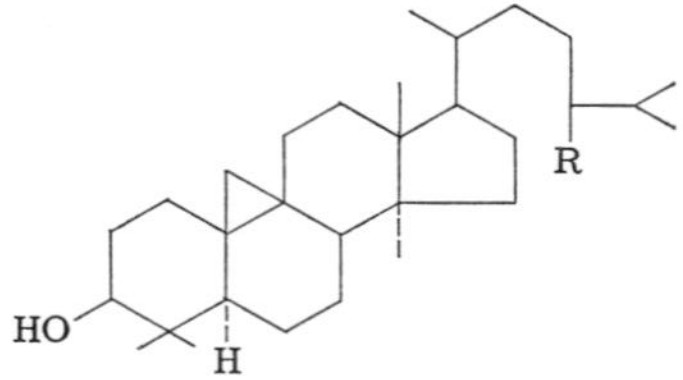

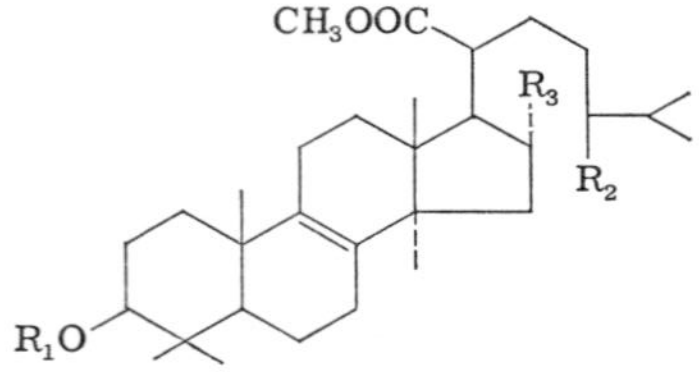

(II) Dammarane group

(15, 16)

(17, 18)

(19)

(15) Betulafolianetriol Tri-TMSi		2.77 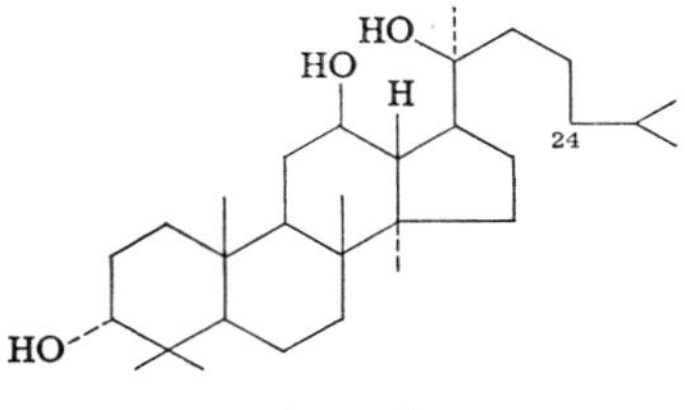
(16) Betulafolienetriol Tri-TMSi	Δ^{24}	3.12
(17) Panaxanol	R = H	2.94 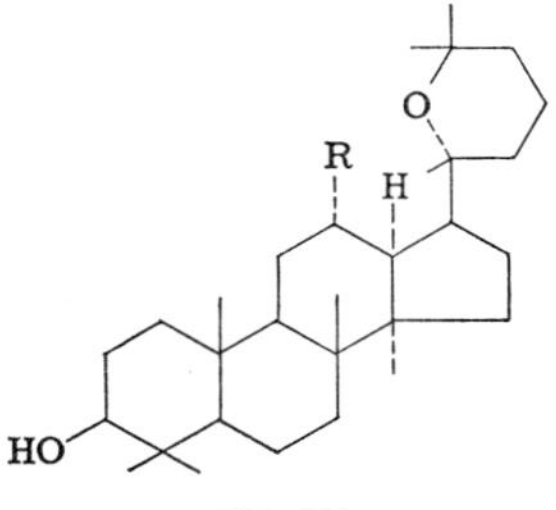
(18) 12-Epipanaxadiol	R = OH	4.37
(19) Isotirucallenol		2.00 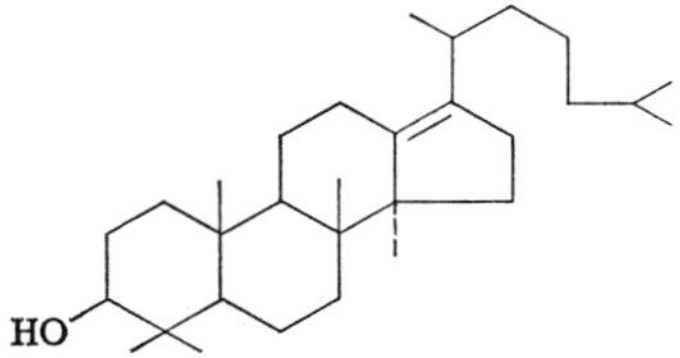

(Continued)

TABLE I (*Continued*)

Compound		1.5% SE-30[b]
(20) Dammaranediol-I	20R-OH	3.82 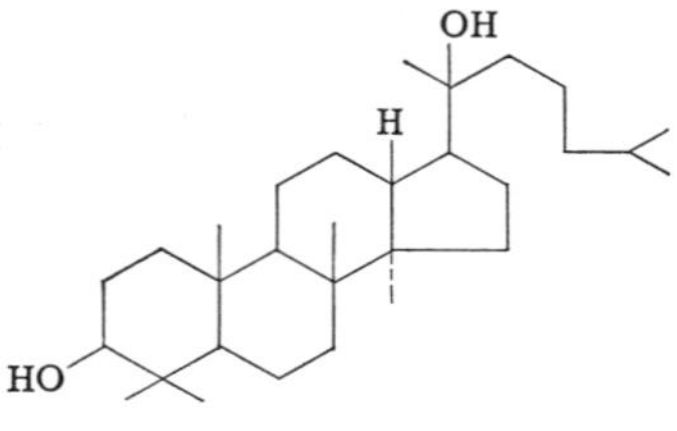
3-acetate		4.94
(21) Dammaranediol-II	20S-OH	3.74
3-acetate		4.86

(20, 21)

[a] N. Ikekawa, S. Natori, and C. Tanaka, unpublished data, 1966.
[b] Column, 1.5% SE-30 on Shimalite W, 80–100 mesh, 180 cm × 4 mm i.d.; column temperature, 248°; carrier gas, N₂, 80 ml/minute. Retention time of cholestane, 4.0 minutes; retention time of cholesterol, 7.2 minutes.
[c] Each peak may correspond to stereoisomers about the C-24 methyl group, but this is not certain.

TABLE II

RELATIVE RETENTION TIMES OF URSANE, OLEANANE AND LUPANE GROUPS[a]

Compound	1.5% SE-30[b]
(III) Ursane group (α-amyrin group)	
(22) Urs-12-ene $R_1 = H, R_2 = CH_3$	1.74 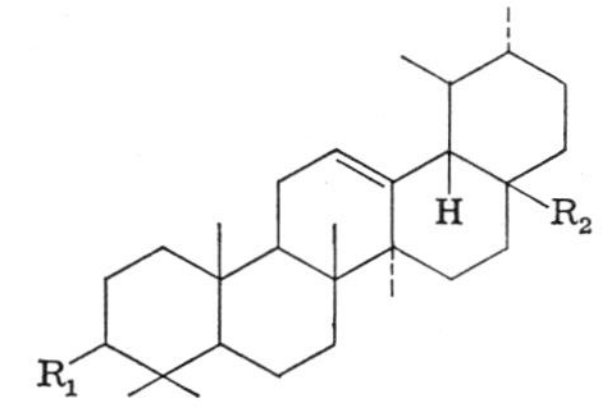
(23) α-Amyrin acetate $R_1 = OH, R_2 = CH_3$	3.65
$R_1 = OAc, R_2 = CH_3$	4.74
(24) Methyl ursolate acetate $R_1 = OH, R_2 = COOCH_3$	5.25
$R_1 = OAc, R_2 = COOCH_3$	6.80
(25) Taraxasterol	5.75 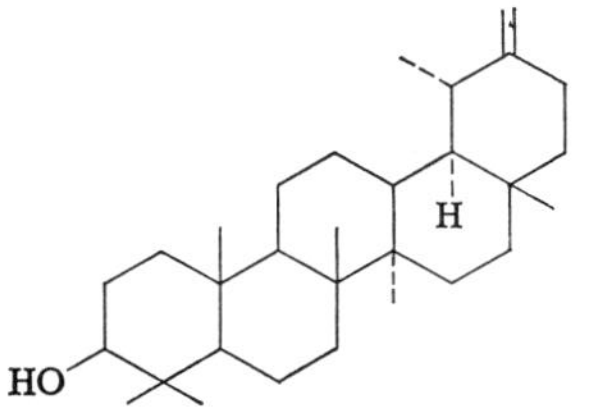

(22–24)

(25)

(*Continued*)

TABLE II (Continued)

(26–38)

	Compound		1.5% SE-30[b]
(IV)	Oleanane group (β-amyrin group)		
(26)	β-Amyrin	$R_1 = OH, R_2 = CH_3$	3.23
(27)	Methyl deoxyoleanolate	$R_1 = H, R_2 = COOCH_3$	2.66
(28)	Methyl oleanolate acetate	$R_1 = OH, R_2 = COOCH_3$ $R_1 = OAc, R_2 = COOCH_3$	4.8 7.35
(29)	Methyl oleanonate	$R_1 = =O, R_2 = COOCH_3$	4.4
(30)	Oleanoyl chloride	$R_1 = OH, R_2 = COCl$	7.4
(31)	Oleanolactone	$R_1 = OH, R_2—C-13 = —COO—$ (non Δ^{12})	3.4
(32)	Erythrodiol Diacetate	$R_1 = OH, R_2 = CH_2OH$ $R_1 = OAc, R_2 = CH_2OAc$	7.32 10.52
(33)	Methyl glycyrrhate	$R_1 = OH, R_2 = CH_3$ C-11 = O, C-20—CH$_3$ $\overset{\displaystyle COOCH_3}{\diagup}$	9.51
(34)	Hederagenin methyl ester	$R_1 = OH, R_2 = COOCH_3$ C-4—CH$_3$ $\diagdown$ CH$_2$OH	8.55
	TMSi		7.37
(35)	Hederatriol diacetate	$R_1 = OAc, R_2 = CH_2OH$ C-4—CH$_3$ $\diagdown$ CH$_2$OAc	9.9
(36)	Primulagenin	$R_1 = OH, R_2 = CH_2OH$ C-16---OH	
	TMSi		6.57
	Triacetate		14.8
(37)	Longispinogenin	$R_1 = OH, R_2 = CH_2OH$ C-16—OH	
	TMSi		5.3
	Triacetate		13.1

(38) Soyasapogenol A

R_1=OH, R_2=CH$_3$
C-21~OH, C-22~OH
C-4—CH$_3$
　　　　CH$_2$OH

6.57

TMSi
(V) Modified oleanane group

(39, 40)

(39) Taraxerol (skimmiol)　　R=OH　　3.14
(40) Taraxerone (skimmione)　R==O　　4.08

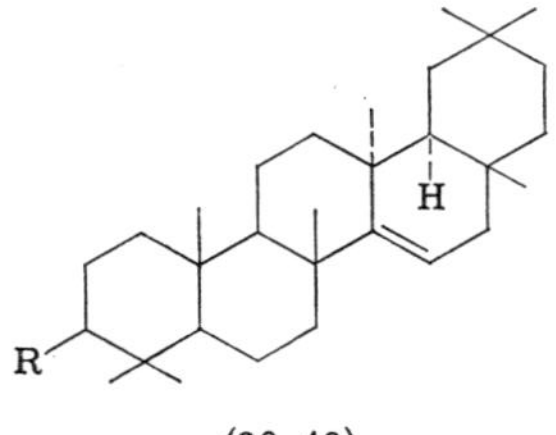

(41)

(41) Friedelin　　5.07

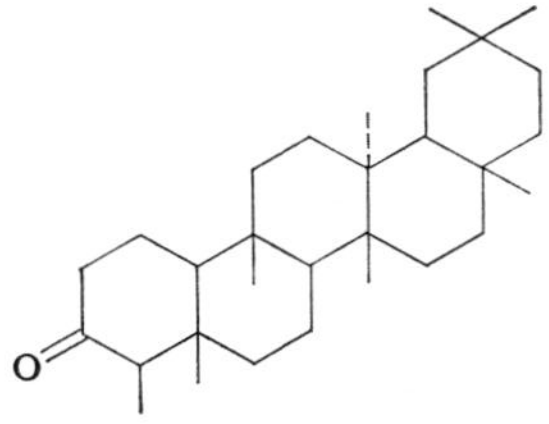

(Continued)

TABLE II (Continued)

(42)

(43–45)

Compound	1.5% SE-30[b]
(42) Dendropanoxide	2.57 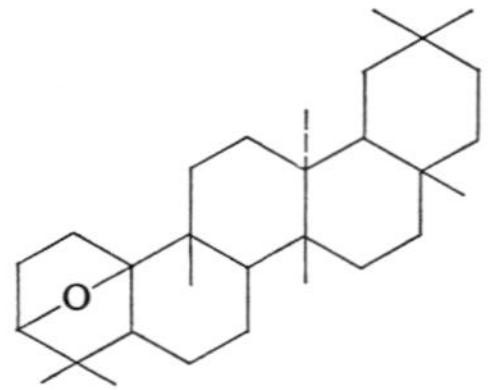
(VI) Lupane group	
(43) α-Lupene $R_1=H, R_2=CH_3$	1.76 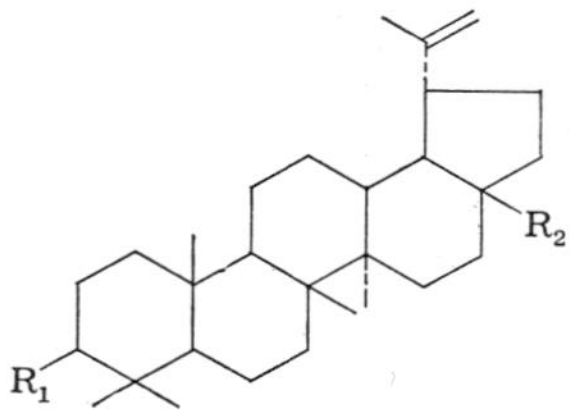
(44) Lupeol $R_1=OH, R_2=CH_3$	3.66
Acetate	4.75
(45) Betulin $R_1=OH, R_2=CH_2OH$	8.93
TMSi	5.36
Diacetate	11.9

[a] N. Ikekawa, S. Natori, H. Itokawa, S. Tobinaga, and M. Matsui, *Chem. Pharm. Bull. (Tokyo)* **13**, 316 (1965).
[b] Column, 1.5% SE-30 on Gas Chrom P, 80–100 mesh, 150 cm × 4 mm i.d.; column temperature, 240°; carrier gas, N₂, 90 ml/minute. Retention time of cholestane, 2.8 minutes; retention time of cholesterol, 5.2 minutes.

TABLE III

RELATIVE RETENTION TIMES OF TRITERPENES OF THE HOPANE-ZEORINANE GROUP (VII) ON SE-30 AND NGS

(46–53)

(54–56)

(57–61)

Compound	2% SE-30[a]	1.5% SE-30[b]	1% NGS[e]
(46) Hopane	2.60	2.68	4.05
(47) 21α-Hopane (moretane)*	2.04	—	—
(48) 17α-Hopane	1.85	—	—
(49) Hop-22(29)-ene (diploptene)	2.63	2.78	5.00
(50) 21α-Hop-22(29)-ene (moretene)	1.98	—	—
(51) Hop-21-ene (hopene a)	2.72	2.87	4.80
(52) Hop-17(21)-ene (hopene I)	1.67	1.73	2.10
(53) Hopa-15,17(21)-diene	1.60	—	—
(54) Neohop-12-ene	2.27	—	—
(55) Neohop-13(18)-ene (hopene II)	1.91	2.01	2.70
(56) Neohopa-11,13(18)-diene	2.05	—	—
(57) Fern-7-ene	2.22	2.34	3.10
(58) Fern-8-ene (isofernene)	1.92	1.96	2.66
(59) Fern-9(11)-ene (fernene)	2.03	2.08	2.96
(60) Ferna-7,9(11)-diene	1.88	1.99	2.60
(61) 5β-Fern-8-ene	2.17	—	—

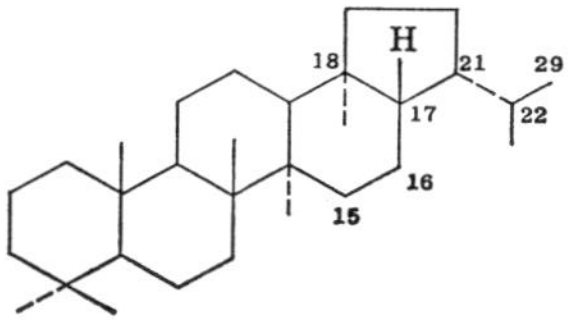

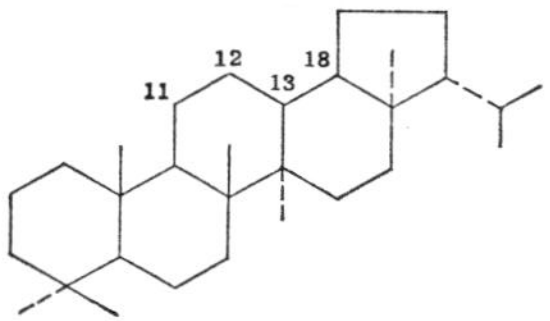

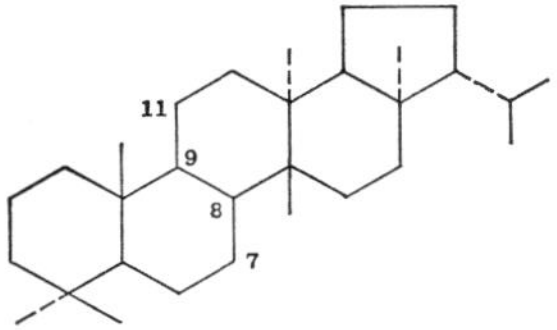

(Continued)

TABLE III (Continued)

Compound	2% SE-30[a]	1.5% SE-30[b]	1% NGS[c]
(62) Adian-5-ene	2.16	2.28	3.30 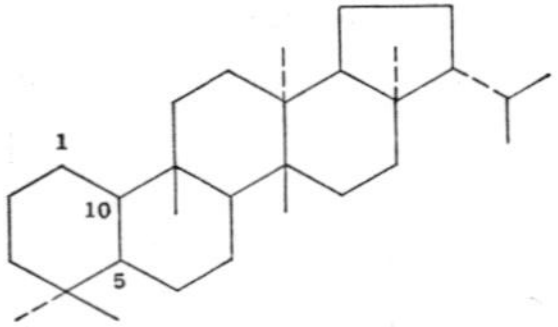
(63) Adiana-1(10),5-diene	2.16	2.29	3.74
(64) Adian-5(10)-ene	2.07	2.22	3.00
(65) Filic-3-ene	2.78	3.06	4.98

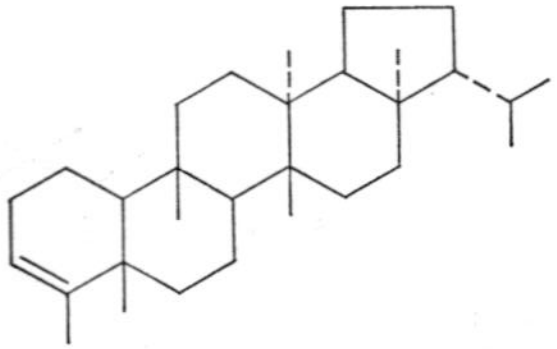

(62–64)

(65)

(66–69)

		a	b	c
(66) Hydroxyhopane (diplopterol) TMSi	$R_1 = R_2 = H$	—	4.30	2.80^d
(67) Hydroxyhopanone	$R_1 = O, R_2 = H$	5.35	—	5.07^d
(68) Zeorin	$R_1 = H, R_2 = \text{----}OH$	—	—	4.96^d
(69) Zeorinone	$R_1 = H, R_2 = O$	—	—	4.42^d
(43) α-Lupene		—	1.76	2.60
(22) Urs-12-ene		—	1.74	2.13

[a] Column, 2% SE-30 on Diatoport F, 80–100 mesh, 200 × 4 mm i.d.; column temperature, 250°; carrier gas, N_2, 60 ml/minute. Retention time of cholestane, 3.0 minutes. Unpublished data of H. Ageta (1966).

[b] Column, 1.5% SE-30 on Gas Chrom P, 80–100 mesh, 150 cm × 4 mm i.d.; column temperature, 225°; carrier gas, N_2, 80 ml/minute. Retention time of cholestane, 7.4 minutes. Data from N. Ikekawa, S. Natori, H. Ageta, K. Iwata, and M. Matsui, *Chem. Pharm. Bull. (Tokyo)* **13**, 320 (1965).

[c] Column, 1% NGS on Gas Chrom P, 80–100 mesh, 150 cm × 4 mm i.d.; column temperature, 219°; carrier gas, N_2, 80 ml/minute. Retention time of cholestane, 3.0 minutes. Data from N. Ikekawa, S. Natori, H. Ageta, K. Iwata, and M. Matsui, *Chem. Pharm. Bull. (Tokyo)* **13**, 320 (1965).

[d] Relative retention times of dehydrated products.

[e] 21—H = α-configuration.

TABLE IV
RELATIVE RETENTION TIMES OF ONOCERANE GROUP (VIII)[a]

Compound		1.5% SE-30[b]	1% NGS[c]
(70) α-Onocerin	R=OH	4.76	—
Diacetate	R=OAc	7.83	—
(71) α-Onoceradiene	R=H	1.34	1.78

(70, 71)

Compound		1.5% SE-30[b]	1% NGS[c]
(72) β-Onocerin	R=OH	6.20	—
Diacetate	R=OAc	9.77	—
(73) β-Onoceradiene	R=H	1.68	1.89

(72, 73)

Compound		1.5% SE-30[b]	1% NGS[c]
(74) γ-Onocerin	R=OH	8.0	—
Diacetate	R=OAc	14.5	—
(75) γ-Onocerene	R=H	2.16	2.90

(74, 75)

Compound	1.5% SE-30[b]	1% NGS[c]
(76) Serratene (Δ14)	2.51	4.00
(77) Isoserratene (Δ13)	2.01	2.78

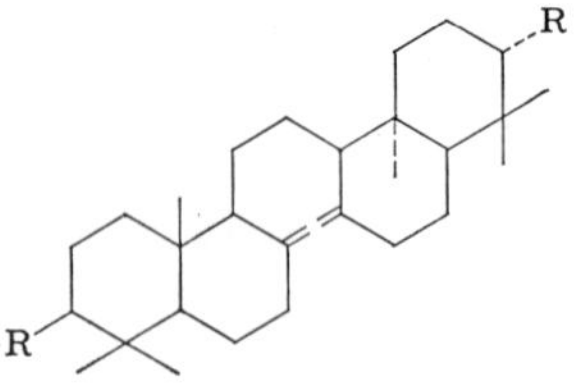

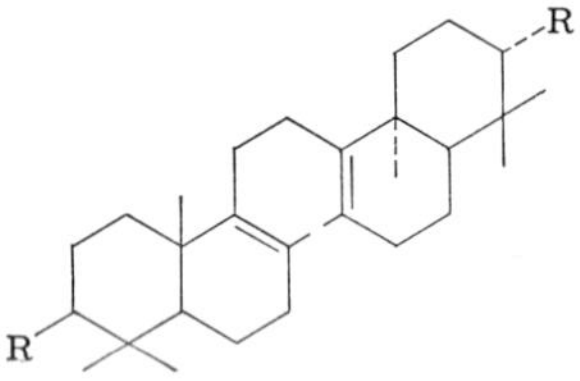

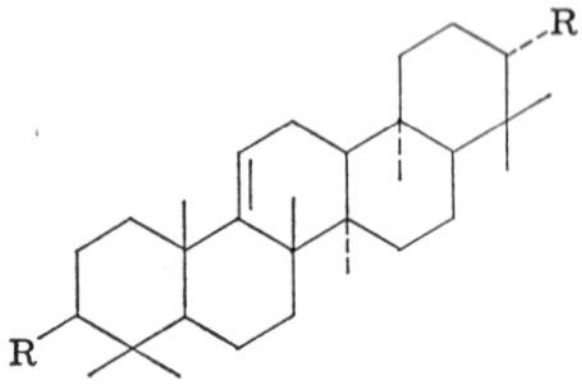

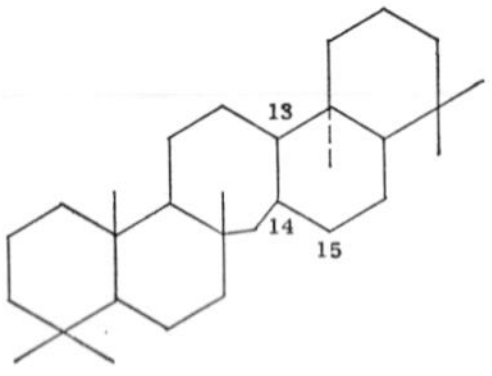

(76, 77)

[a] Data from N. Ikekawa, S. Natori, H. Ageta, K. Iwata, and M. Matsui, *Chem. Pharm. Bull.* (*Tokyo*) **13**, 320 (1965).

[b] Column, 1.5% SE-30 on Anakrom A, 80–100 mesh, 150 cm × 4 mm i.d.; column temperature, 230°; carrier gas, N_2, 90 ml/minute. Retention time of cholestane, 4.4 minutes.

[c] Column, 1% NGS on Gas Chrom P, 80–100 mesh, 150 cm × 4 mm i.d.; column temperature, 220°; carrier gas, N_2, 85 ml/minute. Retention time of cholestane, 2.8 minutes.

TABLE V
RELATIVE RETENTION TIMES OF TRIMETHYLSILYL (TMSi)
DERIVATIVES OF SOME TRITERPENES[a]

Compound and conditions	2% XE-60	1.5% QF-1	1% NGS	1.5% SE-30 (not TMSi derivative)
(23) α-Amyrin TMSi	3.77	3.65	3.96	3.65
(26) β-Amyrin TMSi	3.34	3.12	3.42	3.26
(39) Taraxerol TMSi	3.15	3.12	3.33	3.14
(25) Taraxasterol TMSi	13.0	7.52	—	5.75
Cholestane	1	1	1	1
	(5.7 min)	(1.7 min)	(5.5 min)	(2.8 min)
Column temperature	225°	220°	220°	240°
Column (cm × mm i.d.)	225 × 4	150 × 4	150 × 4	150 × 4
Carrier gas, N₂ (ml/min)	90	90	90	90

[a] Data from N. Ikekawa, S. Natori, H. Itokawa, S. Tobinaga, and M. Matsui, *Chem. Pharm. Bull. (Tokyo)* **13**, 316 (1965).

TABLE VI
RELATIVE RETENTION TIMES OF TRITERPENE METHYL ETHERS[a,b]

Compound	0.5% Apiezon L	1.5% SE-30	1.5% QF-1	1.0% CDMS[e]
Germanicol methyl ether (miliacin)	2.83	2.54	2.76	3.79
δ-Amyrin methyl ether (isomiliacin)	2.80	2.44	2.78	3.79
β-Amyrin methyl ether (isosawamilletin)	2.79	2.45	2.89	3.77
Taraxerol methyl ether (sawamilletin)	2.74	2.45	2.75	3.67
Multiflorenol methyl ether	2.74	2.46	2.83	3.77
α-Amyrin methyl ether	3.20	2.73	3.17	4.25
Bauerenol methyl ether	4.11	3.24	3.42	5.59
Fernenol methyl ether (arundoin)	4.31	3.20	3.52	5.50
Isoarborinol methyl ether (cylindrin)	4.95	3.43	3.47	6.25
Cholestane	1.00	1.00	1.00	1.00
	(11–14 min)	(3–4 min)	(2.2–2.8 min)	(2.5–3 min)
Column temperature	240°	240°	225°	240°

[a] Data from T. A. Bryce, M. Martin-Smith, G. Osske, K. Schreiber, and G. Subramanian, *Tetrahedron* **23**, 1283 (1967).
[b] Solid support, Gas Chrom Z, 100–120 mesh; carrier gas, Ar, 60 ml/min.
[e] CDMS = HiEff-8b.

Related compounds having methyl, hydroxymethyl, and carboxyl groups at C-17 in the ursane, oleanane, and lupane series have been isolated from natural sources, and some correlations of retention times among these compounds are possible. For example, the separation factor of methyl and carbomethoxy compounds (α-amyrin and methyl ursolate) is about 1.5, and that of methyl and hydroxymethyl compounds (β-amyrin and erythrodiol) is about 2.2.

Recently, thirteen hydrocarbons of the hopane series listed in Table III have been either isolated or derived in the studies of triterpenoid constituents of ferns and other plants.[21] Some correlations between positions of double bonds and behavior on GLC have thus been made possible and may be of value in biogenetic studies or in investigations of the acid-catalyzed isomerizations of compounds in this series.

Table III summarizes the retention times observed, related to cholestane, with the liquid phases SE-30 and NGS. Although satisfactory resolution is often obtained on SE-30, NGS gives much better results, and the order of retention times on these nonpolar and polar phases is found to be slightly different, as indicated below.

$$\text{SE-30} \quad \Delta^{17(21)} < \Delta^8 < \Delta^{7,9(11)} < \Delta^{13(18)} < \Delta^{9(11)} < \Delta^{5(10)} < \Delta^5 < \Delta^{1(10),5}$$
$$< \Delta^7 < \Delta^0 < \Delta^{22(29)} < \Delta^{21} < \Delta^3$$

$$\text{NGS} \quad \Delta^{17(21)} < \Delta^{7,9(11)} < \Delta^8 < \Delta^{13(18)} < \Delta^{9(11)} < \Delta^{5(10)} < \Delta^5 < \Delta^7$$
$$< \Delta^{1(10),5} < \Delta^0 < \Delta^{21} < \Delta^3 < \Delta^{22(29)}$$

The compounds having a double bond in the side chain or at C-3 to C-4 show longer retention times than that of hopane, the corresponding saturated hydrocarbon. On the other hand, the compounds having a double bond in the ring system show shorter retention times on both phases. The magnitude of the effect of a double bond depends upon its position in the molecule, but generally tetrasubstituted double bonds in the skeleton, such as $\Delta^{17(21)}$, Δ^8, and $\Delta^{13(18)}$, have been found to shorten the retention time more than the trisubstituted double bond on both phases. It is also interesting that in the case of ferna-7,9(11)-diene the decreasing effect of double bonds at C-7 and C-9 on the retention time is additive. The additivity of the effect of double bonds in increasing the retention time had previously been observed in the case of sterols.[8, 22, 23]

[21] H. Ageta, K. Iwata, and K. Yonezawa, *Chem. Pharm. Bull.* (*Tokyo*) **11**, 407 (1963); H. Ageta, K. Iwata, and S. Natori, *Tetrahedron Letters* p. 1447 (1963); M. N. Galbraith, C. J. Miller, J. W. L. Rawson, E. Ritchie, J. S. Shannon, and W. C. Taylor, *Australian J. Chem.* **18**, 226 (1965); R. T. Aplin, H. R. Arthur, and W. H. Hui, *J. Chem. Soc.* p. 1251 (1966).

[22] K. Tsuda, K. Sakai, and N. Ikekawa, *Chem. Pharm. Bull.* (*Tokyo*) **9**, 835 (1961).

[23] Among the sterols, the correlation between structure and retention time has been investigated by Tsuda *et al.*,[22] Clayton,[8] and Knights.[24]

In the naturally occurring steroids, the number of variations in the stereochemistry of the ring system are few. The *cis*- or *trans*-junction of rings A and B is the commonest single variant. Even here, however, this steric modification is known to influence the contribution of double bonds and other structural changes elsewhere in the molecule. In the polycyclic triterpenes, the range of isomeric possibilities, due to 1–2 shifts of methyl groups and hydrogen atoms during biogenesis is considerably greater, and these steric modifications complicate the comparison of GLC data for related compounds.

However, these data, like others obtained for steroids, indicate that the retention time of a compound is affected by its polarity and molecular volume or shape. Usually the effect of a double bond is to increase the polarity and hence the retention time, but a decrease of retention time following introduction of a double bond may also occur and is probably due to molecular volume changes. The polarity contribution of double bonds within the ring system, especially in tetrasubstituted positions, may be reduced by steric effects, while changes in molecular shape may curtail solute-solvent interaction. On the other hand, double bonds outside the ring system, i.e., the side chain, increase the polarity of a molecule, and hence the retention time.[25]

The difference of orientation of the isopropyl side chain in the hopane series greatly effects the retention time.[26]

Applications

It will be clear that much remains to be done before a satisfactory theoretical understanding of structure-retention time relationships among these compounds is achieved. However, many of the relationships between positions of double bonds and retention times noted for the hopane series may be expected to apply in other series of pentacyclic triterpenes, such as the oleanane, ursane, and lupane groups, and thus may be helpful in structural determination.

A number of examples of the separation of triterpenes of plants have been described. Some of these have involved: rice-bran oil,[15] leaf waxes from Cuban sugar cane,[17, 27] potato plants,[27] rosebay willowherb,[28] vege-

[24] B. A. Knights, *J. Gas Chromatog.* **2**, 160, 338 (1964).
[25] In the steroid series it is established that Δ^9, Δ^{16}, and Δ^{22} bonds decrease the retention time of saturated parent compound, whereas Δ^5, Δ^6, Δ^7, Δ^8, $\Delta^{8(14)}$, Δ^{14}, Δ^{24}, and $\Delta^{24(28)}$ bonds increase it.
[26] Y. Tsuda, K. Isobe, S. Fukushima, H. Ageta, and K. Iwata, *Tetrahedron Letters* p. 23 (1967).
[27] G. Osske and K. Schreiber, *Tetrahedron* **21**, 1559 (1965).
[28] A. T. Glen, W. Lawrie, J. McLean, and M. E. Younes, *Chem. & Ind.* p. 1908 (1965).

table oils,[29] certain grasses,[30] etc. Triterpenes have been isolated from petroleum distillates and oil shale bitumen by Hills *et al.*[5] Castle *et al.*[31] and Rees *et al.*[32] have studied the biosynthesis of plant sterols and triterpenes using gas chromatography.

a) Unsaponifiable fraction

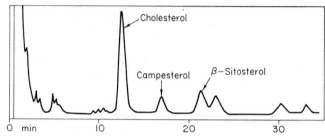

b) Unsaponifiable fraction without sterols

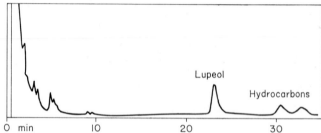

c) Triterpene fraction obtained by silica gel column

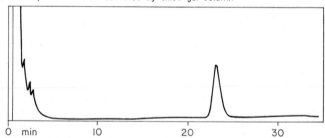

FIG. 1. Gas chromatograms of unsaponifiable fractions from the blood of silkworm larvae. Conditions: 0.75% SE-30; 180 cm × 4 mm i.d.; column temperature, 232°; N_2, 80 ml/min.

[29] E. Fedeli, A. Lanzani, P. Capella, and G. Jacini, *J. Am. Oil Chemists Soc.* **43**, 254 (1966).
[30] K. Nishimoto, M. Ito, S. Natori, and T. Ohmoto, *Tetrahedron Letters* p. 2245 (1965); *Chem. Pharm. Bull. (Tokyo)* **13**, 224 (1965); *ibid.* **14**, 97 (1966); T. Ohmoto, *Japan. J. Pharmacognosy* **20**, 67 (1966).
[31] M. Castle, G. A. Blondin, and W. R. Nes, *J. Am. Chem. Soc.* **85**, 3306 (1963).
[32] H. H. Rees, E. I. Mercer, and T. W. Goodwin, *Biochem. J.* **99**, 726 (1966).

As an example of the application of gas chromatography in the identification and isolation of a polycyclic triterpene in biological fluid, detection of the triterpene lupeol in silkworm blood is described.[20] The larval blood was homogenized and extracted with methanol. After removal of the methanol, the residue was refluxed with ether and the ether extract was saponified with 10% KOH in ethanol. Gas chromatography of the unsaponifiable fraction including sterols, hydrocarbons (C_{33}, C_{34}), and lupeol gave the results shown in Fig. 1a. After removal of the sterols as digitonides from the unsaponifiable fraction, GLC gave the results shown in Fig. 1b. The material was now chromatographed on a silica gel column. The fraction eluted by 10% benzene in hexane was found by GLC to contain the triterpene alcohol as shown in Fig. 1c. The retention time of this substance was identical with that of lupeol, which was isolated in crystalline form from this fraction.

Acknowledgments

The author wishes to express his gratitude and deep appreciation to Dr. R. B. Clayton, Editor, and Professor K. Schreiber, Institut für Kulturpflanzenforschung, Gatersleben, for their review of the manuscript and their contribution of many valuable suggestions. He is also deeply indebted to Drs. M. Shimizu, Daiichi Seiyaku Co., Ltd., S. Natori, National Institute of Hygienic Sciences, H. Ageta, S. Tobinaga, Showa College of Pharmacy, O. Tanaka, Tokyo University, and H. Itokawa, Tokyo College of Pharmacy, for gifts of triterpenes, without which the studies reported here could not have been carried out.

[4] Separation of Steroids by Means of Ion-Exchange Resins

By TOKUICHIRO SEKI

The chromatographic system to be described in this article utilizes ion-exchange resins as the stationary phase for the fractionation of steroids of medium and higher polarity. Aqueous alcohols are used as more polar moving phases, and mixtures of other organic solvents nearly saturated with water are used as less polar moving phases. The sample components distribute themselves between the moving liquid phase and the stationary resin phase with which it is in equilibrium. When one of the polar (aqueous alcoholic) phases is used, the elution sequence of steroids is quite similar to that of reversed-phase partition chromatography. When less polar liquid phases are used, the elution sequence resembles that of normal-phase partition chromatography. Since ethanol and water are used as the moving phase or as the polar components of the less polar moving phase, carboxylic acid type ion-exchange resin, Amberlite IRC-50 (A.G.) must be esterified with alcohol to avoid changes of the sorption properties of the resin during chromatography. Dowex 50W-X4 is used after conditioning.

The chromatographic systems permit separation of various 17-hydroxycorticosteroids,[1,2] 17-ketosteroids,[2,3] and estrogens.[4,5] The particular advantage of this method is that the columns can be used repeatedly and the resolving power does not change after repeated uses. The steroids studied are shown in Table I.

Preparation of Stationary Phase

Dowex 50W-X4. Sodium ion form of the resin (200–400 mesh) is first fractionated in deionized water by the sedimentation method,[6] and the particles that do not sediment within 3 minutes in a 1-liter beaker are collected. The finer resin particles thus obtained are then classified hydraulically according to the method of Hamilton,[7] as modified by Vassiliou and Kunin,[8] and particles of 47–58 μ are obtained. These are transferred to a glass filter and washed successively with 6 N hydrochloric acid (10 volumes), water, 2 N sodium hydroxide (10 volumes), water, 2 N hydrochloric acid (10 volumes), water, and finally 95% ethanol.

Amberlite IRC-50. Commercial Amberlite IRC-50 (A. G.) is converted to the sodium ion form by treatment with 5 volumes of 2 N sodium hydroxide and then is washed with deionized water. The resulting sodium salt of the resin is pulverized in a ball mill without drying. The powdered resin is suspended in deionized water, and preliminary fractionation is performed by the sedimentation method.[6] Then the resin is classified hydraulically,[7,8] and particles of 30–35 μ, 35–40 μ, 40–45 μ, 55–65 μ, and 60–72 μ are collected. These are transferred to a glass filter, washed with 1 N hydrochloric acid until the filtrate is acid, and then washed with 10 volumes of acetone, followed by distilled water, 1 N sodium hydroxide, water, and 1 N hydrochloric acid. Particles of 55–65 μ are refluxed with aqueous acidic alcohol A, those of 60–72 μ with aqueous acidic alcohol B, and those of 30–35 μ, 35–40 μ, and 40–45 μ with acidic alcohol C for 60 hours (Table II). One liter of acidic alcohol is used with 50 ml of H form Amberlite IRC-50, and bumping of the mixture is prevented by

[1] T. Seki and K. Matsumoto, *Endocrinol. Japon.* **13**, 75 (1966).
[2] T. Seki and K. Matsumoto, *J. Chromatog.* **27**, 423 (1967).
[3] T. Seki and K. Matsumoto, *J. Chromatog.* **10**, 400 (1963).
[4] T. Seki, *Nature* **181**, 768 (1958).
[5] K. Matsumoto and T. Seki, *Endocrinol. Japon.* **10**, 136 (1963).
[6] P. Decker and H. Holler, *J. Chromatog.* **7**, 392 (1962).
[7] P. B. Hamilton, *Anal. Chem.* **30**, 914 (1958).
[8] B. Vassiliou and R. Kunin, *Anal. Chem.* **35**, 1328 (1963).

TABLE I
Steroid Nomenclature

Number in the figures	Systematic name	Trivial name
1	11β,17α,21-Trihydroxypregn-4-ene-3,20-dione	Cortisol
2	17α,21-Dihydroxypregn-4-ene-3,11,20-trione	Cortisone
3	3α,11β,17α,21-Tetrahydroxy-5β-pregnan-20-one	Tetrahydrocortisol
4	3α,17α,21-Trihydroxy-5β-pregnane-11,20-dione	Tetrahydrocortisone
5	3α,11β,17α,21-Tetrahydroxy-5α-pregnan-20-one	Allotetrahydrocortisol
6[a]	6β,17α,21-Trihydroxypregn-4-ene-3,20-dione	6β-Hydroxy-11-deoxycortisol
7	17α,21-Dihydroxypregn-4-ene-3,20-dione	11-Deoxycortisol
8	3α,17α,21-Trihydroxy-5β-pregnan-20-one	Tetrahydro-11-deoxycortisol
9	3-Hydroxyestra-1,3,5(10)-trien-17-one	Estrone
10	2-Methoxy-3-hydroxyestra-1,3,5(10)-trien-17-one	2-Methoxyestrone
11	Estra-1,3,5(10)-triene-3,17α-diol	Estradiol-17α
12	Estra-1,3,5(10)-triene-3,17β-diol	Estradiol-17β
13	Estra-1,3,5(10)-triene-3,16α17β-triol	Estriol
14	Estra-1,3,5(10)-triene-3,16β,17β-triol	16-Epiestriol
15	3α,11β-Dihydroxy-5β-androstan-17-one	11β-Hydroxyetiocholanolone
16	3α-Hydroxy-5β-androstane-11,17-dione	11-Ketoetiocholanolone
17	3α,11β-Dihydroxy-5α-androstan-17-one	11β-Hydroxyandrosterone
18	3α-Hydroxy-5α-androstane-11,17-dione	11-Ketoandrosterone
19	11β-Hydroxyandrost-4-ene-3,17-dione	11β-Hydroxyandrostenedione
20	Androst-4-ene-3,11,17-trione	Adrenosterone
21	6β-Hydroxyandrost-4-ene-3,17-dione	6β-Hydroxyandrostenedione
22	3α,17-Dihydroxy-5β-pregnan-20-one	17α-Hydroxypregnanolone
23	3β-Hydroxyandrost-5-en-17-one	Dehydroepiandrosterone
24	3α-Hydroxy-5β-androstan-17-one	Etiocholanolone
25	3β-Hydroxy-5α-androstan-17-one	Epiandrosterone
26	3α-Hydroxy-5α-androstan-17-one	Androsterone
27	Androst-4-ene-3,17-dione	Androstenedione
28	5β-Androstane-3,17-dione	Etiocholanedione
29[a]	5α-Androstane-3,17-dione	Androstanedione
30	5β-Pregnane-3α,17α,20α-triol	Pregnanetriol

[a] Not shown in the figures.

adding two or three pieces of boiling stone (8-mm cube) the edges of which are removed by grinding.

Preparation of the Column

Dowex 50W-X4. The conditioned resin is washed with a mixture of 10 volumes of ethanol–benzene–water (50:100:1 by volume), and then suspended by swirling in about 2 volumes of the same solvent. The

TABLE II

Composition of Aqueous Acidic Alcohols Used for Esterification of Resin

Aqueous acidic alcohol	Composition (by volume)	
A	Methanol–ethanol[a]–2 N HCl	1:3:2
B	Methanol–ethanol[a]–2 N HCl	1:5:2
C	Ethanol[a]–conc. HCl	19:1

[a] 99% ethanol.

suspension is poured into a chromatographic tube through a small funnel fitted with a ground joint and allowed to settle. The next day, the solvent above the column is removed and the column is washed with eluent K or L (Table III). When about 300–500 ml of the solvent have been passed through the column, it is ready for use.

Amberlite IRC-50. The columns used with the aqueous alcoholic eluent are prepared with 55–65 μ particles and 60–72 μ particles which are partially esterified as described above. The former are washed with eluent D, and the latter with eluent E (Table III), and then suspended in 2 volumes of the same eluent. The suspension is poured into a chromatographic tube as described above and packed under an air pressure of 1 atmosphere above atmospheric pressure. After about 200 ml of the eluent have been passed through, it is ready for use.

The column used with eluent F, G, H, I, or J (Table III) is packed in the following way. The resin esterified with acidic alcohol C is washed successively with 10 volumes of ethanol (99%), 10 volumes of a mixture of benzene–ethanol (99.5%)–water (100:50:1 by volume) and 20 volumes of eluent F, H, or I (35–40 μ particles), eluent G (30–35 μ particles), or eluent J (40–45 μ particles). The washed resin is suspended in

TABLE III

Composition of Eluents

Eluent	Composition (by volume)	
D	Methanol–ethanol[a]–water	3:9:8
E	Methanol–ethanol[a]–water	3:15:11
F	Ethanol[b]–benzene–n-hexane–water	50:350:80:3.3
G	Ethanol[b]–benzene–cyclohexane–water	50:350:300:2.3
H	Ethanol[b]–benzene–cyclohexane–water	30:170:200:1.4
I	Ethanol[b]–benzene–cyclohexane–water	30:140:210:1.4
J	Ethanol[b]–benzene–cyclohexane–water	25:50:225:1
K	Ethanol[b]–benzene–n-hexane–water	50:350:80:3
L	Ethanol[b]–benzene–n-hexane–water	20:140:60:1

[a] 99% ethanol.
[b] 99.5% ethanol.

three volumes of the eluent used for the washing and the suspension is poured into a chromatographic tube and allowed to settle. After about 300 ml of the eluent has been passed through the column, it is ready for use.

Quantitative Determination of Steroids

17-Hydroxycorticosteroids. 17-Hydroxycorticosteroids are analyzed by Porter-Silber reagent,[9] prepared by dissolving 50 mg of phenylhydrazine sulfate in a mixture of 21 ml of ethanol (99.5%) and 39 ml of sulfuric acid (158 ml of sulfuric acid to 37 ml of deionized water). Two volumes of the reagent are added to each fraction when eluent D is used. When eluent G, H, or I is used, each fraction is allowed to evaporate at room temperature, the residue is dissolved in 0.5 ml of 60% ethanol, and 1.0 ml of Porter-Silber reagent is then added to each fraction. The mixture is allowed to stand at room temperature for 15 hours, and the optical density finally is measured at 410 mμ.

17-Ketosteroids. Analysis of 17-ketosteroids may be performed by the Epstein method[10] using aqueous Zimmermann reagent after evaporation of the eluent. The eluent is evaporated by placing the test tubes in a suitable rack and heating them in a boiling water bath for about 30 minutes. The residue is dissolved in 0.05 ml of methanol; 0.2 ml of saturated solution of *m*-dinitrobenzene in 5% Hyamine 1622 solution is added, followed by 0.1 ml of 8 N aqueous potassium hydroxide solution with mixing. After the preparation has stood for 30 minutes, 2.0 ml of 5% aqueous Hyamine 1622 solution is added and the optical density is measured at 510 mμ against deionized water.

Estrogens. Estrogens may be measured by their ultraviolet absorption at 280 mμ when eluent E is used. When eluents G, J, K, and L are used, each fraction is allowed to evaporate at room temperature and estrogens are determined either by their ultraviolet absorption at 280 mμ after redissolving in alcohol or fluorimetrically.[11]

Systematic Separation of 17-Hydroxycorticosteroids and 17-Ketosteroids

Separation of 17-hydroxycorticosteroids and 17-ketosteroids is performed on esterified Amberlite IRC-50. As described above, aqueous eluent (eluent D) is used with the resin of a larger particle size (55–65 μ). This system permits the separation of steroids with widely different polarities by a one-step elution and also has provided good separation of the 5α- and 5β-isomers of 3α-hydroxysteroids so far investigated. The

[9] R. H. Silber and C. C. Porter, *Methods Biochem. Anal.* 4, 139 (1957).
[10] E. Epstein, *Clin. Chim. Acta* 7, 735 (1962).
[11] R. W. Bates and H. Cohen, *Federation Proc.* 6, 236 (1947).

TABLE IV

RELATIVE ELUTION VOLUME OF STEROIDS AND DNP-AMINES

Steroids	\multicolumn{6}{Eluent}
	D[a]	F[b]	G[c]	H[c]	I[c]	J[d]
Cortisol	0.50	0.69	—	—	—	—
Cortisone	0.53	0.46	—	—	—	—
Tetrahydrocortisol	0.50	1.00	—	—	—	—
Tetrahydrocortisone	0.50	0.72	—	—	—	—
Allotetrahydrocortisol	0.56	1.05	—	—	—	—
6β-Hydroxy-11-deoxycortisol	0.49	0.91	—	—	—	—
11-Deoxycortisol	0.67	—	1.00	1.00	1.00	—
Tetrahydro-11-deoxycortisol	0.64	—	1.11	—	0.93	—
Estrone	0.63	—	0.34	—	0.31	—
11β-Hydroxyetiocholanolone	0.66	—	0.89	—	0.84	—
11-Ketoetiocholanolone	0.69	—	0.69	—	0.70	—
11β-Hydroxyandrosterone	0.74	—	0.85	0.82	0.82	—
11-Ketoandrosterone	0.74	—	0.64	0.64	0.68	—
11β-Hydroxyandrostenedione	0.67	—	0.62	—	0.64	—
Adrenosterone	0.73	—	0.54	0.59	0.64	—
6β-Hydroxyandrostenedione	0.65	—	0.78	—	—	—
17α-Hydroxypregnanolone	0.73	—	—	0.48	—	—
Dehydroepiandrosterone	0.77	—	—	0.40	—	—
Etiocholanolone	0.88	—	—	—	—	0.76
Epiandrosterone	0.92	—	—	—	—	0.85
Androsterone	1.00	—	—	—	—	—
Androstenedione	0.94	—	—	—	—	1.00
Etiocholanedione	0.88	—	—	—	—	0.57
Androstanedione	1.02	—	—	—	—	—
Pregnanetriol	0.95	—	—	—	—	1.75
N-2,4-Dinitrophenylethylamine (DNP-ethylamine)	—	—	—	—	—	0.57
N-2,4-Dinitrophenylethanolamine (DNP-ethanolamine)	—	0.76	0.71	0.75	—	—
N-2,4-Dinitrophenylisopropylamine (DNP-isopropylamine)	—	—	—	—	—	0.44

[a] The elution volume of androsterone was taken as 1.00.
[b] The elution volume of tetrahydrocortisol was taken as 1.00.
[c] The elution volume of 11-deoxycortisol was taken as 1.00.
[d] The elution volume of androstenedione was taken as 1.00.

separation of 5α- and 5β-isomers of some 3-oxosteroids has also been possible.[2, 12]

The steroids that are poorly resolved, if at all, by the system using aqueous eluent are separated from each other by less polar eluents (eluents F, G, H, I, and J) on esterified Amberlite IRC-50 of smaller particle sizes.

[12] K. Matsumoto and T. Seki, *Endocrinol. Japon.* **9**, 201 (1962).

First Chromatography. Steroids are separated using eluent D. A sample is dissolved in 0.45 ml of a mixture of methanol and ethanol (99%, 1:3 by volume), and 0.35 ml of deionized water is added to the solution with mixing. The solution is then applied to the column and overlayered carefully with the eluent D; elution is performed at 33° under an air pressure of 1 atmosphere above atmospheric pressure. The eluate is collected in fractions of 21 drops using a drop count type automatic fraction collector and a flow rate of 2.5 fractions per hour. The elution pattern is shown in Fig. 1a, and the relative elution volume of some steroids was shown in Table IV.

The elution pattern of steroids is quite reproducible when the temperature of the column is kept constant, and scarcely any change of the elution pattern is observed even in the presence of large amounts of impurities such as are present in a crude urine extract. Since the esterification of the resin is performed with an aqueous acidic alcohol having nearly the same composition as the eluent, there is no danger of deesterification with time, and the column can be used for several years without repacking. The column may be washed once or twice a year with a solvent containing 0.3 N hydrochloric acid instead of deionized water to remove cations adsorbed at the top of the column.

Differences in the elution pattern and/or resolving power have been observed with different lots of Amberlite IRC-50.[2] The greatest variation with different lots of resin is in the position of cortisone relative to those of cortisol and allotetrahydrocortisol. With some lots of Amberlite IRC-50, the temperature of the column must be kept at 33° for cortisone to be eluted just between cortisol and allotetrahydrocortisol, whereas with other lots, the optimum temperature for this is 29°. The elution patterns of Fig. 1 were obtained with a column prepared from resin bought in February, 1964. The elution pattern of the steroids obtained with a column prepared from a resin bought in July, 1965, was somewhat less satisfactory, but the use of a finer resin fraction (45–50 μ) gave satisfactory resolution.

The recovery of steroids from the column is more than 90%[2,12,13] and with the more polar steroids it is almost quantitative. The optimum load of steroids is 50–500 μg/cm² for each component, and milligram amounts of steroids can be chromatographed when there is a good separation of components. The lower quantitative limits for effective separation has not been determined, but is obviously exceedingly low. Submicrogram quantity of steroids, e.g., 0.5 μg of cortisol and 0.2 μg of corticosterone, can be separated and estimated fluorimetrically. Recovery of ³H-labeled steroids and ¹⁴C-labeled steroids is also satisfactory, and scarcely any isotope effect is observed with steroids labeled with ³H or ¹⁴C.

[13] T. Seki, *J. Chromatog.* 2, 667 (1959).

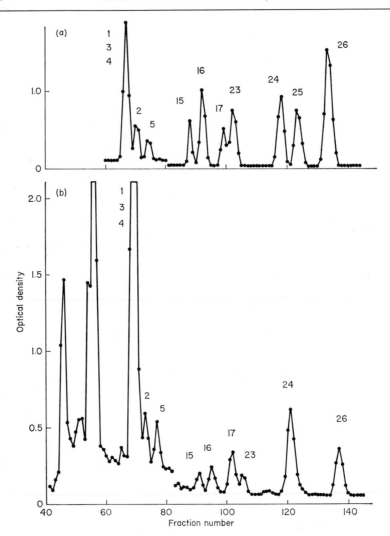

Fig. 1. Elution of 17-hydroxycorticosteroids and 17-ketosteroids. Elution patterns (a) and (b) were obtained in experiments performed under the same conditions. The size of the column was 0.6×131 cm, the moving phase was eluent D, and the temperature of the column was $33°$. The effluent was collected in fractions of 21 drops, and the flow rate was 2.5 fractions per hour. (a) Synthetic mixture. (b) Neutral fraction of β-glucuronidase hydrolyzate of normal human female urine.

Second Chromatography. 17-Hydroxycorticosteroids are separated by eluent F. A sample is dissolved in 0.5 ml of a mixture of ethanol (99.5%)–benzene–water (30:210:1 by volume), and then 0.2 ml of a mixture of *n*-hexane–carbon tetrachloride (5:1 by volume, mixture M) is added with

mixing. This solution is applied to the column and overlayered carefully
with eluent F and elution is performed with the same eluent at 22°. The
eluate is collected in fractions of 20 drops and a flow rate of 1.5 fractions
per hour. The elution pattern is shown in Fig. 2a.

Estrone, 11β-hydroxyetiocholanolone, 11-ketoetiocholanolone, 11-de-
oxycortisol, tetrahydro-11-deoxycortisol, 11β-hydroxyandrostenedione,

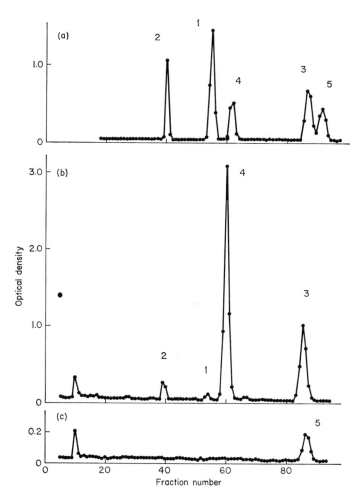

FIG. 2. Elution of 17-hydroxycorticosteroids. Elution patterns (a), (b), and (c)
were obtained in experiments performed under the same conditions. The size of the
column was 0.5 × 71 cm, the moving phase was eluent F, and the temperature was
22°. Effluent was collected in fractions of 20 drops, and the flow rate was 1.5 fractions
per hour. (a) Synthetic mixture of standard samples. (b) Fractions 67–75 (Fig. 1).
(c) Fractions 76–79 (Fig. 1).

and 6β-hydroxyandrostenedione may be separated from each other by eluent G. A sample is dissolved in a 0.3 ml of a mixture of ethanol (99.5%)–benzene–water (50:350:1 by volume) with about 50 μg of DNP-ethanolamine (as marker); then 0.4 ml of a mixture of cyclohexane and carbon tetrachloride (5:1 by volume, mixture N) is mixed with the solution. The preparation is applied to the column and overlayered carefully with eluent G, and elution is then performed with the same eluent. The elution pattern is shown in Fig. 3a.

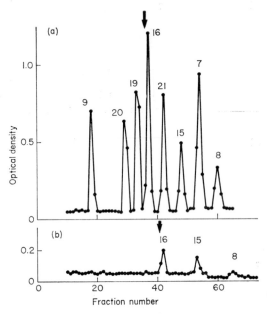

FIG. 3. Elution of 17-ketosteroids and 17-hydroxycorticosteroids. Elution patterns (a) and (b) were obtained in experiments performed under the same conditions. The size of the column was 0.5 × 77 cm, the moving phase was eluent G, and the temperature was 22°C. The arrows in the figure indicate the positions in which DNP-ethanolamine was eluted. The effluent was collected in fractions of 20 drops and the flow rate was 1.5 fractions per hour. (a) Synthetic mixture. (b) Fractions 85-98.

11β-Hydroxyandrosterone, 11-ketoandrosterone, adrenosterone, dehydroepiandrosterone, and 17α-hydroxypregnanolone are separated from each other by eluent H. A sample is dissolved in 0.3 ml of a mixture of ethanol (99.5%)–benzene–water (60:340:1 by volume) with about 50 μg of DNP-ethanolamine, and then 0.4 ml of mixture N is added with mixing. The solution is applied to the column and overlayered carefully with eluent H, and elution is performed with the same eluent. The elution pattern is shown in Fig. 4a.

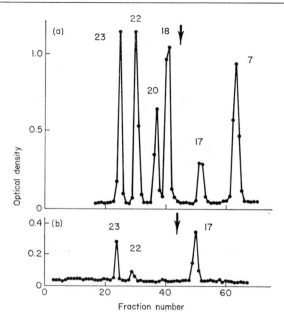

FIG. 4. Elution of 17-ketosteroids. Elution patterns (a) and (b) were obtained by experiments performed under the same conditions. The size of the column was 0.5×61 cm, the moving phase was eluent H, and the temperature was 22°. The arrows in the figure indicate the position in which DNP-ethanolamine was eluted. The effluent was collected in fractions of 20 drops, and the flow rate was 1.5 fractions per hour. (a) Synthetic mixture. (b) Fractions 99–108.

Etiocholanolone, epiandrosterone, androstenedione, etiocholanedione, and pregnanetriol are separated by eluent J. A sample is dissolved in 0.35 ml of a mixture of ethanol (99.5%)–benzene–cyclohexane–water (50:100: 150:1 by volume). Then 0.45 ml of mixture N is mixed with the solution and it is applied to the column, and overlayered carefully with eluent J; elution is performed with the same eluent. The elution pattern is shown in Fig. 5a. The relative elution volume of some steroids with eluent F–J are given in Table IV.

One of the advantages of the system described above is the higher solubility of steroids in eluents F–J than in the moving phase of normal-phase partition chromatography which utilizes Celite or silica gel as the supporting medium of the stationary phase. The maximum load is 5–10 mg/cm² for each component when the components are well separated. The column can be used repeatedly, since the ratio of ethanol to water in the eluents F–J is nearly equal to that in acidic alcohol C. The column can be preserved for many days with the eluent over the column while the flow of the eluent is stopped by inserting a suitable stopper in the

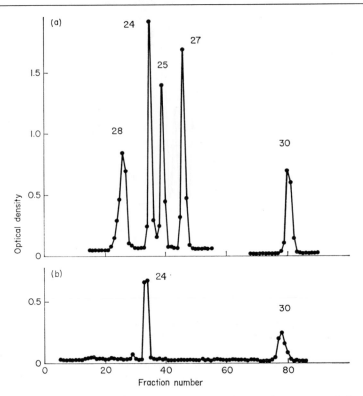

Fɪɢ. 5. Elution of 17-ketosteroids and pregnanetriol. Elution patterns (a) and (b) were obtained by experiments performed under the same conditions. The size of the column was 0.5×84 cm, the moving phase was eluent J, and the temperature was $22°$. Effluent was collected in fractions of 20 drops, and the flow rate was 1.3 fractions per hour. (a) Synthetic mixture. (b) Fractions 118–134.

bottom of the tube. With eluents F–J, the elution volume of steroids is not so constant as with eluent D, and in some cases it varies by about 10%. The most striking decrease in elution volume of steroids or broadening of the width of the peak was observed when a preserved column was used without prewashing with the eluent with which the column was in equilibrium. The reason for such deviations is not clear, they can be prevented by washing the column with 100–200 ml of the eluent before use. The relative elution volumes of steroids have shown little variation with different lots of Amberlite IRC-50 so far studied. The flow rate of the eluent through a column was 3–6 ml/cm²/hour, which was obtained by raising the solvent container 40–60 cm above the column in order to exert a hydrostatic pressure on the column.

Recovery of steroids was about 75–90% with 5–100 μg amount of steroids. With submicrogram amount of steroids, it is a little lower. An isotope effect was observed when cortisol-^{3}H corticosterone-^{3}H, and estrone-^{3}H are separated by eluents F, B and J, respectively, tritiated compounds being eluted slightly more slowly.

Analysis of Urinary Steroids. An example of an analysis of urinary steroids is described. The neutral fraction of a normal female urine (1/5 day's), obtained by hydrolysis with β-glucuronidase, is dissolved and transferred with 15 ml of methanol to a separatory funnel, and 20 ml of *n*-hexane and 1.5 ml of water are added. The less polar lipids are extracted into the *n*-hexane phase by shaking the sample for 5 minutes. The methanol layer is separated, and the *n*-hexane layer is washed with 10 ml of a mixture of methanol and water (9:1 by volume). To effect a preliminary purification, the combined methanolic extract is evaporated to dryness and transferred quantitatively with 5 ml of 60% aqueous ethanol to a small column (0.7 × 2.0 cm) of partially esterified Amberlite IRC-50 (60–80 μ measured in the wet sodium ion form) and overlayered carefully with 10 ml of 70% aqueous ethanol which is allowed to filter through the column. The filtrate is now evaporated to dryness in a rotary evaporator. The residue is separated by eluent D as described above, and each effluent fraction is divided into two equal parts. One is analyzed by Porter-Silber reaction (fractions 41–82) and Zimmermann reaction (fractions 83–145). The other half is used to separate the steroids which do not separate from each other by the first chromatography. In a typical analysis fractions 67–75 and fractions 76–79 were combined separately and chromatographed by eluent F. Eluent G was used to analyze the pooled fractions 85–98, and eluent H was used to analyze the pooled fractions 99–108. Fractions 118–134 were pooled and analyzed with eluent J. The elution patterns of each pooled fraction are shown in Figs. 1b, 2b, 2c, 3b, 4b, and 5b.

Chromatographic Separation of Estrogens

In the chromatographic separation of estrogens described below, the separation of those estrogens that are usually found in human urine and are resistant to alkali treatment is described. Aqueous alcoholic eluent is used to separate estrogens into four fractions[14] (first chromatography). When this system is used for the separation of a phenolic fraction of low-titer urines, it is found that each estrogen fraction is too impure for quantitative estimation of estrogens by fluorimetry.[15] Therefore individual fractions are purified further by second chromatography. Hydrogen ion form Dowex 50W-X4 in equilibrium with eluent K or L is suita-

[14] K. Matsumoto and T. Seki, *Endocrinol. Japon.* **10**, 183 (1963).
[15] R. Nozaki, *Folia Endocrinol. Japon.* **42**, 788 (1966).

ble for the separation of estriol fraction[15] and estradiol fraction, respectively, which are obtained by the first chromatography. The high efficiency of these chromatographic systems may be due to the high content of benzene residues of Dowex 50W-X4 and the presence of an aromatic ring in the estrogen molecule. Separation of estradiol-17β from estradiol-17α is accomplished on esterified Amberlite IRC-50 by means of eluent G. 2-Methoxyestrone can be separated from estrone by eluent J on esterified Amberlite IRC-50.

First Chromatography. In a typical analysis the estrogens were first separated by eluent E. A sample was dissolved in 0.45 ml of a mixture of methanol–ethanol (99%) (1:5 by volume) with about 50 μg of DNP-ethylamine as marker; 0.35 ml of deionized water was added with mixing. The solution was then applied to the column and overlayered carefully

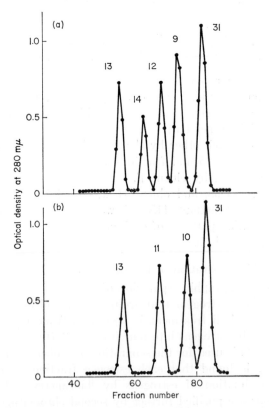

Fig. 6. Elution patterns of standard estrogens. Elution patterns (a) and (b) were obtained from the experiments performed under the same conditions. Size of the column was 0.6 × 112 cm, the moving phase was eluent E, and the temperature of the column was 32°. Effluent was collected in fractions of 20 drops, and the flow rate was 3.5 fractions per hour. (31: DNP-ethylamine.)

with eluent E; elution was performed under an applied air pressure of 1 atmosphere. The eluate was collected in fractions of 21 drops, and the flow rate was 3.5 fractions per hour. Elution patterns are shown in Fig. 6. The elution pattern of estrogens is quite reproducible when the temperature of the column is kept constant. Satisfactory separation of estrogens has been obtained with different lots of Amberlite IRC-50 so far studied, and the column can be used repeatedly. The recovery of estrogens is above 90%.

Second Chromatography. Estriol could be separated from other estrogens by partition chromatography on Dowex 50W-X4. A sample was dissolved in 0.5 ml of a mixture of ethanol (99.5%)–benzene–water (30:210:1 by volume), and 0.2 ml of mixture M was added with mixing. The solution was added to the column and overlayered carefully with eluent K; elution was performed with eluent K. The elution pattern is shown in Fig. 7.

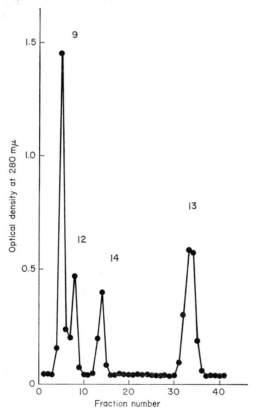

FIG. 7. Elution pattern of estrogens. The size of the column was 0.42 × 44 cm, the moving phase was eluent K, and the temperature was 22°. The effluent was collected in fractions of 20 drops, and the flow rate was 3 fractions per hour.

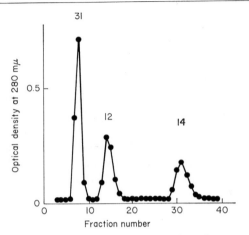

Fig. 8. Elution pattern of estrogens. The size of the column was 0.5×60 cm, the moving phase was eluent L, and the temperature was 22°. The effluent was collected in fractions of 20 drops and the flow rate was 3 fractions per hour. Estrone overlapped with DNP-ethylamine and estradiol-17α overlapped with estradiol-17β (31: DNP-ethylamine).

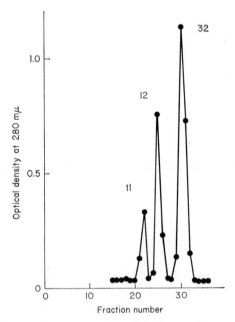

Fig. 9. Elution pattern of estrogens. The size of the column was 0.5×51 cm, the moving phase was eluent G, and the temperature was 22°. The effluent was collected in fractions of 20 drops, and the flow rate was 1.5 fractions per hour. (32: DNP-ethanolamine).

Estradiol-17β was separated from estrone and 16-epiestriol by means of a column of Dowex 50W-X4. A sample was dissolved in 0.5 ml of a mixture of ethanol (99.5%)–benzene–water (50:350:1 by volume) with 50 μg of DNP-ethylamine, and 0.3 ml of mixture M was added with mixing. The solution was added to the column and overlayered carefully with eluent L; elution was performed with the same eluent. The elution pattern is shown in Fig. 8.

Separation of estradiol-17β from estradiol-17α was possible with eluent G using a column of esterified Amberlite IRC-50. A sample was

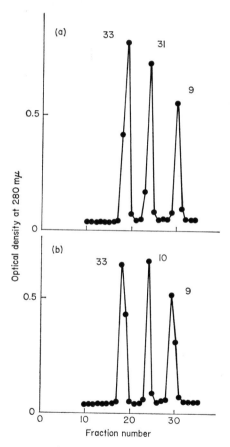

FIG. 10. Elution patterns of estrogens. The size of the column was 0.5 × 84 cm, the moving phase was eluent J, and the temperature was 22°. The effluent was collected in fractions of 20 drops, and the flow rate was 1.5 fractions per hour. The elution patterns (a) and (b) were obtained in separate experiments performed under the same conditions. (31: as in Fig. 8; 33: DNP-isopropylamine.)

chromatographed with DNP-ethanolamine and the elution pattern is shown in Fig. 9.

Estrone was separated from 2-methoxyestrone with eluent J on a column of esterified Amberlite IRC-50. A sample was chromatographed with DNP-ethylamine and/or with DNP-isopropylamine as described above; the elution pattern is shown in Fig. 10. The column of Dowex 50W-X4 could be used repeatedly, and the elution pattern was reproducible. Recovery of estrogens from Dowex column was about 70–80%.

Analysis of Urinary Estrogens. In a typical analysis of urinary estrogens, the phenolic fraction of a normal female urine (1/10 day's urine, preovulatory phase) was obtained by acid hydrolysis and extrac-

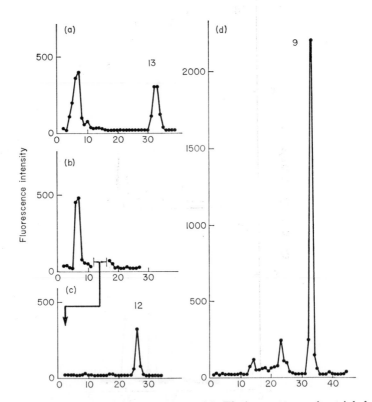

FIG. 11. Elution of urinary estrogens. (a) Elution pattern of estriol fraction chromatographed under the conditions described in Fig. 7. (b) Elution pattern of estradiol fraction chromatographed under the conditions described in Fig. 8. (c) Elution pattern of purified estrogen fraction (fractions 12–16 of the elution pattern b) chromatographed under the conditions described in Fig. 9. (d) Elution pattern of estrone fraction chromatographed under the conditions described in Fig. 10. Abscissas: fluorescence intensity; ordinates: fraction number.

tion according to the Brown method,[16] and purified as described above by filtering the solution in 60% aqueous ethanol through a small column of partially esterified Amberlite IRC-50. The filtrate was evaporated to dryness in a rotary evaporator, and the residue chromatographed as described above, using eluent E. The fractions containing estriol, estradiol-17β, and estrone were pooled separately and purified by chromatography with eluents K, L, G, and J. The position of the fractions containing three estrogens could be determined from the position of the fractions containing DNP-ethylamine. Estimation of 16-epiestriol was not attempted.

The estriol fraction was purified by eluent K. As shown in Fig. 11a, the efficiency of purification was high. The estradiol fraction was first purified by eluent L. The fractions containing estradiol-17β was pooled and further purified by eluent G. The elution patterns were shown in Figs. 11b and 11c. The estrone fraction was purified by eluent J (Fig. 11d).

[16] J. B. Brown, *Biochem. J.* **60**, 185 (1955).

[5] Methods of Analysis in the Biochemistry of Bile Acids

By P. ENEROTH and J. SJÖVALL

The quantitative analysis of mixtures of bile acids in biological materials includes preliminary extraction and purification followed by a chromatographic separation permitting the identification and quantitation of the individual compounds. The concentration and nature of the bile acid mixture as well as the amount and nature of the nonbile acid material in the extract will determine the extraction procedure and the number of purification steps required before the final analysis can be made.

The present chapter describes some procedures that have been satisfactory in work with biological extracts. The chromatographic separation of bile acids is described in the first section. The second section deals with specific applications to biological extracts where extraction and purification procedures are considered. The use of infrared spectroscopy in bile acid analysis will not be discussed. Appropriate references are found in the books by Haslewood[1] and van Belle.[2] The latter book also contains

[1] G. A. D. Haslewood, "Bile Salts." Methuen, London, 1967.
[2] H. van Belle, "Cholesterol, Bile Acids and Atherosclerosis." North Holland Publ., Amsterdam, 1965.

an extensive tabulation of physical constants for a large number of bile acid derivatives.

The term "free bile acid" will be used to mean a substituted 5α- or 5β-cholan-24-oic acid unless otherwise indicated.[3] Bile acids in peptide linkage with glycine or taurine form derivatives that will be referred to as conjugated bile acids.

Preparative Chromatography of Bile Acids

Free bile acids can be suitably prepurified by adsorption chromatography, liquid–liquid partition chromatography, countercurrent distribution,[4] and ion-exchange chromatography (see section on bile acids in blood) before analytical procedures are applied.

Adsorption Chromatography

Silicic acid should be used for free bile acids since more active adsorbents may give incomplete recoveries. Methyl esters or methyl ester acetates of bile acids are usually best separated on aluminum oxide. For specific purposes other adsorbents, e.g., Florisil, may be used.

Different commercial silicic acid preparations give different separations of the bile acids. Silicic acid Mallinckrodt [analytical reagent, 100 mesh (powder), Mallinckrodt Chemical Works, St. Louis, Missouri] and silica gel Davison 923 (Davison Chemical Corp., Baltimore, Maryland) have been most frequently used.

Silicic acid Mallinckrodt is activated for 5 days at 120° and is then rapidly packed in benzene in a chromatography tube. The height of the column should be about four times the diameter to give a suitable solvent flow. The column should provide at least a 25-fold excess of silicic acid with respect to the weight of the material applied. The bile acid mixture is dissolved in benzene (large volumes may sometimes be required depending on the nature of the extract), and the solution is applied onto the column. Nonpolar material is eluted with benzene until the effluent is free from solutes. Free bile acids are then eluted with increasing amounts of acetone, ethyl acetate, or acetic acid in benzene. Table I gives examples of the relative elution order of different bile acids in typical separations of fecal bile acid mixtures. The composition of the solvent required to desorb a bile acid may vary between different runs,

---

[3] The following systematic names are given bile acids referred to by trivial names: cholic acid, 3α,7α,12α-trihydroxy-5β-cholanoic acid; deoxycholic acid, 3α,12α-dihydroxy-5β-cholanoic acid; chenodeoxycholic acid, 3α,7α-dihydroxy-5β-cholanoic acid, ursodeoxycholic acid, 3α,7β-dihydroxy-5β-cholanoic acid; lithocholic acid, 3α-hydroxycholanoic acid.

[4] E. H. Ahrens, Jr., and L. C. Craig, *J. Biol. Chem.* **195**, 763 (1952).

TABLE I

SILICIC ACID CHROMATOGRAPHY OF FECAL BILE ACID FRACTIONS

Acetone % in benzene[a]	Bile acids eluted[b]	Ethyl acetate % in benzene[c]	Bile acids eluted[b]	Ethyl acetate % in benzene[d]	Bile acids eluted[b]
0	Cholanoic	5	3-keto	50	—
2	3-keto; (3β)	10	3β; 3α	55	3α,12α,7-keto; 3α,7α,12-keto
		15	3β,12-keto; (3α)		
4	3β			60	3α,12α,7-keto; 3α,7α,12-keto; (3α,7β,12α)
8	3β; 3α; (3-keto,12α); (3-keto,7α)	20	3β,12-keto; 3α,12-keto		
		22	3α,12-keto	65	3α,7β,12α; 3β,7β,12α; (3β,7α,12α); (3α,7α,12α)
12	3α; 3-keto,12α; 3-keto,7α; (3β,12-keto); (3α,12-keto)	24	3β,12α; 3α,12α		
				70	3β,7α,12α; 3α,7α,12α; (3α,7β,12α); (3β,7β,12α)
16	3β,12-keto; 3α,12-keto; (3β,12α)				
20	3β,12α; 3α,12α			75	3α,7α,12α; 3β,7α,12α
24	3α,12α			80	3α,7α,12α; (3β,7α,12α)
28	3α,12α			100	(3α,7α,12α)

[a] Silicic acid Mallinckrodt, 20-fold excess. The column was eluted with 20 ml of each solvent per gram of silicic acid.

[b] Substituted 5β-cholanoic acids. Greek letters denote configuration of hydroxyl groups on C-3, C-7, or C-12. Figures in parentheses indicate overlapping.

[c] Silicic acid Mallinckrodt, 25-fold excess. The column was eluted with 40 ml of each solvent per gram of silicic acid.

[d] Silicic acid Mallinckrodt, 50-fold excess. The column was eluted with 20 ml of each solvent per gram of silicic acid.

and exact reproducibility can be obtained only if the water content of the silicic acid and the solvents is standardized. However, in preparative work such reproducibility is usually not required since all fractions can be analyzed by thin-layer chromatography (TLC) and gas-liquid chro-

TABLE II

ALUMINUM OXIDE CHROMATOGRAPHY OF MONO- AND DISUBSTITUTED BILE ACID METHYL ESTERS FROM FECES[a]

Solvent	Heptane, % in benzene										Ethyl acetate, % in benzene	
	90	80	70	60	50	40	30	20	10	0	10	20
Compounds eluted[b]	3-keto; (3β)	3β; 3α	3α	3α	3α	3α; 3β,12-keto	3β,12-keto (3α,12-keto)	3α,12-keto	3α,12-keto (3β,12α)	3α,12-keto; (3β,12α)	3β,12α; (3α,12α)	3α,12α

[a] Aluminum oxide Woelm, activity grade V, 15-fold excess, eluted with 2 ml of each solvent per gram of aluminum oxide.
[b] For abbreviations see Table I.

matography (GLC). The use of a standardized group separation is described in the section on fecal bile acids.

For the separation of bile acid methyl esters, Wootton[5] used a 200-fold excess of silica gel Davison 923 in pentane–hexane, 1:1. Ether, 40% in pentane–hexane, eluted methyl lithocholate; methyl deoxycholate was eluted with ether, and methyl cholate with ether–acetone, 1:1. Columns providing such a large excess of adsorbent may permit the resolution of mixtures that are difficult to separate by other preparative methods. Thus, the diacetates of methyl deoxycholate and chenodeoxycholate were partially separated by prolonged elution of the above-mentioned column with 15% ether in pentane–hexane.

Several standardized preparations of aluminum oxide are commercially available. Our experience is limited to the use of Woelm neutral aluminum oxide (Woelm, Eschwege, Germany). The activity of the adsorbent is decreased by addition of water; addition of 3, 6, 10, and 15% (w/w) water yields aluminum oxide of activity grades II, III, IV, and V, respectively. Bile acid methyl esters should preferably be separated on low activity grade aluminum oxide (III–V) to minimize possible ester hydrolysis or other reactions on the adsorbent. Acetone should usually be avoided since it may form the high-boiling mesityl oxide especially on high activity grade aluminum oxide.

Solvent mixtures commonly employed in the aluminum oxide chromatography of bile acid esters include heptane–benzene and ethyl acetate–benzene. Aluminum oxide grade III is usually used for separations of mono- and disubstituted bile acid methyl esters. It is advisable to apply the sample in as nonpolar a mixture of heptane–benzene as possible (e.g., heptane–benzene, 1:1). If the sample is poorly soluble in this mixture it may be dissolved in benzene, and the solution is then diluted to the appropriate concentration with heptane. It is not necessary to keep the volumes at a minimum. Elution with increasing amounts of benzene in heptane followed by ethyl acetate in benzene usually yields methyl monohydroxycholanoates in the interval 80–90% benzene in heptane. Monohydroxy monoketocholanoates are eluted in the interval benzene/heptane 95:5–ethyl acetate/benzene 5:95, followed by dihydroxycholanoates in the interval ethyl acetate–benzene 5:95–25:75. To obtain these results large volumes of each solvent mixture (e.g., 20 ml per gram of aluminum oxide) should be used with a 20-fold excess of adsorbent. Alternatively a 200-fold excess of aluminum oxide may be used; these columns are eluted with smaller volumes (e.g., 2–5 ml/g) of more polar solvent mixtures. For comparison, a chromatography of fecal bile acids on aluminum oxide grade V is shown in Table II.

[5] I. D. P. Wootton, *Biochem. J.* **53**, 85 (1953).

In the choice between separation of free bile acids on silicic acid (Mallinckrodt) or of methyl esters on aluminum oxide (Woelm) the following points may be considered. Aluminum oxide can give better separations of stereoisomers of hydroxycholanoates than silicic acid (e.g., the separation of the methyl esters of deoxycholic and allodeoxycholic acids[6]). Silicic acid should usually be chosen when quantitative yields are of importance. In work with crude biological extracts silicic acid is the most convenient adsorbent since it can be used without prior esterification of the bile acids.

An example of the specific use of Florisil (Floridin Company, Tallahassee, Florida) is the separation of cholic and allocholic acids.[7] Florisil, 100–200 mesh, was washed with hydrochloric acid until the washings were colorless, then with water until neutral, and finally with methanol; it was dried at 280° for 2–3 hours. A 10-g column, 200 × 11 mm, was prepared in benzene for separation of up to 50 mg of bile acids. The bile acids were applied in 0.5 ml of methanol–benzene 1:1, and the column was eluted with 100 ml of benzene. Cholic acid was eluted with 8% methanol in benzene (25-ml fractions were collected). Part of the allocholic acid appeared in the later cholic acid fractions, but the bulk of this acid could be eluted with 15% methanol in benzene.

Partition Chromatography

With the adsorption chromatographic methods described, sometimes it may be difficult to separate bile acids with keto groups from those having an axial hydroxyl group in the corresponding position. A more efficient way to separate such bile acid pairs on a preparative scale is to use reversed-phase partition chromatography.

Siliconized diatomaceous earth, polyethylene powder (Hostalen, Farbwerke, Hoechst, Germany), and methylated Sephadex (Pharmacia, Uppsala, Sweden) have been used to support the stationary phase.[8,9] The following treatments are necessary before these supports can be used. Hyflo Super-Cel (Johns-Manville Co., New York) is a suitable and inexpensive type of diatomaceous earth. It is washed with 6 N hydrochloric acid until a colorless supernatant is obtained. The material is then washed with distilled water until neutral and then with acetone. The dried material is placed in dishes in a desiccator together with a beaker containing 50–100 ml of dimethyldichlorosilane and is left for 2–3 weeks.

[6] H. Danielsson, A. Kallner, and J. Sjövall, J. Biol. Chem. 238, 3846 (1963).
[7] H. J. Karavolas, W. H. Elliott, S. L. Hsia, E. A. Doisy, Jr., J. T. Matschiner, S. A. Thayer, and E. A. Doisy, J. Biol. Chem. 240, 1568 (1965).
[8] J. Sjövall, Methods Biochem. Anal. 12, 97 (1964).
[9] E. Nyström and J. Sjövall, J. Chromatog. 17, 574 (1965).

The support is then washed with ethanol until the filtrate is free from HCl and is dried at 100°. The hydrophobic support thus obtained has been found superior to supports silanized with solutions of dimethyldichlorosilane.

Hostalen is purified by continuous Soxhlet extraction with ethanol until the extract is colorless. The support is dried at temperatures below 75°. The procedure for methylation of Sephadex is described in this series.[10]

TABLE III
SOLVENT SYSTEMS FOR REVERSED-PHASE PARTITION
CHROMATOGRAPHY OF BILE ACIDS

Solvent system	Mobile phase	Amount (ml)	Stationary phase	Amount (ml)	Application
G[a]	Methanol–water	255:45	Heptane	50	Nonpolar bile acid derivatives, e.g., bile acid esters[b]
F2	Methanol–water	180:120	Chloroform–heptane	45:5	Separation of diketo and monohydroxy bile acids[c]
F1	Methanol–water	165:135	Chloroform–heptane	45:5	Separation of monohydroxy-, diketo-, dihydroxy-, and monohydroxymonoketo bile acids[c]
C1	Methanol–water	150:150	Chloroform–isooctanol[d]	15:15	Separation of glycine conjugated bile acids, trihydroxy- and dihydroxymonoketo bile acids[c]
C2	Methanol–water	144:156	Chloroform–isooctanol	15:15	Same as for C1
D	Water	300	n-Butanol	100	Taurine conjugated bile acids[e]

[a] Cannot be used with methylated Sephadex as the support.
[b] A. Norman, *Brit. J. Nutr.* **18**, 173 (1964).
[c] A. Norman and J. Sjövall, *J. Biol. Chem.* **233**, 872 (1958).
[d] 2-Ethyl-hexanol-1.
[e] A. Norman, *Acta Chem. Scand.* **7**, 1413 (1953).

Examples of solvent systems for reversed-phase chromatography of bile acids are given in Table III. After equilibration of the solvent mixture in a separatory funnel a volume of the stationary phase (4 ml/4.5 g of hydrophobic Super-Cel; 3 ml/4.5 g of Hostalen; 6 ml/4.5 g of methylated Sephadex) is added to the support and the mixture is thoroughly homogenized with a spatula for a short time (about half a minute to avoid solvent evaporation). Mobile phase is added, the slurry is ho-

[10] J. Ellingboe, E. Nyström, and J. Sjövall, see Vol. XIV [48].

mogenized and poured into a chromatography tube having a diameter giving columns with a height to diameter ratio of 10–20:1. Air bubbles are removed with a perforated plunger, and the column is allowed to settle by gravity under free solvent flow. Final packing of the top is achieved by light pressure with the plunger. The sample is applied in a small volume of mobile phase; a few drops of stationary phase may be added. Fractions of 2–3 ml (4.5 g columns) are collected with a flow rate of 0.3–0.6 ml/min/cm². The effluent can be monitored by titration or, after evaporation of solvents, by measurement of the UV absorption of sulfuric acid chromogens. The columns should preferably be run at a constant temperature (23 ± 1° has usually been used). A 4.5 g column should not be loaded with more than about 30 mg of a bile acid mixture. With larger samples the column size is proportionally increased. Hostalen columns are sensitive to overloading with resulting peak broadening and bleeding of stationary phase. When high efficiency columns are required, hydrophobic Super-Cel should be used as the support. Alternatively, the more polar methylated Sephadex can be used with solvent systems of the C-type.

Table IV gives the approximate retention volumes of free bile acids

TABLE IV

APPROXIMATE RETENTION VOLUMES OF BILE ACIDS IN REVERSED-PHASE
LIQUID–LIQUID PARTITION CHROMATOGRAPHY
ON HYDROPHOBIC HYFLO SUPER-CEL[a]

	Solvent system			
Bile acid[b]	F2	F1	C1	C2
3α	100[c]	—	—	—
3,12-diketo	120	—	—	—
3α,12-keto	60	90	—	—
3α,12α	35	50	>250	>250
3α,7α	35	50	>250	>250
3α,7β	—	30	—	—
3α,6α	—	30	—	—
3,7,12-triketo	60	90	—	—
3α,7,12-diketo	20	35	—	—
3α,12α,7-keto	Front[d]	Front	65	95
3α,7α,12α	Front	Front	100	130
3α,7β,12α	Front	Front	50	60
3α,6β,7α	Front	Front	80	—
3α,6β,7β	Front	Front	140	—

[a] The values in the table refer to results obtained with a 4.5-g column.
[b] Compound designation, see Table I.
[c] Milliliters of effluent at peak fraction.
[d] Compound eluted with or close to the void volume (12–20 ml).

in different solvent systems using hydrophobic Super-Cel as the support. Retention volumes on Hostalen columns are shorter than on Super-Cel columns, whereas on methylated Sephadex larger retention volumes are obtained.[9] This is probably related to the different capacities of the supports to carry the stationary phase. Phase systems of type C are particularly useful for the separation of polar free bile acids. Small changes in the methanol concentration of the mobile phase cause large changes in the retention volumes. The only drawback with these phase systems is that isooctanol is difficult to evaporate.

In work with bile acid mixtures of biological origin it is usually best to make a separation with a phase system of type F followed by re-chromatography of material eluted with the solvent front using a solvent system of type C. Material retained on the columns can be eluted with ethanol. Because of the poor solubility of nonpolar bile acids (e.g., litho-cholic acid) in methanol–water these acids may give rise to double peaks due to crystallization on the column. These compounds are best chroma-tographed with solvent system F2.

TABLE V

REVERSED-PHASE LIQUID–LIQUID CHROMATOGRAPHY OF CONJUGATED BILE ACIDS ON HYDROPHOBIC HYFLO SUPER-CEL[a]

	Solvent system				
Conjugated bile acid	D	C2	C1	F1	F2
Glycolithocholic	—	>250	>250	80	40
Glycodeoxycholic	—	—	132	24	Front
Glycochenodeoxycholic	—	—	120	24	Front
Glycocholic	>100	44	37	Front	Front
Taurolithocholic	60	—	22	Front	Front
Taurodeoxycholic	30	Front	Front	Front	Front
Taurocholic	Front	Front	Front	Front	Front

[a] For explanations see Table IV.

Reversed-phase partition chromatography can be used for separations of conjugated bile acids (Table V). It should be noted that overlapping occurs between glycine-conjugated bile acids and free bile acids (cf. Tables IV and V).

Straight-phase partition chromatography with aqueous acetic acid as stationary phase has also been successfully applied to bile acid separa-tions.[11-13] These systems are based on the solvent mixtures developed for

[11] J. Sjövall, *Acta Physiol. Scand.* **29**, 232 (1953).
[12] E. H. Mosbach, C. Zomzely, and F. E. Kendall, *Arch. Biochem. Biophys.* **48**, 95 (1954).
[13] J. T. Matschiner, T. A. Mahowald, W. H. Elliott, E. A. Doisy, Jr., S. L. Hsia, and E. A. Doisy, *J. Biol. Chem.* **225**, 771 (1957).

countercurrent distribution of bile acids.[4] Petroleum ether mixed with isopropyl ether or benzene is used as the mobile phase after equilibration with 1/3 of 70% aqueous acetic acid. An advantage with these systems is that they permit the use of nonsilanized support and gradient elution.

Eight milliliters of the acetic acid phase is homogenized with 10 g of Celite 545 (washed as described above for Hyflo Super-Cel but not silanized). After addition of the petroleum ether phase the slurry is packed in a chromatography tube (10 mm diameter) using a perforated plunger. The sample (less than 100 mg of crude bile acids) is dissolved in 0.5 ml of stationary phase which is then added to 0.5 g of Celite. This mixture is applied to the column which is eluted with petroleum ether containing increasing amounts of isopropyl ether[12] or benzene[13] (each solvent is equilibrated with 70% acetic acid). Deoxycholic and hyodeoxycholic acids are eluted in that order with 40% isopropyl ether in petroleum ether and cholic and 3,7,12-triketocholanoic acids in that order with 60% isopropyl ether in petroleum ether. The separation of the latter pair of compounds is typical for these systems in that ketonic bile acids are often more retarded than the corresponding hydroxy acids (see partition coefficients given by Ahrens and Craig[4]). This is in contrast to the behavior in the neutral reversed-phase systems (Table IV). However, the mobilities of keto and hydroxy acids are sufficiently similar to permit a separation of hydroxy and/or keto acids into groups of mono- di-, and trisubstituted cholanoic acids.[11]

Analytical Separations

The term analytical separation has been chosen to indicate separation procedures permitting identification on a microscale of individual bile acids in fractions obtained by preparative chromatography. Chromatographic systems suitable for this purpose and/or for quantitative analysis of individual compounds include paper chromatography, thin-layer chromatography (TLC), glass-paper chromatography, and gas-liquid chromatography (GLC).

Paper Chromatography

Thin-layer and gas chromatographic techniques have to a large extent replaced paper chromatography in the qualitative and quantitative analysis of free bile acids. For conjugated bile acids, particularly in quantitative work, paper chromatography is still the best method for detailed analyses.

Whatman 3 MM is a suitable filter paper with a high capacity. In qualitative work it can be used without pretreatment. For quantitative determinations the following procedure should be used.[14]

[14] J. Sjövall, *Clin. Chim. Acta* **4**, 652 (1959).

The paper is cut in strips 90 × 465 mm. These are cut so that four strips, 15 × 355 mm, separated by a 10 mm space are obtained. Intact paper is left at both ends so that the four strips are held together with 80 mm of paper at the upper end and 30 mm of paper at the lower end. Starting lines are drawn 30 mm from the upper and lower ends of the separate strips for descending and ascending chromatography, respectively.

Six papers are hung in a cylindrical tank (46 × 15 cm) for descending chromatography (3 on each side of the trough) and washed with 400 ml each of ethanol, 0.6 N hydrochloric acid, water, ethanol, isoamyl acetate (150 ml), and ethanol in that order. All solvents should be glass-distilled. The solvents are continuously supplied from a tightly stoppered separatory funnel the tip of which goes down into the trough. The papers are stored in the glass tank and washed with ethanol until used. Immediately before use they are taken out and dried at 100°.

The samples (5–100 μg of bile salt or acid) are supplied to the starting line of three of the lanes (1, 2, and 3). Lane 4 is left for use as blank. Suitable solvent systems and chromatographic conditions are listed in Table VI. The solvents are equilibrated and the nonpolar phase is used as mobile phase. For separation of glycine conjugates the papers are equilibrated in the tanks (dimensions as above) in an atmosphere saturated with both phases (filter paper lining of the walls). Failure to obtain

TABLE VI
SOLVENT SYSTEMS FOR QUANTITATIVE ANALYSIS OF CONJUGATED
BILE ACIDS BY PAPER CHROMATOGRAPHY

Solvent system[a]	Conditions for chromatography	Bile acids[b]	Conditions in 65% sulfuric acid	Abs. max. of chromogen (nm)	Approx. $E_{1cm}^{1\%}$ at abs. max.
i-AA/H/F/W 85:15:70:30	Equilibration 0.5 hr in vapors of mobile phase. Ascending, 20 hr.	TC TCD + TD	60 min, 20°[c] 15 min, 50°	320 305	290 170
E/H/A/W 50:50:70:30	Equilibration 8 hr in vapors of both phases. Descending, 18 hr.	GC GUD GCD GD	60 min, 20°[c] 15 min, 60° 15 min, 60° 10 min, 60°	320 305 305 308	350 210 210 200

[a] i-AA = isoamyl acetate; H = heptane; F = formic acid; W = water; E = ethylene chloride; A = acetic acid. All solvents are redistilled. Ethylene chloride is washed with concentrated sulfuric acid, water, dried over K_2CO_3, and distilled.

[b] T = tauro-; G = glyco-; C = cholic; CD = chenodeoxycholic; D = deoxycholic; UD = ursodeoxycholic. Compounds are listed in order of increasing R_f value.

[c] 3,6,7-Trihydroxycholanoic acids do not give UV absorption under these conditions.

equilibration results in tailing or elongated spots. Too high a flow rate in descending chromatography also results in tailing.

After development, lane 1 is cut out and sprayed with 10–15% phosphomolybdic acid in ethanol.[15] Spots of hydroxy bile acids appear after heating a few minutes at 70–80°. Corresponding zones on lanes 2–4 are cut out with a margin. In descending chromatography, lanes 1 and 3 are both sprayed and the bile acids on lane 2, having mobilities being the mean of those on lanes 1 and 3, are eluted. Lane 4 is used as a blank. One edge (free from bile acid) of the filter paper pieces is clamped between two thick microscopic slides that are obliquely ground on the short sides. The slides are fastened in a vertical position and ethanol, 2 ml in 20 minutes (rate obtained with a capillary 220×0.3 mm), is delivered to the groove formed at the upper end of the slides.[16] The elution is repeated once, allowing the paper piece to dry between the elutions. The eluate is dried at 80°, and 65% (v/v) sulfuric acid (1 ml for 0–50 μg bile acid) is added after cooling. The tubes are thoroughly shaken for 5–10 minutes to dissolve the bile acids. Heating times and wavelengths at which UV absorption of the chromogens are read are given in Table VI. The measurements are made against the identically treated blank zone eluates which should have a low UV absorption (0.01–0.04) when read against sulfuric acid. The amount of bile acid is calculated by comparison with standards (about 20 μg) directly treated with sulfuric acid. Duplicate determinations of, for example, bile acids in bile should not differ by more than 10% of the mean value.

When bile acids in extracts of bile or intestinal contents are analyzed, it is necessary to correct the UV absorption of the zones of taurine conjugated bile acids for nonspecific absorption. A reasonably accurate correction is obtained by measuring the absorption of the eluates from one lane in 80% ethanol. The interfering materials have similar UV absorption in 80% ethanol and sulfuric acid and the value in 80% ethanol is subtracted from that in sulfuric acid to obtain the absorption of the bile acid chromogen.

The sulfuric acid chromogens are sensitive to the presence of heavy metals or peroxides, which give changes in the absorption spectra. The sulfuric acid spectra of standards carried through the procedure should be the same as those of standards directly pipetted into test tubes.

The solvent systems described are useful for analysis of bile acids as well as their derivatives, e.g., methyl esters. In qualitative analysis isopropyl ether may be used instead of ethylene chloride[17] and sometimes

[15] D. Kritchevsky and M. R. Kirk, *Arch. Biochem. Biophys.* **35**, 346 (1952).

[16] J. Sjövall, *Arkiv Kemi* **8**, 317 (1955).

[17] J. Sjövall, *Acta Chem. Scand.* **6**, 1552 (1952); *ibid.* **8**, 339 (1954).

gives larger separation factors. The mobility of the compounds is controlled by changing the proportion of heptane in the mobile phase; increased amounts of heptane decrease the mobility. Temperature and humidity will affect the mobility. Slightly different proportions of heptane may therefore have to be used to obtain the same separation in different laboratories. Equilibration of the papers with stationary phase becomes increasingly important with the higher proportions of heptane; tailing can be prevented by pretreatment of the papers with a 15–20% ether solution of 70% acetic acid.[18] Separation of bile acids with closely similar mobilities is best accomplished by using descending chromatography with overflow in a solvent system giving low mobilities of the compounds. Glycine-conjugated deoxycholic, chenodeoxycholic, ursodeoxycholic, and hyodeoxycholic can be separated (e.g., with the same system given in Table VI), and these types of systems are also useful for separation of some isomeric 3,7,12- and 3,6,7-trihydroxycholanoic acids. In the latter case the method may be used for preparative purposes.

In the isopropyl ether–heptane solvents ketonic acids usually have a lower R_f value than the corresponding hydroxy acids. In most cases axial hydroxyl groups give a higher R_f value than equatorial ones in the corresponding position. A methyl ester has a higher R_f value than the acid; the difference corresponds roughly to the removal of one hydroxyl group from the acid. Glycine conjugation lowers the R_f value somewhat more than the addition of a hydroxyl group, and reduction of the bile acid to a bile alcohol also lowers the R_f value. The influence of the length of the side chain is discussed in the TLC section.

The solvent systems for taurine conjugates (Table VI) can be used for bile alcohol sulfates. Somewhat better separations of these polar compounds have been obtained with the upper phase of n-amyl acetate–heptane–acetic acid–water 85:15:103:47.[19] Descending chromatography for 48 hours with the upper phase of n-amyl acetate (pear oil)–heptane–acetic acid–water 80:20:70:30 is the only method so far reported with which taurodeoxycholic and taurochenodeoxycholic acids can be separated.[8]

Thin-Layer Chromatography

Of the adsorbents available for TLC, silica gel G and H (Merck AG, Darmstadt, Germany), Anasil B (Analytical Engineering Laboratories, Inc., Hamden, Connecticut), and Florisil have been used for the separation of bile acids. For analytical purposes silica gel G has been preferred

[18] I. E. Bush, "The Chromatography of Steroids." Pergamon Press, London, 1961.
[19] R. J. Bridgwater, G. A. D. Haslewood, and A. R. Tammar, *Biochem. J.* **85**, 413 (1962).

by most investigators. Anasil B has proved valuable in the separation of certain mixtures, i.e., some 5α and 5β isomers.[20]

A simple and inexpensive way to prepare chromatoplates is to keep the glass plate pressed against the table with two stainless steel wires 0.3–0.5 mm diameter (or with adhesive tape) running along two sides of the glass plates. The slurry of adsorbent (usually 10 g of silica gel in 19 ml of water) is poured over the center of the plate and is evenly spread with a glass rod resting on the wires. After 30 minutes the plates are dry enough to be activated at 120° for at least 45 minutes.

The bile acids (or derivatives), 10–20 μg dissolved in 5 μl of acetone, methanol, or chloroform–methanol 1:1, are applied to the starting line, 1–1.5 cm from the smoothed edge of the adsorbent layer. The width of the application zone should not exceed 5 mm. For preparative purposes application of the sample as a band by some commercially available applicator is to be preferred to spot by spot application. Development of the plate without saturating the chamber with the solvent vapor often leads to "Randphenomen"; i.e., compounds near the outer parts move faster than those in the center of the plate. This might be overcome by supersaturating the chamber by lining its walls with filter paper drenched in the developer or by reducing the space of the chamber—for example, by using the "sandwich" technique. This refers to a technique where three sides of the plate are lined with cardboard strips, 0.7×7 mm. Another glass plate is placed over the chromatoplate riding on the cardboard strips. The "sandwich" is held together with clips and dipped into the developer in a suitable narrow chamber. It should be noted that supersaturated chambers often give separations and mobilities different from those obtained without supersaturation. Several techniques discussed in more detail in the chapter on TLC of steroids[21] can be successfully applied to the analysis of bile acids. Thus, two-dimensional and multiple one-dimensional development as well as wedge techniques can be very useful.[22]

Some solvent systems used for the separation of *conjugated bile acids* are listed in Table VII. It is possible to separate taurine from glycine conjugates without overlapping between the groups. Nonpolar glycine conjugates overlap with polar free bile acids. None of the systems permits the separation of the glycine or taurine conjugates of chenodeoxy- and deoxycholic acids. Nor is it possible to distinguish between conjugated bile acids isomeric at C-5, i.e., allo and normal bile acids.[23]

[20] A. F. Hofmann, *in* "New Biochemical Separations" (A. T. James and L. J. Morris, eds.), p. 262. Van Nostrand, Princeton, New Jersey, 1964.

[21] B. P. Lisboa, see this volume [1].

[22] J. A. Gregg, *J. Lipid Res.* **7**, 579 (1966).

[23] A. F. Hofmann and E. H. Mosbach, *J. Biol. Chem.* **239**, 2813 (1964).

TABLE VII
SOLVENT SYSTEMS FOR TLC OF CONJUGATED BILE ACIDS

	R_f values in solvent system[b]				
Bile acid[a]	BuOH 10[c] HAc 1 H_2O 1	BuOH 85[d] HAc 10 H_2O 5	i-AAc 4[e] HPr 3 PrOH 2 H_2O 1	TMP 2[f] i-PE 1 HAc 1 i-PrOH 1	EtCl₂ 10[f] HAc 10 H_2O 1
TC	0.08	0.22	0.11	0.06	0.09
TD	0.21	0.41	0.24	—	—
TCD	0.21	0.41	0.24	0.12	0.16
TL	—	—	—	0.16	0.25
GC	0.35	—	0.53	0.22	0.32
GD	0.50	—	0.80	0.39	0.53
GCD	0.50	—	0.79	0.36	0.53
GL	—	—	—	0.55	0.73

[a] See Table VI. L = lithocholic acid.

[b] BuOH = n-butanol; HAc = acetic acid; i-AAc = isoamyl acetate; HPr = propionic acid; PrOH = n-propanol; TMP = 2,2,4-trimethylpentane; i-PE = isopropyl ether; i-PrOH = isopropanol; EtCl₂ = ethylene chloride.

[c] H. Gänshirt, F. W. Koss, and K. Morianz, *Arzneimittel-Forsch.* **10**, 943 (1960).

[d] T. Hoshita, S. Nagayoshi, M. Kouchi, and T. Kazuno, *J. Biochem. (Tokyo)* **56**, 177 (1964).

[e] Mobilities given relative to cholic acid: A. F. Hofmann, *J. Lipid Res.* **3**, 127 (1962).

[f] J. A. Gregg, *J. Lipid Res.* **7**, 579 (1966).

Comprehensive lists of solvent systems and mobilities of free bile acids and bile acid derivatives are given in the chapter on TLC of steroids.[21] The following presentation will be limited to those solvents that have been used in our laboratory and have been found satisfactory for most separation problems in work with biological materials.[24] The solvent system designations given by Lisboa[21] will be used.

For the separation of *trisubstituted bile acids* solvent systems Ij, Im, Io, and Ip (Table VI in Lisboa's[21] chapter) are recommended. Other useful systems have been described by Hofmann[20] (systems IIIb and IIIc) and by Hamilton.[25] *Disubstituted bile acids* are separated with solvent systems Ic, Id, Ij, IIIa, and IIIb. *Monosubstituted bile acids* should be analyzed with systems Ie, If, Ig, and Ij. A few pairs of bile acids in each group cannot be separated with these solvents: 3α,7α-dihydroxy-12-keto- and 3α,12α-dihydroxy-7-ketocholanoic acids; 3,7-diketo- and 3,12-diketocholanoic acids; 7α-hydroxy- and 12α-hydroxy-, as well as 7-keto

[24] P. Eneroth, *J. Lipid Res.* **4**, 11 (1963).

[25] J. G. Hamilton, *Arch. Biochem. Biophys.* **101**, 7 (1963). Solvent composition: isopropyl ether–butanone–acetic acid 100:40:10 or 2,2,4-trimethylpentane–isopropyl ether–acetic acid 100:50:70.

and 12-ketocholanoic acids. However, the three last-mentioned pairs can be separated if analyzed as methyl esters using solvent Ib (Table VI in Lisboa[21]), IIa (Table VII in Lisboa[21]), and IIa, respectively.

The subdivision of the bile acids into groups of mono-, di-, and trisubstituted has been made to simplify the presentation of their TLC behavior. It is evident from Tables VI and VII in Lisboa's chapter[21] that certain members from the three groups overlap. The choice of solvents will depend on the expected nature and complexity of the bile acid mixture. However, when dealing with complex unkown mixtures it is advisable to run several two-dimensional chromatographies (on 27×27 cm chromatoplates) when solvent system Ij is used for development in the first direction. One of the systems recommended for each group is then used in the second direction. The solvent is allowed to ascend 20 cm from the starting point in each direction. On the basis of the results obtained, a suitable solvent can be selected for one-dimensional (single or multiple) development, for example, for preparative purposes.

It may be advantageous to analyze the bile acids as methyl esters. In acidic systems (in contrast to neutral systems) the difference in mobility of an acid and its methyl ester is small. The difference is larger with mono- and disubstituted compounds than with the trisubstituted ones, but the same solvent systems can usually be used both for acids and methyl esters. As an example it may be mentioned that $3\alpha,12\alpha$-dihydroxy-7-ketocholanoic acid and its methyl and ethyl esters have relative mobilities of 1.00, 1.10, and 1.19, respectively, in solvent system Ia (Table VI in Lisboa's chapter[21]).

TABLE VIII

INFLUENCE OF DIFFERENT SIDE CHAINS ON THE MOBILITY OF A BILE ACID
IN PAPER CHROMATOGRAPHY AND TLC[a]

	Solvent[b]	Compound[c]							
		C_{22}	C_{23}	C_{24}	C_{25}	C_{26}	C_{27}	C_{28}	C_{29}
Paper chromatography, R_f values	E/H/A/W 6:4:7:3	0.28	0.38	0.47	0.58	0.68	0.78	0.82	0.85
	i-PE/H/A/W 6:4:7:3	0.10	0.15	0.21	0.29	0.44	0.59	0.72	0.78
TLC, R_f values	CHCl$_3$/EtOAc/ HAc 45:45:10	0.16	0.19	0.24	0.27	0.34	0.40	0.46	0.49

[a] Calculated from K. Morimoto, *J. Biochem.* (*Tokyo*) **55,** 410 (1964).

[b] For abbreviations see Tables VI and VII.

[c] C_{24} denotes $3\alpha,7\alpha,12\alpha$-trihydroxy-5β-cholanoic acid. The other compounds are homologs. C_{28} and C_{29} are C-24-methyl and C-24-ethyl derivatives of the trihydroxy-5β-cholestanoic acid (C_{27}).

Conversion of hydroxyl groups into acetates increases the mobility and usually results in a less efficient separation of isomeric compounds.

Bile acids with more than 5 carbon atoms in the side chain may occur together with C_{24} bile acids.[1] Elongation of the side chain results in an increased mobility of a bile acid (Table VIII).

Substituted 5α-cholanoic (allo) acids may also be found together with the normal 5β-cholanoic acids. It has been noted that the separation of C-5 isomers in many cases is best accomplished with Anasil B as adsorbent.[20] When Silica gel G is used, deoxy- and allodeoxycholic acids are better separated than their methyl esters whereas the reverse is true for the C-5 isomers of cholic and chenodeoxycholic acids. The allo isomers of lithocholic and deoxycholic acids have a higher mobility than the normal isomers. However, when a 7α-hydroxyl group is present, the allo compound is more strongly adsorbed than the normal one (Table IX).

A variety of detection reagents are available (see chapter by Lisboa[21]). Phosphomolybdic acid, 15% in ethanol, is a very sensitive reagent for hydroxy bile acids on silica gel G. After spraying of the plates (immediately after drying at 150°), blue spots appear on a yellow background on heating (may be used even for acetylated bile acids). Concentrated or 50% sulfuric acid may be used as a charring reagent. Ketonic bile acids are difficult to detect with this reagent even after prolonged heating at 200°. Therefore we use a saturated solution of potassium dichromate in 80% sulfuric acid as a universal bile acid detection reagent. With this reagent, excessive heating is not required. Different bile acids often stain differently before charring occurs. Observation of the color development may serve as a guide for subsequent identification work. A reagent designed to give widely different colors with bile acids on TLC consists of anisaldehyde, 0.5 ml, and sulfuric acid, 1 ml, in glacial acetic acid, 50 ml. The reagent is prepared immediately before use, and after spraying the plates are heated at 125° for 10 minutes. The colors are observed in visible and UV light at 366 nm (mμ).[26]

TLC is useful for *small-scale preparative work*. Adsorbents, with or without binder, have to be washed to eliminate interfering materials. Prerunning of the layers in the appropriate solvent is usually less satisfactory. Silica gel G can be purified by refluxing in acetone for 30 minutes. The solvent is filtered off, and the extraction is repeated once with acetone and twice with methanol. The gel is then dried at 80° overnight in a clean oven. Plates with a 0.5–1 mm layer are prepared. The amounts of bile acids that may be applied depend on the separation factors. Not more than 0.2 mg of each bile acid derivative should be applied over a

[26] D. Kritchevsky, D. S. Martak, and G. H. Rothblat, *Anal. Biochem.* **5**, 388 (1963).

TABLE IX

Separation of Pairs of Bile Acids Isomeric at C-5[a]

Pairs of compounds[b]	Configuration at C-5	Relative mobility[c]	R_f of reference compound	Solvent system composition (v/v)		Adsorbent
Propyl cholanoate	5α	0.79 ⎫	—	Heptane	92	Anasil B
	5β	*1.00* ⎭		Diethyl ether	8	
Methyl 3,6-diketo	5α	0.37 ⎫				
	5β	0.66 ⎬				
Methyl 3,12-diketo	5α	0.59 ⎬	0.47	Di-n-butyl ether	85	Anasil B
	5β	0.79 ⎬		Acetone	15	
Methyl 3α	5α	1.16 ⎬				
	5β	*1.00* ⎭				
3-Keto-12α acid	5α	2.48 ⎫				
	5β	2.45 ⎬		Trimethylpentane	5	
$3\alpha,12\alpha$ acid	5α	1.29 ⎬	0.33	Ethyl acetate	25	Silica gel G
	5β	*1.00* ⎬		Acetic acid	0.2	
$3\beta,12\alpha$ acid	5α	1.38 ⎬				
	5β	1.40 ⎭				
Methyl $3\alpha,7\alpha$	5α	0.79 ⎫	0.40	Same		Same
	5β	*1.00* ⎭				
Methyl $3\alpha,12\alpha$	5α	1.26 ⎫	0.45	Di-n-butyl ether	70	Anasil B
	5β	*1.00* ⎭		Acetone	30	
Methyl $3\alpha,7\alpha,12\alpha$	5α	0.72 ⎫	0.28	Cyclohexane	7	Silica gel G
	5β	*1.00* ⎭		Ethyl acetate	23	
				Acetic acid	3.0	
$3\alpha,7\alpha,12\alpha$ acid	5α	0.86 ⎫	0.50	Benzene	30	Silica gel G
	5β	*1.00* ⎭		Isopropanol	10	
				Acetic acid	1	

[a] Calculated from papers by A. F. Hofmann *in* "New Biochemical Separations" (A. T. James and L. J. Morris, eds.), p. 262. Van Nostrand, Princeton, New Jersey, 1964, and by A. Kallner (*Acta Chem. Scand.* **18**, 1502 (1964); *Arkiv Kemi* **26**, 553, 567 (1967)].

[b] For abbreviations, see Table I.

[c] Relative to italicized reference compound.

starting line of 15 cm if the compounds can be just separated in an analytical run. Bile acids separated by 5 cm may be run in 5-mg amounts. The bands are detected by spraying with water if the concentration is 30–50 μg/cm². Smaller amounts are detected with iodine vapors or by spraying with iodine dissolved in ether. Alternatively dyes with known relative mobilities may be cochromatographed with the sample. The bile acid zones are marked and the layer is removed either with a zone extractor[27] or by scraping off the layer with a razor blade. Water-saturated ethyl acetate or methanol have been used to extract bile acids. Recoveries (tested with labeled compounds) of 50–70% have been reported.[28] To minimize the amount of inorganic material frequently extracted with the bile acids, 2% acetic acid in diethyl ether has been used, giving similar recoveries.[28] To obtain better results, bile acids should be converted into methyl esters prior to chromatography. Recoveries of close to 100% are then obtained on elution with methanol.[29]

TLC has been used for quantitative determination of some conjugated and free bile acids in the 15–75 μg range. For details the reader is referred to the studies by Kottke et al.[30] Silica gel G is purified by extraction with 50% sulfuric acid (5–6 hours), water, 6 N HCl (overnight), and water until neutral. After drying and sieving, plates can be prepared with this material. The quantitative determinations are made by scraping the appropriate areas into test tubes and treating this material with 4 ml of 65% sulfuric acid. After centrifugation the absorbancies of the sulfuric acid chromogens are read against a similarly treated blank.

Bile acids (in the range 0.5–4 μg) have also been estimated by direct densitometric reading of absolutely uniform layers stained with phosphomolybdic acid.[31]

The most recent development in quantitative TLC is the use of an enzymatic method for the determination of 3-hydroxy bile acids.[32,33] Individual bile acid spots (revealed with iodine) are eluted with methanol (redistilled from dinitrophenylhydrazine). The enzymatic determination is carried out using 3α- and 3β-hydroxysteroid dehydrogenase from *Pseudomonas testosteroni*.[34] A description of this method as applied to TLC eluates is given by Palmer, this volume [6].

[14]C-Labeled bile acids of relatively high specific activity can be quantitated by densitometric reading of autoradiograms (Kodirex, roentgen film) at 550 nm.[35] Smaller amounts of activity may be determined by liquid scintillation counting after the layer has been scraped into a counting vial containing 2,5-diphenyloxazole (PPO)–p-bis[2-(4-methyl-

[27] J. S. Matthews, V. A. L. Pereda, and P. A. Aquilera, *J. Chromatog.* 9, 331 (1962).
[28] A. Kallner, *Acta Chem. Scand.* 21, 87 (1967).
[29] P. Eneroth, B. Gordon, R. Ryhage, and J. Sjövall, *J. Lipid Res.* 7, 511 (1966).
[30] B. A. Kottke, J. Wollenweber, and C. A. Owen, Jr., *J. Chromatog.* 21, 439; 24, 99 (1966).
[31] G. Semenuk and W. T. Beher, *J. Chromatog.* 21, 27 (1966).
[32] T. Iwata and K. Yamasaki, *J. Biochem.* (*Tokyo*) 56, 424 (1964).
[33] R. H. Palmer and Z. Hruban, *J. Clin. Invest.* 45, 1255 (1966); and R. H. Palmer, personal communication, 1967.
[34] B. Hurlock and P. Talalay, *Endocrinology* 62, 201 (1958).
[35] A. Norman and R. H. Palmer, *J. Lab. Clin. Med.* 63, 986 (1964).

5-phenyloxazolyl)]benzene (POPOP)–Cab-O-Sil (thixotropic gel powder)–toluene in the proportions 0.5:0.03:4.0:100.[36]

Glass Paper Chromatography

This type of chromatography differs from TLC in that glass paper is used as a matrix for the adsorbent. The advantages with this technique are the possibilities of using descending chromatography with overflow or multiple developments and of using adsorbents which do not permit preparation of a thin layer on a glass plate. This may be exemplified with the use of monopotassium phosphate papers.[37] These are prepared by dipping 9×15 cm glass fiber paper 934-AH (H. Reeve Angel and Co. Ltd., London) in $0.1\ M$ KH_2PO_4. Excess solution is wiped off with a glass rod, and the papers are dried over a hot plate. They can be stored wrapped in aluminum foil. Two types of solvent systems are recommended.[38] One consists of 2,2,4-trimethylpentane (100 ml)–butanone (40 ml) to which varying amounts of a butanone–acetic acid, 9:1, mixture can be added, e.g., 5 ml for separation of less polar and 20 ml for polar bile acids. The other system consists of trimethylpentane (100 ml)–isopropyl ether (40 ml) to which varying amounts of acetic acid (e.g., 5–20 ml) are added. The development times are very short (e.g., 10 minutes) and the resolutions are as good as those obtained with TLC. The glass papers are sensitive to overloading; no more than 5 μg of an individual bile acid should be applied if tailing is to be avoided. When the spots are revealed by charring after light spraying with sulfuric acid, it is important that the papers are not contaminated by careless handling. To avoid background staining, the papers may be heated in a furnace at 600° for 15 minutes before the adsorbent is applied.

Gas-Liquid Chromatography

To obtain good results it is necessary to use columns that give minimal tailing of polar methyl hydroxycholanoates. Several factors of importance for the preparation of such columns are still poorly understood, but it is clear that the support and its treatment is of great importance. The procedure described below will usually give satisfactory columns with 2000–3000 theoretical plates for methyl hydroxycholanoates. For further details the review by Horning et al.[39] should be consulted.

[36] P. H. Ekdahl, A. Gottfries, and T. Scherstén, *Scand. J. Clin. Lab. Invest.* **17**, 103 (1965).
[37] J. G. Hamilton and J. W. Dieckert, *Arch. Biochem. Biophys.* **82**, 203 (1959).
[38] J. G. Hamilton, *Arch. Biochem. Biophys.* **101**, 7 (1963).
[39] E. C. Horning, W. J. A. VandenHeuvel, and B. G. Creech, *Methods Biochem. Anal.* **11**, 69 (1963).

Gas-Chrom P (Applied Science Laboratories, State College, Pennsylvania) or Chromosorb W (Johns-Manville Co., New York), 100–120 mesh, 30 g, is treated with concentrated hydrochloric acid (analytical reagent) overnight and then repeatedly washed until the supernatant is colorless. The support is then washed with many portions of glass-distilled water. Fines are removed in this step by decantation. When the washings are neutral, the support is washed with ethanol and then left for 1 hour in pyridine (analytical reagent) and finally washed on a sintered-glass funnel with ethanol. The support is then spread as a thin layer, dried at room temperature and finally at 120° overnight. Silanization can be carried out in liquid or gas phase. In the former case the dried support is poured into 100 ml of a 0.5–1% solution of dimethyldichlorosilane in toluene in a suction flask. Air is removed by intermittent application of vacuum. After 30 minutes the support is poured on a sintered-glass funnel, rinsed with distilled toluene once and then with large volumes of ethanol. The silanized, neutral support is spread on a filter paper, air dried, and finally dried at 100°. For silanization in the gas phase the washed and dried support is placed in a gas wash bottle connected to another gas wash bottle containing dimethyldichlorosilane. The bottles are cooled in dry ice–ethanol, and a slow stream of nitrogen is passed through the system for 3 hours. The support is then rapidly swirled in ethanol and washed with large amounts of this solvent. The support is dried as above. The latter silanization procedure has been found to give the best results in preparation of 3% QF-1 columns.

The inactivated support is coated by the filtration procedure.[39] The support (20–25 g) is added to 100 ml of a 1–3% solution of the stationary phase (SE-30, SE-52, and PhSi in toluene, QF-1 in acetone, XE-60 in ethyl acetate, NGS and Hi-Eff 8 B in chloroform) in a suction flask. Air is removed from the support by intermittent suction. After about half an hour, excess solution is removed by filtration under suction on a sintered-glass funnel. The coated support is spread in a thin layer, dried at room temperature and finally at 100°. When Gas Chrom P is used, the percent of phase on the support is about the same as the concentration in the solution used for coating. Chromosorb W adsorbs about twice this amount.

Glass columns, 2 m × 3–4 mm i.d. are washed with water, methanol, acetone, chloroform, and toluene and are then silanized with a 0.2% solution of dimethyldichlorosilane in toluene. Rinsing with liberal amounts of methanol completes the silanization, and the column is dried. A plug of silanized glass wool is inserted in the outlet of the column. To fill a coiled column the outlet is connected to a water suction pump and a spoonful of packing material is poured into the column. Packing is accomplished by tapping with a pencil covered with thick rubber tubing. The procedure is repeated until the column is filled to a point 5 cm below the flash heater region. The top of the column is protected with a small plug of silanized glass wool. U-shaped columns are packed by filling from both sides and tapping with the pencil. Vibrators should not be used.

SE-30 (temperature limit 300°) and QF-1 (temperature limit 250°) columns are conditioned at 220–240° without carrier gas flow for 48–72 hours and then with carrier gas for 24 hours (polyester and XE-60 columns are not subjected to heating without carrier gas flow). The column can then be connected with the argon or hydrogen flame ionization detector. The column is tested by injection of a mixture of 0.1–2 μg of methyl cholanoate, lithocholate, deoxycholate, chenodeoxycholate, and cholate. Symmetrical peaks with only little tailing of the di- and trihydroxy compounds should be obtained. When broad peaks with a similar degree of tailing are obtained with all compounds the column may be poorly packed and repacking should be tried. If this does not improve the results the stationary phase may be unevenly distributed on the support. When tailing of peaks is related to the polarity of the compounds the inactivation of the support may be incomplete. The following procedure may then improve the results. The detector is disconnected, the column temperature is set at 150° and the carrier gas flow is reduced to a minimum. Hexamethyldisilazane, 100–200 μl, is injected in 10-μl portions and the carrier gas supply is turned off. The column is left overnight and is then conditioned with carrier gas at 220–240° for about 8 hours. The column bleed is then sufficiently reduced to permit connection with the detector and retesting.

Columns which are continually used over extended periods for analysis of biological extracts show a gradually impaired performance. To restore such columns the injection zone is carefully cleaned of carbonaceous material; the glass wool plug and the upper 5 cm of the column packing are replaced with fresh material. The column is then treated with hexamethyldisilazane as described above. This technique may prolong the useful life of a column by several months.

The Separation of Bile Acids by GLC

Separation is determined by the choice of stationary phase and bile acid derivative.

Methyl Esters. Bile acids can be analyzed only if the carboxyl group is converted into an ester—most conveniently by treatment with ethereal diazomethane. This reagent should always be freshly prepared. A simple procedure is as follows. Nitrosomethylurea (stabilized with acetic acid, Fluka AG, Buchs, Switzerland), 0.5 g, is dissolved in 7.5 ml of diethyl ether in a 150 × 18 mm test tube. One milliliter of 50% aqueous KOH is added, and nitrogen is slowly blown through the mixture at room temperature into another test tube containing 15 ml of ether cooled in dry ice–ethanol. Connections are made with glass tubes and cork stoppers. The gas effluent from the second tube is led directly into the exhaust of the hood. The reaction is completed when the mixture in the first tube is almost colorless.

The bile acids are dissolved in a few milliliters of ether–methanol 9:1, and an aliquot of the ethereal solution of diazomethane is added to give a persistent yellow color. After 15–30 minutes at 4–20° the solvents are removed under a stream of nitrogen (under a hood). The residue is dissolved in a suitable volume of acetone, and 1–4 μl is injected into the gas chromatograph.

Trimethylsilyl Ethers (TMS). For complete conversion of axial and equatorial hydroxyl groups at carbon atoms 3, 6, 7, and 12 the bile acid methyl esters (2–5 mg) are dissolved in 0.5 ml of pyridine (dry, refluxed and distilled over BaO), 0.2 ml of hexamethyldisilazane and a few drops of trimethylchlorosilane.[40] After 15–30 minutes the reagents are evaporated under a stream of nitrogen. Under these conditions 3-keto-Δ^4-cholanoates are partly converted into enol TMS ethers. The residue is extracted with 1 ml of dry hexane; insoluble material is removed by centrifugation. If dimethylformamide or dioxane is used instead of pyridine the reaction rates are slower and partial TMS ethers can be prepared selectively. The following conditions are used for conversion of 3α, 3β, 6α, 7β, and 12β hydroxyl groups (the latter three are equatorial) into TMS ethers:[29, 41] 10–100 μg of bile acid methyl ester is dissolved in 0.06 ml of dimethylformamide (purified by refluxing with CaC_2, distilled, and stored over aluminum oxide activity grade I). Hexamethyldisilazane, 0.03 ml, is added and the solution is heated at 50° for 3 hours. The reaction mixture can then be directly analyzed by GLC. 6β, 7α, and 12α hydroxyl groups show essentially no reactivity under these conditions. The conditions described by Briggs and Lipsky may also be used.[42] The sample (1–3 mg) is dissolved in a mixture of reagent grade dioxane, 0.5 ml, hexamethyldisilazane, 0.05 ml, and trimethylchlorosilane, 0.02 ml. The reaction is allowed to proceed at room temperature and is stopped by addition of 2.5 μl of methanol to 50 μl of reaction mixture. An aliquot is used for GLC. With this method 7β and 6α hydroxyl groups can

[40] M. Makita and W. W. Wells, *Anal. Biochem.* **5**, 523 (1963).
[41] J. Sjövall, *in* "Biomedical Applications of Gas Chromatography" (H. Szymanski, ed.), p. 151. Plenum Press, New York, 1964.
[42] T. Briggs and S. R. Lipsky, *Biochim. Biophys. Acta* **97**, 579 (1965).

be distinguished; TMS ethers being formed more slowly with 7β than with 3α or 6α hydroxyl groups.

Trifluoroacetates (TFA).[43-45] Methylated bile acids or their partial TMS ethers, 0.1 μg to 1 mg, are dissolved in 0.2 ml of trifluoroacetic anhydride and allowed to react for 15–20 minutes at 35°. 3-Keto-Δ^4-cholanoates are partly converted into enol trifluoroacetates by means of this procedure. The reagent is removed under a stream of nitrogen and the residue is dissolved in acetone, acetonitrile, or carbon disulfide for GLC analysis. TFA derivatives should be analyzed within 1–2 days since signs of decomposition may appear on storage for more than 48 hours at room temperature.[44, 45]

Trifluoroacetates may undergo thermal decomposition in the chromatograph. The flash heater temperature should only be about 10° higher than that of the column. Particularly sensitive compounds include the TFA of methyl 3,6,7-trihydroxy-, 3-keto-7α-hydroxy-, and 3β,12β-dihydroxycholanoates, but several other compounds have been noted to give two peaks or baseline drift probably due to elimination of trifluoroacetic acid.[46]

Dimethylhydrazones.[47] Bile acid methyl esters, 10–100 μg, are dissolved in 0.1 ml of 1,1-dimethylhydrazine in screw-capped vials. The vials are flushed with nitrogen, closed, and left at room temperature overnight.[29] The reagent is evaporated under a stream of nitrogen, and the residue is dissolved in 2 ml of acetone. After 1 hour the acetone is evaporated and the sample can be analyzed by GLC. Under these conditions dimethylhydrazones are formed only with an unconjugated 3-keto group.[39]

Data which permit the selection of stationary phase and bile acid derivative to suit most separation problems are reported in Tables X–XII. In these tables relative retention times have been used although it is realized that these are subject to variation mainly due to temperature differences and, to a less extent, to differences in column preparation. In our experience the temperature dependence is most pronounced with TMS ethers on Hi-Eff-8B columns. Temperature differences do not affect relative retention times on QF-1 columns to the same extent (cf. Table X).

[43] J. Sjövall, *Acta Chem. Scand.* **16,** 1761 (1962).
[44] A. Kuksis and B. A. Gordon, *Can. J. Biochem. Physiol.* **41,** 1355 (1963).
[45] A. Kuksis, *Methods Biochem. Anal.* **14,** 325 (1966).
[46] J. Sjövall, *in* "The Gas Liquid Chromatography of Steroids" (J. K. Grant, ed.) (*Mem. Soc. Endocrinol.* **16**), p. 243. Cambridge Univ. Press, London and New York, 1967.
[47] W. J. A. VandenHeuvel and E. C. Horning, *Biochim. Biophys. Acta* **74,** 560 (1963).

Bile acid methyl esters with unprotected hydroxyl groups are best separated on QF-1 columns. It is advisable to use 3% columns since it is often difficult to prepare 0.5–1% QF-1 columns which do not give tailing of polar compounds. This difficulty has also been experienced by Okishio and Nair,[48] who therefore developed a column having a mixture of QF-1, SE-30, and NGS as stationary phase. On QF-1 columns epimeric alcohols and the corresponding ketone can be separated. Substituents in the 5β-cholanoate nucleus increase the retention time of the parent compound in the following approximate order: 12α < 12β $\leqq$ 7α < 7β $\leqq$ 6β < 3β < 6α $\leqq$ 3α $\leqq$ 12-keto < 7-keto < 6-keto < 3-keto < 3-keto-Δ^4. Although individual hydroxyl and keto groups often contribute to the retention time with constant logarithmic factors, interactions between two or more substituents in a molecule sometimes makes it difficult to calculate the retention time of a bile acid methyl ester.[41]

QF-1 columns are also most useful for separation of TFA esters. The retention times are not as well predictable as with free hydroxyl groups.

Similar results, but with much smaller separation factors between epimeric alcohols and the corresponding ketones, are obtained with the CNSi and XE-60 phases (silicones with slightly different amounts of cyanoethyl substituents).

Nonselective phases (SE-30 and SE-52) and silicone phases with limited selectivity such as PhSi-20 and PhSi-35 (having 20 and 35 mole %, respectively, of phenyl groups) have not been much used in systematic bile acid analysis. However, some important pairs of bile acid esters can be more easily separated on these phases than on selective ones, e.g., methyl 3α,12α-dihydroxy-12-keto- and 3α,7α-dihydroxy-7-ketocholanoates and methyl 3,12-diketo- and 3,7-diketocholanoates (PhSi-20, Table XI). No other phase permits the separation of the latter pair unless dimethyl-hydrazones are prepared (Table XII).

Allo and normal bile acids are separable on most phases. When methyl 5α- and 5β-cholanoates lacking free or protected hydroxyl groups are analyzed, the 5α-compound has a longer retention time than the corresponding 5β-compound on all phases tested. On QF-1 this is also true in most cases when a 3-hydroxyl group is present, even though the separation factor between a 3α-hydroxy-5β-cholanoate (equatorial OH) and the corresponding 3α-hydroxy-5α-cholanoate (axial OH) may be small. When TMS ethers of cholic and allocholic acids are prepared, the order of elution is reversed; the TMS ether of allocholic acid has the shorter retention time both on QF-1 and SE-30.[23]

The use of GLC in quantitative work is discussed in the sections on bile, serum, and feces.

[48] T. Okishio and P. P. Nair, *Anal. Biochem.* **15**, 360 (1966).

TABLE X

RELATIVE RETENTION TIMES[a] OF DIFFERENT BILE ACID METHYL ESTER DERIVATIVES ON SOME SELECTIVE PHASES

Bile acid methyl ester Substituents[b]:	3% QF-1[c]				0.5-1.0% QF-1[d]		0.5% CNSi[e]		2-3% XE-60[f]				0.5% Hi-Eff-8B[g]		0.4% NGS[h]
	OH	TFA	TMS	p-TMS	OH	TFA	OH	TFA	OH	TFA	TMS	p-TMS	TMS	TMS	TFA
None	—	—	—	—	0.16	—	0.31	0.35	0.09	—	—	—	0.77	—	0.44
3α (e)	0.53	0.47	0.30	—	0.49	0.42	0.37	0.81	0.36	0.80	—	1.02	1.36	—	1.06
(5α)3α (a)	0.52	—	—	—	—	—	—	—	—	—	—	—	—	—	—
3β (a)	0.48	0.46	0.29	—	0.44	0.41	0.34	0.76	—	—	—	—	1.09	—	0.89
(5α)3β (e)	0.58	—	—	—	—	—	—	—	—	—	—	—	—	—	—
7α (a)	0.38	—	0.26	—	0.35	0.26	—	—	0.25	—	—	0.67	—	—	—
7β (e)	0.44	—	0.25	—	0.39	0.32	—	—	—	—	—	—	—	—	—
12α (a)	0.34	—	0.22	—	0.31	0.22	0.24	—	0.23	—	—	0.64	—	—	—
12β (e)	0.37	—	0.21	—	0.34	0.25	0.26	—	—	—	—	—	—	—	—
3-keto	1.00	—	—	—	0.95	—	0.49	—	—	—	—	—	4.26	—	3.39
(5α)3-keto	1.09	—	—	—	—	—	—	—	—	—	—	—	—	—	—
7-keto	—	—	—	—	0.57	—	—	—	—	—	—	—	—	—	—
12-keto	—	—	—	—	0.49	—	0.28	—	—	—	—	—	—	—	—
3α,Δ5	—	—	—	—	—	0.44	0.33	0.77	—	—	—	—	—	—	0.94
3α,Δ6	—	—	—	—	—	0.42	0.37	0.73	—	—	—	—	—	—	0.90
3α,Δ7	—	—	—	—	—	0.45	—	0.78	—	—	—	—	—	—	1.09
3α,Δ7,9	—	—	—	—	—	0.38	—	0.67	—	—	—	—	—	—	0.91
3β,Δ5	—	—	—	—	0.51	0.46	—	—	—	—	—	—	—	—	—
3-keto,Δ4	1.66	—	—	—	—	0.87	0.43	—	—	—	—	—	—	—	—
3-keto,Δ6	—	—	—	—	—	0.89	—	—	—	—	—	—	—	—	—
3-keto,Δ7,9	—	—	—	—	—	—	—	—	—	—	—	—	—	—	—
3α,6α (e,e)	1.48	0.93	0.36	—	1.47	0.98	1.48	1.62	1.55	1.57	—	1.16	1.26	—	1.93
3α,6β (e,a)	—	—	—	—	1.19	0.91	1.35	—	—	—	—	—	—	—	—
3β,6α	—	—	—	—	—	0.96	—	1.59	—	—	—	—	—	—	—

(5α)3β,6α	—	—	—	—	1.18	—	—	—	—	—	—	—	—	—
3β,6β	—	—	—	—	0.88	—	—	—	—	—	—	—	—	—
(5α)3β,6β	—	—	—	—	1.09	—	—	—	—	—	—	—	—	—
3α,7α	1.13	0.83	—	0.64	0.85	—	1.50	1.10	1.39	1.09	0.46	1.08	—	1.50
(5α)3α,7α	1.17	—	—	—	—	1.18	1.53	—	—	—	—	—	—	—
3α,7β	1.22	0.95	0.37	0.58	1.00	1.26	—	1.16	—	1.26	—	1.57	—	2.12
3β,7α	0.96	0.69	—	—	—	1.64	—	—	—	—	—	—	—	—
(5α)3β,7α	1.24	—	—	—	—	1.10	1.77	—	—	—	—	—	—	—
3α,12α	1.00	0.68	—	0.56	0.67	1.18	—	—	—	—	—	—	—	—
(5α)3α,12α	1.07	—	—	—	—	1.00	1.00[i]	1.00	1.00[i]	1.00[i]	0.41	1.00[i]	—	1.00[i]
3α,12β	1.06	0.78	0.30	—	0.78	1.06	—	—	—	—	—	—	—	—
3β,12α	0.88	0.59	—	0.53	0.58	0.91	—	—	—	—	—	—	—	—
(5α)3β,12α	1.15	—	—	—	—	—	—	—	—	—	—	—	—	—
3β,12β	0.96	0.72[j]	0.31	—	—	—	—	—	—	—	—	—	—	—
3α,12α,Δ^4	0.70	—	—	—	—	—	—	—	—	—	—	—	—	—
3β,12α,Δ^4	0.77	—	—	—	—	—	—	—	—	—	—	—	—	—
3α,12α,$\Delta^{8(14)}$	—	—	—	—	—	—	0.82	—	—	—	—	—	—	—
7α,12α	0.73	0.43	—	—	0.58	—	—	0.68	—	0.66	—	—	—	0.90
3α,6-keto	1.73	1.52	1.01	2.61	1.76	1.41	—	—	—	—	—	—	—	—
3α,7-keto	1.57	1.38	0.98	1.76	1.60	—	3.80	—	—	—	—	4.19	—	—
3α,12-keto	1.34	1.23	0.85	1.62	1.54	1.23	—	—	—	—	—	3.97	—	—
3β,12-keto	2.22	1.61[j]	—	1.32	1.28	—	—	—	—	—	—	—	—	—
3-keto,7α	2.29	—	—	2.17	1.76	1.74	—	—	—	—	—	—	—	—
(5α)3-keto,7α	1.84	1.36	—	—	—	—	—	—	—	—	—	—	—	—
3-keto,12α	2.14	—	—	1.83	1.42	1.44	—	—	—	—	—	—	1.35	—
(5α)3-keto,12α	2.98[j]	—	—	—	—	—	—	—	—	—	—	—	—	—
3-keto,7α,Δ^4	3.28	—	—	—	—	—	—	—	—	—	—	—	—	—
3-keto,12α,Δ^4	2.69	—	—	—	—	—	—	—	—	—	—	—	—	—
3,7-diketo	2.71	—	—	2.83	1.61	—	—	—	—	—	—	—	4.47	—
3,12-diketo	3.18	—	—	2.86	1.56	—	—	—	—	—	—	—	4.44	—
(5α)3,12-diketo	—	—	—	—	—	—	—	—	—	—	—	—	—	—

(Continued)

TABLE X (Continued)

Bile acid methyl ester Substituents[b]:	3% QF-1[c]				0.5-1.0% QF-1[d]		0.5% CNS[e]		2-3% XE-60[f]				0.5% Hi-Eff-8B[g]		0.4% NGS[h]
	OH	TFA	TMS	p-TMS	OH	TFA	OH	TFA	OH	TFA	TMS	p-TMS	TMS	TMS	TFA
7,12-diketo	1.41	—	—	—	—	—	3.51	—	—	—	—	—	—	—	—
3α,6α,7α	—	0.87[i]	—	—	2.61	1.24[i]	3.71	1.15[i]	—	—	—	—	1.01	—	—
3α,6β,7α	2.28	—	—	1.24	2.50	0.87[i]	3.12	—	—	—	—	—	0.72	—	—
3α,6α,7β	—	1.32	—	—	2.40	1.60[i]	2.82	2.56	—	—	—	—	2.01	—	—
3α,6β,7β	2.20	—	—	0.72	2.31	1.47	3.41	—	—	—	—	—	1.21	—	—
3α,6β,12α	—	—	—	—	—	—	—	—	—	—	—	—	—	—	—
3α,7α,12α	2.14	1.29	—	1.23	2.33	1.39	3.08	2.27	—	2.20	0.91	1.24	0.75	—	1.94
(5α)3α,7α,12α	2.37	1.37	—	—	—	—	—	—	—	—	—	—	—	—	—
3α,7β,12α	2.17	1.18	—	0.74	2.44	1.33	3.20	—	—	—	—	—	—	—	—
3β,7α,12α	1.81	0.95	—	1.13	1.86	1.17	—	—	—	—	—	—	—	—	—
(5α)3β,7α,12α	2.50	—	—	—	—	—	—	—	—	—	—	—	—	—	—
3β,7β,12α	1.93	1.00	—	0.65	—	—	—	—	—	—	—	—	—	—	—
3α,7α,12-keto	3.33	2.42	—	2.09	4.00	2.80	4.19	6.50	—	—	—	—	3.08	1.00	—
3α,12α,7-keto	3.07	1.99	—	1.88	3.70	2.54	4.07	—	—	—	—	—	3.15	1.00[j]	—
3-keto,7α,12α	3.94	2.44	—	—	4.79	2.79	4.97	—	—	—	—	—	—	—	—
(5α)3-keto,7α,12α	4.50	—	—	—	—	—	—	—	—	—	—	—	—	—	—
3-keto,7α,12α,Δ^4	3.33[i]	—	—	—	—	—	—	—	—	—	—	—	—	—	—
3α,7,12-diketo	—	—	—	—	4.48	4.30	3.98	—	—	—	—	—	3.20	—	—
3,7,12-triketo	5.60	—	—	—	6.33	—	4.55	—	—	—	—	—	—	—	—
(5α)3,7,12-triketo	7.20	—	—	—	—	—	—	—	—	—	—	—	—	—	—

[a] Relative to methyl deoxycholate unless otherwise indicated.

[b] (5α) indicates a 5α-cholanoate, other compounds are substituted 5β-cholanoates. (e) = equatorial; (a) = axial; Δ = double bond; OH = free hydroxyl groups; TFA = trifluoroacetates; TMS = trimethylsilyl ethers; p-TMS = partial TMS where 6β,7α and/or 12α hydroxyl groups are free. For other abbreviations see Table I.

[c] P. Eneroth, B. A. Gordon, R. Ryhage, and J. Sjövall, *J. Lipid Res.* **7**, 511, 524 (1966); A. Kallner, *Arkiv Kemi* **26**, 553 (1967); unpublished data. Column temperature 230–235°.

[d] J. Sjövall, *Acta Chem. Scand.* **16**, 1761 (1962); K. Tsuda, V. Sato, N. Ikekawa, S. Tanaka, H. Higashikuze, and R. Osawa, *Chem. Pharm. Bull. (Tokyo)* **13**, 720 (1965); T. Okishio and P. P. Nair, *Anal. Biochem.* **15**, 360 (1966). Column temperature 210–220°.

[e] D. H. Sandberg, J. Sjövall, K. Sjövall, and D. A. Turner, *J. Lipid Res.* **6**, 182 (1965); Column temperature 225° (OH). K. Tsuda *et al.*, see footnote *d* (TFA).

[f] T. Briggs and S. R. Lipsky, *Biochim. Biophys. Acta* **97**, 579 (1965), (OH), (TMS) and (p-TMS): column temperature 240°; A. Kuksis and B. A. Gordon, *Can. J. Biochem. Physiol.* **41**, 1355 (1963), (TFA): column temperature 200°.

[g] M. Makita and W. W. Wells, *Anal. Biochem.* **5**, 523 (1963). Column temperatures 245° (left column) and 280° (right column).

[h] K. Tsuda *et al.*, see footnote *d*. Column temperature 210°.

[i] These compounds often give two peaks. The retention time of the major peak, which may be a degradation product, is given.

[j] Retention times given relative to this derivative.

TABLE XI
RELATIVE RETENTION TIMES[a] OF DIFFERENT BILE ACID METHYL ESTER DERIVATIVES ON PHASES WITH LIMITED SELECTIVITY

Bile acid methyl ester	0.5–2.3% SE-30[c]			0.75% SE-52[d]		0.5% PhSi-20[e]		0.5% PhSi-35[e]	
Substituents[b]:	OH	OAc	TFA	OH	OAc	OH	OAc	OH	OAc
None	0.29	—	0.71	—	—	0.23	—	0.20	—
3α	0.60	0.83	1.13	0.32	0.72	0.54	0.79	0.49	0.67
3β	—	—	1.08	—	—	—	—	—	—
12α	—	—	—	—	—	0.42	—	0.38	—
3-keto	0.66	—	—	—	—	—	—	—	—
12-keto	—	—	—	—	—	0.45	—	0.42	—
$3\alpha,\Delta^5$	—	—	1.09	—	—	—	—	—	—
$3\alpha,\Delta^6$	—	—	1.01	0.63	—	—	—	—	—
$3\alpha,\Delta^7$	—	—	1.14	—	0.76	—	—	—	—
$3\alpha,\Delta^{7,9}$	—	—	0.98	—	0.66	—	—	—	—
$3\beta,\Delta^5$	—	—	—	—	—	0.64	—	0.58	—
3-keto,Δ^6	0.66	—	—	—	—	—	—	—	—
3-keto,$\Delta^{7,9}$	0.66	—	—	—	—	—	—	—	—
$3\alpha,6\alpha$	1.26	1.81	1.28	1.24	1.55	1.34	1.92	1.30	1.72
$3\beta,6\alpha$	—	—	1.30	—	—	—	—	—	—
$(5\alpha)3\beta,6\alpha$	—	—	1.55	—	—	—	—	—	—
$3\beta,6\beta$	—	—	1.18	—	—	—	—	—	—
$(5\alpha)3\beta,6\beta$	—	—	1.39	—	—	—	—	—	—
$3\alpha,7\alpha$	1.11	1.36	1.17	1.12	1.20	1.16	1.35	1.18	1.29
$3\alpha,7\beta$	1.09	—	1.40	1.22	1.58	1.12	—	1.16	—
$(C_{22})3\alpha,12\alpha$	—	—	—	—	—	—	0.55	—	0.51
$(C_{23})3\alpha,12\alpha$	—	—	—	—	—	—	0.86	—	0.82
$3\alpha,12\alpha$	1.00	1.12	1.00[f]	1.00	1.00[f]	1.00	1.15	1.00	1.14
$3\alpha,12\alpha,\Delta^{8(14)}$	—	—	0.86	1.04	—	—	—	—	—
3α,7-keto	1.08	—	2.00	—	—	1.14	1.49	1.21	1.48
3α,12-keto	1.05	—	—	—	1.20	1.15	1.46	1.12	1.43
3,7-diketo	1.03	—	—	—	—	1.16	—	1.21	—
3,12-diketo	1.06	—	—	—	—	1.27	—	1.25	—
$3\alpha,6\alpha,7\alpha$	2.21	—	—	—	—	2.41	—	2.62	—
$3\alpha,6\beta,7\beta$	—	—	—	—	—	2.20	—	2.24	—
$(C_{22})3\alpha,7\alpha,12\alpha$	—	—	—	—	—	1.03	—	1.17	—
$3\alpha,7\alpha,12\alpha$	1.89	1.60	1.09	1.86	1.44	2.20	—	2.32	—
$(C_{27})3\alpha,7\alpha,12\alpha$	—	—	—	—	—	4.37	—	4.23	—
$3\alpha,7\beta,12\alpha$	—	—	—	—	1.98	2.30	—	2.36	—
$3\alpha,7\alpha$,12-keto	—	—	1.97	—	1.93	2.62	2.57	2.73	2.59
$3\alpha,12\alpha$,7-keto	—	—	—	—	—	2.20	2.24	2.36	2.27
3α,7,12-diketo	1.71	—	—	—	—	2.18	2.52	2.34	2.72
3,7,12-triketo	1.61	—	—	—	—	2.03	—	2.30	—

[a] Relative to methyl deoxycholate unless otherwise indicated.

[b] For abbreviations see Tables X and VIII; OAc = acetates.

[c] W. J. A. VandenHeuvel, J. Sjövall, and E. C. Horning, *Biochim. Biophys. Acta* **48,**

TABLE XII
RETENTION TIMES RELATIVE TO METHYL DEOXYCHOLATE OF
3-DIMETHYLHYDRAZONES (3-DMH) OF BILE ACID METHYL
ESTERS ON 3% QF-1 COLUMN AT 230°

Compound	3-DMH	3-DMH-7-keto	3-DMH-12-keto	3-DMH-7,12-diketo	3-DMH-7α,12α
Retention time	0.50	1.50	1.34	3.32	1.89

Combined Gas Chromatography–Mass Spectrometry

The recently developed gas chromatography–mass spectrometry (GC-MS) instruments[49, 50] have opened new possibilities for analysis of bile acids in biological materials. The technique is based on a separation device between the outlet of the GLC column and the mass spectrometer that will selectively remove most of the carrier gas, thus permitting the ion source to be operated at a pressure of less than 10^{-4} mm Hg. Our experience is limited to the use of the instrument described by Ryhage[49] and the LKB 9000 GC-MS instrument equipped with a Ryhage-Becker molecule separator of the jet type. When used in bile acid and steroid analysis the latter instrument permits the use of the same gas chromatographic conditions as in conventional GLC. The temperature of the molecule separator and the ion source should be kept at least as high as that of the column to prevent condensation of compounds appearing in the effluent. Since the temperature may influence the fragmentation process it should be kept constant to make detailed comparisons between different analyses possible. The energy of the bombarding electrons is kept at 20–23 eV to avoid ionization of the remaining carrier gas (helium, ionization potential 24.8 eV). By amplifying a fraction of the total ion current produced in the ion source, a conventional GLC curve is obtained. Mass spectra can be recorded at any time during the analysis; a suitable scan speed is 5–8 seconds for the mass range 10–700.

[49] R. Ryhage, *Anal. Chem.* **36,** 759 (1964).
[50] J. T. Watson and K. Biemann, *Anal. Chem.* **37,** 844 (1965).

596 (1961); and unpublished: Column temperature 210° (OH). K. Tsuda et al. (see footnote c, Table X): Column temperature 220° (TFA). A. Kuksis and B. A. Gordon, *Can. J. Biochem. Physiol.* **41,** 1355 (1963): Column temperature 195° (OAc).

[d] W. L. Holmes and E. Stack, *Biochim. Biophys. Acta* **56,** 163 (1962): Column temperature 250° (OH). K. Tsuda et al. (see footnote c): Column temperature 260° (OH) and 235° (OAc).

[e] J. Sjövall, C. R. Meloni, and D. A. Turner, *J. Lipid Res.* **2,** 317 (1961): Column temperature 215–200°.

[f] Retention times given relative to this derivative.

The sensitivity in GC-MS analysis of bile acid derivatives is determined by the intensity of the background spectrum, which has to be subtracted from the sample spectrum. For this reason only well conditioned columns with a low percentage of stationary phase should be used. When SE-30 or QF-1 columns are used in work with biological samples, detailed mass spectra are obtained with 0.1–10 μg of bile acid derivatives (depending on retention times and type of derivatives).

Some difficulties in the interpretation of GC-MS data may arise from imperfections in column technology. Thus, tailing on the GLC column of a major component will make the interpretation of the mass spectrum of a following minor compound difficult. This may occur in the analysis of bile acid derivatives with unprotected hydroxyl groups. When thermally labile derivatives (e.g., TFA) are analyzed, degradation products may form on the GLC column and be eluted ahead of the intact molecules. This makes the analysis of mixtures of such compounds difficult.

A discussion of the fragmentation patterns of different bile acid derivatives is beyond the scope of this chapter. For details the reader is referred to reviews[46, 51] and original papers.[29, 52-54]

Chromatographic Identification of Bile Acids in Biological Materials

The following general procedure, based on the combination of TLC and GLC, may be used for complex bile acid mixtures. Crude bile acid methyl esters are first applied to thin-layer plates for preparative subdivision—e.g. into mono-, di-, and trisubstituted compounds—before a more detailed analysis is made. In some cases, for example, in work with fecal bile acids, it is more efficient to make a purification on silicic or aluminum oxide columns before TLC is used. Each group of bile acids can then be further subdivided by preparative TLC. The fractions obtained are analyzed by GLC and TLC. The latter method is of particular value in studies of trihydroxy bile acids, which are less well separated by GLC. The results obtained usually permit a preliminary tentative identification. However, other compounds may have the same chromatographic properties as bile acid methyl esters. A bile acid structure can then usually be established by GLC of the oxidation product. Oxidation is conveniently carried out by dissolving 1–100 μg of material in 1 ml of acetone. After cooling in an ice bath, 10 μl of oxidizing reagent

[51] H. Budzikiewicz, C. Djerassi, and D. H. Williams, in "Structure Elucidation of Natural Products by Mass Spectrometry," Vol. 2. Holden Day, San Francisco, California, 1964.
[52] P. Eneroth, B. A. Gordon, and J. Sjövall, J. Lipid Res. 7, 524 (1966).
[53] S. Bergström, R. Ryhage, and E. Stenhagen, Svensk Kem. Tidskr. 73, 566 (1961).
[54] P. D. G. Dean and R. T. Aplin, Steroids 8, 565 (1966).

(26.72 g of CrO_3 and 23 ml of sulfuric acid diluted to 100 ml with water) is added.[55] The mixture is left in an atmosphere of N_2 for 10 minutes, and the reaction is terminated by addition of 5 ml of water. The reaction product is extracted with ethyl acetate and analyzed on a 3% QF-1 column which specifically retards ketonic compounds. In this way it is possible to identify different methyl ketocholanoates (as such or as the 3-dimethylhydrazones) and to determine the number and position of the original substituents as well as the configuration at C-5. The procedure is not applicable to compounds where the oxidation results in formation of several products, e.g., to 3,6,7-trisubstituted cholanoates. When it is initially believed that the unknown compound is a bile acid methyl ester carrying only keto groups, oxidation should still be carried out to ensure that no peak shift (change in relative retention time after a specific quantitative reaction) occurs.

The number of hydroxyl groups can often be deduced by studying peak shifts on QF-1 columns after trifluoroacetylation of the compound(s) (Table X). The stereochemistry of the hydroxyl groups can be studied with GLC on QF-1 or XE-60 after stereospecific formation of TMS ethers. Conversion of a hydroxyl group into a TMS ether will result in a decreased retention time; the total decrease is related to the number of reacting groups (Table X). Peak shift studies have been more frequently carried out than "spot shift" studies mainly since the derivatives commonly used have been specifically chosen for GLC. However, it is also possible to study changes in mobility (spot shift) on TLC or paper chromatography after formation of derivatives. An excellent description of this technique as used in the paper chromatographic identification of steroids has been published by Bush.[56] It may be mentioned that TFA and TMS derivatives, which are so useful in GLC, can be analyzed also by TLC.

The retention time changes observed when the bile acid methyl esters are subjected to the reactions mentioned give considerable information regarding the structure of the bile acid. When simple mixtures are analyzed, the peak shift technique may often be directly used without preliminary purifications; the results will permit the selection of suitable reference compounds and comparisons can be made with the compounds in the sample. However, the technique cannot be used without preliminary separations if there are too many compounds in a mixture; it is then difficult to establish correspondence between the peaks of the different derivatives. The same difficulty exists when a mixture is analyzed on

---

[55] C. Djerassi, R. R. Engle, and A. Bowers, *J. Org. Chem.* **21**, 1547 (1956).
[56] I. E. Bush, "The Chromatography of Steroids." Pergamon Press, London, 1961.

several columns with different stationary phases. It should therefore be stressed that whenever mixtures containing more than about 5 compounds are analyzed, a preliminary fractionation must be carried out. This is true also when minor components are to be identified; degradation peaks can arise from major components, and these can be mistaken for separate compounds. The predominant compounds should therefore be removed before the minor components are studied. It should also be ascertained (e.g., by analysis of blanks) that interfering compounds are not introduced during the purification procedure—e.g., from insufficiently prepurified silica gel.

Whenever problems arise as to whether a peak is due to a bile acid derivative, or if it is difficult to establish correspondence between the peaks before and after derivative formation, combined GC-MS is used. This technique usually also makes it possible to see whether a peak is due to a degradation product from another compound appearing later in the chromatogram. By GC-MS analysis a molecular weight can usually be obtained for the unprotected methyl cholanoate or its TMS ether. Di- and trifluoroacetates usually do not give a molecular ion peak, and when these derivatives are used the retention time will indicate the probable number of trifluoroacetoxy groups in the molecule. The presence of keto groups is easily established since a characteristic nuclear fragment containing the ketonic oxygen is always found. Positional isomers are distinguished by their mass spectra whereas stereoisomers may not give mass spectral differences. In such cases the complementary use of GLC aids in the identification since the stereoisomers can usually be separated on QF-1 columns.

Quantitative Analysis of Bile Acids in Bile and Small Intestinal Contents

Conjugated bile acids in bile and small intestinal contents may be analyzed directly by paper chromatography. A suitable volume (e.g., 5–100 μl) is then applied to the starting line, and the analysis is carried out as described in the section on paper chromatography (p. 246). When the bile acid concentration is low (<2 meq/liter) proteins may be removed by adding the sample drop by drop to 20 volumes of ethanol. After brief boiling and subsequent cooling, the extract is filtered. The concentrated extract can be used for quantitative paper chromatography or TLC. Total bile acids may be quantitated by enzymatic methods.[32, 33]

For analysis of the small amounts of bile acids present in gallstones,[57] the stones are pulverized and refluxed in chloroform–methanol 1:1 for 1 hour. The extract is filtered and the residue is reextracted with ethanol.

[57] L. J. Schoenfield, J. Sjövall, and K. Sjövall, *J. Lab. Clin. Med.* **68**, 186 (1966).

The combined extracts are evaporated *in vacuo,* and the residue is subjected to a three-stage countercurrent distribution between equal volumes of 70% ethanol and petroleum ether. The 70% ethanol phases are evaporated, and the residue is analyzed for bile acids in the same way as an ethanol extract of bile. Some loss of nonpolar bile acids may occur.

A detailed analysis of individual bile acids present in minor amounts usually requires hydrolysis of the conjugated bile acids followed by quantitation of the free bile acids by GLC. The hydrolytic step is at present a weak point in bile acid analysis which will be eliminated when detailed procedures have been worked out for hydrolysis with the extracellular enzymes produced by strains of *Clostridium.*[58-60]

Many conditions for alkaline hydrolysis of bile acids have been described and claimed to give little destruction of the bile acids. It must be realized, however, that tests with pure free bile acids are not representative for the conditions that exist in a biological extract where the bile acids may form only a minor part of the compounds present. The possibility of artifact formation must therefore always be kept in mind.

The following conditions may be used.[61] An ethanol extract of 0.1–0.5 ml of bile or small intestinal contents is dissolved in about 5 ml of 50% aqueous ethanol containing 15% (w/v) sodium hydroxide. The solution (kept in a Teflon tube in a Parr bomb) is heated at 110° for 10 hours (hydrolysis should not be carried out in glass tubes since losses may occur by adsorption on silicic acid precipitated after acidification). The hydrolyzate is transferred to a separatory funnel, 10 ml of water is added, and the solution is acidified with $4 N$ HCl to about pH 1. It is extracted with 3×20 ml of ethyl acetate. The combined ethyl acetate phases are washed with 5–10 ml portions of water until neutral, each portion being backwashed with 20 ml of ethyl acetate. The ethyl acetate phases are evaporated *in vacuo.* The free bile acids are analyzed by TLC or, after methylation with diazomethane, by GLC.

For quantitative determinations, GLC on QF-1 columns of methyl esters[41, 57] or methyl ester trifluoroacetates[44, 62] or GLC on cyclohexanedimethanol polysuccinate (Hi-Eff 8B) columns of TMS ethers[40] is used. The areas of the GLC peaks are compared with those obtained with appropriate standards, preferably in a linear range.

A stock solution of methyl lithocholate (0.3 mg/ml), methyl deoxycholate (0.8 mg/ml), methyl chenodeoxycholate (0.8 mg/ml), and methyl

[58] A. Norman and R. Grubb, *Acta Pathol. Microbiol. Scand.* **36,** 537 (1955).
[59] A. Norman and O. A. Widström, *Proc. Soc. Exptl. Biol. Med.* **117,** 442 (1964).
[60] P. P. Nair, M. Gordon, and J. Reback, *J. Biol. Chem.* **242,** 7 (1967).
[61] L. Irvin, C. G. Johnston, and J. Kopala, *J. Biol. Chem.* **153,** 439 (1944).
[62] D. H. Sandberg, J. Sjövall, K. Sjövall, and D. A. Turner, *J. Lipid Res.* **6,** 182 (1965).

cholate (3 mg/ml) is prepared in acetone. Dilutions are made 1:2, 1:4, and 1:8. Four microliters of each of the standard solutions is injected on a 3% QF-1 column using a Chaney adaptor for the 10 μl Hamilton syringe to obtain reproducible injections. Suitable dilutions of the unknown sample are prepared, and 4 μl amounts are injected. Calibration curves are prepared from the peak areas of the standards from which the amounts of bile acids in the unknown sample are calculated. If bile acids other than those mentioned are present, appropriate standards are prepared.

This method is sufficiently accurate for most purposes provided that injection errors are minimized by the use of one standard injection volume (e.g., 4 μl, preferably not less). The injection error is small after some practice and can be made smaller by incorporation of a known amount of an internal standard. The use of a dilution series of a standard mixture is advisable since possible day-to-day variations of the relative response of different bile acid methyl esters are then controlled. Small variations may occur as a result of changing properties of the column and detector systems when repeated injections of impure biological samples are made. When trifluoroacetates are determined, the derivatives of the standard samples and the unknown sample should be prepared and analyzed on the same day.

Although factors to correct for the differing responses of various bile acid derivatives have been devised[44, 63] in order to avoid the necessity for individual standards, we have not been able to obtain constant factors valid for different columns.[41, 45] In only one instance the same response has been found with derivatives of different bile acids. Thus Grundy et al.[64] using a hydrogen flame detector and TMS derivatives obtained the same ionization response per unit mass of the parent unsubstituted bile acids. Using a Hi-Eff-8B column and an argon ionization detector Makita and Wells[40] found variable molar responses with different bile acid TMS ethers. It is thus clear that the conditions for quantitative GLC must be individually tested for each column and detector system. When this is done, the accuracy of the GLC quantitations varies with the nature and concentrations of the components and is of the order $\pm5 - \pm10\%$.[63, 65]

Quantitative Determination of Bile Acids in Blood and Liver Homogenates

Total bile acids have been determined by an enzymatic method,[32] but difficulties have been experienced in the application of this method unless

[63] D. K. Bloomfield, *Anal. Chem.* 34, 737 (1962).
[64] S. M. Grundy, E. H. Ahrens, Jr., and T. A. Miettinen, *J. Lipid Res.* 6, 397 (1965).
[65] A. Kuksis, *J. Am. Oil Chemists Soc.*, 42, 276 (1965).

a TLC purification procedure is used.[66] Since we have no personal experience with this technique, it is not described here.

Individual bile acids may be determined by GLC. The procedure described is a slightly simplified version of the published method.[62]

All solvents are redistilled. They are checked for absence of interfering compounds by injection of a concentrate corresponding to 5 ml of solvent into the gas chromatograph. A column of 500 ml of the ion exchanger Amberlyst A 26 (Rohm and Haas, Philadelphia, Pennsylvania) is prepared and washed with 2 liters each of ethanol, hexane, ethanol, water, and 4 N hydrochloric acid. After it has been washed with water until neutral, the column is converted into bicarbonate form with 4 liters of 8% sodium bicarbonate in water. It is then washed with water until netural. The ion exchanger can then be stored in water in dark bottles.

Procedure. A column of 3 ml of ion exchanger is prepared in a chromatography tube, 200 × 7 mm, having a solvent reservoir, 100 ml, at the top. The column is converted into OH⁻ form with 30 ml of 1 N NaOH in water followed by 30 ml of 1 N NaOH in 80% aqueous ethanol. It is then washed with water until the effluent is neutral.

Serum or plasma, 2–5 ml, is diluted with an equal volume of water, and the pH is adjusted to about 11 with 1 M NaOH. The sample is applied to the column and allowed to pass through the resin at a rate of about one drop every 20 seconds. The column is then washed to neutrality with water. The preparation is washed with 20 ml of ethanol, then the bile acids are eluted with 150 ml of 0.2 M ammonium carbonate in 80% aqueous ethanol.[67]

The eluate is evaporated *in vacuo* and transferred with 3 × 2 ml of 15% NaOH in 50% ethanol, to a Teflon container in a steel bomb. Hydrolysis is carried out at 110° for 10 hours. The sample is then transferred to a separatory funnel, the Teflon container being rinsed with 2 × 5 ml of water. After acidification with 4 M HCl, the bile acids are extracted with 3 × 20 ml of ethyl acetate. The pooled extracts are washed with 5–10 ml portions of water until neutral, each portion being reextracted with 20 ml of ethyl acetate.

The pooled ethyl acetate phases are evaporated *in vacuo*. The residue is methylated with freshly distilled diazomethane. Excess reagent and solvents are evaporated *in vacuo*, and the residue is transferred with 5-ml and 2-ml portions of benzene to a column of 1 g of aluminum oxide (Woelm, activity grade IV). After washing with 25 ml of benzene, the bile acids are eluted with 40 ml of acetone–methanol 9:1. This eluate is evaporated to dryness and is transferred with acetone to a 4-ml glass-stoppered centrifuge tube. The solvent is removed under a stream of

[66] R. H. Palmer, personal communication, 1967.
[67] G. W. Kuron and D. M. Tennent, *Federation Proc.* **20,** 268 (1961).

nitrogen, the residue is dissolved in 0.2 ml of trifluoroacetic anhydride, and the tube is stoppered and heated for 15 minutes at 35°. Appropriate standard mixtures of bile acid methyl esters are also converted into trifluoroacetates.

The trifluoroacetates are analyzed on 1% QF-1 or XE-60 columns or on a mixed QF-1/SE-30/NGS column.[48] Conditions giving a linear peak area:mass ratio are established for the range 0.02–0.5 μg. This can be done with an ^{90}Sr argon ionization detector or a flame ionization detector. The injection volume, 4 μl, is kept constant for samples and standard (see section on analysis of bile acids in bile). To avoid large solvent fronts, acetonitrile and carbon disulfide are used with argon ionization and flame ionization detectors, respectively.

When used for determinations of deoxycholic, chenodeoxycholic, and cholic acids in normal human serum a loss of 15–40% occurs, mainly during hydrolysis. Free bile acids—for example, present in rat portal blood[68]—are determined with the same procedure with omission of the hydrolytic step.

Lithocholic and other nonpolar bile acids, if present, are partly lost during aluminum oxide chromatography. If such acids are to be determined, the sample is applied to the alumina column in hexane–benzene 9:1. After washing with 20 ml of this solvent, methyl lithocholate may be eluted with benzene, 35 ml; di- and trihydroxycholanoates being eluted with acetone–methanol 9:1.[69]

Okishio and Nair have described the use of GLC for determination of bile acids in liver tissue.[69] Homogenates or subcellular fractions are lyophilized and then refluxed for 30 minutes with 100 ml of 95% ethanol containing 0.1% ammonium hydroxide. The residue is reextracted with the same volume of fresh solvent and filtered. The material left on the filter paper is washed with 50 ml of solvent. After evaporation of the combined extracts *in vacuo*, the residue is dissolved in 3–7 ml of 0.1 M NaOH and transferred to a 5-ml Amberlyst A 26 column in the OH⁻ form with the addition of 35 ml of water. The column is washed with 20-ml portions of ethanol, ethanol–ethylene chloride 1:1, and 80% ethanol. Elution of bile acids and subsequent steps are carried out as described above.

Quantitative Determination of Bile Acids in Feces

In connection with studies of cholesterol metabolism much work has been devoted to the problem of quantitating fecal bile acids. The complexity of the fecal lipid mixture makes the development of a generally

[68] T. Cronholm and J. Sjövall, *European J. Biochem.* **2**, 375 (1967).
[69] T. Okishio and P. P. Nair, *Biochemistry* **5**, 3662 (1966).

applicable method difficult. Older methods based on titrimetric or spectrophotometric determinations of bile acids after simple extraction procedures suffer from lack of specificity. Recent improvements in chromatographic techniques as indicated in preceding sections have made it possible to determine fecal bile acids with much increased accuracy. At present, the final quantitative determination of bile acids in feces from different species is preferably made by GLC. Irrespective of whether the bile acids are measured individually or as a group, a fecal extract has to be prepared, saponified, and subjected to a preliminary chromatographic purification before GLC can be used. To the initial extract internal standard(s) may be added to correct for specific and nonspecific losses during workup. Specific losses, for example, of the least polar or most polar bile acids, may occur during group purification. To check that such losses do not occur, the purification procedures can be tested by addition of ^{14}C-labeled (New England Nuclear Corp., Boston, Massachusetts) 3-ketocholanoic (obtained by oxidation of lithocholic acid as described in section on chromatographic identification) and/or cholic acids to the initial extract. Radiopurity is checked before use by TLC-autoradiography. The excretion products of radioactive bile acids or cholesterol given to the subjects studied may also serve as internal standards. The choice of internal standard to correct for nonspecific losses depends on the TLC conditions used for the quantitation since the retention time of this standard must be different from those of the bile acids to be determined. Alternatively labeled deoxycholic acid (a major fecal bile acid) may be used.

A prerequisite for the quantitative GLC methods is that the composition of the fecal bile acid mixture is known. By application of the procedures described in the section on chromatographic identifications a number of isomeric mono-, di- and trisubstituted cholanoic acids in human feces can be identified (Table XIII).[29, 52] Some of the 3β-hydroxycholanoic acids occur as esters with unidentified acid(s).[70] Normally, conjugated bile acids are present only in small amounts. In pathological conditions this amount may be significantly increased. Other bile acids, for example, those carrying a hydroxyl group at C-6, may be found in animal feces. The three methods described below for fecal bile acid quantitation have been developed for analysis of human feces. However, it must be remembered that they are not necessarily directly applicable to all experimental conditions. When dietary changes are introduced, the composition both of bile acids and nonbile acid contaminants may change and influence the analytical results. Under such circumstances and when the methods are used for analysis of bile acids in animal feces,

[70] A. Norman, *Brit. J. Nutr.* **18**, 173 (1964).

TABLE XIII

BILE ACIDS PRESENT IN HUMAN FECES[a]

Unsubstituted	Monosubstituted[c]	Disubstituted	Trisubstituted
Unsaturated cholanoic acid[b]	3-keto	3,12-diketo; $3\alpha,7\alpha$	$3\alpha,7\alpha,12$-keto
(not characterized)	3β	3-keto,7α; $3\alpha,7\beta$	$3\alpha,12\alpha,7$-keto
	3α	3-keto,12α; $3\beta,7\alpha$	$3\alpha,7\alpha,12\alpha$
		$3\alpha,7$-keto; $3\alpha,12\alpha$	$3\alpha,7\beta,12\alpha$
		$3\alpha,12$-keto; $3\alpha,12\beta$	$3\beta,7\alpha,12\alpha$
		$3\beta,12$-keto; $3\beta,12\alpha$	$3\beta,7\beta,12\alpha$
		$3\beta,12\beta$	$(5\alpha)3\alpha,7\alpha,12\alpha$

[a] P. Eneroth, B. A. Gordon, R. Ryhage, and J. Sjövall *J. Lipid Res.* **7**, 511, 524 (1966).
[b] A. Norman and R. H. Palmer *J. Lab. Clin. Med.* **63**, 986 (1964).
[c] For abbreviations, see Table X.

the specificity of the method should be tested by GLC and GC-MS analysis of the purified samples.

Method 1.[71] Feces are homogenized in chloroform–methanol 1:1 and are continuously extracted with hot chloroform–methanol 1:1 for 48 hours in a Soxhlet extractor. An aliquot (100–200 ml) corresponding to one-tenth of a daily portion is withdrawn into a 500-ml round-bottomed flask with a ground joint made to fit a liquid–liquid extractor of the upward displacement type.[72] The solution is concentrated *in vacuo* to about 20–50 ml (until foaming begins). Peroxide-free dioxane (4 ml per 100 g of solids in the extract) and $4 M$ KOH (2 ml/100 mg) are added and the mixture is refluxed for 3 hours. The hydrolyzate is neutralized, and a volume of the solvent equal to that of the dioxane added is distilled off. The concentrated hydrolyzate is acidified to pH 2–3 with $6 N$ HCl and the extractor is assembled. The solution is extracted with peroxide-free diethyl ether for 16 hours with magnetic stirring. The ether extract is washed 3 times with 0.05 volumes of a $0.08 M$ citrate–phosphate buffer, pH 5.8, and 3 times with 0.1 volumes of water. Each of the washings is passed through a second separatory funnel containing diethyl ether. The combined ether extracts are taken to dryness *in vacuo* and weighed. The residue is dissolved in benzene and an aliquot (100 mg is always sufficient) is applied to a silicic acid column in benzene providing a 25-fold excess of silicic acid (Mallinckrodt, activated for at least 5 days at 120–130°). Four fractions are collected: (I) benzene 10 ml/g silicic acid; (II) benzene–acetic acid 99:1, 30 ml/g silicic acid;

[71] P. Eneroth, K. Hellström, and J. Sjövall, *Acta Chem. Scand.* in press; and unpublished results from this laboratory.
[72] L. Kahn and C. H. Wayman, *Anal. Chem.* **36**, 1340 (1964).

(III) benzene–acetic acid 3:1, 40 ml/g silicic acid; (IV) chloroform–methanol 1:1, 10 ml/g silicic acid. The first two fractions contain mainly fatty acids and sterols. If present, cholenoic acid(s)[35] can be found in these fractions. Fraction III contains mono- and disubstituted bile acids whereas the predominant part of the trisubstituted bile acids are found in fraction IV. Fractions III and IV are evaporated, and from each fraction two samples are withdrawn and methylated with freshly distilled diazomethane. One of the two samples is trifluoroacetylated. The methyl esters and the trifluoroacetates are quantitated by GLC on QF-1 columns as described in the sections on analysis of bile and serum. The quantitation of methyl esters as well as trifluoroacetates permits an estimation of the accuracy with which individual bile acids are determined. Furthermore, this technique increases the possibility of detecting interfering non-bile acid compounds.

It is apparent from Table XIII that all bile acids cannot be determined individually with a single GLC method. However, 3α-hydroxy-, 3β-hydroxy-, 3α,12α-dihydroxy-, 3β,12α-dihydroxy-, 3α-hydroxy-12-keto-, and 3β-hydroxy-12-keto-5β-cholanoic acids have usually been found to be the predominant bile acids in a number of subjects on six different diets including both solid and formula type diets.[71] Several isomeric disubstituted cholanoic acids (see Table XIII) occurring in minor amounts will be included in the quantitation of the above-mentioned compounds. The collection of trisubstituted cholanoic acids in a separate silicic acid fraction facilitates their GLC determination, particularly when they are present in small amounts, as is usually the case. If the GLC analysis indicates an unusual bile acid composition the amounts of purified bile acids available are sufficient for extensive chromatographic identifications when the amounts of feces indicated (1/10 daily portion) have been worked up.

It should be mentioned that the method described is not applicable to the analysis of conjugated bile acids. The more vigorous conditions required to hydrolyze conjugated bile acids have been avoided for reasons given in the section on bile analysis.

To reduce the number of bile acids that have to be determined and yet obtain some information on the amounts of metabolites of the primary hepatic bile acids (chenodeoxycholic and cholic acids) the purified fecal bile acid may be oxidized with the oxidizing reagent described in the section on chromatographic identification. In this way methyl 3-keto- and 3,7-diketocholanoates (from chenodeoxycholic acid metabolites) and methyl 3,12-diketo- and 3,7,12-triketocholanoates (from cholic acid metabolites) are obtained. A precise quantitative evaluation of this method is still lacking.

Method 2.[64, 73] Feces are homogenized in an equal volume of water; 1 g of homogenate is mixed with 20 ml of 1 *M* NaOH in 90% aqueous ethanol. Deoxycholic acid-24-^{14}C is added as internal standard, and the mixture is refluxed for 1 hour. Ten milliliters of water is added, and neutral steroids are removed by extraction with 3×50 ml of petroleum ether (phases separated by centrifugation for 5 minutes at 1000 *g*). To the lower aqueous phase 2 ml of 10 *M* NaOH is added and hydrolysis (of esters and conjugated bile acids) is made at 2 atm for 3 hours in a pressure cooker. The pH is then adjusted to 2 (with HCl), and 75 ml of chloroform–methanol 2:1 is added. After the preparation has been shaken and centrifuged as above, the lower phase is removed. The procedure is repeated twice with 50-ml portions of chloroform, and the combined lower phases are taken to dryness *in vacuo.*

The residue is partially dissolved in 25 ml of heptane–propionic acid 99:1 and is applied to a column of 5 g of Florisil (60–100 mesh, activated overnight at 200°). Elution with 50 ml of the same solvent removes fatty acids. The material that is not dissolved in heptane–propionic acid is dissolved in benzene–methanol–acetic acid 80:20:1 and applied to the column in as small a volume as possible (including rinsing). The column is then eluted with 10 ml of diethyl ether–acetic acid 9:1. To this effluent 25 ml of ethyl acetate and 15 ml of water are added, and the solution is swirled. The lower phase is removed and extracted with 20 ml of ethyl acetate. The combined organic phases are taken to dryness, and the residue is methylated with 4 ml of 5% HCl in superdry methanol (overnight, or 2 hours at 100°). After evaporation of the solvent *in vacuo*, the residue is purified by TLC (see below). The Florisil purification step may be omitted when the bile acid concentration exceeds 200 μg per gram of homogenate (as is usually the case) and when larger quantities of bile acids are not required for further characterization.

When Florisil chromatography is not performed, one-fifth of the chloroform phase (obtained after extraction of the acidified saponification mixture) is methylated in 4 ml of methanol–HCl. After evaporation the residue is applied to a 0.5 mm silica gel H chromatoplate, 20×20 cm, which has been prerun in methanol–acetic acid 90:10 and subsequently activated at 120° for 1 hour. Methyl cholate is run as reference compound. The plate is first developed in benzene and exposed to iodine vapors. Fatty acid methyl esters appear as a band at R_f 0.65. A line is drawn below this zone, and the plate is developed with isooctane–isopropanol–acetic acid 120:40:1 to this line. Iodine vapor reveals the methyl cholate. An area of silica gel from methyl cholate to the line drawn below

[73] We are grateful to Dr. E. H. Ahrens, Jr., for providing valuable information regarding this method.

the fatty acid methyl ester zone is collected in a vacuum aspirator[74] and is eluted with 25 ml of methanol. To this solution is added a known amount of 5α-cholestane (internal standard for GLC, usually 0.25 mg), and the solvent is evaporated. The residue is dissolved in 10.0 ml of ethyl acetate, and 3 ml is removed for radioactivity determination. The remaining solution is taken to dryness in a small glass vial, and 100 μl of pyridine–hexamethyldisilazane–trimethylchlorosilane 9:3:1 is added. After 1 hour an aliquot, 1–3 μl, is analyzed by GLC on 1–2% SE-30 or DC-560 for determination of total fecal bile acids as a group. All peaks eluted after cholestane are considered as bile acid derivatives and their total area is determined. Bile acid peak areas are proportional to the weight of the parent bile acid provided that a hydrogen flame ionization detector is used and that the columns are satsifactory. It is advisable to check this relationship for each individual column and GLC-instrument with different bile acid TMS ethers. The amount of bile acids in the sample is calculated from the total bile acid peak area, the cholestane peak area, and the amount of cholestane added. Corrections for losses are made on the basis of the recovery of the radioactive deoxycholic acid. The specificity of this method has so far been proved for subjects on liquid formula diets.

Method 3.[75] Lyophilized feces (approximately 1–5% of a daily portion) is transferred to a glass-stoppered 50-ml centrifuge tube and suspended in 25 ml of 2 M NaOH in 50% ethanol. The mixture is then extracted with 4 successive portions of 12.5 ml of petroleum ether–diethyl ether 1:1. Each upper phase is siphoned off after shaking and centrifugation (800 g, 5–10 minutes). The pooled petroleum ether–ether phases are washed once with 10 ml of water. The combined aqueous phases are gently heated to remove traces of diethyl ether and are then autoclaved at 120° for 4 hours. The hydrolyzate is acidified with 6 M HCl to give a pH of 2–3 and is extracted with 4 successive portions of 15 ml of diethyl ether by means of the procedure described above. To help phase separation, small volumes of ethanol may be added. The combined ether extracts are washed once with 10 ml of water and are then taken to dryness. The residue is dissolved in minimal amounts of dry methanol and methylated with ethereal diazomethane. The solution is taken to dryness under a stream of nitrogen, and the residue is subjected to preparative TLC as described under method 2 with the exception that the bands are located by spraying with a 0.04% solution of 2,7-dichlorofluorescein in ethanol

[74] B. Goldrich and J. Hirsch, *J. Lipid Res.* **4**, 482 (1963).
[75] S. S. Ali, A. Kuksis, and J. M. R. Beveridge, *Can. J. Biochem.* **44**, 957, 1377 (1966). We are grateful to Dr. A. Kuksis for providing valuable information regarding details of this method.

followed by viewing under UV light. Extraction of the adsorbent is made with chloroform–methanol 3:1. The bile acid methyl esters in this extract are then quantitated on QF-1 columns before and after conversion into trifluoroacetates. With this procedure individual fecal bile acids can be determined as described under method 1.

[6] The Enzymatic Assay of Bile Acids and Related 3α-Hydroxysteroids; Its Application to Serum and Other Biological Fluids

By ROBERT H. PALMER

Introduction

The utilization of induced bacterial enzymes for the estimation of steroids was introduced and developed by Talalay and co-workers[1-5] and others.[6,7] 3α-Hydroxysteroid dehydrogenase (EC 1.1.1.50),[8-13] from *Pseudomonas testosteroni,* forms the basis for a highly sensitive and convenient assay for the determination of 19, 21, and 24 carbon steroids with 3α-hydroxyl groups. The enzyme is a catalyst for the reaction

$$3\alpha\text{-Hydroxysteroid} + NAD^+ \rightleftarrows 3\text{-ketosteroid} + H^+ + NADH$$

At alkaline pH and with ketone trapping agents, oxidation of the steroid alcohol is practically quantitative, and the NADH produced can be measured by spectrophotometric or fluorometric techniques. The sensitivity of the steroid assay is limited only by the methods of measuring NADH, or about 10^{-9} mole (<0.5 μg of steroid) with the spectrophotometer and 10^{-11} mole (<5 mμg of steroid) with the fluorometer.

The structural specificity of the enzyme and the molecular stoi-

[1] P. Talalay, M. M. Dobson, and D. F. Tapley, *Nature* **170**, 620 (1952).
[2] P. Talalay and M. M. Dobson, *J. Biol. Chem.* **205**, 823 (1953).
[3] B. Hurlock and P. Talalay, *Proc. Soc. Exptl. Biol. Med.* **93**, 560 (1956).
[4] B. Hurlock and P. Talalay, *J. Biol. Chem.* **227**, 37 (1957).
[5] B. Hurlock and P. Talalay, *Endocrinology* **62**, 201 (1958).
[6] R. S. Stempfel, Jr. and J. B. Sidbury, Jr., *J. Clin. Endocrinol.* **24**, 367 (1964).
[7] T. Iwata and K. Yamasaki, *J. Biochem.* **56**, 424 (1964).
[8] P. I. Marcus and P. Talalay, *J. Biol. Chem.* **218**, 661 (1956).
[9] P. Talalay and P. I. Marcus, *J. Biol. Chem.* **218**, 675 (1956).
[10] S. Delin and J. Porath, *Biochim. Biophys. Acta* **67**, 197 (1963).
[11] S. Delin, P. G. Squire, and J. Porath, *Biochim. Biophys. Acta* **89**, 398 (1964).
[12] P. G. Squire, S. Delin, and J. Porath, *Biochim. Biophys. Acta* **89**, 409 (1964).
[13] J. Boyer, D. N. Baron, and P. Talalay, *Biochemistry* **4**, 1825 (1965).

chiometry of the reactions provide significant advantages over conventional methods for the quantitative estimation of bile acids and steroids which utilize color reactions or absorption spectra that may vary with different steroids; for example, lithocholic acid is not detected by methods of bile acid analysis involving sulfuric acid absorption spectra[14-17] or the Pettenkofer reaction,[18] and may be grossly underestimated by others.[19] In comparison with gas chromatographic techniques, problems of derivative formation, variations in detector response with steroids of different molecular weight, and difficulties in peak area measurements are all obviated. The method is particularly convenient for the estimation of bile acids, since preliminary hydrolysis of amino acid conjugates is not necessary. Steroids conjugated at the C-3 position, 3-ketosteroids, and 3β-hydroxysteroids are not normally assayed by the method, but can be determined by prior appropriate hydrolytic treatment, reversing the enzyme reaction or using β-hydroxysteroid dehydrogenase.[4-6]

For the estimation of total 3α-hydroxysteroids many biological fluids, such as bile, intestinal contents, spinal fluid, and extracts of feces, can be assayed directly; certain other extracts, e.g., plasma,[20] require preliminary purification. The sensitivity of the method permits chromatographic separation and estimation of individual steroids when desired. The general method to be described, which has been used and progressively modified over the past eight years, consists of a simple and reliable working procedure for obtaining α-enzyme, spectrophotometric and fluorometric techniques using the enzyme to estimate bile acids and related 3α-hydroxysteroids, and a method for preparing a serum extract suitable for enzymatic analysis of individual free and conjugated bile acids.

Preparation of 3α-Hydroxysteroid Dehydrogenase (α-Enzyme)

Highly purified enzyme preparations (700,000 units/mg protein)[11,13] and a crude extract (3000–4000 units/mg protein) suitable for the analysis of relatively pure steroids[6] have been described. An account has also

[14] E. H. Mosbach, H. J. Kalinsky, E. Halpern, and F. E. Kendall, *Arch. Biochem. Biophys.* **51**, 402 (1954).

[15] J. Sjövall, *Arkiv Kemi* **8**, 317 (1955).

[16] D. Rudman and F. E. Kendall, *J. Clin. Invest.* **36**, 530 (1957).

[17] J. B. Carey, Jr., *J. Clin. Invest.* **37**, 1494 (1958).

[18] J. L. Irwin, C. G. Johnston, and J. Kopala, *J. Biol. Chem.* **153**, 439 (1944).

[19] A. P. Wysocki, O. W. Portman, and G. V. Mann, *Arch. Biochem. Biophys.* **59**, 213 (1955).

[20] Iwata and Yamasaki[7] have described the enzymatic assay of an acetone-alcohol plasma extract. In our hands, however, preliminary purification has always been necessary for reliable enzyme assays.

been given in an earlier volume in this series.[21] The present method is comparatively simple and results in a preparation with a specific activity of 10,000–15,000 units/mg protein, an advantage over the crude extract in the analysis of biological fluids and extracts. It is also relatively free of 3(and 17)-β-hydroxysteroid dehydrogenase (EC 1.1.1.51), or β-enzyme, an analogous dehydrogenase that is a catalyst for the oxidation of 3β-, 16β-, and 17β-hydroxyl groups.[1–6, 8–13] The β-enzyme present in crude extracts can also be isolated by the procedure described below, and the two enzymes can be useful in the identification and measurement of steroid isomers.[4–6, 13]

Assay of Enzyme Activity

Reagents

Sodium pyrophosphate buffer, 0.1 M, pH 9.5
Methanol, doubly redistilled from 2,4-dinitrophenylhydrazine[6]
Standard α- and β-hydroxysteroids:
Androsterone, 0.75 mg/ml in methanol (α-activity)
Testosterone or 17β-estradiol, 0.75 mg/ml in methanol (β-activity)
NAD+, 5 mM, kept on ice and stored frozen

Assay Procedure.[8] The following modification, at room temperature, is adequate for locating enzyme activity and estimating its amount. The system consists of 0.1 ml of NAD+, 0.02 ml of steroid, and 2.88 ml of pyrophosphate buffer in a 10-mm light path spectrophotometer cuvette. The reaction is started by adding 20 μl of enzyme solution and stirring rapidly. At 30 seconds and every 15 seconds thereafter, optical density measurements at 340 mμ are taken against a control cuvette containing all components except steroid. Velocities are calculated during the first 90–120 seconds, when zero-order kinetics are approximated. One unit of α- or β-enzyme activity produces an increment in optical density ($\log_{10} I_0/I$) of 0.001 per minute.

Enzyme Purification

Reagents

Water, distilled and deionized
Glassware, rinsed with a solution containing 2 g of EDTA (ethylenediaminetetraacetate disodium salt) and 1.35 g of sodium bicarbonate per liter
Pseudomonas testosteroni (ATCC 11996),[8] obtained commercially[22]

[21] P. Talalay, see Vol. V, p. 512.
[22] Nutritional Biochemicals Corp., Sigma Chemical Co., or Worthington Biochemical Corp.

Sörensen's phosphate buffer, 0.03 M, pH 7.2: 216 ml of 0.2 M Na_2HPO_4, 84 ml of 0.2 M KH_2PO_4, 600 mg of EDTA, water to 2 liters

Acetone, redistilled

Protamine sulfate, 20 mg/ml in 0.001 M EDTA. The protamine sulfate is purified by dissolving in EDTA (2 g/liter) and precipitating with ethanol. The process is repeated, and the viscous precipitate is washed with ethanol and dried *in vacuo*.

Procedure. Five grams of *P. testosteroni* are suspended in 50 ml of phosphate buffer at 4° overnight. All subsequent operations are carried out at 4° or less. Twenty-five milliliter portions of the suspension are added to 125-ml portions of acetone, precooled to −20° in two 250-ml Teflon centrifuge bottles. After rapid stirring, the suspensions are centrifuged 10 minutes at 2000 g, −20°. The supernatants are decanted and the precipitates are resuspended briefly in 125 ml of acetone at −20° and centrifuged again. The pellets are quickly drained, combined, and suspended in 100 ml of buffer.

Protamine sulfate, 8 ml (equal to one-fifth of the protein content) is added to precipitate nucleic acids. The material is stirred, allowed to stand for 1 hour, and centrifuged for 20 minutes at 10,000 g, 4°. The supernatant is decanted and concentrated by acetone precipitation: it is divided among four 250-ml Teflon centrifuge bottles each containing 150 ml of acetone at −20°; the bottles are balanced rapidly and centrifuged 30 minutes at 10,000 g, −20°. The supernatants are discarded, and the pellets are suspended and bottles washed in three successive 5-ml portions of water; a final centrifugation at 4° removes insoluble material.

The enzyme solution (14.5 ml) is placed on a 40 × 400 mm column of Sephadex G-100, made up and run with phosphate buffer. The fractions (8–9 ml) are assayed for α- and β-activity, and appropriate fractions are combined (Fig. 1). The preparations are concentrated by freeze-drying to about 25,000 units/ml. Contamination of α-enzyme with β-enzyme is about 1.5%; alcohol dehydrogenase activity is also present.

The enzyme can be stored indefinitely at −20°, but it loses activity at room temperature or when repeatedly frozen and thawed at acid pH; this can be prevented by 20–25% glycerol.[13] We have not found this to be necessary if the enzyme is kept on ice or frozen when not in actual use.

Enzymatic Assay of 3α-Hydroxysteroids

Spectrophotometric Assay

Principle. NADH, produced stoichiometrically during the oxidation of steroid alcohols, is measured by its absorption of light at 340 mμ. The

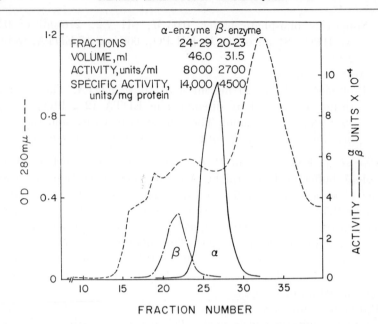

FIG. 1. Separation of α- and β-enzymes from 5 g of *Pseudomonas testosteroni* with a 40×400 mm column of Sephadex G-100 in $0.03\,M$ phosphate buffer, pH 7.3.

reaction goes to completion because of the alkaline pH and the use of ketone trapping agents. Since the extinction coefficient of NADH is known, the amount of steroid present can be calculated without the use of internal standards.

Reagents—Spectrophotometric Assay

Sodium pyrophosphate buffer, $0.1\,M$, pH 9.5
Methanol, doubly redistilled from 2,4-dinitrophenylhydrazine[6]
NAD$^+$, 5 mM in pyrophosphate buffer; prepared weekly, kept on ice, stored frozen
Hydrazine hydrate, $1\,M$, pH 9.5. 5.0 ml hydrazine hydrate (100%) and 1.5 ml $2\,N$ H$_2$SO$_4$ are made up to 100 ml
Sodium lauryl sulfate (95%), $0.1\,M$
Girard T reagent, $0.1\,M$ in pyrophosphate buffer

Procedure—Spectrophotometric Assay. The assay is essentially that described by Hurlock and Talalay.[5] The system consists of 0.01–0.2 ml of steroid in water or methanol,[23] 0.1 ml of NAD$^+$, 1.0 ml of hydrazine, and pyrophosphate buffer to 3.0 ml in a 10-mm light path spectrophotom-

[23] Up to 0.2 ml of methanol can be incorporated in the 3.0-ml assay system without interfering with enzyme activity.

eter cuvette; a similar control cuvette and a reagent blank (without steroid) are also prepared. After initial readings of the optical density at 340 mμ,[24] 500 units of α-enzyme in 0.020–0.025 ml is added, with stirring, to the reaction cuvette and the reagent blank. Readings are taken periodically (every 5 minutes) until the reaction comes to completion, usually in 10–20 minutes. After correcting for the addition of enzyme, the increase in optical density is used to calculate the amount of NADH produced and therefore steroid oxidized. The molar extinction coefficient of NADH at 340 mμ is 6220; in a 3.0 ml system, the micromoles of steroid present equal the change in optical density divided by 2.07.[25]

Comment. For relatively pure samples $\geqslant 10^{-8}$ mole, the spectrophotometric assay is reliable and convenient; no internal standard is necessary. However, nonspecific increases in optical density may occur with highly impure samples, due in part, presumably, to precipitation of hydrophobic materials in the extract[24] or from uncharacterized reactions between the extract and NAD+. These can often be avoided by using the fluorometric assay.

Fluorometric Assay

Principle. NADH is measured by virtue of its native fluorescence at the alkaline pH of the enzymatic reaction. The sensitivity can be increased by using smaller volumes, or by using the strong alkali method for measuring NADH.[26] The latter is less convenient, however.

Reagents—Fluoromertic Assay

Buffer, NAD+, Girard T reagent, methanol: as in the spectrophotometric assay

Reaction mixture: 0.5 ml NAD+ and 0.5 ml Girard T reagent made to 50 ml in pyrophosphate buffer. Store frozen in 10-ml portions.

[24] These should remain constant for several readings over 10–15 minutes. Occasionally, particularly in extracts of high lipid content, a nonspecific and progressive increase in optical density occurs, associated at times with visual turbidity in the cuvette. This can often be remedied by incorporating 0.1 ml of 0.1 M sodium lauryl sulfate and somewhat more enzyme in the system, or the substitution of Girard T reagent, which may help solubilize hydrophobic ketones, for hydrazine as the trapping agent.

[25] Thionicotinamide-adenine dinucleotide has a greater equilibrium constant with α-enzyme than NAD+ (6.0×10^{-8} *vs* 9.2×10^{-9}) and can be used to advantage in this assay.[13] Its molar extinction coefficient at 395 mμ is 11,300, so it also permits a 2-fold increase in sensitivity. The use of microcuvettes, with a proportionate reduction in reagents, allows a further 10-fold increase in sensitivity. However, the fluorometric method is a more convenient way to increase the sensitivity and has other advantages.

[26] O. H. Lowry, N. R. Roberts, and J. H. Kapphahn, *J. Biol. Chem.* **224**, 1047 (1957).

Quinine standards

Stock solution, $10^{-3}\,M$ quinine sulfate in $0.1\,M$ H_2SO_4. Keep refrigerated.

Working standard, $3.33 \times 10^{-8}\,M$ quinine sulfate in $0.1\,M$ H_2SO_4. Must be made up daily from stock solution.[27]

Enzyme. 0.1 ml α-enzyme (25,000 units/ml), 5.0 mg bovine serum albumin, 0.4 ml pyrophosphate buffer. Keep on ice, store frozen.

Standard hydroxysteroid. $12.5 \times 10^{-5}\,M$ deoxycholic acid in methanol.[28]

Procedure—Fluorometric Assay. The procedure is essentially that of Lowry *et al.*[26] The Farrand model A fluorometer with an incident filter transmitting at 365 mμ (Corning Glass No. 5860) and an emergent filter complex transmitting at 470 mμ (Corning Glass Nos. 4308, 5562, and 3387) is highly satisfactory. It is convenient to set the sensitivity so that the working standard gives a full-scale deflection using aperture 5. It is important to remove all tubes from the machine between readings, so that they may remain at room temperature.

Control and reaction tubes are made up with 10^{-8} to 10^{-9} mole steroid in water or methanol (total volume of methanol <0.15 ml), and 2.0 ml of reaction mixture. A standard tube, containing 1.25×10^{-9} mole deoxycholic acid standard, and a corresponding control tube are also prepared. After initial readings, 50–500 units of α-enzyme in 0.02 ml are added and the tubes mixed thoroughly with a Vortex mixer.[29] Readings are taken until the reaction comes to completion, at which time an internal standard of 1.25–2.50×10^{-9} mole of deoxycholic acid in 0.02 ml of methanol is added to the assay cuvette. The reaction is again followed to completion, and the amount of steroid originally present is calculated from the changes in fluorescence resulting from the sample and the internal standard.

Comment—Fluorometric Assay. The fluorometric assay has been extremely useful and reliable provided that certain precautions as to fluorometric technique are observed.[26, 30–32] The increased sensitivity has obvious advantages; substances in plasma extracts that interfere with

---

[27] Quinine fluorescence deteriorates in dilute solutions.

[28] Deoxycholic acid, recrystallized, was chosen because of its water solubility and because it assayed 100% pure with α-enzyme. Any other pure 3α-hydroxysteroid can be used.

[29] Covers must be held on the tubes to prevent inadvertent changes in volume.

[30] D. J. R. Laurence, Vol. IV [7].

[31] O. H. Lowry, Vol. IV [17].

[32] D. Glick, "Quantitative Chemical Techniques of Histo- and Cytochemistry," Vol. 1, p. 361. Wiley, New York, 1961.

the enzyme reaction can be diluted out, solubility difficulties largely are avoided, and smaller amounts of enzyme are used. The use of internal standards minimizes problems of quenching or enhancement of fluorescence.

Preparation of Serum Samples for Enzymatic Analysis of Bile Acids

Principle. The objective is to recover quantitatively all bile salts, with polarities ranging from lithocholate to taurocholate, in an extract suitable for enzyme analysis. Serum proteins are precipitated with ethanol, the filtrate is subjected to mild hydrolysis to destroy complex lipids, and less polar substances, such as fatty acids and cholesterol, are removed by preparative thin-layer chromatography. Areas corresponding to free bile acids, conjugated bile acids, or both are eluted and assayed. Individual bile acids are separated by thin-layer chromatography prior to analysis when desired.

Procedure. Five milliliters of serum or plasma are dropped with stirring into 100 ml absolute ethanol in a 250-ml centrifuge bottle. After heating 15 minutes at 78°, the suspension is centrifuged at 2000 rpm, 4°. The supernatant is decanted, filtered, and taken to dryness with a rotary vacuum evaporator.[33] The residue is dissolved in methanol (redistilled from 2,4-dinitrophenylhydrazine) and transferred to a small (10–25 ml) pear-shaped flask, from which it can be quantitatively applied to a preparative thin-layer plate.[34] Preparative TLC is performed using a 1-mm layer of silica gel H[35] on a 20 × 20 cm plate and phase system S9A (trimethylpentane, 40; isopropyl alcohol, 10; acetic acid, 1.0). Reference spots of lithocholic and cholic acid (20 μg) are applied on either side of the sample. After chromatography, the reference spots are visualized with iodine vapor. The silica gel in areas corresponding to free bile acids (lithocholic through cholic) and conjugated bile acids (behind cholic through origin) is removed with a razor blade and transferred to 250-ml round-bottom flasks. The silica is moistened with water and refluxed with chloroform–methanol (redistilled from 2,4 dinitrophenylhydrazine) 1:1 for 1 hour. After cooling and filtering, the filtrate is taken to dryness. Total free and conjugated bile acids can be assayed directly with the fluorometric method described above, or the individual bile acids can be separated by TLC and then assayed. Conjugated bile acids are hydrolyzed in 20 ml of water and 5 ml of 10 N sodium hydroxide in a Teflon bottle by autoclaving at 15 pounds pressure overnight (16 hours). After

---

[33] Buchler Instruments, Fort Lee, New Jersey.
[34] Applied Science Laboratories, Inc., State College, Pennsylvania, make an excellent sample streaker for applying large volumes to a thin-layer plate.
[35] Brinkmann Instruments, Inc., Westbury, New York.

acidification to pH 1 with concentrated HCl, the hydrolyzed bile acids are extracted with three 75-ml portions of ether. Free bile acids and hydrolyzed conjugates can then be separated on a 0.5-mm TLC plate of silica gel H using a variety of systems.[36,37]

Comment. The measurement of bile acids in serum and other biological fluids has presented numerous difficulties. Endogenous bile acids vary widely in polarity and solubility characteristics, so that more polar compounds (e.g., taurocholate or sulfate esters) are extracted with difficulty from aqueous solutions, while significant losses of less polar compounds (e.g., lithocholic acid) occur during liquid–liquid extraction of neutral lipids and fatty acids with hexane or ether. Most methods have accepted the disadvantages of a preliminary hydrolysis of bile acid conjugates, with variable destruction and losses of different bile acids, so that the free bile acids can be extracted from aqueous solutions; in more recent methods hydrolysis has been necessary for gas–liquid chromatography. The enzymatic method does not require hydrolysis for the measurement of total bile acids, but no satisfactory way of separating individual conjugated bile acids is available that does not require hydrolysis.

It is probable that the enzymatic analysis could be used with any method resulting in an extract clean enough for GLC analysis, such as that described by Grundy *et al.*[38] and Sandberg *et al.*[39]; however the present method is relatively convenient, does not require strong hydrolytic procedures, and gives a good recovery of bile acids.[40] The use of strong ion-exchange procedures, with resins[39] or salts soluble in organic solvents,[41] may give good recoveries and an extract suitable for enzymatic analysis; this would provide a considerable increase in convenience, and we are currently investigating this possibility.

[36] P. Eneroth, *J. Lipid Res.* **4**, 11 (1963).
[37] A. F. Hofmann, *in* "New Biochemical Separations" (A. T. James and L. J. Morris, eds.), p. 261. Van Nostrand, Princeton, New Jersey, 1964.
[38] S. M. Grundy, E. H. Ahrens, Jr., and T. A. Miettinen, *J. Lipid Res.* **6**, 397 (1965).
[39] D. H. Sandberg, J. Sjövall, K. Sjövall, and D. A. Turner, *J. Lipid Res.* **6**, 182 (1965).
[40] In three determinations, lithocholic acid-24-C^{14} added to plasma was recovered from the free bile acid area of the thin-layer plate in yields of 86.0, 91.6, and 92.4%; recoveries of cholic acid were comparable.
[41] A. F. Hofmann, *J. Lipid Res.* **8**, 55 (1967).

[7] The Gas–Liquid Chromatography of Retinol (Vitamin A) and Related Compounds

By Percy E. Dunagin, Jr., and James Allen Olson

Principle

Conjugated unsaturated compounds, such as retinol, are readily isomerized, polymerized, and otherwise transformed when subjected to high temperatures, acid, and reactive surfaces.[1,2] By careful control of the experimental conditions, however, six derivatives of retinol have been separated in good yield by gas–liquid chromatography. Diatomaceous earth supports are treated with base to remove acidic sites, and all surfaces in the column itself are covered with a silanizing material. An antioxidant is used to reduce destruction of the more labile compounds. Other parameters are adjusted in order that compounds may be eluted with short retention times and at low temperatures.

Preparation of Derivatives of Retinol

All-*trans* retinol, all-*trans* retinal, 9-*cis* retinal, 11-*cis* retinal, 13-*cis* retinal, all-*trans* retinoic acid, and all-*trans* retinyl acetate were obtained from Distillation Products Industries (Rochester, New York). 13-*cis*-Retinoic acid was a generous gift from Hoffmann-La Roche, Inc., Basel, Switzerland.

Anhydroretinol is prepared by the dehydration of 100 mg of crystalline all-*trans* retinol in 1 liter of HCl–ethanol 1:40 for 12 minutes at room temperature (24°) in the dark. The reaction mixture is then neutralized with 500 ml of 10% KOH. The products are extracted with 100 ml of hexane, and the hexane layer is washed with H_2O and dried over Na_2SO_4. While unreacted retinol remains on a column of 50 g of chromatographic Al_2O_3, grade II, the anhydroretinol band is eluted with 0.5% acetone in hexane. The ultraviolet spectrum of the product (maxima are at 346, 365, and 388 mμ) is similar to that reported by Shantz *et al.*[3]

Methyl retinyl ether, presumably the all-*trans* isomer, is prepared by treating the lithium derivative of all-*trans* retinol with dimethyl sulfate.[4] The ether, which is eluted with 1% acetone in hexane from a column

[1] E. E. Royals, "Advanced Organic Chemistry." Prentice-Hall, Englewood Cliffs, New Jersey, 1958.
[2] W. J. Zubyk and A. Z. Conner, *Anal. Chem.* **32**, 912 (1960).
[3] E. M. Shantz, J. D. Cawley, and D. E. Norris, *J. Am. Chem. Soc.* **65**, 901 (1943).
[4] A. R. Hanze, T. W. Conger, E. C. Wise, and D. I. Weisblat, *J. Am. Chem. Soc.* **70**, 1253 (1948).

containing 50 g of alumina, grade II, is readily separated from unreacted retinol.

All-*trans* methyl retinoate is prepared by treating all-*trans* retinoic acid with an excess of diazomethane in ether. Diazomethane is released from *N*-methyl-*N*-nitroso-*p*-toluenesulfonamide (Diazald, Aldrich Chemical Co., Inc., Milwaukee, Wisconsin) by treatment with ethanolic NaOH or KOH, and is distilled into cold ether in an apparatus similar to that used by Lipsky and Landowne.[5] The ethereal solution of methyl retinoate is evaporated to dryness under N_2 in the dark, and the resulting methyl retinoate is crystallized twice from methanol–H_2O 5:1. Its melting point and ultraviolet and infrared spectra agree with reported values.[6,7] The methyl ester of 13-*cis* retinoic acid was prepared in an analogous manner, but was not crystallized.

In determining the amount of retinoic acid in a complex mixture, the preparation is dissolved in ether in the presence, if need be, of a small amount of ethanol. Ethereal diazomethane is added until the evolution of N_2 ceases, which usually takes less than 5 minutes, or until the initial solution takes on the yellow color of excess diazomethane. Retinoic acid is completely methylated by this procedure, with no apparent *cis-trans* isomerization or other chemical transformation. The solvents are evaporated under a N_2 stream at room temperature, and the residue is dissolved in 0.1–0.5 ml of hexane for GLC analysis. All procedures are best carried out in darkness to prevent isomerization and destructive reactions.

Chromatographic Apparatus

We routinely use a Research Specialties Co. Model 600 gas chromatograph equipped with a column oven designed to hold vertically positioned U-shaped columns and with a Minneapolis-Honeywell 5-mV recorder. The detector is housed in a separate oven connected to the column exit by a stainless steel tube 50 cm in length.

Sample Delivery. The sample dissolved in hexane (retinol dissolves better in ethyl ether) is introduced into the top of the column by one of two methods: (1) A capillary containing the sample is placed into the carrier gas stream through a standard vapor locking system. The sample solution is vaporized on a block maintained at 75–100° above the column temperature and passed to the column through a stainless steel tube 15 cm in length. By this method 0.5 μl to 2 μl can be reproducibly injected.

[5] S. R. Lipsky and R. A. Landowne, Vol. VI, p. 513.
[6] K. R. Farrar, J. C. Hamlet, H. B. Henbest, and E. R. H. Jones, *J. Chem. Soc.* p. 2657 (1952).
[7] C. D. Robeson, J. D. Cawley, L. Weisler, M. H. Stern, C. C. Eddinger, and A. J. Chechak, *J. Am. Chem. Soc.* **77**, 4111 (1955).

(2) The sample solution in a microsyringe is injected through a silicon rubber septum directly into a 3-cm deep plug of glass wool covering the column packing. Amounts in excess of 2 μl are introduced more accurately onto the column by this method, and destruction on the hot metallic surfaces of the vaporizer and tubing is minimized.

Column Preparation. A diatomaceous earth support is washed first with acid, then with base, dried, treated with dimethyldichlorosilane and finally coated with 1–3% SE-30. The prepared packing, on Gas Chrom P (silanized), is commercially available from Applied Science Laboratories Inc. (State College, Pennsylvania). U-Shaped glass columns, 0.55–1.4 m long with an inside diameter of 4 mm, and glass wool plugs are siliconized before use. The packing material is poured into each arm of the glass column with gentle tapping to permit settling.

Detector. An argon triode ionization detector is used when the separated components are collected *in toto* for spectral and radioactivity analyses. Retinal and methyl retinoate can also be detected by an "electron capture" detector, which does not respond to most hydrocarbons. When it is desirable to detect other compounds also, the electron capture detector is usually connected in parallel with a flame ionization detector by means of a Swagelok "T" at the nitrogen effluent stream from the column. Both detectors respond to as little as 5×10^{-9} g.

Conditions for Gas Chromatography of Retinol Derivatives

Short Columns. Retinal, anhydroretinol and methyl retinoate can be chromatographed with no obvious destruction on a "fresh" 55-cm column containing 1% SE-30 coated on 60–80 mesh Gas Chrom P at a column temperature of 150° and with a flow rate of 150 ml/minute (Fig. 1). After the column is aged by passing carrier gas through it for 2–3 days under normal operating conditions, methyl retinyl ether can also be chromatographed without appreciable destruction (Fig. 1).

On the other hand, retinol and retinyl acetate are completely converted under the above conditions to a compound appearing in the anhydroretinol area and having the ultraviolet absorption spectrum of anhydroretinol. Dehydration and deacetylation are significantly reduced, however, by pretreating the column with β-carotene. After the column temperature is raised to 250° and the vaporizer temperature to 300°, 50–60 μl of a solution of β-carotene (5 mg/ml) in ether is injected into the vaporizer. After 2 or 3 hours, the temperature is reduced by 100° to normal values. At a flow rate of 150 ml/minute, either retinol, with a retention time of 8.5 minutes or retinyl acetate, with a retention time of 16 minutes, accounts for 50–70% of the total peak area. In either case, the remaining 30–50% appears in the anhydroretinol area of the chro-

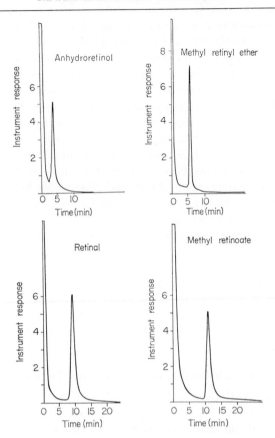

Fig. 1. Gas chromatography of individual retinol derivatives. Approximately 1 μg of each derivative was chromatographed on a column of 60–80 mesh Gas Chrom P (silanized) coated with 1% SE-30. Column length, 55 cm; column temperature, 150°; argon flow rate, 150 ml/minute. Reprinted from P. E. Dunagin and J. A. Olson [*Anal. Chem.* **36,** 756 (1964)]. Copyright 1964 by the American Chemical Society and reprinted by permission of the copyright owner.

matogram. When the flow rate is increased to 880 ml/minute, however, retinol and retinyl acetate account for 94% of the total area (Fig. 2). The protective effect of β-carotene lasts for 2 or 3 days. Its action in minimizing the dehydration of retinol is approximately the same whether it is injected through the hot metallic vaporizer, directly onto the column, or through the septum of the on-column injection system. Presumably β-carotene reacts with active sites on the column proper, and not solely in the vaporizer.

Hydroquinone, which is a well-known inhibitor of free-radical-initiated oxidations, also significantly reduces the dehydration of retinol on

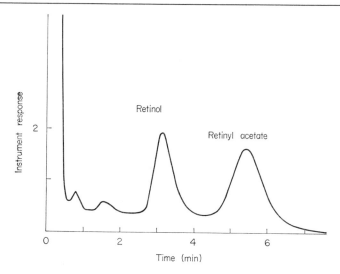

FIG. 2. Gas chromatography of retinol and retinyl acetate. Approximately 5 μg of each compound was separated on a column of silanized Gas Chrom P coated with 1% SE-30 and treated with β-carotene as described in the text. Column length, 55 cm; column temperature, 150°; argon flow rate, 880 ml/minute. Reprinted from Dunagin and Olson [*Anal. Chem.* **36**, 756 (1964)]. Copyright 1964 by the American Chemical Society and reprinted by permission of the copyright owner.

columns packed with glass beads coated with 0.1% SE-30, but not on columns packed with diatomaceous earth supports. In this case the protection lasts only for a few hours. Columns packed with 10% polyvinyl-pyrrolidone (PVP) on Chromasorb W also dehydrate retinol, but to a lesser degree than untreated SE-30 columns. However, the greatly increased retention time of retinol and retinyl acetate on PVP-coated columns is disadvantageous.

Although the formation of anhydroretinol is probably catalyzed by acidic sites on unneutralized supports,[8] hydroquinone and β-carotene presumably inhibit a free-radical-induced reaction on columns treated with base. In any event, NH₄OH and pyridine injected through the vaporizer do not reduce the destruction of either retinol or retinyl acetate.

Although mild conditions are necessary for the chromatography of methyl retinyl ether, retinol, and retinyl acetate, other derivatives are more stable. By use of the 55-cm column mentioned above with a reduced flow rate, retinal can be chromatographed at 160°, methyl retinoate at 175°, and anhydroretinol at >200° without obvious alterations in the chromatographic pattern.[9]

[8] T. Ninomiya, K. Kidokoro, M. Horiguchi, and N. Higosaki, *Bitamin* **27**, 349 (1963).
[9] P. E. Dunagin, Jr., and J. A. Olson, *Anal. Chem.* **36**, 756 (1964).

Long Columns. The efficient separation of retinal and methyl retinoate is of particular interest because of their importance in retinol metabolism.[10] After considerable experimentation, 3% SE-30 coated on a base washed and silanized support, such as Gas Chrom P, was chosen for the separation of retinal and methyl retinoate. This packing is favored for several reasons: (1) Methyl retinoate can be chromatographed with no apparent destruction at temperatures as high as 230° on this packing. (2) The heavier loading with liquid phase increases the efficiency for methyl retinoate by reducing tailing of the peak. As expected, the efficiency of the column can be further enhanced by increasing the column length and by optimizing the flow rate. Defects in other packing materials are as follows: (1) On polar coatings such as QF 1 (Dow Corning), XE-60 (General Electric Corp.), and ethylene glycol adipate polyester (Applied Science Laboratories), retinal and methyl retinoate separate better than on SE-30, but the retention time of retinal was excessively long. (2) Retinal and methyl retinoate, when separated on silanized supports which have been washed only with acid, can be chromatographed satisfactorily only after the column has been aged for 2 or 3 days. (3) Both retinal and methyl retinoate give extraneous peaks on the polyester-coated support even after many days of aging, evidently owing to excess acid in the coating itself.

Ultimately an efficiency of 1300 theoretical plates[11] is obtained with a glass column 1.4 m long which contains 100–120 mesh silanized Gas Chrom P coated with 3% SE-30. The column temperature is 180°, the vaporizer temperature is 230°, and the argon flow rate is 50 ml/minute. An on-column injection system is employed. Both the inside of the glass column as well as the glass wool plugs are siliconized. This column is excellent for separating the methyl ester of retinoic acid from tissue lipids, and conceivably can be used equally well for anhydroretinol. However, all-*trans* retinal is partially destroyed under these conditions.

Collection of Eluted Samples

Collection on Al_2O_3 at the Detector Exit. Retinal and methyl retinoate are collected in 94% yield at the conventional detector output port in a right-angle glass collection tube (7 mm o.d.) which contains a 4 mm deep layer of 100-mesh Al_2O_3 over a small glass wool plug. The collected compound is then eluted with a minimum of 0.5 ml of ethanol.

Collection in Capillary Tubing. A sleeve made from AWG No. 15

[10] P. E. Dunagin, Jr., R. D. Zachman, and J. A. Olson, *Biochim. Biophys. Acta* **124**, 71 (1966).
[11] Theoretical plates $= 16 \ (y/x)^2$ where $x =$ peak width at the base and $y =$ retention distance.

Teflon tubing is connected snugly to the outlet tube ($\frac{1}{16}$ inch o.d.) from the detector or directly to the exit tube ($\frac{1}{16}$ inch o.d.) from the column. Before the injection or during the analysis of the sample, a disposable, inexpensive 300-mm length of capillary tubing (1.5 mm o.d.) is rapidly inserted into the Teflon sleeve and is changed as often as desired. The effluent material which condenses on the inside wall of the capillary is eluted with as little as 0.1 ml of solvent introduced from a syringe with a hypodermic needle (No. 22). The efficiency of collecting methyl retinoate with this method is 80%.

Unstable derivatives such as methyl retinyl ether, retinol, and retinyl acetate can be collected at the column exit in capillaries with little destruction. Although in our laboratory the flow to the detector is usually interrupted during sample collection, use of a stream splitter would allow continuous monitoring of a small part of the effluent gas while the remainder was passed through the capillary collection tube.

Automatic Collection on Anthracene. The exit tubing ($\frac{1}{16}$ inch) from the argon ionization detector is connected to the inlet tubing ($\frac{1}{16}$ inch) of an automatic gas fraction collector (Packard Instrument Co.) by means of AWG No. 15 Teflon tubing. Components of the effluent stream are condensed in anthracene-filled cartridges, and the entire cartridge is placed into a glass vial and counted in a liquid scintillation spectrometer.

Ultraviolet Spectroscopy

A Zeiss PMQ II single-beam spectrophotometer with microcuvette adaptors is used to analyze about 0.1 μg of eluted compound in ≥ 0.25 ml of solution. For measuring the spectra of such small volumes in a Perkin Elmer Model 4000 double-beam spectrophotometer, a blackened shield having an orifice 1 mm in diameter is placed into each light path. By use of a calibrated 1-cm microcuvette with a curved bottom, which is centered in the reduced light path, the spectrum of as little as 0.04 μg of a retinol derivative in 0.10 ml of solution can be accurately determined. This same solution is then quantitatively washed into a counting vial with 1 ml of ethanol for radioactive counting or is used for other studies.

Spectra of Chromatographed Retinol Derivatives. Each of the six studied retinol derivatives can be chromatographed separately as virtually a single peak on short columns at 150° (Figs. 1 and 2). In order to ascertain whether these compounds are transformed during analysis, the spectrum of the collected peak is compared with the spectrum of the solution initially injected. When methyl retinoate and anhydroretinol are chromatographed at 150° and collected at the detector exit at 165°, the spectra of the eluted compound and of the initial solution are identical (Fig. 3). When all-*trans* retinal is collected at the detector exit, however,

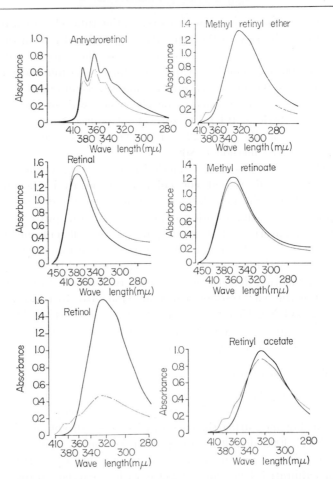

FIG. 3. Spectra of retinol derivatives before and after gas chromatography. ——, Spectra of solutions applied to gas chromatography; ...., spectra of collected compounds. Spectra of retinyl acetate reprinted from P. E. Dunagin and J. A. Olson [*Anal. Chem.* **36,** **756** (1964)]. Copyright 1964 by the American Chemical Society and reprinted by permission of the copyright owner.

the absorption maximum shifts slightly to a lower wavelength, which is indicative of the formation of *cis* isomers. The product collected at the detector exit after chromatography of methyl retinyl ether, retinyl acetate or retinol is largely anhydroretinol. When these compounds are collected at the column exit, however, only a small amount of anhydroretinol is present (Fig. 3).

These spectrophotometric studies point up certain necessary precautions in handling these compounds: (1) Although anhydroretinol and methyl retinoate can be collected without transformation at the detector exit (165°), retinal is better collected at the column exit to minimize isomerization. (2) Methyl retinyl ether, retinol, and retinyl acetate are converted in part to anhydroretinol even when collected at the column exit (150°), but are largely dehydrated when collected at the detector exit. (3) Since destruction of components after separation on the column affects neither the response of the argon detector to mass nor the accuracy of radioactivity detection methods, quantitative measurements based on peak dimensions are still meaningful, even when the compound is transformed in the detector. Destruction in the vaporizer and on the column in these cases is known to be minimal.

Separation of Cis-Trans Isomers of Methyl Retinoate and of Retinal

All-*trans* methyl retinoate appears as a single unaltered peak upon chromatography on a short column (55 cm) at 150°. When chromatographed at 180° on a 1.4 m column of 3% SE-30 on Gas Chrom P, however, 2–3% of the total peak area appears in an earlier component. Furthermore, when methyl retinoate is isomerized by exposing an ethanolic solution of the all-*trans* isomer to UV light for 24 hours, three peaks—components A, B, and C—are obtained by GLC analysis (Fig. 4). The UV spectra of all three components are similar, except that components A and B absorb maximally at shorter wavelengths than does all-*trans* methyl retinoate. Individual peaks can be collected from columns run at 150°, dissolved in ethanol and rechromatographed at 150° without isomerization. When components are collected and rerun at 180°, however, considerable isomerization takes place, as shown in Table I. If the effluent gas is also passed through the detector, the extent of isomerization is even greater.

Of the six known *cis-trans* isomers of retinol, only four of the corresponding isomers of methyl retinoate have been synthesized and characterized.[7] The retention time and λ_{max} of known isomers and of the three GLC components are summarized in Table II. Component C is mainly, if not solely, the all-*trans* isomer. If the 13- *cis* isomer, which has essentially the same retention time as the all-*trans* isomer, is also present in component C, the amount present is not sufficient to shift the absorption maximum to a longer wavelength. The *cis* isomers present in components A and B have not been precisely identified. In view of its relative stability, however, component B may be the 9-*cis* or the 9,13-*cis* isomer. Of these two, the λ_{max} of the 9-*cis* isomer approximates that of component B. The least stable component, component A, seemingly has an absorption maximum lower than any of the defined isomers. Since the 11-*cis* or 11,13-*cis* isomers would probably be rapidly isomerized at higher

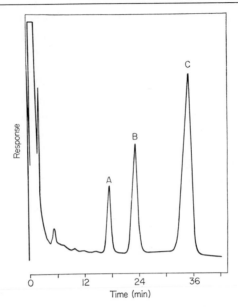

FIG. 4. Gas chromatography of the methyl esters of an isomerate of retinoic acid. Separation was performed on a 1.4 m column of 100–200-mesh Gas Chrom P (silanized) coated with 3% SE-30. Column temperature, 180°; argon flow rate, 50 ml/minute. Reproduced from P. E. Dunagin *et al.* [*Biochim. Biophys. Acta* **124**, 71–85 (1966) Fig. 1].

temperatures, however, their presence in component A is questionable. Further studies are necessary to establish these assignments. Quite possibly the application of more refined methods, such as capillary columns, might ultimately allow the separation of all six isomers of methyl retinoate at lower temperatures and with little destruction.

Isomers of retinal have also been partially differentiated by GLC.

TABLE I

ISOMERIZATION OF METHYL RETINOATE DURING CHROMATOGRAPHY AT 180°

Initial component	Retention time[a] (min)	Percent distribution after rechromatography		
		A	B	C
A	13	56.4	19.5	24.1
B	17	0	89.3	10.7
C	25	0	1.4	98.6

[a] Conditions: 1.4 m column of 3% SE-30 on Gas Chrom P; column temperature, 180°; argon flow rate, 100 ml/minute. Samples were applied by on-column injection and collected directly from the column exit port.

TABLE II
PROPERTIES OF *Cis-Trans* ISOMERS OF METHYL RETINOATE

Isomer or component	Relative retention time[a] (cm)	λ_{max}[b] (mμ)
all-*trans*	15.0	354
13-*cis*	14.8	359
9-*cis*	—	348
9,13-*cis*	—	352
A	7.6	344
B	10.1	347
C	15.0	354

[a] Conditions: Column temperature, 180°; flow rate, 50 ml argon per minute; 1.4 m column with 3% SE-30 on Gas Chrom P (silanized).

[b] C. D. Robeson, J. D. Cawley, L. Weisler, M. H. Stern, C. C. Eddinger, and A. J. Chechak, *J. Am. Chem. Soc.* **77**, 4111 (1955).

Best results are obtained on a 1 m column of 3% SE-30 on 100–120-mesh Gas Chrom P (silanized), which is operated at 150° with a flow rate of 200 ml of argon per minute. Samples are delivered with the on-column injection system, and collected immediately from the column effluent. Under these conditions the retention time of given isomers of retinal are as follows: 9-*cis*, 38.3 minutes; 11-*cis*, 40.4 minutes; 13-*cis*, 41.4 minutes; and all-*trans*, 43.2 minutes. As in the case of methyl retinoate, the *cis* isomers are eluted before the all-*trans* isomer on nonpolar columns.

The peaks obtained with various isomers of retinal, however, are broad at 150°, which indicates that an appreciable amount of isomerization takes place. From recent data on the temperature-dependent rate of isomerization of 11-*cis* retinal to the all-*trans* isomer,[12] we have calculated that the half-life, $t_{1/2}$, for this conversion at 150° is 3.7 minutes in heptane and 45.5 minutes in propanol. Similarly, the half-life for the conversion of all-*trans* retinal to the 11-*cis* isomer at 150° is 18.1 minutes in heptane, and in propanol is probably greater than 200 minutes. Since nonpolar SE-30 columns are used in the present studies, the heptane values might better approximate gas phase relationships on the column. On the other hand, the 9-*cis* and 9,13-*cis* isomers of retinal should be somewhat more stable than the 11-*cis* compound.

Determination of Retinoic Acid in Tissues and in Biological Fluids

Retinol and retinal cannot be readily separated from biological tissues by GLC because of the limited efficiency of the chromatography column and of the presence of large amounts of interfering tissue lipids. On the

[12] R. Hubbard, *J. Biol. Chem.* **241**, 1814 (1966).

other hand, retinoic acid can be analyzed effectively, especially when the nonacidic tissue lipids are eliminated by anion-exchange chromatography prior to gas chromatographic analysis.

Procedure. After the administration of radioactive retinoic acid to a 200–300 g rat, the liver or small intestine is homogenized with 20 volumes of $CHCl_3$–MeOH 2:1, and filtered. The tissue filtrates, or a methanolic solution (diluted 4:1) of a 24-hour bile sample, are placed separately on a column (2 cm i.d.) containing 25 ml (wet volume) of the anion exchange resin Bio-Rad AG2-X8 in the acetate form. The nonacidic tissue lipids are eluted with methanol, usually 100 ml, until the eluate

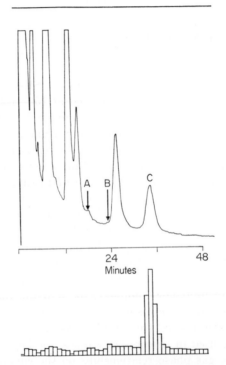

FIG. 5. Gas chromatographic analysis of a crude fraction of rat bile collected for 8 hours after the administration of radioactive all-*trans*-(6,7-^{14}C)-retinoic acid. Nonacidic components of bile were removed with methanol from a column of Bio-Rad AG2-X8 in the acetate form. The fraction eluted with 1% acetic acid in methanol was methylated with diazomethane and analyzed on a 1.4 m column of 3% SE-30 coated on 100–120-mesh Gas Chrom P (silanized). Column temperature, 180°; flow rate, 50 ml/minute. The upper trace shows the response of the argon detector, and the lower trace shows the radioactivity collected automatically in anthracene cartridges. Reproduced from P. E. Dunagin *et al.* [*Biochim. Biophys. Acta* **124**, 71–85 (1966), Fig. 5].

becomes colorless. The fraction containing retinoic acid is then eluted easily with 1% acetic acid in methanol. A gradient elution system may also be used.[10] In either case the acidic eluting solvent is evaporated at 20° or less under a nitrogen stream in the dark. The residue is treated with diazomethane, the solvent is evaporated under N_2 and the esterified residue is dissolved in 0.1–0.5 ml of hexane. Aliquots of 5–20 μl are analyzed by gas chromatography for the presence of radioactive retinoic acid (Fig. 5). The recovery of crystalline retinoic acid from the anion exchange column is 85–90%. In the presence of methanol, however, some trans-esterification of retinoic acid esters occurs on the ion-exchange resin.

Acknowledgment

The authors wish to acknowledge the important contributions of John D. Hendrix of the University of Florida College of Medicine to studies on the separation of isomers of methyl retinoate and retinal by GLC. Supported by grants-in-aid from the U.S. Public Health Service (5-R01-AM-01278) and from the National Science Foundation (GB-2164).

Section II
Special Synthetic Methods

[8] Synthesis of Isotopically Labeled Steroids of High Specific Activity

By N. K. Chaudhuri and Marcel Gut

I. Introduction

With the increasing interest in metabolic and biosynthetic studies of both natural and synthetic steroidal hormones and sterols, the demand for isotopically labeled steroids of highest specific activity has greatly increased in recent years. This has led to the development of various special methods for their preparation. The description of these methods are scattered throughout the literature and have been reviewed[1] from time to time. The last review in this field appeared in 1958.[1] It is proposed to discuss in the present review those methods and modifications which have been published since then with special emphasis on the description of the most *practical* preparations of those labeled key steroids which are either commercially not available or too expensive to buy in large amounts. More specifically, this review excludes all procedures that do not furnish highest specific activities, e.g., biosyntheses, exchange methods and introduction of label in the form of tritide. The mentioned examples do not exhaust the list of syntheses of radioactive steroids. Different combinations of similar steps will lead to further examples. In view of the great variety of needs for different labeled substrates it will be realized that in the last analysis the specific details of a synthesis will often have to be tailor-made.

II. Purity of Labeled Steroids[2]

During the past few years, the criteria of purity of labeled compounds have become increasingly stringent. Improved methodology in all phases of chromatography, a better understanding of the pitfalls in detecting high specific activity contaminants, and a growing awareness of the dangers of radiation decomposition have alerted users of labeled compounds to the need for determining whether or not high standards of purity are actually being achieved and maintained during storage.

The determination of the purity of labeled steroids illustrates most of the problems one might expect to encounter in this work. Subtle dif-

[1] G. H. Twombly, The synthesis and metabolism of radioactively labeled steroids, *Vitamins Hormones* **9**, 237 (1951). K. Bloch, Synthesis and degradation of labeled steroids, this series, Vol. IV, p. 732; A. Murray, III, and D. L. Williams, "Organic Synthesis with Isotopes." Wiley, (Interscience), New York, 1958.

[2] L. E. Geller and T. F. Sullivan, *Atomlight* **53**, 10 (1966).

ferences in the structure of steroids, such as substituent isomers, double-bond isomers, etc., require a full appreciation of the hazards of employing analytical methods that are not sensitive enough to detect compounds very similar in structure to the authentic one. The cautious investigator, therefore, is content only after a combination of pertinent physical and chromatographic data indicates that the labeled steroid contains no unreasonable contaminants. The use of naive terms, such as "chromatographically pure," is meaningless unless qualified and demonstrated to be applicable to the detection of likely impurities.

The paper chromatographic systems reported in the literature are primarily designed for the separation of steroid isomers and closely related compounds isolated from natural sources. These standard, well-accepted systems are frequently unsuccessful for the separation of particular isomers and by-products resulting from synthesis or from radiation decomposition. This situation necessitates the use of a wide variety of partition chromatography systems and cognate systems, such as those containing t-butanol, dioxane, or acetic acid. Reverse-phase systems are routinely employed for steroids of low and moderate polarity. Frequently, modification of existing systems or development of entirely new systems is required. While no change in basic chromatographic technique is necessary, an awareness of the delicate nature of some of the steroids is essential for good results.

The choice of methods for the demonstration of purity of labeled steroids at the time of synthesis involve the use of suitable Bush (for the composition of various Bush systems see Appendix) and Zaffaroni type paper chromatographic systems, thin–layer and gas–liquid chromatography, ultraviolet spectrophotometry, and isotope dilution analysis. Steroids of high specific activity are diluted with cold carrier prior to running a chromatogram in order to diminish the apparent impurities at the origin resulting from irreversible adsorption. The addition of cold carrier does not always prevent spurious results, and occasionally derivatives are prepared that offer better stability during equilibration in Bush systems. Compounds, such as the corticoids and aldosterone, decompose very rapidly when ultramicro amounts are spotted and allowed to equilibrate for hours in the presence of oxygen. Therefore, they are equilibrated for only a brief period of time before chromatography is initiated. Two-dimensional chromatography is frequently employed in an attempt to answer the annoying question of whether the impurities existed before, or were formed during, chromatography.

Purity demonstrated by a single peak on a radiochromatogram does not necessarily mean that impurities do not exist; but only that all the radioactivity present runs as a single entity. Further examination of the state of purity of a steroid is made by employing the technique of isotope dilution analysis.

For the isotope dilution test, 350–600 mg of carrier steroid is employed. Three samples of the initial dilution, three of the crystallized product, and three of the residue of the mother liquors are taken for assay. Each sample is counted in duplicate by liquid scintillation counting. For the analysis to be acceptable, the change in the specific activity of the crystallized compound and the residue from the mother liquor, from that of the starting material should not vary by more than 2–2.5%, which is consistent with the experimental error of the procedure employed.

For those steroids that are prone to cocrystallization with other steroids (isomorphism), such as the sterols and estrogens, use is made of derivative-isotope dilution analysis. For example, isotope dilution analysis is first performed in the usual manner on estradiol-6,7-^{3}H and a sufficient amount of that crystallized product is converted to its acetate. This new product is crystallized, and three samples of it are assayed as before. If the isotope dilution result indicates impurities in excess of the statistical experimental error, then new chromatographic systems are sought that will resolve the contaminants. A purification is effected and the compound is again submitted for isotope dilution analysis.

In addition, the majority of labeled steroids are examined in the ultraviolet and/or by gas chromatography in order to ensure an accurate determination of the specific activity. These measurements provide additional data on both chemical and radiochemical purity.

It is recognized that the state of purity of any labeled compound changes with the passage of time because of the effect of the radiation. In order to minimize the rate of radiation decomposition, labeled steroids are stored in solution in the refrigerator at temperatures above the freezing point of the solvent. Benzene and benzene-ethanol mixtures are the preferred solvents. Some labeled steroids decompose at a rate of approximately 5% every 6 months in solution, while others begin to noticeably deteriorate in a shorter period of time. Labeled steroids have to be checked routinely by chromatography before use. When impurities of more than a few percent are demonstrated, the steroids have to be repurified.

The especially sensitive, high specific activity steroids such as the tritium-labeled corticoids, estrogens, glucuronides, and sulfates have been observed to undergo radiation decomposition at a rapid rate. For example, chromatographic analysis of estradiol-6,7-^{3}H, at a specific activity of 30 curies (C) per millimole, indicates that approximately 5–10% decomposition occurs in only 4 weeks. In a recent publication, Geller and

Silberman,[2a] conclude that nonradioactive colored matter is often the cause of decomposition. Removal of these tars extended the stability of tritiated estrogens.

The above approach to quality control is not considered a panacea for all problems arising in the purity examination of steroids. It is believed, however, that the detection of impurities by these procedures is reasonably sensitive. The chance that a steroid (product) is both isomorphic (cocrystallizes) and isopolar (cochromatographs) with another steroid (impurity) is very unlikely. The combination of various procedures employed should preclude the presence of anything but small amounts of contaminants.

III. Steroids Labeled with ^{14}C

With very few exceptions, e.g., 21-^{14}C, organic synthesis has almost exclusively been designed to place the carbon-14 label at C-4. For metabolic studies, 4-^{14}C-labeled steroids are especially suited since the label forms part of a ring and can therefore be lost only by ring fission.

1. *19-Nortestosterone-4-^{14}C, Estradiol-4-^{14}C, Estrone-4-^{14}C, and Equilenin-4-^{14}C*[3] *(Fig. 1)*

19-Nortestosterone Acetate-4-^{14}C. To a solution of 2 millimoles of methyl magnesium iodide-^{14}C (4 mC) in 15 ml of ether and cooled to —20° was added, dropwise over a period of 30 minutes, a solution of 700 mg (2.2 millimoles) of 17β-acetoxy-5-hydroxy-3,5-seco-4-norestr-5-en-3-oic acid 3,5-lactone[4] in 60 ml of ether–benzene (1:1), care being taken to exclude all moisture. Then the mixture was allowed to warm up to 25°; after 6 hours it was decomposed with an ice-cold saturated ammonium chloride solution and the product was extracted with benzene. The benzene layer was washed with dilute sodium carbonate solution and with water, dried and evaporated to dryness *in vacuo*. The residue was dissolved in 15 ml of glacial acetic acid, and 1.5 ml of concentrated hydrochloric acid was added to it. The solution was kept for 48 hours at room temperature in an atmosphere of nitrogen. After removal of the volatile acids *in vacuo*, the remaining sirup was dissolved in benzene and washed with sodium bicarbonate solution and with water. The benzene layer was dried, and the solvent was distilled off, leaving a partially crystalline residue which was chromatographed. The benzene–ether fractions gave,

[2a] L. E. Geller and N. Silberman, *Atomlight* **60**, 11 (1967).

[3] Compare M. Uskoković and M. Gut, *J. Org. Chem.* **22**, 996 (1957); S. Kushinsky, *J. Biol. Chem.* **230**, 31 (1958).

[4] J. A. Hartman, A. J. Tomasewski, and A. S. Dreiding, *J. Am. Chem. Soc.* **78**, 5662 (1956).

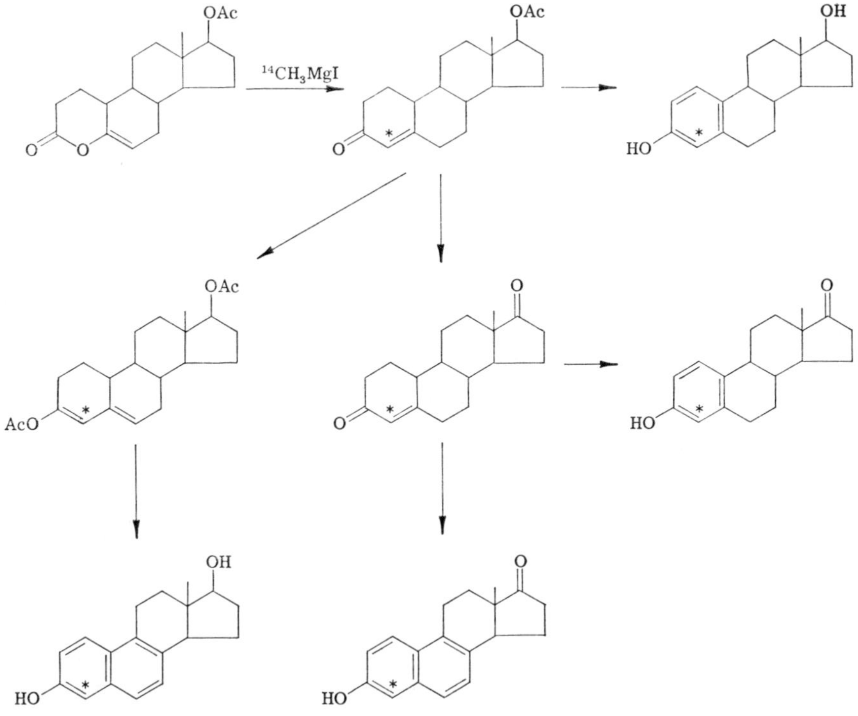

FIG. 1.

after recrystallization from ether–pentane, 298 mg of 19-nortestosterone acetate-4-^{14}C, m.p. 91–93°, λ_{max}^{MeOH} 240 mμ, ϵ 17,700. The specific activity was 2 mC/millimole.

19-Nortestosterone-4-^{14}C. A solution of 100 mg of 19-nortestosterone acetate-4-^{14}C in 100 ml of 1 N methanolic potassium hydroxide solution was left under a nitrogen atmosphere overnight at room temperature. After neutralization with a 50% aqueous acetic acid, the methanol was removed under reduced pressure and the concentrate, after addition of more water, was extracted with methylene chloride. The extract was washed with water, dried, and evaporated *in vacuo.* Chromatography on silica gel eluted with benzene–ethyl acetate mixtures gave 81.7 mg of 19-nortestosterone-4-^{14}C, m.p. 121–123°, λ_{max}^{MeOH} 240 mμ, ϵ 16,000.

19-Norandrost-4-ene-3,17-dione-4-^{14}C. To a solution of 50 mg of 19-nortestosterone-4-^{14}C in 10 ml of 90% acetic acid was added a solution of 25 mg of chromic oxide in 1 ml of 90% acetic acid. The combined solution was left at room temperature for 15 minutes, then the excess reagent was removed by the addition of 4 drops of methanol. After the preparation had stood for 30 minutes, the solvents were removed *in vacuo,* the residue was extracted with benzene, and the solution was washed with a saturated sodium bicarbonate solution and with water. The washed solution was dried then evaporated to dryness, and the residue was chromatographed on silica gel. The eluate obtained with benzene–ether mixtures gave, after recrystallization from ether–pentane, 46 mg of 19-norandrost-4-ene-3,17-dione-4-^{14}C,[5] m.p. 168–170°, λ_{max}^{MeOH} 240 mμ, ϵ 16,500.

3,17β-Diacetoxyestra-3,5-diene-4-^{14}C. A solution of 316 mg of 19-nortestosterone acetate-4-^{14}C in 20 ml of acetic anhydride–acetyl chloride (1:3) was refluxed under nitrogen for 2 hours. The solvents were evaporated off *in vacuo*, and the residue was recrystallized from ethanol containing a drop of pyridine to give 200 mg of 3,17β-diacetoxyestra-3,5-diene-4-^{14}C,[6] m.p. 133–156°, λ_{max}^{MeOH} 234 mμ, ϵ 20,200.

17-Dihydroequilenin-4-^{14}C. Refluxing a mixture of 320 mg of 3,17β-diacetoxyestra-3,5-diene-4-^{14}C in 20 ml of acetic acid containing 200 mg of freshly sublimed selenium dioxide under nitrogen for 15 minutes gave 235 mg of impure 17-dihydroequilenin 3,17-diacetate-4-^{14}C, m.p. 112–119°. This crude product was subjected to reductive hydrolysis with lithium aluminum hydride. After the usual work-up the crude product was chromatographed on a Celite partition column using the system[7] 0.4 N sodium hydroxide for stationary phase and benzene for mobile phase. Thereby was eluted 188 mg of 17-dihydroequilenin-4-^{14}C, m.p. 238–241°, $[\alpha]^{23}$ + 53° (dioxane) and λ_{max}^{MeOH} 230 mμ (ϵ 7200), 271 mμ (ϵ 6500), 283 mμ (ϵ 7500), 292 mμ (ϵ 5000), 328 mμ (shoulder, ϵ 2800), 341 mμ (ϵ 3900).

Equilenin-4-^{14}C. A solution of 300 mg of 19-norandrost-4-ene-3,17-dione-4-^{14}C in 15 ml of acetyl chloride and 5 ml of acetic anhydride was refluxed for 2 hours and then the solvents were removed under reduced pressure. The residue of crude 3-acetoxy-19-norandrosta-3,5-dien-17-one-4-^{14}C was dissolved in 20 ml of acetic acid; 200 mg of selenium dioxide was added, and the mixture was refluxed for 15 minutes. After the usual work-up, followed by hydrolysis of the acetate with 1 N methanolic potassium hydroxide solution, there was obtained 235 mg of crude equilenin-4-^{14}C. Purification on a Celite column using the system of Haenny *et al.*[7] gave 98 mg of pure equilenin-4-^{14}C, m.p. 250–253°.

Estradiol-4-^{14}C and Estrone-4-^{14}C. Medium: *Corynebacterium simplex* (ATCC 6946) was grown in a buffered yeast extract medium (pH 7.0) consisting of the following materials: $K_2HPO_4 \cdot 3 H_2O$, 13.3 g; KH_2PO_4, 4.0 g; yeast extract, 1.0 g; distilled water, 1000 ml. The stock culture medium consisted of a growth medium to which 2% agar was

[5] A. L. Wilds and N. A. Nelson, *J. Am. Chem. Soc.* 75, 5366 (1953).
[6] J. A. Hartman, *J. Am. Chem. Soc.* 77, 5151 (1955).
[7] E. O. Haenny, J. Carol, and D. Banes, *J. Am. Pharm. Assoc.* 42, 167 (1953).

added as a solidifying agent. Growth from the stock slant was transferred to 100 ml of growth medium and incubated at 30° for 24 hours.

Conversion. For steroid conversion, 1 liter of growth medium in a 5-liter round-bottomed flask was inoculated with a 1 or 2% cell suspension from a culture which had previously been passed through three serial transfers. The incubation was conducted at 28° with constant aeration through a tube constricted to about 3 mm, at a maximal rate permissible without splattering of the broth on the rubber stopper at the top (air was sterilized by passage through cotton, concentrated sulfuric acid, and sterile distilled water). After 24 hours of growth, the steroid, 250 mg per liter of broth, dissolved in absolute methanol (not more than 1 ml of methanol per 100 ml of broth), was added dropwise; the aeration was continued for the duration of the incubation. At the end of the incubation the steroid was extracted with three 1-liter portions of methylene chloride.

According to the above procedure, 250 mg of 19-nortestosterone-4-^{14}C was incubated for 5 hours. The resultant extract contained a mixture that was separated by partition chromatography on Celite 545 (solvent system, 70% aqueous methanol and various mixtures of *n*-hexane–benzene). Crystallization of the residue from the 10% benzene fractions gave 3.5 mg (2%) of a substance, m.p. 170–173°, whose infrared absorption spectrum (CS_2) showed maxima at 5.77 μ (five-membered ring ketone) and 5.97 μ (conjugated ketone), but none due to a hydroxyl group. These facts are consistent with the assignment of its structure as 19-nor-Δ^4-androstene-3,17-dione-4-^{14}C, reported[5] m.p. 170–171°; the parent compound with the 17-hydroxyl was oxidized to a ketone. Crystallization of the residue from the 20% benzene fractions gave 33.2 mg (22%) of a substance, m.p. 258–262°, whose infrared spectrum was found to be identical with that of authentic estrone. Crystallization of the residue from the 50% benzene fractions gave 80.5 mg (52%) of a substance, m.p. 174–180°, whose infrared spectrum (after acetylation to increase its solubility in CS_2) was found to be identical with that of authentic estradiol diacetate (prepared from estradiol).

19-Nortestosterone acetate (250 mg) was incubated as above except that a 24-hour incubation period was used. Direct crystallization of the residue of the extract gave a 77% yield of estradiol-17β monoacetate, m.p. 218–222° (reported[8] m.p. 217–219°). Incubation of 74.6 mg and 74.7 mg of 19-nortestosterone acetate and 19-nortestosterone acetate-4-^{14}C, respectively, in 500 ml of broth with a 2% cell suspension gave 77 and 79% yields of estradiol-17β monoacetate-4-^{14}C, respectively. The overall

[8] C. Djerassi, G. Rosenkranz, J. Romo, S. Kaufmann, and J. Pataki, *J. Am. Chem. Soc.* **72**, 4534 (1950).

yield of estradiol-17β monoacetate-4-^{14}C was 28% (based on the ^{14}C used). A sample of estradiol-17β monoacetate-4-[14]C (6.9 mg) was hydrolyzed[9] with $KHCO_3$ and resulted in a 98% yield of estradiol-4-[14]C (6.9 mg, m.p. 178–179°, and 1.3 mg, m.p. 176–178°). The infrared spectrum of the estradiol-4-^{14}C was identical with that of authentic estradiol (m.p. 178–179°), λ_{max} 280 mμ, log ϵ 3.27 ($\pm$10%) (reported[10] log ϵ 3.31). Incubation of 19-nortestosterone acetate for only 5 hours resulted in a mixture containing estradiol-17β monoacetate (approximately 25%) and starting material.

2. *17β-Hydroxy-5β-androstane-3-one-4-[14]C*[11,12] (*Fig. 2*)

To a solution of 30 mg of testosterone-4-^{14}C in 10 ml of absolute ethanol was added 30 mg of 10% palladium-on-charcoal and a solution of 8 mg of potassium hydroxide in 1 ml water; the mixture was hydrogenated at a pressure of 45 pounds for 2 hours. Then the solution was

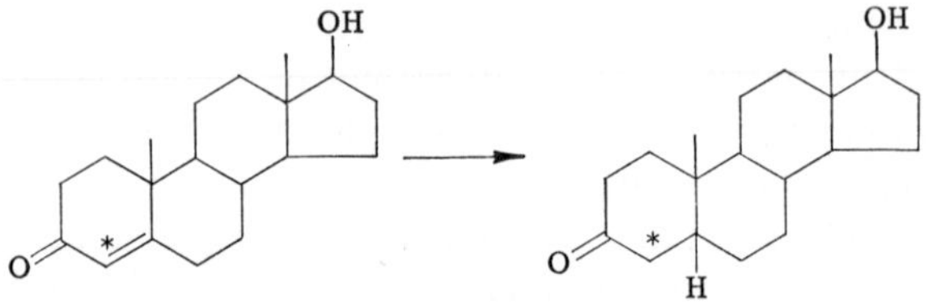

FIG. 2.

filtered, and the filtrate was neutralized with acetic acid, diluted with water, and extracted with methylene chloride. The solvent was removed *in vacuo* and the residue was chromatographed on a Bush B$_3$ Celite partition column to give 17β-hydroxy-5β-androstane-3-one-4-^{14}C in a 50% yield.

3. *17β-Hydroxy-5α-androstane-3-one-4-[14]C* (*Fig. 3*)[13]

A solution of 290mg of testosterone-4-^{14}C in 3 ml of dioxane and 3 ml of ether was added dropwise with magnetic stirring to a solution of 35 mg of lithium in 30 ml liquid ammonia over a period of 10 minutes. An addi-

[9] F. Sondheimer, S. Kaufmann, J. Romo, H. Martinez, and G. Rosenkranz, *J. Am. Chem. Soc.* **75**, 4712 (1953).

[10] M. R. Ehrenstein, A. R. Johnson, P. C. Olmstead, V. I. Vivian, and M. A. Wagner, *J. Org. Chem.* **15**, 264 (1950).

[11] Compare R. B. Gabbard and A. Segaloff, *J. Org. Chem.* **27**, 655 (1962).

[12] In identical fashion 17α-methyltestosterone-4-^{14}C was reduced to 17β-hydroxy-17α-methyl-5β-androstan-3-one-4-[14]C.

[13] Compare F. L. Weisenborn and H. E. Applegate, *J. Am. Chem. Soc.* **81**, 1960 (1959).

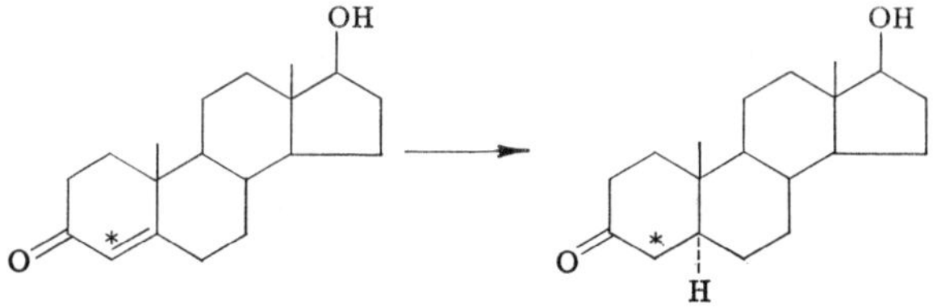

FIG. 3.

tional 10 mg of lithium was added, and the stirring was continued for an additional 30 minutes while the blue color persisted. The solution was then neutralized by the addition of 800 mg of ammonium chloride and the ammonia was allowed to evaporate. The residue was extracted with methylene chloride; the extract was washed with water and dried; and the solvent was removed *in vacuo*. The residue was chromatographed on a Bush B₃ Celite partition column to give 241 mg of 17β-hydroxy-5α-androstan-3-one-4-¹⁴C, identical with a nonradioactive authentic standard.

4. *Cortisol-4-¹⁴C (Fig. 4)*[14]

17α,20;20,21-Bismethylenedioxy-3,5-seco-4-norpregnane-5,11-dion-3-oic acid. A solution of 5 g of 17α,20;20,21-bismethylenedioxypregn-4-ene-3,11-dione[15] in a mixture of 200 ml of methylene chloride and 20 ml of ethyl acetate was ozonized at −78° until the color of the solution had a blue tint (1.5–2.5 hours). The solvent was then removed *in vacuo*, and the amorphous residue was dissolved in 20 ml of acetic acid. After the addition of 5 ml of 30% hydrogen peroxide, the solution was left overnight at room temperature. The solution was then concentrated *in vacuo* to a small volume of approximately 25 ml, 100 ml of water was added and extracted with a benzene–ether mixture. The organic phase was extracted with 1 N sodium hydroxide solution, and the alkaline extracts were acidified by pouring them into an ice cold concentrated hydrochloric acid mixture. The desired keto acid separated as a sirup and was extracted with methylene chloride. The extract was washed with water and dried, and the solvent was evaporated to leave a residue which, upon chromatography on silica gel, gave 3.1 g of 17α,20;20,21-bismethylene-dioxy-3,5-seco-4-norpregnane-5,11-dion-3-oic acid, m.p. 176–177°.

5-Hydroxy-17α,20;20,21-bismethylenedioxy-3,5-seco-4-norpregn-5-en-11-on-3-oic Acid 3,5-Lactone. To a solution of 3 g of 17α,20;20,21-bis-methylenedioxy-3,5-seco-4-norpregnane-5,11-dion-3-oic acid in 150 ml of

---

[14] Compare U.S. Patent 2,951,074, August 30, 1960.

[15] R. E. Beyler, R. M. Moriarty, F. Hoffman, and L. H. Sarett, *J. Am. Chem. Soc.* 80, 1517 (1958).

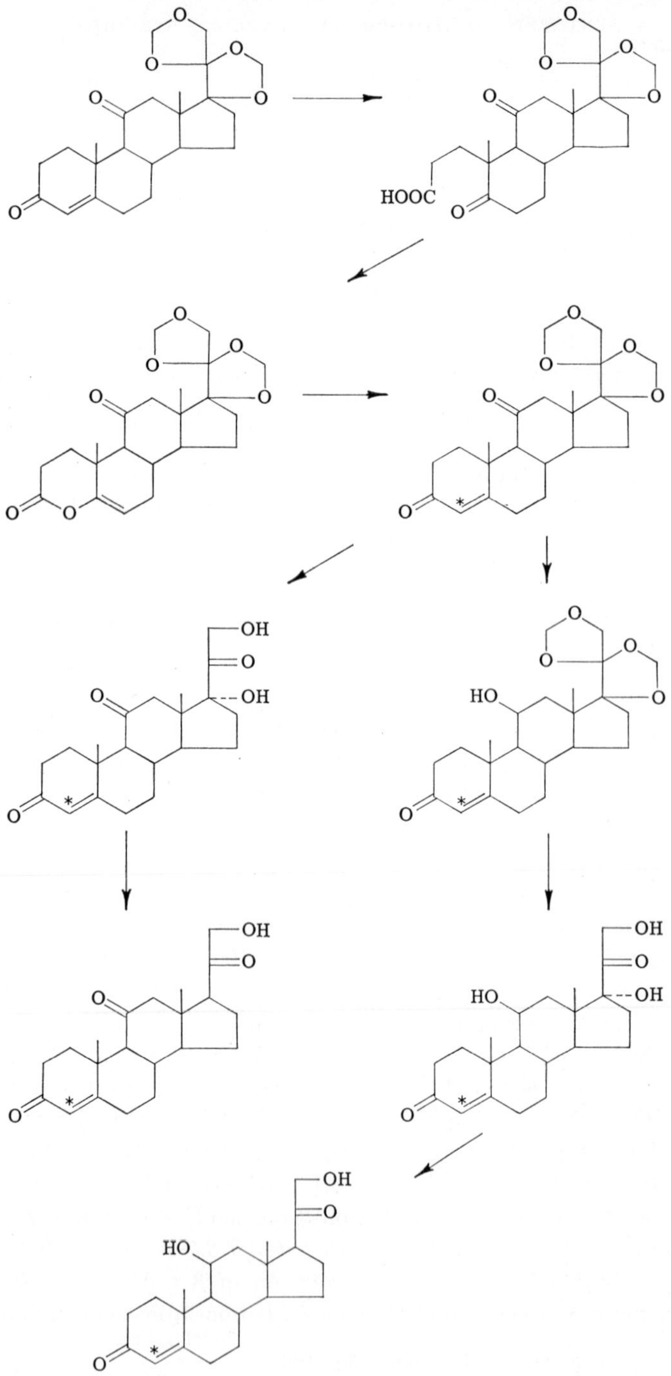

Fig. 4.

acetic anhydride was added 10 g of anhydrous sodium acetate; the mixture was refluxed for 2 hours. The solvent was removed *in vacuo*, and the residue was dissolved in ethyl acetate. This solution was washed with water and with 2 N sodium carbonate solution, then dried and concentrated. After recrystallization from acetone-ether there was obtained 1.5 g of lactone, m.p. 240–242°.

17α,20;20,21-Bismethylenedioxypregn-4-ene-3,11-dione-4-[14]C. To a magnetically stirred solution of 2 millimoles of methyl magnesium iodide-^{14}C in 15 ml of ether which was cooled to 0° to −10°, was added dropwise a solution of 2.4 millimoles of the above lactone in 20 ml of anhydrous benzene and 10 ml of anhydrous ether. After 1 hour of stirring at 0°, the mixture was left at room temperature overnight and then added to an ice cold saturated solution of ammonium chloride. The mixture was extracted with methylene chloride, the extract was evaporated to dryness, and the residue was dissolved in 100 ml of methanol to which 30 ml of water containing 10 g of potassium hydroxide was added. The mixture was allowed to stand for 18 hours at room temperature, 20 ml of acetic acid was added, and the methanol was evaporated *in vacuo*. The residual aqueous layer was extracted with methylene chloride, dried, and evaporated. The sirupy residue was chromatographed on a silica gel column whereby, from the ethyl acetate–benzene eluates, there was obtained the 17α,20;20,21-bismethylenedioxy derivative of cortisone-4-^{14}C in a yield of 52%, calculated on the methyl iodide-[14]C.

11β-Hydroxy-17α,20;20,21-bismethylenedioxypregn-4-en-3-one-4-[14]C. A solution of 600 mg of the methylenedioxy derivative of cortisone-4-^{14}C in 20 ml of anhydrous tetrahydrofuran was added dropwise to a boiling solution of 600 mg of lithium aluminum hydride in 20 ml of tetrahydrofuran, the resulting solution was refluxed for 2 hours. After the preparation an excess of ethyl acetate was added dropwise, followed by a saturated solution of sodium sulfate to a point when the suspended inorganic salts separated as a cake from the clear supernatant. The solvent was removed *in vacuo* from the filtered organic phase, and the residue was dissolved in 10 ml of dioxane to which 400 mg of 2,3-dichloro-5,6-dicyanobenzoquinone was added. After the solution had stood overnight at room temperature, the solvent was removed *in vacuo* and the residue was chromatographed over silica gel. The ethyl acetate–benzene fraction furnished, after recrystallization from ether, 553 mg of the bismethylenedioxy derivative of cortisol-4-^{14}C, m.p. 220–223°.

Cortisol-4-[14]C. A solution of 120 mg of the bismethylenedioxy derivative of cortisol-4-^{14}C in 10 ml of 50% acetic acid was heated for 2.5 hours on a steam bath, and then the solution was neutralized, diluted with water, and extracted with methylene chloride. The solvent was evaporated *in vacuo* and the residue was chromatographed on a Bush B₅ Celite partition column. The yield of pure cortisol-4-^{14}C amounted to 61 mg, identical in its chromatographic behavior with the nonradioactive compound.

5. *Cortisone-4-^{14}C*

A solution of 120 mg of the bismethylenedioxy derivative of cortisone-4-^{14}C was heated in 25 ml of 50% acetic acid on a steam bath for 4 hours. The work-up, identical to the one described for cortisol-4-^{14}C above, gave, after chromatography on a Bush B₅ Celite partition chromatography, 28 mg of cortisone-4-^{14}C and 17 mg of starting material.

6. *Corticosterone-4-^{14}C*

To 2 g of zinc dust was added a solution of 200 mg of cortisol-4-^{14}C in 100 ml of 50% acetic acid, the mixture was refluxed for 2 hours. The hot solution was decanted, and the zinc was washed with methanol. The acetic acid was neutralized, then the methanol was evaporated off *in vacuo* and the aqueous phase was extracted with methylene chloride. The solvent was removed *in vacuo,* and the residue was chromatographed on a Bush B₂ Celite partition column to give 167 mg of corticosterone-4-^{14}C.

7. *11-Dehydrocorticosterone-4-^{14}C*

To a solution of 250 mg of cortisone-4-^{14}C in 100 ml of acetic acid was added 3 g of zinc dust; the mixture was boiled for 2 hours. The hot solution was filtered, and the zinc was washed with hot methanol. The filtrate was neutralized and the methanol evaporated *in vacuo*. The aqueous phase was extracted with methylene chloride, the extract was dried, and the solvent was evaporated. The residue was chromatographed on a Bush B₂ Celite partition column to give 209 mg of 11-dehydrocorticosterone-4-^{14}C, in all respects identical with the nonradioactive material.

8. *17α,21-Dihydroxypregn-4-ene-3,20-dione-4-^{14}C* (*Fig. 5*)

A solution of 5 g of 17α,20;20,21-bismethylenedioxypregn-4-en-3-one[16] in 100 ml of methylene chloride and 10 ml of ethyl acetate was treated with ozone at −70°, as described above for the bismethylenedioxy derivative of cortisone. After working up the reaction in a similar fashion, the crude keto acid was refluxed with acetic anhydride containing 9% anhydrous sodium acetate. The reaction mixture was worked up as described for the lactone of the bismethylenedioxy derivative of cortisone to give

[16] R. E. Beyler, F. Hoffman, R. M. Moriarty, and L. H. Sarett, *J. Org. Chem.* **26**, 2421 (1961).

1.3 g of 17α,20;20,21-bismethylenedioxy-5-hydroxy-3,5-seco-4-norpregn-5-en-3-oic acid 3,5-lactone, m.p. 185–187°.

This lactone was allowed to react with methyl magnesium iodide-¹⁴C in the same fashion as described above for the preparation of cortisone-4-¹⁴C. After the usual work-up there was obtained, in a yield of 47%, the bismethylenedioxy derivative of 17α,21-dihydroxypregn-4-ene-3,20-dione, identical with the nonradioactive compound.

F_IG. 5.

The bismethylenedioxy protective grouping was hydrolyzed by heating the material on a steambath for 4 hours with 50% acetic acid. After the usual work-up and chromatography on a Bush B₂ Celite partition column, there was obtained the desired 17α,21-dihydroxypregn-4-ene-3,20-dione-4-¹⁴C in a yield of 22%. In addition, 35% of the starting material could be recovered.

9. Deoxycorticosterone-4-^{14}C

A solution of 100 mg of 17α,21-dihydroxypregn-4-ene-3,20-dione-4-^{14}C in 100 ml of 50% acetic acid was treated with zinc dust as described for the transformation of cortisol-4-^{14}C into 11β,21-dihydroxypregn-4-ene-3,20-dione-4-^{14}C. After a similar work-up, the crude product was chromatographed on a Bush B₃ Celite partition column to give deoxycorticosterone-4-^{14}C in 77% yield.

10. 3β,17α,21-Trihydroxypregn-5-en-20-one-4-^{14}C (Fig. 6)

A solution of 100 mg of the bismethylenedioxy derivative of 17α,21-dihydroxypregn-4-ene-3,20-dione-4-^{14}C (see footnote 16) in 20 ml of acetyl chloride was heated under reflux for 1 hour and then the solvent

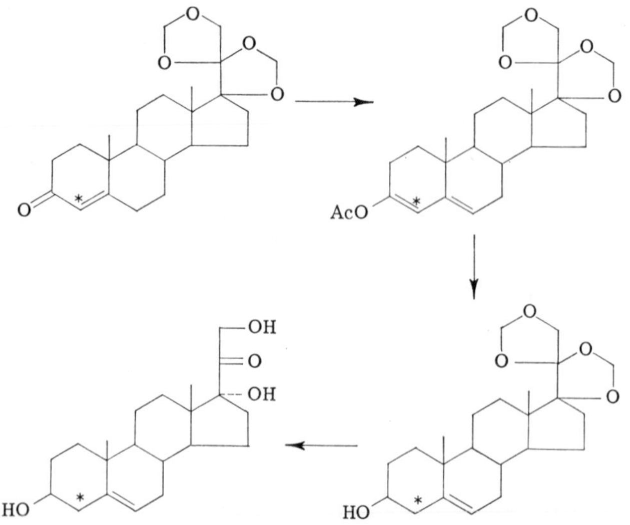

FIG. 6.

was evaporated off in vacuo. The residue was recrystallized from benzene–ether, m.p. 159–162°. However, this purification is not essential for the following step, and the crude residue can directly be treated with a solution of 300 mg of sodium borohydride in 95% ethanol. After the solution had stood overnight at room temperature, the ethanol was removed in vacuo, water was added to the residue and the aqueous phase was extracted with methylene chloride. The solvent was evaporated off in vacuo, the residue was dissolved in 25 ml of 50% acetic acid and the solution was heated on the steambath for 2½ hours. Then the solution was neutralized, diluted with water, and extracted with methylene chloride. The solvent was evaporated *in vacuo* and the residue chromatographed on a Bush B$_2$ Celite partition column, yielding 23 mg of pure $3\beta,17\alpha,21$-trihydroxypregn-5-en-20-one-4-^{14}C, m.p. 230–232°.

11. *17α-Hydroxyprogesterone-4-^{14}C Caproate and 17α-Hydroxyprogesterone-4-^{14}C (Fig. 7)*[17]

17α-Hydroxy-3,5-seco-4-norpregnane-5,20-dion-3-oic Acid 17-Caproate. A solution of 20 g of 17α-hydroxyprogesterone caproate (m.p. 119–121°, $[\alpha]_D + 60°$ [CHCl$_3$])[17] in 80 ml of chloroform, 120 ml of ethyl acetate and 10 ml of acetic acid was cooled to −75°; within 7 hours 1.2 equivalents of ozone was introduced. After warming up to 0°, 15 ml of a 30% hydrogen peroxide solution was added and stirred overnight at 6°. Then the solution was concentrated *in vacuo* at 25° and the residue was extracted with ether. The ether solution was extracted exhaustively with a solution of sodium bicarbonate. The bicarbonate extracts were immedi-

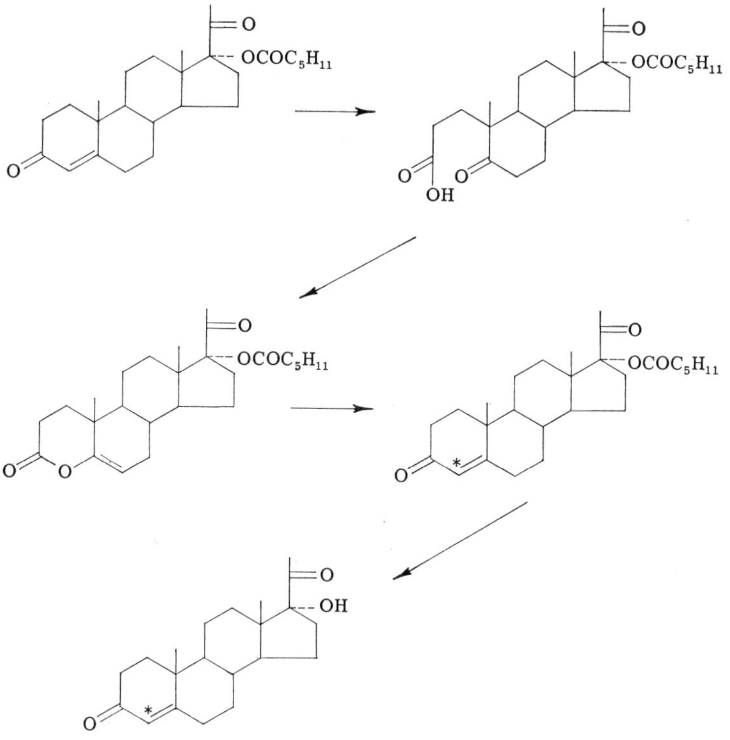

FIG. 7.

[17] P. E. Schulze, Z. *Naturforsch.* **136**, 409 (1958).

ately acidified and extracted with ether; the ether solution was dried and concentrated. The neutral portion which remained in the first ether solution was evaporated to dryness, dissolved in 300 ml of acetic acid, and treated with a solution of 9 g of periodic acid in 20 ml of water. After 2 hours the solution was concentrated *in vacuo* and worked up as described above. Both acidic portions were combined to give 17.6 g of a colorless oil that did not crystallize.

5,17α-Dihydroxy-3,5-seco-4-norpregn-5-en-3-oic   Acid   3,5-Lactone Caproate. A solution of 17.6 g of 17α-hydroxy-3,5-seco-4-norpregnane-5,20-dion-3-oic acid 17-caproate in 80 ml of acetic anhydride and 80 ml of acetyl chloride was heated under reflux for 48 hours. Then the solution was concentrated *in vacuo* and extracted with ether; the ether extract was washed with a solution of sodium bicarbonate and with water and saline, and was dried and evaporated. The residue was chromatographed on 350 g of silica gel. The substance was applied in a benzene solution and eluted with benzene–methylene chloride mixtures (75:25). The combined crystalline fractions, recrystallized from isopropyl ether, gave a yield of 7 g of lactone, m.p. 124–136°; $[\alpha]_D - 73°$ ($CHCl_3$). Analysis: Calculated: C, 72.6; H, 8.9. Found: C, 72.65; H, 9.15.

17α-Hydroxyprogesterone Caproate-4-[14]C. A Grignard solution, prepared from 710 mg of methyl iodide-[14]C (1 mC/millimole), 150 mg of magnesium, and 5 ml of ether was added dropwise, while stirring under nitrogen, within 1 hour at room temperature to a solution of 2.16 g of the lactone in 800 ml of ether (distilled over lithium aluminum hydride under nitrogen). After 12 hours of stirring, a saturated ammonium chloride solution was added, the aqueous phase was extracted several times with ether, and the ether extract was washed with sodium bicarbonate and sodium chloride solution and finally evaporated to dryness. The oily residue was dissolved in 40 ml of acetic acid, and 4 ml of concentrated hydrochloric acid was added. The solution was kept for 48 hours under nitrogen at 35°. The solution was concentrated *in vacuo* and then extracted with ether; the ether solution was washed with bicarbonate and sodium chloride solution, dried, and evaporated. The oily residue was dissolved in a small amount of benzene and chromatographed on 35 g of aluminum oxide (Woelm, acidic). The eluates with benzene–methylene chloride mixtures 60:40 afforded crystalline fractions which were recrystallized from isopropyl ether to yield 770 mg of 17α-hydroxyprogesterone caproate-4-^{14}C, m.p. 118–120°. This material was identical with an authentic sample in its melting point, rotation, ultraviolet, and infrared spectra.

17α-Hydroxyprogesterone-4-[14]C. A solution of 100 mg of 17α-hydroxyprogesterone caproate-4-^{14}C in 50 ml of a 0.5% methanolic solution of potassium hydroxide was refluxed for 1 hour. After cooling, the solution was neutralized with a 50% acetic acid solution, the methanol was removed *in vacuo*, and the concentrate was extracted with methylene chloride. The extract was washed with a 2 N sodium carbonate solution and dried, and the solvent was evaporated *in vacuo*. The residue was chromatographed on a Bush B$_3$ Celite partition column to give 53 mg of 17α-hydroxyprogesterone-4-^{14}C.

12. *3β-Hydroxyandrost-5-en-17-one-4-^{14}C and 3β-Hydroxypregn-5-en-20-one-4-^{14}C (Fig. 8)* [18]

3β-Hydroxyandrost-5-en-17-one-4-^{14}C. A solution of 100 mg of testosterone-4-^{14}C in 10 ml of acetyl chloride and 2 ml of acetic anhydride was refluxed for 2 hours, and then the solvents were removed under reduced pressure. The 103-mg residue, which consisted mainly of 3,17β-diacetoxyandrosta-3,5-diene-4-^{14}C, was dissolved in 50 ml of 95% ethanol containing 2 drops of pyridine. To this solution was added a solution of 250 mg of sodium borohydride in 25 ml of 95% ethanol, and the combined solutions were left at room temperature overnight. Then 30 ml of a 1 N hydrochloric acid solution was added, and most of the ethanol was evaporated off *in vacuo*. The concentrate was extracted with methylene chloride, the extract was washed with water and dried, and the solvent was evaporated under reduced pressure. The residue was dissolved in 25 ml of 90% ethanol and combined with a hot solution of 500 mg of digitonin in 20 ml of 90% ethanol. After the preparation had stood for 20 hours at room temperature, the resulting digitonide precipitate was centrifuged, then washed with cold 90% ethanol and with ether. The digitonide was dried under reduced pressure, then dissolved in 20 ml acetic acid; a solution of 250 mg of chromic oxide in 2 ml water was added with stirring and the mixture was let stand for 30 minutes at room temperature. The excess chromic acid was destroyed with a few drops of methanol. The solvents were removed *in vacuo*, 80 ml of water was added to the residue, and the insoluble oxidized digitonide was then centrifuged, washed with water, and dried. The dried residue was dissolved in a minimal volume of pyridine and heated on a steam bath for 1.5 hours. Then 80 ml of ether was added, and the solid was centrifuged and washed with ether. The combined ether washings were washed with

---

[18] For procedure, compare D. Kupfer, E. Forchielli, and R. I. Dorfman, *J. Am. Chem. Soc.* **82**, 1257 (1960). Other procedures, when checked by the reviewers, seemed more time consuming, without providing a substantially superior yield: M. Uskoković, R. I. Dorfman, and M. Gut, *J. Org. Chem.* **23**, 1947 (1958); M. Gut and M. Uskoković, *ibid.* **24**, 673 (1959); P. N. Rao and L. R. Axelrod, *ibid.* **26**, 1607 (1961); Z. T. Glazer and M. Gut, *ibid.* 4725 (1961).

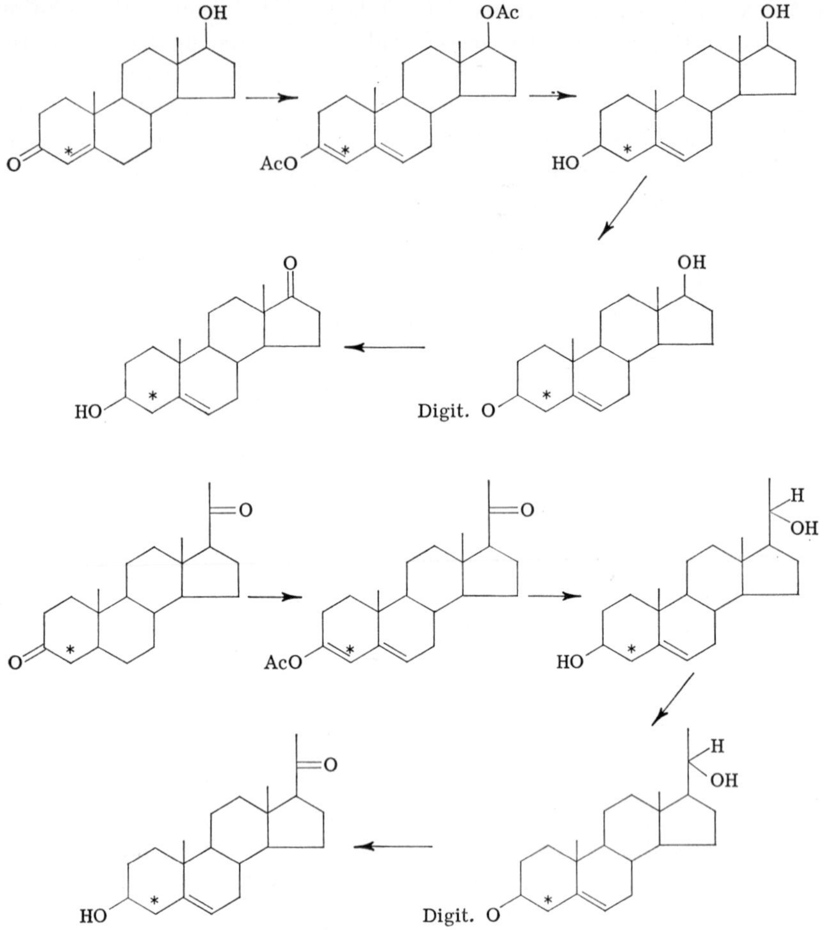

FIG. 8.

0.1 N sodium hydroxide solution and with water, dried, and evaporated. The residue was chromatographed on a Bush A Celite partition column whereby 18 mg of pure 3β-hydroxyandrost-5-en-17-one-4-^{14}C, identical with authentic material, could be isolated.

3β-Hydroxypregn-5-en-20-one-4-[14]C. This compound was prepared from progesterone-4-^{14}C in strict analogy to the above procedure for the conversion of testosterone-4-^{14}C to 3β-hydroxyandrost-5-en-17-one-4-[14]C. The yields amounted to 15–20%, calculated on progesterone-4-[14]C.

13. *3β,21-Dihydroxypregn-5-en-20-one-4-[14]C (Fig. 9)*[19]

To a solution of 125 mg of 3β-hydroxypregn-5-en-20-one-4-^{14}C in 5 ml of anhydrous benzene and 0.3 ml of methanol, was added 1 ml of

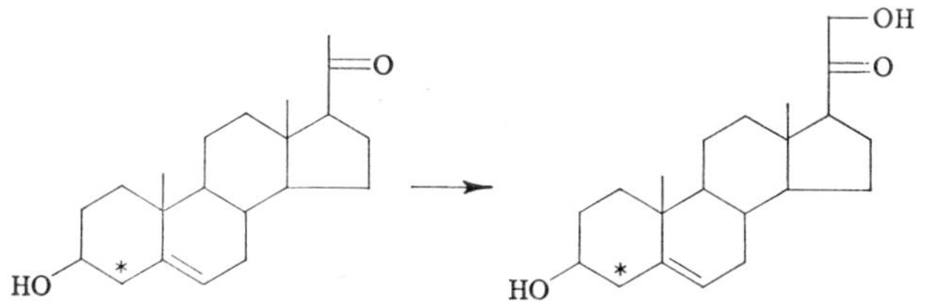

Fig. 9.

boron trifluoride–ether complex followed by 200 mg of lead tetraacetate; the solution was stirred for 4 hours at room temperature. The crude product was isolated by extraction with benzene, the solvent was removed *in vacuo*, and the residue was hydrolyzed with a methanolic solution of potassium hydroxide. The crude hydrolysis product was chromatographed on a Bush B$_2$ Celite partition column to yield 51 mg of 3β,21-dihydroxy-pregn-5-en-20-one-4-^{14}C, m.p. 159–162°.

IV. Steroids Labeled with ³H

A. Introduction

There are many basic differences between carbon-14 and tritium other than their chemical nature, and most of these will make tritium the label of choice for the selection of tagged steroids.

The price ratio of four dollars per millicurie for carbon-14, compared with two dollars per curie for tritium, will become even larger when two steroids, labeled with carbon-14 and with tritium, respectively, are being compared.

The difference in half-life of carbon-14 (5568 years) and that of tritium (12.5 years) expresses itself in their respective specific activities of 58 C/millimole for tritium and 55 mC/millimole for carbon-14. This fact is of particular importance where, in a given biological system, a steroid with counts but without measurable weight has to be introduced.

The most frequently encountered disadvantages of high specific activity tritium labeling over carbon-14 are occasionally observed isotope fractionation[20] of the tritiated steroid or its derivatives from the non-labeled analog which does preclude the measurement of constant isotope rates over the chromatographic peaks, and the rather limited shelf life, which is due to the effect of the radiation and makes continuous quality control imperative. Since the preparation of steroids by biosynthesis and

---

[19] Compare J. D. Cocker, H. B. Henbest, G. H. Phillipps, G. P. Slater, and D. A. Thomas, *J. Chem. Soc.* **1965**, 6 (1965).

[20] V. Cejka and E. M. Venneman, *Atomlight* **55**, 5 (1966) and references therein.

also by catalytic exchange lead to relatively low specific activities, this review will consider only "specific" labeling by either catalytic tritiation of a double bond or catalytic tritiation of a halosteroid, whereby specific activities of over 100 mC/millimole and 30 mC/millimole, respectively, can be obtained.

B. Syntheses of Steroids-1,2-³H

The required starting material for the preparation of 3-keto-Δ⁴-steroids-1,2-³H can readily be synthesized by the dehydrogenation[21] of the commercially available 3-keto-Δ⁴-analog. The selective tritiation of the 1,2-double bond can be carried out by either homogeneous or heterogeneous catalysis.

Selective Reduction by Homogeneous Catalytic Hydrogenation

Recently, Djerassi and Gutzwiller[22] have used the method of homogeneous rhodium catalysis for deuterium labeling of steroids by the reduction of double bonds in the various positions of the steroid ring. Their preliminary report indicates that the method has great promise for selective reductions and may also be used for tritium labeling.

Preparation of the Catalyst Solution (0.005 M). To a solution of 197 mg (0.75 millimole) of triphenylphosphine in 25 ml benzene was added an equal volume of ethanol (or methanol) and 48 mg (0.125 millimole) of μ-dichlorotetraethylenedirhodium.[23]

The resulting yellow solution of tris(triphenylphosphine) chlororhodium was filtered through cotton and used directly for the subsequent hydrogenation. The corresponding iodo catalyst (brownish black solution) was prepared by simply adding 1 equivalent of sodium iodide to the solution of the chloro complex.

Typical Reduction Procedure. A 100-mg sample of a steroid was dissolved in 20–50 ml of 0.005 M catalyst solution and stirred at room temperature and atmospheric pressure in a tightly stoppered flask which had first been evacuated and filled with deuterium (tritium may be used for tritium labeling). After several hours, the flask was evacuated to remove the labeling gas, and the solution was evaporated to dryness; the residue was taken up in a mixture of petroleum ether and methylene chloride and filtered through alumina. Evaporation of the solvent gave the product.

Results. Unhindered Δ¹-, Δ²-, and Δ³-cholestenes were reduced by either the chloro or iodo catalyst in nearly quantitative yield in 1.5–2 hours. More highly substituted olefins such as Δ⁴-, Δ¹⁴-, and Δ⁸⁽¹⁴⁾ steroids

[21] D. Burn, D. N. Kirk, and V. Petrow, *Proc. Chem. Soc.* **1960**, 14.
[22] C. Djerassi and J. Gutzwiller, *J. Am. Chem. Soc.* **88**, 4537 (1966).
[23] R. Cramer, *Inorg. Chem.* **1**, 722 (1962).

were not reduced during this time. More hindered disubstituted olefins such as Δ^{11}-steroids are reduced only to the extent of 13% in 36 hours.

The reduction of α,β-unsaturated ketones[24] is particularly interesting. Δ^1-3-Keto-5α-steroids were readily reduced to the saturated ketones in 6–16 hours, while Δ^4-3-ketones or Δ^5-7-keto-3β-acetates were recovered almost unchanged. Using the chloro catalyst and periods of 16–72 hours, $\Delta^{1,4}$-androstadiene-3,17-dione and $\Delta^{4,6}$-androstadiene-3,17-dione were converted in 75–85% yield into the Δ^4-3-ketones, the rest of the material being the completely saturated diketones. Δ^4-3-Ketones labeled with deuterium or tritium at position 1 may, therefore, now be prepared in high yields by the homogeneous catalytic reduction of $\Delta^{1,4}$-dien-3-ones instead of the heterogeneous catalytic hydrogenations using palladium, which usually give a large number of reduction products and poor yields of Δ^4-3-ketones.

One further difference was noted with homogeneous rhodium catalysis. Whereas palladium-catalyzed tritiation and deuteration of $\Delta^{1,4}$-androstadiene-3,17-dione proceed from the β-face to give ultimately labeled Δ^4-androstene-3,17-dione in about 15% overall yield, rhodium-catalyzed (chloro catalyst was used) deuteration of the above dienone occurred from the α-face to give Δ^4-androstene-3,17-dione-1α,2α-D_2 in 85% yield.

These reviewers prefer to prepare the "homogeneous" catalyst as a solid:[25] A solution of 300 mg of triphenylphosphine and of 50 mg of trichlororhodium hydrate in 20 ml of ethanol is refluxed in a nitrogen atmosphere. The desired tris(triphenylphosphine) chlororhodium precipitates during heating, and the crystallizate can be collected by filtration. The catalyst can be stored at room temperature without the exclusion of air.

The advantage of this preparation over the catalyst solution is that tritiation can be carried out in dioxane solution, which will minimize the exchange of tritium as compared to the use of hydroxylic solvents.

Selective Reduction by Heterogeneous Catalytic Tritiation (Fig. 10)

Cortisol-1,2-$^3H^{26}$ (70% 1β,2β).[27] A solution of 360 mg (1 millimole) of prednisolone in 60 ml of dioxane to which 300 mg of 5% palladized charcoal had been added was reduced with 5 C of tritium, diluted with

---

[24] In the reduction of α,β-unsaturated ketones, a 1:1 ethanol–benzene solution is preferred to 1:1 methanol–benzene because the latter gives the dimethyl ketal of the saturated ketone.

[25] J. Biellmann and H. Liesenfelt, *Compt. Rend. Acad. Sci.* 263, 251 (1966).

[26] Compare P. A. Osinski and H. Vanderhaeghe, *Rec. Trav. Chim.* 79, 216 (1960).

[27] For configuration of isotope see H. J. Brodie, M. Hayano, and M. Gut, *J. Am. Chem. Soc.* 84, 3766 (1962). See also H. J. Brodie, K. Raab, and M. Gut, *Proc. 2nd Intern. Congr. Hormonal Steroids, Intern. Congr. Ser.* 111, 119. Excerpta Medica, Amsterdam, 1966.

carrier hydrogen. After the uptake of 1 millimole of hydrogen-tritium mixture, the reaction was worked up as usual and the crude reaction product was chromatographed on a Bush B₅ Celite partition column. This furnished 74 mg cortisol-1,2-^{3}H (specific activity 835 mC/millimole) and 33 mg dihydrocortisol-1,2,4,5-^{3}H (specific activity 1557 mC/millimole).

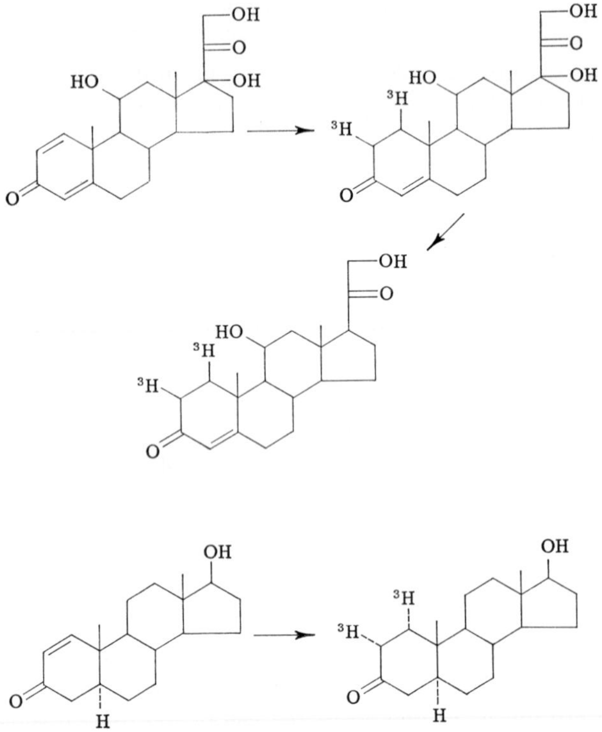

Fig. 10.

Corticosterone-1,2-^{3}H.[25] To a solution of 30 mg of cortisol-1,2-^{3}H (specific activity 412 mC/millimole) in 6 ml of 50% acetic acid was added 500 mg of zinc dust and the mixture was refluxed for 75 minutes. After the usual work-up,[28] the material was chromatographed on a Bush B₂ Celite partition column to give 11.5 mg (0.033 millimole) of corticosterone-1,2-^{3}H (specific activity 414 mC/millimole). The product was identified by acetylation and by its spectrum in concentrated sulfuric acid.

[28] Exactly as described for the conversion of cortisol-4-^{14}C to corticosterone-4-^{14}C on page 316.

5α-Androstane-3,17-dione-1-³H   (*90%   1α*).[29]   5α-Androst-1-ene-3,17-
dione in dioxane solution was catalytically tritiated over 10% palladium-
on-charcoal, as described for the preparation of aldosterone-1,2-³H.[30] The
residue was dissolved in 90% acetic acid and reoxidized with chromic
acid, followed by several equilibrations with a 1 N methanolic potassium
hydroxide solution. After the usual work-up, the crude product was
chromatographed on a Bush A Celite partition column to yield pure 5α-
androstane-3,17-dione-1-³H in a yield of 30%, calculated on tritium.

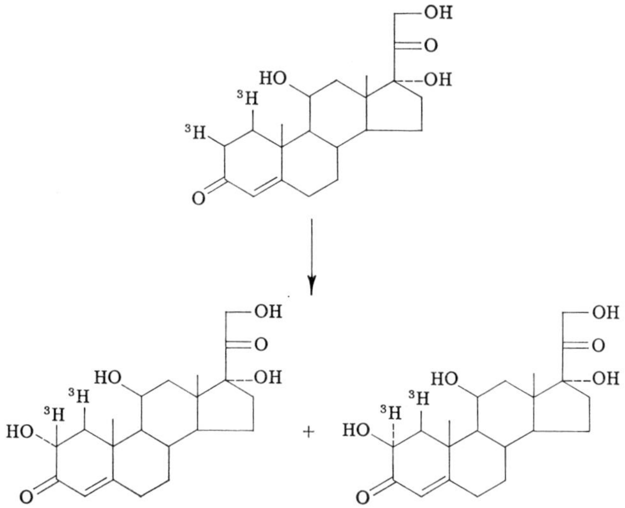

FIG. 11.

*2α-Hydroxycortisol-1,2-³H   and   2β-Hydroxycortisol-1,2-³H   (Fig.
11).*[31, 32] 1,2-Tritiated cortisol 21-acetate (approximately 1 C/millimole)
was heated at 95° in 4.0 ml of glacial acetic acid containing 0.05 ml of
acetic anhydride and 300 mg of lead tetraacetate for 1.5 hours (until a
negative coloration was obtained with KI-starch paper). The cooled
reaction mixture was taken up in methylene chloride, shaken with sufficient concentrated sodium hydroxide to neutralize 90% of the acetic
acid, washed with sodium bicarbonate solution and water, and evaporated
to dryness. The residue was dissolved in 16 ml of freshly distilled

[29] For configuration of isotope, see H. J. Ringold, M. Gut, M. Hayano, and A.
Turner, *Tetrahedron Letters* 18, 835 (1962).
[30] See page 330.
[31] Compare S. Burstein, H. L. Kimball, and M. Gut, *Steroids* 8, 789 (1966). The label
of this material at other positions than C-1 and C-2 amounts to less than 8%.
[32] 2α- and 2β-Hydroxycortisol-4-¹⁴C has been prepared in a similar fashion from
cortisol-4-¹⁴C.

methanol which had previously been boiled while nitrogen was passed through the solvent. To the solution 0.6 ml of 1 M methanolic potassium hydroxide (prepared from the same methanol) was added, thoroughly mixed, and left at 30° under nitrogen for 10 minutes. Then 0.06 ml of water (oxygen free) was added, and the solution was kept for an additional 2 minutes. The solution was made slightly acidic with 1 M acetic acid, and the methanol was evaporated under reduced pressure and extracted with a large amount of ethyl acetate. The ethyl acetate extract was washed with a saturated sodium bicarbonate solution and water and finally evaporated *in vacuo*. The residue was then chromatographed in the system ethylene chloride–methanol–water 200:17:33 on Celite 545 on a 5 × 100 cm column collecting 50-ml fractions. The 2α- and 2β-hydroxy isomers which were recovered in fractions 48–67 were acetylated with acetic anhydride and pyridine and chromatographed on Celite 545 on a 2 × 60 cm column in the system benzene–Skellysolve C–methanol–water 5:5:4:1 collecting 10-ml fractions. Fractions 42–58 contained the tritiated 2β-hydroxycortisol 2,21-diacetate, and the tritiated 2α-hydroxycortisol 2,21-diacetate was in fractions 62–81. The diacetates were each rechromatographed in the system benzene–Skellysolve C–methanol–water 5:5:4:1 and methanolized to the free steroid as described previously. The free steroids were then rechromatographed in the system ethylene chloride–methanol–water 200:17:33 and measured by ultraviolet spectrophotometry; 6 mg of 2α- and 10 mg of 2β-hydroxycortisol were obtained with specific activities of 945 and 893 mC/millimole, respectively.

The 2α- and 2β-hydroxycortisol exhibited single chromatographic peaks after chromatography in the systems ethylene chloride–methanol–water 200:17:33 and benzene–*n*-butanol–methanol–water 10:1:3, indicating a better than 95% purity. The radiochemical homogeneity of the 2α- and 2β-hydroxycortisol was further checked by subjecting aliquots of the materials, following dilution with respective carrier, to chromatographic separation as the free steroids and as acetates and to degradation experiments directed at the determination of the fraction of alkali-exchangeable [3]H.

6β-Hydroxycortisol-1,2-[3]H (Fig. 12).[31, 33] A mixture of 21 mg of 6β-hydroxycortisol 6,21-diacetate[32] and 20 mg of 2,3-dichloro-5,6-dicyano-benzoquinone in 3.0 ml of dioxane was heated under reflux for 19 hours. The mixture was evaporated to dryness, and the residue was dissolved in methylene chloride and washed with alkali and water. The reaction product was chromatographed on a thin-layer silica gel G in the system chloroform containing 4% methanol. The major product (10 mg), which

[33] Compare S. Bernstein and R. Littell, *J. Org. Chem.* **25**, 313 (1960).

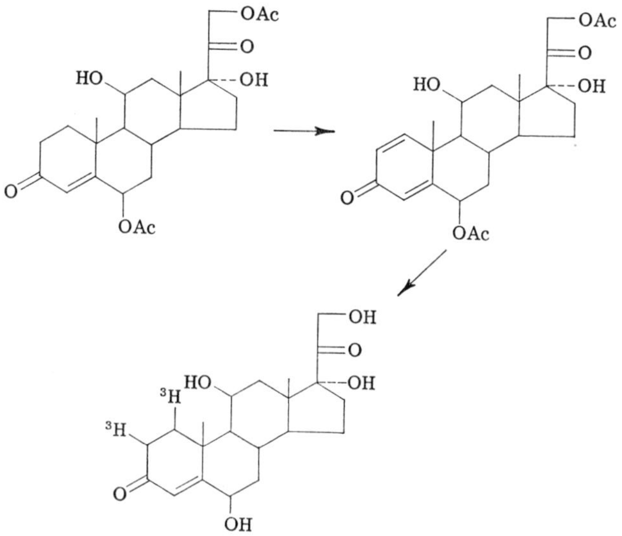

Fig. 12.

separated from the starting material, exhibited a maximum absorption at 243 mμ in methanol and showed no soda fluorescence indicating a 3-oxo-$\Delta^{1,4}$-diene. Without further characterization, the material (7.6 mg) was catalytically hydrogenated with tritium gas in 3.0 ml of dioxane over 10 mg of 10% palladium-on-charcoal. The mixture was allowed to absorb approximately 1 equivalent of tritium (approximately 2.2 C).

The tritiation product was chromatographed on thin-layer silica gel in the system chloroform containing 4% methanol. The zone corresponding to 6β-hydroxycortisol 6,21-diacetate was eluted and hydrolyzed to the free steroid by methanolysis for 20 minutes; the product was chromatographed on thin layer in the system chloroform containing 15% ethanol. The zone corresponding to 6β-hydroxycortisol was chromatographed on paper in the system ethylene chloride–methanol–water 200:17:33 (for 64 hours), and in the system benzene–n-butanol–methanol–water 10:1:3:3, and finally in the system chloroform containing 4% methanol on thin-layer silica gel. The undiluted material (188 μg) exhibited a spectrum in sulfuric acid which was identical with that of authentic 6β-hydroxycortisol. By the Porter-Silber reaction, the specific activity of the material was 13.8 C/millimole. Small portions of the tritiated 6β-hydroxycortisol chromatographed as single zones on paper (inseparable from admixed authentic 6β-hydroxycortisol) in the systems ethylene chloride–methanol–water 200:17:33, benzene–n-butanol–methanol–water 10:1:3:3, and benzene–methanol–water 2:1:1.

Testosterone-1-³H (83%-β) (Fig. 13). Testosterone-1,2-³H was pre-

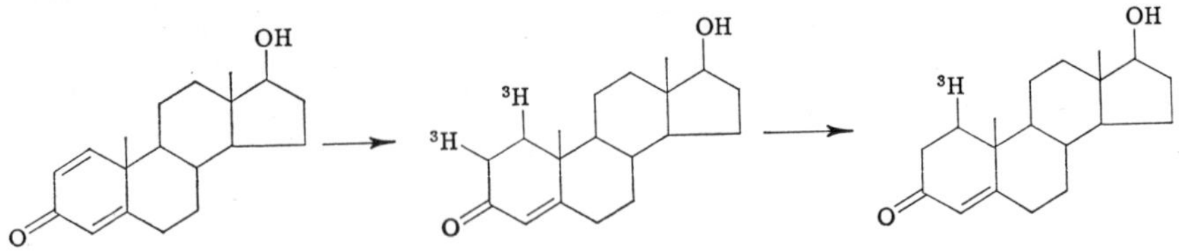

FIG. 13.

pared by reduction of 17β-hydroxyandrosta-1,4-dien-3-one with carrier-free tritium on 5% palladium-on-charcoal catalyst.[26] The product was chromatographed on a Bush B$_2$ Celite partition column. The eluates which contained the material having the mobility of testosterone (120 mC/millimole) were combined, the solvents were evaporated under reduced pressure, and the residue was dissolved in 2% (w/v) potassium hydroxide solution in 500 ml of 50% methanol–water and refluxed under nitrogen until no further exchange of tritium with solvent was noted. This was determined by periodically withdrawing aliquots into scintillation vials, acidifying with hydrochloric acid, evaporating to dryness, and then adding scintillation fluid and counting. The testosterone, now tritiated essentially at C-1 was recovered from the acidified reaction mixture by extraction with benzene after most of the methanol was evaporated under reduced pressure. The dried benzene extract was concentrated and chromatographed on a Bush B$_2$ Celite partition column. The eluted testosterone-1-^{3}H chromatographed as one radioactive zone with the mobility of testosterone in the paper systems ligroin (60–90°)–propylene glycol and benzene–ligroin–methanol–water (4:1:4:1). A portion was diluted with authentic testosterone and crystallized from benzene-hexane to constant specific activity of 16,760 dpm/micromole. Refluxing this material for 2 hours with potassium hydroxide in aqueous methanol as indicated above gave, on isolation and recrystallization, material of constant specific activity of 17,100 dpm/micromole, indicating complete exchange.

Aldosterone-1,2-^{3}H (Fig. 14)[34, 35]

Into a 25-ml three-neck flask (one neck having a break seal, one connected to the vacum line, and a third serving for introduction of the starting material and the reagent) was put a solution of 70 mg of Δ^1-aldosterone 21-acetate,[36] m.p. 192–194°, in 8 ml of dioxane, and 30 mg of

[34] K. R. Laumas and M. Gut, *J. Org. Chem.* 27, 314 (1962).
[35] Purification of the solvents and washing of the Celite were carried out as described by C. Flood, D. S. Layne, S. Ramcharan, E. Rossipal, J. F. Tait, and S. A. S. Tait, *Acta Endocrinol.* 36, 265 (1961). In all chromatographic separations only the peak fractions were combined and counted.
[36] E. Vischer, J. Schmidlin, and A. Wettstein, *Experientia* 12, 50 (1956).

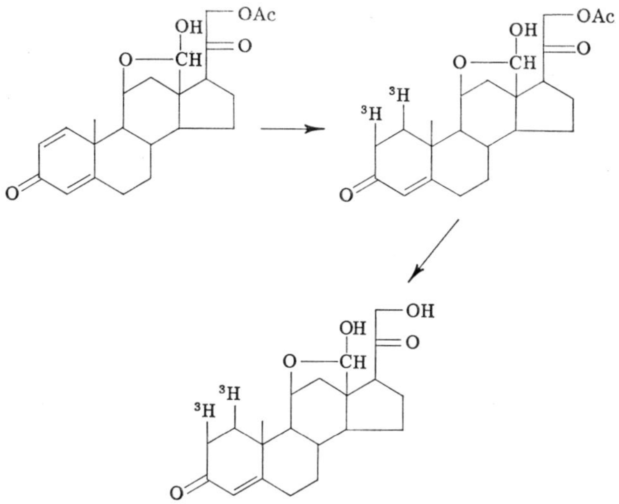

FIG. 14.

5% palladium-on-charcoal. Then one neck was closed off; the flask was connected to the vacuum line, degassed, and evacuated; and 6 ml of carrier-free tritium ($\simeq$15 C) was transferred by means of a Toepler pump. Then the flask was sealed off, removed from the line, and shaken at 25° for 18 hours. The flask was connected to the line through the break seal, the break seal was broken, traces of tritium were toeplered out, and the solvent was pumped off. The residue was dissolved in 50 ml of methanol, the catalyst was filtered off and washed with methanol, and the filtrate was concentrated, whereafter the distilled solvent was replaced with a fresh one. Finally the solution was reduced to a small volume, and the radioactivity was measured on an aliquot. There was a total activity of 10 ± 0.5 C in the mixture.

The above mixture was divided into three portions. A typical run using one portion of the material is described below. One portion was taken down to dryness and transferred to 1 g of Celite 545 with a small quantity of methanol. The Celite was dried with a gentle stream of nitrogen, 0.5 ml of the stationary phase was added, and the mixture was packed on top of a 60 cm-long column. The column contained 28 g of Celite 545 and 14 ml of the stationary phase. The mobile phase was benzene–Skellysolve C (1:1 v/v) and stationary phase was methanol–water (4:1 v/v). The mobile phase was run through the column, and 5-ml fractions were collected with an automatic fraction collector. From every fraction 5 μl was taken out and added to 10 ml of a specially prepared quenched scintillation liquid in a glass vial. The quenched scintillation liquid contained one part of the usual scintillation liquid

[4 g of diphenyloxazole and 100 mg of 1,4-bis-(2,5-phenyloxazolyl)ben-zene, per liter of toluene] and 10 parts of absolute ethanol. The vials were counted in a Tri-Carb liquid scintillation counter at a high voltage setting giving an efficiency of 16.4% for tritium. In this quenched solution 1 μC of tritium gave about 1000 cpm. A solution of 20 μC progesterone-7-^{3}H in the quenched scintillation liquid was used as standard.

The first few fractions from the solvent front did not show any ultraviolet absorbance at 240 mμ, characteristic of Δ^4-3-ketones, and were considered to contain the saturated 3-ketones of aldosterone-21-acetate-1,2-^{3}H. Aliquots from the second peak fractions did not give any alkali fluorescence and contained Δ^1-5α-aldosterone-21-acetate-4,5-^{3}H. The first and second peak fractions were not further investigated. The third peak comprising fractions 19, 20, 21, and 22 were pooled and identified as aldosterone-21-acetate-1,2-^{3}H by paper chromatographic comparison with an authentic sample in a Bush B$_2$ system [mobile phase toluene–Skellysolve C (2:1 v/v); stationary phase methanol–water (4:1 v/v)]. It gave a color with the blue tetrazolium reagent and alkali fluorescence on heating. The radiochemical purity of the substance was tested by running about 1 μC of the radioactive material alone with carrier aldosterone-21-monoacetate in the Bush B$_2$ system. Carrier aldosterone-21-acetate was added and scanned in a Vanguard automatic windowless flow chromatogram scanner (Model 800 autoscanner, efficiency about 3%). It showed a single sharp peak corresponding to the marked spot due to aldosterone-21-monoacetate, and thus was found to be pure.

In order to clean the column before reuse, 500 ml of the mobile phase was allowed to run through, and 1 g of Celite was removed from the top. The second batch of the aldosterone-21-monoacetate-1,2-^{3}H was put on the column as described above. From every three runs, four peak fractions containing the 1,2-^{3}H-aldosterone-21-monoacetate were pooled, and the total activity was about 1500 mC. The specific activity of the substance was determined and found to be 100 μC/μg.

Hydrolysis of the Aldosterone-21-monoacetate-1,2-^{3}H. (i) Enzymatic hydrolysis. The solution of 10 ml of 0.05 M monohydrogen sodium orthophosphate was prepared, and the pH was adjusted to 7.0 with a few drops of acetic acid. Then 40 mg of wheat germ lipase (Worthington Biochemical Corp., Freehold, New Jersey) was dissolved in that solution and warmed to 37° in a constant temperature oven. The solution of 200 mC of aldosterone-21-monoacetate-1,2-^{3}H in 0.4 ml of propylene glycol was added to the enzyme solution in four portions with approximate 0.5-hour intervals between additions. Incubation was continued for a total period of 4 hours. The incubation mixture was extracted with 4 $\times$ 15 ml of methylene chloride. The combined solvent extract was washed with 2 ml of water and taken down to dryness. It was further purified using a partition column made up with 28 g of Celite 545 in 14 ml of the stationary phase.

The mobile phase (benzene) of the solvent system was equilibrated against the stationary phase [methanol–water (1:1 v/v)]. Fractions of 5 ml were collected and assayed for tritium using 5 μl from every fraction dissolved in the quenched scintillation liquid. The peak fractions 9, 10, and 11 containing aldosterone-1,2-^{3}H were pooled and their identity with standard aldosterone confirmed by paper chromatography, blue tetrazolium reaction, and soda fluorescence. The radiochemical purity of free compound and diacetate, prepared by acetylation with acetic anhydride and pyridine was determined as follows. Chromatography of a mixture of 1 μC of the aldosterone-1,2-^{3}H and inert aldosterone (2 μg) in a Bush B_5 system gave a spot due to inert aldosterone, which was marked by pencil as observed in the ultraviolet light, and a strip (1.25 inch wide) of the chromatogram was scanned in a Vanguard automatic chromatogram scanner. It showed only one peak corresponding to the inert aldosterone; yield 100 mC. The specific activity of the aldosterone-1,2-^{3}H was found to be 100 μC/μg.

A mixture of about 1 μC of the aldosterone-1,2-^{3}H and 3 μg of inert aldosterone was acetylated with 0.15 ml of acetic anhydride and 0.3 ml of pyridine. The diacetate was run in the Bush B_3 system and scanned in a chromatogram scanner. Again only one sharp peak corresponding to the aldosterone diacetate was observed, thereby establishing the radiochemical purity of the aldosterone-1,2-^{3}H.

(ii) ALKALINE HYDROLYSIS. Aldosterone 21-monoacetate-1,2-^{3}H (220 mC) was hydrolyzed according to the method of Simpson et al.[37] in a sealed tube. The reaction mixture was extracted with 4 × 15 ml methylene chloride; the total extract was washed with 2 ml of water and evaporated to dryness. It was purified by column chromatography, and the fractions containing aldosterone-1,2-^{3}H were identified. The radiochemical purity[38] of the product was established as described above in the case of aldosterone-1,2-^{3}H obtained from the enzymatic hydrolysis; yield 66 mC. The specific activity was found to be 98 μC/μg.

In a similar fashion, using the 1-dehydro analog as starting material, the following steroids were prepared in our laboratories:

cholestenone-1,2-^{3}H
cortisone-1,2-^{3}H
21-hydroxypregn-4-ene-3,11,20-trione-1,2-^{3}H

[37] S. A. Simpson, J. F. Tait, A. Wettstein, R. Neher, J. Von Euw, O. Schindler, and T. Reichstein, Helv. Chim. Acta 37, 1163 (1954).
[38] It should be noted that no 17-isoaldosterone-1,2-^{3}H could be detected (chromatography of a mixture of authentic aldosterone and its 17-isomer showed feasibility of its separation with a Bush B_5 system on paper, as well as on a Celite column).

17-methyltestosterone-1,2-³H
21-hydroxypregn-4-ene-3,20-dione-1,2-³H
17α,21-dihydroxypregn-4-ene-3,20-dione-1,2-³H
progesterone-1,2-³H
17α-hydroxypregn-4-ene-3,20-dione-1,2-³H
androst-4-ene-3,17-dione-1,2-³H

C. Synthesis of Steroids-6,7-³H and 7-³H

The starting material for the labeled steroids with a 3-keto-Δ^4 grouping can be prepared from 3-keto-Δ^4 steroids by dehydrogenation[39] with chloranil to give $\Delta^{4,6}$-3-ketosteroids. The starting material for the C_{18}-steroids (6-dehydroestrone and 6-dehydroestradiol) can be prepared[3] by bromination, followed by dehydrobromination of 19-norandrost-4-ene-3,17-dione and 19-nortestosterone acetate, respectively. For the preparation of 3β-hydroxy-Δ^5-steroids-7-³H the respective 3β-acetoxy-7α-bromo-Δ^5-steroid serves as starting material. However, the purification of this starting material is very difficult, due to its instability. It is therefore preferable to take the crude mixture, as obtained from the bromination, and to tritiate it in that state. Although this tritiation usually gives a mixture of compounds with varying contents of label at C-4, C-6, and C-7 whereby a minimum of 80% of the label is found at C-7 (see table).[40]

TRITIUM CONTENT OF VARIOUS STEROIDS AT POSITIONS C-3, C-4,
C-6, and C-7

	Percent of total tritium			
Compounds	C-3, C-4, and C-6	C-7	7α	7β
Cholesterol-7α-³H	10.0	86	26	60
Dehydroepiandrosterone-7α-³H				
Lot A	9.4	89	45	44
Lot B	16.0	84	82	2
Testosterone-7α-³H	—	100	80	20
Pregnenolone-7α-³H	13.0	87	42	45

Progesterone-7-³H and Androst-4-ene-3,17-Dione-7-³H (Fig. 15)[41]

Progesterone-6,7-³H. To 43.4 mg (0.139 millimole) of pregna-4,6-diene-3,20-dione,[42] dried in a reaction vessel of about 7 ml capacity by

[39] E. J. Agnello and G. D. Laubach, *J. Am. Chem. Soc.* **82**, 4293 (1960).
[40] L. E. Geller, A. Sanke, F. Flynn, and N. Silberman, *Atomlight* **62**, 11 (1967).
[41] W. H. Pearlman, *J. Biol. Chem.* **236**, 700 (1961).
[42] A. Wettstein, *Helv. Chim. Acta* **23**, 388 (1940); cf. footnote 39.

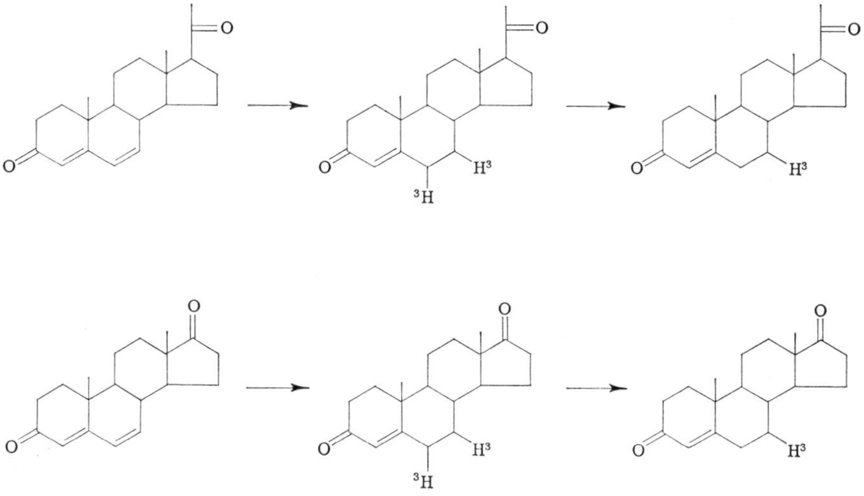

FIG. 15.

evaporation from toluene solution, were added 60 mg of 10% palladium-on-charcoal and 4 ml of toluene, freshly distilled over calcium hydride. The reaction vessel was evacuated after freezing its contents with solid carbon dioxide–ethanol mixture. Tritium gas (1.77 C in 0.76 ml) and hydrogen (3.39 ml) were introduced into the system at 24° with a Toepler pump, which also served as a manometer. The gas pressure in the system was adjusted to atmospheric by introducing about 5.4 ml of nitrogen dried over magnesium perchlorate; this permitted direct measurement of the tritium-hydrogen uptake. The reaction mixture, after thawing to the equilibration temperature (24°) was agitated with a magnetic stirrer; during the first 2½ hours, 3.57 ml of gas were consumed, and after an additional 16 hours, 0.60 ml. Since the amount of catalyst employed consumed 0.77 ml of gas as determined in blank experiments, 3.40 ml (0.139 millimole) of tritium-hydrogen apparently were taken up by the steroid. The catalyst was removed from the reaction mixture by centrifugation and washed well with toluene. The toluene extract on evaporation gave 42.0 mg of a white crystalline product (radioassay 355 mC) which contained 29.6 mg of progesterone, estimated by measuring the optical density at 240 mμ in ethanol, and less than 0.07 mg of $\Delta^{4,6}$-pregnadiene-3,20-dione, estimated similarly at 284 mμ. This product was purified by reverse phase partition column chromatography on siliconized Celite with the solvent system toluene–70% methanol in a manner essentially that described in detail by Contractor and Pearlman.[43]

[43] S. F. Contractor and W. H. Pearlman, *Biochem. J.* **76**, 36 (1960).

Progesterone-7-³H. To a mixture of 27.3 mg of progesterone-6,7-³H above and 24.6 mg of progesterone (total 51.9 mg, specific activity, 2.39 mC/mg), were added 15 ml of cold 5% KOH in 90% methanol. The reaction mixture was refluxed under nitrogen for 2 hours, chilled, acidified to Congo red with 2 N HCl, and diluted with 50 ml of water. Extraction was made once with 50 ml, and twice with 25 ml, of ether; the ether extracts were combined and washed with 10 ml of 1% NaOH and then four times with 10 ml of water. Evaporation of the ether extract gave 50.7 mg of a yellow semicrystalline product. It was purified by chromatography on aluminum oxide as described above except that twice as much of the aluminum oxide and eluting solvents were employed. There was thus obtained 41.3 mg of white crystalline material which on repeated crystallization from ether–petroleum ether furnished 19.5 mg of progesterone-7-³H, m.p. 129–130°, λ_{max} (ethanol) 240 mμ, ϵ 16,800, specific activity, 1.67 mC/mg. Thus, about 70% of the radioactivity in the steroid molecule was retained after alkali treatment.

Δ⁴-Androstene-3,17-dione-6,7-³H. The toluene solution of 57.5 mg (0.202 millimole) of androsta-4,6-diene-3,17-dione[44] was treated under the conditions described above with 5.90 ml of tritium-hydrogen containing 2.58 C of tritium in the presence of 82 mg of 10% palladium-on-charcoal at 24°; 5.64 ml of gas was consumed over a period of 7 hours, of which 4.59 ml (0.188 millimole) was consumed by the steroid, if allowance is made for the catalyst. The reaction mixture, after removal of the catalyst, furnished 54.5 mg of a white crystalline product, radioassay 340 mC, which was estimated to contain 44.5 mg of Δ⁴-androstene-3,17-dione and 7.2 mg of Δ⁴,⁶-androstadiene-3,17-dione according to optical density measurements made at 240 mμ and 282 mμ, respectively, in ethanol solution. Of this reaction product, 53.8 mg were subjected to reverse-phase partition column chromatography on 25 g of siliconized Celite with the solvent system toluene–70% methanol, under conditions identical with those described above. In this solvent system, values of K for Δ⁴-androstene-3,17-dione, 5α-androstane-3,17-dione, and 5β-androstane-3,17-dione are 8.1, 17, and 17, respectively, hence permitting a complete separation of the Δ⁴-dione from the small amount of epimeric saturated diones which are presumably formed as end products in the tritiation. The optical density at 240 mμ and at 282 mμ in ethanol is plotted against the eluate fraction number. A slight contamination of Δ⁴-androstene-3,17-dione with unreacted starting material is apparent. The specific activity of the peak fraction 37 was 6.03 mC per milligram of estimated Δ⁴-androstene-3,17-dione. Eluate fractions 32–40 were

44 L. Ruzicka and W. Bosshard, *Helv. Chim. Acta* **20**, 328 (1937); cf. C. Djerassi, G. Rosenkranz, J. Romo, S. Kaufmann, and J. Pataki, *J. Am. Chem. Soc.* **72**, 4534 (1950).

pooled, and the residue (43.3 mg) was estimated to contain 35.9 mg of Δ^4-androstene-3,17-dione-6,7-^{3}H and 4.1 mg of $\Delta^{4,6}$-androstadiene-3,17-dione.

Δ^4-Androstene-3,17-dione-7-^{3}H. The eluate residue above containing Δ^4-androstene-3,17-dione-6,7-^{3}H was treated with methanolic KOH in a manner identical with that employed in removing labile tritium from progesterone-6,7-^{3}H. There was thus recovered 41.6 mg of a slightly yellowish crystalline product which was subsequently purified by reverse-phase partition column chromatography with the solvent system toluene–70% methanol, exactly as described above, and with similar results. The specific activity of the peak fraction 32 was 5.00 mC per milligram of estimated Δ^4-androstene-3,17-dione. Eluate fractions 26–37 were pooled to furnish 36.4 mg of a crystalline residue which on repeated crystallization from acetone–heptane gave 28.8 mg of Δ^4-androstene-3,17-dione-7-^{3}H, m.p. 175–176°, λ_{max}(ethanol) 240 mμ, ϵ 14,800, specific activity, 4.05 mC/mg. This preparation appeared to be free from contamination with $\Delta^{4,6}$-androstadiene-3,17-dione according to ultraviolet analysis. A preponderant amount of the radioactivity of the impure Δ^4-androstene-3,17-dione-6,7-^{3}H was thus retained after alkali treatment.

Estrone-6,7-^{3}H and Estradiol-6,7-^{3}H (Fig. 16)[45]

Estrone-6,7-^{3}H. Into a 25-ml three-neck flask (one neck having a break seal, one connected to the vacuum line, and a third serving for

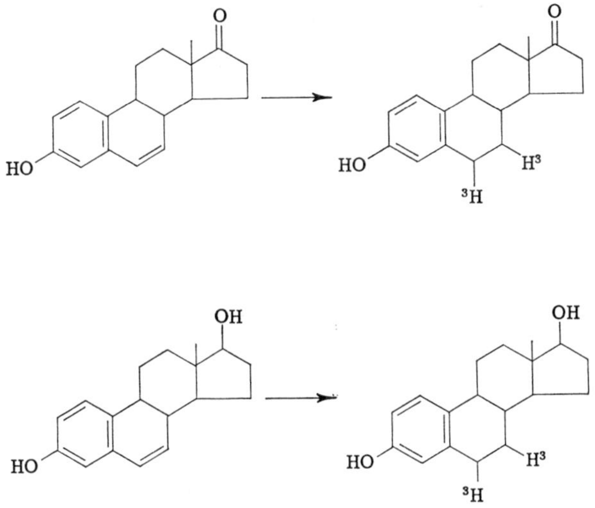

Fig. 16.

[45] Compare W. H. Pearlman and M. R. J. Pearlman, *J. Am. Chem. Soc.* **72**, 5781, (1950); M. Uskoković and M. Gut, *J. Org. Chem.* **22**, 996 (1957); and V. J. O'Donnell and W. H. Pearlman, *Biochem. J.* **69**, 38P (1958).

introduction of the starting material, the solvent, and the reagent) was put a solution of 68 mg of Δ^6-estrone[46] in 8 ml of dioxane and 50 mg of 10% palladium-on-charcoal. Then one neck was closed off; the flask was connected to the vacuum line, degassed, and evacuated; and 6 ml of carrier-free tritium ($\simeq$15 C) was transferred by means of a Toepler pump. Then the flask was sealed off, removed from the vacuum line, and shaken at 25° for 6 hours. The flask was connected to the vacuum line through the break seal, the break seal was broken, traces of tritium were toeplered out, and the solvent was pumped off. The residue was dissolved in 50 ml of a 1:1 mixture of acetone–methanol, the catalyst was filtered off and washed with methanol, and the filtrate was evaporated to dryness and immediately chromatographed on a Bush B$_2$ Celite partition column. From the eluates there was obtained 57 mg of estrone-6,7-^{3}H, with a specific activity of 155 mC/mg.

Estradiol-6,7-^{3}H. The tritiation of Δ^6-estradiol[45] was carried out as described for the tritiation of Δ^6-estrone to estrone-6,7-^{3}H. Similar yields and slightly lower specific activities were obtained in this preparation.

Synthesis of 17β-Hydroxy-17α-ethynylestr-5(10)-en-3-one-6,7-^{3}H, 17β-Hydroxy-17α-ethynylestr-4-en-3-one-6,7-^{3}H, and 17α-Ethynylestr-4-ene-3β,17β-diol Diaceate-6,7-^{3}H from Estradiol-17β-6,7-^{3}H (Fig. 17)[47]

Estradiol 3-Methyl Ether-6,7-^{3}H. To a solution of 200 mg of estradiol-6,7-^{3}H in 16 ml of absolute ethanol were added 300 mg of anhydrous potassium carbonate and 1.0 ml of methyl iodide. The reaction mixture was heated at reflux temperature under nitrogen atmosphere for 5 hours; after 2.5 hours of heating, an additional 1 ml of methyl iodide was added. After the 5-hour heat treatment, the mixture was cooled to room temperature and water and methylene chloride were added. The water layer was separated and the methylene chloride solution was washed with a 1 N potassium hydroxide solution, then washed to neutrality with water and finally with brine. The solution was dried over sodium sulfate, filtered, and evaporated to dryness *in vacuo*. The residual oil crystallized on standing, yielding about 197 mg of estradiol-3-methyl ether-6,7-^{3}H.

The 1 N potassium hydroxide wash was neutralized with acetic acid and extracted with methylene chloride. The extract was washed with

[46] S. Kaufmann, J. Pataki, G. Rosenkranz, J. Romo, and C. Djerassi, *J. Am. Chem. Soc.* **72**, 4531 (1950).

[47] This sequence of reactions has been submitted by Mr. R. T. Nicholson, Searle Chemicals, Inc. and has been carried out with isotope by Dr. K. I. H. Williams, Worcester Foundation for Experimental Biology.

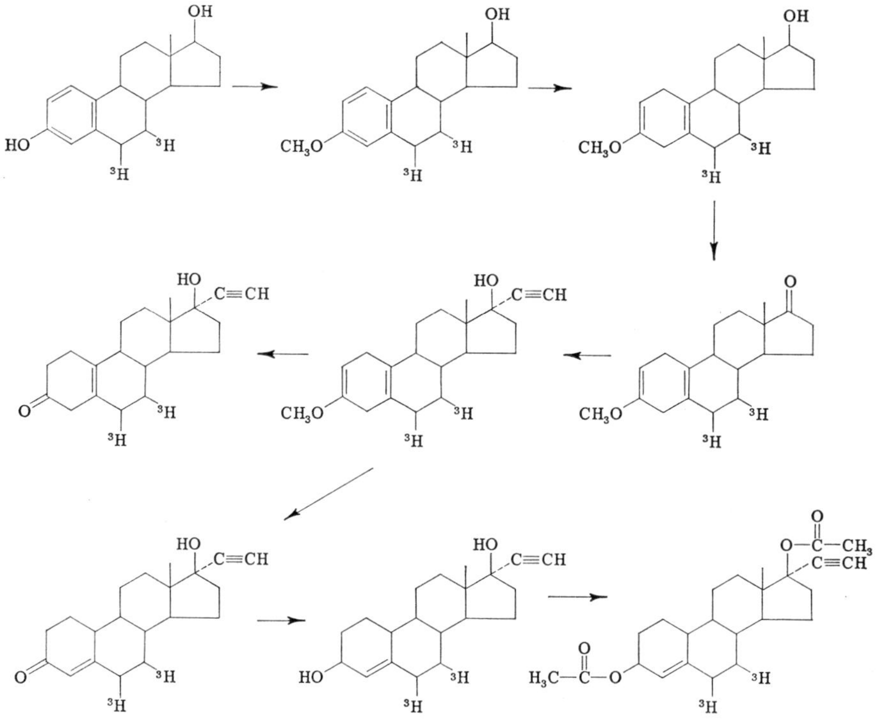

FIG. 17.

water and dried over sodium sulfate. Filtration and evaporation produced 7 mg of crystalline estradiol-6,7-³H, m.p. 175–178°.

3-Methoxyestra-2,5(10)-dien-17β-ol-6,7-³H. To a 50-ml three-neck flask fitted with a dry ice condenser and previously flushed with nitrogen and protected from moisture were added 197 mg of estradiol 3-methyl ether, 5.75 ml of dry tetrahydrofuran, and 3.85 ml of isopropanol. The solution was stirred magnetically and cooled with a dry ice–acetone bath while 15 ml of ammonia was distilled into the mixture. Small pieces from 0.57 g sodium metal, freshly cut and washed in toluene and pentane, were added rapidly. After the addition, the blue color lasted for between 45 and 90 minutes. When the mixture became milky white, 3.85 ml of methanol was added dropwise and the ammonia was allowed to evaporate with the aid of a jet of nitrogen. Water was added, and the mixture was extracted with benzene. The benzene extract was washed to neutrality with water and dried over sodium sulfate. Filtration and evaporation to dryness yielded 3-methoxyestra-2,5(10)-dien-17β-ol-6,7-³H as a crude crystalline residue weighing about 197 mg, m.p. 114–118°.

3-Methoxyestra-2,5(10)-dien-17-one-6,7-³H. A solution of 191 mg of

17β-hydroxy-3-methoxyestra-2,5(10)-diene-6,7-^{3}H in 25 ml of toluene and 0.90 ml of freshly distilled cyclohexanone was heated at reflux under a nitrogen atmosphere, and about 5 ml of toluene was removed by distillation in order to dry the system. Then 181.4 mg of aluminum isopropoxide in 4 ml of dry toluene was added to the refluxing mixture, and the whole was heated at reflux for 1 hour. The mixture was cooled to about 50°, and 0.19 g of potassium sodium tartrate in 9 ml water was added. The mixture was steam distilled until the distillate was clear (about 20–30 minutes). The resulting suspended solid was chilled in an ice bath, removed by filtration, and washed with water. The solid was air-dried briefly, and then dissolved in tetrahydrofuran and filtered to remove inorganic salts. Evaporation under nitrogen and final drying *in vacuo* yielded 181 mg of crystalline 3-methoxyestra-2,5(10)-dien-17-one-6,7-^{3}H, m.p. 122–138°.

17α-Ethynyl-3-methoxyestra-2,5(10)-dien-17β-ol-6,7-[3]H. Small pieces from 600 mg of potassium metal, cut under mineral oil and washed quickly in toluene and weighed into a beaker containing *t*-amyl alcohol, were added rapidly (under a nitrogen atmosphere) to vigorously stirred and refluxing *t*-amyl alcohol (10 ml). Complete solution was achieved in 30 minutes to 1 hour. The mixture was cooled to 0–5°, 10 ml dry ether was added, and acetylene was bubbled through a water scrubber, a blank, two sulfuric acid scrubbers, a blank, and then over the surface of the stirred ether-potassium amylate mixture for 45 minutes. 3-Methoxyestra-2,5(10)-dien-17-one-6,7-[3]H (176.5 mg) was added, with stirring, in one portion as a solid, and acetylene addition was continued for 3 more hours. The mixture was allowed to stand in a refrigerator overnight. After addition of 5–10 ml of water, the mixture was steam distilled to remove the *t*-amyl alcohol. The pot residue was extracted with ethyl acetate, washed with 5% aqueous sodium bicarbonate, water, and brine and dried over sodium sulfate. The solution was filtered and divided into two parts. One part was converted to 17β-hydroxy-17α-ethynylestr-5-en-3-one-6,7-^{3}H; the second part was carried on three steps to 17α-ethynylestr-4-ene-3β,17β-diol-6,7-[3]H. The total weight of the two parts was 163 mg, m.p. 158–176°.

17α-Ethynyl-17β-hydroxyestr-5(10)-en-3-one-6,7-[3]H. To the solution of 81 mg of 17α-ethynyl-3-methoxyestra-2,5(10)-dien-17β-ol-6,7-^{3}H in 9 ml of acetic acid at 25° under nitrogen was added 1 ml of water. The mixture was stirred briefly and allowed to stand for 2 hours. Water was then added and the mixture was extracted with ethyl acetate. The extract was washed with 5% aqueous sodium bicarbonate, water, and brine; dried over sodium sulfate; filtered; and evaporated to dryness. The residual oil weighed 70 mg. The product was purified by column chromatography using 15 g of Davison 950 silica gel (60–200 mesh) in a column with 10 mm i.d. The product was obtained with 5% ethyl acetate in benzene, 41 mg, m.p. 165–175°.

17α-Ethynyl-17β-hydroxyestr-4-en-3-one-6,7-³H. To a solution of 82 mg of 17α-ethynyl-3-methoxyestra-2,5(10)-dien-17β-ol-6,7-³H in 7.2 ml methanol in a nitrogen atmosphere were added 0.48 ml of water and 0.24 ml of concentrated hydrochloric acid. The mixture was stirred at room temperature for 2 hours, then diluted with water and extracted with ethyl acetate. The extract was washed with 5% aqueous sodium bicarbonate, water, and brine. After drying over sodium sulfate, filtering, and evaporating to dryness, the crude crystalline residue of 17α-ethynyl-17β-hydroxyestr-4-en-3-one-6,7-³H weighed 75 mg, m.p. 190–204°. Purification by column chromatography, using the same conditions as for the norethynodrel column, gave 51 mg of 17α-ethynyl-17β-hydroxyestr-4-en-3-one, m.p. 205–210° which came off with 10% ethyl acetate in benzene.

17α-Ethynylestr-4-ene-3β,17β-diol-6,7-³H. A solution of 150 mg of lithium aluminum tri-*t*-butoxyhydride in 6 ml of tetrahydrofuran (THF, previously distilled from Grignard reagent) was cooled to 0–5° while 51 mg of 17α-ethynyl-17β-hydroxyestr-4-en-3-one-6,7-³H in 3 ml of dry, purified tetrahydrofuran was added dropwise at a rapid rate with stirring. Stirring was continued for 3½ hours, the ice bath being allowed to melt and the reaction mixture to warm to room temperature. The reaction mixture was then poured into a stirred mixture of 90 ml of ice and water containing 2.4 ml acetic acid. After 15 minutes of stirring, the crystalline precipitate was removed by filtration and washed with water. The damp cake was dissolved in a small amount of acetone and filtered through a bed of filter-aid to remove inorganic salts. The filtrate was diluted with water to give crystalline 17α-ethynylestr-4-ene-3β,17β-diol-6,7-³H; weight was 37 mg, m.p. 142–143°.

3β,17β-Dihydroxy-17α-ethynylestr-4-ene-6,7-³H Diacetate. A mixture of 37 mg of 17α-ethynylestr-4-ene-3β,17β-diol-6,7-³H in 1 ml of pyridine and 0.5 ml acetic anhydride was heated on a steam bath at 96–97° in a nitrogen atmosphere for 18 hours. The mixture was cooled to room temperature, then poured slowly into 30 ml of ice and water with stirring. The precipitate was removed by filtration 20–30 minutes later. The tacky precipitate was dissolved in ether and washed with 5% aqueous hydrochloric acid, 5% aqueous sodium bicarbonate, water, and brine. After drying over sodium sulfate, filtering and evaporating to dryness, the oily residue was crystallized from 2 ml methanol by adding water until the supernatant remained (ca. 1 ml) turbid. Filtration of the crystals gave 33 mg of 17α-ethynylestr-4-ene-3β,17β-diol-6,7-³H diacetate, m.p. 126–130°.

Pregnenolone-7-³H and Progesterone-7-³H (Fig. 18)[48]

7-Bromo-3β-hydroxypregn-5-en-20-one Acetate. This compound was prepared exactly as described[49] and the crude reaction product was washed ten times with small amounts of petroleum ether, dried, and used within 18 hours.

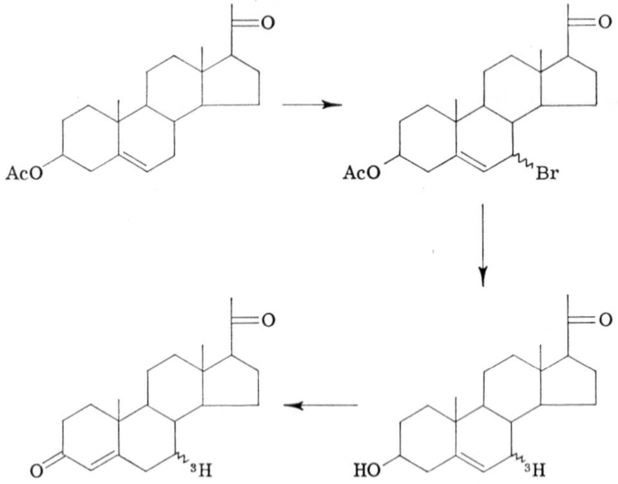

FIG. 18.

3β-Hydroxypregn-5-en-20-one-7-³H. To a 50-ml two-neck flask, one neck equipped with a break-seal, containing a suspension of 1.0 g of 5% palladium-on-calcium carbonate in 11 ml of ethyl acetate, 500 mg of 7-bromo-3β-hydroxypregn-5-en-20-one acetate were added, and then 40 ml of tritium (~100 C) was introduced with the help of a Toepler pump while the flask was cooled with liquid nitrogen. The flask was sealed off and then shaken for 1½ hours. It was then connected to the line and frozen; the break-seal was broken and the excess tritium was toeplered out. The mixture was filtered, the solvent was evaporated, and the residue was equilibrated with a 1% methanolic potassium hydroxide solution. The equilibrated product was purified on an absorption-type silica gel column; the fractions with 10% ethyl acetate in benzene furnished 210 mg of pregnenolone-7-³H with an activity of 9.03 C, m.p. 186–190°.

Progesterone-7-³H. 3β-Hydroxypregn-5-en-20-one-7-³H (240 mg, 9.6 C) was oxidized with 10 ml of cyclohexanone and 250 mg of aluminum

[48] M. Gut and M. Uskoković, *Naturwissenschaften* **47**, 40 (1960).
[49] R. Antonucci, S. Bernstein, D. Giancola, and K. J. Sax, *J. Org. Chem.* **16**, 1126 (1951).

isopropoxide under the usual Oppenauer conditions. After the addition
of 2 ml of a saturated sodium sulfate solution, the mixture was left overnight, then anhydrous sodium sulfate was added and the crude product
was chromatographed on a silica gel column. The benzene eluates contained practically all the cyclohexanone and cyclohexanol of the reaction mixture. The eluate with 10% ethyl acetate in benzene furnished
191 mg of crude progesterone which was rechromatographed on a Celite
column with the solvent system methanol–water (4:1, v/v) petroleum
ether. The fractions 7–18 furnished **174** mg of progesterone-7-³H, m.p.
119–121°, and an ultraviolet maximum $\epsilon_{240\,m\mu}$(EtOH) 16,000. The specific
activity was 31 mC/mg. The fractions 1–7 contained less than 5% of the
total radioactivity of progesterone.[50]

In similar fashion the following steroids were prepared in our laboratories:

3β-hydroxyandrost-5-en-17-one-7-³H
3β,17β-dihydroxypregn-5-en-20-one-7-³H
3β,21-dihydroxypregn-5-en-20-one-7-³H
3β,17α,21-trihydroxypregn-5-en-20-one-7-³H
cholesterol-7-³H
17α,20α-dihydroxycholesterol-7-³H
17α,20β-dihydroxycholesterol-7-³H
20α,22R-dihydroxycholesterol-7-³H

D. Synthesis of Steroids-19-³H (Fig. 19)

Androst-4-ene-3,17-dione-19-³H.[51] To a solution of **27** mg of 5β,19-
cycloandrostane-3,17-dione[52] in 0.5 ml of dioxane was added by distillation 4 mg of tritiated water (13 C) and 0.05 ml of concentrated hydro-

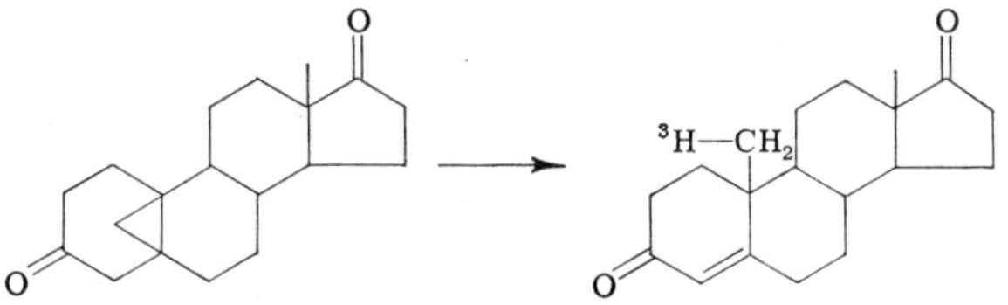

FIG. 19.

[50] Therefore, if the Oppenauer oxidation product had contained any 5α-pregnane-
3,20-dione (arising from tritium addition on the Δ⁵-double bond), it could not have
amounted to more than 5%.
[51] S. Rakhit and M. Gut, *J. Am. Chem. Soc.* **86**, 1432 (1964); compare C. Djerassi
and M. A. Kielczewski, *Steroids* **2**, 125 (1963).
[52] J. J. Bonet, H. Wehrli, and K. Schaffner, *Helv. Chim. Acta* **45**, 2615 (1962);
L. H. Knox, E. Velarde, and A. D. Cross, *J. Am. Chem. Soc.* **85**, 2533 (1963).

chloric acid. The flask was sealed, then the solution was heated on a steam bath for 24 hours and kept an additional 24 hours at 25°. The break-seal of the flask was then opened, the solvents were distilled off, and the residue was equilibrated with 2.5% potassium hydroxide in methanol–water (1:2 v/v).

After addition of water, the mixture was extracted with methylene chloride, and the extract was dried over anhydrous sodium sulfate, and evaporated. The crude mixture was finally chromatographed on a Celite partition column using the Bush A system, which yielded 11 mg of androst-4-ene-3,17-dione-19-³H with a specific activity of 20 μC/μg. The radiochemical purity was tested by chromatographing about 1 μC of the material with 5 μg of authentic androst-4-ene-3,17-dione on paper on the Bush A system. Scanning in a Vanguard automatic windowless flow chromatogram scanner (Model 800 autoscanner) gave a single sharp peak corresponding to the marked spot due to the carrier and found by inspection with the ultraviolet lamp.

In a second experiment, 20 mg of 5β,19-cycloandrostane-3,17-dione was dissolved in a flask containing 1 ml of dioxane and 0.1 mg of sodium hydroxide. The flask was connected to the vacuum line, degassed, and evacuated, and 4 mg of tritiated water (13 C) was distilled into it. Then the flask was sealed off, removed from the line, heated on a steam bath for 6 hours, and left at room temperature for 18 hours. The break-seal was opened, the solvents were distilled off on the vacuum line, and water was added to the residue and worked up exactly as described above. After chromatography, 12 mg of androst-4-ene-3,17-dione-19-³H with a specific activity of 30 mC/mg was isolated.

E. Synthesis of a Steroid-5α,6α,7α-³H (Fig. 20)

5α-Androstane-3,17-dione-5α,6α,7α-³H.[53] 3β-Hydroxyandrost-5-ene-3, 17-dione in dioxane was reduced with carrier-free tritium and 10% palladium-on-charcoal. The product was repeatedly dissolved in methanol and concentrated to remove labile tritium and was then chromatographed on paper in ligroin–propylene glycol for 10 hours. A radioscan showed that approximately 90% of the radioactivity had the mobility of 3β-hydroxy-5α-androstan-17-one. This zone was eluted and used in the following experiment.

[53] H. J. Brodie, S. Baba, M. Gut, and M. Hayano, *Steroids* 6, 659 (1965). This example is offered only in order to point out that the saturation of an isolated double bond does not necessarily incorporate tritium only at the double bond but that other positions might be implicated. In all these cases where the position and configuration of the label are of importance, a thorough analysis is required. These considerations hold also for the steroids labeled by the tritiation of 7-bromosteroids (see Table, page 334).

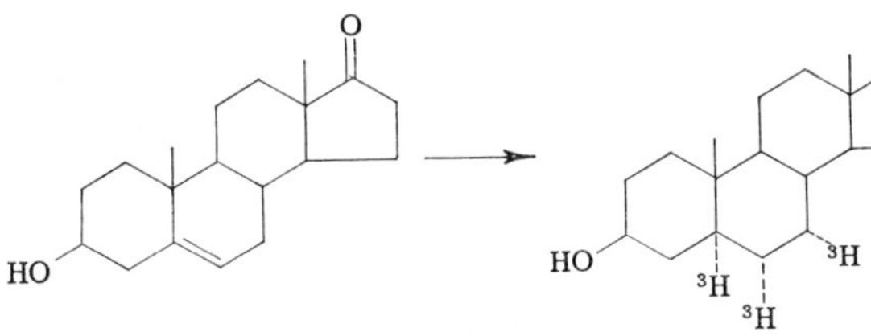

FIG. 20.

A portion of the radioactive material which had the mobility of 3β-hydroxy-5α-androstan-17-one was diluted with 1.965 g of authentic material and was recrystallized from benzene–hexane to constant specific activity (5.57×10^4 cpm/micromole). After the steroid was dissolved in 50 ml of acetone and cooled to 5°, 2 ml of 8 N chromic acid in sulfuric acid was added with stirring after the method of Bowden et al.[54] The mixture was stirred for 30 minutes at 5° and 30 minutes at room temperature, then it was poured onto ice and the aqueous layer was extracted with ether. After drying the ether solution over sodium sulfate, filtration and evaporation of the solvent yielded 1.92 g of product. This was recrystallized to constant specific activity (5.53×10^4 cpm/micromole), with benzene–hexane; melting point, infrared spectrum, and mobility on thin-layer plate (benzene–ethyl acetate 9:1, v/v) were identical with authentic 5α-androstanedione. A portion of the product was refluxed with 2% potassium hydroxide in 50% methanol–water for 3 hours in a nitrogen atmosphere. After neutralization with hydrochloric acid and evaporation of methanol under reduced pressure, the aqueous mixture was extracted with ethyl acetate. The ethyl acetate solution was dried with sodium sulfate, and the solvent was removed by evaporation. The residue was recrystallized as above, and the specific activity was determined (5.40×10^4 cpm/micromole).

Appendix

PARTITION CHROMATOGRAPHIC SYSTEMS APPLIED TO STEROIDS[a]

	Stationary phase	Mobile phase	Supporting material
A	80% aqueous methanol	Skellysolve C	Celite
B₃	80% aqueous methanol	Skellysolve C–Benzene 2:1	Celite
B₂	80% aqueous methanol	Toluene–Skellysolve C 2:1	Celite
B₅	80% aqueous methanol	Benzene or toluene	Celite

[a] I. E. Bush, Biochem. J. **50**, 370 (1952).

[54] K. Bowden, I. M. Heilbron, E. R. H. Jones, and B. E. L. Weedon, J. Chem. Soc. **1946**, 39 (1946).

Acknowledgment

Financial support by grant AM-03419 from the National Institute of Arthritis and Metabolic Diseases, National Institutes of Health is gratefully acknowledged.

[9] Synthesis of Labeled Squalene and Squalene 2,3-Oxide

By R. G. NADEAU and R. P. HANZLIK

Introduction

The role of squalene as the acyclic precursor of lanosterol and cholesterol has been established for many years.[1] Recent work[2,3] has shown that squalene 2,3-oxide is an intermediate in the enzymatic cyclization of squalene to lanosterol, and in consequence this aspect of steroid and terpenoid biogenesis has become the center of renewed interest. Improved methods for the chemical synthesis of labeled squalene and squalene 2,3-oxide are therefore described here.[4] The procedures described are for tritium labeling, but could be equally applied to labeling with ^{14}C.

Squalene-4-^{3}H

In general terms, the synthesis starts with thiourea-clathrate purified squalene (I),[9] which is converted to squalene 2,3-oxide (III). The oxide, in three almost quantitative steps, is degraded to squalene trisnor-aldehyde,[10] which is then labeled in the α-position via an acid-catalyzed exchange reaction with tritiated water. Surprisingly, it has been shown by using a deuterium exchange model reaction that no double bond isomeri-

[1] T. T. Tchen and K. Bloch, *J. Biol. Chem.* **226**, 921, 931 (1957). See also T. T. Tchen, Vol. V, p. 489.

[2] J. D. Willett, K. B. Sharpless, K. E. Lord, E. E. van Tamelen, and R. B. Clayton, *J. Biol. Chem.* **242**, 4182 (1967).

[3] P. D. G. Dean, P. R. Ortiz de Montellano, K. Bloch, and E. J. Corey, *J. Biol. Chem.* **242**, 3014 (1967).

[4] The enzymatic method for squalene synthesis (this volume, p. 442), although valuable for small-scale preparations, is expensive and not convenient for large applications. Similarly, previously described chemical methods[5-8] lack simplicity and the radiochemical yield is less than satisfactory.

[5] D. W. Dicker and M. C. Whiting, *J. Chem. Soc.* p. 1994 (1958).

[6] J. W. Cornforth, R. H. Cornforth, and K. K. Mathew, *J. Chem. Soc.* p. 2539 (1959).

[7] R. E. Wolff and L. Pichat, *Compt. Rend. Acad. Sci.* **246**, 1868 (1958).

[8] R. Maudgal, T. T. Tchen, and K. Bloch, *J. Am. Chem. Soc.* **80**, 2589 (1958).

[9] N. Nicolaides and F. Laves, *J. Am. Chem. Soc.* **76**, 2596 (1954).

[10] E. E. van Tamelen, K. B. Sharpless, J. D. Willett, R. B. Clayton, and A. L. Burlingame, *J. Am. Chem. Soc.* **89**, 3920 (1967).

zation or other aberrations of the aldehyde molecule occur during the exchange reaction. The aldehyde requires no further purification and stores well in dilute hexane solution under nitrogen.

Tritium-labeled squalene results from subjecting squalene trisnor-aldehyde-α-³H to a Wittig reaction using isopropyltriphenylphosphonium iodide.

Outline

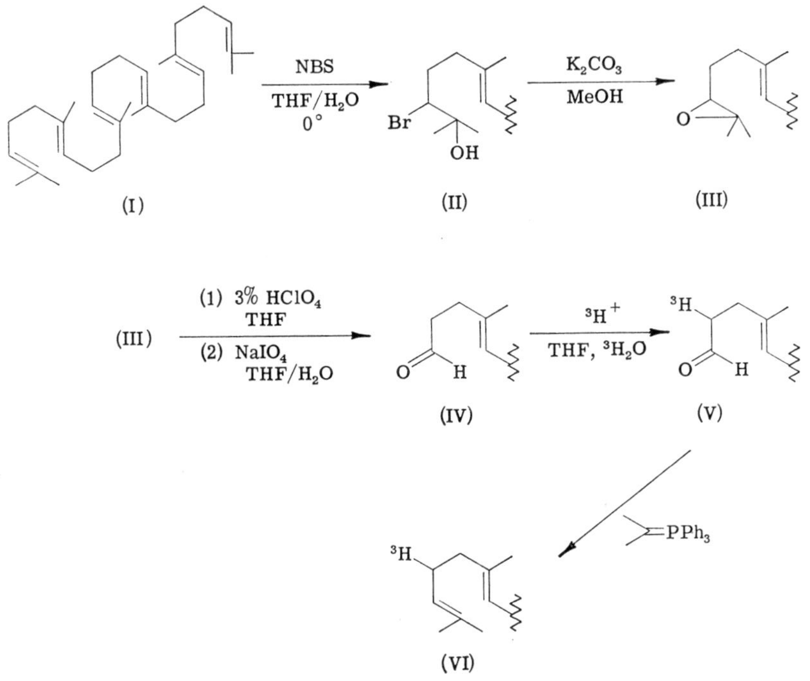

Procedure

Step 1. 2-Hydroxy-3-bromosqualene. Pure squalene (I), 22.5 g (0.055 mole) is dissolved in 1500 ml of tetrahydrofuran (THF). The solution is stirred under nitrogen, cooled to 0°, and water is added until the solution becomes cloudy. Then a small amount of THF is added to clear the solution. Next, 10.8 g (0.061 mole) of N-bromosuccinimide is added in small portions over a period of 10 minutes. Stirring at 0° is continued for 20 more minutes.

The crude product is isolated by evaporation of about half the solvent, addition of water and extraction with hexane. This gives 27.2 g of crude reaction mixture.

Step 2. Squalene 2,3-Oxide (III). The crude bromohydrin (II) described above, 25.0 g, is stirred for 3 hours at room temperature with 10.0 g of K_2CO_3 in 200 ml of methanol under nitrogen. The crude oxide, 20.2 g, is isolated by addition of water and extraction with hexane.

Pure squalene 2,3-oxide is obtained by column chromatography using 1000 g of deactivated (10% H_2O) silica gel. The column is initially eluted with hexane until all the nonpolar impurities are collected. Then the solvent polarity is gradually increased with anhydrous ether, beginning with 2% ether–hexane. The oxide is readily eluted with 7% ether–hexane. The oxide fractions are combined to give a total weight of 7.46 g of pure compound. Homogeneity is confirmed by carbon-hydrogen analysis, infrared and nuclear magnetic resonance spectra, and thin-layer chromatography (TLC). The overall yield, based on squalene, is 32%.

Step 3. Trisnorsqualene Aldehyde (IV). The oxide, 7.00 g, is dissolved in 70 ml of ethylene glycol dimethyl ether, and to this solution is added 7.0 ml of 3% $HClO_4$. After 2 hours of stirring at room temperature, TLC (silica gel, 20% EtOAc–hexane) shows complete disappearance of oxide (R_f 0.65) and appearance of a new compound, glycol (R_f 0.10). Routine work-up gives 7.05 g squalene-2,3-glycol which is homogeneous on TLC. The yield is 97%.

The above glycol, 7.00 g, is then dissolved in 100 ml of THF and water is added to the saturation point. While the solution is being stirred at room temperature under nitrogen, 10.3 g (3:1 molar excess) $NaIO_4$ is added. The reaction is monitored by TLC and is complete in 1 hour. The yield of pure aldehyde is 6.21 g (98% of theory).

Step 4. Trisnorsqualene Aldehyde-α-³H (V). A stock solution of $^3H \oplus$ is made by cautiously adding 225 mg of PCl_5 to 1.0 ml of tritiated water (1 C/ml) at ice bath temperature. The solution is ready for use when the PCl_5 is completely dissolved.

To 50 mg of trisnorsqualene aldehyde under nitrogen in 1.0 ml anhydrous THF is added 50 μl $^3H \oplus$ stock solution. The container is well sealed and stored in the dark at room temperature for 48 hours. Hexane (5 ml) is now added to the reaction mixture, precipitated water is removed by pipette, and anhydrous K_2CO_3 is added to the hexane solution to remove traces of water. The hexane is then filtered through K_2CO_3 and evaporated under a stream of nitrogen at 50°. The resulting product is homogeneous on GLC and has a specific activity in the range 50,000–70,000 dpm/μg.

Step 5. Squalene-4-³H (VI). To 2.24 g (5.2 millimoles) of isopropyltriphenyl phosphonium iodide suspended in 25 ml of anhydrous THF at 0° under nitrogen is added 2.26 ml of 2.3 M MeLi. Next a solution of 0.50 g (1.3 millimoles) of trisnorsqualene aldehyde-α-³H (20,000 dpm/μg) in 30 ml of anhydrous THF is added. The mixture is stirred for 2 hours at 0°, then for 6 hours at room temperature.

The reaction mixture is filtered under gravity, and the filtrates are evaporated to small volume. The residual material is chromatographed on a column of grade I alumina from which it is eluted with hexane. In this way is isolated 384 mg (70%) squalene with specific activity of 18,100 dpm/μg.

Conversion of Trisnorsqualene Aldehyde-α-³H (V) to Squalene 2,3-Oxide-4-³H (VII)

Diphenyl sulfide, isopropyl iodide, and silver fluoroborate react to produce silver iodide and diphenyl isopropyl sulfonium fluoroborate, which is an isolable crystalline salt. Treatment of the sulfonium salt with phenyl lithium generates a sulfur ylide which reacts with aldehydes to give epoxides.

Procedure[11]

Step 1. Salt Formation. Diphenyl sulfide (7.45 g, 40.0 millimoles) and isopropyl iodide (5.50 g, 32.4 millimoles) are stirred under nitrogen and cooled in an ice H_2O bath. Carefully dried (48 hours at 68° and 0.01 mm Hg) silver fluoroborate[14] (6.38 g, 32.4 millimoles) is added in small portions, care being taken to exclude moisture. Almost immediately, yellow silver iodide begins to precipitate and the reaction mixture thickens. The ice bath is removed and the mixture is stirred for 2 hours at room temperature under nitrogen.

Methylene chloride is added and the mixture is filtered to remove silver iodide, which is formed quantitatively. The methylene chloride is evaporated *in vacuo* at 35°, and the resulting oil is washed with several portions of dry ether, the ether and excess diphenyl sulfide being decanted away from the insoluble sulfonium salt. The resulting crude salt is recrystallized from methylene chloride–ether solutions until it is pure white. Percentage yield of purified salt (m.p. 117–120°)[15] = 55% based on isopropyl iodide.

---

[11] This synthetic method is based on the work of V. Franzen *et al.*[12] Recently, an alternate synthesis of diphenyl sulfonium isopropylide has appeared.[13] We have not tested this alternate method since it is more complicated, involving sequential generation of the ethylide, methylation, generation of the isopropylide, and reaction with the aldehyde, all *in situ*. Furthermore, our procedure consistently gives respectable (80%) yields of epoxide, and the reaction is very clean with no complications caused by excess base.

[12] V. Franzen, H. Schmidt, and C. Mertz, *Chem. Ber.* 94, 2942 (1961).

[13] E. J. Corey and M. Jautelot, *Tetrahedron Letters* 24, 2325 (1967).

[14] Silver perchlorate (AgClO₄) can be used, but the perchlorate salt decomposes when dried and may be a potential explosion hazard.

[15] Carbon and hydrogen analyses and NMR data consistent with all the proposed structures have been obtained.

Step 2. Generation of the Sulfur Ylide and Reaction with Aldehyde V. All glassware must be oven-dried for this reaction. The tetrahydrofuran for use as solvent is refluxed over lithium aluminum hydride and is then distilled into the reaction vessel under a stream of dry nitrogen. The reaction is conveniently carried out in a small two-necked flask. One neck is equipped with a nitrogen inlet and outlet and the other first is used for collection of the solvent from the distillation apparatus and afterward is closed with a rubber septum or serum cap through which other reactants can be introduced by injection.[16]

The proper amount of salt, about 2.0–2.5 moles per mole of aldehyde, is suspended in 10 ml of tetrahydrofuran and cooled with stirring under nitrogen in a dry ice–acetone bath. Slightly less than an equivalent amount of phenyl lithium[17] is added by syringe to the suspended salt. The solution becomes yellow indicating formation of the ylide.[18] After 15 minutes there is only a trace of solid material (unreacted salt). The aldehyde (V) is dissolved in tetrahydrofuran to give an approximately 10% solution and is added dropwise from a syringe. The yellow color will lighten (but should not disappear if moisture has been excluded) indicating the presence of excess ylide. After 15 minutes the dry ice bath is removed and stirring is continued. While the reaction mixture warms to room temperature the remaining yellow color disappears. The reaction mixture is extracted with hexane and water. The hexane is dried with sodium sulfate and evaporated, giving a mixture of diphenyl sulfide and epoxide (VII). This mixture is dissolved in methanol and treated with potassium borohydride to reduce any unreacted aldehyde to the alcohol, which separates more cleanly than the aldehyde from the epoxide.[19] The mixture of diphenyl sulfide and epoxide separate by preparative TLC on unactivated silica gel with 5% ethyl acetate in hexane.

The epoxide (VII) from the TLC (usually in 75–85% yield) is purified via its thiourea clathrate once before use in incubations, by adding dropwise a 10% solution of epoxide in benzene to 20–25 volumes of a saturated solution of thiourea in methanol while the mixture is being stirred rapidly. When addition is complete, the solution, which may now contain solid

---

[16] For larger-scale preparations (several hundreds of milligrams) a 100-ml three-neck flask with a 25-ml dropping funnel is convenient. Column chromatography on a 100:1 excess of $SiO_2 \cdot 15\%$ H_2O may be more convenient than TLC.

[17] t-BuOK and MeLi are unsatisfactory bases for generation of the ylide.

[18] Dilute solutions of the ylide are yellow whereas more concentrated solutions are orange colored.

[19] Usually, no alcohol is produced when the ylide reaction mixture is treated with KBH_4 if excess ylide is present; however, trace amounts of several compounds more polar than the epoxide always result from the KBH_4 treatment. They are easily separable from the epoxide by TLC.

particles, is cooled in a refrigerator for 2 hours and then in a freezer
($-15°$) for 2 hours. The precipitate is filtered[20] as rapidly as possible to
prevent warming and is *not* washed. It can be decomposed by distribution
between water and hexane in a separatory funnel, copious quantities of
water being used to wash the hexane. The hexane can be dried briefly
over sodium sulfate and evaporated. Recovery is between 50 and 70% of
the material recovered from TLC.

[20] A 15-ml glass-stoppered centrifuge tube is a very convenient vessel for forming
thiourea clathrates on a 10-mg scale. The tube is centrifuged and the mother
liquors are decanted, leaving the clathrate to be decomposed in the centrifuge tube.

[10] The Preparation and Properties of Steroid Sulfate Esters

By R. Clifton Jenkins and Eugene C. Sandberg

Principle. In 1936, Sobel *et al.*[1] outlined three reactions which have
since formed the basis for most of the chemical methods employed in
the synthesis of sulfurylated steroids. The reactions appear to be equally
applicable to primary and secondary steroid alcohols, and also to the
sulfurylation of phenolic steroids. The active sulfurylating agent in each
instance is a pyridine-sulfur trioxide complex which is fully reactive only
under anhydrous conditions. The reactions pragmatically represent three
different chemical means by which this active agent may be formed.

Joseph *et al.*[2] have recently reported a method for steroid sulfurylation that utilizes sulfamic acid as the sulfurylating agent (Procedure 4).
The major advantage of this technique lies in the fact that anhydrous
conditions are not required.

Methods

Procedure 1

$$ClSO_3H + 3C_5H_5N \rightarrow (C_5H_5N)_2 \cdot SO_3 + C_5H_5N \cdot HCl$$

$$ROH + (C_5H_5N)_2 \cdot SO_3 \xrightarrow{C_5H_5N} ROSO_3^- \, C_5H_5NH^+ + C_5H_5N$$

Freshly distilled chlorosulfonic acid is added drop by drop to a 6-fold
molar excess of dry, freshly distilled pyridine in a glass-stoppered vessel
immersed in an ice bath. Spattering normally occurs with this technique
and, while this can be partially alleviated by introducing the chloro-

[1] A. E. Sobel, I. J. Drekter, and S. Natelson, *J. Biol. Chem.* **115**, 381 (1936).
[2] J. P. Joseph, J. P. Dusza, and S. Bernstein, *Steroids* **7**, 577 (1966).

sulfonic acid beneath the surface of the pyridine, safety precautions are essential regardless of the technique employed.

The resulting gel is liquefied by gentle heating (50–60°), and the steroid, previously dissolved in a minimal quantity of dry, freshly distilled pyridine, is added to a 2- to 4-fold stoichiometric excess of the liquefied pyridine-sulfur trioxide. The reaction is allowed to proceed overnight at room temperature in the closed vessel and is terminated by the addition of a quantity of dilute aqueous ammonium, sodium, or potassium hydroxide sufficient to double the reaction volume and to bring the mixture to pH 8–9. The sulfurylated steroid is extracted into n-butanol (5 volumes $\times$ 2). In general, a yield of 70–80% is to be anticipated.

This has been the most widely applied method for the synthesis of both gross and radioactive tracer quantities of steroid sulfates.[3-14] In addition to its common utilization for the esterification of mono-hydroxylated steroid compounds, the method has also been used to form disulfated compounds[13] and to synthesize certain monoesterified derivatives of dihydroxy steroids as well. For example, in the synthesis of estradiol 17β-monosulfate, Kirdani[9] utilized the pyridine-chlorosulfonic acid method to sulfurylate estradiol 3-benzoate. The benzoate radical was then removed by selective hydrolysis in 5% sodium hydroxide in methanol.

Procedure 2

$$HOSO_3H + 2C_5H_5N \rightarrow HOSO_3H \cdot (C_5H_5N)_2$$

$$ROH + HOSO_3H \cdot (C_5H_5N)_2 \xrightarrow{C_5H_5N, Ac_2O} ROSO_3^- C_5H_5NH^+ + C_5H_5N + H_2O$$

This method of synthesis involves the preparation of pyridine sulfate (pyridine–sulfuric acid complex) and the subsequent utilization of this substance for sulfurylation in the presence of an internal dehydrating

[3] S. Burstein and S. Lieberman, *J. Am. Chem. Soc.* **80**, 5235 (1958).

[4] A. Butenandt and H. Hofstetter, *Z. Physiol. Chem.* **259**, 222 (1939).

[5] H. I. Calvin and S. Lieberman, *Biochemistry* **3**, 259 (1964).

[6] W. R. Dixon, V. Vincent, and N. Kase, *Steroids* **6**, 757 (1965).

[7] G. A. Grant and W. L. Glen, *J. Am. Chem. Soc.* **71**, 2255 (1949).

[8] G. W. Holden, I. Levi, and R. Bromley, *J. Am. Chem. Soc.* **71**, 3844 (1949).

[9] R. Kirdani, *Steroids* **6**, 845 (1965).

[10] J. M. McKenna and J. K. Norymberski, *J. Chem. Soc.* p. 3889 (1957).

[11] J. R. Pasqualini, "Contribution à l'Étude Biochimique des Corticosteroïdes," pp. 71–78. R. Foulon, Paris, 1962.

[12] J. E. Plager, *J. Clin. Invest.* **44**, 1234 (1965).

[13] E. Wallace and N. Silberman, *J. Biol. Chem.* **239**, 2809 (1964).

[14] E. C. Sandberg and R. C. Jenkins, unpublished observations, 1964.

agent. In the absence of the dehydrating agent only a negligible degree of sulfurylation occurs, demonstrating the intense inhibitory effect of water in this and other methods in which pyridine–sulfur trioxide is used as the sulfurylating material.

Concentrated sulfuric acid is added drop by drop, with swirling, to a 3-fold molar excess of dry, freshly distilled pyridine dissolved in 10–20 ml of chloroform. The resulting precipitate is washed several times with chloroform and dried under vacuum. It may then be used directly or following storage under desiccation. Pyridine sulfate is extremely hygroscopic, and due haste must be exercised in weighing and in transferring this material to the reaction vessel.

A 1.0–2.0 M solution of pyridine sulfate in pyridine is prepared. A 10% volume of acetic anhydride is added, and the mixture is stirred for 30 minutes. The steroid, previously dissolved in a minimal quantity of dry, freshly distilled pyridine, is then added to a 3- to 4-fold molar excess of the pyridine sulfate. The mixture is stirred overnight at room temperature. The reaction is terminated by the addition of a sufficient volume of 10% sodium bicarbonate (or dilute sodium or potassium hydroxide) to bring the reaction mixture to pH 8–9. Unreacted steroid and any steroid acetate formed during the reaction are extracted into four volumes of diethyl ether.[15] The aqueous phase is brought to a pH of 10–11 with 2 N sodium or potassium hydroxide and the sulfurylated steroid is extracted into n-butanol (5 volumes $\times$ 2). The yield is generally in the range of 70–80%.

This method has been found to be effective in the sulfurylation of both mono- and dihydroxy C-18 steroid[9,16] and C-19 steroid[14] compounds. Levitz[16] has demonstrated that the concentration of acetic anhydride in this reaction is influential in that a 2-fold molar excess of this material relative to pyridine sulfate (1.0 M pyridine sulfate in pyridine plus 20% acetic anhydride by volume) will yield the steroid acetate as the primary product.

Procedure 3

$$SO_3 + 2C_5H_5N \rightarrow (C_5H_5N)_2{\cdot}SO_3$$

$$ROH + (C_5H_5N)_2{\cdot}SO_3 \xrightarrow[C_6H_6]{C_5H_5N, Ac_2O} ROSO_3^- \, C_5N_5NH^+ + C_5H_5N$$

The pyridine-sulfur trioxide complex utilized in this procedure is pre-

---

[15] Steroid sulfates of relatively low polarity, such as cholesterol sulfate, are moderately soluble in ether. When such compounds are being synthesized, extraction of the reaction mixture with this solvent should be omitted or a solvent of lower polarity should be employed.

[16] M. Levitz, *Steroids* **1**, 117 (1963).

pared by the direct reaction of solid sulfur trioxide with a 3-fold molar excess of pyridine in chloroform. This reaction is rather violent and should be carried out in an ice bath with constant stirring. Equal weights of dried complex and steroid are then refluxed for 20 minutes in a volume of benzene–pyridine–acetic anhydride (10:1:1, v:v:v) sufficient to afford approximately a 7% solution of pyridine–sulfur trioxide (w:v). After cooling, the sulfurylated product is precipitated by adding four volumes of petroleum ether to the reaction mixture. Yields are described as being essentially quantitative.

Although Sobel and Spoerri[17] demonstrated the usefulness of this reaction in the preparation of gross quantities of the sulfate esters of dibromocholesterol, cholesterol, ergosterol, and lanosterol, the method has not often been used in the preparation of steroid sulfates inasmuch as sulfur trioxide is difficult to handle.

A reportedly effective modification of this method involves the use of sulfur trioxide methylamine. Equal weights of steroid and sulfur trioxide methylamine are dissolved in a minimal quantity of pyridine, and the mixture is heated on a steam bath for 10–20 minutes. The resulting gel is evaporated under nitrogen and redissolved in aqueous sodium hydroxide (pH 8–9). The sulfate is recovered by extraction into n-butanol (5 volumes $\times$ 2). We are unaware of data relating to the yield from this procedure and have had no experience with its use.

Procedure 4

$$H_2NSO_3H + C_5H_5N \xrightarrow{\text{heat, } N_2} H_2NSO_3H \cdot C_5H_5N$$

$$ROH + H_2NSO_3H \cdot C_5H_5N \xrightarrow{\text{heat, } N_2, C_5H_5N} ROSO_3^- NH_4^+ + C_5H_5N$$

Pulverized sulfamic acid is mixed with a 5- to 10-fold molar excess of pyridine, and the resulting suspension is vigorously agitated in a nitrogen atmosphere at 85–95°. The steroid is then added to a 3- to 4-fold molar excess of the sulfamic acid and agitation in a nitrogen atmosphere at the above temperature is continued for 1–2 hours. The reaction mixture is cooled and then dissolved in a minimal volume of methanol. An excess of aqueous ammonium hydroxide is added, and the inorganic solids are removed by filtration. The sulfurylated steroid may be extracted into n-butanol (5 volumes $\times$ 2). A lengthier method for the formation and extraction of the sodium or potassium salt of the steroid sulfate has been included by Joseph et al.[2] in their description of this procedure.

These authors have employed the technique to synthesize gross

[17] A. E. Sobel and P. E. Spoerri, *J. Am. Chem. Soc.* **63**, 1259 (1941).

quantities of the sulfate esters of certain C-18, C-19, and C-21 mono-hydroxylated steroid compounds and also to synthesize the disulfate of 17α-ethynylestradiol. Yields are reported to vary from 20 to 65%.

Purification

Methods for the purification of steroid sulfates recovered from the butanol extract of the reaction mixtures include column,[18-21] paper,[5, 22, 23] and thin-layer[24-26] chromatography. Crystallization may be accomplished from such solvent combinations as methanol–diethyl ether, methanol–water, acetone–water, and methanol–acetone.[27]

Properties

Data that relate to a limited number of the physical properties of certain steroid sulfate esters are summarized in the table.

Sulfurylated steroid compounds are readily hydrolyzed in the presence of heat and/or acid. While a number of these compounds appear to be quite stable under ordinary conditions, many undergo spontaneous degeneration if left exposed to air, light, and average room temperatures. To obviate this effect, storage of crystalline material in a vacuum desiccator within an unlighted refrigerator is advised. When dissolved in absolute ethanol and kept under refrigeration, most of these compounds are reasonably stable for months. However, a modest degree of spontaneous hydrolysis (usually less than 5%) should be anticipated. The product of hydrolysis may be readily removed by partitioning the stored material between diethyl ether and water before use. A less polar solvent than ether must be utilized with sulfurylated steroids of relatively low polarity in order to prevent an excessive transfer of the steroid sulfate to the organic phase. For example, hexane has been found to be more appropriate than ether when dealing with cholesterol sulfate.

[18] H. I. Calvin, K. D. Roberts, C. Weiss, L. Bandi, J. J. Cos, and S. Lieberman, *Anal. Biochem.* 15, 426 (1966).
[19] H. I. Calvin, R. L. Vande Wiele, and S. Lieberman, *Biochemistry* 2, 648 (1963).
[20] N. M. Drayer, K. D. Roberts, L. Bandi, and S. Lieberman, *J. Biol. Chem.* 239, PC3112 (1964).
[21] N. M. Drayer and S. Lieberman, *Biochem. Biophys. Res. Commun.* 18, 126 (1965).
[22] W. R. Dixon, J. G. Phillips, and N. Kase, *Steroids* 6, 81 (1965).
[23] J. J. Schneider and M. L. Lewbart, *Recent Progr. Hormone Res.* 15, 201 (1959).
[24] H. Moser, A. B. Moser, and J. C. Orr, *Biochim. Biophys. Acta* 116, 146 (1966).
[25] G. W. Oertel, M. C. Tornero, and K. Groot, *J. Chromatog.* 14, 509 (1964).
[26] F. S. Wusteman, K. S. Dodgson, A. G. Lloyd, F. A. Rose, and N. Tudball, *J. Chromatog.* 16, 334 (1964).
[27] D. W. Killinger and S. Solomon, *J. Clin. Endocrinol. Metab.* 25, 290 (1965).

PHYSICAL PROPERTIES OF CERTAIN STEROID SULFATES

Parent compound	Derivative	Salt[a]	Mol. wt.[b]	Melting point (°C)	Optical rotation (α_D)	Infrared spectrum	References[c]
Androsterone	3α-Sulfate	K+	408.6	184–185	—		3
		Na+	392.5	140	—		a
		Na+	—	212–214	+78° (H₂O)	+	b
Isoandrosterone	3β-Sulfate	K+	408.6	223	—		3
		Na+	392.5	157–160	—		a
Etiocholanolone	3α-Sulfate	Na+	392.5	159–161	—		a
		Na+	—	150–151	—		12
Cholestanol	3β-Sulfate	K+	506.8	234–235(d)	—		10
		C₅H₅NH+	547.8	165–169(d)	+17° (CHCl₃)		10
Dehydroisoandrosterone	3β-Sulfate	K+	406.6	219–223	—		3
		Na+	390.5	169–172	—	+	a, c
		Na+	—	182–186	—		d
		NH₄+	385.5	206–207	+19°	+ +	2, c
		C₅H₅NH+	447.6	194–195	—	+ +	c, e
16α-Hydroxy-dehydroisoandrosterone	3β-Sulfate	K+	422.6		—	+	c
Δ⁵-Androstenediol	3β-Sulfate	Na+	392.5	222–225	—		6
		NH₄+	387.5	228–231	—		13
	3β,17β-Disulfate	NH₄+	—	210–213	—	+	c
		NH₄+	484.6	239–242	—		13
Pregnenolone	3β-Sulfate	K+	434.6	210–212	—		2
		Na+	418.5	195–197	—		a
		NH₄+	413.6	198–201	—		19
		C₅H₅NH+	475.6	174–177	—		f

Compound	Ester	Cation	M.W.	M.p.	$[\alpha]$		Ref.
Δ⁵-Pregnene-3β,20α-diol	3β-Sulfate	NH_4^+	415.6	196–198	—	+	f
17α-Hydroxypregnenolone	3β-Sulfate	NH_4^+	429.6	195–198	—	+	5, f
		$C_5H_5NH^+$	491.6	192–195	—	+	5
Δ⁵-Pregnene-3β,17α,20α-triol	3β-Sulfate	NH_4^+	431.6	219–222	—	+	c
Cholesterol	3β-Sulfate	K^+	504.8	222–224(d)	−21°	+	2, 24
		Na^+	488.7	184–185(d)		+	20
		Na^+	—	194–195	−19°		2
		NH_4^+	483.7	214–215	−24°		2
		NH_4^+	—	197–200		+	21
		$C_5H_5NH^+$	545.8	179(d)	+23.8° ($CHCl_3$)	+	17, 20
Estrone	3-Sulfate	K^+	388.5	219–220	+111°	+	2
		Na^+	372.4	228–230(d)	+110° ($CHCl_3$)	+	4, b
		NH_4^+	367.5	225–227(d)	—		14
		$C_5H_5NH^+$	429.5	173–175(d)	+84.1° ($CHCl_3$)		4
2-Methoxyestrone	3-Sulfate	K^+	418.5	>260	—	+	2
Equilin	3-Sulfate	Na^+	370.4		+218° (H_2O)		7
Equilenin	3-Sulfate	Na^+	368.4		+70° (H_2O)		7
Estradiol	3-Sulfate	Na^+	374.4	207–212(d)	+41° (H_2O)	+	9
		NH_4^+	369.5	185–188(d)	—		14
	17β-Sulfate	Na^+	374.4	188–190(d)	+20° (H_2O)	+	9
		NH_4^+	369.5	188–191(d)	—		14
	3,17β-Disulfate	Na^+	476.5	186–191(d)	+33° (H_2O)	+	9
17α-Ethynylestradiol	3,17β-Disulfate	K^+	532.7	165–170	−2° (DMSO)		2
		Na^+	500.5	130	−6°		2
17α-Ethynylestradiol, 3-methyl ether	17β-Sulfate	K^+	428.6	130	—		2

(Continued)

PHYSICAL PROPERTIES OF CERTAIN STEROID SULFATES (*Continued*)

Parent compound	Derivative	Salt[a]	Mol. wt.[b]	Melting point (°C)	Optical rotation (α_D)	Infrared spectrum	References[c]
Testosterone	17β-Sulfate	K+	406.6	260	+58°		2
		Na+	390.5	215	+74.5° (H_2O)		8
		$C_5H_5NH^+$... NH4+	385.5	218–221	—		6
		NH4+	—	201–203	+70°		2
Deoxycorticosterone	21-Sulfate	$C_5H_5NH^+$	489.6	185–188	+107° (EtOH-95%)	+	11
Corticosterone	21-Sulfate	K+	464.6	178–180	—	+	11
11-Dehydrocorticosterone	21-Sulfate	$C_5H_5NH^+$	503.6	175–176	+142° (MeOH)	+	11
Cortisone	21-Sulfate	$C_5H_5NH^+$	519.6	195–198	+136° (MeOH)	++	11
		Na+	462.6		—	+	11
11-Deoxycortisol	21-Sulfate	$C_5H_5NH^+$	505.6	176–178	+89° (MeOH)	+	11

[a] Steroid sulfates have been prepared as salts of a variety of cations by treating the sulfated compound with an aqueous solution of the desired cation (see text footnote 17).

[b] Water of hydration frequently associated with steroid sulfates has been disregarded in calculating the molecular weights listed in this table.

[c] The numbers refer to the corresponding text footnotes and to the following sources:

(a) R. Vihko, *Acta Endocrinol.* **52**, Suppl. 109, 27 (1966).
(b) W. Neudert and H. Röpke, "Atlas of Steroid Spectra." Springer, New York, 1965.
(c) K. D. Roberts, L. Bandi, H. I. Calvin, W. D. Drucker, and S. Lieberman, *Biochemistry* **3**, 1983 (1964).
(d) L. Segal, B. Segal, and W. R. Nes, *J. Biol. Chem.* **235**, 3108 (1960).
(e) N. B. Talbot, J. Ryan, and J. K. Wolfe, *J. Biol. Chem.* **148**, 593 (1943).
(f) H. I. Calvin and S. Lieberman, *J. Clin. Endocrinol. Metab.* **26**, 402 (1966).

[11] Chemical Syntheses of Substrates of Sterol Biosynthesis

By R. H. CORNFORTH and G. POPJÁK

General Introduction

The enzymes of sterol biosynthesis in yeast and several of the intermediates of sterol biosynthesis derived from mevalonate have been described in these volumes by Tchen.[1] In this article and article [12] we present the preparation of the enzymes of sterol biosynthesis from liver, the enzymatic and/or chemical synthesis of all the intermediates of squalene biosynthesis from mevalonate up to and including farnesyl pyrophosphate. All the methods to be described are those developed and used in our laboratories and those of which we have personal experience; hence we are not attempting a comprehensive review.

This article describes chemical syntheses of substrates of sterol biosynthesis:[2]

(1) syntheses suitable for the preparation of 1-^{14}C- and 2-^{14}C-labeled mevalonolactone;

(2) syntheses of unlabeled and of 3′,4-^{13}C$_2$-labeled mevalonolactone;

(3) synthesis of $R(-)$mevalonolactone (natural isomer);

(4) synthesis of 5-D$_2$- and 4-^{14}C-labeled mevalonolactone;

(5) synthesis of mevalonates labeled stereospecifically at C-4 with hydrogen isotopes;

(6) preparation of mevalonates labeled stereospecifically at C-2 with hydrogen isotopes;

(7) synthesis of methyl and ethyl farnesoates; procedure suitable for labeling with ^{14}C at C-2;

(8) synthesis of isopentenyl pyrophosphate;

---

[1] T. T. Tchen, see Vol. V [66], p. 489; Vol. VI [75], p. 505.

[2] The following abbreviations and terminology will be used in articles [11] and [12]: Mvald = mevaldic acid; MVA, mevalonic acid; 5-P-MVA, 5-phosphomevalonate; 5-PP-MVA, 5-pyrophosphomevalonate; I-PP, isoprenyl (= isopentenyl = 3-methylbut-3-en-1-yl) pyrophosphate; Mp-PP, monoprenyl (= 3,3-dimethylallyl) pyrophosphate; Dp-PP, *trans*-diprenyl (= geranyl) pyrophosphate; Tp-PP, *trans-trans*-triprenyl (= *trans-trans*-farnesyl) pyrophosphate; ATP and ADP, adenosine tri- and diphosphate; NAD$^+$ and NADH, nicotinamide adenine dinucleotide coenzyme, oxidized and reduced forms; NADP$^+$ and NADPH, nicotinamide adenine dinucleotide phosphate coenzyme, oxidized and reduced forms; P$_i$ and PP$_i$, inorganic orthophosphate and pyrophosphate, respectively.

For descriptions of absolute configuration around an asymmetric center we will use the R and S convention introduced by R. S. Cahn, C. K. Ingold, and V. Prelog, *Experientia* **12**, 81 (1956).

(9) syntheses of prenyl pyrophosphates (3,3-dimethylallyl, geranyl, and farnesyl pyrophosphate).

I. Synthesis of Unlabeled and 1-^{14}C- and 2-^{14}C-Labeled Mevalonolactone by the Reformatsky Reaction on 4-Acetoxybutan-2-one

Principle. 4-Acetoxybutan-2-one is condensed with methyl bromoacetate and zinc; the resulting methyl 5-acetoxy-3-hydroxy-3-methylpentanoate is hydrolyzed by alkali to mevalonic acid, which is isolated as the lactone by distillation.

Procedure. The procedure described here is that reported by Cornforth *et al.*[3] and used by the Radiochemical Centre, Amersham, Bucks, England. Alternative experimental conditions were reported by Hoffman *et al.*[4] 4-Acetoxybutan-2-one may be prepared by acetylation of 4-hydroxybutan-2-one[3,4] by addition of acetic acid to methylvinyl ketone[5] (described here) or by LiAlH$_4$ reduction of the ethylene ketal of ethyl acetoacetate, followed by acetylation of the alcohol and hydrolysis of the ketal group.[6] 4-Hydroxybutan-2-one is now available commercially (Eastman Organic Chemicals).

4-Acetoxybutan-2-one. Methylvinyl ketone (35 g), acetic acid (150 ml), and water (1 drop) are refluxed overnight. Fractionation through a Widmer column gives unchanged ketone and acid and 4-acetoxybutan-2-one (28.6 g; 44%), b.p. 78–84° at 15 mm; this is freed from traces of acetic acid by heating at 35–40° at 0.2 mm for 1.5 hours. It is best stored at low temperature (4°) to avoid fission of acetic acid.

Methyl 5-Acetoxy-3-hydroxy-3-methylpentanoate. The apparatus for the Reformatsky reaction is oven-dried at 120° and moisture is excluded during the reaction. 4-Acetoxybutanone (3.0 g; 23.1 millimoles) and methyl bromoacetate (3.5 g; 22.9 millimoles) are dissolved in dry ether (15 ml) and added to zinc wool (3.0 g; 2-fold excess; cut into short lengths and activated by washing with 2% HCl, ethanol, acetone, and dry ether and heating to 100° *in vacuo* with a crystal of iodine, or by the method of Fieser and Johnson[7]). The mixture is stirred and refluxed for 4 hours, during which time a gummy complex separates. (The reaction

[3] J. W. Cornforth, R. H. Cornforth, G. Popják, and I. Youhotsky Gore, *Biochem. J.* **69**, 146 (1958).

[4] C. H. Hoffman, A. F. Wagner, A. N. Wilson, E. Walton, C. H. Shunk, D. E. Wolf, F. W. Holly, and K. Folkers, *J. Am. Chem. Soc.* **79**, 2316 (1957).

[5] J. W. Cornforth, R. H. Cornforth, A. Pelter, M. G. Horning, and G. Popják, *Tetrahedron* **5**, 311 (1959).

[6] L. Canonica, L. Gaudenzi, G. Jommi, U. Valcavi, *Gazz. Chim. Ital.* **91**, 1400 (1961).

[7] L. F. Fieser and W. S. Johnson, *J. Am. Chem. Soc.* **62**, 576, (1940).

usually becomes exothermic for a time after a period varying from 20 minutes to 1.5 hours). After cooling, the complex is decomposed with acetic acid (1.3 ml) in water (10 ml) with vigorous stirring. The aqueous layer is saturated with ammonium chloride and extracted four times with ether. The ethereal solution is dried over anhydrous MgSO₄ and evaporated; the residue is distilled, giving 2.9 g (62%) of methyl 5-acetoxy-3-hydroxy-3-methylpentanoate, b.p. 80–82° at 0.1 mm.

3-Hydroxy-3-methylpentano-5-lactone (Mevalonolactone). Methyl 5-acetoxy-3-hydroxy-3-methylpentanoate (2.9 g; 14.2 millimoles) is allowed to stand at room temperature for 40 hours with an excess of methanolic 1 N KOH solution (65 ml); a solution of dry HCl in methanol, equivalent to the KOH used, is added, the precipitated potassium chloride is filtered after chilling, and the filtrate is evaporated. The residue is extracted with chloroform, and the solution is filtered, evaporated, and distilled, giving 1.95 g, b.p. 115° at 0.1 mm; redistillation gives a small forerun and 1.6 g (86%) of mevalonolactone, b.p. 110° at 0.1 mm.

A paper chromatogram of the hydroxamate, developed with fresh butanol–acetic acid–water (4:1:5, by volume) and sprayed with ferric chloride solution gives a single spot with an R_f of 0.57.

1-¹⁴C- and 2-¹⁴C-labeled mevalonolactones are made exactly as described above using the appropriately labeled methyl bromoacetates in the Reformatsky reaction. The synthesis has been satisfactorily performed on one-third of the scale given; there is no reason why the scale should not be further reduced.

We have found the reactions to be reproducible, though the yield in the Reformatsky reaction is sometimes less.

The Radiochemical Centre also makes mevalonolactone-2-T₂ by the same process.

II. Synthesis of Mevalonolactone by the Reaction of Ketene with 4-Acetoxybutan-2-one

Principle. 4-Acetoxybutan-2-one reacts with ketene in the presence of boron trifluoride to give a β-lactone which, without isolation, is hydrolyzed by alkali to mevalonic acid. Provided a ketene generator is available, this is the most convenient method for the preparation of mevalonolactone on a large scale.

Procedure.[5] 4-Acetoxybutan-2-one (26 g; 0.2 mole) and dry ether (20 ml) are cooled in a bath at −30°, and boron trifluoride etherate (0.6 ml) is added. Gaseous ketene[8] (slightly over one equivalent) is passed

---

[8] J. W. Williams and C. D. Hurd, *J. Org. Chem.* **5**, 122 (1940); W. E. Hanford and J. C. Sauer, *Org. Reactions* **3**, 108 (1947).

and the mixture is then left at $-25°$ for 2 hours; some crystallization usually occurs. The mixture is added, with the aid of more ether, to 1 N methanolic KOH (600 ml) cooled to $-15°$. The solution is kept over-night at room temperature, then neutralized to phenol red by a solution of dry hydrogen chloride in methanol and evaporated at low pressure. The residue, after washing with ether and chloroform several times to remove neutral material, is treated with methanol and then with suf-ficient hydrogen chloride in methanol to bring the total acid added to 592.4 meq (7.6 ml of N alkali is consumed by 0.6 ml of boron trifluoride etherate). Potassium chloride is removed by filtration, the filtrate is evaporated at low pressure, and the residue is extracted with chloroform. The filtered chloroform solution is evaporated, and the residue is distilled in high vacuum; after removal of low-boiling material the lactone is distilled rapidly by wide-path distillation, i.e., in an apparatus with wide-bore glass connections between flask, receiver, and pump, in order to minimize the risk of dehydration to $\alpha\beta$-unsaturated lactone during pro-longed distillation. The boiling point varies with the conditions of dis-tillation between 100° and 114° at 0.001 to 0.01 mm. The product con-tains a small amount of 3-hydroxyisovaleric acid formed from acetone carried over in the ketene gas stream; it is therefore treated with ethereal diazomethane until the yellow color of the reagent persists and is then redistilled. The methyl 3-hydroxyisovalerate is easily separated (b.p. 70° at 10 mm). The yield of redistilled mevalonic lactone is 23.3 g (90%). Consistently good yields have been obtained in this process.

This method has also been applied on a small scale to the synthesis of mevalonolactone-3′,4-^{13}C$_2$.[5]

III. Synthesis of R-Mevalonolactone from Coriandrol [(+)-Linalool]

Principle. (+)-Linalool is hydrated by Brown's[9] method of reaction with diborane followed by oxidation with alkaline peroxide to give 3,7-dimethyloctane-1,3,6-triol. Acid-catalyzed reaction of the triol with acetaldehyde gives 2,4-dimethyl-4-(3-hydroxy-4-methylpentyl)-1,3-di-oxan, which is oxidized with concentrated aqueous chromium trioxide in pyridine to 2,4-dimethyl-4-(3-oxo-4-methylpentyl)-1,3-dioxan. Conden-sation of this ketone with methyl formate and sodium methoxide gives the hydroxymethylene ketone, which is immediately oxidized with aque-ous methanolic sodium periodate; the acetal group is hydrolyzed during isolation, and the resulting R-mevalonic acid is isolated as the lactone.

Procedure.[10] The S-(+)-linalool ($[\alpha]_D^{20} + 16.9°$) required for this

[9] H. C. Brown and B. C. Subba Rao, *J. Org. Chem.* **22**, 1136 (1957).
[10] R. H. Cornforth, J. W. Cornforth, and G. Popják, *Tetrahedron* **18**, 1351 (1962).

synthesis was obtained by fractionation of oil of *Melaleuca quinque-nervia*, (Cav.), S. T. Blake Variety 'A,' the Australian broad-leafed tea tree.

3,7-Dimethyloctane-1,3,6-triol. A solution of lithium borohydride[11] (10 g, containing 8.8 g of $LiBH_4$, 0.4 mole, by iodometric assay[12]) in anhydrous tetrahydrofuran (300 ml) is cooled in ice-water and stirred during addition (1.5 hours) of a solution of boron trifluoride etherate (71.8 g, 0.51 mole, freshly distilled from calcium hydride at 40 mm pressure) in tetrahydrofuran (150 ml). (+)-Linalool (52 g, 0.34 mole) in tetrahydrofuran (25 ml) is added during the next hour and the mixture is left overnight at room temperature. Stirring is resumed and excess of diborane is destroyed by addition of methanol (30 ml). The mixture is then cooled in ice and a 3 N solution (270 ml) of sodium hydroxide in 50% (v/v) aqueous methanol is added. Hydrogen peroxide (30%, 84 ml) is introduced dropwise during 1 hour, the internal temperature rising to 27–28°. The solvents are removed by distillation from a steam bath followed by steam distillation. The residue is saturated with sodium chloride and extracted several times with ether. The ethereal solution is washed with saturated aqueous sodium chloride, dried, and evaporated. Distillation gives (i) 19.0 g, b.p. $< 100°$ at 0.001 mm and (ii) 38.0 g (59% yield) of 3,7-dimethyloctane-1,3,6-triol, a very viscous colorless oil, b.p. 114–120° at 0.001 mm.

2,4-Dimethyl-4-(3-hydroxy-4-methylpentyl)-1,3-dioxan. A mixture of 3,7-dimethyloctane-1,3,6-triol (77 g, 0.405 mole), dry ether (45 ml) and toluene-*p*-sulfonic acid (100 mg) is stirred during gradual addition of acetaldehyde (25 ml, 0.447 mole). The solution is allowed to stand for 2 days; it is then diluted with more ether, washed with saturated solutions of sodium bicarbonate and of sodium chloride, and dried. Evaporation and distillation give a forerun (7.9 g) and then 2,4-dimethyl-4-(3-hydroxy-4-methylpentyl)-1,3-dioxan (71.4 g; 81.6% yield), a colorless liquid, b.p. 85–90° at 0.01 mm, 144–145° at 17 mm; $n_D^{22.5}$ 1.4590.

2,4-Dimethyl-4-(3-oxo-4-methylpentyl)-1,3-dioxan. Chromium trioxide (50 g; 0.5 mole) is dissolved in water (30 ml), with warming on a steam bath to complete solution. The cooled solution is added gradually with stirring to pyridine (anhydrous, laboratory reagent grade; 500 ml) cooled in ice-water; 2,4-dimethyl-4-(3-hydroxy-4-methylpentyl)-1,3-dioxan (35.7 g, 0.165 mole) in pyridine (100 ml) is then added and the mixture is allowed to stand for 90 hours. It is then poured into water and filtered through a Celite pad. The residue is washed well with ether

[11] A suspension of sodium borohydride (but not potassium borohydride) may be used instead.
[12] D. A. Lyttle, E. H. Jensen, and W. A. Struck, *Anal. Chem.* **24,** 1843 (1952).

and the filtrate is extracted four times with ether. The combined ethereal solution is dried and evaporated, and the residue is distilled to give 2,4-dimethyl-4-(3-oxo-4-methylpentyl)-1,3-dioxan as a pale yellow oil (31.1 g, 88% yield; b.p. 64° at 0.005 mm; 146–150° at 19 mm; n_D^{21} 1.4506).

R-(−)-*Mevalonolactone.* Sodium (6.2 g; 0.27 mole) is dissolved in dry methanol, and the solution is evaporated to dryness finally at 200° for 3 hours at 10 mm pressure; the residual dry sodium methoxide is cooled to −15° under nitrogen and methyl formate (180 ml; dried over calcium chloride and distilled) is added with stirring. 2,4-Dimethyl-4-(3-oxo-4-methylpentyl)-1,3-dioxan (29 g; 0.136 mole) in methyl formate (50 ml) is then added during 30 minutes. The mixture is left standing at room temperature overnight; it is then cooled in ice water, and acetic acid (16 ml) in ether (50 ml) is added. The mixture is washed with saturated aqueous sodium chloride, and the solvents are removed at room temperature. The residue is dissolved in ether and washed with saturated sodium bicarbonate solution; the hydroxymethylene ketone is then extracted from the ether with ice-cold 0.5 N sodium hydroxide (270 ml). The solution is immediately acidified with acetic acid (7.7 ml) and extracted with ether. Evaporation of the dried ethereal extract at low pressure leaves the crude hydroxymethylene ketone (20.6 g). A solution of sodium metaperiodate (60 g; 0.28 mole) in water (425 ml), and methanol (95 ml) is added; the mixture is shaken intermittently and kept at a pH of about 5 by addition of 2 N sodium hydroxide as required (total, 55 ml). Considerable heat is evolved and an acid sodium iodate crystallizes out. The mixture is left overnight; after filtration the crystals are washed with methanol. Sulfur dioxide is passed into the filtrate until the iodine first liberated is reduced. Powdered silver sulfate (26 g; 0.083 mole) is added, and the mixture is shaken until precipitation of silver iodide is complete; sodium chloride is then added to remove silver ions from solution. After filtration from silver salts the solution is concentrated to 500 ml by distillation, first at atmospheric pressure for 1 hour to ensure complete hydrolysis of the acetal, then at reduced pressure. It is then extracted continuously with chloroform for 24 hours. The combined chloroform extracts from two batches give on wide-path distillation (cf. page 362) 15.6 g of the lactone, b.p. 100–108° at 0.005 mm. Extraction for a further 48 hours gives an additional 1.65 g of lactone. Crystallization from a mixture of methylene chloride (20 ml) and dry ether (30 ml) at −70° gives R-(−)-mevalonolactone (12.0 g; 34.1% yield) as highly hygroscopic crystals, m.p. 19–20° with some previous sintering, $[\alpha]_D^{20}$ −23.0° ($c = 6$ in ethanol).

The purity of the product is demonstrated, after conversion to the sodium salt, by its substantially complete utilization by mevalonate kinase. The optical purity of the product, of course, corresponds to the optical purity of the starting (+)-linalool. The overall yield of pure crystallized mevalonolactone from linalool is 14.5%.

IV. Synthesis of 5-D_2- and 4-^{14}C-Labeled Mevalonolactone

Principle. 4,4-Dimethoxybutan-2-one, by Reformatsky condensation with methyl bromoacetate and zinc, gives methyl 5,5-dimethoxy-3-hydroxy-3-methylpentanoate. The ester group is reduced by lithium aluminum hydride, and the resulting alcohol is acetylated; the acetal group is hydrolyzed and oxidized by peroxyformic acid, and distillation of the product gives mevalonolactone. When lithium aluminum deuteride is used for reduction of the ester, the product is mevalonolactone-5-D_2,[13, 14] and when methyl bromoacetate-2-^{14}C is used in the Reformatsky reaction the product is mevalonolactone-4-^{14}C.[5] The method could equally well be adapted to the preparation of 5-T_2-, 5-^{14}C-, or 4-T_2-labeled mevalonolactone.

Procedure. The reaction conditions given here incorporate improvements on those described by us for our original preparations of 4-^{14}C- and 5-D_2-labeled mevalonolactone. The procedure is described for unlabeled materials but can readily be scaled-down, say, to 8–10 millimoles, for radioactive preparations.

Methyl 5,5-Dimethoxy-3-hydroxy-3-methylpentanoate

Note. The Reformatsky reaction on 4,4-dimethoxybutan-2-one has proved rather capricious. Shunk *et al.*,[15, 16] using granular zinc and ethyl bromoacetate, reported yields of 26–30%, and Eggerer and Lynen,[17] using zinc dust and ethyl bromoacetate, had yields of 30–33% on an 0.2 M scale. On a scale of 10–30 millimoles we frequently obtained yields of 50–58%, but on a larger scale (50–300 millimoles) the yields varied from 25–42%. In our earlier experiments zinc "needles" were used and the reaction proceeded without the aid of a catalyst; this form of zinc is no longer available and we use British Drug Houses zinc "wool," cut into short lengths and activated by the method of Fieser and Johnson;[7] if a trace of mercuric chloride is added the reaction always starts within 10 minutes.

[13] G. Popják, D. S. Goodman, J. W. Cornforth, R. H. Cornforth, and R. Ryhage, *J. Biol. Chem.* **236**, 1934 (1961).
[14] J. W. Cornforth, R. H. Cornforth, C. Donninger, G. Popják, G. Ryback, and G. J. Schroepper, Jr., *Proc. Roy. Soc.* **B163**, 436 (1966).
[15] C. H. Shunk, B. O. Linn, and K. Folkers, U.S. Patent 3,014,963 (1961).
[16] *Chem. Abstr.* **56**, 12745g (1962).
[17] H. Eggerer and F. Lynen, *Ann. Chem.* **608**, 71 (1957).

The normal procedure in larger-scale Reformatsky reactions is to use only a small portion of the reactants at the start, then to add the remainder gradually as the rate of reaction allows. We have found that much better yields are obtained when all the reactants are mixed at the beginning, as is the custom in small-scale experiments.

All reagents should be pure and dry and the apparatus should be dried in an oven at 120°. Dimethoxybutanone of high quality is available commercially (Fluka; Aldrich Chemical Company) and is redistilled before use. We have also prepared it by the method of Royals and Brannock.[18] Methyl bromoacetate is redistilled. Ether is sodium dried.

Activated zinc (13 g; 2 equivalents), mercuric chloride (90 mg), 4,4-dimethoxybutan-2-one (13.2 g; 0.1 mole), methyl bromoacetate (15.3 g; 0.1 mole) and dry ether (100 ml) are placed in a 500-ml two-necked flask fitted with a mechanical stirrer, efficient reflux condenser, and drying tube. The mixture is stirred and gently refluxed until a vigorous reaction sets in (5–10 minutes). The source of heat is removed until the reaction subsides; stirring and refluxing are then continued for 2 hours. The reaction mixture is cooled and decomposed with a chilled solution of acetic acid (7 ml) in water (70 ml) with vigorous stirring for 10 minutes. The aqueous layer is then saturated with ammonium chloride. The ether layer is separated and the aqueous layer is extracted three times with ether (25 ml). The combined ethereal extract is washed with 5-ml portions of saturated potassium bicarbonate solution until free of acid, and dried ($MgSO_4$). The ether is evaporated and the residue is distilled under reduced pressure; the lower boiling material is removed in water-pump vacuum (bath up to 100°) and methyl 5,5-dimethoxy-3-hydroxy-3-methylpentanoate then distills at 60–62° at 0.04 mm. The yields in two separate experiments were 12.4 g (60%) and 11.6 g (56%). On an 0.2 M scale a 45% yield was obtained.

1-Acetoxy-5,5-dimethoxy-3-methylpentan-3-ol. Lithium aluminum hydride (2.3 g; 0.06 mole) is stirred and refluxed in dry ether (60 ml) for 1 hour; the solution is cooled to room temperature and a solution of methyl 5,5-dimethoxy-3-hydroxy-3-methylpentanoate (10.3 g; 0.05 mole) in dry ether (30 ml) is added at a rate to maintain steady refluxing. The mixture is stirred and refluxed for 2 hours, then cooled to room temperature. Water or saturated potassium bicarbonate solution is cautiously added dropwise until the alumina coagulates (5–6 ml); the alumina is filtered and washed well with ether. After evaporation of the ether the residue is acetylated with acetic anhydride (7 ml) in pyridine (10 ml) for two hours at room temperature. The excess acetic anhydride and

[18] E. E. Royals and K. C. Brannock, *J. Am. Chem. Soc.* **75**, 2050 (1953).

pyridine are evaporated in high vacuum (from a bath at 40°) and condensed in a cold trap at −70°. The residue is dissolved in ether and washed with a small volume of saturated sodium bicarbonate solution; the ether solution is dried and evaporated and the product is distilled in high vacuum. The yield is 8.8 g (80%), boiling at 80–82° at 0.02 mm.

Mevalonolactone. 1-Acetoxy-5,5-dimethoxy-3-methylpentan-3-ol (1 g) is mixed with 98% formic acid (2 ml) and 30% hydrogen peroxide (1 ml) and left at room temperature for 2 hours; there is a small rise in temperature. The solution is then warmed to 75–80° for 30 minutes, when a test portion gives only a slight reaction with a solution of 2,4-dinitrophenylhydrazine in 2 N HCl. Water (8 ml) is added and the solution is refluxed until the peroxyformic acid is destroyed; a test with starch iodide paper is very slight after 2 hours. The solution is evaporated on a rotary evaporator in water-pump vacuum from a bath at 30°, and the residue is evaporated 5 or 6 times with 20 ml methanol to remove the formic acid as completely as possible. A solution of the residue (∼0.6 g) in chloroform is dried and evaporated; the product is distilled, giving mevalonolactone (0.51 g; 86%) boiling at 110° at 0.02 mm. The I.R. spectrum is identical with that of an authentic specimen. $E_{1cm}^{1\%}$ at 215 mμ is 590, indicating the presence of about 8% of the $\alpha\beta$-unsaturated lactone (ε 8000 at 215 mμ).

In another experiment on a 5 g scale the reaction mixture was left 4 hours at room temperature, water (40 ml) was added and the solution was refluxed for 4 hours in a bath at 110°. A 74% yield of mevalonolactone containing only 2.8% of unsaturated lactone was obtained; the lower boiling material was more completely separated during distillation. GLC analysis on a 5-foot 3% QF1 column (0.25″ O.D.) at 110° indicated 97–98% purity.

V. Syntheses of 4R and 4S-Mevalonates-4-*H_1[19]

Principle. Reformatsky condensation of methylvinyl ketone with methyl bromoacetate and zinc gives methyl 3-hydroxy-3-methylpent-4-enoate, which is converted to a mixture of the methyl esters of the *cis*- and *trans*-5-acetoxy-3-methylpent-3-enoic acids by treatment with phosphorus tribromide and pyridine followed by reaction with anhydrous potassium acetate in acetone.[20] The *cis* and *trans* unsaturated hydroxy acids produced by the very mild alkaline hydrolysis of the acetoxy esters are readily separated by distillation, since the *cis* isomer forms a lactone boiling at a much lower temperature than the *trans* hydroxy acid. Each

---

[19] For notation of stereochemical configurations see footnote 2. *H can be either deuterium or tritium.

[20] Method of L. Ruzicka and G. Firmenich, *Helv. Chim. Acta* **22**, 392 (1939).

isomer is then converted to its diphenylmethylamide and epoxidized with perbenzoic acid. The racemic mixture of epoxides derived from each isomer is then reduced with lithium borodeuteride, or lithium borotritide, to give a racemic mixture of 4-D_1-, or 4-T_1-labeled mevalonic diphenylmethylamides, from which the free mevalonic acids are obtained by alkaline hydrolysis. Scheme 1 shows that the specimen of mevalonate

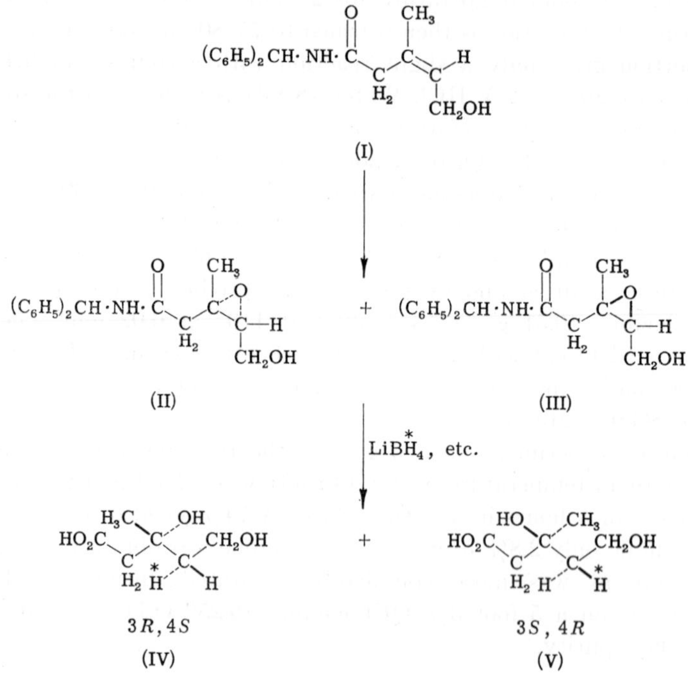

SCHEME 1. Derivation of (3R,4S)-(IV) and of (3S,4R)-(V) mevalonic acid-4-*H_1 from the diphenylmethylamide of cis-5-hydroxy-3-methylpent-3-enoic acid (I) through the epoxides (II) and (III). In enzymatic reactions only the 3R-isomer (IV) of mevalonate is utilized. *H in the formulas can be either deuterium or tritium.

derived from the amide of cis-5-hydroxy-3-methylpent-3-enoic acid (I) contains the 3R,4S + 3S,4R enantiomers of the isotopically labeled substance (IV and V, respectively). The amide of the trans-5-hydroxy-3-methylpent-3-enoic acid (VI), on the other hand, gives a mixture of the 3R,4R and 3S,4S enantiomers (Scheme 2). Since in biosynthetic experiments, on account of the absolute stereospecificity of mevalonate kinase,[10] only the 3R-isomer is utilized, the specimen of the mevalonate-4-*H_1 derived from the amide of the cis-acid (I) is equivalent to (4S)-meva-

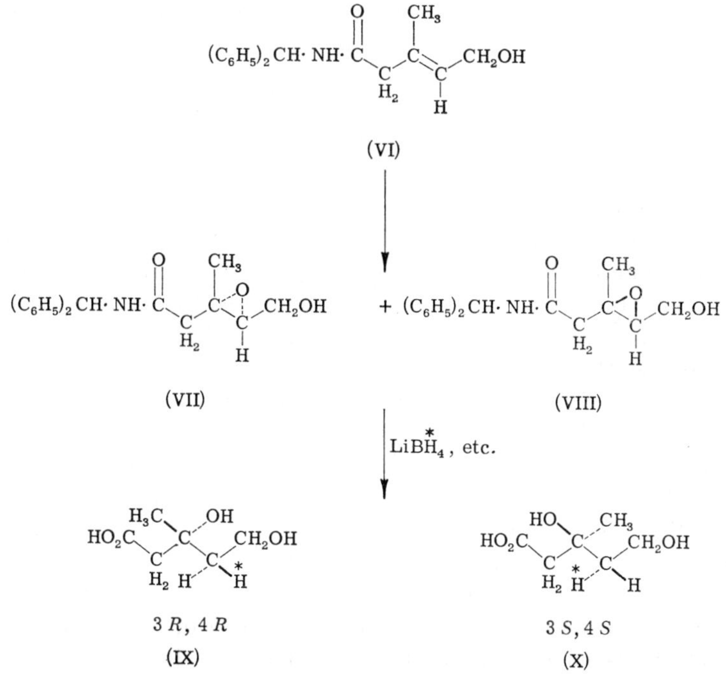

SCHEME 2. Derivation of $(3R,4R)$- (IX), and of $(3S,4S)$- (X) mevalonic acid-4-*H_1 from the diphenylmethylamide of *trans*-5-hydroxy-3-methylpent-3-enoic acid (VI) through the epoxides (VII) and (VIII). In enzymatic reactions only the $3R$-isomer (IX) of mevalonate is utilized. *H in the formulas can be either deuterium or tritium.

lonate-4-*H_1 (IV) and the one obtained from the amide of the *trans*-acid (VI) is equivalent to $(4R)$-mevalonate-4-*H_1 (IX).

Procedure. See Cornforth *et al.*[21, 22]

Methyl 3-hydroxy-3-methylpent-4-enoate. All apparatus is oven-dried at 120° and moisture is excluded during the reaction. Zinc wool is cut into short lengths and activated by the method of Fieser and Johnson.[7] Commercial methylvinyl ketone is stirred with calcium chloride for 3 hours, filtered, and dried over anhydrous calcium sulfate (Drierite) overnight; after filtration it is distilled (b.p. 37° at 145 mm), a forerun being discarded; its infrared spectrum should exhibit only a small hydroxyl peak.

[21] J. W. Cornforth, R. H. Cornforth, C. Donninger, and G. Popják, *Proc. Roy. Soc.* **B163**, 492 (1966).
[22] J. W. Cornforth, R. H. Cornforth, and K. K. Mathew, *J. Chem. Soc.* p. 112 (1959).

Activated zinc wool (45 g; 0.69 g-atom), mercuric chloride (0.4 g), and dry benzene (212 ml) are placed in a 2-liter 3-necked flask equipped with a mechanical stirrer, efficient condenser, and dropping funnel; 60 ml of a solution of methylvinyl ketone (43 g; 0.61 mole) and methyl bromoacetate (85 g; 0.56 mole) in dry benzene (212 ml) are added and the mixture is refluxed and stirred (bath temperature 85–90°) until a vigorous reaction sets in (about 10 minutes). The remainder of the solution is then added at a rate sufficient to maintain vigorous refluxing (about 30 minutes; bath maintained at 85°). The mixture is stirred and refluxed for a further hour, cooled and decomposed by acetic acid (41 ml; 0.35 mole) in water (350 ml) with vigorous stirring. The layers are separated and the aqueous layer is saturated with ammonium chloride and extracted several times with ether. The combined benzene–ether solution is washed with saturated sodium bicarbonate solution (which is reextracted once with ether) and dried ($MgSO_4$). The solvents are evaporated on a rotary evaporator in water pump vacuum from a water-bath at 25°. The residue is distilled at 0.2 mm (b.p. 28–30°), the distillate being collected in a receiver cooled to −30°; a polymeric residue remains. The distillate is redistilled at water pump vacuum, yielding 47.7 g (60% calculated on methyl bromoacetate[23]) of methyl 3-hydroxy-3-methylpent-4-enoate, b.p. 66–68° at 13.5 mm. It contains a little methyl bromoacetate, which may be removed through a fractionating column, but is sufficiently pure for the next operation.

Methyl 5-Acetoxy-3-methylpent-3-enoate. A solution of methyl 3-hydroxy-3-methylpent-4-enoate (84 g; 0.58 mole) and pyridine (anhydrous laboratory reagent grade; 14 ml) in light petroleum (b.p. 40–60°; 170 ml) is cooled in an ice-salt mixture and stirred mechanically during addition of a solution of redistilled phosphorus tribromide (25 ml) in light petroleum (90 ml) during 30 minutes, with exclusion of moisture. The mixture is stirred for 2 hours longer, the temperature being allowed to rise gradually to +10°. It is then cooled to 0° and decomposed by rapid addition of ice and water (about 300 ml) with vigorous stirring. Stirring is continued for 30 minutes; the petroleum layer is separated, washed with water and sodium bicarbonate solution, dried and evaporated at room temperature and low pressure. The crude bromide is shaken with finely divided anhydrous potassium acetate (330 g)[24] in dry acetone (2 liters)[25] for 2–3 days. After filtration the acetone is evaporated and the residue is dissolved in ether; the ether solution is washed with sodium

---

[23] Over a number of runs, the yield varied from 48 to 60%.

[24] Laboratory reagent grade may be used directly from freshly opened bottles; material which has absorbed moisture should be fused and powdered.

[25] Analytical grade reagent is sufficiently dry.

bicarbonate solution, dried, and distilled. It gives methyl 3-methylpenta-2,4-dienoate (27.0 g; b.p. 58–60° at 15 mm), an intermediate fraction (5.1 g; b.p. 38–40° at 0.4 mm), and methyl 5-acetoxy-3-methylpent-3-enoate (40.0 g; 37%; b.p. 70–72° at 0.02 mm). The product consists of about 30% *cis* and 70% *trans* isomer.

3-Methylpent-3-eno-5-lactone and 5-hydroxy-3-methyl-trans-pent-3-enoic acid.[26] Methyl 5-acetoxy-3-methylpent-3-enoate (22.2 g; 0.12 mole) is hydrolyzed with 0.1 N NaOH solution (2660 ml) for 24 hours at room temperature. The solution is neutralized to phenol red with 1 N H_2SO_4 and concentrated to about 50 ml on a rotary evaporator; it is then acidified with 10 N H_2SO_4, saturated with ammonium chloride and extracted five times (or by continuous extraction) with ether. The dried solution is evaporated and the residue is distilled. Most of the acetic acid is removed in water pump vacuum (bath up to 80°) and the product is then distilled rapidly[27] by wide-path distillation at 0.01 mm. The lower fraction (6.2 g; b.p. about 40°; bath 80–100°) is collected in a receiver cooled to −70°; the bath is then preheated to 140–150° for distillation of the *trans*-hydroxy acid (7.4 g; b.p. about 120°). For purification the lactone is dissolved in ether, washed free of acid with sodium bicarbonate solution, and distilled; the yield of purified 3-methylpent-3-enolactone (b.p. 112° at 14 mm) is 2.7 g. The acid is crystallized from dry ether at −70°, yielding 7.1 g 5-hydroxy-3-methyl-*trans*-pent-3-enoic acid (m.p. 56–57°). If the starting material is assumed to be 30% *cis* and 70% *trans*, the yields are 69% and 65%, respectively.

N-Diphenylmethyl-(5-hydroxy-3-methyl-cis-pent-3-en-)amide. 3-Methylpent-3-eno-5-lactone (2.7 g; 0.024 mole) is refluxed with diphenylmethylamine (4.4 ml; 0.026 mole) in a mixture of cyclohexane (10 ml) and benzene (5 ml) for 1 hour; the product crystallizes on cooling. A first crop of 4.3 g (m.p. 129–131°) and a second crop of 1.1 g (m.p. 133–134°) are obtained; total yield 5.4 g (76%).[28]

N-Diphenylmethyl-(5-hydroxy-3-methyl-trans-pent-3-en-)amide. Dicyclohexylcarbodiimide (5.54 g; 0.027 mole) is dissolved in methylene

---

[26] Both these substances polymerize on storing.

[27] This is necessary to minimize polymerization of the hydroxy acid. The distillation flask, receiver, and pump should have wide-bore connections; the authors used glass tubes (12 mm internal diameter) with spherical joints to connect pump and receiver.

[28] In other hands this preparation once yielded a product of m.p. 109–112° which could not be satisfactorily purified by recrystallization; it was dissolved in chloroform, washed with dilute H_3PO_4, saturated NaCl, and saturated $NaHCO_3$ solutions; after drying and evaporation of the chloroform the residue gave a satisfactory product when crystallized from benzene. No claim is made for reproducibility of the yield quoted above.

chloride in a roomy flask (500 ml) and diphenylmethylamine (5.0 ml; 0.029 mole) is added. The solution is heated to boiling, and a boiling solution of 5-hydroxy-3-methyl-*trans*-pent-3-enoic acid (3.5 g; 0.027 mole) in methylene chloride (36 ml) is added rapidly with shaking, a further 10 ml of methylene chloride being used for rinsing. A vigorous exothermic reaction takes place and dicyclohexylurea begins to separate immediately. The flask is shaken intermittently until the reaction subsides. After several hours the dicyclohexylurea (4.6 g; 77%; m.p. 236°) is filtered and washed with methylene chloride. The methylene chloride is distilled off from a steam bath, and the residue is crystallized from benzene (150 ml). The yield of N-diphenylmethyl-(5-hydroxy-3-methyl-pent-3-en-)amide is 6.1 g (77%) m.p. 141–142°. On a larger scale (5.8 g of acid) a 72% yield was obtained. On some occasions, when the reaction mixture was allowed to stand overnight, some amide separated along with the dicyclohexylurea; it can be extracted with hot methylene chloride.

The yield quoted here is reproducible provided the reaction conditions are observed strictly. If the solutions are not hot when mixed, the diphenylmethylamine salt of the acid may separate along with the dicyclohexylurea.

N-Diphenylmethyl-cis-3,4-epoxy-5-hydroxy-3-methylpentanamide. A solution of N-diphenylmethyl-(5-hydroxy-3-methyl-*cis*-pent-3-en-)amide (3.5 g) in chloroform (35 ml) is cooled in ice and stirred magnetically during addition of a benzene solution of perbenzoic acid[29] (1.1 molar equivalents). The solution is allowed to stand overnight at +5° and is then washed free of acid with 1 N NaOH; after two washings with saturated sodium chloride solution, it is dried and evaporated and the residue is crystallized from benzene, giving 3.4 g (89.6%) m.p. 147–148°.

N-Diphenylmethyl-trans-3,4-epoxy-5-hydroxy-3-methylpentanamide. A solution of N-diphenylmethyl-(5-hydroxy-3-methyl-*trans*-pent-3-en-)amide(7.0 g) in chloroform (70 ml) is treated as described for the *cis* isomer. The residue remaining after evaporation of the solvents is a viscous oil; a seed is obtained by rubbing a sample with light petroleum until it crystallizes. A solution of the remainder in hot 1-chlorobutane (50 ml) is cooled slowly in a large bath of hot water and seeded. The product separates as a mixture of oil and crystals and gradually crystallizes completely; it is collected and dried in a vacuum desiccator, finally at 0.2 mm pressure. The yield of N-diphenylmethyl-5-hydroxy-3,4-*trans*-epoxy-3-methylpentanamide is 6.6 g (89.4%), m.p. 83–85°, softening from 81°.

[29] Prepared by the method of L. S. Silbert, E. Siegel, and D. Swern, *Org. Syntheses* **43**, 93 (1963).

Lithium borodeuteride reduction of cis- and trans-N-diphenylmethyl-3,4-epoxy-3-methylpentanamides. When lithium borodeuteride reacts with alcohol groups there is an exchange of isotopic hydrogen on the boron atom; as a result, the atom-% excess deuterium at position 4 of the mevalonic diphenylmethylamides is considerably lower than that of the starting lithium borodeuteride. It has been established that the diphenyl-methylamido group reacts only sluggishly with lithium borohydride; the theoretical amount of the latter required for reaction with the primary alcohol group and reduction of the epoxide group is 0.5 molar equivalent. When 0.8 molar equivalent of lithium borodeuteride is used the resulting N-diphenylmethyl mevalonamide contains 72.5 atom-% excess D at position 4; with 1.5 molar equivalents this rises to 83% and with 2.5 molar equivalents to 87%. By carrying out an exchange of the primary alcoholic H against D_2O in tetrahydrofuran prior to reduction with 0.8 molar equivalent lithium borodeuteride a product containing 85 atom-% excess D can be obtained; by reducing the acetate instead of the free alcohol the atom-% excess D can be raised to 94%.

General Method of Reduction. To a solution of N-diphenylmethyl-3,4-epoxy-3-methylpentanamide in anhydrous tetrahydrofuran (approximately 1 ml per millimole) is added the chosen excess of lithium borodeuteride in tetrahydrofuran solution (approximately $0.7 M$ is a convenient dilution). The mixture is refluxed for 1.5 hours; excess of methanol (approximately 1 ml per millimole of lithium borodeuteride used) is added, and refluxing is continued for 10–15 minutes. The solution is then diluted with water and acidified with dilute phosphoric acid. The solvents are largely removed by boiling off (from a steam bath) or on a rotary evaporator, depending on the scale of the experiment. The product may be isolated in one of two ways: (1) it is collected by filtration, washed free of acid with water, and dried in a vacuum desiccator, or (2) it is extracted into chloroform and the chloroform solution is washed with saturated sodium chloride and sodium bicarbonate solutions, dried, and evaporated; the former method gives a crude product of higher purity, but in lower yield. It can be purified by crystallization from benzene (0.5 ml per 100 mg); pure N-diphenylmethyl mevalonamide melts at 101°; melting points of 97° and over are satisfactory. Yields of purified product are approximately 75% when 1.5 molar equivalents of lithium borodeuteride are used for the reduction. The yield from the *cis* epoxide is generally lower than that from the *trans* and the difference is especially marked when small excesses of reducing agent are used. The reaction has been performed on scales varying from 78 mg to 7 g.

Method of Hydrolysis to Mevalonic Acid. N-Diphenylmethyl meva-lonamide is refluxed for 2–3 hours with a 2.5 N solution of NaOH in 50%

aqueous methanol (1 ml per 50 mg); aqueous NaOH may be used, in which case the mixture should be stirred vigorously to ensure complete hydrolysis. The mixture is cooled and extracted three times with ether; the aqueous layer is neutralized and the methanol and dissolved ether are removed by evaporation on a steam bath or at low pressure. The aqueous solution, adjusted to pH 7.6, may be used directly in enzymatic experiments after assay for mevalonate content with mevalonate kinase (cf. this volume, article [12]).

For isolation of mevalonic lactone the solution is acidified and extracted continuously with chloroform for 24 hours; after evaporation of the chloroform the residue is heated to 100° at 12 mm for 10 minutes. The lactone thus obtained is entirely satisfactory for enzymatic work without further purification.

Crude mevalonic diphenylmethylamide may be hydrolyzed directly. Thus, N-diphenylmethyl-trans-3,4-epoxy-5-hydroxy-3-methylpentanamide (7 g; 22.51 millimoles) was reduced with lithium borodeuteride (18.24 millimoles); the crude mevalonic diphenylmethylamide (6.59 g), isolated by chloroform extraction, was hydrolyzed as described above to give 2.26 g (77%) of mevalonic lactone-4-D_1, which was used without distillation for conversion to mevalonic lactone-2-D_1 (see p. 376).

Mass Spectrometry of 4-D_1-labeled Mevalonic Lactones. The molecular ion regions of the mass spectra of mevalonic lactones and diphenylmethylamides can rarely be used for calculation of the proportion of deuterated molecules because of variable ion-molecule reactions. Using an MS.2 (A.E.I. Limited) mass spectrometer, the $[M - CH_3]^+$ region was found satisfactory.[30]

Lithium Borotritide Reduction of cis- and trans-N-Diphenylmethyl-3,4-epoxy-3-methylpentanamides. In our first reductions of N-diphenylmethyl *cis-* and *trans-*3,4-epoxy-5-hydroxy-3-methylpentanamides with lithium borotritide[20] a device was adopted, in order to obtain efficient incorporation of tritium, which has since been shown to have been based on a false assumption. It was assumed that both the alcoholic-OH and amide-NH groups would react with lithium borohydride before reduction of the epoxide group; therefore, 0.5 molar equivalent of nonradioactive lithium borohydride was added first to react with these groups; highly radioactive lithium borotritide (less than 0.25 molar equivalent) was then allowed to react, and finally more unlabeled borohydride was added to complete the reduction. It is now known that the diphenylmethylamido group reacts only very sluggishly with lithium borohydride and that reduction of the epoxide group takes place concomitantly with reaction

[30] J. W. Cornforth, R. H. Cornforth, G. Popják, and L. Yengoyan, *J. Biol. Chem.* **241**, 3970 (1966).

of the primary alcohol group; therefore, it is not possible in this instance
to use a "sandwich" technique. Nor is it possible to allow a deficiency
of lithium borotritide to react to completion before finishing the reduction
with unlabeled borohydride, for if the epoxides are refluxed with less
than sufficient lithium borohydride to ensure fairly rapid and complete
reduction, isomerization occurs leading to inferior products that cannot
easily be purified. Therefore, in order to minimize the isotope effect as far
as possible, the minimum quantities of lithium borotritide required to
give chemically satisfactory products have been used; these quantities
are 0.55 molar equivalent for the *trans* and 0.76 molar equivalent for the
cis epoxide.

The preparations described here have been carried out in collabora-
tion with the Radiochemical Centre; the quantities of radioactivity used
are accordingly much higher than would normally be employed in a
laboratory synthesis for biosynthetic work.[31]

Lithium Borotritide Solutions. Solid lithium borotritide is very slow
to dissolve in tetrahydrofuran, and a considerable residue remains undissolved. A solution provided by the Radiochemical Centre contains 1000
mC/ml, the specific activity of the solid lithium borotritide being $\sim$ 1500
mC/millimole by combustion. Calculated on radioactivity the solution
is approximately 0.67 M; however, by chemical estimation[12] (on 0.2 ml)
it is only 0.325 M; therefore, the specific activity of the dissolved lithium
borotritide is $\sim$ 3000 mC/millimole.[32]

*Mixture of 3R,4S- and 3S,4R-N-Diphenylmethyl-4-$^3H_1$ Mevalonam-
ides.* To N-diphenylmethyl-*trans*-3,4-epoxy-5-hydroxy-3-methylpentan-
amide (146 mg; 0.47 millimole) was added 0.325 M lithium borotritide
solution in tetrahydrofuran (0.8 ml; 0.26 millimole containing 750 mC);
the ampule was rinsed out into the reaction vessel with 2 $\times$ 0.4 ml of
anhydrous tetrahydrofuran. The mixture was refluxed for 1.5 hours;
methanol (0.2 ml) was added and refluxing was continued for 10 minutes.
Water (2 ml) was then added and the mixture was acidified with dilute
phosphoric acid; the solvents were very carefully boiled off from a steam
bath, and more water was added. The product solidified on cooling and
rubbing and was collected, washed with water until the washings were
neutral, and dried. The yield of crude product was 116 mg (78%) with
specific activity of 1264 μC/mg. The crude product (115 mg) was diluted
with unlabeled N-diphenylmethylmevalonamide (230 mg) and crystallized from benzene (1.8 ml) to give 303 mg (m.p. 98.5–100° on a Kofler

---

[31] Mevalonates labeled stereospecifically with tritium at C-4 and C-2 are available
from The Radiochemical Centre, White Lion Road, Amersham, Bucks, England.
[32] Tetrahydrofuran solutions of lithium borotritide of known molarity as well as
radioactivity are available from The Radiochemical Centre.

block; unlabeled pure specimen had m.p. 99–100.5°); the specific activity
was 427 μC/mg (134 mC/millimole).

Mixture of 3R,4R- and 3S,4S-N-Diphenylmethyl-4-$^3H_1$ Mevalonamides. N-Diphenylmethyl-*cis*-3,4-epoxy-5-hydroxy-3-methylpentanamide
(133 mg; 0.428 millimole) was reduced in a similar manner with 0.325 M
lithium borotritide solution (1 ml, containing 1000 mC). The yield of
crude product was 89.6 mg (67%), having a specific activity of 1124
μC/mg. After dilution of 89 mg with 178 mg of unlabeled material and
crystallization from benzene (1.4 ml), 246 mg was obtained having m.p.
99–100° (Kofler block) and specific activity 369 μC/mg (115 mC/millimole).

Hydrolyses are carried out as described for the corresponding deuterated N-diphenylmethyl mevalonamides.

VI. Preparation of 2R- and 2S-Mevalonates-2-D_1
from 4R- and 4S-Mevalonates-4-D_1

Principle. The stereospecifically 4-D_1-labeled mevalonic lactones,
prepared as described in Section V are converted, via the silver mevalonates to the methyl esters. These are oxidized by zinc permanganate
in acetone to the corresponding monomethyl 3-hydroxy-3-methylglutarates which, after conversion to their lithium salts, are reduced exclusively at the ester group by lithium borohydride. As a result of this
interconversion of carbinol and carboxyl groups the labeled hydrogen
atoms are at position 2 of the new mevalonates and the configuration
around C-3 has suffered inversion. Thus, the mixtures of enantiomorphs
behaving as 3R,4R- and 3R,4S-mevalonates-4-D_1, respectively, in enzymatic reactions become mixtures of enantiomorphs behaving as 2R,3R-
and 2S,3S-mevalonates-2-D_1, respectively.

Procedure.[30] The procedure is described for unlabeled material.

Methyl Mevalonate. Mevalonic lactone (1.03 g) is titrated at 40–50°
with standardized barium hydroxide solution (phenolphthalein indicator);
14.46 ml of 0.53 N (97% of the theoretical) is used. The solution of
barium mevalonate is added to an exact equivalent of silver sulfate
(1.195 g of 99%) dissolved in boiling water (120 ml). The barium sulfate
is filtered off through a pad of Celite and washed with water. The filtrate
is evaporated to dryness on a rotary evaporator with slight warming and
the residue is finally dried at 50–60° and 0.02 mm pressure for 4 hours.
The dried silver mevalonate (1.95 g) is triturated under anhydrous
carbon tetrachloride (30 ml); the resulting suspension is stirred vigorously
with a mechanical stirrer during addition of a solution of methyl iodide
(3 ml) in dry carbon tetrachloride (10 ml) over a period of 30 minutes.
The mixture is stirred for a further 2 hours; the silver iodide is removed

by filtration through a Celite pad and is washed with carbon tetrachloride. The filtrate is evaporated at room temperature and low pressure, finally at 0.02 mm, leaving methyl mevalonate (1.07 g; 84% yield;
n_D^{20} 1.4525). Methyl mevalonate cannot be distilled and gradually cyclizes
to mevalonic lactone even at room temperature. If strictly anhydrous
conditions are not observed during this preparation the product always
contains some mevalonic lactone. The presence of mevalonic lactone is
indicated by a rise in refractive index (the lactone has n_D^{20} 1.4740) and by
the appearance of characteristic peaks in the infrared spectrum.

Monomethyl 3-Hydroxy-3-methylglutarate. Freshly prepared methyl
mevalonate (1.07 g; 6.6 millimole) in analytical grade acetone (100 ml)
is cooled in ice water and stirred magnetically while $Zn(MnO_4)_2 \cdot 6 \ H_2O$
(2.23 g; 5.4 millimoles) is added in small portions during 20 minutes.
Stirring is continued for 15 minutes, and excess permanganate is then
destroyed by careful addition of 30% hydrogen peroxide solution. The
sludge of manganese dioxide and zinc salts is separated by centrifuging
and washed twice with acetone; evaporation of the acetone leaves an oil
(175 mg) containing mevalonic lactone. The sludge is suspended in water,
and the manganese dioxide is dissolved by careful addition of 30%
hydrogen peroxide, the pH being kept at 3–5 by addition of $4 N$ sulfuric
acid. The solution is then acidified to pH 2.5 and extracted continuously
with chloroform for 24 hours. Evaporation of the solvent on a rotary
evaporator, finally at 0.02 mm pressure with slight warming, leaves
crude monomethyl 3-hydroxy-3-methylglutarate (684 mg). This is chromatographed on a 100 g Celite 0.05 N sulfuric acid column (8 ml of
acid per 10 g of Celite) with chloroform (25 ml fractions) as eluent.
Monomethyl 3-hydroxy-3-methylglutarate (525 mg; 3 millimoles) is
isolated from fractions 19 to 36.

Mevalonic Lactone from Monomethyl 3-Hydroxy-3-methylglutarate.
Monomethyl 3-hydroxy-3-methylglutarate (528 mg; 3 millimoles) is
neutralized with 0.2 N lithium hydroxide solution (14.88 ml); the solution is evaporated at low pressure on a rotary evaporator, finally at
80–90° and 0.02 mm pressure for 4 hours. The partly crystalline lithium
salt is dissolved in anhydrous tetrahydrofuran (40 ml) by stirring overnight at room temperature, or more rapidly by warming. A 1 M solution
of lithium borohydride in anhydrous ether or tetrahydrofuran (5 ml) is
added, and the mixture is refluxed for 2 hours. Methanol (8 ml) is
added, and refluxing is continued for 30 minutes; water (20 ml) is then
added, and the organic solvents are removed on a rotary evaporator. The
aqueous solution is brought to pH 8 and extracted overnight with chloroform in order to remove most of the boric acid. It is then acidified to
pH 2.5 and extracted with chloroform for 24 hours. After removal of

the solvent, the residue is evaporated five times with methanol to remove traces of boric acid; it is then heated to 100° at 12 mm pressure for 15 minutes. The residual mevalonic lactone (372 mg; 95%) gives an infrared spectrum identical with that of an authentic specimen.

The overall yield from mevalonic lactone to mevalonic lactone is 36%. Yields of up to 45% have been obtained. The crude 4R-mevalonic lactone-4-D$_1$, whose preparation was described in Section V, gave a 41% yield of 2R-mevalonic lactone-2-D$_1$; assay with mevalonate kinase showed this to be 99% pure.

VII. Synthesis of Methyl and Ethyl Farnesoates from *Trans*-Geranyl Acetone

Principle. Trans-geranyl acetone is condensed with diethyl methoxy- or ethoxy-carbonylmethylphosphonate carbanion[33] to give a mixture of *cis-trans* and *trans-trans* farnesoic esters, which are separated by gas liquid chromatography.

Procedure. Trans-geranyl acetone is made according to Isler *et al.*[34] The phosphonates are prepared by means of the Michaelis-Arbuzov reaction.[35] Equivalent quantities of triethyl phosphite and methyl or ethyl bromoacetate are mixed and heated, under an air condenser, to 130°; alkyl bromide begins to evolve below 100°. The mixture is heated at 130° for a further 3–5 hours and distilled. Diethyl methoxycarbonylmethylphosphonate boils at 132° at 9 mm, and diethyl ethoxycarbonylmethylphosphonate boils at 140° at 9 mm. Tetrahydrofuran is dried by stirring with powdered calcium hydride for 2 days; lithium aluminium hydride is then added until no further reaction takes place. The tetrahydrofuran is freshly distilled as required.

Ethyl Farnesoate. A slurry of sodium hydride (50% in oil; 1.2 g) in dry tetrahydrofuran (40 ml) is cooled to 20° and stirred magnetically in an atmosphere of nitrogen during addition of diethyl ethoxycarbonylmethylphosphonate (5.6 g; 0.025 mole); 5 ml of tetrahydrofuran is used for washing in. Stirring is continued at room temperature until solution occurs and hydrogen is no longer being evolved (30–60 minutes). *Trans*-geranyl acetone (4.875 g; 0.025 mole) is now added dropwise, another 5 ml of tetrahydrofuran being used for washing in; there is a slight rise in temperature (from 20 to 24.5°) which is maintained for some hours. The

[33] H. Pommer, *Angew. Chem.* **72**, 811, 911 (1960); S. Trippett and D. M. Walker, *Chem. & Ind.* p. 990 (1961); W. S. Wadsworth and W. D. Emmons, *J. Am. Chem. Soc.* **83**, 1733 (1961).

[34] O. Isler, R. Rüegg, L. Chopard-dit-Jean, H. Wagner, and K. Bernhard, *Helv. Chim. Acta* **39**, 897 (1956).

[35] G. M. Kosolapoff, "Organophosphorus Compounds," 1st ed., Chap. 7. Wiley, New York, 1950.

reaction mixture is allowed to stand at room temperature for 2–3 days, during which time a gelatinous yellow mass separates. Most of the solvent is evaporated on a rotary evaporator at room temperature; water is added and the product is extracted with light petroleum (b.p. 40–60°). After evaporation of the solvent the residue is distilled giving 6.23 g of product, b.p. 95–100°/0.01–0.02 mm. Gas-liquid chromatographic analysis on a 6-foot 10% Apiezon-L column at 185° shows this to have the composition: 9% geranyl acetone, 18% ethyl *cis-trans* farnesoate, 73% ethyl *trans-trans*-farnesoate. Therefore the ethyl farnesoate is 80% *trans-trans*-isomer and is produced in 86% yield.

Methyl Farnesoate-2-14C. The procedure described above has not been applied to the preparation of 14C-labeled farnesoate, but would obviously be the method of choice. The experimental conditions described here should therefore be modified to take advantage of the improved procedure.

Redistilled triethyl phosphite (339 mg; 2.04 millimoles) is weighed into a suitable reaction vessel,[36] and methyl bromoacetate-2-14C (281 mg; 1.84 millimole containing 4 mC) is distilled into it by vacuum line technique. The mixture is heated at 130° for 5 hours and excess triethyl phosphite is removed on the vacuum line with slight warming. To the residue (383 mg) is added *trans*-geranyl acetone (355 mg; 1.83 millimole) and dry dimethylformamide (2 ml). The solution is cooled in ice water and stirred under dry nitrogen during addition of a solution of dry sodium methoxide (from evaporation, finally at 200° at 12 mm, of 2 ml of a 0.1 M solution in methanol) in dimethylformamide (8 ml). The mixture is kept for 60 hours at room temperature; it is then diluted with an equal volume of water and extracted several times with light petroleum (b.p. 40–60°). The solution is concentrated to small volume by a current of nitrogen and then warmed to 50° at 12 mm pressure for a short time. The residue (370 mg) analyzed by gas-liquid radiochromatography[37] on an ethylene glycol–succinate polyester column at 172° has the following composition: geranyl acetone, 49.5%; methyl *cis-trans*-farnesoate, 13.5%; *trans-trans*-farnesoate, 37%. Ninety percent of the radioactivity is associated with the farnesoates.

VIII. Preparation of *Trans-Trans*-Farnesol

Principle. The esters of farnesoic acid (cf. Section VII) are resolved into the *cis-trans* and *trans-trans*-components by preparative gas-liquid

---

[36] The authors used a tube (with ground glass joint for attachment to the vacuum line), around the upper portion of which a small lead condenser coil could be placed during the period of heating.

[37] G. Popják, A. E. Lowe, D. Moore, L. Brown, and F. A. Smith, *J. Lipid Res.* **1**, 29 (1959); G. Popják, A. E. Lowe, and D. Moore, *ibid.* **3**, 364 (1962).

chromatography;[38] the *trans-trans*-isomer is then reduced with LiAlH$_4$ to the alcohol. Reduction with lithium aluminium tritide, or deuteride, gives 1-^{3}H$_2$- and 1-D$_2$-labeled *trans-trans*-farnesol.

Preparative Gas-Liquid Chromatography (GLC) of Farnesoates. Suitable column packing for the separation of methyl or ethyl farnesoates is Celite 545, or Chromosorb (80–120 mesh), coated with 10% Apiezon L vacuum grease,[39] or with 10–15% Carbowax 20M (Union Carbide Corporation). We prefer the Apiezon column. The operating temperature is 190–195°. We have used 180 × 2 cm and 180 × 8 mm columns with equally good results: satisfactory resolution of the *cis-trans* and *trans-trans* isomers is obtained on the larger column with a load of 50 mg; the capacity of the smaller column (used in an Aerograph Autoprep instrument) is about 20 mg.

The vapors of farnesoates, as well as of farnesol, are difficult to condense. With the Aerograph Autoprep instrument, when the collecting tubes are immersed in melting ice, only 50–60% of the farnesoates applied to the column are trapped; the rest escapes as aerosol. When unlabeled material is fractionated and ample supplies are available, this loss is of no consequence, but with isotopically (e.g., ^{14}C) labeled material it cannot be tolerated.

When radioactive specimens are fractionated by GLC it is not permissible to allow at any time the passage of vapors from the column without an efficient collector being attached to the exit of the apparatus. For the quantitative collection of vapors of methyl farnesoate-2-^{14}C, the simple device illustrated in Fig. 1 was found adequate, as the vapors could be directed with the turning of the two-way tap into either of two U-traps, packed with cotton wool moistened with light petroleum (b.p. 40–60°) and which had previously been extracted thoroughly with fat solvents in a Soxhlet apparatus. The methyl *trans-trans*-farnesoate-2-^{14}C is collected in one trap and the *cis-trans*-isomer and impurities are condensed in the other. For the extraction of the substance(s) trapped in the U-tubes, the cotton wool is transferred with forceps to a straight tube fitted with a fine sintered disk. First, enough light petroleum (b.p. 40–60°) is poured over the cotton wool just to cover it and then a further 50 ml of the solvent is delivered from a burette (ungreased stopcock!) at a rate corresponding to the rate of flow through the sinter. The U-trap is rinsed with a few milliliters of light petroleum which is combined with

---

[38] It is preferable to effect the separation of the geometric isomers on the esters rather than after reduction to farnesol because the separation factor for *cis-trans* and *trans-trans* farnesoate is 1.2 as against 1.1 for the two farnesols (see footnote 39).

[39] G. Popják and R. H. Cornforth, *J. Chromatog.* 4, 214 (1960).

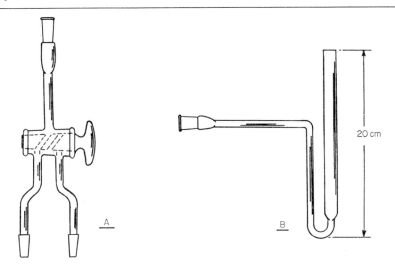

FIG. 1. Two-way tap (A) and U-trap (B) packed with defatted cotton wool moistened with petroleum (b.p. 40–60°) for the condensation of methyl farnesoate in preparative gas-liquid chromatography [G. Popják, J. W. Cornforth, R. H. Cornforth, R. Ryhage, and D. S. Goodman, *J. Biol. Chem.* **237**, 56 (1962)]. Reproduced by the permission of the Editor, *Journal of Biological Chemistry.*

the eluate of the cotton wool. The solvent is removed on a rotary evaporator. When the bath temperature does not exceed 40°, methyl farnesoate (and farnesol) is recovered without perceptible losses under water-pump vacuum.

A further precaution taken against contamination of the environment with radioactive material is that the load of the GLC column is reduced: in the case of methyl farnesoate-2-^{14}C and using the collector illustrated in Fig. 1, a 10-mg load is the largest that can be trapped quantitatively.[40]

By preparative GLC specimens of *trans-trans*-farnesoate of 95–98% purity are readily obtained by a single fractionation; the 2–5% impurity remaining is the *cis-trans* isomer.

Reduction of Methyl trans-trans-Farnesoate to Farnesol. The reduction of farnesoates to farnesol with LiAlH$_4$ is carried out at −30°. The procedure[41] is illustrated with the preparation of 1-D$_2$, 2-^{14}C-labeled *trans-trans*-farnesol.

[40] When the object is the preparation of *trans-trans*-farnesol-2-^{14}C it is preferable to synthesize a small quantity of the labeled farnesoate with a high specific activity, resolve the *cis-trans* and *trans-trans* isomers on small scale, and then dilute the labeled *trans-trans*-farnesoate with the unlabeled material to the desired specific activity.

[41] G. Popják, J. W. Cornforth, R. H. Cornforth, R. Ryhage, and D. S. Goodman, *J. Biol. Chem.* **237**, 56 (1962).

Methyl *trans-trans*-farnesoate-2-^{14}C (506 mg; specific activity 0.02 μC per micromole; 95% *trans-trans* isomer), dissolved in 5 ml of dry ether, is added slowly over 30 minutes to a stirred solution of LiAlD$_4$ in ether kept at $-30°$ (422 mg LiAlD$_4$, 99.2% D, in 50 ml ether).[42] After addition of all the farnesoate to the LiAlD$_4$ and washing in the contents of the dropping funnel with 5 ml of dry ether, the reaction mixture is stirred for a further 90 minutes at $-30°$. After the usual working up, the yield of farnesol is 425 mg (95% *trans-trans*-isomer by GLC; $> 98\%$ farnesol-1-D$_2$ by mass spectrometry).

In another preparation, 577 mg of methyl *trans-trans*-farnesoate-2-^{14}C (specific activity 284 μC per micromole) was reduced similarly with 127 mg of LiAlD$_4$ and gave 510 mg of *trans-trans*-farnesol.

IX. Chemical Synthesis of Isopentenyl (3-Methylbut-3-en-1-yl) Pyrophosphate

The method described here[43] gives a higher yield than any previously published procedure.[44–47] Isopentenol is first phosphorylated to the monophosphate with POCl$_3$ and pyridine;[47] the monophosphate is then converted via the phosphoromorpholidate[48] to the pyrophosphate.

Reagents

(1) Isopentenol, prepared via the Grignard reagent in tetrahydrofuran from β-methylallyl chloride.[46] Isopentenol-1-^{14}C is made by the carbonation of the Grignard reagent with ^{14}CO$_2$

(2) Pyridine, analytical grade reagent, refluxed over barium oxide, distilled and then stored over sticks of calcium hydride

(3) Morpholine, dried and distilled as the pyridine above

(4) Dicyclohexylcarbodiimide

(5) POCl$_3$, freshly distilled

(6) Lithium hydroxide

(7) *tert*-Butanol

(8) Orthophosphoric acid, 88%

(9) Tri-*n*-butylamine

[42] The seemingly large excess of deuteride was used because we had reason to believe that the preparation used was not entirely satisfactory.

[43] C. Donninger and G. Popják, *Biochem. J.* **105**, 545 (1967).

[44] F. Lynen, H. Eggerer, U. Henning, and I. Kessel, *Angew. Chem.* **70**, 738 (1958).

[45] C. Yuan and K. Bloch, *J. Biol. Chem.* **234**, 2605 (1959).

[46] H. Eggerer and F. Lynen, *Ann. Chem.* **630**, 58 (1960).

[47] C. D. Foote and F. Wold, *Biochemistry* **2**, 1254 (1963).

[48] J. G. Moffatt and H. G. Khorana, *J. Am. Chem. Soc.* **83**, 649 (1961).

Method

Dilithium Isopentenyl Phosphate. The procedure to be described on a large scale with unlabeled isopentenol is equally applicable with the isotopically labeled alcohol on a millimolar scale.

Freshly distilled $POCl_3$ (4.1 ml; 45 millimole) and dry ether (6 ml) are introduced into a 50-ml round-bottom flask fitted with a dropping funnel and provided with facilities for magnetic stirring. The mixture is cooled in an ice-salt bath to $-10°$ to $-15°$. Isopentenol (2.58 g; 30 millimoles), dry pyridine (3.63 ml; 45 millimoles), and dry ether (6 ml), mixed in the dropping funnel, are added drop by drop to the cooled solution of $POCl_3$ with continuous stirring. After the addition of the alcohol (30 minutes) the flask is stoppered and left at $-20°$ overnight. The reaction mixture is filtered next morning through a sintered-glass plate, in order to remove the precipitate of pyridine hydrochloride; the filtrate is collected on 10 g of crushed ice. The filter is washed with the ether-rinsing of the reaction flask. Lithium hydroxide (4.32 g dissolved in 50 ml of water) is then added to the ether-ice mixture. The ether is removed on a rotary evaporator and the pH of the aqueous phase is adjusted to 12 with ammonia. The inorganic trilithium phosphate, which precipitates, is centrifuged down and is washed with ice-cold water. The supernatant is combined with the washings of the lithium phosphate precipitate, and the pH of the solution is adjusted to 8 with HCl; the preparation is then lyophilized.

The dry residue is washed with and suspended in ethanol–ether (1:2, v/v) and transferred with the solvent to a centrifuge tube. The insoluble material, dilithium isopentenyl monophosphate, is centrifuged down, washed once more with ethanol-ether and twice with ether and is finally dried *in vacuo* over P_2O_5. Yield of product is 2.71 g (51%), which gives analysis correct for $C_5H_9O_4PLi_2$.

Isopentenyl Phosphoromorpholidate and Trilithium Isopentenyl Pyrophosphate. Dilithium isopentenyl phosphate (890 mg; 5 millimoles) is dissolved in a little water and is added on to a 5×1 cm Dowex-50 column (H⁺ form): the isopentenyl monophosphoric acid is washed off the column with 50 ml of water. An equal volume of *tert*-butanol (50 ml) and an excess of morpholine (1.7 ml; 20 millimoles) are added to the column eluate and the mixture is brought to a gentle reflux in a flask fitted with a dropping funnel and condenser. Dicyclohexylcarbodiimide (4.12 g; 20 millimoles) dissolved in *tert*-butanol (20 ml) is added through the dropping funnel over 1 hour to the mixture, which is then refluxed gently for 4 hours and can be left then at room temperature overnight.

The dicyclohexylurea formed is filtered off and washed with *tert*-butanol. The combined filtrates are concentrated *in vacuo* until nearly all the butanol is driven off. The remaining solution is extracted with ether and then evaporated to dryness *in vacuo*. A few milliliters of methanol are added to the residue and distilled off *in vacuo;* this is repeated three times. The same process is repeated with ether.

The product, isopentenyl phosphoromorpholidate, a yellow gum, is dried, without further purification, by two evaporations of its solution in dry pyridine (20 ml) under high vacuum. Dry N_2-gas is led into the flask during the addition of fresh pyridine.

The phosphorylating reagent is prepared separately by dissolving 88% orthophosphoric acid (1.02 ml; 15 millimoles) and tri-*n*-butylamine (3.6 ml; 15 millimoles) in dry pyridine (10 ml) and by distilling off the pyridine *in vacuo*. Fresh pyridine (10 ml) is added to the residue and is distilled off again. The butylammonium phosphate is then dissolved in fresh pyridine and the solution is added to the dry isopentenyl phosphoromorpholidate. After mixing, the pyridine is removed *in vacuo,* and the dry residue is dissolved in fresh pyridine (10 ml). After 2 days at room temperature (about 20°) the pyridine is distilled off *in vacuo*. The gummy residue is dissolved in water, the solution is transferred to a stoppered centrifuge tube, and lithium hydroxide is added to pH 12. The mixture is extracted with ether. The emulsion, which invariably forms, is broken by centrifuging, which separates four layers: a precipitate of trilithium phosphate, an aqueous layer, an interphase precipitate, and an ether layer. The ether layer is siphoned off and discarded. The interphase precipitate is decanted with the aqueous layer and the precipitate of trilithium phosphate is washed with a little cold water. The combined washings and the aqueous phase are then evaporated to dryness under reduced pressure. The solid residue is extracted thoroughly with ethanol–ether (1:2, v/v), and the insoluble material is collected by centrifuging. The pellet of trilithium isopentenyl pyrophosphate is washed with ether and dried *in vacuo* over P_2O_5. The yield (1.1 g) is 83% from isopentenyl monophosphate. The product gives a slightly low analysis for P (21.8% instead of 23.4%) on account of a small contamination with isopentenyl monophosphate.

Analysis of Isopentenyl Pyrophosphate. High-voltage electrophoresis (e.g., at 8 kV and 30 mamp for 15–20 minutes) on paper in pyridine–acetic acid–water, pH 6.1 (100 ml of pyridine and 8 ml of acetic acid diluted with water to 2.5 liters) separates isopentenyl monophosphate, isopentenyl phosphoromorpholidate, and isopentenyl pyrophosphate. The electrophoretic mobilities of these three substances on Whatman No. 1 paper relative to the mobility of inorganic orthophosphate are: 0.75, 0.57, and 1.06.

Chromatography on a column of silica gel (Whatman SG-31, 100–200

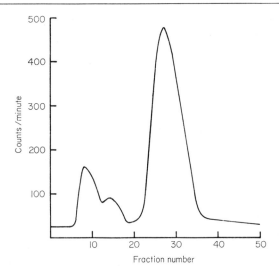

Fig. 2. Purification of ¹⁴C-labeled isopentenyl pyrophosphate synthesized by the method of C. Donninger and G. Popják [*Biochem. J.* **105**, 545 (1967)] on a silicic acid column. The largest component is isopentenyl pyrophosphate (5-ml fractions).

mesh), with n-propanol–ammonia (sp. gr. 0.88)–water (6:3:1, by volume), is highly suitable for the final purification of isopentenyl pyrophosphate. Figure 2 shows the chromatography of a specimen of isopentenyl-1-¹⁴C pyrophosphate, synthesized by the method described, on such a column. The same solvent system, or isopropanol–ammonia–water, can be used in thin-layer chromatography of isopentenyl pyrophosphate; on silica gel H plates, 0.25 mm thick, isopentenyl pyrophosphate has an R_f of 0.15–0.20 with these two solvents.

For paper chromatography of isopentenyl pyrophosphate see Vol. V [66].

Xa. Synthesis of Prenyl Pyrophosphates

The method is based on the procedure of Cramer and Böhm[49] as modified by Popják *et al.*[41] The overall reactions of the process are:

$$RCH_2OH + HO \cdot PO_3^{2-} + CCl_3CN \rightarrow RCH_2\text{-}OPO_3^{2-} + CCl_3CONH_2$$
$$RCH_2\text{-}OPO_3^{2-} + H_2PO_4 + CCl_3CN \rightarrow RCH_2\text{-}O\text{-}P_2O_6^{3-} + CCl_3CONH_2$$

This method is equally applicable for the phosphorylation of isopentenol, 3,3-dimethylallyl alcohol, geraniol, and farnesol. For the preparation of isopentenyl pyrophosphate we prefer another method (cf. Section IX). In spite of the poor yield (about 15%) with the allylic

[49] F. Cramer and W. Böhm, *Angew. Chem.* **71**, 775 (1959).

alcohols, the method coupled with isolation procedures improved since originally described, gives reliable and pure preparations.

Reagents

    (1) Di-triethylammonium phosphate. The crystalline salt is prepared by dissolving 85–100% orthophosphoric acid in acetonitrile and adding to it 2 equivalents of redistilled triethylamine: 20 g of H_3PO_4 is dissolved in 100 ml of acetonitrile to which 41.3 g of redistilled triethylamine is added. Di-triethylamine phosphate crystallizes overnight at room temperature. The crystals are collected on a sintered funnel, washed briefly with acetonitrile, and kept in a desiccator over silica gel

    (2) Acetonitrile, redistilled

    (3) Trichloroacetonitrile, redistilled

    (4) 3,3-Dimethylallyl alcohol; geraniol, *trans-trans*-farnesol[50]

Procedure. To 1 millimole of the alcohol to be phosphorylated are added 6 millimoles of trichloroacetonitrile in a flask fitted with a stirrer and dropping funnel. Di-triethylammonium phosphate, 2.4 millimoles (~0.7 g), dissolved in acetonitrile (20 ml) is then added slowly through the dropping funnel to the solution of the alcohol at room temperature over a period of 3–4 hours, the mixture being stirred continuously. After standing for 2 hours more at room temperature the reaction mixture is diluted with 100 ml of ether and transferred to a separating funnel; it is then extracted three times with 25–30 ml of 0.1 N aq. ammonia.[51] The aqueous extracts are combined and extracted three times with 50–100 ml of ether[51] and then concentrated to a few milliliters on a rotary evaporator (bath temperature about 50°). The aqueous concentrate contains the prenyl mono- and pyrophosphates and also unreacted phosphorylating reagent. The manner of isolation of the pyrophosphate depends on the alcohol phosphorylated.

Isolation of 3,3-Dimethylallyl Pyrophosphate or of Geranyl Pyrophosphate. When either 3,3-dimethylallyl alcohol or geraniol is phosphorylated, the aqueous concentrate obtained in the previous step is applied to a column of silica gel (Whatman SG-31; 60 × 1.3 cm) equilibrated with *n*-propanol–conc. ammonia (sp. gr. 0.88)–water (6:3:1, by volume)[52] and the column is eluted with the same solvent system.

---

[50] 3,3-Dimethylallyl alcohol is prepared by the reduction of 3-methylcrotonic acid with $LiAlH_4$. Pure geraniol is available commercially (Meranol; A. Boake Roberts & Co., London). For preparation of *trans-trans*-farnesol see Section VIII.

[51] Obstinate emulsions form during these extractions; the phases can be separated satisfactorily only by centrifuging.

[52] H. Plieninger and H. Immel, *Chem. Ber.* **98**, 414 (1965).

When 3-ml fractions are collected the monophosphates of 3,3-dimethyl-allyl alcohol and of geraniol appear between fractions 7 and 25, dimethyl-allyl pyrophosphate between fractions 34 and 45, and geranyl pyrophosphate, in a somewhat tailing band, between fractions 31 and 68.[53] This column is suitable also for the purification of isopentenyl pyrophosphate (cf. Fig. 2), which is eluted from it with about the same volume of solvent as is dimethylallyl pyrophosphate. The solution of the allyl pyrophosphates is concentrated to a few milliliters (2–4 ml), made alkaline with a drop of ammonia and an equal volume of acetone and a little ethanol (about one-eighth to one-tenth of the volume of acetone) is added. The triammonium allyl pyrophosphate (dimethylallyl or geranyl pyrophosphate) crystallizes out as a white solid; it is recrystallized twice from 50% aqueous acetone. The yield of the recrystallized product is about 15%.[53]

   Isolation of Farnesyl Pyrophosphate. The aqueous concentrate obtained from the phosphorylation reaction is treated with XAD-2 resin as described for the purification of the enzymatically synthesized farnesyl pyrophosphate (cf. this volume, article [12]). When the farnesol phosphorylated is labeled with ^{14}C the adsorption of farnesyl monophosphate and pyrophosphate to the resin may be followed conveniently by the disappearance of radioactivity from the supernatant of the resin. When the adsorption is complete the resin is collected on a sintered-glass filter and washed repeatedly with 0.01 N aqueous ammonia. The farnesyl monophosphate and pyrophosphate are then eluted from the resin with methanol containing 0.01 N ammonia. The methanol extract, concentrated if necessary to a few milliliters, is then applied on to a DEAE-cellulose column which is then eluted with methanol containing 0.08 M ammonium formate as described for the purification of farnesyl pyrophosphate from enzyme incubations (cf. this volume, article [12]). From a column of 12 × 2 cm DEAE-cellulose (Whatman DE-11, or Bio-Rad Cellex D, capacity 0.99 mEq/g) farnesyl monophosphate is eluted between 100 and 250 ml and the pyrophosphate between 300 and 450 ml of effluent. The pooled fractions of the pyrophosphate are concentrated on a rotary evaporator, the methanol being gradually replaced by 0.01 N aqueous ammonia. When all the methanol has been driven off and the ammonium formate begins to crystallize, the concentration is stopped, a little 0.01 N aq. ammonia is added just sufficient to dissolve the ammonium formate and the farnesyl pyrophosphate is separated from the ammonium formate by a second adsorption on to and elution from XAD-2 resin, as described for the purification of enzymatically synthesized farnesyl pyrophosphate. In syntheses ranging from 0.5 to 2 millimolar scale the yields varied

[53] P. W. Holloway and G. Popják, *Biochem. J.* **104,** 57 (1967).

between 10 and 15%. Slightly alkaline solutions (pH 8–9) of farnesyl pyrophosphate are stable at −20° for many months.

Xb. Some Properties of Prenyl Pyrophosphates

The prenyl pyrophosphates, in common with other substances in which an anionic, or potentially anionic, group is in the α-position to a double bond, are electrophilic reagents and can be expected to lose the pyrophosphate group readily. The electrophilic character of the prenyl pyrophosphates accounts, no doubt, for their particular reactivity in biosynthetic reactions as well as for their instability to acid. It was this latter property, at least in part, which led to the recognition of the biosynthesis of farnesyl pyrophosphate in yeast[44] as well as in liver extracts.[54] The dependence of the hydrolysis of farnesyl pyrophosphate on pH in a buffered medium is shown in Fig. 3, which indicates that the undissociated

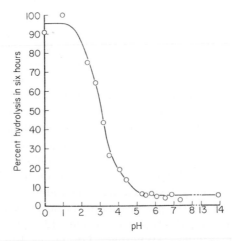

Fig. 3. Percentage hydrolysis of farnesyl pyrophosphate in 6 hours as a function of pH in buffered solutions at 22° [D. S. Goodman and G. Popják, *J. Lipid Res.* **1**, 286 (1960)]. Reproduced by the permission of the Editor, *Journal of Lipid Research*.

form of the farnesyl pyrophosphoric acid is the unstable species.[54] At any pH the hydrolysis at room temperature goes to completion, but the rate varies inversely with pH. The rate of hydrolysis of farnesyl pyrophosphate at pH 3.19 and at 22° is shown in Fig. 4. At pH 4 about 5 days are needed for complete hydrolysis. It may be predicted that if periods of exposure to acid shorter than 6 hours had been used for the hydrolysis

[54] G. Popják, *Tetrahedron Letters* **19**, 19 (1959); D. S. Goodman and G. Popják, *J. Lipid Res.* **1**, 286 (1960).

of farnesyl pyrophosphate, the curve shown in Fig. 3 would have been shifted to the left to a limiting position with its mid point at pH 1.9, corresponding to the pK_1 for the dissociation of a single hydrogen ion from pyrophosphoric acid.

The products of acid catalyzed hydrolysis of farnesyl pyrophosphate are inorganic pyrophosphate,[44, 54] nerolidol, and *trans-trans*-farnesol,[54] the two alcohols appearing in a ratio of about 4:1. The only view compatible with

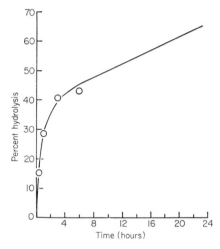

FIG. 4. Rate of hydrolysis of farnesyl pyrophosphate in 0.1 M phthalate buffer, pH 3.19, at 22° [D. S. Goodman and G. Popják, *J. Lipid Res.* **1**, 286 (1960)]. Reproduced by the permission of the Editor, *Journal of Lipid Research*.

the formation of nerolidol from farnesyl pyrophosphate at an acid pH is that the carbon-oxygen bond of the ester is cleaved, that the pyrophosphate moiety of the molecule extracts an electron from the organic portion and leaves an anion resulting in an allylic carbonium ion which is partially stabilized by resonance between two mesomeric forms, (a) and (b). These carbonium ions in the presence of water then accept a hydroxyl ion and form the primary and tertiary alcohol.

$$\underset{\text{(a)}}{\overset{\overset{\displaystyle CH_3}{|}}{-C}=CH\overset{(+)}{-}CH_2} \leftrightarrow \underset{\text{(b)}}{\overset{\overset{\displaystyle CH_3}{|}}{-C}\underset{(+)}{-}CH=CH_2}$$

In contrast to its instability in acid, farnesyl pyrophosphate is fairly stable in neutral and alkaline solutions. In a solution of pH 8 frozen at −20° only 2% hydrolysis was observed in 3 months. At 100° in 0.1 N

KOH farnesyl pyrophosphate is hydrolyzed to the extent of 15% in 1 hour and the product of hydrolysis is principally farnesol. Thus during alkaline hydrolysis the oxygen-phosphorus bond is the one that is broken.

3,3-Dimethylallyl and geranyl pyrophosphate have properties similar to those of farnesyl pyrophosphate. In the presence of acid these give, in addition to inorganic pyrophosphate, methylvinylcarbinol (2-methylbut-3-en-2-ol) and 3,3-dimethylallyl alcohol, linalool, and geraniol, respectively. Detailed information about the rates of hydrolysis of the monoprenyl pyrophosphate is not available, but the half-life of geranyl pyrophosphate in $0.1\ N$ HCl was reported to be 1.71 minutes at $25°$.[55]

All the assays used so far for the enzymatic synthesis of the prenyl pyrophosphates from ^{14}C-labeled mevalonate, or from ^{14}C-labeled isopentenyl pyrophosphate, depend on their instability to acid and on the resulting liberation of the primary and tertiary ^{14}C-labeled alcohols, extractable from the incubations either with ether or with light petroleum (cf. this volume, article [12]).

[55] H. Eggerer, *Chem. Ber.* **94**, 174 (1961).

Section III

Enzyme Systems of Terpenoid and Sterol Biosynthesis and Sterol Ester Formation

[12] Enzymes of Sterol Biosynthesis in Liver and Intermediates of Sterol Biosynthesis

By G. POPJÁK

Introduction

Chemical syntheses of substrates of sterol biosynthesis are presented in this volume, article [11].[1]

This article is divided into two sections:

I. Individual enzymes of sterol biosynthesis and the enzymatic preparation of substrates derived from mevalonate.

Although mevaldate reductase is believed not to be involved directly in sterol biosynthesis, the preparation of this enzyme will be included since it is most useful for the preparation of mevalonate labeled stereospecifically at C-5 with a hydrogen isotope.

Thus in Section I we will deal with the enzymes that catalyze reactions (1) to (7).[2]

$$MVald + NADH \text{ (or NADPH)} + H^+ \rightarrow MVA + NAD^+ \text{ (or NADP}^+) \tag{1}$$

$$MVA + ATP \rightarrow 5\text{-P-MVA} + ADP \tag{2}$$

$$5\text{-P-MVA} + ATP \rightleftarrows 5\text{-PP-MVA} + ADP \tag{3}$$

$$5\text{-PP-MVA} + ATP \rightarrow I\text{-PP} + CO_2 + P_i + ADP \tag{4}$$

$$I\text{-PP} \rightleftarrows Mp\text{-PP} \tag{5}$$

$$Mp\text{-PP} + I\text{-PP} \rightarrow Dp\text{-PP} + PP_i \tag{6}$$

$$Dp\text{-PP} + I\text{-PP} \rightarrow Tp\text{-PP} + PP_i \tag{7}$$

II. Multienzyme systems of sterol biosynthesis:

(1) enzyme system synthesizing squalene and sterols from mevalonate;

(2) enzyme system synthesizing farnesyl pyrophosphate and isopentenyl pyrophosphate from mevalonate;

(3) enzyme system synthesizing squalene from farnesyl pyrophosphate;

(4) a system reconstituted from soluble enzymes and microsomes for the synthesis of sterols and squalene.

The synthesis of squalene from farnesyl pyrophosphate is included in

---

[1] For general introduction to articles [11] and [12], see this volume, article [11].

[2] The abbreviations and terminology used are given in footnote 2 of article [11], this volume.

Section II because there is now evidence indicating that a hitherto unrecognized intermediate exists between farnesyl pyrophosphate and squalene.[3]

I. Individual Enzymes of Sterol Biosynthesis from Liver

A. MEVALONATE:NAD AND MEVALONATE:NADP OXIDOREDUCTASES (=MEVALDATE REDUCTASES: EC 1.1.1.32 AND 1.1.1.33)

The mevaldate reductases, although believed not to be involved in the normal pathway of sterol biosynthesis, are useful for the preparation of mevalonate labeled stereospecifically at C-5 with either tritium or deuterium.[4]

Reaction Catalyzed. Mevaldate reductase from pig liver or rat liver catalyzes the stereospecific transfer of the R hydrogen[5] from C-4 of the dihydronicotinamide ring of NADH or NADPH to the R position at C-5 of mevaldate:[4]

$$(1)$$

Although the enzyme is stereospecific in respect to the hydrogen transfer, it shows no stereospecificity for the substrate: the product of the reduction of racemic $(3RS)$ mevaldate is the optically inactive $3RS$-mevalonate.[4]

Assay

The reaction of mevaldate reduction is followed spectrophotometrically by the decrease of absorption of NADH or NADPH at 340 mμ and 25°. The assay mixture consists of: 2 ml of $0.4\ M$ potassium phosphate buffer, pH 6.0; 0.2 ml of 6 mM NADH (or NADPH) dissolved in $0.02\ M$ $KHCO_3$; and enzyme solution and water to 5.9 ml. The reaction mixture is divided between two spectrophotometer cells of 1-cm light paths: 2.9 ml is measured into the test cell and the remainder is added to the reference cell. The reaction is started by the addition of 5 micromoles of potassium mevaldate (0.1 ml) to the test cell.

[3] H. C. Rilling, *J. Biol. Chem.* **241**, 3233 (1966).
[4] C. Donninger and G. Popják, *Proc. Roy. Soc.* **B163**, 465 (1966).
[5] Mevaldate reductase is an "A" side specific oxidoreductase. For definition of the "A" and "B" side of the dihydronicotinamide ring of NADH or NADPH in terms of absolute configuration, see J. W. Cornforth, R. H. Cornforth, C. Donninger, G. Popják, G. Ryback, and G. J. Schroepfer, Jr., *Proc. Roy. Soc.* **B163**, 436 (1966).

Units. One enzyme unit is that amount of enzyme which catalyzes the oxidation of NADH or NADPH, under the conditions of the assay, at the initial rate of 1 micromole per minute. Specific activities are expressed as enzyme units per milligram of protein.

Preparation of Substrate Solution for Mevaldate Reductase Assay. A solution of potassium mevaldate can be prepared most conveniently from N,N'-dibenzylethylenediammonium-bis-(5,5'-dimethoxy-3-hydroxy-3-methyl pentanoate) (=DBED salt of mevaldate dimethylacetal)[6] according to Knauss *et al.*[7] An example is given for the preparation of approximately $0.05\,M$ solution of potassium mevaldate: 62.6 mg of the DBED salt is dissolved in 1.8 ml of water and 0.2 ml of $5\,N$ $(NH_4)OH$ is added, which precipitates dibenzylethylenediamine. The precipitate is extracted with ether and then the residual ether and excess ammonia are removed by a stream of N_2. Of this solution, 0.5 ml is acidified to $0.5\,N$ HCl by the addition of $5\,N$ HCl, and after 15 minutes at room temperature it is neutralized with KOH (bromothymol blue) and made up to 1 ml. The solution of mevaldate is prepared freshly before use and is stored on ice.

Purification of Mevaldate Reductase from Pig Liver

The method is based on the procedure of Schlesinger and Coon.[4,8]

Fresh pig liver, chilled in ice, is finely chopped or minced in a meat grinder. The mince is extracted, in 500-g portions, by stirring for 30 minutes with 1 liter of ice-cold $0.1\,M$ potassium phosphate buffer, pH 7.4. The extract, filtered through several layers of gauze, is centrifuged for 20 minutes at 10,000 g at 0°; the supernatant is the initial extract for further fractionation.

First Ammonium Sulfate Precipitation. The initial extract is fractionated with ammonium sulfate and the fraction precipitating between 35% and 70% saturation with ammonium sulfate (F_{35}^{70}) is collected by centrifuging at 10,000 g for 20 minutes.[9] The F_{35}^{70} precipitate is press-dried between several layers of filter paper and can be stored at −20° for several months. The enzyme is purified from this F_{35}^{70} fraction in small batches.

Calcium Phosphate Gel Treatment. The F_{35}^{70} precipitate obtained in

---

[6] Obtainable from Mann Research Laboratories, New York.

[7] H. J. Knauss, J. D. Brodie, and J. W. Porter, *J. Lipid Res.* **3**, 197 (1962).

[8] M. J. Schlesinger and M. J. Coon, *J. Biol. Chem.* **236**, 2421 (1961); C. Donninger, Ph.D. Thesis, Univ. of London (1964).

[9] The small protein fraction precipitating between 0 and 35% saturation with ammonium sulfate is discarded or it may be preserved as a crude preparation of mevalonate kinase.

the previous step is dissolved in 0.02 M KHCO$_3$ (10 ml of KHCO$_3$ per 2 g damp-dry precipitate) and is dialyzed for 3–4 hours against 4 liters of the same fluid. The protein content of the dialyzed solution is usually 30–40 mg/ml. The dialyzed solution is treated with calcium phosphate gel[10] in a protein/gel ratio of 2:1, and the mixture is stirred for 10 minutes; the gel is then removed by centrifuging and discarded.

Second Ammonium Sulfate Precipitation. Solid ammonium sulfate is added to the supernatant from the previous step to 40% saturation, the pH of the mixture being maintained between 7.8 and 8 (glass electrode) by the dropwise addition of 5 N NH$_4$OH. The small dark brown precipitate is removed by centrifuging and is discarded. The supernatant is brought to 70% saturation with ammonium sulfate; the precipitate is collected by centrifuging and kept for the next step.

DEAE-Cellulose Treatment. The precipitate obtained in the previous step is dissolved in 0.02 M Tris-HCl buffer, pH 8.0, containing 1 mM EDTA and 1 mM cysteine, and is dialyzed against several liters of the same buffer. Seven hundred milligrams of DEAE-cellulose, equilibrated

TABLE I

PURIFICATION OF MEVALDATE REDUCTASE FROM PIG LIVER ACCORDING TO SCHLESINGER AND COON (1961)[a,b]

Preparation	Volume (ml)	Protein (mg/ ml)	Units (per ml)	Total protein (mg)	Total units	Specific activity	Recovery (%)
1st Ammonium sulfate precipitate dissolved in 0.02 M KHCO$_3$, after dialysis	115	35.2	1.45	4048	166.8	0.04	100
Supernatant after calcium phosphate gel treatment	138	18.0	0.97	2490	133.0	0.05	90
2nd Ammonium sulfate precipitate	40	32.4	3.00	1296	120	0.09	74
Supernatant after DEAE-cellulose treatment	71	16.6	1.54	1178	109	0.09	66
Supernatant after bentonite treatment	19	24.0	4.64	456	88	0.20	53

[a] M. J. Schlesinger and M. J. Coon, *J. Biol. Chem.* **236**, 2421 (1961).

[b] Data from C. Donninger, Ph. D. Thesis, Univ. of London, 1964.

[10] Prepared according to K. K. Tsuboi and P. B. Hudson, *J. Biol. Chem.* **224**, 880 (1957). The suspension of the gel usually contains 30–35 mg of dry gel per milliliter.

with 5 mM Tris-HCl buffer, pH 8.0, is added to 15-ml portions of the dialyzed protein solution. After 15 minutes of stirring, the cellulose is removed by centrifuging and is discarded.

Third Ammonium Sulfate Precipitation. Solid ammonium sulfate is added to the supernatant from the previous step to a concentration of 50% saturation; the precipitate formed is sedimented by centrifuging and discarded. The concentration of ammonium sulfate in the supernatant is now raised to 70% saturation, and the white precipitate formed is collected by centrifuging and dissolved in 10 mM Tris-HCl buffer, pH 8.0, and dialyzed against the same solution for 3–4 hours.

Bentonite Treatment. The solution from the previous step is treated with bentonite in a bentonite/protein ratio of 2:1, by the addition of solid bentonite and continuous stirring for 10 minutes. The bentonite is then removed by centrifuging at 15,000 g for 30 minutes at 0°.

The supernatant contains the enzyme which is stable at $-20°$ for at least 4 months. Preparations with a specific activity of about 0.2 are obtained by the procedure with an approximately 50% recovery of the enzyme. Further details of the purification are given in Table I.

The initial reaction rate of the enzyme from pig liver with NADH as coenzyme is about three times faster than with NADPH.

Purification of Mevaldate Reductase from Rat Liver

The method to be described, excepting minor modifications, is that of Knauss, Brodie, and Porter.[7] The assay conditions for this enzyme are the same as for the enzyme prepared from pig liver, except that NADPH (instead of NADH) is used as coenzyme.

Initial Extract. The starting material for the fractionation is the F_{30}^{60} enzyme preparation described on page 444.

Calcium Phosphate Gel Treatment. The preparation of F_{30}^{60} enzymes, dissolved in 0.02 M KHCO$_3$ and dialyzed against the same solution for 4 hours, is treated with calcium phosphate gel[10] in a protein/gel ratio of 1:1. The mixture is stirred for 10 minutes, and then the gel is removed by centrifuging and is discarded. The supernatant is treated once again in the same way and the gel, after sedimentation in the centrifuge, is discarded.

Ammonium Sulfate Precipitation. The supernatant, after the second calcium phosphate gel treatment, is brought to 60% saturation by the gradual addition of solid ammonium sulfate. The resulting precipitate is collected by centrifuging (15,000 g for 20 minutes at 0°) and is dissolved in 0.02 M KHCO$_3$ to give a pale yellow solution. The specific activity of mevaldate reductase in this preparation is usually about 0.1 and is stable at $-20°$ for at least 3 months. The rat liver enzyme, like the one from pig liver, is active with both NADH and NADPH as coenzyme, but the reaction rate is about three times faster with NADPH than with NADH; the reverse is true for the enzyme from pig liver. The steps of purification are summarized in Table II.

TABLE II

PURIFICATION OF MEVALDATE REDUCTASE FROM RAT LIVER[a]

Preparation	Protein (mg/ ml)	Volume (ml)	Units (per ml)	Total protein (mg)	Total units	Specific activity	Recovery (%)
Initial extract (F_{30}^{60}- enzymes) dialyzed	14.0	40	0.165	560	6.6	0.013	100
Supernatant after 2nd calcium phosphate gel treatment	2.4	49	0.085	118	4.15	0.035	63
Ammonium sulfate precipitate dissolved in 0.02 M KHCO$_3$	14.5	3	1.35	43.5	4.07	0.093	61.5

[a] Data from C. Donninger, Ph. D. Thesis, University of London, 1964.

Preparation of $5R$-5-$^3H_1$- and $5R$-5-D_1-Labeled Mevalonate with Mevaldate Reductase

The mevaldate reductases, purified either from pig liver or from rat liver, display the same stereospecificity whether NADH or NADPH is used as coenzyme in the reaction: only the R-hydrogen at C-4 of the dihydronicotinamide ring is transferred to mevaldate. Since the preparation of stereospecifically labeled NADH is simpler and cheaper than that of NADPH, we recommend the use of $4R$-4-$^3H_1$- or $4R$-4-$^2H_1$-labeled NADH in conjunction with the pig liver enzyme for the preparation of the stereospecifically labeled mevalonates.[11] Although the reaction rate

[11] Since it is known that only the R-hydrogen at C-4 of the dihydronicotinamide ring of NADH or NADPH is transferred to mevaldate (cf. footnote 4), stereospecifically labeled coenzymes are not necessarily needed in the reaction; reduction of NAD$^+$ or of NADP$^+$ with dithionite in ^{3}HHO or in D_2O should give reduced coenzymes with sufficiently high isotope content for the preparation of the stereospecifically labeled mevalonates; we have not tried, however, the use of such coenzymes with mevaldate reductase. S. Chaykin and L. Meissner [*Biochem. Biophys. Res. Commun.* **14**, 233 (1964)] and S. Chaykin, K. Chakraverty, L. King, and J. G. Watson [*Biochim. Biophys. Acta* **124**, 1 (1966)] reported that reduction of NAD$^+$ with tritium-labeled sodium borohydride gives a mixture of 1,2-, 1,4-, and 1,6-$^3H_2$-NADH. Alcohol dehydrogenase utilized only the 1,4-$^3H_2$-component of the mixture, and the stereospecificity of the enzyme was fully maintained. The observations of Chaykin *et al.* thus support the suggestion that a stereospecifically labeled coenzyme is not needed for the preparation of a stereospecifically labeled product.

of mevaldate reductase from rat liver is slower with NADH than with NADPH, the rat liver enzyme can be used quite satisfactorily with the labeled NADH also.

We shall describe in detail the preparation of mevalonate-5R-5-D_1 the method being equally applicable for making the tritium-labeled specimen.

Preparation of NADH-4R-4-D_1

NAD^+ is reduced with ethanol-1-D_2 and yeast alcohol:NAD oxido-reductase,[5, 12] and the product is isolated as the barium salt.[13]

For large-scale work, i.e., for making deuterio-NADH in gram quantities, we have found the following reaction mixture satisfactory: 0.2 M Tris (pH not adjusted), 10 mM NAD^+, 60–80 mM ethanol-1-D_2, and 8–10 mg of yeast alcohol dehydrogenase per 100 ml. The reaction, at room temperature, is followed by withdrawing 10-μl samples and diluting them to 3 ml with 0.02 M Tris (pH not adjusted) and measuring the light extinction at 340 mμ. Using the above concentrations of reagents, the reaction, in 200–500 ml incubations, is completed within 25 minutes when the NADH is isolated as the barium salt, as described by Rafter and Colowick.[13] The yield is usually 80% or better. Solutions of the NADH are prepared by dissolving the barium salt in water and by the addition of a small excess of 0.5 M K_2SO_4; the barium sulfate is centrifuged down and washed with a little water, the washing being combined with the first supernatant. The ultraviolet spectrum of the preparation usually gives an E_{260}:E_{340} ratio of 2.5.

Reduction of Mevaldate to Mevalonate on a Large Scale with 4R-4-D_1-Labeled NADH

The reaction mixture contains 0.15 M potassium phosphate buffer, pH 6, 25 mM potassium mevaldate, 10 mM NADH-4R-4-D_1, and 50 units of pig liver mevaldate reductase (specific activity 0.2) per 100 ml of reaction mixture. The reaction, at 25°, is followed by withdrawing 10-μl samples, diluting these to 3 ml with 0.02 M potassium phosphate buffer, pH 6.0, and measuring the light extinction at 340 mμ. When all the NADH becomes oxidized the reaction mixture is acidified to pH 2.0 with 5 N HCl and is extracted with chloroform in a continuous extractor for 2 days.

The amount of mevalonate formed is determined by assaying a portion of the chloroform extract with mevalonate kinase. The chloroform

---

[12] H. F. Fisher, E. E. Conn, B. Vennesland, and F. H. Westheimer, *J. Biol. Chem.* **202**, 687 (1953).

[13] G. W. Rafter and S. P. Colowick, Vol. III, p. 887.

extract is reduced to a convenient volume, e.g., to 100 ml from a 100-ml incubation, and an aliquot (1 ml) is evaporated to dryness. The residue is dissolved in 0.3 ml of 1 N KOH and 0.2 ml of water; the solution is kept at 50° for 15 minutes, when it is brought to pH 7.0 with 1 N HCl and is made up to 2 ml with water; 0.1–0.2 ml samples of this solution are then assayed with mevalonate kinase (see p. 402).

Based on the amount of NADH oxidized, the yield of mevalonate is usually 80–90%. However, since the mevalonate formed by the reduction of racemic (3RS) mevaldate is 3RS-mevalonate, and since mevalonate kinase is specific for 3R-mevalonate, the enzymatic assay shows only one-half of this amount.

Purification of Mevalonate Formed by the Reduction of Mevaldate

When mevaldate is reduced with deuterio-NADH it is convenient to label the mevalonate-5R-5-D$_1$ formed with 2-[14]C- or 4-^{14}C-labeled mevalonate; this will assist in following the purification of mevalonate as well as the calculation of yields of products when the deuterio-mevalonate is used as substrate in subsequent experiments. After the amount of mevalonate in the chloroform extract of mevaldate-reductase incubations has been determined with mevalonate kinase, enough ^{14}C-labeled mevalonic lactone is added to the chloroform extract to give a specific activity of 0.005–0.010 μC per micromole.[14] The chloroform extract, which contains the mevalonate mostly in the form of its lactone, is concentrated on a rotary evaporator to a small volume, e.g., to 5 ml in the case of an extract derived from a 100 ml incubation, and is applied to a Celite (or Hyflo Supercel) 0.5 N H$_2$SO$_4$ column,[15] which is then eluted with chloroform equilibrated with 0.5 N H$_2$SO$_4$.

The Celite for this column is prepared as follows: 200 g of Celite are stirred with 1 liter of distilled water and the particles are allowed to sediment for 1 minute; the supernatant—containing fine particles—is decanted and the process is repeated three times. The sediment of the coarse particles is collected on a Büchner funnel and dried at 100° overnight and then baked at 300° for 3 hours. The baked Celite is then extracted several times with concentrated HCl until the acid remains colorless. The Celite is then washed with water until neutral and dried at 100° overnight.

---

[14] It is preferable to use ^{14}C-labeled mevalonic lactone of high specific activity in order to reduce the dilution of the deuterated species with nondeuterated molecules to a minimum. Mevalonic lactone-2-^{14}C with specific activities of 3–5 μC per micromole is available from The Radiochemical Centre, White Lion Road, Amersham, Bucks, England.

[15] Cf. Fig. 2 in J. W. Cornforth, G. D. Hunter, and G. Popják, *Biochem. J.* **54**, 597 (1953).

A column, suitable for the purification of up to 3–4 millimoles of mevalonic lactone, is prepared by mixing thoroughly 20 g of the acid-washed Celite with 16 ml of $0.5\,N$ H_2SO_4 and by suspending this stationary phase in chloroform and packing it into a 2.5-cm diameter tube. The resulting column is 14–15 cm high. The chloroform used for the elution is prepared by shaking 1 liter of the solvent in a separating funnel with 20 ml of $0.5\,N$ H_2SO_4 and, when the phases have separated, by filtering the chloroform through filter paper.

When 10 ml fractions are collected, mevalonic lactone is eluted from a column of 14.5×2.5 cm between fractions 17 and 33 with the peak at fraction 22 or 23. A small unidentified impurity (? dehydration product of mevalonic lactone), invariably present, is eluted in fraction 5; mevaldic acid is retained on the column.

For smaller amounts of mevalonic lactone up to a few hundred micromoles, a column made from 10 g of Celite and 8 ml of $0.5\,N$ H_2SO_4 and packed to a size of 13.5×1.5 cm, is sufficient. When 4 ml fractions are being collected the small impurity referred to is eluted between fractions 6 and 8, and mevalonic lactone between fractions 25 and 45 with the peak at fraction 34 or 35.

The fractions containing the mevalonic lactone are pooled and evaporated to dryness on a rotary evaporator. The residue is suitable for analysis by mass spectrometry and for preparation of substrate-solution for assay with mevalonate kinase and for other enzyme experiments (cf. p. 403).

When the NADH-4R-4-D_1 is made by the reduction of NAD^+ with 95–98% ethanol-1-D_2 and yeast alcohol dehydrogenase as described, and this NADH is used for the reduction of mevaldate, the resulting mevalonic lactone contains 80–82% monodeuterated molecules.[4]

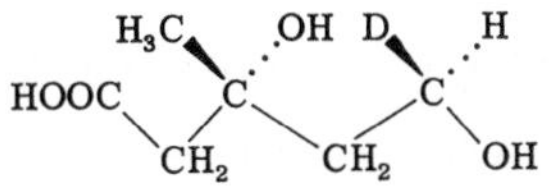

(I)

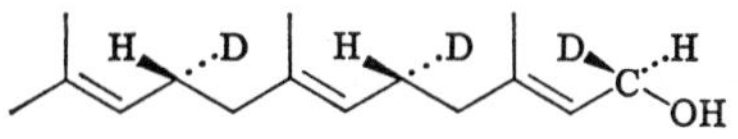

(II)

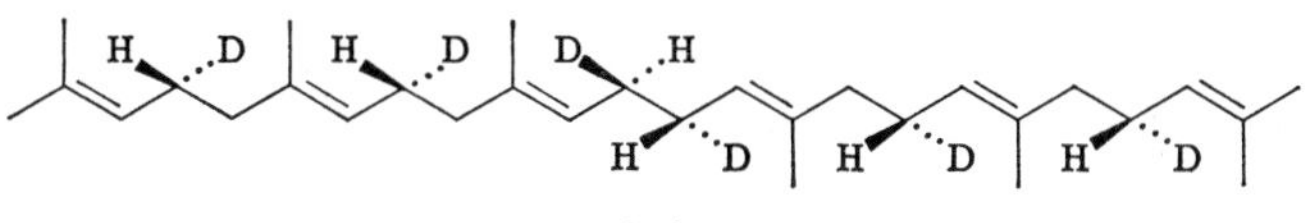

(III)

The absolute configuration of the enzymatically reactive enantiomer in this preparation of mevalonate-5-D_1 is $3R,5R$ as shown in formula (I)[4] and of the farnesol and squalene biosynthesized from it as shown in formulas (II) and (III).[4, 16]

B-1. ATP:MEVALONATE 5-PHOSPHOTRANSFERASE
(EC 2.7.1.36, MEVALONATE KINASE)

Mevalonate kinase is probably the best characterized among all the enzymes of sterol biosynthesis, as its purification offers the least difficulty. Since its first isolation by Tchen[17] from yeast extracts, it has been identified in mammalian as well as in plant tissues.[18-20] Liver is probably its most abundant source. Irrespective of its source, the enzyme catalyzes the reaction (2):

$$(2)$$

It was proved conclusively by the use of liver mevalonate kinase and synthetic R and S mevalonates as substrates that this enzyme was stereo-specific for the R-enantiomer[21] and that the product of the reaction is the levorotatory R-mevalonate 5-phosphate,[22] $[\alpha]_{589}^{24}$ $-6.1°$ in water.

Assay of Mevalonate Kinase

a. *Spectrophotometric Assay*

This is basically the same assay as that described by Tchen[22a] for yeast mevalonate kinase but modified to meet the optimum conditions for the liver enzyme. The assay measures the amount of ADP formed in the kinase reaction (2) by coupling it with the pyruvate kinase (8) and lactate dehydrogenase reactions (9):

$$\text{MVA} + \text{ATP} \rightarrow \text{5-P-MVA} + \text{ADP} \qquad (2)$$
$$\text{ADP} + \text{PEP} \rightarrow \text{ATP} + \text{pyruvate} \qquad (8)$$
$$\text{pyruvate} + \text{NADH} + \text{H}^+ \rightarrow \text{lactate} + \text{NAD}^+ \qquad (9)$$

Sum: MVA + PEP + NADH + H$^+$ → 5-P-MVA + lactate + NAD$^+$

[16] J. W. Cornforth, R. H. Cornforth, C. Donninger, and G. Popják, *Proc. Roy. Soc.* **B163**, 492 (1966).

[17] T. T. Tchen, *J. Biol. Chem.* **233**, 1100 (1958).

[18] H. R. Levy and G. Popják, *Biochem. J.* **75**, 417 (1960).

[19] K. Markley and E. Smallman, *Biochim. Biophys. Acta* **47**, 327 (1961).

[20] W. D. Loomis and J. Battaile, *Biochim. Biophys. Acta* **67**, 54 (1963).

[21] J. W. Cornforth, R. H. Cornforth, and G. Popják, *Tetrahedron* **18**, 1351 (1962).

[22] R. H. Cornforth, K. Fletcher, H. Hellig, and G. Popják, *Nature* **185**, 923 (1960).

[22a] T. T. Tchen, see Vol. V [66], p. 489; Vol. VI [75], p. 505.

Reagents

(i) Tris-HCl buffer, 0.3 M, pH 7.3–7.4

(ii) MgCl$_2$, 0.3 M solution

(iii) Potassium *RS*-mevalonate, 0.1 M solution. Crystalline (or distilled) mevalonic lactone, 130 mg, is dissolved ın 5.5 ml of 0.2 N KOH and the solution is heated at 50° for 15 minutes (or at 37° for 30 minutes) ; after the solution has been cooled to room temperature its pH is adjusted to 7.3 with 0.1 N HCl and its volume is made up to 10 ml with water.

(iv) ATP solution, 0.20 M, neutralized to pH 7. The ATP must be free of ADP; a contamination of ATP with more than about 0.2% of ADP is not acceptable, or if present may necessitate an increase in the concentrations of phosphoenolpyruvate and of NADH in the assay system.

(v) NADH, 6 mM, dissolved in 0.01 M KHCO$_3$

(vi) Sodium phosphoenolpyruvate, 50 mM solution (PEP)

(vii) Lactate dehydrogenase (EC 1.1.1.27), 250 μg/ml, and pyruvate kinase (EC 2.7.1.40), 1 mg/ml, dissolved together in 1% bovine serum albumin (LDH/PK solution).

(viii) Cysteine hydrochloride, 0.6 M solution not neutralized

(ix) KOH, 1 N, for the neutralization of cysteine

(x) KF, 1 M solution. This is needed only in assays of the "initial extract" and of the "protamine supernatant" in order to inhibit ATPase.

(xi) Mevalonate kinase solution. For assays of highly purified preparations it is convenient to dilute a concentrated stock solution of the enzyme with 1% bovine serum albumin to a level of about 0.2–0.5 kinase units per milliliter.

(xii) ADP, 2.5 mM solution, neutralized to pH 7. This is needed for testing the assay system.

Assay Procedure. For an assay a reaction mixture is made up from the described reagents and water as follows:

1. Water		2.64 ml
2. Tris		2.00 ml
3. MgCl$_2$		0.10 ml
4. KF		0.06 ml
5. ATP		0.20 ml
6. PEP		0.10 ml
7. NADH		0.40 ml
8. Cysteine		0.10 ml
9. KOH		0.10 ml
10. LDH/PK		0.20 ml
11. Kinase		0.10 ml
Total Volume		6.00 ml

Before the assay of the enzyme is made this reaction mixture is tested with ADP: 2.9 ml are pipetted into two spectrophotometer cells of 1-cm light path; 0.1 ml of water is added to the reference cell and 0.1 ml of the ADP solution (xii) into the test cell. The latter must result in the oxidation within 30 seconds of 0.25 micromole of NADH; this corresponds to a change in the optical density between the reference and test cell of 0.517 at 340 mμ, since the oxidation of 0.0483 micromole of NADH in 3 ml viewed in a spectrophotometer cell with a 1-cm light path at 340 mμ gives a change in optical density of 0.1.

If the magnitude of the change in the optical density is as expected, but the reaction takes longer than 30 seconds to completion, then either the amount of lactate dehydrogenase or the amount of pyruvate kinase used is inadequate. If the reaction comes to a standstill before a difference of 0.517 in the optical density between the test and reference cell is attained, the test system is inadequate in phosphoenolpyruvate and/or NADH. If any of these deficiencies are observed, they should be corrected before progressing to the assay of mevalonate kinase.

For the assay of mevalonate kinase 2.9 ml of the prescribed reaction mixture are pipetted into one spectrophotometer cell (the test cell) with a 1-cm light path, and 2.9 ml are added to another. When a double-beam recording spectrophotometer is available it is most convenient to place the "test cell" in the reference position of the instrument and to add 0.1 ml water to the "reference cell" (placed in the test position), and then to start the reaction by stirring into the test cell 0.1 ml of the mevalonate solution and to record the rate of change in the optical density at 340 mμ for a few minutes. The amount of NADH oxidized is equivalent to the amount of MVA phosphorylated.

This assay is of only limited value with the "initial extracts" of liver, because these contain enzymes that strongly interfere with the assay: (a) NADH oxidase which may consume the NADH added to the reaction mixture before the reaction proper with mevalonate could be started; (b) ATPase which may generate ADP and cause further consumption of NADH; and (c) some dehydrogenase whose substrate is ADP, or adenine released from it, and which causes a reduction of NAD$^+$ generated in the reaction mixture.

When a double-beam recording difference spectrophotometer is used the interfering reactions (a) and (b) are readily compensated for, since the reference cell contains all ingredients except mevalonate. However, there is no compensation for the interference by reaction (c), since the ADP generated by the mevalonate kinase reaction acts as an additional substrate for the dehydrogenase which reduces NAD$^+$.

In order to obviate interference by reactions (a) and (b), it is our custom, when assaying "initial extracts" or the "protamine supernatant," to work near the spectrophotometer and as fast as possible: there must be a minimum delay between the addition of the test enzyme solution (LDH/PK solution) to the reaction mixture, the pipetting of the assay solution (2.9 ml) into the test cell, and the addition of the mevalonate solution to the test cell. As there is no compensation for interfering reaction (c) we prefer to use very small amounts of enzyme (no more than 0.01 unit) when assaying initial extracts. The interfering reaction (c) is not present in preparations of the "protamine supernatant" and in later fractions. Because of these circumstances it is not unusual to find that the spectrophotometric assay will give a higher number of mevalonate kinase units in the protamine supernatant than in the initial extract.

For none of the fractions beyond the "protamine supernatant" need these precautions be taken, and the KF can be omitted from the reaction mixture.

When several assays have to be made in one day, we find it convenient to make a large volume of the part of the reaction mixture comprising reactants 1 to 7 inclusive. For example, for five assays we make the following "cocktail":

1. Water	13.2 ml
2. Tris	10.0 ml
3. $MgCl_2$	0.5 ml
4. KF	0.3 ml
5. ATP	1.0 ml
6. PEP	0.5 ml
7. NADH	2.0 ml
Total Volume	27.5 ml

To 5.5 ml of this "cocktail" we add 0.1 ml each of the cysteine solution and of the 1 N KOH, 0.2 ml of the LDH/PK, and 0.1 ml of the mevalonate kinase solution; 2.9 ml of this mixture is pipetted into the reference and test cuvettes; 0.1 ml of water is added to the reference cell, and 0.1 ml of the mevalonate solution is added to the test cell to start the reaction.

The purified preparations of mevalonate kinase (after the calcium phosphate gel treatment) are ideal to determine the concentration of mevalonate in unknown solutions. For this purpose the same reaction mixture is prepared as for the assay of the enzyme itself, except that 0.5–1.0 unit of mevalonate kinase is used per assay and that the amount of mevalonate added to the test cell is about 0.25 micromole or less. With a purified enzyme, under such conditions the reaction comes to completion within a few minutes and from the change in the optical density between the reference and test cell at 340 mμ the concentration of R-mevalonate in an unknown solution can be determined with great accuracy.

b. *Radiochromatographic Assay*

When the substrate in the mevalonate kinase reaction is labeled with ^{14}C, the 5-phosphomevalonate-^{14}C formed may be determined by chromatography of the deproteinized reaction mixture on paper[22a, 23] or on thin-layer plates (G. Popják, unpublished), followed by the scanning of the chromatograms for radioactivity.

For this purpose the mevalonate kinase incubation mixture is deproteinized by the addition of an equal volume of ice-cold 10% (w/v) perchloric acid, and the precipitated protein is centrifuged down. The protein-free supernatant is then neutralized with KOH and the insoluble $KClO_4$ is centrifuged off at $0°$. The supernatant is lyophilized and the residue is dissolved in a little water, e.g., in 0.5 ml from a 3-ml reaction mixture. Suitable aliquots of this solution are applied to strips of Whatman No. 1 paper or onto thin-layer plates of silica gel G.

For the development of the paper strips, isobutyric acid–conc. ammonia (sp. gr. 0.88)–water mixture[23] (66:3:30 by volume) is very satisfactory as it gives highly reproducible results.[24] For the development of 30 cm-long strips by the descending technique, 8–12 hours are needed. The dried strips are then scanned for radioactivity, preferably with an automatic device. 5-Phosphomevalonate has an R_f of 0.37, mevalonic acid an R_f of 0.66, and mevalonic lactone an R_f of 0.75. When the radioactivity is recorded in conjunction with a linear rate-meter, the areas under the peaks are proportional to the amount of each component and hence the fraction of mevalonate converted into 5-phosphomevalonate may be calculated.

In this paper-chromatographic system mevalonate always appears as an equilibrium mixture of the acid and of the lactone, the amount of the acid being slightly larger than that of the lactone. ADP has an R_f in this system identical with that of 5-phosphomevalonate, whereas ATP migrates more slowly with an R_f of 0.2–0.25.

The thin-layer plates are developed with *n*-propanol–conc. ammonia–1% EDTA (6:3:1 by volume):[25] 5-phosphomevalonate just perceptibly moves away from the origin (R_f 0.05), but mevalonate migrates as a single component with an R_f of 0.43.

Unit of Activity. One unit of enzyme activity is defined as that amount of enzyme which, when assayed by the spectrophotometric method, phosphorylates mevalonate at an initial rate of 1 micromole per minute at $25°$. Specific activities are expressed as units of enzyme activity

---

[23] A. de Waard and G. Popják, *Biochem. J.* **73**, 410 (1959).
[24] For other paper chromatographic systems see Vol. V [66], p. 491.
[25] H. Plieninger and H. Immel, *Chem. Ber.* **98**, 414 (1965).

per milligram protein. These units are sixty times larger than those defined for the similar enzyme from yeast by Tchen.[22a]

Purification of the Enzyme

The method of purifying mevalonate kinase from pig liver, to be described, is based on the procedure of Levy and Popják;[18] it gives reproducible preparations of the enzyme which phosphorylates mevalonate at an initial rate of about 1 micromole min⁻¹ mg⁻¹.

Reagents

    (a) "Sucrose solution," 0.35 M, containing 35 mM KHCO$_3$ and 1 mM EDTA

    (b) Ammonium sulfate recrystallized from a saturated aqueous solution neutralized with NH$_4$OH and containing 1 mM EDTA

    (c) "Dilute potassium phosphate buffer:" 0.02 M, pH 7.5, containing 1 mM EDTA

    (d) Protamine sulfate solution containing 10 mg/ml. Protamine sulfate made from salmon sperm is the preferred preparation.

    (e) Calcium phosphate gel;[10] the preparation should contain 30–35 mg dry weight per milliliter of suspension

    (f) Potassium phosphate buffers, pH 7.5: 0.01 M, 0.02 M, and 0.05 M

Method

Initial Extract. Pig liver, obtained as soon as possible after slaughter of the animal and chilled in ice, is finely chopped or minced in a meat grinder. Each 500-g batch of the mince is extracted with 1 liter of the "sucrose solution," reagent (a), by gentle stirring for 90 seconds. The mixture is filtered through two layers of cheesecloth. The volume of the filtrate is usually equal to the volume of the extracting fluid used and contains 60–80 mg protein per milliliter, with a specific activity of 6 to 9 × 10⁻³ units per milligram (cf. Table III).

Protamine Precipitation. Of the protamine solution, reagent (d), 176 ml is added to 1 liter of initial extract to give a final concentration of 1.5 mg/ml. After 2 minutes of stirring at 0° the mixture is centrifuged at 2000 g for 30 minutes to give an almost clear "protamine supernatant" without any loss of mevalonate kinase units, but a solution containing one-third to one-fourth of the amount of protein of the initial extract with a corresponding increase in the specific activity.

First Ammonium Sulfate Precipitation. Finely powdered ammonium sulfate is added to the "protamine supernatant" to a concentration of 85% saturation at 0°; the mixture is stirred gently until the salt dissolves and then the precipitate is allowed to settle overnight. The reddish brown precipitate is collected by centrifuging and is dissolved in the "dilute potassium buffer," reagent (c): 400 ml is a suitable volume for the precipitate obtained from 1 liter of the "protamine supernatant." A small insoluble residue is centrifuged off.

At this stage there is usually a complete recovery of enzyme units with a slightly increased specific activity (e.g., an increase from 0.03 to 0.04).

TABLE III

PURIFICATION OF MEVALONATE KINASE FROM PIG LIVER

Preparation	Volume (ml)	Total protein[a] (g)	Enzyme units	Specific activity
Initial extract	1000	60–80	360–720	6–9×10^{-3}
Protamine supernatant	920	18–24	360–720[b]	2–3×10^{-2}
Second ammonium sulfate fraction (30–45% saturation) after dialysis	120	3.45–4.2	340–630	0.1–0.15
Enzyme after calcium phosphate gel treatment				
First eluate	100	0.35–0.45	140–270	0.4–0.6
Second eluate	100	0.17–0.22	100–220	0.6–1.0

[a] Protein determination by the biuret method, Vol. III [73].

[b] Occasionally it happens, when the enzyme is assayed by the spectrophotometric method, that there is an apparent increase in the total enzyme units in the protamine supernatant as compared to the initial extract. Owing to interfering reactions the spectrophotometric method may give too low values for the enzyme units contained in the initial extract (cf. pp. 404, 405).

Second Ammonium Sulfate Precipitation. The concentration of ammonium sulfate in the solution of the first ammonium sulfate precipitate is increased to 30% saturation, and the small precipitate formed is removed by centrifuging at 12,000 g for 30 minutes. More ammonium sulfate is added to the supernatant to a concentration of 45% saturation. The precipitate, which contains about 80% of the mevalonate kinase from the previous step, is collected by centrifuging at 12,000 g for 30 minutes and dissolved in the "dilute potassium phosphate buffer," reagent (c); 100 ml is sufficient for the precipitate derived originally from 1 liter of the "protamine supernatant." The solution is then dialyzed against 3 changes of 0.01 M potassium phosphate buffer, pH 7.5, for 4 hours. In this step a 2.5 to 3-fold purification is achieved.

Calcium Phosphate Gel Treatment. Twelve milliliters of the calcium phosphate gel suspension, reagent (e), is added to 120 ml of the dialyzed solution obtained in the previous step and the mixture is stirred for 10 minutes at 0°. The gel is centrifuged off and discarded. A further 60–65 ml of gel is added to the supernatant, and the mixture is stirred for 30 minutes at 0°. Usually 90–95% of the mevalonate kinase is adsorbed onto the second gel, which is collected by centrifuging; the supernatant, almost free of mevalonate kinase, is discarded.

The gel is washed successively with 50 ml of potassium phosphate buffers, pH 7.5: once with the 0.01 M and twice with the 0.02 M buffer. These washings elute mostly inert protein. The mevalonate kinase is eluted by four washings with 50 ml each of 0.05 M potassium phosphate buffer: the first two and the last two washings are combined separately (first and second eluates, cf. Table III) as the specific activity of the protein in the third and fourth washings is nearly twice as high as in the first two.

For storage purposes the enzyme from the gel eluates may be precipitated with ammonium sulfate (80% saturation), the precipitate collected by centrifuging and the supernatant poured off. The precipitated enzyme keeps well for many months at −20°. When needed, solutions are made in 0.02 M potassium phosphate buffer, pH 7.5, and dialyzed before use against the same solution for 4 hours. The solutions of the enzyme may be kept frozen at −20° for several weeks with very little deterioration.

The enzyme preparations, after the gel treatment, are free of ATPase, of phosphomevalonate kinase, and of pyrophosphomevalonate decarboxylase and have specific activities ranging from 0.4 to 1.0. They are suitable for the assay of unknown solutions of mevalonate and for the preparation of 5-phosphomevalonate.

Further Purification of Mevalonate Kinase. Occasionally, particularly when very large batches of enzyme are being processed, the specific activity of the protein eluted from the calcium phosphate gel is not higher than 0.4–0.6. The specific activity of such preparations can be further improved by refractionating the eluates with ammonium sulfate. The fraction precipitating between 35% and 50% saturation is dissolved in a volume of 0.01 M potassium phosphate buffer, pH 7.5, to give a solution of 30 mg protein per milliliter (usually 10 ml of buffer for 2 g wet precipitate) and is dialyzed against the same buffer for 4 hours. The dialyzed solution (20–25 mg of protein per milliliter) is treated with bentonite in a protein:bentonite ratio of 1:1 by the slow addition of solid bentonite to the stirred protein solution. The stirring is continued for 10 minutes after all the bentonite has been added; the bentonite is then centrifuged off at 15,000 g for 30 minutes and the supernatant is subjected again to the fractionation with calcium phosphate gel.

By this further purification process mevalonate kinase preparations with a specific activity of greater than 1 are invariably obtained.

Properties of Liver Mevalonate Kinase

The liver enzyme, like the one from yeast,[22a] is activated by divalent metal ions, Mg^{++}, Ca^{++}, or Mn^{++}. Mg^{++}, at a concentration of 3 mM, is the preferred cation; concentrations higher than 4 mM are inhibitory, unless the concentration of ATP is increased in the reaction mixture (see below).

The reaction rates in the presence of 5 mM Ca^{++} are about 60% of those observed at similar concentrations of Mg^{++}; Mn^{++} ions are inferior to Mg^{++} or Ca^{++} as activators of this enzyme. The activation of mevalonate kinase by Ca^{++} is an unusual property and contrasts with the behavior of the yeast enzyme, which is inactive with Ca^{++}.

Liver mevalonate kinase, unlike the enzyme from yeast, is highly sensitive to acid. Hence the step of adjusting the pH of the protein solution to 5.1, as described in the purification of yeast mevalonate kinase by Tchen[22a] cannot be practiced at any stage of the purification of the liver enzyme. Exposure to pH 5.1 for 5 minutes results in an irreversible and total inactivation of liver mevalonate kinase preparations.

Purified preparations of liver mevalonate kinase are almost inactive without cysteine or glutathione: a concentration of 10 mM of either substance in the reaction mixture gives about the maximum effect. Ascorbic acid cannot substitute for cysteine or glutathione.

Mevalonate kinase cannot phosphorylate mevalonic lactone and is specific for the R-enantiomorph of mevalonate. The K_m for R-mevalonate is 50 μM; maximum reaction rates are observed at a concentration of about 1 mM of the natural substrate (2 mM of the synthetic RS-mevalonate). The unnatural S-enantiomer of mevalonate is not inhibitory up to a concentration of at least 5 mM. The pH optimum of the reaction is 7.3.

Adenosine triphosphate and inosine triphosphate are equally effective phosphate donors with liver mevalonate kinase. The concentration of ATP which gives maximum reaction rates varies with the concentration of Mg^{++} ions in the reaction mixture; the optimum ratio of ATP to Mg^{++} (mM/mM) is 1.5–2. Thus the optimum concentration of ATP with 3.3 mM Mg^{++} is 5.5 mM and with 5 mM Mg^{++} it is 7–8 mM. When the enzyme is assayed by the spectrophotometric method, concentrations of ATP greater than 8 mM cannot be used on account of the inhibition of the pyruvate kinase reaction.

B-2. ENZYMATIC PREPARATION OF $(-)(R)$5-PHOSPHOMEVALONATE FROM (RS)MEVALONATE

Although the chemical synthesis of 5-P-MVA has been described, the yield of the process is very poor and its product is the racemic RS-5-

phosphomevalonate of which only the R-enantiomer is reactive enzymatically. The enzymatic method we now describe, based on the procedures of de Waard and Popják[23] and of Levy and Popják,[18] gives the barium salt of (R)5-phosphomevalonate in 80–85% yield and has been used by us on a 1–6 millimolar scale with equally good results. For preparations on a small scale (micromoles) the isolation procedure described by Tchen (Vol. VI [75], p. 506) is preferred.

Principle. The $(-)(R)$5-phosphomevalonate formed enzymatically from (RS)-mevalonate is isolated as the ethanol-insoluble barium salt. The unreacted S-enantiomer of mevalonic acid is converted into the lactone; as this does not hydrolyze readily in the cold at a mildly alkaline pH and does not form an ethanol-insoluble barium salt, it remains in solution when the barium 5-phosphomevalonate is precipitated. The adenine nucleotides (ADP, ATP) are separated from 5-phosphomevalonic acid by precipitation with 80% ethanol, in which 5-phosphomevalonic acid is soluble.

Reagents

> Mevalonate kinase from liver purified to a specific activity of 0.6–1.0 (see p. 407)
>
> A 0.5 M solution of potassium RS-mevalonate labeled with ^{14}C. (RS)-Mevalonic lactone (see pp. 360, 361), 642 mg (4.9 millimoles), is dissolved in 1 N KOH (5.5 ml) and kept at 37° for 30 minutes; the solution is then titrated to pH 7.4 with 1 N HCl. One milliliter of a similarly prepared 0.1 M solution of potassium (RS)-mevalonate-2-^{14}C, containing conveniently 100 μC of ^{14}C, is then added and the volume of the mixture is made up to 10 ml to give a 0.5 M solution of (RS)-mevalonate
>
> ATP, crystalline disodium salt
> Tris(hydroxymethyl)aminomethane
> $MgCl_2 \cdot 6 H_2O$
> Cysteine hydrochloride
> 5 N HCl
> Acetic acid
> 1 M BaCl$_2$
> Ethanol

Procedure. A preparation of 5-P-MVA from 3 millimoles of (RS)-mevalonate in an incubation of 600 ml is described.

ATP, 3.046 g (4.5 millimoles); $MgCl_2 \cdot 6 H_2O$, 0.61 g (3 millimoles); cysteine hydrochloride, 0.939 g (6 millimoles); and tris(hydroxymethyl)-aminomethane, 7.26 g (60 millimoles) are dissolved in 200 ml of water and the solution is titrated with 5 N HCl to pH 7.4 (glass electrode). Six milliliters of the 0.5 M potassium (RS)-mevalonate (3 millimoles) are added and the volume of the mixture made up to 590 ml with water. After this reaction mixture has been warmed up to 37°, 10 ml of a dialyzed solution of liver mevalonate kinase (300 units) is added and the incubation is continued at 37°. With the amount of substrate and enzyme used here, the reaction is completed in 30–35 minutes. This is checked by adding 0.1 ml of the incubation solution to 2.9 ml of the spectrophotometric reaction mixture used for assaying mevalonate kinase, but from which the mevalonate kinase has been omitted (see p. 403), and by measuring the change in optical density at 340 mμ: this should be 0.5–0.517.

At the end of the incubation the reaction mixture is acidified with acetic acid to pH 4 and heated to about 60° for 2 minutes; it is cooled in ice and poured into 2.5 liters of ice-cold ethanol.

The sticky precipitate (ADP, ATP, and protein) is centrifuged down. The supernatant, containing the 5-phosphomevalonic acid and unreacted mevalonic acid, is decanted and saved. The precipitate is taken up in 20 ml of water and reprecipitated by the addition of 80 ml of ethanol; the process is repeated three times. All the supernatants are combined and evaporated to dryness on a rotary evaporator (bath temperature 50°). The residue is finally dried in high vacuum at 80° for 30 minutes in order to complete the lactonization of free mevalonic acid.

First Barium Precipitation of 5-Phosphomevalonate. The following operations are carried out at ice temperature. The dry residue from the previous step is dissolved in 50 ml of ice water and 10 ml of 1 M BaCl$_2$ are added. A small precipitate may be formed at this stage, particularly if the enzyme preparation used was not dialyzed adequately to remove all the (NH$_4$)$_2$SO$_4$ from it, and is removed in a refrigerated centrifuge. The supernatant, combined with the water washings of the precipitate, is then made alkaline (faint pink) to phenolphthalein with NH$_4$OH, and ice-cold ethanol is added to it to a concentration of 65%. The copious flocculent precipitate is collected after 1–2 hours (or after standing in the cold room overnight) by centrifuging. The supernatant (which contains most of the unreacted mevalonic acid lactone) is discarded. The precipitate is washed four times with 80% ethanol; the washings—after sedimentation of the precipitate by centrifuging—are discarded.

The precipitate (first barium precipitate) contains all the 5-phosphomevalonate, bare traces of unreacted mevalonate, but is contaminated with adenine nucleotides. The latter are removed in the next step.

Second Barium Precipitation of 5-Phosphomevalonate. The first barium precipitate—without previous drying—is dissolved in 17 ml of water and 3 ml of 1 N HCl; the volume of the solution is now about 30 ml. Ice-cold ethanol (120 ml) is added to a concentration of 80% and precipitates the remaining adenine nucleotides. After 2–3 hours at ice temperature (or overnight in the cold room) the sediment is centrifuged off and washed three times with small volumes of 80% ethanol.

The supernatant and the washings combined are concentrated on a rotary evaporator (bath temperature 50°) to about 5 ml. The thick solution is transferred into a centrifuge tube with water (total volume 10 ml) and is made alkaline (phenolphthalein pink) with NH_4OH; 1 ml of 1 M $BaCl_2$ and 40 ml of ethanol are added. The white precipitate which forms on addition of ethanol is allowed to settle overnight in the cold room.

The precipitate is collected by centrifuging and is washed three times with 80% ethanol (10 ml each), once with dry ethanol, and three times with acetone. The preparation, barium 5-phosphomevalonate, is finally dried at 70° in an oven.

The yield is usually 80–85% of the theoretical. The preparation is free of unreacted mevalonate and of adenine nucleotides, but is contaminated with small amounts of $BaCO_3$; it usually contains 2.2 micromoles of 5-phosphomevalonate per milligram of barium salt (theory: 2.35 micromoles per milligram).

In the particular example given from our laboratory notebooks, from 3 millimoles of (RS)-mevalonate 1.234 millimoles of (R)5-phosphomevalonate were obtained (based on assay of ^{14}C); the weight of the barium salt was 623 mg.

Preparation of Potassium 5-Phosphomevalonate from Barium Salt. One hundred milligrams of the barium salt is dissolved in 0.6 ml of 0.5 N HCl, and 0.6 ml of 0.5 M K_2SO_4 is added. The precipitated $BaSO_4$ is centrifuged down and is washed twice with 0.3 ml of water. The supernatants are combined and neutralized to pH 7 (0.038 ml of 40% KOH), and the volume is made up to 2 ml to give an approximately 0.1 M solution of potassium 5-phosphomevalonate containing also some KCl. The preparation may be assayed enzymatically with purified phosphomevalonate kinase (see p. 407), by counting of ^{14}C and by examination on paper chromatograms or thin-layer plates (see p. 406; also Vol. V [66], p. 491).

C-1. PHOSPHOMEVALONATE KINASE (ATP: 5-PHOSPHOMEVALONATE PHOSPHOTRANSFERASE; EC 2.7.4.2)

This enzyme catalyzes the reversible reaction (3):

$$5\text{-P-MVA} + \text{ATP} \rightleftarrows 5\text{-PP-MVA} + \text{ADP} \tag{3}$$

Its purification from yeast autolyzate has been described by Tchen.[22a]

The method of isolation from liver extracts described here is based on the work of Levy and Popják,[18] and of Hellig and Popják,[26] and gives preparations of about the same specific activity as those obtained from yeast.

Assays of Phosphomevalonate Kinase

The methods of assay for this enzyme are identical with those described for mevalonate kinase (cf. p. 402) except that the substrate is 5-P-MVA.

a. *Spectrophotometric Assay*

The method measures the amount of ADP formed in the reaction. The reagents needed are the same as those described for the spectrophotometric assay of mevalonate kinase except that the buffer should have a pH of 7.3 and that in the preparation of the 6.0-ml assay mixture the volume of the $0.2\,M$ ATP is reduced to 0.1 ml and the volume of the water is correspondingly increased to 2.74 ml. The concentration of the solution of the substrate R-$(-)$-5-phosphomevalonate[27] should be 30–50 mM. As with the mevalonate kinase assay, 2.9 ml of the reaction mixture is pipetted into the reference and test cuvettes of 1-cm light path; 0.1 ml of water is added to the former and 0.1 ml of the substrate solution is stirred into the latter to initiate the reaction, which is followed by measuring the change in optical density at 340 mμ for a few minutes. A change in optical density of 0.1 corresponds to the oxidation of 0.0483 micromole of NADH, which is equal to the amount of ADP formed and of 5-phosphomevalonate phosphorylated.

The comments about the spectrophotometric assay made in conjunction with the assay of mevalonate kinase are equally applicable here.

b. *Radiochromatographic Assay*

For this purpose a 1-ml incubation mixture of the following composition is the most suitable: 50 mM Tris-HCl buffer, pH 7.3; 5 mM MgCl$_2$; 3.6 mM ATP; 1 mM R-$(-)$-5-phosphomevalonate-^{14}C (specific activity 0.2–1.0 μC per micromole); 10 mM KF (omitted with purified enzyme preparations); 10 mM cysteine (neutralized with KOH) and 0.05–0.1 unit of phosphomevalonate kinase. After suitable incubation periods the reaction mixture is treated as described for the radiochromatographic assay of mevalonate kinase. Aliquots of the deproteinized mixture are applied to Whatman No. 1 paper strips, which are then developed by the

[26] H. Hellig and G. Popják, *J. Lipid Res.* **2**, 235 (1961).
[27] The preparation of R-(−)-5-phosphomevalonate is described on page 410.

descending technique with the isobutyric acid–conc. aq. ammonia (specific gravity 0.88)–water (66:3:30 by volume) solvent system. The radioactive spots, 5-PP-MVA and 5-P-MVA, are located on the strip by a suitable scanning device. 5-PP-MVA has an R_f of 0.25–0.27, and 5-P-MVA an R_f of 0.37–0.38 in this system.

Unit of Activity. One unit of enzyme activity is defined as that amount of enzyme which when assayed, under the optimum conditions, by the spectrophotometric method, phosphorylates 5-P-MVA at the initial rate of 1 micromole per minute at 25°. The specific activity of the enzyme is expressed as units per milligram of protein.

Purification

The starting material for the isolation of phosphomevalonate kinase is the "protamine supernatant" of pig liver extracts, which also serves for the isolation of mevalonate kinase (cf. p. 407).

Ammonium Sulfate Fractionation. The purification of phosphomevalonate kinase from 3.14 liters of protamine supernatant (obtained from 1.6 kg of pig liver) will be described as an illustration of the method: ammonium sulfate (720 g) is added to 30% saturation, and 1 hour after addition of the salt the precipitate is centrifuged off at 2000 g for 30 minutes. This fraction (fraction I) contains only mevalonate kinase (specific activity 0.08–0.10). To the supernatant (3.35 liters) ammonium sulfate (355 g) is added to 48% saturation; the precipitate (fraction II) is collected by centrifuging and is dissolved in 300 ml of 20 mM potassium phosphate buffer, pH 7.6, containing 1 mM EDTA, and is dialyzed for 4 hours against two changes of 10 mM potassium phosphate buffer, pH 7.6, containing 1 mM EDTA. This dialyzed fraction II (525 ml) usually contains mevalonate kinase and phosphomevalonate kinase in a ratio of 2:1 and traces of pyrophosphomevalonate decarboxylase.

Calcium Phosphate Gel Treatment. A suspension of calcium phosphate gel[10] (450 ml; 30 mg dry weight per milliliter) is added to the dialyzed fraction II (525 ml); after gentle stirring for 30 minutes the gel, which adsorbs nearly all the mevalonate kinase, is centrifuged off. Approximately one-half of the units of phosphomevalonate kinase remain in the supernatant (720 ml).

Acid Treatment and Second Ammonium Sulfate Fractionation. The supernatant from the gel treatment is acidified in 30–50 ml batches to pH 5.2 with 2 N acetic acid. The precipitate is centrifuged off and is discarded. The pH of the supernatant is adjusted to 7.4 with 1 N KOH. The preparation contains about one-third of the phosphomevalonate kinase units originally present in the "protamine supernatant"; it is free of mevalonate kinase (which is inactivated at pH 5.2), but contains just detectable traces of pyrophosphomevalonate decarboxylase (about 3% of total enzyme activity). The latter can be removed by a repeated ammonium sulfate fractionation: the protein precipitated between 30 and 45% saturation contains the kinase. The precipitate (about 1.4 g protein) is collected by centrifuging at 10,000–12,000 g for 20 minutes and is dissolved in 30 ml of 10 mM phosphate buffer, pH 7.4; it is dialyzed against two changes of the same buffer. There is some loss (20%) of enzyme units by this reprecipitation, without gain in specific activity, but the decarboxylase is removed by it completely. Solutions of phosphomevalonate kinase (10–20 mg protein per milliliter) kept frozen at —20° lose about 20% of their activity in 6 months.

The steps of purification are summarized in Table IV.

TABLE IV

PURIFICATION OF PHOSPHOMEVALONATE KINASE FROM "PROTAMINE SUPERNATANT" OF PIG LIVER EXTRACTS[a]

Fraction	Total protein (g)	Specific activity	Total enzyme units
Protamine supernatant	78.5	0.005	415
Fraction II[b]	18.9	0.016	305
Calcium phosphate-gel supernatant	4.17	0.037	146
Gel supernatant after acid treatment	1.67	0.078	130

[a] Preparation from 1.6 kg of pig liver, 3.14 liters.

[b] Precipitated between 30 and 48% saturation with ammonium sulfate.

Properties of Liver Phosphomevalonate Kinase

ATP is the only nucleotide triphosphate that can act as a phosphate donor with this enzyme, which also requires Mg^{++} ions (in preference to Mn^{++}) for activation. Maximum reaction rates are observed in the presence of 3.5 mM ATP and 5 mM $MgCl_2$. The K_m value for 5-phosphomevalonate is 0.3 mM. Unlike the similar enzyme from yeast,[22a] which is equally active in the pH range 5.3–10, the liver enzyme has a distinct pH optimum of 7.3. Zn^{++} ions, which activate the yeast enzyme almost as effectively as Mg^{++}, are severely inhibitory to the liver enzyme, which is also inhibited by N-ethylmaleimide (90% inhibition at 5 mM), p-chloromercuribenzoate (100% inhibition at 0.1 mM) and by o-iodosobenzoate (90% inhibition at 5 mM). Iodoacetamide is only slightly inhibitory (9% inhibition at 5 mM).

Although phosphomevalonate kinase appears to be a sulfhydryl enzyme, it is not as sensitive to molecular oxygen as is liver mevalonate kinase, and cysteine or glutathione are not required absolutely for its activation, although cysteine increases slightly the activity of the enzyme.

The phosphomevalonate kinase reaction is freely reversible; when 5-PP-MVA is incubated with ADP and enzyme, 5-P-MVA and ATP are formed.[26] The equilibrium constant

$$K_{eq} = \frac{[\text{5-PP-MVA}][\text{ADP}]}{[\text{5-P-MVA}][\text{ATP}]}$$

at pH 7.3 and 37°, as determined from the forward reaction, is close to unity (0.7–1.1).

The phosphomevalonate kinase reaction can be driven to completion in the forward direction by coupling it with another enzyme which regenerates continuously ATP from the ADP; pyruvate kinase is ideally suited for this purpose and is used for the preparation of 5-PP-MVA.

C-2. Preparation of (−)(R)5-Pyrophosphomevalonate (5-PP-MVA)

A method for the small-scale preparation of 5-PP-MVA with yeast phosphomevalonate kinase has been described in a previous volume by Tchen (Vol. VI [75], p. 507). The procedure used in our laboratory is suitable for a relatively large-scale preparation of this substance.

Procedure

5-P-MVA is converted completely into 5-PP-MVA in the following incubation mixture: Tris-HCl buffer, pH 7.3, 50 mM; ATP, 3.6 mM; MgCl$_2$, 5 mM; cysteine (neutralized with KOH), 10 mM; 5-P-MVA, 1 mM; phosphoenol pyruvate, 1.5–2.0 mM; pyruvate kinase, 35 μg/ml, and phosphomevalonate kinase, 0.1 unit/ml. After 1 hour of incubation at 37° the pH of the reaction mixture is adjusted to 6 with 1 N HCl; ethanol is added to 75% and the mixture is warmed quickly to 60° and kept at that temperature for 90 seconds. The mixture is then cooled in ice and the precipitate (protein and ATP) is sedimented by centrifuging. The supernatant is decanted and saved, and the precipitate is washed three times with 75% ethanol. The washings combined with the first supernatant are then concentrated on a rotary evaporator until all the ethanol has been removed and the volume of the solution reduced to 10–15 ml for each 100 ml of the original incubation mixture. The contents of the evaporation flask are transferred quantitatively with a little water to a centrifuge tube and the pH of the solution is adjusted to 8.5 with KOH. One molar BaCl$_2$, 0.2 ml for each 100 ml of the original incubation mixture, is added to the concentrate and precipitates much of the remaining ATP. After the barium salt of ATP has been removed by centrifuging, the same amount of BaCl$_2$ and ethanol to a concentration of 50% are added to the supernatant in order to precipitate the barium salt of 5-PP-MVA, which is allowed to settle out overnight in the cold room and is collected by centrifuging and washed with 75% and then with absolute ethanol.

The barium salt of 5-PP-MVA is dissolved in a minimum amount of 0.1 N HCl and the barium is precipitated by the addition of a Na_2SO_4 solution. After removal of the $BaSO_4$ by centrifuging, the pH of the supernatant is adjusted to 6.9 with KOH. At this stage about 75% of the 5-PP-MVA synthesized is found in the preparation which is still contaminated with small amounts of ATP (about 0.3–0.5 micromole of ATP per micromole of 5-PP-MVA).

The contaminating ATP can be removed by any of the following three procedures: (i) paper chromatography; (ii) chromatography on ion-exchange resin; and (iii) differential adsorption of ATP onto charcoal.

Procedure (*i*). This is suitable for small-scale work involving a few micromoles of 5-PP-MVA. The preparation is chromatographed on Whatman No. 1 paper with *tert*-butanol–formic acid–water (40:10:16 by volume) (cf. Vol. VI [75], p. 507); 5-PP-MVA and ATP have R_f values of 0.33 and 0.1, respectively, in this system. 5-PP-MVA is eluted from the appropriate portion of the dried paper with dilute aqueous ammonia; the eluate is then freeze-dried.

Procedure (*ii*). This procedure is suitable for separation of several hundred micromoles of 5-PP-MVA.[23] The solution of 5-PP-MVA obtained from its barium salt is applied to a 10 × 1 cm column of Amberlite IRA-400 resin (chloride form) (Rohm and Haas Co., Philadelphia, Pennsylvania), which is then eluted according to the gradient elution technique of Hurlbert *et al.*:[28] the mixing flask for the eluent contains initially 100 ml of water and the reservoir 50 ml of 0.015 N HCl; the contents of the reservoir are changed to 0.2 M (100 ml), 0.5 M (50 ml), and 1 M (100 ml) KCl in 0.015 N HCl after the following volumes of effluent have been collected: 25, 100, and 150 ml; 5-PP-MVA is eluted between 140 and 190 ml of effluent. The combined fractions containing the 5-PP-MVA are lyophilized and the dry residue is extracted with 90% methanol; the extract is clarified by centrifuging, and the solvent is removed on a rotary evaporator. The residue, 5-PP-MVA, uncontaminated by ATP, is dissolved in dilute aqueous ammonia.

Procedure (*iii*). This is less elaborate than procedure (ii); it gives equally good preparations of 5-PP-MVA, but may result in a 20% loss of the substance. The solution of the 5-PP-MVA, derived from the barium precipitation, is diluted to a concentration of about 1.5 m*M* and

[28] R. B. Hurlbert, H. Schmitz, A. F. Brumm, and V. R. Potter, *J. Biol. Chem.* **209**, 23 (1954).

then acid-washed charcoal (Nuchar C190, N; West Virginia Pulp and Paper Co., Tyrone, Pennsylvania), 100 mg for each 10 ml, is stirred into the ice-cold solution and removed by filtration through sintered glass after 30 minutes. The charcoal is washed on the filter with small volumes of water; the filtrate and washings are combined and lyophilized giving 5-PP-MVA uncontaminated by ATP.

The overall yield, through the barium precipitation and procedure (iii), for the final purification is about 50%. The properties of 5-PP-MVA have been described by Tchen (Vol. VI [75], p. 507 and ff.); it should be added only that acid solutions (even at pH 6.0) of 5-PP-MVA are unstable when kept frozen: 50% hydrolysis to 5-P-MVA may occur in 3 months. Alkaline solutions deteriorate more slowly (3–5% hydrolysis in 6 months). When large amounts of 5-PP-MVA are made it is best to keep the specimen in the form of the dry barium salt and prepare solutions when required.

5-PP-MVA, like 5-P-MVA, is levorotatory, although accurate values for its rotation have not yet been obtained.

D. ATP:5-Pyrophosphomevalonate Carboxy-lyase (Dehydrating) (Pyrophosphomevalonate Decarboxylase, EC 4.1.1.33)

Reaction Catalyzed. It was shown by Bloch et al.,[29] with the enzyme isolated from yeast, that it catalyzed the following reaction:

$$5\text{-PP-MVA} + \text{ATP} \rightarrow \text{I-PP} + CO_2 + \text{ADP} + P_i \qquad (4)$$

This was confirmed also for the enzyme isolated from liver.[30] The oxygen of the 3-OH group of 5-PP-MVA is found after decarboxylation in the inorganic orthophosphate liberated;[31] it has been suggested, therefore, that the 3-phospho-5-pyrophosphomevalonate might be an intermediate in the reaction. No evidence could be obtained so far for the existence of this intermediate. It was also found with the liver decarboxylase that the appearance of CO_2 and of ADP was stoichiometric and synchronous in the pH range of 5.1–8.9.[32] It is consistent with all the experimental observations recorded with both the yeast and liver decarboxylase that the role of ATP in the reaction is not that of a phosphate donor, but of an acceptor of the electrons of the 3-OH group[31] with the resulting displacement of ADP. This is shown in Fig. 1, which also illustrates that the

[29] K. Bloch, S. Chaykin, A. H. Phillips, and A. de Waard, *J. Biol. Chem.* **234**, 2595 (1959).
[30] H. Hellig and G. Popják, *Biochem. J.* **80**, 47P (1961).
[31] M. G. Lindberg, C. Yuan, A. de Waard, and K. Bloch, *Biochemistry* **1**, 182 (1962).
[32] H. Hellig, Ph.D. Thesis, Univ. of London, 1962.

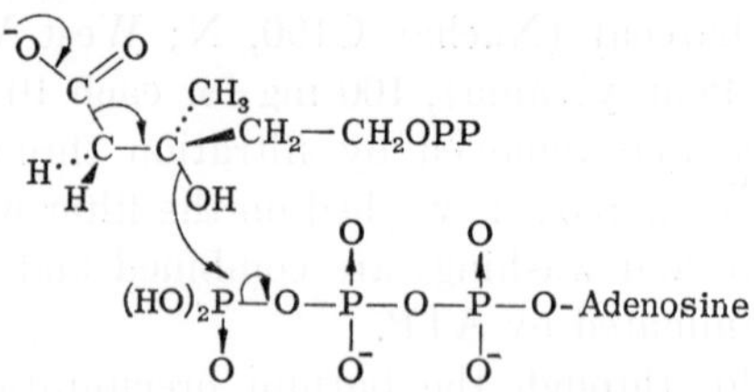

FIG. 1. Probable mechanism of 5-pyrophosphomevalonate decarboxylase reaction. ATP acts as an acceptor of the electrons of the bond between the hydroxyl group and C-3 of 5-pyrophosphomevalonate with the consequent displacement of ADP from ATP. [M. Lindberg, C. Yuan, A. de Waard, and K. Bloch, *Biochemistry* **1**, 182 (1962)].

The scheme also shows that the loss of the carboxyl and the 3-hydroxyl group occurs by a *trans*-elimination of these groups [J. W. Cornforth, R. H. Cornforth, G. Popják and L. Yengoyan, *J. Biol. Chem.* **241**, 3970 (1966)]. Reproduced by permission of the Editor, *Journal of Biological Chemistry*.

decarboxylation proceeds by a *trans*-elimination of the carboxyl and 3-hydroxyl groups.[33]

Assay of Pyrophosphomevalonate Decarboxylase

a. Spectrophotometric Assay

In this assay, as for mevalonate kinase and phosphomevalonate kinase, the amount of ADP produced in the reaction is measured by the coupled pyruvate kinase–lactate dehydrogenase system. This assay can be used at the pH optimum of the liver decarboxylase (5.1) for only short periods (5–10 minutes) because of the gradual destruction of NADH in the assay system and because of a gradually increasing turbidity of the enzyme solution at this pH. The assay is satisfactory at pH 7.4, but the observed enzymatic activity is 2.5 times less than the maximum found at pH 5.1.

For assays at pH 7.4 the reaction mixture (5.9 ml) contains 100 mM Tris-HCl buffer, 2 mM MnCl$_2$, 0.2 mM MgCl$_2$ (for the activation of pyruvate kinase), 3 mM ATP, 0.3 mM phosphoenol pyruvate, 0.2 mM NADH, 200 μg lactate dehydrogenase, 52 μg pyruvate kinase, 10 mM NaF, 10 mM cysteine (freshly neutralized with KOH), and pyrophospho-mevalonate decarboxylase (0.005–0.01 unit per milliliter). A 2.9-ml amount of this reaction mixture is pipetted into 2 spectrophotometer cells of 1-cm light path; 0.1 ml of water is added to the reference cell and 0.1

[33] J. W. Cornforth, R. H. Cornforth, G. Popják, and L. Yengoyan, *J. Biol. Chem.* **241**, 3970 (1966).

ml of a 0.6 mM solution of 5-PP-MVA (cf. p. 417) is stirred into the test cell to initiate the reaction, which is followed for a few minutes at 340 mμ.

For assays at pH 5.1 the Tris-HCl buffer is replaced by 100 mM sodium acetate buffer; the other ingredients are the same as described above.

b. Assay by Measuring Release of $^{14}CO_2$ from 5-Pyrophosphomevalonate-1-^{14}C

These assays are carried out in 3-ml Warburg flasks; the center well contains 0.2 ml of 40% KOH and frilled filter paper in the usual way and the side arm 0.2 ml of 5 N HCl. The 1-ml incubation mixture in the main compartment consists of 100 mM sodium acetate buffer, pH 5.1, 3 mM MnCl$_2$, 3 mM ATP, 10 mM NaF, 10 mM cysteine (neutralized), and about 0.01 unit of decarboxylase. The substrate, 1-^{14}C-labeled $(-)(R)$5-pyrophosphomevalonate[34] (suitable specific activity 0.2 μC per micromole or higher) is added last to a concentration of 0.1 mM; the flask is stoppered immediately. The mixture is incubated at 25° for 5 minutes when the acid is tipped from the side arm into the main compartment and the incubation is continued with shaking for 2 hours more. The contents of the center well are then transferred with a little water to a centrifuge tube containing 1.5 ml of 20% KOH to which 150 micromoles of Na$_2$CO$_3$ have been added. After several hours, or overnight, at room temperature the piece of filter paper is removed from the centrifuge tube, rinsed thoroughly with hot water and the CO$_3^{2-}$ is precipitated by the addition of 0.2 ml of 1 M BaCl$_2$. The precipitate of BaCO$_3$ is then collected for ^{14}C-counting.

This assay, although slow, is most useful for studying the properties of the enzyme, as it allows the changing of the buffer and the concentrations of the various reactants. Sodium acetate, potassium phosphate, and Tris-HCl buffers—in their appropriate pH ranges—are suitable for this enzyme.

c. Combined Assay

The assays described under a and b may be combined in a specially constructed spectrophotometer cell shown in Fig. 2, and using the spectrophotometric assay system with 5-pyrophosphomevalonate-1-^{14}C as substrate. After the amount of ADP formed has been measured for a few minutes the reaction is stopped by the injection of 0.3 ml of 0.1 M EDTA and 0.2 ml of octanol through the side inlet. The two needles shown in

---

[34] Prepared from R($-$)-5-phosphomevalonate-1-^{14}C with phosphomevalonate kinase (cf. pp. 410, 413).

Fig. 2 are then inserted through the rubber caps, 0.5 ml of $5\,N$ HCl is injected through the side inlet and the CO_2 is driven out of the reaction mixture with slow bubbles of N_2 gas and trapped in a 40% KOH solution in a "wash-bottle." The KOH in the side arm and in the "wash-bottle" are combined and the CO_3^{2-} is precipitated with $BaCl_2$ as in assay (b), and its ^{14}C-content is determined.

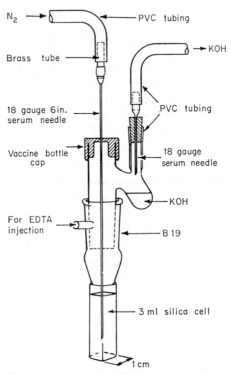

FIG. 2. Modified spectrophotometer cell for the assay of 5-pyrophosphomevalonate decarboxylase by measurement of the amount of ADP formed and $^{14}CO_2$ released from the substrate. (H. Hellig, Ph.D. Thesis, University of London, 1962.)

It is found by this method that the amount of ADP formed and CO_2 evolved in the decarboxylase reaction are equivalent (micromole/micromole).

Unit of Enzyme Activity. One unit of the decarboxylase is defined as that amount of enzyme which under the conditions of assay and 25° generates 1 micromole of ADP or of CO_2 per minute. Enzyme specific activities are expressed as units per milligram of protein. The enzymatic activity observed at pH 7.4 has to be multiplied by a factor of 2.5 to give the activity at the optimum pH of 5.1.

Purification of Pyrophosphomevalonate Decarboxylase
from Liver Extracts

So far this enzyme, among the enzymes acting on mevalonate and its derivatives, has proved to be the most difficult one to purify and to obtain in quantity from liver extracts, mainly on account of the instability of even partially purified preparations. The method to be described for the purification gives preparations free of other enzymes acting on mevalonate derivatives and of ATPase, but its specific activity is inferior to the specific activity of the similar enzyme prepared from yeast extracts.[22a]

Method

The starting material for the purification is the "protamine supernatant" of pig liver extracts as described for mevalonate kinase and phosphomevalonate kinase (p. 407).

The decarboxylase is contained in the protein fraction precipitating between 30 and 65% saturation with ammonium sulfate (pH 7.2–7.4).

The precipitated protein is collected by centrifuging at 10,000–12,000 g for 20 minutes and dissolved in 10 mM potassium phosphate buffer, pH 7.4, containing 1 mM EDTA to give a protein concentration of 40–50 mg/ml. The solution is dialyzed for 4 hours against the same buffer (Fraction I, Table V).

Treatment with Alumina-Cγ Gel. The dialyzed solution obtained in the previous step is diluted with distilled water (neutralized with ammonia) to a protein concentration of 10 mg/ml and a suspension of alumina-Cγ gel (85 mg solid per milliliter) is stirred into it to give a gel:protein ratio of 1:1. Nearly all the protein and all the decarboxylase is adsorbed onto the gel, which is collected by centrifuging; the supernatant is discarded. The gel is washed three times with a volume of 20 mM potassium phosphate buffer, pH 6.9, equal to the volume of the packed gel. The washings are discarded. The decarboxylase is eluted by five similar washings with 100 mM potassium phosphate buffer, pH 7.1 (Fraction II; Table V).

Second Ammonium Sulfate Fractionation. The gel eluate is fractionated between the limits of 45 and 65% saturation with ammonium sulfate by the use of a saturated solution of the salt (prepared from ammonium sulfate recrystallized from 1 mM EDTA and neutralized with ammonia), the pH of which has been adjusted to 8.0 with ammonia. Fourteen milliliters of this saturated solution is added to 20 ml of the eluate with gentle stirring. After 30 minutes the precipitate is sedimented at 10,000 g for 20 minutes. The supernatant is treated with 24 ml of the saturated salt solution and the precipitate is collected after 1 hour by centrifuging as above. The supernatant is discarded and the precipitate dissolved in a volume of 10 mM potassium phosphate buffer, pH 7.2, containing 1 mM cysteine, to give a protein concentration of about 50 mg/ml. The solution is then desalted in 3 ml portions by gel filtration through a 12 × 2 cm Sephadex G-25 column by the method of Flodin,[35] the 10 mM phosphate buffer being used for elution. The protein is recovered without loss of quantity or of enzymatic activity with a 2-fold dilution of concentration (Fraction III; Table V). Removal of the ammonium sulfate by dialysis results in serious losses of enzymatic activity.

Chromatography on DEAE-Cellulose Column. Portions (4–5 ml) of the desalted preparation of the decarboxylase (about 100 mg of protein) are applied to a 19 × 1.5 cm DEAE-cellulose column (Whatman DE-50), equilibrated with 1 mM potassium phosphate buffer, pH 6.9, which is then developed according to the gradient elution technique of Hurlbert *et al.*[28] The mixing flask contains initially 150 ml of 1 mM potassium phosphate buffer, pH 6.9, and the reservoir 150 ml of the same buffer + 20 mM NaCl; after 150 ml of effluent has been collected the solution in the reservoir is changed to 300 ml of the buffer fortified with 200 mM NaCl. When 5-ml fractions are collected the decarboxylase with the highest specific activity appears in fractions numbers 62–72 (Fraction IV; Table V).

For the preservation of the enzyme the fractions of the highest specific activity eluted from the DEAE column are pooled and the protein is

TABLE V

PURIFICATION OF PYROPHOSPHOMEVALONATE DECARBOXYLASE
FROM PIG LIVER EXTRACT

Fraction	Total protein (g)	Total enzyme units[a]	Specific activity[a]
I. First ammonium sulfate fraction from "protamine supernatant"[b]	17.90	32	0.0018
II. Eluate from alumina-C$_\gamma$ gel	6.25	25	0.0041
III. Second ammonium sulfate fraction after desalting	2.74	17.3	0.0063
IV. Eluate from DEAE-cellulose column	0.432	10.8	0.025

[a] Determined by the spectrophotometric assay at pH 7.4. The activity at the optimum pH of 5.1 is 2.5 times greater.
[b] Preparation from 1.7 kg of liver.

[35] P. Flodin, *J. Chromatog.* **5**, 103 (1961).

precipitated by 65% saturation with ammonium sulfate. The precipitate, collected by centrifuging, is dissolved in the minimum volume of 20 mM potassium phosphate buffer, pH 7.4, containing 1 mM 2-mercaptoethanol and is stored at −20°.

The purification steps are summarized in Table V.

Properties of Pyrophosphomevalonate Decarboxylase

This is the least stable of the liver enzymes acting on mevalonate and its derivatives: storage of the purified preparation at −20° may result in the loss of almost all the activity in 4 months.

The decarboxylase from liver, in contrast to the similar enzyme from yeast, has a distinct pH optimum at 5.1; the observed activities at pH 4 and 7.4 are 40% of the maximum. ATP is the specific coenzyme of the liver decarboxylase; maximum activity is observed at a concentration of 2–3 mM. Mn^{++}, Mg^{++}, and Co^{++} ions are nearly equally effective in activating the enzyme at 3 mM concentration, though Mn^{++} is slightly superior to the other two cations. The K_m value for 5-pyrophosphomevalonate is about 0.5 μM. Cysteine (5–10 mM) and ascorbate (5–10 mM) stimulate the enzyme (10–20%). EDTA (10 mM) causes complete inhibition, and p-chloromercuribenzoate (0.1 mM) a 60% inhibition which is reversible by 1 mM 2,3-dimercaptopropanol or by cysteine. The following substances are without effect: potassium phosphate (100 mM), HCO$_3^-$ (10 mM) KCN (5 mM), Na$_3$ASO$_3$ (5 mM), N-ethylmaleimide (5 mM), iodoacetamide (5 mM), lipoic acid (10 mM), thiamine pyrophosphate (10 mM), avidin (0.6 mg/ml).

E. ISOPENTENYL PYROPHOSPHATE ISOMERASE (EC 5.3.3.2) AND GERANYL TRANSFERASE (EC 2.5.1.1)

These two enzymes will be treated together because they are obtained from the same extract of liver, and because they are assayed by similar methods.

Isopentenyl Pyrophosphate Isomerase (Prenyl Isomerase). This enzyme was recognized first by Agranoff et al.[36] in yeast extracts as catalyzing the reversible reaction (5):

$$\text{(5)}$$

Until recently only indirect proof existed for the presence of this

---

[36] B. W. Agranoff, H. Eggerer, U. Henning, and F. Lynen, *J. Biol. Chem.* **235**, 326 (1960).

enzyme in liver: in the multienzyme system which synthesizes farnesyl pyrophosphate from mevalonate and ATP (p. 443) addition of 5 mM, iodoacetamide stopped the reactions at isopentenyl pyrophosphate.[33] Although inhibition of the isomerase from yeast by iodoacetamide is one of the most outstanding characteristics of this enzyme, the accumulation of I-PP in the multienzyme system poisoned with iodoacetamide is not an entirely convincing proof for the existence of this enzyme, because the geranyl transferase from liver is also inhibited by this reagent.

Recently the partial purification of prenyl isomerase from pig liver has been reported by Shah *et al.*,[37] and the enzyme has been obtained also in our laboratory[38] free from other enzymes acting on mevalonate or its derivatives.

The elimination of hydrogen from C-2 of isopentenyl pyrophosphate in reaction (5) is a stereospecific process: in enzyme systems synthesizing *trans*-polyprenyls H_R is eliminated,[16] in *cis*-polyprenyl synthesis (rubber) H_S is lost[39] as shown in reactions (6) and (7):

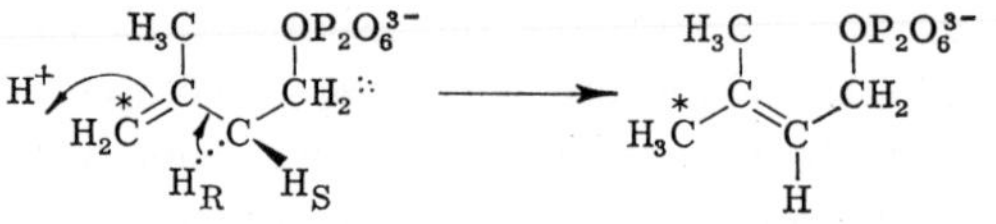

$$\tag{6}$$

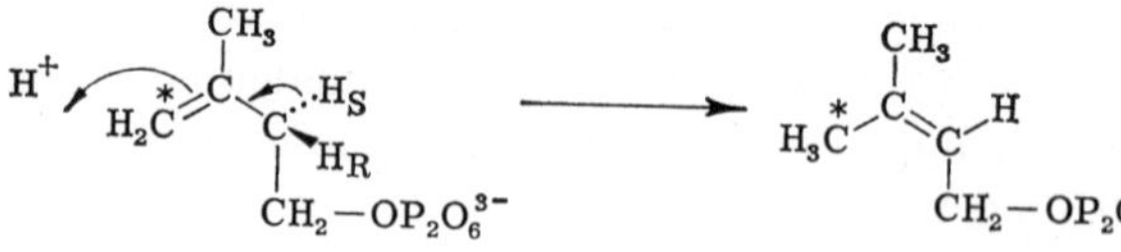

$$\tag{7}$$

Geranyl Transferase (trans-Diprenyl Transferase), EC 2.5.1.1. This enzyme was first recognized in yeast autolyzates by Lynen *et al.*,[40] who reported a 20-fold purification. Benedict, Kett, and Porter[41] described recently a 15-fold purification of geranyl transferase from pig liver. The method to be described here gives a 100-fold purification of the enzyme from pig liver and is based on the work of Holloway and Popják.[38, 42]

Reaction Catalyzed. The enzyme preparations purified 100-fold from pig liver catalyze two reactions, (8) and (9):

[37] D. H. Shah, W. W. Cleland, and J. W. Porter, *J. Biol. Chem.* **240**, 1946 (1965).
[38] P. Holloway and G. Popják, *Biochem. J.* **104**, 57 (1967); and *Biochem. J.* **106**, 835 (1968).
[39] B. L. Archer, D. Barnard, E. G. Cockbain, J. W. Cornforth, R. H. Cornforth, and G. Popják, *Proc. Roy. Soc.* **B163**, 519 (1966).
[40] F. Lynen, B. W. Agranoff, H. Eggerer, U. Henning, and E. M. Möslein, *Angew. Chem.* **71**, 657 (1959).
[41] C. R. Benedict, J. Kett, and J. W. Porter, *Arch. Biochem. Biophys.* **110**, 611 (1965).
[42] P. W. Holloway and G. Popják, *Biochem. J.* **100**, 61P (1966).

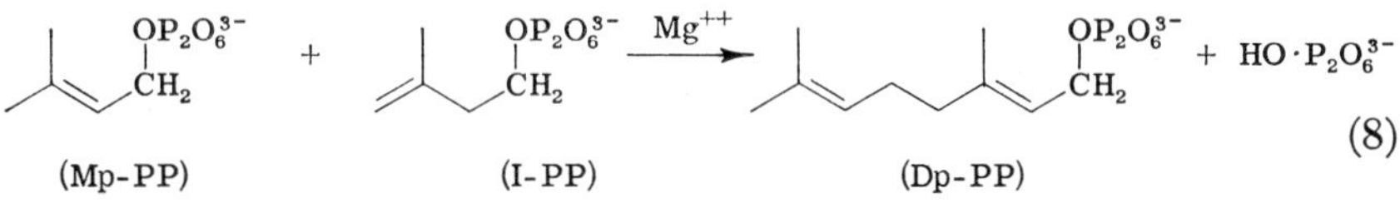

$$\text{(Mp-PP)} \quad + \quad \text{(I-PP)} \quad \xrightarrow{\text{Mg}^{++}} \quad \text{(Dp-PP)} \quad + \text{ HO} \cdot P_2O_6^{3-} \tag{8}$$

$$\text{(Dp-PP)} \quad + \quad \text{(I-PP)}$$

$$\xrightarrow{\text{Mg}^{++}} \quad \text{(Tp-PP)} \quad + \quad \text{HO} \cdot P_2O_6^{3-} \tag{9}$$

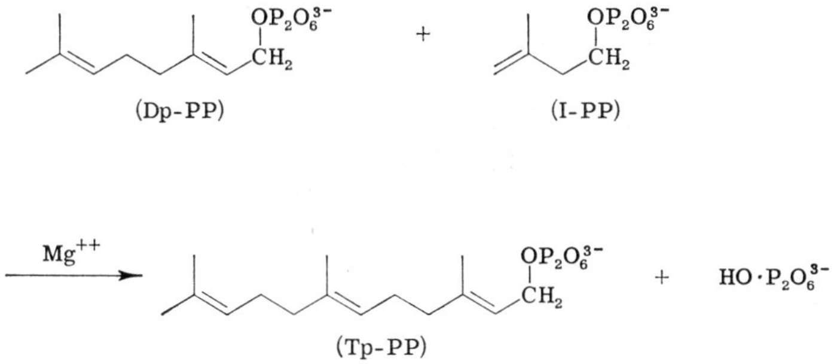

It is very probable that one and the same protein catalyzes both of these reactions because (a) irrespective of whether the substrates offered to the enzyme are Mp-PP + I-PP, or Dp-PP + I-PP, the product of the reaction is *trans-trans*-farnesyl pyrophosphate (Tp-PP), and because (b) the ratio of the catalytic activities of the various enzyme preparations

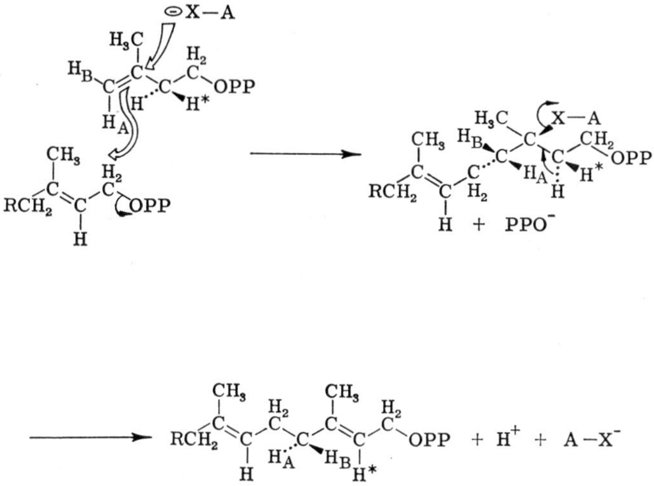

SCHEME 1. Stereochemistry of *trans*-prenyl transferase reaction. [J. W. Cornforth, R. H. Cornforth, G. Popják and L. Yengoyan, *J. Biol. Chem.* **241**, 3970 (1966)]. Reproduced by the permission of the Editor, Journal of Biological Chemistry.

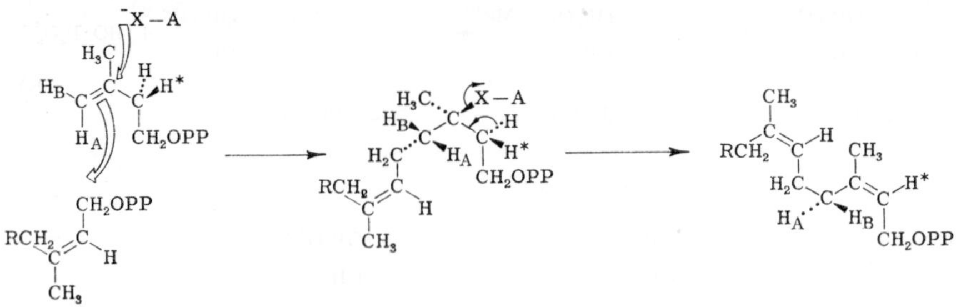

SCHEME 2. Predicted steric course of *cis*-prenyl transferase reaction as, for example, in the biosynthesis of rubber. [J. W. Cornforth, R. H. Cornforth, G. Popják, and L. Yengoyan, *J. Biol. Chem.* **241**, 3970 (1966)]. Reproduced by the permission of the Editor, *Journal of Biological Chemistry*.

with Mp-PP + I-PP and with Dp-PP + I-PP remain the same throughout the purification procedure.

Reactions (8) and (9) describe the overall process correctly, but not completely. The formation of the carbon-to-carbon bonds on each addition of a prenyl residue to the double bond of I-PP is a stereospecific process and occurs probably in two steps: first the allyl residue and an as yet undefined nucleophilic, X^-, group are added to the double bond in a *trans*-manner; this is followed by a *trans*-elimination of the nucleophile and of H_R from C-2 of I-PP. The formation of the carbon-to-carbon bond occurs with an inversion of configuration at C-1 of the prenyl pyrophosphate and the prenyl residue is added on to that side of the double bond of I-PP on which the groups $—CH_2CH_2—OP_2O_6{}^{3-}$, $—CH_3$, and $=CH_2$ appear in a clockwise order.[4, 16, 33] The stereochemical consequences of these reactions are summarized in Scheme 1. This particular stereochemistry applies only to the biosynthesis of *trans*-polyprenyl substances; during the synthesis of *cis*-polyprenyls (e.g., rubber) the hydrogen eliminated from C-2 of -IPP is H_S.[39] The probable steric course of a *cis*-prenyl transferase reaction is shown in Scheme 2: this is similar to the *trans*-reaction except for a different orientation of substrates on the enzyme.

Assays of Prenyl Isomerase and of Diprenyl Transferase

Principle of the Assays. The principle of the assays for both enzymes is the instability of the prenyl pyrophosphate in an acid medium (cf. p. 388). Thus Mp-PP, the product of the isomerase reaction, gives by acid hydrolysis methylvinyl carbinol (2-hydroxy-2-methylbut-3-ene) and 3,3-dimethylallyl alcohol, and Tp-PP, the product of the geranyl transferase reaction, gives nerolidol and farnesol in a ratio of about 4:1 in addition to inorganic pyrophosphate. Thus using ^{14}C-labeled I-PP of known specific activity as substrate either alone (in the isomerase reaction), or in combination with Mp-PP, or with Dp-PP (in the transferase reaction), the amount of product formed is deduced from the amount of radioactivity extractable with appropriate solvents from the reaction mixture after acid hydrolysis.

Assay of Prenyl Isomerase

METHOD 1. ASSAY COUPLED WITH GERANYL TRANSFERASE

The principle of this assay is that the Mp-PP formed in the reaction in the presence of excess I-PP and diprenyl transferase is converted into farnesyl pyrophosphate (Tp-PP).

This assay is most suitable for crude enzyme preparations which contain both isomerase and transferase (cf. Table VI; S_{45}, F_{50}, F_{60}), but is at best a compromise since the assay has to be done near the pH optimum of the transferase (7.9) rather than that of the isomerase (pH 6.0); hence the units of enzyme activity obtained by the method are less than the values one might obtain at the pH optimum of the isomerase.

One-milliliter reaction mixtures containing 100 micromoles of Tris-HCl buffer, pH 7.9, 5 micromoles of $MgCl_2$, 2 micromoles of $MnCl_2$, 50 millimicromoles of ^{14}C-labeled isopentenyl pyrophosphate (suitable specific activity 0.1–0.5 μC per micromole or higher) and 0.1 mg, or less, of protein are incubated at 37° for 5 minutes. The reaction is stopped by the addition of 1 ml of 80% ethanol, containing 2 N HCl. Geraniol, nerolidol, and farnesol, about 1 mg each, are added to the mixture, which after 30 minutes at 37° is made alkaline (pH 10) with 10 N NaOH and

TABLE VI
DISTRIBUTION OF ISOPENTENYL PYROPHOSPHATE ISOMERASE AND PRENYL TRANSFERASE
IN AMMONIUM SULFATE FRACTIONS OF PIG LIVER EXTRACT

Ammonium sulfate fraction	Volume (ml)	Total protein (mg)	Isomerase[a]		Transferase	
			Specific activity	Total enzyme units	Specific activity	Total enzyme units
Unfractionated S_{45} from 800 g liver	920	20,240	0.3	6072	3.5	70,840
0–40 (F_{40})	110	3,740	0	0	1.0	3,740
40–50 (F_{50})	99	2,970	0.3	891	9.0	26,730
50–60 (F_{60})	110	3,520	0.8	2816	8.5	29,920
60–70 (F_{70})	97	2,328	1.2	2794	0	0

[a] Assayed by the coupled assay system (see above).

TABLE VII

PURIFICATION OF ISOPENTENYL PYROPHOSPHATE ISOMERASE[a]

| | | | Isomerase | |
Fraction	Volume (ml)	Total protein (mg)	Specific activity	Total enzyme units
Unfractionated S_{45}[b]	400	6400	0.3	1920
F-70[c]	20	600	1.2	720
G-200-2[c]	25	43	6.2	266

[a] Data of P. W. Holloway and G. Popják, *Biochem. J.* **106,** 835 (1968).
[b] Assayed by the coupled system (p. 429).
[c] Assayed by the direct method at pH 7.9 and with Mg^{++} instead of Mn^{++}. At the optimum pH of 6 and with Mn^{++} the activity of the enzyme is four times higher.

is extracted three times with light petroleum (b.p. 30–40°). The extract is dried over anhydrous $MgSO_4$ and aliquots are taken for counting of radioactivity.

For comparative purposes, in order to follow the progress of the purification of the isomerase, this assay can be extended to preparations free of geranyl transferase (cf. Tables VI and VII; F_{70}, G-200-2) by the addition of purified diprenyl transferase to the assay mixture: 10 units of purified transferase (G-200-1, Table VIII) are then added to the reaction mixture described above for each unit of isomerase to be assayed.

Unit of Activity. One unit of isomerase activity, by the coupled assay, is expressed as one-third of the millimicromoles of ^{14}C-labeled isopentenyl pyrophosphate converted into ^{14}C-labeled farnesyl pyrophosphate per minute in the presence of an excess of transferase.

METHOD 2. DIRECT ASSAY

For preparations of the isomerase free of transferase (F_{70}, G-200-2, cf. Tables VI and VII) the following assay is used.

TABLE VIII

PURIFICATION OF *trans*-DIPRENYL TRANSFERASE FROM PIG LIVER EXTRACT[a]

Fraction	Total protein (mg)	Specific activity	Total enzyme units
S_{45}[b]	20,800	1.5	31,200
F_{50}	1,710	7.5	12.800
FH	1,450	7.5	10,900
G-200-1	100	39.0	3,900
DEAE-eluate	11	152.0	1,670

[a] Data of P. W. Holloway and G. Popják, *Biochem. J.* **104,** 57 (1967).
[b] Derived from 800 g liver.

One-milliliter reaction mixtures containing 100 micromoles of sodium acid maleate buffer, pH 6, 2 micromoles of $MnCl_2$, 0.5 micromole of 2-mercaptoethanol, 50 millimicromoles of ^{14}C-labeled isopentenyl pyrophosphate (suitable specific activity 0.1–0.5 μC per micromole, or higher) and 1–5 units of isomerase are incubated for 5 minutes at 37°. The reaction is stopped by the addition of 0.1 ml of 5 N HCl, and 1 mg of carrier methylvinyl carbinol[43] is added. After 30 minutes at 37° the mixture is made alkaline (pH 10) with 10 N NaOH, it is saturated with NaCl, and extracted four times with 1 ml of light petroleum (b.p. 30–40°). The combined extracts, after drying over $MgSO_4$, are analyzed for [14]C. The amount of the product formed is calculated from the ^{14}C-content of the petroleum extract and from the known specific activity of the substrate used.

Unit of Isomerase Activity. The "direct assay" applied to preparations free of diprenyl transferase gives the true value of isomerase activity. One unit of activity is that amount of enzyme which converts 1 millimicromole of I-PP into Mp-PP per minute under the conditions of the assay. Specific activities of the preparations are expressed as units of enzyme activity per milligram of protein.

The direct method of assay on purified preparations gives at least five times higher values than does the "coupled assay" at pH 7.9.

Analysis of the Product of Isomerase Reaction. Incubations of the purified isomerase, G-200-2, are set up as for the "direct assay," but using more enzyme, e.g., 10 units per milliliter. At the end of the chosen incubation period the pH of the reaction mixture is adjusted with KOH to 9.5 and 0.1 ml of a 1% solution of intestinal alkaline phosphatase in 0.05 M $KHCO_3$ and 0.05 ml of 0.1 M $MgCl_2$ are added for each milliliter of the incubation, which is continued for 4 hours more at 37°. At the end, the mixture is saturated with NaCl; after addition of carrier isopentenol, 3,3-dimethylallyl alcohol and farnesol, 1 mg each, the alcohols are extracted at ice temperature in the cold room with ethyl chloride: three extractions (1 ml of ethyl chloride per milliliter of incubation) are sufficient. After addition of 20–50 μl of benzene, the ethyl chloride is evaporated off at room temperature and the benzene solution is analyzed by gas-liquid radiochromatography.[44]

For large incubations (10–100 ml) the above procedure is cumbersome; the following technique is recommended instead. At the end of the incubation with the alkaline phosphatase the reaction mixture is transferred quantitatively with the minimum amount of water to a distilla-

---

[43] Prepared by boiling 0.5 ml of 3,3-dimethylallyl alcohol with 1 drop of conc. HCl; almost complete isomerization of the primary to the tertiary alcohol occurs.

[44] G. Popják, A. E. Lowe, D. Moore, L. Brown, and F. A. Smith, *J. Lipid Res.* **1**, 29 (1959); G. Popják, A. E. Lowe, and D. Moore, *J. Lipid Res.* **3**, 364 (1962).

tion flask which has a volume three times larger than the volume of the reaction mixture. Solid ammonium sulfate[45] is now added to the flask to saturate the solution. After assembly of the distillation apparatus and packing of the receiver in crushed ice, the distillation flask is heated very cautiously with a microburner for about 2 minutes, during which time the protein is coagulated and the mixture is brought to boiling point. Thereafter the heating of the distillation flask with the microburner is regulated by hand to produce a smooth distillation. All the isopentenol and dimethylallyl alcohol liberated by the action of phosphatase are collected in the first 5 ml of distillate. The alcohols are then extracted, after saturation of the solution with NaCl, with ethyl chloride as described above.[33]

For the gas-liquid radiochromatographic analysis of the extract a 9 foot $\times$ 4 mm column packed with 10% "Carbowax 20M" (Union Carbide Corporation, New York) on Celite or Chromosorb is most suitable. Two analyses are made: one at 80° and a second at 195°. The analysis is made at the higher temperature in order to check whether the farnesol (which is retained on the column at 80°) contains any radioactivity or not. Presence of radioactivity in the farnesol indicates contamination of the isomerase with prenyl transferase.

For the analysis of the products of isomerase reaction in incubations of 5 ml and larger, gas-liquid radiochromatography is not needed. Mass detectors are now available in conjunction with gas-liquid chromatography (GLC), such as the argon- or flame-ionization detectors, which respond to a few micrograms, or even less, of a substance and hence obviate the need to use isotopically labeled I-PP in the study of the isomerase reaction. After the distillation of the alcohols liberated by the phosphatase and after their extraction by ethyl chloride as described above, but without added carriers, a few microliters (5–10) of pure octanol are added to the extract and the ethyl chloride is evaporated off at room temperature. A few microliters of the residual solution of the prenols in octanol are then applied to the gas-liquid chromatography column (Carbowax 20M, at 80–85°), connected to one of the high-sensitivity ionization detectors. Isopentenol emerges from the column 5–10 minutes after the application (depending on the gas-pressure and gas flow used) followed immediately by dimethylallyl alcohol. Octanol has a retention time about twice that of dimethylallyl alcohol. The use of specially pure octanol as the final solvent for the prenols is most important: reagent and analytical grade octanols contain impurities which

---

[45] Other salts, e.g., MgSO₄ or NaCl, are unsuitable. The ammonia liberated at the alkaline pH (9.5) of the reaction mixture suppresses the frothing during distillation.

have retention volumes very close to those of isopentenol and dimethyl-allyl alcohol and hence interfere with the analysis. The octanol needed for such experiments is purified first by preparative GLC on a Carbowax 20M column at 85°, or by careful fractional distillation.

Assay of Diprenyl Transferase

The assay is similar to the coupled assay described for prenyl isomerase.

Method. One-milliliter reaction mixtures containing 100 micromoles of Tris-HCl buffer, pH 7.9, 5 micromoles of $MgCl_2$, 50 millimicromoles of dimethylallyl pyrophosphate (or 50 millimicromoles of geranyl pyrophosphate), 50 millimicromoles of ^{14}C-labeled isopentenyl pyrophosphate (0.1–0.5 μC per micromole) and 0.1 mg, or less, of protein are incubated for 5 minutes at 37°. The reaction is stopped by the addition of 1 ml of 80% ethanol containing 2 N HCl. Geraniol, nerolidol, and farnesol, 1 mg each, are added to the mixture, which after 30 minutes at 37° is made alkaline (pH 10) with 10 N NaOH and is extracted three times with light petroleum (b.p. 30–40°). The combined extracts are dried over $MgSO_4$, made up to 5 ml with the solvent, and an aliquot is assayed for radio-activity.

During the purification of the enzyme control assays of the same composition as described above are also set up except that the dimethyl-allyl pyrophosphate or geranyl pyrophosphate is omitted. When the preparations of transferase are free of isomerase there is no radioactivity extractable with petroleum from the control assays after acidification.

Unit of Activity. One unit of the transferase is defined as that amount of enzyme which in the presence of dimethylallyl or geranyl pyrophosphate converts 1 millimicromole of ^{14}C-labeled isopentenyl pyrophosphate into farnesyl pyrophosphate in 1 minute. The specific activity is defined as units of enzyme per milligram of protein.

Identification of Products of Geranyl Transferase Reaction. The products of the transferase reaction are identified from incubations larger than used for the standard assay. A 4-ml reaction mixture containing 400 micromoles of Tris-HCl buffer, pH 7.9, 20 micromoles of $MgCl_2$, 0.5 micromoles of either dimethylallyl pyrophosphate or of geranyl pyro-phosphate, 0.5 micromoles of ^{14}C-labeled isopentenyl pyrophosphate (specific activity 0.1 μC per micromole), and 4 mg of purified transferase (specific activity 80–150) is incubated at 37° for 10 minutes, when the pH of the mixture is adjusted to 9.5 with 1 N NaOH and 2 mg of in-testinal alkaline phosphatase (EC 3.1.3.1) contained in 0.2 ml of 0.05 M $KHCO_3$ is added. After 4 hours at 37° linalool, geraniol, nerolidol, and farnesol, 1 mg of each, or less, are added to the incubation, which is extracted three times with light petroleum (b.p. 30–40°). The combined extracts are dried over $MgSO_4$ and concentrated to a small volume; portions of the extract are then analyzed by gas-liquid radiochromatography[44] at 185–195° on a 9 foot $\times$ 4 mm column packed with Celite or Chromosorb (100–120 mesh) coated with 10% "Carbowax 20M" (Union Carbide Corporation, New York). The usual finding, when purified transferase (G-200-1; DEAE-eluate; cf. Table VIII) is used, is that 90% of the radioactivity in the petroleum extract is associated with *trans-trans*-farnesol, irrespective of whether dimethylallyl pyrophosphate + isopentenyl pyrophosphate-^{14}C, or geranyl pyrophosphate + isopentenyl pyrophosphate-^{14}C were the substrates in the reaction. The remaining 10% emerges from the column in two poorly resolved fractions between linalool and geraniol[45a] but no radioactivity is seen with geraniol even when dimethylallyl pyrophosphate + isopentenyl pyrophosphate-^{14}C are the substrates.

When acid hydrolysis, instead of alkaline phosphatase, is used to release the alcohol(s) from the pyrophosphate esters—as in the standard assay—90% of the radioactivity of the petroleum extracts is associated with nerolidol + *trans-trans*-farnesol in a ratio of 4:1 (cf. p. 389), and the remainder in the same two components appears between linalool and geraniol found after hydrolysis with phosphatase.

Preparation of Prenyl Isomerase and *trans*-Diprenyl Transferase from Pig Liver

Initial Extract. Fresh pig liver, chilled on ice, is cut into small pieces and is homogenized in 100-g lots in a Waring blendor for 30 seconds with 200 ml of a $0.25\,M$ sucrose solution containing 25 mM $KHCO_3$ and 1 mM EDTA. Alternatively the liver is minced first in a meat-grinder and is extracted with 2 volumes of the above medium by stirring for 10 minutes; the suspension is filtered through several layers of cheesecloth. Whichever method is used for extraction, the subsequent treatment of the preparation is the same and yields similar results. The homogenates, or crude extracts, are centrifuged successively at 600 g for 15 minutes, at 10,000 g for 20 minutes, and finally at 45,000 g for 3 hours, the sediment after each centrifuging being discarded. The last supernatant, S_{45}, filtered through cotton wool to remove floating fat, is the source of the two enzymes.

Ammonium Sulfate Fractionation. The S_{45} preparation is fractionated

---

[45a] Since the time that this chapter was submitted these two fractions were identified as isomers of farnesene; they result from the thermal dehydration of farnesol in the GLC apparatus especially when the temperature of the injection port is 200° or higher.

with ammonium sulfate between the following limits of saturation: 0–40, 40–50, 50–60, and 60–70%. The pH of the preparation is kept at 7.5 with ammonium hydroxide during the fractionation. The precipitated proteins are collected by centrifuging at 10,000–12,000 g for 20 minutes, and dissolved in a small volume of 10 mM Tris-HCl buffer, pH 7.7, to give solutions of 30–50 mg protein per milliliter, which are dialyzed against the same buffer for 3 hours.

The fraction precipitating between 40 and 50% saturation (F_{50}) is almost free of isomerase and contains about one-third of the transferase units of the S_{45} and is used for the further purification of this enzyme. The fraction precipitating between 60 and 70% saturation (F_{70}) is not infrequently free of transferase and may contain as much as one-half of the isomerase units of the S_{45}; the F_{70} is used for the further purification of the isomerase.

The distribution of isomerase and transferase enzymes among the various ammonium sulfate fractions of the S_{45} is shown in Table VI.

Further Purification of Prenyl Isomerase.[46] The pellet of F_{70} proteins (Table VI) after centrifugation is transferred directly into a dialysis sack and is dialyzed against 10 mM Tris-HCl buffer, pH 7.7, containing 1 mM mercaptoethanol for 6 hours, the dialysis fluid being changed twice. The solution is centrifuged at 40,000 g for 30 minutes, and the clear red supernatant (30–40 mg protein per milliliter) is applied in 10-ml portions onto a Sephadex G-200 column (36 $\times$ 2.5 cm) equilibrated with 10 mM Tris-HCl buffer, pH 7.7, containing 1 mM mercaptoethanol. The column, which has a void volume of 67 ml, is eluted with the same buffer. The isomerase has a V_e of 170 ml. A red-colored substance (hemoglobin) is eluted just before the isomerase and acts, conveniently, as a marker: the 20-ml of effluent immediately after this contains the isomerase. EDTA is added to this solution of isomerase to a concentration of 0.1 mM; the solution is then concentrated by "dialysis" against solid sucrose overnight. The limp dialysis sack is then tightened and the sucrose is removed by dialysis against 10 mM Tris-HCl buffer, pH 7.7, containing 1 mM mercaptoethanol but no EDTA. If EDTA is not added to the isomerase solution before concentrating it by "dialysis" against solid sucrose, part of the enzyme activity is lost.

This preparation (G-200-2) of the isomerase usually represents a twenty-fold purification from the initial S_{45} extract (Table VII) and is entirely free of prenyl transferase: the only product of its action on isopentenyl pyrophosphate is dimethylallyl pyrophosphate. The solutions of the enzyme are kept frozen at $-20°$.

[46] This procedure is based on the work of P. W. Holloway and G. Popják, *Biochem. J*. **106**, 835 (1968).

Further Purification of Diprenyl Transferase. The pH of the dialyzed F_{50} (Table VI) solution is adjusted to 5 with $1 N$ acetic acid, and the copious white precipitate (mostly glycogen and a little protein) is removed by centrifugation at 2000 g for 15 minutes; the pH of the supernatant is brought to 7.5 with $1 N$ KOH. A further small precipitate, formed on neutralization, is removed by centrifugation at 105,000 g for 30 minutes. The supernatant (FH) retains its transferase activity for several months at $-20°$.

Samples of FH solution (8 ml containing about 400 mg protein) are further fractionated on a Sephadex G-200 column (36 $\times$ 2.5 cm) equilibrated with 10 mM Tris-HCl buffer, pH 7.7. The column, which has a void volume of 67 ml, is eluted with the buffer used for equilibration. The transferase has the same elution volume (130 ml) as crystalline bovine serum albumin; the protein fraction eluted between 130 and 146 ml has the highest specific activity in respect of the transferase. The eluate (16 ml) is added slowly to 32 ml of a stirred saturated solution of ammonium sulfate, pH 7.7, and after 30 minutes the precipitate is collected by centrifuging it at 5700 g for 20 minutes; the supernatant is discarded. The pellet of precipitated protein may be preserved at $-20°$ without deterioration for many months; it is reconstituted by dissolving in 10 mM Tris-HCl buffer, pH 7.7 and dialysis against the same buffer for 3 hours. When 2 g of the damp centrifuged precipitate is dissolved in 10 ml of buffer and the solution is dialyzed, the resulting solution usually has a protein content of 20–30 mg/ml. Such solutions (G-200-1) are free of prenyl isomerase and represent a 20- to 25-fold purified enzyme.

Further purification of the transferase from the G-200-1 solution has so far been achieved only on a small scale on a DEAE-cellulose column (Whatman DE-11), 10 $\times$ 1.5 cm equilibrated with 10 mM Tris-HCl buffer, pH 7.7, containing 1 mM 2-mercaptoethanol. One milliliter of the G-200-1 solution is applied to the column, which is then eluted with the buffer of equilibration with successive stepwise increases in NaCl content of 0, 100, and 150 mM. The buffers must be made up so that, irrespective of the NaCl content, the solutions contain 1 mM of mercaptoethanol; without it the transferase is inactivated. When 40 ml of each buffer is run through the column, the protein fraction eluted with the 150 mM NaCl contains the transferase with the highest specific activity (Table VIII).

Because the transferase in dilute solution is unstable in the presence of high salt concentration (0.2–0.5 M NaCl), the eluate from the DEAE-cellulose column should be treated as the eluate from the Sephadex G-200 column (see above). This salt inactivation is prevented by mercaptoethanol, or by the joint addition of geranyl pyrophosphate plus Mg^{++} to concentrations of 20 μM and 2 mM, respectively.

Properties of Prenyl Isomerase and Diprenyl Transferase

Properties of Prenyl Isomerase.[47] The enzyme from liver has a distinct pH optimum at 6; at pH 5 and 7.5 its activity is 40% and at pH 7.9 it is only 25% of the maximum. It is specifically activated by Mn^{++} rather than Mg^{++}; this property, however, does not appear until after dialysis of the enzyme against EDTA and repeated dialysis against a medium free of EDTA. This suggests that Mn^{++} is firmly bound to the native enzyme. The full activation of the enzyme by Mn^{++} ions is attained at a concentration of 2 mM when its activity is 4–5 times higher than in the presence of Mg^{++} ions of similar or higher concentration. The K_m value for isopentenyl pyrophosphate is 4 μM. The equilibrium constant $K_{eq} =$ [Mp-PP]/[I-PP] at 37° and pH 6 is about 9.[40, 46]

Properties of Diprenyl Transferase. The preparations of the transferase are equally active with dimethylallyl pyrophosphate + isopentenyl pyrophosphate, and with geranyl pyrophosphate + isopentenyl pyrophosphate as substrates, at all stages of purification. Whichever pair of substrates is used with the enzyme, the product of the reaction is farnesyl pyrophosphate. The pH optimum of the enzyme is 7.9 and requires Mg^{++} ions as activator in preference to Mn^{++} at an optimum concentration of 1–2 mM. The K_m for geranyl pyrophosphate is 4.3 ± 1.7 μM and for isopentenyl pyrophosphate 1.45 ± 0.12 μM.[48]

Phosphate ions at a concentration of 0.1 M, even at the optimum pH, inhibit geranyl transferase to the extent of 80–90%. Sulfhydryl reagents are also inhibitory: 80% inhibition is observed after pre-

---

[47] We have reason to believe that a previous description of the properties of the enzyme by Shah *et al.*[37] is incorrect both in respect to pH optimum and metal ion requirement.

[48] Benedict *et al.*[41] reported on a 15-fold purified "farnesyl synthetase," which they claimed was active only with geranyl pyrophosphate + isopentenyl pyrophosphate as substrates, and K_m values of 4 and 2 μM for geranyl and isopentenyl pyrophosphate, respectively. In a more recent publication from the same laboratory [J. K. Dorsey, J. A. Dorsey, and J. W. Porter, *J. Biol. Chem.* **241**, 5353 (1966)] the preparation of 200-fold purified "geranyl synthetase" is reported which appears to have properties similar to those of the enzyme we have described here: it is active with dimethylallyl pyrophosphate + isopentenyl pyrophosphate, or with geranyl pyrophosphate + isopentenyl pyrophosphate. They report K_m values of 1.25 and 2.2 μM for isopentenyl pyrophosphate and dimethylallyl pyrophosphate, respectively.

The discrepancies between the reports of Benedict *et al.* and Dorsey *et al.*, we believe, are due to the fact that the former workers did all their incubations in 0.05 M PO_4^{2-} buffer and in the presence of iodoacetamide without realizing that both inhibited severely the transferase.

incubation of the enzyme at 37° for 5 minutes with 2 mM iodoacetamide, 5 μM N-ethylmaleimide, or 0.5 μM p-hydroxymercuribenzoate. The inhibition by the last reagent is completely reversed by 1 mM 2-mercaptoethanol.

II. Multienzyme Systems of Sterol Biosynthesis

The multienzyme systems of liver have been extensively used for studying various phases of sterol biosynthesis and for the preparation of isotopically labeled sterols, squalene, farnesyl pyrophosphate, isopentenyl pyrophosphate, etc. Their description here is a departure from the former practice adopted in these volumes of describing the preparation and properties only of purified enzymes. There are still several unsolved problems of sterol biosynthesis for which these multienzyme systems should prove useful; moreover, they offer simple means for the preparation of substances labeled with a variety of isotopes.

A. Enzyme System for the Synthesis of Sterols from Mevalonate

Homogenates are prepared according to Bucher and McGarrahan[49] either from the liver of young rats (100–150 g) or from fresh pig liver. Rat liver is the preferred tissue when the object of the experiment is the preparation of isotopically labeled sterols.

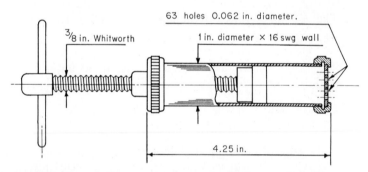

Fig. 3. A tissue press, made of stainless steel, for the preparation of liver pulp free of connective tissue, bile ducts, and blood vessels.

The animals are killed either by decapitation or by dislocation of the neck; the liver (weighing usually 4–5 g) is excised and packed in crushed ice. When the organ is thoroughly chilled, it is blotted on filter paper, all extraneous tissues are trimmed off, and the organ is put through a stainless steel press shown in Fig. 3. The liver pulp thus produced is free of connective tissues, blood vessels, and bile ducts, which are retained on the

[49] N. L. R. Bucher and K. McGarrahan, *J. Biol. Chem.* **222**, 1 (1956).

perforated disk of the press. The pulp is collected in a tared beaker containing some ice-cold $0.1\ M$ potassium phosphate (or Tris HCl) buffer, pH 7.5, and enriched with 5 mM MgCl$_2$, 2 mM MnCl$_2$, and 30 mM nicotinamide. After all the tissue pulp has been collected and weighed, enough buffer is added to make 2.5 volumes per unit weight of pulp. The suspension is then homogenized in a smooth-walled glass homogenizer fitted with a Teflon piston, the diameter of which is 0.4 mm less than that of the glass tube. The piston is driven by a 0.1 h.p. electric motor geared down to 200–300 rpm. In order to avoid warming up of the preparation, the homogenizer is conveniently immersed in a plastic cylinder packed with melting ice and which can be moved by hand up and down together with the homogenizer tube. The homogenizing is continued for 2 minutes with about 15 up and down strokes of the homogenizer per minute.

The homogenate is centrifuged first at 700–800 g for 20 minutes at 0° and the supernatant, filtered through several layers of gauze, is spun again at 10,000 g for 30 minutes, also at 0°. The supernatant (S$_{10}$) after the second centrifuging is decanted carefully from the loosely packed mitochondrial pellet through several layers of gauze or cotton wool moistened with a little buffer. The S$_{10}$ preparation contains all the enzymes and cofactors needed for the biosynthesis of sterols, or squalene, either from acetate or from mevalonate, but the system is 30–50 times more efficient with mevalonate than with acetate as substrate.

For maximum conversion of mevalonate into squalene or sterols, the S$_{10}$ preparations have to be fortified with ATP and NADPH (or NADH). The following additions are optimal: 3 mM ATP, 1 mM NADP, 3 mM glucose 6-phosphate [or 1 mM NAD and 6 mM ethanol; if available 3,3-dimethylallyl alcohol (3-methylbut-2-en-1-ol) can suitably replace ethanol], and 2 mM RS-mevalonate, labeled with an isotope appropriate for the experiment (e.g., with ^{14}C at C-2).

For sterol biosynthesis the S$_{10}$ preparations, supplemented as above, are incubated aerobically at 30–37° in stoppered flasks with shaking at 100–120 oscillations per minute. For maximum conversion of mevalonate into sterols a 3-hour incubation is needed, the addition of ATP, NADP, and glucose 6-phosphate (or of NAD and ethanol) being repeated at the end of the first and second hour. Under these conditions usually 0.3–0.7 micromoles of mevalonate are converted into sterols per milliliter of S$_{10}$ incubated. Since six molecules of mevalonate are needed for the synthesis of one of sterol, this represents a *de novo* synthesis of 0.05–0.11 micromole of sterol; calculated as cholesterol (but see later) this is equivalent to 19–40 μg. The S$_{10}$ preparations contain 250–300 μg of endogenous sterol per ml, most of which is cholesterol. It follows that the newly synthesized sterols will be diluted around 10-fold with endogenous material. Since only about one-half of the sterols synthesized is cholesterol (see below), the newly synthesized labeled cholesterol represents only about 5% of the total sterol extracted from the preparations. These values may serve as a useful guide in predicting the isotope content of biosynthesized material.

For the analysis of the S_{10}-incubations the following procedure proves useful when the interest is in total sterols.

Analysis of S_{10}-Incubations for Total Sterol

At the end of the reaction time ethanol and KOH (either as solid pellets or as a 40% solution) are added to final concentrations of 50% and $2\,N$, respectively. The mixture is heated at 70° under N_2 for 1 hour after which it is extracted three times with light petroleum (b.p. 40–60°). The combined petroleum extracts, washed with water and dried over anhydrous $MgSO_4$, are concentrated to a few milliliters and applied to an alumina[50] column (approximately 1 g of alumina per milligram of total unsaponifiable material to be fractionated). The column is eluted first with light petroleum (b.p. 40–60°), 5 ml per gram of alumina, in order to collect squalene,[51] and then with acetone–ether (1:1, v/v), 5 ml per gram of alumina, in order to remove the sterols and some other polar substances (e.g., farnesol) synthesized from mevalonate.

The acetone-ether eluate is concentrated to a small volume in a centrifuge tube and the sterols are precipitated by the addition of a 0.5% solution of digitonin in 90% ethanol. In the case of a 10-ml incubation of S_{10} the acetone-ether eluate is most suitably concentrated to 0.5 ml to which 5 ml of the digitonin solution and 6 drops of $1\,N$ HCl are added; the mixture is heated to about 70° for 10 minutes and then left at room temperature for several hours, or overnight. The precipitate is centrifuged down, washed with ethanol–ether (1:1, v/v), acetone–ether (1:1, v/v), and acetone (2 ml each). A 10 ml S_{10} incubation usually yields 12–15 mg of digitonides. For further analysis these are dissolved in acetic acid (1 ml per 10 mg digitonide) and heated (100°) for 5 minutes; the solution is then cooled and treated with 5 volumes of ether. The precipitate (digitonin) is centrifuged and washed twice with ether. The combined ethereal solution, washed with water twice and with $1\,M$ $NaHCO_3$ until neutral,

---

[50] The alumina is prepared from Woelm Grade I alumina by treating it first with methyl formate either by passing methyl formate through a column of the alumina until evolution of heat ceases, or by stirring the alumina with methyl formate. The alumina is then washed with methanol and, after filtering off the methanol, it is dried at 110° and kept in a vacuum desiccator over silica gel.

[51] Five to 10% of the mevalonate utilized in the S_{10} preparations under aerobic conditions appears as squalene.

is dried with $MgSO_4$; the solvent is evaporated off leaving crude sterol; usual yield: 2 mg sterol from 10 mg digitonide.

The further analysis of this crude sterol is illustrated by an excerpt from our laboratory notebook: 23.7 mg of crude sterols, obtained by the above procedures from incubations of S_{10} with mevalonate-2-^{14}C and containing 27 μC of ^{14}C, were dissolved in 2 ml of benzene and applied to a 80 × 1 cm column of silicic acid-Hyflo Super Cel (2:1, w/w) suspended in benzene.[52] The column was eluted with pure benzene, 4.9 ml fractions being collected. Figure 4 shows the elution pattern; only 46% of the

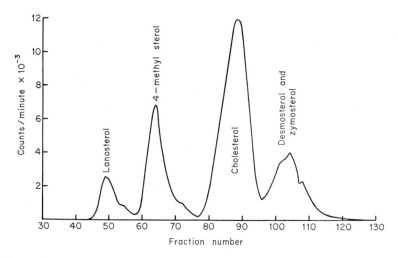

FIG. 4. Chromatogram of ^{14}C-labeled sterols, biosynthesized from mevalonate-2-^{14}C by a rat liver homogenate. For details see text. (G. Popják, unpublished.)

total radioactivity was associated with cholesterol, the remainder was found in lanosterol, 4-methylsterol, and in other sterols. The combined fractions of cholesterol, when evaporated under N_2, left a slightly colored crystalline solid, 23 mg, and had a specific activity of 0.81 μC/mg of 0.312 μC per micromole of cholesterol. Since the mevalonate-2-^{14}C used in the biosynthesis had a specific activity of 0.825 μC per micromole, and since cholesterol contains 5 positions derived from C-2 of mevalonate, the newly synthesized cholesterol must have had a specific activity of 0.825 × 5 = 4.125 μC per micromole. Hence the biosynthetic cholesterol was diluted with endogenous material in the ratio of 4.125:0.312. This example is typical of what may be expected of the biosynthetic potential of the Bucher-McGarrahan homogenate of rat liver. Although the specific

[52] I. D. Frantz, Jr., E. Dulit, and A. G. Davidson, *J. Biol. Chem.* **226**, 139 (1957).

activity of the cholesterol obtained is relatively low, the specific activity of the other sterols—as these occur in the liver only in minute amounts—is very high.[52a]

B. ENZYME SYSTEM FOR SYNTHESIS OF SQUALENE FROM MEVALONATE

The S_{10} preparation described in the preceding section, when incubated anaerobically, synthesizes squalene from mevalonate as efficiently as it forms sterols in the presence of O_2. However, when squalene is the product aimed at it is not necessary to use rat liver, nor to make the homogenates by the Bucher-McGarrahan technique. Pig liver, obtained freshly from the slaughterhouse, chilled in ice, and minced in a meat-grinder, is a very good source of the enzyme system. The liver mince is suspended in 2.5 volumes of 0.1 M Tris-HCl buffer, pH 7.5, fortified with 5 mM $MgCl_2$, 2 mM $MnCl_2$, and 30 mM nicotinamide and is homogenized for 30 seconds in a Waring blendor. An S_{10} preparation is then obtained by centrifuging as described in the previous section.

For maximum yield of squalene, the same supplementation is used as described for sterol biosynthesis, but before the addition of mevalonate the S_{10} preparation (supplemented with cofactors) is degassed on a water pump and saturated with N_2. Suction with the water pump is continued until the homogenate begins to froth, when N_2 is let in from a side arm; the process is repeated five times. The solution of substrate is then added to the preparation under a stream of N_2; the flask is then sealed and incubated at 30–37° with shaking.

When the highest yield of squalene is desired, the same procedure is followed as for sterol biosynthesis, but the cofactors should, of course, be added at the hourly intervals under a blanket of N_2.

For the isolation of squalene the unsaponifiable material is extracted, after alkaline hydrolysis, and the squalene separated from the sterols by chromatography on alumina, as described in the previous section (see p. 440). The petrol eluate of the alumina column contains, in our experience, no other substance biosynthesized from mevalonate than squalene, but is frequently contaminated with paraffinic hydrocarbons.

The total yield of squalene from mevalonate-2-^{14}C may be confidently deduced from the ^{14}C-content of this petrol eluate, but when pure squalene is needed, as for mass-spectrometric analysis of the substance for its content of a stable isotope, or for chemical degradations, purification through the thiourea clathrate is recommended.

When the amount of biosynthetic squalene is too small for isolation

---

[52a] For the analysis of sterols from small incubations thin-layer chromatography is more suitable.

as the clathrate, and/or there is no objection against dilution of isotope content, a suitable quantity (a few milligrams) of carrier squalene is added to the petrol eluate and the solvent is evaporated off. The residue is dissolved in a little benzene (0.5–1.0 ml) and a saturated solution of thiourea in methanol (2–3 ml) is added. The long needles of thiourea-squalene adduct begin to form immediately and are collected after a few hours by centrifuging. The crystals are washed with a little light petroleum, then decomposed with water and the squalene is extracted with light petroleum.

We have prepared squalene on several occasions by these procedures in quantities of 100 mg or higher. The squalene obtained with the S_{10} preparations from variously labeled mevalonates is never diluted by more than 5–10% of endogenous material.

Products Other than Squalene and Sterols Formed from Mevalonate in S_{10} Preparations

Squalene and sterols are not the only products formed from mevalonate in S_{10} preparations of liver. The farnesyl pyrophosphate synthesized from mevalonate in the liver homogenates is metabolized by an alternative pathway not leading to squalene and sterols. The microsomes, which are the particulate components of S_{10} preparations, contain a phosphatase[53] which hydrolyzes farnesyl pyrophosphate to free farnesol and inorganic phosphate at a pH optimum of 7 even in the absence of a divalent metal ion. In turn, the liberated farnesol is attacked by liver alcohol dehydrogenase and an aldehyde dehydrogenase, both present in the nonparticulate fraction of the S_{10} preparations of liver, and is thus converted to farnesoic acid.[54]

C. Enzyme System Synthesizing Farnesyl Pyrophosphate from Mevalonate

The S_{10} preparations, described in Sections II,A and II,B, made either from rat liver by the technique of Bucher and McGarrahan[49] or from pig liver by homogenizing in a Waring blendor, are centrifuged at 100,000 g for 1 hour or at 45,000 g for 3 hours. The sediment consists of two layers: a transparent pellet of glycogen and an opaque pinkish-brown layer of microsomes. The supernatant is the source of the enzymes synthesizing farnesyl pyrophosphate from mevalonate and is removed most

[53] J. Christophe and G. Popják, *J. Lipid Res.* **2**, 244 (1961).
[54] The formation of farnesoic acid from farnesol (or from farnesyl pyrophosphate) has not been reported from laboratories studying the sterol-synthesizing enzyme systems of yeast. This is because yeast alcohol dehydrogenase, in contrast to liver alcohol dehydrogenase, is inert toward farnesol.

carefully with a Pasteur pipette without disturbing the microsomal pellet. If the supernatant is decanted, its contamination with microsomes is almost unavoidable and will result in enzyme preparations which will carry the biosynthesis from mevalonate as far as squalene. It is best to leave a 3–4 mm layer of the supernatant over the microsomes in order to avoid the possibility of this contamination.

All the enzymes required to catalyze the synthesis of farnesyl pyrophosphate from mevalonate are present in that fraction of the supernatant, S_{100}, which precipitates between 30 and 60% saturation with ammonium sulfate. This fraction, F_{30}^{60}, is conveniently obtained by the addition of the appropriate amounts of the finely powdered salt to the S_{100}, the pH of the mixture being kept at 7–7.5 by the addition of NH_4OH.

The protein fraction which precipitates between 30 and 60% saturation with ammonium sulfate is collected by centrifuging at 12,000 g for 30 minutes. The sediment is press-dried between several layers of filter paper to a crumbly consistency and can be preserved at $-20°$ without loss of activity for many months. The enzymes are reconstituted by dissolving the precipitate in 0.02 M $KHCO_3$ and by dialysis against several changes of the same fluid for 3–5 hours. The dialyzed solution is clarified by centrifuging at 10,000 g for 20 minutes; it contains 30–50 mg protein per milliliter. The dialyzed solution can be kept frozen at $-20°$ for several weeks without deterioration of its enzymatic activity.

For work on a large scale, for example, for the preparation of 1–2 liters, or more, of the soluble supernatant of liver homogenates, or when a high-speed centrifuge of large capacity is not available, preparations of the soluble protein fraction of liver homogenates, equivalent to the S_{100} fraction, and of microsomes may be obtained (1) by adjusting the pH of S_{10} fractions of homogenates to 6.5 with 1 N acetic acid, and (2) by the addition of protamine sulfate (derived from salmon sperm) to a final concentration of 1 mg/ml. Thirty minutes after thorough mixing with the protamine the preparation is centrifuged at 2000 g for 30 minutes at 0°. The clear supernatant has the same properties as the S_{100} preparation (see above) and the sediment (microsomes and some precipitated soluble RNA) the unimpaired properties (as far as squalene and sterol biosynthesis are concerned) of microsomes sedimented by high-speed centrifugation.[55] The F_{30}^{60} enzyme fraction is obtained from the

---

[55] The fraction of protein precipitating between 0 and 30% saturation with ammonium sulfate, F_0^{30}, is collected by centrifugation at 12,000 g for 30 minutes and may be saved as a preparation of mevalonate kinase, which usually has a specific activity as high as 0.1 (cf. page 406 for definition) and is free of all other enzymes acting on mevalonate or on its derivatives (G. Popják, unpublished). The enzyme is reconstituted by dissolving 1–2 g of the wet sludge in 10 ml of 0.02 M $KHCO_3$, and

supernatant of S_{10} preparations after protamine treatment in the same way as the corresponding fraction of the S_{100} preparations.

For the biosynthesis of farnesyl pyrophosphate with the reconstituted F_{30}^{60} enzymes, the following reactants are needed:[56] 0.1 M Tris-HCl buffer, pH 7.5, 5 mM MgCl$_2$, 2 mM MnCl$_2$, 7 mM ATP, 30 mM glutathione (this is needed only when several-months-old preparations of F_{30}^{60} enzymes are used; cysteine can replace glutathione), 2 mM RS-mevalonate (labeled, e.g., with ^{14}C at C-2 or C-4) and F_{30}^{60} protein, 10–15 mg/ml. Usually 50–80% of the enzymatically reactive R-mevalonate is converted into farnesyl pyrophosphate during a 2-hour incubation at 37°.

Isolation of Farnesyl Pyrophosphate from Incubations of F_{30}^{60} Enzymes with Mevalonate

The isolation of farnesyl pyrophosphate biosynthesized from mevalonate in incubations of F_{30}^{60} enzymes can be done by the procedure of collidine extraction, etc., as described originally,[56] but recently we have developed a more efficient procedure which gives better yields and purer preparations than the earlier method (G. Popják, unpublished).

The method to be described depends on (a) the solubility of farnesyl pyrophosphate in ethanol and methanol; (b) on its stability at pH values above 7; (c) on its solubility in dilute aqueous alkali; (d) on its absorption from an aqueous solution by a lipophilic resin from which it can be eluted quantitatively with methanol; and (e) on a final purification step by chromatography on a DEAE-cellulose column. At all steps of the purification the pH of solutions containing farnesyl pyrophosphate should be kept above 7 by the addition of drops of conc. ammonia.

Method. At the end of the reaction time of the F_{30}^{60} incubation ethanol is added to a concentration of 70%. The mixture is heated quickly to 60°, kept at that temperature for 5 minutes, then cooled in ice. The precipitate is centrifuged down and washed three times with 80% ethanol. The first supernatant and the ethanol washings are combined and concentrated on a rotary evaporator (bath temperature 50°) under water pump vacuum to a small volume convenient for further handling (e.g., to 20–40 ml after a 1-liter incubation), by which time all the ethanol has been removed. The turbid solution is diluted to 100 ml with 0.01 N NH$_4$OH and 10 ml of 0.25 M EDTA is added for each liter of the incubation. This solution is

---

by dialyzing the solution against the same for 3–5 hours. This crude preparation of mevalonate kinase is most useful for assaying solutions of mevalonate by the spectrophotometric technique (cf. pp. 402–405) or for the preparation of small amounts of 5-phosphomevalonate (cf. page 410).

[56] G. Popják, *Tetrahedron Letters* **19**, 19 (1959); D. S. Goodman and G. Popják, *J. Lipid Res.* **1**, 286 (1960).

now treated with Amberlite XAD-2 polystyrene resin[57] which absorbs all the farnesyl pyrophosphate and some of the unused mevalonate from the solution. When the expected yield of farnesyl pyrophosphate is 100–200 micromoles, 20 g of the damp resin are added to the solution, which is then stirred overnight. The resin is then collected by filtration and is washed with 0.01 M aqueous ammonia, which removes inorganic salts, some— but not all—of the unused mevalonate and phosphorylated forms of mevalonate. The resin is then washed with methanol containing 0.01 M ammonia which elutes the farnesyl pyrophosphate and the remainder of mevalonate. For 20 g of resin, four washings of 10 ml each are needed. The methanol eluate is concentrated to a small volume (2–5 ml) on a rotary evaporator and applied to a DEAE-cellulose column[3] (18 × 1.2 cm) equilibrated with methanol containing 0.08 M ammonium formate. The column is eluted with the same solvent:[58] unused mevalonate appears first, followed by farnesyl pyrophosphate (Fig. 5). The fractions containing the farnesyl pyrophosphate are combined and concentrated on a rotary evaporator until some ammonium formate begins to crystallize out; a few milliliters of 0.01 M aqueous ammonia are now added and the evaporation continued. The process of evaporation and addition of aqueous ammonia is repeated until all the methanol has been driven off. The aqueous, slightly alkaline, solution of farnesyl pyrophosphate and ammonium formate is treated once again with the XAD-2 resin, which absorbs the farnesyl pyrophosphate but not the ammonium formate. The latter is washed out with 0.01 M aqueous ammonia, and the farnesyl pyrophosphate is then eluted with methanol containing 0.01 M ammonia. The methanolic solution of purified farnesyl pyrophosphate is concentrated once again on a rotary evaporator, the methanol being gradually replaced by dilute (0.01 M) aqueous ammonia. The solution of farnesyl pyrophosphate is stabilized by the addition of 1 M Tris-HCl buffer, pH 8,

---

[57] Product of Rohm and Haas, Philadelphia, Pennsylvania. The resin is prepared by washing it alternately several times with methanol containing 0.01 M (NH₄)OH and with 0.01 M aq. (NH₄)OH, the last washing being aqueous. At each treatment the resin is stirred for about 15 minutes with the particular solvent and is then filtered by suction on a coarse sintered filter funnel.

The resin XAD-1 is unsuitable because it absorbs farnesyl pyrophosphate almost irreversibly.

[58] Rilling[3] recommends elution with methanol containing 0.08 M ammonium formate and a linear gradient of ammonia from 0 to 0.14 M in 800 ml. We find that the gradient is not required. The elution pattern observed by us is different from that reported by Rilling: we find that farnesyl pyrophosphate is eluted from an 18 × 1.2 cm column between about 100 and 210 ml of effluent when the solvent is made up by diluting 16 ml of 5 M aqueous ammonium formate (pH 7.0) to 1 liter with dry methanol. We have used both Whatman DE-11 and BioRad Cellex-D (capacity 0.99 meq/g) celluloses with identical results.

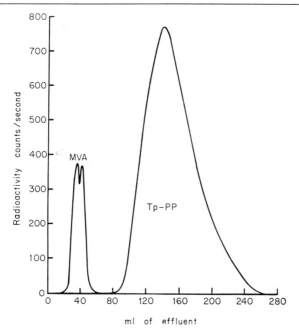

Fig. 5. Purification of farnesyl pyrophosphate, biosynthesized from mevalonate-2-^{14}C by soluble (F_{30}^{60}) liver enzymes, on DEAE-cellulose column. The first doubly spiked fraction contains unused mevalonate and some phosphorylated derivatives of mevalonate. The main component is radiochemically pure *trans-trans*-farnesyl pyrophosphate. (G. Popják, unpublished.)

to a final concentration of 0.1 M and may be preserved at $-20°$ for several weeks without hydrolysis; it is suitable for direct addition to buffered enzymatic incubations, as for the biosynthesis of squalene by liver microsomes (see p. 450).

The method described gives farnesyl pyrophosphate in a nearly quantitative yield from enzymatic incubations. The purification procedure may be followed most conveniently by the use of two thin-layer chromatographic systems: (a) on silica gel G plates and (b) on Eastman Kodak precoated sheets (Chromatogram Sheet, Type K.301, R2), both developed with n-propanol–conc. ammonia–water (6:3:1 by volume). On silica G plates farnesyl pyrophosphate (and inorganic phosphate) stay at the origin, but mevalonate migrates with an R_f of 0.8. On the Eastman Kodak sheets inorganic phosphate moves with an R_f of less than 0.1, farnesyl pyrophosphate with an R_f of about 0.6, while mevalonate moves with the solvent front.[38]

The use of mevalonate labeled with ^{14}C at a convenient position (e.g., at C-2 or at C-4) in the incubation is a great help, for it enables one to calculate yields readily and also to follow the purification procedure by the scanning of thin-layer chromatographic plates for the position of radioactive spots in addition to visualizing the spots by exposure to iodine vapor.

Isolation of Farnesol from F_{30}^{60} Incubations

When the product of incubations of F_{30}^{60} enzymes with mevalonate is farnesyl pyrophosphate, and the main interest is not in the pyrophosphate but in the farnesol, it is not necessary to isolate the farnesyl pyrophosphate first and then hydrolyze it in order to obtain the free alcohol. At the end of the reaction time of the incubation the pH of the mixture is adjusted to pH 9.5 (using a glass electrode) by the addition of 1 N KOH, then 1 mg of intestinal alkaline phosphatase (dissolved in 0.02 M $KHCO_3$) is added for each milliliter of F_{30}^{60} incubations and the mixture is kept at 37° for a further 3–4 hours, when it is extracted four times with light petroleum (b.p. 40–60°). The petroleum extract, which contains farnesol and may also contain some other allylic alcohols (3,3-dimethylallyl alcohol and geraniol) derived from mevalonate, is dried over $MgSO_4$ and is then concentrated to a small volume or evaporated to dryness on a rotary evaporator or by distillation of the solvent. A solution of farnesol in ether or petroleum may be safely evaporated to dryness under waterpump vacuum without significant loss if the bath temperature does not exceed about 50°.

In isotopic experiments, when dilution of the biosynthetic material with the unlabeled substance causes no interference with the object of the experiment, a few milligrams of unlabeled farnesol may be added to the incubation mixture before extraction with petroleum. In experiments in which mevalonate labeled with a stable isotope is used as a substrate and the farnesol biosynthesized has to be obtained undiluted and in a pure form for further analysis (e.g., by mass spectrometry, or by chemical degradation[4, 16, 33]) one of three methods may be used for the purification of farnesol, depending on the scale of the experiment.[59]

On a small scale, when micrograms up to 1–2 mg of farnesol are to be purified, thin-layer chromatography is the most suitable procedure. On plates coated with a 250 μm thick absorbent, made from 30 g of "Kieselgel G nach Stahl" and 60 ml of 12.5% aqueous silver nitrate, and developed with ethyl acetate containing 0.2% acetic acid, trans-trans farnesol has an R_f of 0.23.[16] About 100 μg of farnesol can be purified on such plates,

---

[59] The purification is necessary because the petroleum extract of the F_{30}^{60} incubations, after hydrolysis with intestinal alkaline phosphatase, contains, besides farnesol, some lipids that interfere with the mass spectrometry and/or with the chemical degradations.

measuring 5×20 cm, when the solution of crude farnesol is applied in a band. After development of the plate, the band containing the farnesol is scraped off into a small stoppered flask and the farnesol is extracted by shaking with 15 ml of diethyl ether. The ether solution is then extracted with 5 ml of 3.5% aq. KCl solution and the aqueous layer is extracted with ether. The ether extracts are combined, dried over $MgSO_4$, and evaporated to dryness. The farnesol so obtained is pure according to mass spectrometric analysis.

When purification of more than 1–2 mg of farnesol is needed, chromatography on a silicic acid-Celite column,[60] developed with benzene as the sole solvent, is the most suitable. The column is prepared by mixing 2 parts of silicic acid with 1 of Celite (by weight) and suspending the mixed adsorbent in benzene. A column of 80×1 cm is then packed by gravity. The solution of crude farnesol, dissolved in 2 ml of benzene, is applied to the column, which is then developed with benzene as the sole solvent. Farnesol emerges from such a column after about 900 ml of solvent has passed through it; it needs 2–3 days for the completion of the run. This is the same chromatographic system as described for the separation of sterols (cf. Section II,A).

A third method for the purification of crude farnesol preparations in milligram quantities can be considered if serious losses in final yield are of no consequence. Gas-liquid chromatography offers an excellent method for the purification of farnesol if it were not for the fact that vapors of farnesol emerging from a preparative gas-liquid chromatographic column are among the most difficult ones in our experience to trap quantitatively.

For the analysis, or purification, of farnesol by gas-liquid chromatography polar stationary phases of the polyester or ether type are the most suitable. We have had experience with only two: ethylene glycol–succinate polyester[61] and "Carbowax 20M" (product of Union Carbide Corporation, New York). The latter is superior to the ethylene glycol–succinate stationary phase for the separation of prenols and particularly for the separation of the geometric isomers of farnesol. Apart from giving a better resolution, freshly prepared columns of "Carbowax 20M" can be used immediately, whereas columns coated with ethylene glycol–succinate polyester have to be conditioned at 180° for at least 1 week before they become suitable for the analysis of farnesol: *cis-trans* and *trans-trans*-farnesol are retained completely on freshly prepared columns of ethylene glycol–succinate. For routine use we recommend columns of 10% "Carbowax 20M" on Celite or Chromosorb (80–120 mesh), 6–9 feet

[60] G. J. Schroepfer, Jr., and I. Youhotsky Gore, *J. Lipid Res.* **4**, 266 (1963).
[61] G. Popják and R. H. Cornforth, *J. Chromatog.* **4**, 214 (1960).

long at 195°, as most suitable for analytical and preparative purposes. Such columns separate clearly dimethylallyl alcohol, linalool, geraniol, nerolidol, *cis-trans-* and *trans-trans*-farnesol.

D. Enzyme System for the Synthesis of Isopentenyl Pyrophosphate from Mevalonate

The F_{30}^{60}-enzyme preparations, described for the synthesis of farnesyl pyrophosphate (p. 444), are also suitable for the preparation of isopentenyl pyrophosphate in micromolar amounts from isotopically labeled mevalonate (e.g., from $2\text{-}^{14}C\text{-}$ or $4\text{-}^{14}C$-labeled mevalonate).

Method. Iodoacetamide is added to the reaction mixture described on page 445 for the enzymatic synthesis of farnesyl pyrophosphate from mevalonate to a final concentration of 5 mM. After 2 hours at 37°, during which time the reaction mixture becomes very turbid, 1 volume of ice-cold 10% perchloric acid is added to the incubation mixture which has also been cooled in ice. The precipitated protein is centrifuged down. The decanted supernatant is neutralized with KOH and the potassium perchlorate which crystallizes at 0° is removed by centrifuging. The supernatant is lyophilized and the residue is dissolved in 0.1 M aqueous ammonia (0.3 ml for each ml of the original reaction mixture). One-tenth milliliter of this solution is applied in a line to a 5 cm-wide strip of Whatman No. 3 MM paper for development by high-voltage electrophoresis in a pyridine–acetic acid–water buffer, pH 6.1 (100:8:2400, by volume). After 30 minutes of exposure to a voltage gradient of 72V $\times$ cm^{-1}, three radioactive bands are found on the paper with the following electrophoretic mobilities: (i) 0.109, (ii) 0.127, and (iii) 0.146 mm $\times$ min$^{-1} \times$ V$^{-1} \times$ cm^{-1}. The fastest-moving component is isopentenyl pyrophosphate; it is eluted from the paper with 0.1 M aqueous ammonia.[42]

This method is suitable for the preparation of isopentenyl pyrophosphate on a micromolar scale; the yields are 30–40%.

E. Enzyme System Synthesizing Squalene from Farnesyl Pyrophosphate

Reaction Catalyzed

The reaction scheme (p. 451) summarizes the overall process of the biosynthesis of squalene from two molecules of farnesyl pyrophosphate: it implies that one hydrogen atom, H_S, at C-1 of one of the two farnesyl residues exchanges with H_S (tritium in the reaction written) at C-4 of the dihydronicotinamide ring of NADH or NADPH, that this hydrogen exchange proceeds as if the steric configuration around the carbon atom involved in the exchange had not changed, and that the configuration

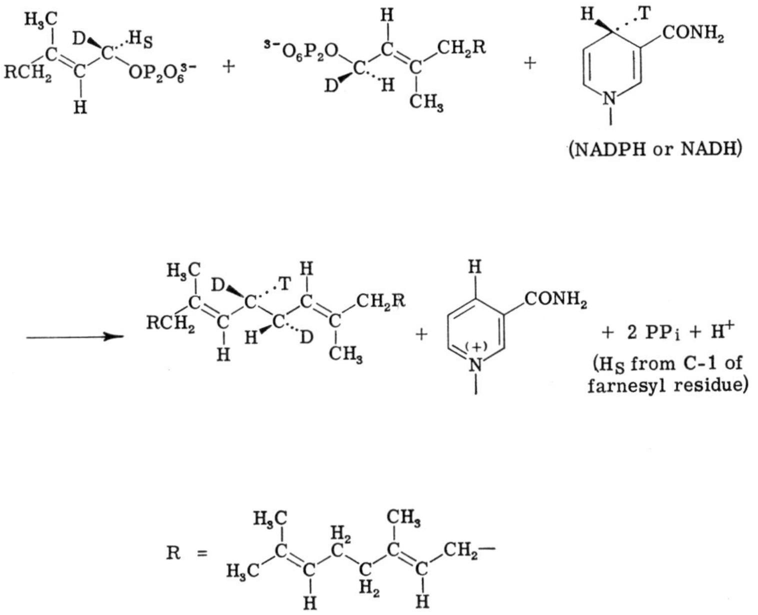

(NADPH or NADH)

+ 2 PPᵢ + H⁺
(Hₛ from C-1 of
farnesyl residue)

at C-1 of the second farnesyl residue not involved in the hydrogen exchange was inverted.[4, 16, 62–66] It is to be noted that a proton is not incorporated into squalene during its synthesis from farnesyl pyrophosphate.

The enzymes catalyzing the synthesis of squalene from farnesyl pyrophosphate are attached to microsomal particles.[56, 67]

[62] G. Popják, D. S. Goodman, J. W. Cornforth, R. H. Cornforth, and R. Ryhage, J. Biol. Chem. 236, 1934 (1961).

[63] G. Popják, J. W. Cornforth, R. H. Cornforth, R. Ryhage, and D. S. Goodman, J. Biol. Chem. 237, 56 (1962).

[64] G. Popják, G. J. Schroepfer, Jr., and J. W. Cornforth, Biochem. Biophys. Res. Commun. 6, 438 (1961/62).

[65] J. W. Cornforth, R. H. Cornforth, C. Donninger, G. Popják, G. Ryback, and G. J. Schroepfer, Jr., Proc. Roy. Soc. B163, 436 (1966).

[66] G. Popják and J. W. Cornforth, Biochem. J. 101, 553 (1966).

[67] G. Krishna, H. W. Whitlock, Jr., D. H. Feldbruegge, and J. W. Porter [Arch. Biochem. Biophys. 114, 200 (1966)] claimed to have solubilized the microsomal squalene synthetase. However, the resulting enzyme preparations were very feeble and their specific activities were reported to be much lower than the specific activities of the starting microsomes. Since there is now strong evidence suggesting the existence of an intermediate between farnesyl pyrophosphate and squalene (cf. footnote 3), the "squalene synthetase" must be treated as a multienzyme system consisting of at least two enzymes.

Preparation of System

Reagents

(1) Farnesyl pyrophosphate, 10 mM solution, labeled suitably with an isotope (e.g., ^{14}C, ^2H, or ^3H). For preparation of farnesyl pyrophosphate see p. 385

(2) Potassium phosphate buffer, 1 M, pH 7.4

(3) MgCl$_2$, 0.1 M

(4) NADPH, 10 mM

(5) Nicotinamide, 1 M

(6) Bovine serum albumin, 1% solution

(7) NaF, or KF, 1 M

(8) Microsomes, prepared from rat liver or pig liver (see below)

Preparation of Microsomes. The supernatant of rat liver or pig liver homogenates centrifuged at 10,000 g (see p. 439) is centrifuged at 105,000 g for 1 hour or at 45,000 g for 3 hours. The soluble supernatant is pipetted off the sediment, which consists of two layers: an upper layer of microsomes and a lower transparent pellet of glycogen. One to 2 ml of a 0.1 M potassium phosphate buffer, pH 7.4, containing 5 mM MgCl$_2$ and 30 mM nicotinamide, are added to each centrifuge tube; the tubes are then shaken vigorously by hand resulting in the separation of the microsomal layer from the glycogen pellet, which remains firmly attached to the tube. The gelatinous microsomal layers, partly suspended in the buffer, are poured into a smooth-walled glass homogenizer of the type used for making the Bucher-McGarrahan homogenate (see p. 439). Buffer, sufficient to bring the volume of the microsomal suspension to one-half of that of the original homogenate, is added to the microsomes, which are now dispersed in the buffer by a few strokes of the loosely fitting piston. The microsomes are sedimented once more by centrifuging at 105,000 g for 30 minutes. The supernatant is discarded, and the microsomes are suspended by the use of the homogenizer in a volume of the phosphate buffer equal to one-tenth of the volume of the homogenate from which the microsomes were separated.

Such concentrated suspensions of washed microsomes may be preserved at $-20°$ for several months without serious impairment of their ability to synthesize squalene from farnesyl pyrophosphate, provided that after a storage for 1 month, or longer, the microsomes are sedimented, washed and resuspended in fresh buffer.

Incubations. Each milliliter of the reaction mixture for the synthesis of squalene from farnesyl pyrophosphate contains 100 micromoles of potassium phosphate buffer, pH 7.4, 5 micromoles of MgCl$_2$, 1 micromole of NADPH (or NADH), 30 micromoles of nicotinamide, 10 micromoles of NaF (or KF), 2 mg of bovine serum albumin, 0.03–0.1 micromole of ^{14}C-labeled farnesyl pyrophosphate, and 0.05–0.1 ml of microsome suspension. This mixture is incubated at 30–37° anaerobically for 1 hour, after which it is saponified; the squalene is extracted from it and purified as described on p. 442. The yield of squalene is calculated from the ^{14}C-content of the purified extract and the known specific activity of ^{14}C-labeled farnesyl pyrophosphate used.

Properties of the Microsomal Squalene Synthetase System[56]

The system has a distinct pH optimum at 7.4; its metal ion requirements are equally satisfied by Mg^{++}, Mn^{++}, or Co^{++}. Either NADPH or NADH is essential for the synthesis of squalene, NADPH being the preferred coenzyme.

The system as described above usually converts 20–40 millimicromoles of farnesyl pyrophosphate into squalene per milliliter in 1 hour. If Tris-HCl buffer, pH 7.4, is used instead of phosphate, and the F^- ions and the serum albumin are omitted, the yield of squalene is usually very poor on account of the rapid hydrolysis of farnesyl pyrophosphate by microsomal phosphatase; this phosphatase is most powerfully inhibited by serum albumin and to a lesser extent by phosphate and fluoride ions.[53]

The squalene synthetase system itself is inhibited by p-chloromercuribenzoate (100% inhibition at 1 mM; reversed by 5 mM reduced glutathione), N-ethylmaleimide (about 50% inhibition at 1 mM, and 75% inhibition at 3 mM), by Cu^{++} (100% inhibition at 1 mM) and Ca^{++} ions (about 60% inhibition at 1 mM), and by EDTA (complete inhibition at 10–20 mM). Iodoacetamide has no effect even at a concentration of 10 mM.

Squalene is not formed (or only in minute amounts) when NADPH and NADH are omitted from the incubations of microsomes with farnesyl pyrophosphate; instead a new substance appears (G. Popják, unpublished) which is in all probability the dephosphorylated form of intermediate "X" between farnesyl pyrophosphate and squalene found recently by Rilling[3] in butanol extracts of incubations of yeast subcellular particles with farnesyl pyrophosphate.

F. Reconstituted Enzyme System for Synthesis of Squalene and Sterols from Mevalonate

The enzyme system synthesizing squalene and sterols from mevalonate and ATP can be reconstituted from the dialyzed F_{30}^{60} soluble enzymes, which synthesize farnesyl pyrophosphate (cf. p. 444), and from the suspension of washed microsomes (cf. p. 452), by combining 1 ml of the soluble enzymes (30–50 mg protein) with 0.2 ml of the washed microsome suspension in a 3 ml reaction mixture which also contains 0.1 M Tris-HCl buffer, pH 7.4 (or 0.05 M potassium phosphate buffer, pH 7.4), 5 mM $MgCl_2$, 2 mM $MnCl_2$, 5 mM ATP, 1 mM NADPH, 1 mM NADH, 1–2 mM (RS)-mevalonate-2-^{14}C 10 mM KF and 30 mM nicotinamide.

This reconstituted system, when incubated under N_2, synthesizes squalene almost as efficiently as a fresh 3-ml S_{10} preparation (cf. p. 442). In the presence of O_2 the system is almost inactive: neither squalene nor sterols are synthesized. Full activation is, however, obtained by adding to the reaction mixture either ascorbate, cysteine, or reduced glutathione to a final concentration of 10 mM.[68]

[68] G. Popják, L. Gosselin, I. Youhotsky Gore, and R. G. Gould, *Biochem. J.* **69**, 238 (1958).

[13] Enzymatic Synthesis of Phytoene in Tomato

By FIROZE B. JUNGALWALA and JOHN W. PORTER

$$\text{Isopentenyl-PP}^1 + \text{dimethylallyl-PP} \rightarrow \text{geranyl-PP} \tag{1}$$

$$\text{Geranyl-PP} + \text{isopentenyl-PP} \rightarrow \text{farnesyl-PP} \tag{2}$$

$$\text{Farnesyl-PP} + \text{isopentenyl-PP} \rightarrow \text{geranylgeranyl-PP} \tag{3}$$

$$2 \text{ Geranylgeranyl-PP} \rightarrow \text{phytoene} \tag{4}$$

An enzyme system present in the chromoplasts of tomato fruit converts either isopentenyl pyrophosphate or farnesyl plus isopentenyl pyrophosphates to phytoene.[2] Isopentenyl pyrophosphate is required for the conversion of farnesyl pyrophosphate to phytoene. Concomitantly with the synthesis of phytoene from isopentenyl pyrophosphate, acid-labile compounds accumulate that can be cleaved with bacterial alkaline phosphatase to yield, primarily, geranylgeraniol. Hence it is assumed that geranylgeranyl pyrophosphate is an intermediate and that two molecules of this compound condense to form phytoene.

Assay Method

General Procedure. The enzymatic synthesis of phytoene is determined by assay for the incorporation of radioactivity of isopentenyl-PP-4-^{14}C. The biosynthesized phytoene is extracted from the incubation

[1] The following abbreviation is used: PP, pyrophosphate. For example, isopentenyl-PP = isopentenyl pyrophosphate.
[2] F. B. Jungalwala and J. W. Porter, *Arch. Biochem. Biophys.* **119**, 209 (1967).

mixture with petroleum ether and then separated from other compounds by chromatography on aluminum oxide. The incorporation of radio-activity into phytoene is confirmed by hydrogenation to lycopersane followed by gas–liquid chromatography of this product.

Reagents

Glycylglycine buffer, 1.0 M, pH 8.0
Borate buffer, 0.5 M, pH 8.0
$MgCl_2$, 0.1 M
$MnCl_2$, 0.1 M
Dithiothreitol, 0.1 M
Isopentenyl-PP-4-^{14}C, 0.001 M

Isopentenyl-PP-4-^{14}C is biosynthesized from *dl*-mevalonic-2-^{14}C acid by a soluble pig liver enzyme system as described by Shah, Cleland, and Porter,[3] and then it is purified according to the method of Bloch *et al.*[4]

Incubation System. The incubation mixture for the biosynthesis of phytoene contains isopentenyl-PP-4-^{14}C, 24.0 millimicromoles and 100,000 cpm; $MnCl_2$, 2.0 micromoles; $MgCl_2$, 15.0 micromoles; glycylglycine or borate buffer, pH 8.0, 100.0 micromoles; dithiothreitol, 10.0 micromoles, and enzyme protein 1–3 mg, in a final volume of 1 ml. Incubation is carried out for 4 hours at 20°, with nitrogen as the gas phase.

Isolation of Phytoene. Enzyme activity in the incubation mixture is stopped by heat denaturation of the protein at 70° for 3 minutes. Two volumes of absolute ethanol are added and then lipids are extracted with five or six 7.0-ml aliquots each of petroleum ether (b.p. 40–60°). (Saponi-fication prior to extraction of phytoene is not necessary.) Known amounts of carrier phytoene (2–3 mg) are added, and the petroleum ether extract is dried over anhydrous sodium sulfate. The volume of the solution is reduced to 10.0 ml by evaporation of solvent at room temperature under a stream of nitrogen.

The extract is chromatographed on a 1.8 × 10 cm column of aluminum oxide (Merck & Co., suitable for chromatographic adsorption). Develop-ment of the chromatogram is effected by successive additions of 50 ml of petroleum ether, 50 ml of 0.75%, 50 ml of 1.5%, and 80 ml of 4% diethyl-ether in petroleum ether. (The latter solvent is purified by passage through silica gel before use.[2]) Ten-milliliter eluate fractions are col-lected. Squalene (if present) begins to appear in the eluate at the change-over to 0.75% diethylether. Phytoene appears near the end of the addi-

[3] D. H. Shah, W. W. Cleland, and J. W. Porter, *J. Biol. Chem.* **240**, 1946 (1965).
[4] K. Bloch, S. Chaykin, A. H. Phillips, and A. DeWaard, *J. Biol. Chem.* **234**, 2595 (1959).

tion of 1.5% of diethylether in petroleum ether. Slight adjustments are made in the concentration of diethylether in petroleum ether whenever a new lot of alumina is used. This is necessary because there is variation in the adsorptive properties of different lots of aluminum oxide.

One-milliliter aliquots of the fractions (10 ml) collected from the column are assayed for radioactivity in a diphenyloxazole–dimethyl-*p*-bis-2',5'-phenyloxazoyl)benzene–toluene solution in a liquid scintillation spectrometer. Ultraviolet spectral determinations[5] are also made on an aliquot of each eluate fraction, and those fractions which have nearly identical specific radioactivities are combined.

Enzyme Activity. Total enzyme activity is expressed as millimicromoles of isopentenyl-PP incorporated into phytoene under the described conditions of the incubation. Specific activity is expressed as millimicromoles of isopentenyl-PP incorporated per milligram of protein per hour.

Purity of Phytoene. Purity of the isolated phytoene can be checked by gas–liquid chromatography. The chromatographically separated phytoene is catalytically hydrogenated in the presence of platinum catalyst to lycopersane.[6] The resultant lycopersane is then chromatographed on a 6 ft × 6 mm column of 5% SE-30 on Chromosorb W at 270° with an argon gas flow rate of 100 ml per minute.[2] Effluent gases are trapped at short time intervals on glass wool plugs in a fraction collector. The contents of each collection tube are then flushed into a scintillation counting vial with phosphor solution and a determination of radioactivity is made. The retention time for lycopersane is about 7 minutes; 85–90% of the injected radioactive material is recovered from the chromatogram and usually more than 90% of the radioactivity emerging is associated with the lycopersane peak.

Assay for Isopentenyl-PP Isomerase and Subsequent Condensing Enzyme Activities

Acid Hydrolysis

Reagents. The reagents required are the same as those described earlier for the assay of the phytoene-synthesizing enzyme system. In addition 10.0 N HCl is required.

Procedure. The incubation system is the same as that described earlier for the synthesis of phytoene. However, the incubation is carried out for a shorter period of time (1 hour or less) and usually with a smaller amount of protein. (The amount of phytoene synthesized during this time is relatively small compared to the amount of terpenyl-PP

[5] F. B. Jungalwala and J. W. Porter, *Arch. Biochem. Biophys.* **110**, 291 (1965).
[6] D. G. Anderson and J. W. Porter, *Arch. Biochem. Biophys.* **97**, 509 (1962).

formed.[2]) Phytoene and other nonsaponifiable lipids are extracted as described earlier. Then 100 μl of 10 N HCl is added to the remaining incubation mixture and the solution is kept at room temperature for 20 minutes. A layer of petroleum ether (5 ml) over the reaction mixture prevents the loss of liberated volatile terpenols. The reaction products are extracted three times with petroleum ether (5 ml) and dried over anhydrous sodium sulfate; an aliquot is assayed for radioactivity. Carrier farnesol, nerolidol, geranylgeraniol, and geranyllinalool are added to the petroleum ether extracts and then assays are made for the individual products of the acid hydrolysis by gas–liquid chromatography of an aliquot of this extract.[7] Separations of the terpenols are made with a Barber-Colman Model 10 gas chromatograph on a 6 ft $\times$ 6 mm column of 20% butane-diol-succinate on Chromosorb W at 190° with an argon flow rate of 120 ml per minute. The retention times for nerolidol, farnesol geranyllinalool, and all-*trans* geranylgeraniol are 4, 9, 17, and 43 minutes, respectively. Ten carbon atom terpenols, if present, emerge with the solvent front. All samples are trapped on glass wool plugs in collection tubes surrounded with dry ice.

Enzymatic Hydrolysis

Reagents. The reagents required are the same as those described earlier for the assay of the phytoene synthesizing enzyme system, except for the following:

Tris buffer, 1.0 M, pH 8.0
Bacterial alkaline phosphatase (Worthington Biochemical Corp., Freehold, New Jersey)

Procedure. The incubation system is the same as that described for the assay of the enzyme system synthesizing phytoene, and incubation of the mixture is carried out for 1 hour. After the extraction of phytoene and other lipids, the remaining mixture is evaporated to dryness at 40° in a rotary evaporator. To this residue are added Tris buffer, pH 8.0, 50 micromoles; Mg[++], 6 micromoles; bacterial alkaline phosphatase, 4 mg of protein, and distilled water to make 1 ml. This mixture is incubated for 3–4 hours at 38° and then for 16–18 hours at room temperature.[8] Two volumes of ethanol are added, and the liberated terpenols are extracted with petroleum ether. The extract is assayed for radioactivity, and aliquots are assayed for individual terpenols by gas–liquid chromatography[7] as described in the preceding section.

---

[7] See reference cited in footnote 2 for sources of carrier compounds and details of the chromatographic procedure.
[8] D. L. Nandi and J. W. Porter, *Arch. Biochem. Biophys.* **105,** 7 (1964).

Preparation of Tomato Enzyme System

Partial Purification. All operations are carried out at 4° unless otherwise stated.

Tomato plastids are prepared from fresh semiripened tomato parenchyma tissue (about 2.0 kg). The tissue is homogenized for 5 seconds at full speed in a Waring blendor with an equal volume of 0.2 M Tris-HCl buffer, pH 8.2, containing EDTA, 0.001 M, and 2-mercaptoethanol, 0.001 M. The homogenate is then filtered twice through four layers of a cheesecloth–glass wool mat to remove cell debris. On the first filtration residual solution is squeezed from the pulp by hand. The plastids are then collected by centrifugation at 3500 g for 35 minutes in an International centrifuge (large head). The plastids are washed twice with potassium phosphate buffer, 0.1 M and pH 6.5 (four times the volume of plastids) containing 2-mercaptoethanol, 0.001 M, and then they are recovered by centrifugation. The plastids are resuspended in phosphate buffer, 0.1 M and pH 6.5, and twice the volume of plastids, and added slowly to about 40 volumes of acetone at −20°, with stirring. The mixture is stirred for 15 minutes, rapidly filtered in the cold, and the residue is washed with acetone at −20°. The residue is dried under vacuum at 0° for 6–8 hours. The dried acetone powder is then extracted with phosphate buffer, 0.1 M, pH 6.8, containing 2-mercaptoethanol, 0.001 M (500 mg of powder in 10 ml of buffer) and centrifuged at 105,000 g for 45 minutes. The precipitate is reextracted as above and the extracts are combined.

An alternative method of enzyme preparation involves the extraction of tomato plastids with phosphate buffer. This extraction is achieved through homogenization of plastids with glass beads in a micro-Waring blendor.[2]

Enzyme protein in the supernatant solution is fractionated with a saturated solution of ammonium sulfate adjusted to pH 7.0. The protein precipitating between 25 and 40%[9] of saturation with ammonium sulfate is dissolved in phosphate buffer, 0.1 M and pH 6.8. This solution is dialyzed against 6 liters of 0.05 M phosphate buffer, pH 6.8 and containing 0.001 M 2-mercaptoethanol, for 6 hours with a change of buffer at 3 hours. The dialyzed protein is stored at −20° under nitrogen in the

---

[9] Recent experiments (Feldbruegge, Shah, and Porter) have shown that the protein precipitated between 40 and 60% of saturation has greater activity for the formation of geranylgeranyl pyrophosphate from isopentenyl pyrophosphate than the 25–40% ammonium sulfate precipitate. In addition the former enzyme fraction has practically no activity for the conversion of geranylgeranyl pyrophosphate to phytoene.

presence of $0.01\ M$ dithiothreitol. Enzyme activity is retained for at least 2 weeks.

About an 8-fold purification in enzyme activity is obtained by this procedure (see the table).

PARTIAL PURIFICATION OF PHYTOENE-SYNTHESIZING ENZYME SYSTEM

Purification step	Total protein (mg)	Total activity (mμmoles/ hour)[a]	Specific activity (mμmoles/ mg/hour)[a]	Recovery (%)	Purification (-fold)
Plastids	484	47.36	0.097	—	—
Acetone powder extract	230	37.89	0.165	80	1.6
Ammonium sulfate, 25–40% saturation	43.4	32.68	0.753	69	7.7

[a] Millimicromoles of isopentenyl pyrophosphate-4-^{14}C incorporated into phytoene.

Enzyme systems synthesizing geranylgeranyl pyrophosphate from isopentenyl pyrophosphate have also been obtained from carrot roots[8] and *Micrococcus lysodeikticus*.[10]

Properties of Enzyme System

Cofactor and Substrate Requirements. Mn^{++} and Mg^{++} ions are required for the synthesis of phytoene; however, these ions appear to partially substitute for each other in the formation of terpenyl pyrophosphates. A thiol compound (preferably dithiothreitol) is required for maximum synthesis of phytoene, but pyridine nucleotides are not required. Farnesyl-PP-4,8,12-^{14}C[11] is also incorporated into phytoene by this enzyme system, provided that nonradioactive isopentenyl-PP is included in the reaction mixture.

Optimum pH. The enzyme appears to have a double pH optimum: pH 6.0 in phosphate or imidazole buffer and pH 8.0 in glycylglycine or borate buffer.[2] The incorporation of radioactivity into phytoene is considerably higher at pH 8.0 than at pH 6.0, but Tris buffer is inhibitory at pH 8.0.

Activators and Inhibitors. Dithiothreitol increases the activity of the enzyme system forming phytoene whereas sulfhydryl inhibitors,

---

[10] A. A. Kandutsch, H. Paulus, E. Levin, and K. Bloch, *J. Biol. Chem.* **239**, 2507 (1964).

[11] D. G. Anderson, M. S. Rice, and J. W. Porter, *Biochem. Biophys. Res. Commun.* **3**, 591 (1960).

p-hydroxymercuribenzoate, N-ethylmaleimide, and iodoacetamide are inhibitory.

Substrate Affinity. The K_m value for isopentenyl-PP is $1.33 \times 10^{-5}\,M$.

Acknowledgment

The experimental work that forms the basis for these methods was supported by grants A-1383 from the National Institute of Arthritis and Metabolic Diseases of the National Institutes of Health, U.S. Public Health Service, and CB-3306 from the National Science Foundation.

[14] Enzymatic Synthesis of β-Carotene in *Phycomyces*

By Henry Yokoyama and C. O. Chichester

$$\text{Mevalonic acid-2-}^{14}\text{C} \rightarrow \beta\text{-carotene-}^{14}\text{C} + CO_2$$

The above reaction is catalyzed by a multienzyme system that can be extracted from *Phycomyces blakesleeanus*. The individual enzymes participating in the reaction have not been isolated or studied in detail from this source. Methods that can be used to obtain cell-free extracts that carry out the above reaction are described below.

Reagents

Mevalonic acid, $0.05\,M$. Sufficient radioactive mevalonic acid is added to carrier mevalonic acid to make a total of 37 mg of substrate which is dissolved in 5 ml of distilled water (15,000 cpm per micromole). Mevalonic acid is added as either the potassium salt or N,N'-dibenzylethylenediamine salt.

Glutathione, $0.002\,M$

Nicotinamide, $0.02\,M$

Tris-HCl buffer, pH 7.8

NADH, $0.015\,M$

NADP+, $0.015\,M$

NADPH, $0.015\,M$

Cu++, $0.02\,M$

Mn++, $0.2\,M$

Protein (gelatin): 40 mg of protein per 5 ml of incubation mixture

Nonradioactive carrier pigment. Eight- to 10-day-old mycelium of *P. blakesleeanus* is disintegrated in the Waring blendor and centrifuged at about 15,000 g for 10 minutes. The supernatant is discarded and pellets are used as carriers.

Preparation of Cell-Free Extract

The (—) strain of *P. blakesleeanus* NRRL 1554 which normally synthesizes more β-carotene than the (+) strain is used as the source of enzymes. *P. blakesleeanus* is grown aerobically under illumination in the manner described previously.[1] Four- to 5-day-old mycelial mats of *P. blakesleeanus* are cooled to approximately 4°, cut into small pieces, and placed in a cold homogenizing medium containing 0.2 M Tris-HCl buffer, pH 7.8, 0.002 M glutathione, and 0.02 M nicotinamide. The mixture is ground for 30 seconds in a very loosely fitting Potter-Elvehjem homogenizer containing 2 volumes of the suspending medium. The homogenizer is powered by a cone-driven stirring motor operating at approximately 500 rpm. This treatment is insufficient to disrupt all the tissues; the unbroken cells and tissue debris are eliminated by centrifugation for 5–10 minutes at 4000 g at 4°. Since prolonged storage in the cold is not feasible, the supernatant is used immediately after centrifugation.

Reaction Mixture. Each flask contains 4 ml of the homogenate and 40 mg of protein. Substrate (5 μM mevalonic acid-2-^{14}C) and cofactors (10.0 μM ATP, 1.5 μM NADH, 1.5 μM NADP+, 1.5 μM NADPH, 1.0 μM Cu++, and 10.0 μM Mn++) neutralized to pH 7.0 and brought to a volume of 1.0 ml are then added to each flask. There should be a large interfacial area and then liquid layer (approximately 5 mm) to allow adequate diffusion of gases throughout the homogenate. The incubation period is 20 hours at room temperature. Incubations are stopped by the addition of 3 volumes of ethanol. After the addition of nonradioactive carrier pigment, the reaction mixture is heated for 1 minute at 100° and centrifuged for 5 minutes at 10,000 g. The supernatant is discarded. Carotene and related polyenes are extracted from the pellet with acetone and methanol. β-Carotene is isolated in the crystalline state[1] and counted[2] in the manner described previously.

Properties

The homogenate also converts β-hydroxy-β-methylglutarate into β-carotene. However, for maximal incorporation the system requires CoA. The two reactions are inhibited by iodoacetamide (10^{-4} M).

[1] C. O. Chichester, T. Nakayama, G. Mackinney, and T. W. Goodwin, *J. Biol. Chem.* **214**, 515 (1955).
[2] H. Yokoyama, T. O. M. Nakayama, and C. O. Chichester, *J. Biol. Chem.* **237**, 681 (1962).

[15] The Conversion of all-*Trans* β-Carotene into Retinal[1-3]

By DeWitt S. Goodman[3a] and James Allen Olson

β-Carotene can replace the growth requirement of rats for vitamin A[4,5] and is converted into vitamin A in lymph fistula rats and pigs,[6,7] rat intestinal loops and slices,[8,9] and isolated perfused rat livers.[10] Recently a soluble enzyme which cleaves β-carotene into retinal was prepared from rat intestinal mucosa[11] and shortly thereafter from rat liver.[12] This enzyme has tentatively been designated as β-carotene 15,15′-oxygenase.

Assay Method

Principle. Radioactive β-carotene is cleaved by the enzyme to form radioactive retinal, which may be separated from the substrate by

[1] Retinal, retinol, anhydroretinol, and retinyl ester are used in place of retinene, vitamin A alcohol, anhydro vitamin A, and vitamin A ester, respectively, in accordance with recommended rules of nomenclature. The term, vitamin A, is used generically in this chapter, and includes any or all compounds of this class.

[2] IUPAC Commission on the Nomenclature of Biological Chemistry, *J. Am. Chem. Soc.* **82**, 5581 (1960).

[3] IUPAC-IUB Commission on Biochemical Nomenclature, *Biochim. Biophys. Acta* **107**, 1 (1965).

[3a] Career Scientist of the Health Research Council of the City of New York under Contract I-399.

[4] H. von Euler, B. von Euler, and H. Hellstrom, *Biochem. Z.* **203**, 370 (1928).

[5] T. Moore, *Biochem. J.* **23**, 803 (1929); **24**, 692 (1930).

[6] S. Y. Thompson, R. Braude, M. E. Coates, A. T. Cowie, J. Ganguly, and S. K. Kon, *Brit. J. Nutr.* **4**, 398 (1950).

[7] H. Wagner, F. Wyler, G. Rinde, and K. Bernhard, *Helv. Physiol. Pharmacol. Acta* **18**, 438 (1960).

[8] J. A. Olson, *J. Biol. Chem.* **236**, 349 (1961).

[9] J. A. Olson, *J. Lipid Res.* **5**, 402 (1964).

[10] R. D. Zachman and J. A. Olson, *J. Biol. Chem.* **238**, 541 (1963).

[11] D. S. Goodman and H. S. Huang, *Science* **149**, 879 (1965).

[12] J. A. Olson and O. Hayaishi, *Proc. Natl. Acad. Sci. U.S.* **54**, 1364 (1965).

chromatography on an alumina column or on a thin-layer plate of silica gel. The isolated product may be further characterized by the formation of several derivatives.

Synthesis of β-carotene-^{14}C. Radioactive β-carotene is synthesized by a modification of the procedure of Lilly *et al.*[13] A culture of *Phycomyces blakesleeanus* was obtained from the Centraalbureau Schimmelcultures, Baarn, Holland. This mold sporulates well on a medium of 5 g of puffed wheat (Quaker Oats Company), 0.2 ml of glycerol, 5 mg of asparagine, and 5 ml of water. When many aerial hyphae are present, spores are harvested in distilled water and 1 ml of the spore suspension, containing about 10^6 spores, is used to inoculate each 100 ml of Lilly's medium in a stationary Roux bottle at room temperature under constant illumination with white light. Each liter of the medium contains D-glucose, 10 g; potassium acetate, 0.72 g; ammonium sulfate, 1 g; monobasic potassium acid phosphate, 1 g; $MgSO_4 \cdot 7 H_2O$, 0.5 g; $FeCl_3$, 0.1 mg; $ZnSO_4$, 0.1 mg; $MnCl_2$, 0.05 mg; $CaCl_2$, 10 mg; and thiamine, 0.1 mg. The pH of the medium is adjusted to 5.7–5.8 before autoclaving and becomes very slightly more acidic during sterilization. When pigmentation begins in 3–4 days, 1 ml of acetate-1-^{14}C solution, containing 2.5×10^7 cpm in 1.4 mg, is added daily to each Roux bottle for 4 successive days. On the tenth day the mold is filtered with suction, ground in a mortar with anhydrous Na_2SO_4 and sand, and left overnight under distilled diethyl ether. The ether extract is filtered, evaporated to a small volume, and saponified under nitrogen with methanolic KOH. The unsaponifiable fraction is extracted with *n*-hexane, washed with water, dried over Na_2SO_4, concentrated to a small volume, and chromatographed on 10 g of deactivated alumina containing 5–6% H_2O. β-Carotene is eluted with 30–70 ml of hexane in 5-ml fractions, and the solvent is evaporated under nitrogen from the major carotene-containing fractions. The labeled β-carotene is dissolved in a small amount of benzene, and 2 volumes of methanol are added to induce crystallization. Usually three crystallizations are required to obtain a product with constant specific activity. About 0.5 mg of pure radioactive β-carotene ($E_{1cm}^{1\%} = 2600$ at 450 mμ in hexane) with a specific radioactivity of 1000–2000 cpm/μg is obtained from each Roux bottle. The overall conversion of radioactive acetate to β-carotene is about 0.5–1%. The three-times recrystallized compound is 95–98% pure as judged either by two-dimensional chromatography on thin-layer plates of silica gel G developed with acetone–*n*-hexane 1:19 (v/v) as one solvent and ether–hexane 1:1 as the other, or by one-dimensional chromatography with benzene–hexane 1:19.

---

[13] V. G. Lilly, H. L. Barnett, R. F. Krause, and F. J. Lotspeich, *Mycologia* **50**, 862 (1958).

Specially prepared silicic acid columns may also be used for the purification of labeled β-carotene.[14] Mallinckrodt 100-mesh silicic acid, 115 g, is suspended in 600 ml of water containing 15 g of NaOH. After filtration and washing with water, the silicic acid is dried at 45° for 48 hours. The column is prepared by suspending 25 g of silicic acid in low-boiling petroleum ether and by packing the column under positive pressure to a volume of 2.5 cm diameter × 17.2 cm. Oxygen and other active sites in the column are eliminated by passing 2 mg of nonradioactive β-carotene in hexane through the column before the radioactive material is added. Thereafter, the extract containing radioactive carotene in petroleum ether is placed on the column and is eluted with petroleum ether at a rate of 5 ml/min. The major orange band is collected, evaporated to dryness under nitrogen, and washed with 50 ml of methanol. After washing the sample in a similar fashion 8 times, the resultant crystalline all-*trans* β-carotene has an $E_{1\,cm}^{1\%}$ value of 2600 at 450 mμ.

Hexane solutions of β-carotene-^{14}C are kept in the dark in low-actinic glass vessels at $-20°$. β-Carotene-^{14}C with a specific radioactivity of 236,000 dpm/μg has been prepared by using acetate with a higher specific radioactivity.[15] Tritiated acetate may also be used for the preparation of tritium-labeled β-carotene.

Storage of Carotenoids and Retinal Derivatives. Compounds of this class are best stored in the crystalline state under a dry, inert gas and in a dark, cold place. Solutions of all these compounds will deteriorate more or less rapidly, depending on the conditions of storage. Although antioxidants such as α-tocopherol reduce the deterioration rate, solutions of radioactive carotene should be tested periodically and purified just before use by passage through a small column of deactivated alumina. When stored in 10% ethanol in hexane in the dark, all-*trans* retinal is fairly stable but does isomerize slowly. Retinol in solution may be protected somewhat by the presence of α-tocopherol and a trace of ammonia.

Incubation Conditions. In order to demonstrate and study the cleavage reaction, incubations should be carried out in Erlenmeyer flasks of amber glass, so as to exclude light from the mixture. Maximal activity of the mucosal enzyme is achieved with a 2-ml incubation mixture containing the following components: F_{20}^{45} (defined below) equivalent to 1 ml of the soluble supernatant fraction of a homogenate of intestinal mucosa (containing about 7 mg of protein); plus 200 micromoles of potassium phosphate buffer pH 7.7; plus 30 micromoles of nicotinamide; plus 10 micromoles of glutathione; plus 12 micromoles of sodium glycocholate;

[14] F. J. Lotspeich, R. F. Krause, V. G. Lilly, and H. L. Barnett, *Proc. Soc. Exptl. Biol. Med.* **114**, 444 (1963).
[15] H. S. Huang and D. S. Goodman, *J. Biol. Chem.* **240**, 2839 (1965).

plus 400 μg of egg lecithin added as an aqueous emulsion (see below); plus 0.2 mg of α-tocopherol added in 20 μl of acetone. The substrate labeled β-carotene (usually 0.8–1 μg) is added to this mixture in solution in 50 μl of acetone, by means of a 50-μl syringe. The incubation is carried out at 37° with room air as the gas phase.

Of the ingredients just listed, the phosphate buffer serves to provide the optimal pH for the reaction, the glutathione helps to keep the enzyme's —SH groups in a reduced state, and the glycocholate plus lecithin provide the detergent mixture necessary for the *in vitro* reaction (see below). The function of the nicotinamide is not clear, since no other cofactors (in particular, no nicotinamide-containing nucleotides) are necessary for the reaction. Omission of nicotinamide, however, usually results in a slightly (approximately 10–20%) reduced yield of retinal. Tocopherol is added as an antioxidant, and does not take part in the reaction. Omission of tocopherol usually is associated with a lower recovery of total radioactivity, but does not affect the yield of the reaction (expressed as the percent of recovered radioactivity found as retinal). Molecular oxygen—an essential component of the reaction—is provided by using room air as the gas phase. The yield of retinal obtained with this system is usually close to 50% after a 40-minute incubation.

Extraction. At the end of the incubation the mixture is extracted with 20 volumes of chloroform–methanol (2:1, v/v), containing 8 μg each of carrier β-carotene, retinal, and retinoic acid, plus 20 μg of retinol and 20 μg of a mixture of retinyl esters. Five volumes of 0.01 N H_2SO_4 are then added, and the lower, chloroform, phase is collected and evaporated. The total lipid extracts so obtained are then chromatographed on columns of alumina. Alternatively, four volumes of distilled acetone may be added to the incubation mixture, followed by four volumes of ether and four volumes of water. The acetone contains 30 μg of carrier β-carotene, retinal, and retinol plus 50 μg of α-tocopherol. Solutions are swirled in a separatory funnel, and the aqueous layer is reextracted with 20 ml of acetone–ether 1:1. The combined lipid extract is dried over anhydrous sodium sulfate, reduced to a small volume in a rotoevaporator under nitrogen at less than 55° in the dark, and transferred to a small tube with ether. The aqueous emulsion is broken with anhydrous sodium sulfate. The ether extract, after drying over sodium sulfate, is evaporated to a small volume and then chromatographed on a thin-layer plate of buffered silica gel G.

Alumina Column Chromatography. β-Carotene, retinyl esters, retinal, retinol, and retinoic acid can easily be separated from each other by chromatography on columns of alumina (Woelm, neutral Al_2O_3) of activity grade III. Quantitative separations are obtained with columns of 1 cm diameter containing 2 g or more of the alumina. The columns are packed by adding the alumina as a slurry in *n*-hexane, and the sample is applied to the column as a solution in *n*-hexane. The maximum load applied to the column is 10 mg of total lipid per gram of alumina. Six fractions are eluted from each column. The order of elution, and volume of eluent per 5 g of alumina are: fraction 1, β-carotene (20 ml of *n*-hexane); fraction 2, retinyl esters (20 ml of benzene–hexane 3:17, v/v); fraction 3, retinal (20 ml of benzene–hexane 1:1); fraction 4, retinol (50 ml of benzene); fraction 5, more polar, but nonacidic, compounds (20 ml of methanol); and fraction 6, acids including retinoic acid (20 ml of methanol–25% acetic acid 3:1). The efficiency of separation is such that less than 3% of any vitamin A derivative is eluted in a fraction other than the one expected. After chromatography, the separated fractions may be assayed for radioactivity or subjected to other analyses.

Thin-Layer Chromatography (TLC) of β-Carotene and Retinol Derivatives. Thin-layer plates (20 cm × 20 cm) of buffered silica gel G are prepared by mixing 2.02 volumes of 0.05 M potassium phosphate buffer pH 8.0 with each gram of silica gel G (Merck-Darmstadt). The slurry is plated at a thickness of 0.2–0.4 mm with a Desaga spreading apparatus, and the plates are dried for 1 hour at 110° and stored over calcium chloride until used. The TLC developing chambers are lined with Whatman 3 MM filter paper and closed tightly to reduce solvent evaporation during chromatography. All operations are conducted in a photographic darkroom under brown light. For two-dimensional chromatography, a sample containing less than 4 mg of lipid and about 20 μg of α-tocopherol is spotted under nitrogen and immediately developed in acetone–hexane 1:49. When the solvent is well advanced, the plate is removed and dried under nitrogen in a plastic bag equipped with inlet and outlet stopcocks. The plate is then turned 90 degrees and developed in ether–hexane 1:1. For one-dimensional chromatography, acetone–hexane 1:19 or ether–hexane 1:1 is usually employed. After development, compounds are located by brief exposure to a 360 mμ ultraviolet light, and are scraped into 1.5 ml of ethanol–hexane 1:9 which also contains 10 μg of α-tocopherol and a trace of ammonia. The gel suspensions are stirred or shaken and then are left for 5–10 minutes. Aliquots can then be taken for radioassay and for spectrophotometric analyses. When proper precautions are taken, the recovery of all carotene and retinol derivatives upon two-dimensional TLC is greater than 80%. When TLC is conducted without particular care, yields as low as 20–40% are obtained. Even under poor conditions of TLC analysis, however, retinal is somewhat more stable than β-carotene or retinol. The chromatographic behavior of β-carotene and retinol derivatives in various solvent systems is given in the table.

CHROMATOGRAPHIC BEHAVIOR OF β-CAROTENE AND RETINOL DERIVATIVES ON
THIN-LAYER PLATES OF SILICA GEL G

Developing solvent	β-Carotene	Long-chain retinyl esters[a]	Retinyl acetate	Retinal	Retinol
Acetone–hexane, 1:99	0.71	0.50	0.30	0.13	0.03
Acetone–hexane, 1:49	0.75	0.57	0.40	0.25	0.05
Acetone–hexane, 1:19	0.76	—	0.47	0.29	0.07
Ether–hexane, 3:7	0.80	—	0.58	0.39	0.16
Ether–hexane, 1:1	0.80	0.65	0.62	0.50	0.25

[a] From natural sources.

Thin-layer plates of alumina,[16] either with or without $CaSO_4$ binder, can also be employed for the separation of carotene and retinol derivatives. Useful developing solvents with deactivated alumina (7% H_2O) are 12% benzene in hexane, or 10% ether in hexane. The resolution of components on alumina plates is, however, somewhat poorer than on silica gel plates.

Chemical Characterization of Retinal. The retinal eluates from alumina columns or from thin-layer plates are cautiously evaporated to dryness in the dark under nitrogen, and immediately dissolved in 0.3 ml of ethanol. About 1 mg of solid $NaBH_4$ is added, and the tube is left at room temperature for 1–2 hours in the dark. The mixture is separated on a thin-layer plate with ether–hexane 1:1 as the developing solvent. The retinol spot, detected by its fluorescence upon brief exposure to ultraviolet light (360 mμ), is scraped from the thin-layer plate and dissolved in 1.5 ml of 10% ethanol in hexane. The optical density at 328 mμ is measured and an aliquot is counted.

The retinol thus formed can be further characterized by dehydration to anhydroretinol in an acid medium.[17] The retinol solution is evaporated to dryness and quickly dissolved in 0.3 ml of absolute ethanol. Thereafter, 0.01 ml of 1 N HCl in ethanol is added, and the ratio of optical density at 371 mμ to 325 mμ is followed. When the ratio reaches a maximum of 1.7–2.0, the reaction is stopped by the addition of 0.011 ml of 1 N ammonia in ethanol. The mixture is spotted on thin-layer plates of buffered silica gel G and developed with ether–hexane 1:1. Anhydroretinol has an R_f of 0.75 in this solvent, and quenches under ultraviolet light. The amount of each compound present is calculated from its ultraviolet absorption, using the following $E_{1cm}^{1\%}$ values: retinal, 1530 at 381 mμ;

[16] J. Davidek and J. Blattna, *J. Chromatog.* **7**, 204 (1962).
[17] K. Harashima, H. Okazaki, and H. Aoki, *J. Vitaminol. (Kyoto)* **7**, 150 (1961).

retinol, 1832 at 328 mμ; and anhydroretinol, 3650 at 371 mμ. Aliquots of each solution are counted under standard conditions, and specific radioactivities are calculated.

Alternatively, retinal may be diluted with additional carrier and reacted with semicarbazide hydrochloride.[8, 11, 18, 19] The retinal (0.5–5 mg) is dissolved in 3.0 ml of 90% ethanol which contains 22 mg sodium acetate and 15 mg semicarbazide hydrochloride. After warming briefly, the solution is left overnight in the dark at room temperature and then is extracted with hexane. The hexane is evaporated under nitrogen and the residue is crystallized from ethanol or from ether–hexane mixtures. The melting point of the recrystallized retinal semicarbazone is 189–190°, and its $E_{1cm}^{1\%}$ value is 2082 at 385 mμ.[18] If it fails to crystallize, the retinal semicarbazone, which consists of *syn* and *anti* isomers, may be isolated from unreacted retinal by chromatography on deactivated alumina (7% H_2O). Retinal is eluted with 2% acetone, whereas the semicarbazones are eluted with 12% acetone in hexane.[8]

Gas–Liquid Chromatography of Retinal. The product of the reaction which takes place during the *in vitro* incubation of ^{14}C-labeled β-carotene with intestinal homogenate fractions has been identified as retinal by gas–liquid chromatography.[20] Approximately 15 μg of pure unlabeled retinal is added to the extracted incubation mixture containing 1–2 μg of newly formed labeled retinal. The extract is chromatographed on a column of alumina (see above) and the retinal fraction (fraction 3) is subjected to gas–liquid chromatography on a helical glass column packed with 1% SE-30 on gas Chrom P of mesh 110–120. Good results are obtained with a column 45 cm long with an internal diameter of 4 mm, using an argon flow rate of 180 ml/min at a pressure of 20 psi. The column temperature should be 154° and the flash-heater temperature 10–20° higher. At the end of the column the effluent stream is split, so that a measured fraction of it enters the mass detector (e.g., an argon ionization detector), whereas most of it is collected at an exit port at 1-minute intervals for assay of radioactivity. In this system retinal emerges as a symmetrical peak with a retention time of approximately 9.5 minutes, whereas retinol, retinyl acetate, and their degradation products emerge earlier, with retention times of about 4.5 minutes. In an experiment carried out with these conditions, about 85% of the recovered radioactivity emerged from the column as a single peak which was coincident with

[18] S. Ball, T. W. Goodwin, and R. A. Morton, *Biochem. J.* **42**, 516 (1948).
[19] C. D. Robeson, W. P. Blum, J. M. Dieterle, J. D. Cawley, and J. G. Baxter, *J. Am. Chem. Soc.* **77**, 4120 (1955).
[20] D. S. Goodman, H. S. Huang, M. Kanai, and T. Shiratori, *J. Biol. Chem.* **242**, 3543 (1967).

the mass peak of retinal. Since, in this experiment, there was a quantitative recovery of radioactivity from the column, this confirmed the identification of the radioactive product of the reaction as retinal. Procedures for the isolation of various derivatives of vitamin A by gas–liquid chromatography are also treated elsewhere in this volume.[21]

Separation of Different Retinyl Esters. Studies on the conversion of β-carotene into vitamin A with lymph fistula animals[6,7,15] or with isolated organs or tissues[8-10] have resulted in the appearance of retinyl esters as the major chemical form of newly formed vitamin A. The composition of the newly formed retinyl ester mixture can be determined by separating the different retinyl esters by a combination of two different kinds of chromatographic procedures. First, retinyl esters are separated on the basis of the degree of unsaturation of the fatty acid component, by thin-layer chromatography on alumina gel G impregnated with silver nitrate.[15] Good results are obtained with thin layers (thickness 200 μ) prepared from a slurry of 25 g of alumina gel G plus 55 ml of 0.7% $AgNO_3$ in water. The ascending solvent is benzene–hexane 1:1. The chromatographic jar is flushed with nitrogen and covered to exclude light during the procedure. Chromatography of a mixture of retinyl palmitate, oleate, and linoleate by this procedure results in the wide separation of three bands, comprising the saturated ester (palmitate, R_f about 0.9), the monounsaturated ester (oleate, R_f about 0.6), and the diunsaturated ester (linoleate, R_f about 0.35). The retinyl esters can be localized by their fluorescence under UV light, and can then be scraped onto filter funnels, and eluted with chloroform.

Retinyl esters of the same degree of unsaturation can be separated according to fatty acid chain length by reversed phase paper chromatography.[15] Strips 14 cm wide, prepared by cutting sheets of Whatman No. 1 filter paper across the fiber, are impregnated with a solution of 10% of DC 200 silicone fluid (viscosity, 10 centistokes at 25°) (Dow-Corning Corporation) in ethyl ether, and are air dried. Descending chromatography is carried out in tanks lined with Whatman No. 3 filter paper previously saturated with methanol–chloroform–n-butanol–water (6:2:1:1). The sample, containing approximately 50 μg of retinyl esters, is applied as a spot to the paper, and the chromatogram is then developed with acetonitrile–methanol–water–glacial acetic acid (30:70:5:1). The separated esters are visualized directly under UV light, and specific areas are cut out and eluted with boiling ethyl ether. This procedure will separate retinyl myristate, palmitate, and stearate from each other; or will separate retinyl palmitoleate from retinyl oleate. In this system, however, retinyl oleate will chromatograph together with retinyl palmitate, and

[21] P. E. Dunagin, Jr., and J. A. Olson, this volume, p. 289.

retinyl linoleate together with retinyl myristate. In order to separate completely a mixture of retinyl esters, the esters should first be separated according to their degree of fatty acid unsaturation by thin layer chromatography, and then the individual TLC bands separated according to fatty acid chain length by reversed-phase paper chromatography.

All chromatographic systems should be standardized with pure reference compounds before being used in experimental situations. Pure β-carotene, retinyl acetate, retinal, retinol, and retinoic acid can be obtained in gram quantities from commercial sources. Pure retinyl esters (e.g., palmitate, stearate, oleate, linoleate) can be synthesized by permitting retinol to react with the appropriate acyl chloride. To 1.2 millimole of retinol in a 50-ml round-bottom flask are added 0.3 ml of pyridine and 1.4 millimole of acyl chloride. The flask is flushed with nitrogen, briefly warmed to 50–60°, and intermittently shaken for 1–2 hours in the dark. The contents of the flask are then extracted with n-hexane (50 ml), and the hexane solution is washed with 50 ml each of 0.1 N NaOH in 50% ethanol, 0.1 N HCl, 0.03 N NaOH in 50% ethanol, and water. After evaporation of the hexane, the oily residue is chromatographed on a column of alumina grade III as described above. The yields of retinyl ester are usually in the range of 70–90%.

Enzyme Preparation and Purification

Considerable information is available concerning the properties of the β-carotene cleavage enzyme present in rat intestinal mucosa[11, 20] and liver.[12] In order to prepare enzymatically active homogenates of intestinal mucosa, rats are fasted overnight, and the proximal one-third to one-half of the small intestine is then removed and flushed with ice-cold isotonic saline. The intestine is cut lengthwise and rinsed with isotonic saline, and the mucosa is scraped from the underlying muscularis with a spatula. The mucosal scrapings are homogenized in a Potter-Elvehjem homogenizer with a loose-fitting Teflon pestle, in 0.1 M potassium phosphate buffer, pH 7.7, containing 30 mM nicotinamide and 4 mM MgCl$_2$. One milliliter of buffer is used for each 10 cm of intestine. The homogenate is centrifuged at 2000 g for 20 minutes to provide a supernatant solution which usually contains about 20 mg protein per milliliter. Further centrifugation of this supernatant solution at 104,000 g for 60 minutes results in the separation of soluble and particulate fractions. The soluble fraction contains the cleavage enzyme, and usually contains about 15 mg of protein per milliliter.

In order to prepare an active homogenate fraction from rat liver, the organ is removed, washed quickly in cold saline, and homogenized in 4–5 volumes of cold 0.15 M Tris buffer, pH 8.2, containing 0.01 M cysteine and 0.01 M nicotinamide, for 5 minutes in a motor-driven, loose-fitting glass homogenizer. The homogenate is centrifuged in the cold at 13,000 rpm for 10 minutes, and the resultant supernatant solution is centrifuged at 105,000 g at 4° for 60 minutes. After each centrifugation, the floating lipid layer is dabbed up with a small piece of absorbent paper. The resulting high speed supernatant solution, which is a clear pinkish color, contains the enzyme.

The intestinal enzyme can be further purified by precipitation with ammonium sulfate between 20 and 45% saturation. In order to achieve this, 14.1 g of finely ground solid ammonium sulfate is added slowly with stirring to each 100 ml of the iced soluble fraction in order to produce a solution 20% saturated in ammonium sulfate. The mixture is stirred for 20 minutes, then is centrifuged at 10,000 g; the precipitate is discarded. To the supernatant solution is then added 17.7 g of ammonium sulfate per original 100 ml, to achieve 45% saturation. This mixture is stirred at ice temperature for 20–30 minutes, then again centrifuged. The precipitate contains the cleavage enzyme; it can be stored at −20° for several weeks with only a minimal loss of enzymatic activity. Before use the precipitate is dissolved in a small volume of 0.01 M potassium phosphate buffer pH 7.7, and dialyzed against a large excess of this buffer for several hours in order to remove ammonium sulfate. The resultant solution, designated the F_{20}^{45}, usually contains 40–45% of the protein in the 104,000 g supernatant from which it was derived.

Significant further purification of the cleavage enzyme has not yet been achieved. Chromatography of a portion of mucosal F_{20}^{45} on a column of Sephadex G-200 results in the recovery of some enzymatic activity in that portion of the eluate corresponding to a probable molecular weight in the range of 100,000–200,000. Only very slight purification (in terms of activity per milligram of protein) is obtained however, apparently because the enzyme is partly inactivated during chromatography.

Properties

Distribution in Nature. The carotene cleavage system is found in the rat intestine[11] and liver,[12] perhaps in kidney,[12] but not in microorganisms. In both the liver and the intestinal mucosa the enzyme is localized in the supernatant fraction after centrifugation at 105,000 g for 1 hour.

Dispersion of β-Carotene. Studies with the mucosal enzyme[20] have extensively surveyed the effects of the addition of a variety of detergents and related compounds on cleavage enzyme activity. Comparable information is not available for the liver enzyme. The following discussion of the effects of detergents and of substrate dispersion hence refers only to the properties of the mucosal enzyme.

An appropriate detergent or detergent–lipid combination is required in order to effect the *in vitro* conversion of β-carotene to retinal with the mucosal enzyme.[20] In the incubation mixture described above for the assay of cleavage enzyme activity, this requirement is satisfied by the addition of sodium glycocholate plus egg lecithin to the mixture. Omission of glycocholate from this mixture results in complete loss of enzymatic activity. Omission of lecithin produces a large decrease in the activity of the system to about 25% of that observed with glycocholate and lecithin. The lecithin requirement is nonspecific, and comparable maximal reaction rates have been obtained by adding, in the presence of glycocholate, either 600 μg of lysolecithin, 600 μg of sphingomyelin, or washed cell particles equivalent to 1 ml of homogenate. The cleavage rate was stimulated to a lesser degree by replacing lecithin with 400 μg of either oleyl acid phosphate, sodium dodecyl sulfate, hexadecyl pyridinium chloride, or Tween 80, or with monoolein, palmitic acid, or linoleic acid. Several other lipids tested, both phospholipids and nonphospholipids, were without effect.

In order to achieve maximal stimulation, the lecithin (or other detergent or lipid) is emulsified in water by first dissolving it in a very small volume (e.g., a few microliters) of methanol, followed by the addition of the desired amount of water, and finally by thoroughly mixing the resultant emulsion on a vortex stirrer.

The bile salt requirement is also nonspecific, and comparable yields of retinal can be obtained with sodium glycocholate, cholate, deoxycholate, glycodeoxycholate, taurodeoxycholate, glycochenodeoxycholate, and taurochenodeoxycholate. In the presence of egg lecithin (400 μg), all these bile salts produce comparable, maximal reaction rates. In the absence of egg lecithin (i.e., with the bile salt alone), all the bile salts produce yields approximately 20–30% of those obtained with added lecithin.

Several synthetic detergents, of which the most effective is sodium dodecyl sulfate, can substitute for the bile salt. The addition of 2–6 mg of sodium dodecyl sulfate to the incubation mixture stimulates the reaction almost as much as does the addition of glycocholate plus lecithin. Addition of sodium dodecyl sulfate plus lecithin stimulates the reaction to its maximal rate, i.e., to the rate observed with glycocholate (or other bile salt) plus lecithin.

The role played by the detergent, or detergent–lipid combination, in activating the cleavage reaction has not been defined. It is probable that these components serve to provide a proper dispersion of the substrate β-carotene for combination with the enzyme. It is also possible that these components activate the enzyme directly, although the lack of specificity of the detergent requirement makes this unlikely. Further information is lacking about the physical or chemical characteristics of a "proper" dispersion, as compared to an inactive one.

In all the experiments described in this section, the substrate β-carotene was added to the incubation mixture in solution in 50 μl of acetone. The labeled β-carotene can also be added as a clear micellar solution in one of the Tweens or in sodium dodecyl sulfate. Regardless of the manner in which β-carotene is initially added, however, maximal rates with the mucosal enzyme are achieved only by using an appropriate combination of detergents, or of detergent and lipid.

Stability. The intestinal enzyme is completely inactivated by heating at 64° for 55 seconds. The rat liver enzyme may be frozen and thawed repeatedly without loss of activity, provided that a high concentration of ammonium sulfate is not present. Aqueous acetone 9:1 destroys the liver enzyme at room temperature.

pH Optimum. There is a relatively narrow optimal pH range for the intestinal enzyme, with maximal yield obtained at approximately pH 7.7, and with a considerable decrease in yield occurring as pH is increased above 8.0 or decreased below 7.5.

Effect of Incubation Time. With the 2 ml of incubation mixture described above, and containing the intestinal enzyme, the reaction rate is constant for approximately 30 minutes and then decreases progressively until 90 minutes, beyond which no further reaction occurs. The yield obtained at 90 minutes is usually about 1.5 times that obtained at 30 minutes.

Effect of Enzyme Concentration. The reaction rate is directly proportional to the intestinal enzyme concentration in the range of 0–7 mg of F_{20}^{45} protein per 2 ml of incubation mixture. Increasing the enzyme concentration above 8 mg of F_{20}^{45} protein in the 2-ml incubation mixture results in only a very slight further increase in the rate of the reaction.

Effect of β-Carotene Concentration. For both the intestinal and liver enzymes, the reaction rate is directly proportional to the amount of added substrate up to approximately 1.8 μg of β-carotene added to a 2-ml incubation mixture. Adding β-carotene in amounts greater than 2 μg results in a progressive decline in the percent yield for any given time interval. When $1/v$ and $1/s$ are plotted in the usual fashion, a straight line results. For the intestinal enzyme a V_{max} of 8.3×10^{-9} mole of retinal formed per hour (for 7 mg of F_{20}^{45} protein) and a K_m of $3.3 \times 10^{-6} M$ were calculated. Because of the physical state of the substrate, which appears to be presented to the enzyme in a micellar solution of bile salt plus lecithin, it is difficult to ascribe precise meaning to the value for the Michaelis constant.

Involvement of SH Groups. Omission of glutathione from the incubation mixture results in a yield of retinal approximately two-thirds that obtained with added GSH. Equivalent stimulation is obtained by adding GSH in amounts of from 5 to 20 μmole per 2 ml. Cysteine and mercaptoethanol are equally effective. The reaction is powerfully inhibited by -SH group inhibitors, including N-ethylmaleimide, iodoacetamide, sodium arsenite, p-OH-mercuribenzoate, and silver ions. These results suggest the involvement of one or more sulfhydryl groups of the enzyme in this reaction.

Effect of Chelators. The reaction is strongly inhibited by α,α'-dipyridyl and by 1,10-phenanthroline, two compounds known to be effective chelators for ferrous iron. The extent of inhibition increases with increasing concentration of inhibitor. Addition of 1 micromole of inhibitor to a 2-ml incubation mixture containing intestinal F_{20}^{45} as enzyme produces approximately 85% inhibition with o-phenanthroline, and approximately 55% inhibition with α,α'-dipyridyl. The intestinal and liver enzymes behave similarly.

Mechanism. The reaction seemingly proceeds by the addition of 1 mole of molecular oxygen across the 15,15' double bond of β-carotene to yield an unstable four-membered peroxide ring, which then collapses to yield two molecules of retinal.[11,12,22] Molecular oxygen is absolutely required for catalysis by both the liver and intestinal enzymes, and the inhibitory effect of α,α'-dipyridyl and o-phenanthroline, suggest the involvement of a bound metal ion in the enzyme. Since cyanide does not inhibit the reaction at low concentrations, ferric ion in a heme complex is apparently not present. Stoichiometrically, approximately one molecule of carotene disappears for each two molecules of retinal which appear.[11] Furthermore, β-apo carotenals or other products are not detected in the incubation mixture.[12] The lack of a requirement for reduced pyridine nucleotides as well as the retention of enzyme activity after prolonged dialysis suggest that the cleavage enzyme is a dioxygenase, not a cofactor-dependent monooxygenase. Furthermore, experiments with doubly labeled β-carotene, uniformly labeled with ^{14}C throughout the molecule but specifically labeled with ^{3}H at the C-15 and C-15' positions indicate that there is complete retention of the hydrogen atoms attached to the two central carbon atoms of β-carotene during the conversion of β-carotene into vitamin A.[22] These findings are all consistent with the mechanism outlined above.

Physiological Implications

The conversion of β-carotene to retinal is an important reaction, since it converts the plant product β-carotene into a product necessary for the

[22] D. S. Goodman, H. S. Huang, and T. Shiratori, *J. Biol. Chem.* **241**, 1929 (1966).

growth and life of the animal organism. Throughout the course of evolution, most, if not all, of the vitamin A utilized by animals, was ultimately derived from this reaction. In recent years, however, vitamin A has been synthesized chemically by several procedures, and chemical synthesis is now the main commercial source of vitamin A.

The conversion of β-carotene to vitamin A mainly takes place in the intestinal mucosa, during the absorption of dietary β-carotene. In the rat, virtually no β-carotene is absorbed unchanged beyond the intestinal mucosa.[15] Instead, the β-carotene is apparently absorbed into the mucosal cells and there converted to retinal. The retinal is then reduced to retinol, by means of a soluble enzyme in the mucosal homogenate, which requires a reduced pyridine nucleotide as cofactor.[23] The retinol is next esterified with long-chain fatty acids, and the newly formed retinyl esters are transported via the intestinal lymphatics, mainly in association with lymph chylomicrons, eventually to enter the vascular compartment. The main form in which retinol compounds enter the blood and tissues is thus in the form of retinyl esters. In the rat, 90% of the radioactivity absorbed into the lymph, after the feeding of either labeled retinol or labeled β-carotene, is found as labeled retinyl esters.[15] Man differs from the rat in that the human intestine is able to absorb a small amount of unchanged dietary β-carotene into the lymph.[24] The quantitative difference in the ability of these two species to absorb intact β-carotene is, however, not very great, since most of the radioactivity absorbed into human lymph after feeding labeled β-carotene is also found in retinyl esters, but not in unchanged β-carotene. In both man[23] and the rat[15] the composition of the lymph retinyl esters is remarkably constant, regardless of the fatty acid composition of the diet and regardless of whether the retinyl esters are derived from preformed dietary vitamin A or from β-carotene. In all instances, retinyl palmitate is the predominant ester in lymph, and the saturated esters retinyl palmitate plus stearate comprise approximately three-fourths of the lymph retinyl esters. After entering the vascular compartment, the newly formed retinyl esters are mainly taken up by the liver, where they undergo turnover, and where vitamin A is mainly stored, in ester form, in the body.

[23] N. H. Fidge and D. S. Goodman, *Federation Proc.* **26**, 849 (1967).
[24] D. S. Goodman, R. Blomstrand, B. Werner, H. S. Huang, and T. Shiratori, *J. Clin. Invest.* **45**, 1615 (1966).

[16] Rubber Transferase from *Hevea brasiliensis* Latex

By B. L. ARCHER and E. G. COCKBAIN

$$-[CH_2\!-\!\overset{\overset{\displaystyle CH_3}{|}}{C}\!=\!CH\!-\!CH_2]_n\!-\!PP + CH_2\!=\!\overset{\overset{\displaystyle CH_3}{|}}{C}\!-\!CH_2\!-\!CH_2\!-\!PP \rightarrow$$

Rubber pyrophosphate   isopentenyl pyrophosphate

$$[CH_2\!-\!\overset{\overset{\displaystyle CH_3}{|}}{C}\!=\!CH\!-\!CH_2]_{n+1}\!-\!PP + PP$$

Rubber transferase catalyzes the incorporation of isopentenyl-PP into rubber (cis-1,4-polyisoprene). By analogy with the reaction between geranyl-PP and isopentenyl-PP to form farnesyl-PP,[1,2] it is believed that rubber transferase catalyzes the transfer of *cis*-1,4-polyisoprenyl-PP to isopentenyl-PP with the elimination of inorganic pyrophosphate, the reaction occurring at the surface of the rubber particles in *Hevea* latex.[3] The enzyme, which is distributed between the aqueous phase of the latex and the surface of the rubber particles,[4] has been partially purified.

Assay Method

Principle. The activities of rubber transferase preparations are compared by measuring the rates of incorporation of isopentenyl-PP-1-^{14}C into rubber. A suspension of washed rubber particles prepared from fresh *Hevea* latex serves as the acceptor for isopentenyl-PP in the assay. Essential cofactors for the incorporation are magnesium and glutathione (or cysteine). Different preparations of washed rubber particles, prepared by the same procedure, usually show different rates of incorporation of isopentenyl-PP into rubber in the presence of the same amounts of rubber transferase and cofactors. This variability is most likely due to variations in the size distribution of the rubber particles and/or the number of rubber molecules with terminal pyrophosphate groups at the surface of the particles. No method of determining the latter figure is available, nor has it been possible so far to establish assay conditions in which the rate of incorporation of isopentenyl-PP is zero order with respect to the rubber content of the suspension. Consequently, the assay method described

---

[1] G. Popják, *Tetrahedron Letters* **19**, 19 (1959).
[2] F. Lynen, H. Eggerer, U. Henning, and I. Kessel, *Angew. Chem.* **70**, 738 (1958).
[3] B. L. Archer, B. G. Audley, E. G. Cockbain, and G. P. McSweeney, *Biochem. J.* **89**, 565 (1963).
[4] A. I. McMullen and G. P. McSweeney, *Biochem. J.* **101**, 42 (1966).

below is suitable only for comparing the relative transferase contents of different preparations, using the same suspension of washed rubber particles.

The rate of incorporation of isopentenyl-PP into the washed rubber does not increase linearly with transferase concentration but obeys the relationship[5]

$$\frac{1}{R} = \frac{A}{[P]} + B \tag{1}$$

where R is the rate of incorporation of isopentenyl-PP, $[P]$ is the bulk concentration of the protein expressed on the volume of aqueous phase in the rubber suspension, and A and B are constants. Plots of $1/R$ against $1/[P]$ for different preparations of rubber transferase give straight lines from which the corresponding values of A are obtained. It can readily be shown from Eq. (1) that the transferase activity per unit weight of a preparation is inversely proportional to the value of A for that preparation.

Reagents

Tris-HCl buffer: The solution is $0.05\,M$ with respect to tris(hydroxymethyl)aminomethane and $0.0268\,M$ with respect to HCl; pH $= 8.2$ at $20°$

Cofactor solution: 19.2 mg of glutathione is dissolved in 0.5 ml of 2.8% (w/v) Tris in water and 0.5 ml of $0.0625\,M$ MgSO$_4$ added

Isopentenyl-PP-1-^{14}C: 2.2 mg of the lithium salt is dissolved in water and diluted to 1 ml; 30 μl of the solution contains 0.25 micromole of isopentenyl-PP (approximately 1400 dps)

EDTA: $0.2\,M$ ethylenediaminetetraacetic acid (sodium salt) adjusted to pH 8.0 with NaOH

Trichloroacetic acid: 1% (w/v) solution in pure toluene

Acetic acid–ethanol: 5% v/v acetic acid in ethanol

Preparation of Washed Rubber Particles. All operations are carried out at 0–5°. Latex is obtained from *Hevea brasiliensis* trees in commercial production, by tapping into clean vessels cooled in ice. The latex is centrifuged at 2000 g for 10 minutes to remove the sedimentable non-rubber bottom fraction particles.[6] Then 60 ml of the upper fraction is diluted with 30 ml of Tris-HCl buffer and centrifuged at 7000 g for 1

---

[5] B. L. Archer, *Proc. Nat. Rubber Producers' Res. Assoc. Jubilee Conf. Cambridge, 1964* p. 101. Maclaren, London, 1965.

[6] L. N. S. Homans, J. W. van Dalfsen, and G. E. van Gils, *Nature* **161**, 177 (1948).

hour. The lower layer containing the finer particles is separated and recentrifuged at 150,000 g for 1 hour in a swing-out rotor (e.g., Spinco SW 39L). The layer of rubber particles is carefully redispersed in Tris-buffer and diluted to 60 ml. The centrifugation and redispersion are repeated twice more, any coagulated rubber being removed at each stage. The final volume of rubber dispersion should be about 10 ml, and it should contain about 1 g of rubber. It is stable for up to 2 days at 0°.

Assay Procedure. A number of 100-mg samples of the suspension of washed rubber are weighed into $2 \times \frac{1}{2}$-inch rimless tubes; cofactor solution (20 μl) is added, followed by the enzyme preparation to be assayed. Each preparation is used in a range of concentrations varying by a factor of about fifty. The mixtures are each diluted to 220 μl with 0.05 M Tris buffer, and 30 μl of IPP-^{14}C solution is added. Incubations are carried out for 1 hour at 30° and terminated by the addition of 0.25 ml of 0.2 M EDTA. The contents of the tubes are dried in a stream of air or oxygen at 70° while the tubes are rotated at about 100 rpm with their axes horizontal. When dry, the rubber in the tubes is immersed in acetic acid–ethanol for 1 hour in order to produce a completely nondispersible film which is then washed several times with water. The rubber films are then digested at 100° for 1 hour in N KOH in order to remove hydrolyzable impurities, washed with water and extracted with boiling ethanol continuously for 16 hours to ensure removal of low molecular weight isoprenoid contaminants. After superficial drying, each sample of rubber is dissolved in 2 ml of trichloroacetic acid solution in toluene, scintillator is added to give a total volume of 6 ml, and the solutions are counted by a standard scintillation technique. The above purification procedure has been checked by ozonolysis of the rubber and recrystallization of the resulting levulinic acid as the dinitrophenylhydrazone. No retention of nonrubber isoprenoid contaminants could be detected. After correction for the residual biosynthetic activity of the washed rubber in the absence of added enzyme, the reciprocal of the rate of incorporation of ^{14}C into the rubber is plotted against the reciprocal of the protein concentration for each of the enzyme preparations and their relative purities are determined from the values of $1/A$ [Eq. (1)]. Protein concentrations are determined by the Folin-Ciocalteau method or by ultraviolet absorptiometry at 260 and 280 mμ.[7] Provided the incorporation of ^{14}C into rubber does not exceed 30%, the uptake *vs* time graphs are linear for the first hour. As the observed enzymatic activities are dependent on the sample of washed rubber used, it is not possible to define absolute units of rubber transferase activity. The method appears to be satisfactory for both unpurified and partially purified samples of the transferase enzyme.

[7] E. Layne, see Vol. III [73].

Enzyme Purification

All operations are carried out at 0–4°.

Step 1. Preparation of Serum. Latex is collected, and the bottom fraction is removed as outlined above. The upper fraction is then centrifuged at 60,000 g to remove most of the rubber, and the resulting, almost clear, aqueous serum is freeze-dried. Then 100 g of the dried solids (ex 1600 ml of serum) is dissolved in oxygen-free water, diluted to 300 ml, and centrifuged at 105,000 g for 90 minutes to remove most of the remaining rubber.

Step 2. Acidification. The pH of the solution is lowered to 5.4 by the gradual addition of approximately 35 ml of 0.83 M H_3PO_4. The precipitate of inactive protein is centrifuged off, and the supernatant is rapidly neutralized to pH 7.0 with 0.1 N Tris.

Step 3. Ammonium Sulfate Fractionation. Saturated ammonium sulfate containing 0.01 M magnesium sulfate and adjusted to pH 7.0 is then added stepwise to give fractions precipitating at 25, 50, and 60% of saturation. The mixtures are each left to stand for 30 minutes before centrifugation. The first fraction normally retains the last of the very small rubber particles and leaves a crystal clear supernatant. The last fraction precipitated between 50 and 60% of saturation contains the active enzyme, which, after washing with 65% saturated ammonium sulfate solution, is dissolved in 15 ml of 0.01 M magnesium sulfate and dialyzed overnight against a large excess of 0.2 M Tris-HCl buffer at pH 8.2 containing 0.001 M MgSO$_4$.

Step 4. Gel Filtration Chromatography on Sephadex G-100. The solution containing the enzyme is applied to a 140 × 2 cm diameter Sephadex G-100 column which has been equilibrated at 0–4° with 0.2 M Tris-HCl buffer containing 0.001 M MgSO$_4$, and the enzyme eluted with the same solvent. The process is followed by ultraviolet spectrophotometry and the first (inactive) protein peak appearing at 135 ml is discarded. The second peak at 185 ml is collected, concentrated by freeze-drying, and then dialyzed against 0.05 M phosphate buffer pH 8.0 containing 0.001 M MgSO$_4$ and 0.005 M cysteine.

Step 5. Chromatography on DEAE-Sephadex A-50. The resulting solution from the above dialysis is applied to a 20 × 2 cm diameter column of the anion exchanger which has been equilibrated against the above 0.05 M phosphate buffer. After the application of the enzyme the column is washed with 145 ml of the same buffer and then eluted with a solution which is 0.1 M in NaCl, 0.05 M in phosphate buffer pH 8.0, 0.001 M in MgSO$_4$, and 0.005 M in cysteine. The active fraction is obtained between 100 and 260 ml after the change to the sodium chloride solution.

Properties

After the above procedure the enzyme has been purified about 350 times, based on the weight of protein present, but the product is still contaminated with dimethylallyltransferase (EC 2.5.1.1).

Stability. Preparations of rubber transferase are generally unstable when purified. The enzyme is partially denatured within 10 minutes at 40° and is completely inactivated at 60°.

Molecular Weight. An approximate value of 60,000 for the molecular weight of rubber transferase has been obtained using the gel-filtration technique of Whitaker.[8]

pH Optimum. The incorporation of isopentenyl-^{14}C-PP proceeds maximally in the pH range 6.8–7.5.

Cofactors. Magnesium and sulfhydryl compounds are essential for full activity of the rubber transferase enzyme. Maximum incorporation rates have been observed at 5 mM for both compounds. Excess glutathione is not inhibitory. Magnesium cannot be replaced by manganese, cobalt, or iron. Ascorbic acid is not a cofactor.

Inhibitors. The enzyme is strongly inhibited by EDTA (0.05 M), or by $5 \times 10^{-5} M$ *p*-chloromercuribenzoate, iodoacetamide, or *N*-ethylmaleimide provided endogenous sulfhydryl compounds are first removed. Farnesyl pyrophosphate also inhibits, but neryl pyrophosphate is acceleratory.

Stereochemical Specificity. Rubber transferase is strictly stereospecific and yields exclusively *cis*-polyisoprene from isopentenyl-PP.[9]

[8] J. R. Whitaker, *Anal. Chem.* 35, 1950 (1963).
[9] B. L. Archer, D. Barnard, E. G. Cockbain, J. W. Cornforth, Rita H. Cornforth, and G. Popják, *Proc. Roy. Soc.* B163, 519 (1966).

[17] Enzymatic Synthesis of (−)-Kaurene and Related Diterpenes

By CHARLES A. WEST and CHRISTEN D. UPPER

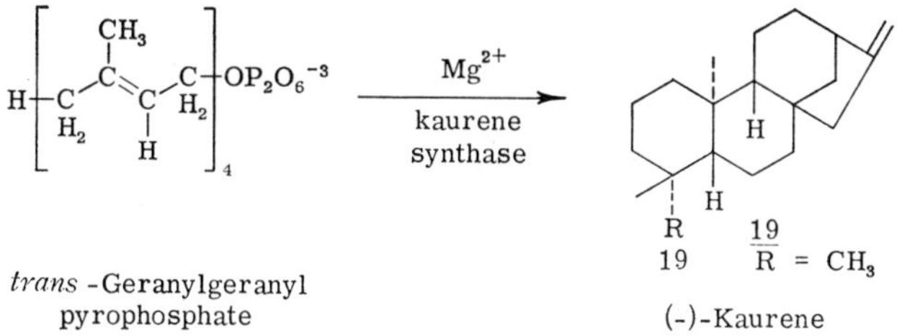

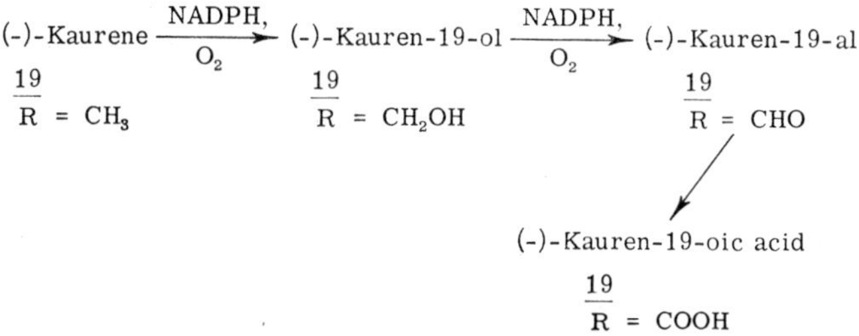

$$\left[H\!-\!\underset{H_2}{C}\!\overset{CH_3}{\underset{}{\overset{}{C}}}\!\!=\!\!\underset{H}{C}\!-\!\underset{H_2}{C}\!-\!OP_2O_6{}^{-3} \right]_4 \xrightarrow[\substack{\text{kaurene}\\\text{synthase}}]{Mg^{2+}}$$

trans-Geranylgeranyl
pyrophosphate

R
19
19   R = CH₃ → $\overline{\dfrac{19}{R}} = CH_3$

(-)-Kaurene

$$(\text{-})\text{-Kaurene} \xrightarrow[O_2]{\text{NADPH,}} (\text{-})\text{-Kauren-19-ol} \xrightarrow[O_2]{\text{NADPH,}} (\text{-})\text{-Kauren-19-al}$$

$\dfrac{19}{R} = CH_3$ $\qquad$ $\dfrac{19}{R} = CH_2OH$ $\qquad$ $\dfrac{19}{R} = CHO$

(-)-Kauren-19-oic acid

$\dfrac{19}{R} = COOH$

(−)-Kaurene is one of a relatively large group of cyclic diterpenes that occur naturally. It has been isolated from higher plants[1] and culture filtrates of the fungus *Fusarium moniliforme* Sheld.[2] Interest recently has been focused on this particular diterpene and the related metabolites shown above because of their intermediary role in the biosynthesis of the plant growth-regulating gibberellins.[3] This article summarizes some of the characteristics of cell-free enzymatic systems from higher plants which catalyze the conversion of *trans*-geranylgeranyl pyrophosphate to (−)-kaurene (kaurene synthase) and the further oxidation of (−)-kaurene to (−)-kauren-19-oic acid.

Kaurene Synthase

Assay Methods

trans-Geranylgeranyl Pyrophosphate as Substrate[4]

Principle. The (−)-kaurene-¹⁴C synthesized from *trans*-geranylgeranyl-2-¹⁴C pyrophosphate in the presence of enzyme and MgCl₂ is extracted and measured by radioassay after chromatographic purification.

Reagents

trans-Geranylgeranyl-2-^{14}C pyrophosphate, 0.70 mM (approximately 2.5×10^5 cpm/micromole)

$MgCl_2$, 25 mM

Tris-glutarate, 1 M, pH 6.5

Procedure. At least two different levels of the enzyme solution are tested by the following procedure. Enzyme solution (0.01–0.20 ml), 0.010 ml $MgCl_2$, 0.010 ml Tris-glutarate, and sufficient water to make the total volume 0.25 ml after addition of substrate (below) are mixed (a conical centrifuge tube is convenient for these small volumes) and placed in a 30° water bath for 3–5 minutes to allow thermal equilibration. The reaction is initiated by the addition of 0.030 ml of *trans*-geranylgeranyl-2-^{14}C pyrophosphate (approximately 5×10^3 cpm) and allowed to proceed at 30° for 30 minutes. The reaction is stopped by heating the tube in a boiling water bath for 4 minutes. The precipitate is separated by centrifugation and extracted twice with 0.25-ml portions of acetone. The combined acetone extracts and aqueous supernatant fraction are extracted twice with 1 ml portions of benzene and the combined benzene extracts are washed with 0.5 ml water. The organic phase is concentrated in a stream of nitrogen to a small volume of residual solution which is applied to the origin of a silica gel G thin-layer chromatographic plate. The plate is developed to 9 cm with *n*-hexane (see the table for details of chromatographic separations). The region between 5 and 9 cm from the origin (or just the kaurene band which is revealed with I_2 vapors if carrier kaurene is cochromatographed with the extract) is subjected to radioassay. It is most convenient to scrape the silica gel region to be assayed directly into a toluene solution of scintillator [e.g., 40 mg of diphenyloxazole (PPO) and 0.50 mg *p*-bis[2′-(5-phenyloxazolyl)]-benzene (POPOP) in 10 ml of toluene] for assay by liquid scintillation spectrometry. The silica gel does not change the counting efficiency for kaurene-[14]C.

Definition of Unit and Specific Activity. The amount of kaurene formed up to 3 nanomoles per 30 minutes of incubation is a linear function of the amount of enzyme present; however, this straight line does not always pass through the origin in a plot of kaurene formed *vs* enzyme

---

[1] J. H. Hosking, *Rec. Trav. Chim.* **47**, 578 (1928); L. H. Briggs and R. W. Cawley, *J. Chem. Soc.* p. 1888 (1948).

[2] B. E. Cross, R. H. B. Galt, J. R. Hanson, P. J. Curtis, J. F. Grove, and A. Morrison, *J. Chem. Soc.* p. 2937 (1963).

[3] J. E. Graebe, D. T. Dennis, C. D. Upper, and C. A. West, *J. Biol. Chem.* **240**, 1847 (1965).

[4] C. D. Upper and C. A. West, *J. Biol. Chem.* **242**, 3285 (1967).

concentration. Thus, the units of enzyme activity are calculated from the slope of the line, or as the difference in the amount of kaurene formed in the two tubes of a two-point assay divided by the difference in enzyme concentration between these two tubes. A unit of activity is defined as the amount of enzyme that will catalyze the formation of 1 nanomole of kaurene in 1 hour in this system. The specific activity is expressed in terms of the number of units of enzyme activity per milligram of protein.

Identification of (−)-*Kaurene as the Product Formed.* As indicated in the table, the triterpene hydrocarbon squalene and the C_{40} hydrocarbon lycopersene and the more unsaturated carotenes such as β-carotene are well resolved from kaurene by thin-layer chromatography in system 1 as used in the assay described. However, many diterpene hydrocarbons such as kaurane, isokaurene, pimaradiene, trachylobane, stachene, and phyllocladene are not well resolved from kaurene in this system. There-fore, it is necessary to apply additional criteria of identification of kaurene as the product in a given system under investigation, at least initially. Thin-layer chromatography in system 2 (see the table) on 3% $AgNO_3$–silica gel G resolves most of the diterpenes tested. Only the stereoisomers phyllocladene and kaurene of the various closely related diterpenes tested could not be distinguished in this system. Thus, careful cochromatography of the ^{14}C-labeled lipid eluted from the region where kaurene runs in system 1 mixed with authentic kaurene[5] should be per-formed as a further test of the presence of kaurene as a product. If a mixture of products including kaurene is formed, thin-layer chromatog-raphy in system 2 can be added as an additional step in the assay. Cocrystallization from ethanol of suspected kaurene-^{14}C with authentic kaurene to constant specific radioactivity was employed in the identifica-tion of kaurene as the product in *Echinocystis macrocarpa.*[3] The con-version of kaurene-^{14}C to gibberellic acid-^{14}C was demonstrated in cul-tures of *Gibberella fujikuroi* in this case also.[3]

Mevalonate as Substrate[3]

Principle. A procedure which utilizes mevalonate-2-^{14}C as substrate is more convenient in screening for the kaurene-synthesizing activity of plant extracts. The formation of kaurene depends on the presence in the extract of mevalonate kinase, phosphomevalonate kinase, pyrophospho-

---

[5] Authentic (−)-kaurene has been isolated from *Gibberella fujikuroi* cultures by Cross *et al.*[2] and from the leaf oils of the New Zealand kauri (*Agathis australis* Salisb.) and *Podocarpus macrophyllus* Don [see L. H. Briggs, B. F. Cain, R. C. Cambie, B. R. Davis, P. S. Rutledge, and J. K. Wilmshurst, *J. Chem. Soc.* p. 1345 (1963) for a discussion of the isolation of (−)-kaurene and closely related com-pounds from plant sources]. (±)-Kaurene also has been synthesized [R. A. Bell, R. E. Ireland, and R. A. Partyka, *J. Org. Chem.* **27**, 3741 (1962)].

MOBILITIES OF KAURENE AND RELATED MATERIALS IN THIN-LAYER CHROMATOGRAPHY[a]

Compound	System 1 R_K	System 2 R_K	System 3 R_G
Diterpene hydrocarbons			
Kaurane	1.18	1.22	—
Trachylobane	1.18	1.22	—
Kaurene	(1.00)	(1.00)	—
Phyllocladene	—	1.00	—
Pimaradiene	—	0.96	—
Atiserene	—	0.91	—
Isopimaradiene	—	0.88	—
Sandaracopimaradiene	1.00	0.79	—
Isophyllocladene	—	0.69	—
Isokaurene	1.00	0.41	—
Stachene	1.00	0.40	—
Isoatiserene	—	0.31	—
Other isoprenoid hydrocarbons			
Squalene	0.39	—	—
Lycopersene	0.17	—	—
β-Carotene	0.11	—	—
Other diterpene derivatives			
Kaurenal	—	—	2.16
Kaurenol	—	—	1.12
Geranylgeraniol	—	—	(1.00)
Kaurenoic acid	—	—	0.40

[a] Plates are prepared by spreading a well-dispersed slurry of 1 part of adsorbent in 2 parts of water (or a 1.5% solution of $AgNO_3$ where indicated) to a thickness of 0.25 mm and are activated by heating in an oven at 110–120° for at least 30 minutes. Reference compounds are detected on developed silica gel G plates by one of the following procedures: (a) exposure to I_2 vapor, (b) phosphomolybdate spray and heating, (c) 2',7'-dichlorofluorescein spray, or (d) exposure of developed plates containing 0.1% sodium fluoresceinate in the adsorbent to bromine vapor. Detection methods (b) and (c) are also suitable for $AgNO_3$-silica gel G plates.

R_K: mobility relative to kaurene as a reference standard. The R_f varies from 0.70 to 0.90 in system 1 and from 0.55 to 0.70 in system 2 depending on such variables as the degree of saturation of the developing chamber with the developing solvent. R_G: mobility relative to geranylgeraniol as a reference standard. The R_f of geranylgeraniol in system 3 varies from 0.40 to 0.45 depending on such variables as the degree of saturation of the developing chamber with the developing solvent. System 1: Silica gel G developed with n-hexane. System 2: 3% $AgNO_3$ in silica gel G developed with n-hexane–benzene 7:3. System 3: Silica gel G developed with benzene–ethyl acetate 9:1.

mevalonate decarboxylase, isopentenyl pyrophosphate isomerase, and a dimethylallyl transferase catalyzing geranylgeranyl pyrophosphate synthesis as well as kaurene synthase. These required activities have been found together in extracts of some plants in the soluble fraction remaining after centrifugation at 105,000 g for an hour. DL-Mevalonate-2-^{14}C is incubated with the extract in the presence of ATP and $MgCl_2$. The kaurene-^{14}C formed is extracted from the incubation mixture and measured by radioassay after chromatographic purification.

Reagents

ATP, 20 mM, neutralized to pH 7

$MgCl_2$, 20 mM

Mevalonic acid-2-^{14}C, N,N'-dibenzylethylenediamine salt (approximately 3 μC/μmole), 0.20 mM

Procedure. To each of two tubes are added 0.050 ml of ATP, 0.050 ml of $MgCl_2$, 0.10 ml of mevalonate-2-^{14}C and sufficient water so that the final volume in each tube will be 1.0 ml after addition of the enzyme as described below. (For some extracts it may be necessary to add Tris buffer to bring the final pH after addition of the extract to the desired value of 7.) The mixture is preincubated at 30° for 3–5 minutes, and the reaction is started by addition to each tube of 0.30 to 0.80 ml of enzyme solution which has been preequilibrated at 30°. Five-tenths milliliter of acetone is added immediately to one tube, which serves as the zero time control. The other tube is incubated for 1 hour at 30° after which time 0.50 ml of acetone is added to stop the reaction. The aqueous acetone mixture in each tube is extracted 3 times with 2-ml portions of benzene. The combined benzene extracts are washed with 1 ml of water and then are assayed for kaurene-^{14}C as described in the kaurene synthase assay. The total radioactivity found in the kaurene region from the incubation mixture is corrected for any radioactivity found in that region in the zero time control. The additional criteria described above for the identification of the product as kaurene should be applied here, at least initially.

Preparations

Kaurene Synthase from Echinocystis macrocarpa Endosperm.[4] *Echinocystis macrocarpa* Greene (also classified as *Marah macrocarpa* Greene and referred to more commonly as wild cucumber) grows wild in the coastal mountain sage, scrub, and chapparal areas of Southern California. Fruit can be obtained in the general period of January through April and the fruit or seed can be stored frozen for long periods without appreciable loss of kaurene synthase activity. These plants have apparently not been cultivated and fruit are not available from any commercial source. The endosperm from immature seeds of this species was originally chosen for study because of its relatively high content of gibberellin-like substances.[6]

[6] M. R. Corcoran and B. O. Phinney, *Physiol. Plantarum* **15**, 252 (1962).

Closely related species can also be found in other parts of the United States and Canada (e.g., *E. lobata* in the eastern and central United States and Canada and *E. oreganus* along the northern Pacific coast). However, the enzymatic properties of seed extracts from these species have not been examined.

Seeds of immature fruit of *E. macrocarpa* in which the embryo and cotyledons do not completely surround the endosperm are cut open and the endosperm is removed with a spatula, leaving behind as completely as possible the embryo and any nucellus present. The pooled endosperm (10 ml from approximately 20 seeds) is gently homogenized by hand in a glass homogenizer and passed through glass wool to remove cell debris. This extract is centrifuged for 1–2 hours at 105,000 g at 0°. All the kaurene-oxidizing activity and almost all the phosphatase activity is sedimented under these conditions, whereas the supernatant solution contains the kaurene synthase activity.

The endosperm homogenate, which has a protein concentration of 1–2 mg/ml, shows a specific activity of 50–100 units of kaurene synthase activity per milligram of protein. The 105,000 g supernatant fraction has up to twice the specific activity of the crude extract with almost 100% of the total activity of the crude extract. No extensive purification of the activity has been possible to date because of the greatly increased lability of the activity after steps such as ammonium sulfate fractionation or Sephadex G-100 or DEAE-Sephadex A-50 chromatography.

Crude Extract from Castor Bean Seedlings Capable of Converting Mevalonate-^{14}C to Kaurene and Other Diterpenes.[7] Seeds of castor bean (*Ricinus communis*) (Baker Hybrid 66) from which the seed coats have been removed are germinated in darkness for 60–72 hours at 30–32°. Seedlings (minus roots) are mixed with 0.050 M Tris-bicarbonate buffer adjusted to pH 7.3 and containing 0.01 M 2-mercaptoethanol in a ratio of 1.5 ml buffer per gram fresh weight and "Polyclar AT" (water-insoluble polyvinylpyrrolidone supplied by General Aniline and Film Corp., New York) in a ratio of 0.25 g polymer per gram fresh weight. This mixture is homogenized (Virtis "23" homogenizer for 2 minutes at three-quarters maximum speed) and the homogenate is squeezed through 4 layers of cheesecloth yielding a filtrate of pH 6.7–6.8. This filtrate is centrifuged at 105,000 g for 60 minutes at 0°. One milliliter of the supernatant solution from this centrifugation contains 2.5–3 mg of protein and is capable of converting 5–10 nanomoles of mevalonate to diterpene hydrocarbons, including kaurene, per hour when incubated with 1 mM $MgCl_2$, 1 mM $MnCl_2$, 2 mM ATP, and 0.05 mM DL-mevalonic acid-2-^{14}C, N,N'-dibenzylethylenediamine salt, at 30°.

[7] D. Robinson and C. A. West, to be published.

Substrate. trans-Geranylgeranyl-2-^{14}C pyrophosphate has been synthesized chemically.[4] Farnesylacetone is condensed with the carbanion generated from methyl diethyl phosphonoacetate-2-^{14}C by the method of Wadsworth and Emmons.[8] The methyl *trans*-geranylgeranoate-2-^{14}C is resolved from methyl *cis*-geranylgeranoate and unreacted ketone by alumina column chromatography, and reduced with LiAlH$_4$ to *trans*-geranylgeraniol-2-^{14}C. This alcohol is phosphorylated by the method of Cramer and Böhm,[9] and the pyrophosphate ester is separated from the resulting mixture by either countercurrent distribution or DEAE-cellulose column chromatography. *trans*-Geranylgeranyl-^{14}C pyrophosphate can also be purified from incubation mixtures of *Echinocystis macrocarpa* endosperm with mevalonate-2-^{14}C, ATP, and MgCl$_2$.[10] It has also been reported as a biosynthetic product of *Micrococcus lysodeikticus*,[11] yeast,[12] carrot,[13] and pig liver.[13]

Properties

Occurrence. Kaurene synthase activity has been observed in *E. macrocarpa* endosperm in germinating castor bean seedlings, and in the fungus *F. moniliforme* Sheld, which produces gibberellins. Apparent activity is present also in extracts of normal and dwarf maize (*Zea mays*) seedlings, although the identification of kaurene as the product in this case was not as rigorous. If kaurene is an intermediate in gibberellin synthesis and gibberellins are ubiquitous constituents of flowering plants as believed, then kaurene synthase must also be of very widespread occurrence. The assay for kaurene synthase may prove to be a useful screening procedure for estimating the occurrence and development of gibberellin-synthesizing activity in various plants and plant tissues.

In the *E. macrocarpa* endosperm preparations only geranylgeraniol, kaurene, and metabolites formed from kaurene are observed as benzene-extractable products of mevalonate and geranylgeranyl pyrophosphate metabolism. However, in extracts of germinating castor bean seedlings, squalene and at least four other diterpenoid hydrocarbons are formed from mevalonate in addition to kaurene. This points to the necessity of careful identification of the product as kaurene before routine application of the assay described.

[8] W. S. Wadsworth, Jr., and W. D. Emmons, *J. Am. Chem. Soc.* **226**, 921 (1957); see also this volume, p. 385.
[9] F. Cramer and W. Böhm, *Angew. Chem.* **71**, 775 (1959).
[10] M. O. Oster and C. A. West, *Arch. Biochem. Biophys.* **127**, 112 (1968).
[11] A. A. Kandutsch, H. Paulus, E. Levin, and K. Bloch, *J. Biol. Chem.* **239**, 2507 (1964).
[12] K. Kirschner, dissertation, University of Munich, Munich, Germany, 1961.
[13] D. L. Nandi and J. W. Porter, *Arch. Biochem. Biophys.* **105**, 7 (1964).

Stability. The kaurene synthase activity of *E. macrocarpa* endosperm disappears only slowly over a period of months when stored at −20° as intact or lyophilized endosperm, but inactivation occurs in a matter of hours once the endosperm has been subjected to purification procedures. Efforts to stabilize the activity have not been successful to date, and this has precluded extensive purification.

Activators and Inhibitors. The kaurene synthase of *E. macrocarpa* is inactivated on dialysis or by treatment with EDTA. Activity is partially restored by addition of 1–5 mM $MgCl_2$ or 0.2–0.3 mM $CoCl_2$, but either not very effectively or not at all by $ZnCl_2$, $CdCl_2$, $FeCl_2$, $CuCl_2$, and $CaCl_2$. Neither ascorbate, 2-mercaptoethanol, nor 1,4-dithioerythritol affects the activity. No evidence for other cofactors has been found. The reaction proceeds equally well under aerobic or anaerobic conditions.

The reaction is almost completely inhibited by 2′-isopropyl-4′-(trimethylammonium   chloride)-5′-methylphenyl   piperidine-1-carboxylate (Amo 1618), tributyl-2,4-dichlorobenzylammonium chloride (Phosfon S), and tributyl-2,4-dichlorobenzylphosphonium chloride (Phosfon) at concentrations of 100 μg of inhibitor per milliliter, and to lesser extents at lower inhibitor concentrations.[14] It is postulated that the *in vivo* action of these and related plant growth retardants may be due, at least in part, to the inhibition of gibberellin synthesis at this site. *o*-Phenanthroline at 0.1 mM is also a potent inhibitor of kaurene synthase activity.

pH Optimum. Kaurene synthase activity shows a relatively sharp pH optimum at 6.55.

Kinetic Properties. No evidence has been found to date for a requirement of more than one protein fraction in the conversion of geranylgeranyl pyrophosphate to kaurene, although purification work has not been very extensive. Careful measurements of the kinetic parameters of kaurene synthase also await more extensive purification of the enzyme.

Conversion of Kaurene to Kauren-19-ol, Kauren-19-al, and Kauren-19-oic Acid

Assay Method[15]

The aerobic incubation of suspensions of kaurene-^{14}C with a microsomal pellet prepared from the endosperm of *Echinocystis macrocarpa* seed (or with the whole endosperm homogenate) in the presence of NADPH leads to the formation of a mixture of kauren-19-ol-^{14}C, kauren-19-al-^{14}C, and kauren-19-oic acid-^{14}C and some other labeled, but un-

[14] D. T. Dennis, C. D. Upper, and C. A. West, *Plant Physiol.* **40**, 948 (1965).
[15] D. T. Dennis and C. A. West, *J. Biol. Chem.* **242**, 3293 (1967).

identified, acids. The reactions have been demonstrated to occur in the sequence kaurene to kauren-19-ol to kauren-19-al to kauren-19-oic acid to unidentified acids, but the enzymes responsible for the individual steps have not been fractionated and characterized. A typical experiment to demonstrate these interconversions is described below.

Mixed in a 25-ml Erlenmeyer flask are the following: 0.50 ml of a microsomal suspension in 0.1 M potassium phosphate buffer, pH 7.0, 0.10 ml of 10 mM MgCl$_2$, and 0.10 ml of 1 mM NADPH made up to a total volume of 1.00 ml with water. Kaurene-[14]C (1.27 nanomole; approximately 17,500 dpm) dissolved in 0.10 ml of ethanol is added to start the reaction. This mixture is incubated under air with shaking for 30 minutes in a water bath at 30°. The reaction is stopped by heating the tube for 3 minutes in a boiling water bath. The precipitate is separated by centrifugation and extracted three times with a total volume of 2 ml of acetone. The combined acetone extracts and aqueous supernatant fraction is extracted three times with 1-ml portions of benzene. The combined benzene extracts are evaporated in a stream of nitrogen to a small volume of residual solution which is transferred to the origin of a silica gel G thin layer plate. The plate is developed to 15 cm with n-hexane, a mark is made across the plate 10 cm from the origin, and the plate is then redeveloped in the same direction with benzene–ethyl acetate (9:1, v/v) to this mark. The gel from indicated regions is scraped from the plate as follows for radioassay: 10–15 cm for kaurene-^{14}C, 8–10 cm for kauren-19-al-^{14}C, 4–8 cm for kauren-19-ol-^{14}C, and origin to 4 cm for the mixture of kauren-19-oic acid-^{14}C and labeled, unidentified acids. Radioassay of these fractions is most conveniently performed by liquid scintillation spectrometry of the gel scraped directly into toluene solution of a scintillator. A scan of the plate for radioactive zones before removal of the samples for radioassay is useful. Careful chromatography and location of radioactive zones is required for radioassay of kauren-19-oic acid-^{14}C and the unidentified, labeled acid fraction as separate entities.

Preparations

A microsomal pellet prepared from *E. macrocarpa* endosperm by differential centrifugation catalyzes the oxidation of kaurene. One volume of endosperm is mixed with 2 volumes of 0.25 M sucrose in a glass homogenizer by hand. This mixture is centrifuged at 10,000 g for 20 minutes at 0°, and the supernatant fraction is then centrifuged again at 105,000 g for 60 minutes at 0°. The pellet obtained from the latter centrifugation is resuspended in a volume of 0.1 M potassium phosphate buffer, pH 7.0, equivalent to the original volume of endosperm taken.

Kaurene-^{14}C prepared biosynthetically as a product of mevalonate-

2-^{14}C metabolism in the endosperm serves as the substrate. Kaurene-17-^{14}C can also be prepared chemically from unlabeled kaurene.[16]

Properties

These reactions have been studied only in preparations from *E. macrocarpa* endosperm.[15] All three activities are localized in the microsomal pellet. No resolution of fractions capable of catalyzing the individual reactions has been achieved. The oxidation of kaurene to kauren-19-ol and of kaurene-19-ol to kaurene-19-al both require air as well as a reduced pyridine nucleotide, with NADPH more effective than NADH. No one of the substances NADP+, NAD+, or FAD serves as a coenzyme under either aerobic or anaerobic conditions. The conversion of kauren-19-ol to kauren-19-al is inhibited by NADP+, *p*-chloromercuribenzoate, and β-diethylaminoethyl diphenylvalerate dihydrochloride (SKF 525-A). These properties suggest that both these reactions may be catalyzed by mixed function oxidases. Similar studies for the kauren-19-al to kauren-19-oic acid conversion have not been made to date.

[16] B. E. Cross, R. H. B. Galt, J. R. Hanson, P. J. Curtis, J. F. Grove, and A. Morrison, *J. Chem. Soc.* p. 295 (1964).

[18] Geranylgeranyl Pyrophosphate Synthetase of *Micrococcus lysodeikticus*

By ANDREW A. KANDUTSCH

$$\text{Dimethylallyl-PP} + \text{isopentenyl-PP} \xrightarrow{\text{Mg}^{++}} \text{geranyl-PP}$$

$$\text{Geranyl-PP} + \text{isopentenyl-PP} \xrightarrow{\text{Mg}^{++}} \text{farnesyl-PP}$$

$$\text{Farnesyl-PP} + \text{isopentenyl-PP} \xrightarrow{\text{Mg}^{++}} \text{geranylgeranyl-PP}$$

Purified geranylgeranyl-PP synthetase preparations obtained thus far catalyze reactions between isopentenyl-PP and C_5, C_{10}, and C_{15} allyl pyrophosphates. It is, however, uncertain whether all three reactions are catalyzed by a single enzyme or by two or three different enzymes.[1]

Assay Method

Principle. The assay method determines the extent to which ^{14}C-labeled isopentenyl-PP reacts with C_5, C_{10}, and C_{15} allyl pyrophosphates

[1] A. A. Kandutsch, H. Paulus, E. Levin, and K. Bloch, *J. Biol. Chem.* **239**, 2507 (1964).

to form ^{14}C-labeled compounds which, after acid hydrolysis, are extractable into petroleum ether. Isopentenyl-PP is resistant to treatment with acid whereas allyl pyrophosphates yield a mixture of products soluble in nonpolar solvents and consisting of the corresponding primary alcohols, as well as isomeric alcohols and degradation products (hydrocarbons). Geranylgeranyl-PP is the major product of the reaction regardless of the chain length of the allyl pyrophosphate substrate. However, ^{14}C-labeled reaction products found after treatment with bacterial phosphatase include minor amounts of geranyllinalool and nonpolar compounds that may contain more than 20 carbon atoms. Geranylgeranyl-PP, formed by the reaction of isopentenyl-PP-^{14}C with dimethylallyl-PP incorporates 3 ^{14}C atoms per molecule; with geranyl-PP, 2 ^{14}C atoms per molecule; and with farnesyl-PP, 1 ^{14}C atom per molecule. Measurement of the extent of the reaction between isopentenyl-PP and farnesyl-PP may be an adequate routine assay for the enzyme. However, the differential lability of enzyme activities toward C_5, C_{10}, and C_{15} allyl pyrophosphates makes advisable an occasional assay with each of these substrates.

Incorporation of isopentenyl-PP-^{14}C into the extractable fraction in the absence of any added allyl pyrophosphate indicates contamination of the enzyme preparation by isopentenyl-PP isomerase. This enzyme may be present in crude preparations, but it is absent from purified enzyme.

Reagents

Isopentenyl-PP-^{14}C. Isopentenyl-PP-1-^{14}C is synthesized from isopentenol-1-^{14}C by the method of Yuan and Bloch.[2] Methods for the biosynthesis of isopentenyl-PP-^{14}C from mevalonic acid-2-^{14}C and its purification are described in a paper by Nandi and Porter.[3]

Allyl pyrophosphates. Dimethylallyl alcohol, geraniol, and farnesol are pyrophosphorylated by the method described by Cramer, Rittersdorf, and Böhm.[4] Slight modifications of the procedure have been described.[1,5,6]

Solutions

Potassium phosphate buffer, 1 M, pH 7.4

MgCl$_2$, 0.1 M

[2] C. Yuan and K. Bloch, *J. Biol. Chem.* **234**, 2605 (1959).
[3] D. L. Nandi and J. W. Porter, *Arch. Biochem. Biophys.* **105**, 7 (1964).
[4] F. Cramer, W. Rittersdorf, and W. Böhm, *Ann. Chem.* **654**, 180 (1962).
[5] G. Popják, J. W. Cornforth, R. H. Cornforth, R. Ryhage, and D. S. Goodman, *J. Biol. Chem.* **237**, 56 (1962).
[6] C. R. Childs, Jr., and K. Bloch, *J. Biol. Chem.* **237**, 62 (1962).

Isopentenyl-PP-^{14}C, 0.001 M
Farnesyl-PP, 0.00016 M
Geranyl-PP, 0.00016 M
Dimethylallyl-PP, 0.001 M
Trichloroacetic acid, 50% (w/v)
KCl, 2 M
NaOH, 5 M
Skelly-solve B (petroleum ether, boiling range 60–70°), distilled
Enzyme in dilute phosphate or Tris buffer

Specific Activity. No unit of enzymatic activity has been defined. Specific activity has been defined in terms of the amount of isopentenyl-PP-^{14}C reacting with farnesyl-PP, to give ^{14}C-labeled compounds which, after treatment with acid, are soluble in petroleum ether. Since 30–47% of the ^{14}C-labeled products found after enzymatic hydrolysis of pyrophosphate esters were compounds other than geranylgeraniol, a more precise definition of activity does not seem justified.

Procedure. In a 10-ml test tube in an ice bath are placed 0.1 ml of phosphate buffer, and 0.05 ml each of MgCl$_2$, isopentenyl-PP-^{14}C, and allyl pyrophosphate (farnesyl-PP, geraniol-PP, or dimethylallyl-PP). Water and enzyme are added to give a final volume of 0.5 ml and the mixture is incubated for 20–30 minutes at 37°. The reaction is then stopped by the addition of 0.5 ml of 50% (w/v) trichloroacetic acid, and the products are hydrolyzed at 60° for 10 minutes. To the ice-cold solution, 0.3 ml of 5 N sodium hydroxide and 2.5 ml of 2 M potassium chloride are added (final pH, 8–9), and the mixtures are extracted with 10 ml of Skelly-solve B. An aliquot of the solvent layer is removed, evaporated to dryness, and counted in a liquid scintillation counter.

The relationship between enzyme concentration and the rate of the reaction may be nonlinear, and estimates of the specific activities of enzyme preparations should be made only after activity has been determined over a wide range of enzyme concentrations.

Purification Procedure

All steps are carried out at 2–4° unless otherwise indicated. Mechanical stirring is used during all additions.

1. *Preparation of Crude Extract.* Twenty grams of spray-dried *Micrococcus lysodeikticus* cells are suspended in 400 ml of 0.05 M potassium phosphate buffer, pH 7.4. Crystalline egg white lysozyme (400 mg), obtained from Armour Pharmaceutical Company, Chicago, is added to the suspension and the mixture is stirred manually for 15–20 minutes at room temperature. Crystalline deoxyribonuclease (20 mg), obtained from

Worthington Biochemical Corp., Freehold, New Jersey, is added to the resulting thick gel, and stirring is continued for 15 minutes at the same temperature. The viscosity of the mixture is further reduced by the addition of 120 ml of 0.05 M phosphate buffer, pH 7.4, the diluted mixture is centrifuged at 60,000 g for 1 hour, and the sediment is discarded.

2. *Ammonium Sulfate Precipitation.* Solid ammonium sulfate is added to the supernatant layer, and the fraction precipitating between 30 and 50% saturation is collected and dissolved in approximately 5 ml of 0.02 M Tris buffer pH 7.4. After the removal of sulfate by passage through a column of Sephadex G-50 (void volume, 15 ml) equilibrated with 0.02 M Tris buffer pH 7.4, the eluted protein is diluted to a concentration of 10 mg/ml with the same buffer and a second fractionation with ammonium sulfate is carried out. The fraction precipitating between 35 and 50% saturation with ammonium sulfate is collected, dissolved in a few milliliters of 0.01 M potassium phosphate buffer pH 6.4, and passed through a Sephadex G-50 column (void volume, 15 ml) with the same buffer.

3. *Chromatography on Hydroxylapatite.* An aliquot of the gel eluate containing approximately 200 mg of protein is chromatographed on a hydroxylapatite column, 1.7 × 14 cm. Elution of the enzyme from the column is accomplished with a linear potassium phosphate buffer gradient at pH 6.4. Two hundred milliliters of 0.01 M buffer is placed in the mixing flask and 0.06 M buffer is placed in the reservoir. Approximately 40% of the protein is eluted in forty 10-ml fractions. Protein concentrations in the collected fractions are determined from absorbancy measurements at 260 and 280 mμ.

4. *Ammonium Sulfate Precipitation.* Fractions containing the major part of the enzyme activity are pooled, and the protein precipitating between 35 and 45% saturation with ammonium sulfate is collected and

RECOVERIES OF PROTEIN AND ACTIVITY AT STAGES OF PURIFICATION

| | | Recovery | |
Fraction	Total protein (mg)	Protein (%)	Activity (%)
Crude extract	1750	—	100
Second (NH$_4$)$_2$SO$_4$ precipitate, 0.35–0.50 saturation	285	16.3	—
Hydroxylapatite column	62	3.5	—
Third (NH$_4$)$_2$SO$_4$ precipitate, 0.35–0.45 saturation	29	1.7	29[a]

[a] The increase in specific activity from crude extract to purified enzyme was approximately 20-fold (see text).

dissolved in a few milliliters of $0.05 M$ phosphate buffer pH 7.4. At this stage the purified enzyme is stable on storage for several months at $-20°$, but at room temperature approximately 50% of the activity with farnesyl-PP was lost after 24 hours.

The table shows the results of a purification procedure. The increase in specific activity of the purified preparation over that for the crude extract, estimated from the highest specific activities attained with the two preparations, was 26-fold when the allyl pyrophosphate substrate was dimethylallyl-PP, 18-fold when the substrate was geranyl-PP, and 20-fold when the substrate was farnesyl-PP. These differences in the degree of purification obtained as measured by reactions with the three allyl-pyrophosphate substrates were not considered significant.

Properties

Activators. Magnesium ions are an absolute requirement for the reaction between isopentenyl-PP and C_5, C_{10}, or C_{15} allyl-pyrophosphate. The apparent K_m value for Mg^{++} is approximately $2 \times 10^{-3} M$ with farnesyl-PP or geranyl-PP and 7×10^{-3} with dimethylallyl-PP.

Optimum pH. The reaction rates with farnesyl-PP or geranyl-PP as substrate are optimal over the pH range 6.5–8. The optimal pH range with dimethylallyl-PP as substrate is somewhat narrower, the highest rate of reaction occurring at pH 7.5.

Affinities for Substrates. Apparent K_m values found for dimethylallyl-PP, geranyl-PP, and farnesyl-PP were 2.9×10^{-5}, 3.1×10^{-6}, and $3.8 \times 10^{-6} M$, respectively. Since the farnesyl-PP preparation used to obtain a K_m value contained approximately 30% of the inactive *cis,trans* isomer, the K_m value for this substrate is probably one-third lower than the value found.

Physical and Chemical Properties. A purified enzyme preparation obtained by this procedure migrated as a single band in electrophoresis on cellulose acetate strips in barbital buffer at pH 8.6. It was excluded from Sephadex G-100 gel, but was eluted from a column of Sephadex G-200 gel as a single symmetrical band after passage through the column of approximately 2 void volumes of buffer. Sedimentation velocity analysis indicated the presence of a minor (4.7% of total protein) component heavier than the major component. The sedimentation coefficient $(S_{20,w})$ of the major component in $0.05 M$ phosphate buffer was 1.54.

Activities of the enzyme preparation with dimethylallyl-PP, geranyl-PP, and farnesyl-PP substrates are affected differentially by heat and by urea. Activity with farnesyl-PP is the most stable to these agents and activity with dimethylallyl-PP is the least stable.

[19] Enzymatic Cyclization of Squalene 2,3-Oxide

By Peter D. G. Dean

2,3-Oxidosqualene Cyclase of Hog Liver

The failure of attempts to isolate "squalene oxidocyclase" by several workers during the last decade[1,2] may be rationalized as a result of evidence that has recently come from two laboratories,[3,4] that squalene epoxide (2,3-oxidosqualene) is an intermediate in cholesterol biosynthesis. More recently, the enzyme which cyclizes 2,3-oxidosqualene has been solubilized and partially purified from hog liver in this laboratory.[5]

Extraction, Solubilization, and Partial Purification of 2,3-Oxidosqualene Cyclase from Hog Liver

Hog liver (4500 g), cooled in ice immediately after removal from the animal, is dissected and cartilaginous material is discarded. The liver is then minced and homogenized at 0° in a Gifford-Wood colloid mill with (11 liters) 0.1 M potassium phosphate buffer, pH 7.4. Nuclei plus cell debris and mitochondria are separately precipitated by centrifugation at 600 g and 12,000 g, respectively. The supernatant is centrifuged in a Spinco zonal (model L 4) centrifuge to collect microsomal material.

The microsomal sediment is scraped from the rotor and rapidly frozen in test tubes (in 10-g batches) by immersing the tubes in liquid nitrogen. The supernatant is discarded although it contains some activity—the amount depends on the extent of the original blending of the liver. The activity is not stable on storage and decreases rapidly on freezing and thawing.

The microsomal material (10 g, ca. 1 g protein) is suspended in 0.1 M phosphate buffer, pH 7.4 (50 ml) using a Potter-Elvehjem homogenizer. The suspension is treated with a cold solution of 3% (w/v, i.e., 0.077 M) sodium deoxycholate (DCA) (25 ml). The mixture is well dispersed using the same homogenizer over 15 minutes. After this time a cold solution of

[1] T. T. Tchen and K. Bloch, *J. Am. Chem. Soc.* **77**, 6085 (1955); T. T. Tchen and K. Bloch, *J. Biol. Chem.* **226**, 921 (1957).

[2] D. S. Goodman, *J. Biol. Chem.* **236**, 2429 (1961).

[3] E. J. Corey, W. E. Russey, and P. R. Ortiz de Montellano, *J. Am. Chem. Soc.* **88**, 4750 (1966).

[4] E. E. van Tamelen, J. D. Willett, R. B. Clayton, and K. E. Lord, *J. Am. Chem. Soc.* **88**, 4757 (1966).

[5] P. D. G. Dean, P. R. Ortiz de Montellano, K. Bloch, and E. J. Corey, *J. Biol. Chem.* **242**, 3014 (1967).

0.077 M calcium chloride (26 ml) is added and the precipitate of calcium deoxycholate is removed by centrifugation at 37,000 g for 20 minutes at 0°.

The exposure to deoxycholate should not be in excess of 15 minutes. The amount of deoxycholate is not so critical, provided that sufficient calcium ions are added; this is in contrast to the pea seedling enzyme (see below).

The supernatant obtained after the removal of calcium deoxycholate is centrifuged at 105,000 g for 6 hours and the upper 85% of the supernatant is very carefully removed from a layer of more viscous supernatant (plus the pellet). This upper supernatant is brought to 30% saturation (at 4°) with ammonium sulfate, and the precipitated protein is collected by centrifugation at 37,000 g for 20 minutes at 0°. The precipitate is dissolved in 0.1 M phosphate buffer (pH 7.4) (30 ml) and further centrifuged to remove undissolved protein. The supernatant is removed by pipette and is then cautiously extracted with cold peroxide-free ether (3 × 30 ml). The ether extracts contain approximately a weight of lipid equal to that of the protein as determined by a biuret reaction; the nucleic acid content at this stage is about 3%.

The extracted aqueous solution is freed from dissolved ether by gassing with nitrogen at 0°. The resulting solution is then fractionated with ammonium sulfate: The protein which precipitates between 10 and 20% contains much of the activity and may be stored[6] (but not frozen) until required. At this stage the recovery of protein is approximately 6% and the specific activity between 20 and 50 times greater than that of the microsomes; the specific activities are listed in Table I.

TABLE I
SUMMARY OF PURIFICATION PROCEDURE FOR 2,3-OXIDOSQUALENE
CYCLASE FROM HOG LIVER

Fraction	Protein (g)	Specific activity[a]	Total activity	Yield %
Microsomes	1.1	4.2	4,600	—
After DCA and calcium chloride	1.0	47.5	14,300	100
High-speed supernatant	0.36	132.0	9,400	66
0–30% (NH₄)₂SO₄	0.10	156	9,000	63
Ether extraction; 10–20% (NH₄)₂SO₄	0.06	250	8,700	61

[a] Millimicromoles per hour per milligram of protein.

Properties of the Enzyme

2,3-Oxidosqualene cyclase is unaffected by the addition of any of the following reagents (at $10^{-3} M$): EDTA, ascorbate, fluoride, iodoacet-

[6] The activity is stable for about 6 days in this state.

amide, glutathione, cyanide, nicotinamide, thiocyanate, azide, calcium chloride. N-Ethylmaleimide ($10^{-3}\,M$) inhibited the enzyme by 20%; the enzyme is also completely inactivated by organic solvents such as acetone, alcohols, etc., even at very low temperatures.

It may be shown that while the cyclizing enzyme is markedly inhibited by the imino analog of the substrate, squalene epoxidase (which converts squalene to the 2,3-epoxide) is hardly affected. This affords a means of assaying both enzymes independently. The sulfur analog of the substrate (for the cyclase) is a much less effective inhibitor ($K_m = K_i = 10^{-4}\,M$).

Formation of Lanosterol from 2,3-Oxidosqualene by a Soluble Enzyme from Hog Liver

Principle. The overall reaction is shown in Eq. (1).

$$\tag{1}$$

The assay is based on the ready separation of product from substrate by thin-layer chromatography.

Since it has not yet been possible to isolate the biologically reactive isomer of 2,3-oxidosqualene, the *dl*-mixture is used in the assay; the inert isomer serves as an internal control for recovery and for counting efficiency estimations.

Reagents. The substrate is prepared from squalene via the bromohydrin (see this volume, p. 346). An amount equivalent to 66 millimicromoles per incubation is emulsified with Tween 80 (Atlas Powder Co.) by evaporating to dryness the 2,3-oxidosqualene with a 4-fold excess (by weight) of Tween (20 μg). Buffer previously boiled and allowed to cool to 40° (0.5 ml per incubation) is then added and the mixture well agitated with a vortex mixer.

Silica gel H for thin-layer chromatography (Merck) is slurried in chloroform (20 g per 85 ml) and poured onto glass plates (20 × 5 cm). These are rapidly tilted to give an even layer and allowed to dry at room temperature. Plates made in this way are easier to scrape and have more uniform activities than those made from aqueous slurries (with subsequent baking), especially with the nonpolar solvent systems used in this assay.

Incubations are performed anaerobically with dialyzed enzymes, in order to reduce epoxidase activity as well as the activities of other oxidative enzymes.

Procedure. Incubations are carried out in a series of test tubes equipped with a side arm (connected to a manifold) so that each tube may be freed from air to the same extent. In each tube 0.2–1.0 ml of the enzyme solutions is placed, and the tubes are then sealed with septum caps. The tubes are evacuated and gassed with pure nitrogen (4–5 times) and allowed to reach 37° in a shaking bath.

The substrate, emulsified as described above, is injected into each tube and the tubes are shaken at 37° for 15 minutes.

Termination and Extraction of Radioactivity. The reaction is terminated by the addition of 1 N methanolic KOH (2 ml) and the tubes are shaken at 37° for 10 minutes.

The hydrolyzate is extracted three times with petroleum ether (3 ml) and once with petroleum ether–ether 1:1 (5 ml). Emulsions and gels may be broken with methanol and centrifugation at low speeds.

The extracts are passed through a column of anhydrous magnesium sulfate (5 × 0.5 cm). The dried extracts are evaporated in a stream of nitrogen to about 50 μl. The solution is applied to thin-layer plates (20 × 5 cm) of silica gel; each tube is washed at least 4–5 times. The washings are concentrated in the tube and applied to the same plate. In this way 80–95% of the radioactivity is chromatographed.

The plates are placed in a tank with 3% ethyl acetate-benzene at the bottom and removed when the solvent has moved 10 cm from the points of application.

The plates are allowed to dry at room temperature, and areas corresponding to 2,3-oxidosqualene (R_f, 0.45) are removed and placed in a counting vial. Scintillation fluid (phenyloxazole-phenyl in toluene, 4 g/l) (15 ml) is then poured into each vial and the samples are counted after 30 minutes (after this time most of the radioactivity is in the toluene).

The product may be further purified by carrier crystallization (methanol–benzene) with pure lanosterol or chromatographed in the gas–liquid system described by Corey, Russey, and Ortiz de Montellano.[3]

2,3-Oxidosqualene Cyclase in the Pea Seedling (*Pisum sativum*)

Mevalonic acid[7-9] and squalene[10,11] are both converted to tetra- and pentacyclic triterpenes in higher plants.

[7] D. J. Baisted, E. Capstack, and W. R. Nes, *Biochemistry* **1**, 537 (1962).
[8] H. H. Rees, E. I. Mercer, and T. W. Goodwin, *Biochem. J.* **96**, 30P (1965).
[9] R. Aexel, S. Evans, M. Kelley, and R. T. Nicholas, *Phytochemistry* **6**, 511 (1967).
[10] R. D. Bennett and E. Heftmann, *Phytochemistry* **4**, 475 (1965).
[11] E. Capstack, N. Rosin, G. A. Blondin, and W. R. Nes, *J. Biol. Chem.* **240**, 3258 (1965).

The cell-free preparation from pea seedling used by Capstack *et al.* (24 hours' germination)[11] accumulates radioactive β-amyrin from 1,5,9,16,20,24-^{14}C-labeled squalene (as an emulsion with Tween 20 and water). β-Sitosterol is not formed from squalene in this preparation.[11]

With the cell-free homogenate described by Capstack *et al.*,[11] it has been shown[12] that 2,3-oxidosqualene is converted to β-amyrin (approximately 35% yield) under anaerobic conditions.

The solubilization and partial purification of the 2,3-oxidosqualene cyclase from *Pisum sativum*[12] closely follows the general procedures used in this laboratory for the extraction of the lanosterol-producing enzyme from hog liver,[5] as described in the preceding section.

Extraction, Solubilization, and Partial Purification of 2,3-Oxidosqualene Cyclase from the Pea Seedling

The seeds (Burpee's blue bantam peas) (280 seeds weighing 76 g) are half immersed in water in petri dishes and allowed to germinate for 24 hours at room temperature.

The peas are then added to cold 0.1 M potassium phosphate buffer (pH 7.4) (200 ml) and homogenized in a Waring blendor for 90 seconds.[13] The homogenate is poured through cheesecloth to remove large particles and coagulated lipids. The filtrate is centrifuged at 20,000 g for 30 minutes at 0° to remove nuclear material, mitochondrial fractions, and unbroken cells. The supernatant (A) is decanted and further centrifuged at 105,000 g for 90 minutes at 2°. The supernatant is first freed of floating lipid by means of a Pasteur pipette attached to a water pump. The remaining supernatant (which is discarded) is cautiously removed from the microsomal pellet.

The latter is resuspended in cold 0.1 M phosphate buffer, pH 7.4 (25 ml), and the suspension is treated with a cold solution of 3% (w/v, i.e., 0.077 M) sodium deoxycholate (6.3 ml). The suspension is occasionally dispersed using a Potter-Elvehjem homogenizer over 12–15 minutes. After this time a cold solution of 0.077 M calcium chloride (6.4 ml) is added and the resulting suspension of calcium deoxycholate is removed by centrifugation at 37,000 g for 20 minutes at 0°.

The exposure to sodium deoxycholate, both in amount and time, is critical (overtreatment in either case leads to intractable gels).

The supernatant obtained after removal of calcium deoxycholate is brought to 30% saturation (at 5°) with ammonium sulfate and the suspension so formed is centrifuged at 37,000 g for 20 minutes at 0°.

The precipitate is taken up in 0.1 M phosphate buffer, pH 7.4 (30 ml)

[12] E. J. Corey, and P. R. Ortiz de Montellano, *J. Am. Chem. Soc.* **89**, 3362 (1967).

[13] Glutathione, sucrose, and magnesium salts, added by Capstack *et al.*[11] were not used in this preparation.

with the aid of a Teflon-in-glass homogenizer (Potter-Elvehjem type). Undissolved protein (B) is removed by centrifugation at 105,000 g for 3 hours at 2°. The supernatant (C) is removed by pipette and stored at 0° until used.

The activities of this enzyme preparation (C), the original homogenate (A) and the high speed pellet (resuspended) (B) obtained in the last step above, are compared with that of the hog liver enzyme (D) in Table II. It is seen that the specific activity increases about 12-fold when this procedure is used although even these values never reach those obtained with the mammalian liver enzyme.

TABLE II

COMPARISON OF SQUALENE 2,3-OXIDE β-AMYRIN CYCLASE
AND LANOSTEROL CYCLASE ACTIVITIES

Sample	Protein (mg)	Specific activity[a]
Microsome (A)	202	4.5×10^{-4}
Precipitated after 3 hours' centrifugation (B)	18	10.5×10^{-4}
Supernatnant after 3 hours' centrifugation (C)	12	51×10^{-4}
Hog liver soluble enzyme (D)	—	90

[a] Millimicromoles per hour per milligram of protein.

The product of incubation of the soluble fraction (C) with 2,3-oxidosqualene is almost entirely β-amyrin; proof may be found elsewhere.[12]

Properties of the Enzyme

The enzyme is effectively inhibited by the imino analog of squalene oxide. Other properties are less well defined than in the case of the hog liver enzyme.

Formation of β-Amyrin from 2,3-Oxidosqualene by a Soluble Enzyme from the Pea Seedling

Principle. The overall reaction is shown in Eq. (2).

$$(2)$$

The assay is based on the ready separation of product from substrate by thin-layer chromatography.

As with the assay for the hog liver enzyme, the *dl*-mixture of the substrate is used in the assay.

Reagents. The substrate is the same as that for hog liver enzyme; an amount equivalent to 8 millimicromoles per incubation is emulsified as described in the section on reagents for the hog liver enzyme; the thin-layer plates are also identical to those described in the same section.

Procedure. Incubations are carried out in an identical manner to those described in the "procedure" section for the hog liver enzyme, except that the volumes of enzyme solutions added were 2–5 ml and incubation tubes were shaken at 30° for 15 hours.

Termination and Extraction of Radioactivity. The incubations are terminated by the addition of $5 N$ methanolic KOH (2–4 ml) and the tubes are shaken at 30° overnight to complete saponification (without this treatment, the organic phases during extraction appear as gels; if the mixture is acidified, the gels are not formed so easily, but the unchanged substrate is converted to polar compounds which are not easily separated from the product β-amyrin). The extraction of the hydrolyzate and the separation of the product from the substrate is identical to the procedure described for the hog liver enzyme except that the R_f of the product, β-amyrin, is 0.17.

The purification and identification of the product by crystallization with carrier β-amyrin (from ethanol–water) and the gas chromatographic analysis have been described elsewhere.[12]

[20] Δ⁷-Sterol Δ⁵-Dehydrogenase and Δ⁵,⁷-Sterol Δ⁷-Reductase of Rat Liver

By Mary E. Dempsey

A complex enzyme system isolated from rat liver catalyzes the conversion of naturally occurring 3β-hydroxysteroids containing 27 carbon atoms to cholesterol, by the reaction pathways outlined in Fig. 1.[1-3] Although the enzymes involved in these conversions are as yet poorly

[1] M. E. Dempsey, *J. Biol. Chem.* **240**, 4176 (1965).

[2] M. E. Dempsey, J. D. Seaton, G. J. Schroepfer, Jr., and R. W. Trockman, *J. Biol. Chem.* **239**, 1381 (1964).

[3] For information on the enzymes catalyzing early steps in sterol biosynthesis, see G. Popják, this volume [12].

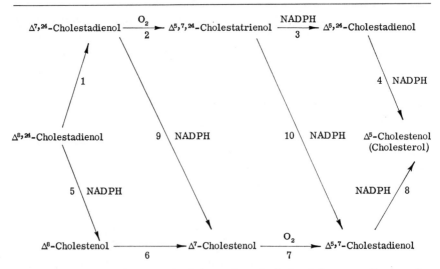

FIG. 1. Pathways of enzymatic cholesterol synthesis involving naturally occurring 3β-hydroxysteroids containing 27 carbon atoms. Reaction steps 2 and 7 are catalyzed by Δ⁷-sterol Δ⁵-dehydrogenase; steps 3 and 8 by Δ⁵,⁷-sterol Δ⁷-reductase; steps 1 and 6 by Δ⁸-sterol Δ⁷-isomerase; steps 4, 5, 9, and 10 by Δ²⁴-sterol Δ²⁴-reductase. Known cofactor requirements for each reaction are indicated. From M. E. Dempsey, *J. Biol. Chem.* **240**, 4176 (1965). Reproduced by permission of the *Journal of Biological Chemistry.*

characterized, procedures for their assay have been devised and some progress has been made in the study of their distribution in liver fractions. Two of these enzymes, Δ⁷-sterol Δ⁵-dehydrogenase and Δ⁵,⁷-sterol Δ⁷-reductase, have been partially purified. Methods of assay and the preparation and properties of the enzyme system, in particular the Δ⁵-dehydrogenase and Δ⁷-reductase activities, are described in this chapter.

Assay Methods

Δ⁵,⁷-STEROL Δ⁷-DEHYDROGENASE (Reaction Steps 2 and 7, Fig. 1)

This activity may be assayed by an isotopic method involving conversion of the labeled Δ⁷-cholestenol[4] to Δ⁵,⁷-cholestadienol and the

---

[4] Systematic names of the sterols referred to in the text by their abbreviated names are: Δ⁷-cholestenol, 5α-cholest-7-en-3β-ol; Δ⁵,⁷-cholestadienol, cholesta-5,7-dien-3β-ol; Δ⁷,²⁴-cholestadienol, 5α-cholesta-7,24-dien-3β-ol; Δ⁵,⁷,²⁴-cholestatrienol, cholesta-5,7,24-trien-3β-ol; Δ⁵,²⁴-cholestadienol, cholesta-5,24-dien-3β-ol; Δ⁸,²⁴-cholestadienol, 5α-cholesta-8,24-dien-3β-ol; Δ⁸-cholestenol, 5α-cholest-8-en-3β-ol; epiperoxide of Δ⁵,⁷,²⁴-cholestatrienol, 5α,8α-epiperoxy-cholesta-6,24-dien-3β-ol; epiperoxide of Δ⁵,⁷-cholestadienol, 5α,8α-epiperoxy-cholest-6-en-3β-ol; 4α-methyl-Δ⁷-cholestenol, 4α-methyl-5α-cholest-7-en-3β-ol; 4,4,14α-trimethyl-Δ⁸-cholestenol, 4,4′,14α-trimethyl-5α-cholest-8-en-3β-ol.

photooxidation of the latter to its $5\alpha,8\alpha$-epiperoxide after dilution with unlabeled dienol. Alternatively, a spectrophotometric technique may be used, in which the $\Delta^{5,7}$-dienol is determined by its absorption at 281.5 mμ. Both methods are sensitive and specific. The spectrophotometric method has the advantage of speed and does not require a labeled Δ^7-sterol as substrate. However, when low levels of enzyme activity are involved, this method is subject to inaccuracies due to unidentified endogenous materials absorbing in the 282 mμ region.

Isotopic Assay[1]

Reagents

Δ^7-Cholestenol-4-^{14}C, 0.26 mM $(9.4 \times 10^4$ cpm per micromole) in propylene glycol solution[5,6]

Enzyme system in 0.1 M potassium phosphate buffer, pH 7.3, e.g., fraction B after Sephadex treatment (15 mg protein per ml)[7]

Oxygen

Nitrogen

Petroleum ether (b.p. 30–60°)

Eosin-Y, 0.4% (w/v), water- and alcohol-soluble (National Allied Chemical and Dye Corporation), in 95% (v/v) ethanol.

$\Delta^{5,7}$-Cholestadienol, 2.1 mM (Mann Research Laboratories) in 95% (v/v) ethanol[8]

Silicic acid, 100 mesh, analytical reagent, suitable for chromatographic analysis (Mallinckrodt Chemical Works)

Super-Cel (Johns-Manville Corporation)

[5] G. J. Schroepfer, Jr., and I. D. Frantz, Jr., *J. Biol. Chem.* **236**, 3137 (1961).

[6] An alternate technique for preparation of labeled Δ^7-cholestenol is to reduce Δ^7-cholesten-3-one with tritium-labeled lithium aluminum hydride to yield the 3α-^{3}H-sterol or to label the hydrogens of carbons 2 and 4 by alkaline isomerization of the ketone in ^{3}H$_2$O followed by reduction of the ketone and isolation of the 3β-hydroxysteroid; compare M. Lindberg, F. Gautschi, and K. Bloch, *J. Biol. Chem.* **238**, 1661 (1963).

[7] A combination (total volume, 2 ml) of fraction A with either fraction C or fractions D_1 or D_2 of the enzyme system (cf. Table I) may be used in place of fraction B, if the following conditions are met: Fraction A has been treated with Sephadex to remove endogenous NADPH, as described in the text for fraction B, and the ratio of fraction A protein to that of either fraction C or D_1 or D_2 in the incubation is 3:1 or greater. The latter requirement ensures maximal activation of the dehydrogenase by fraction A. Instead of Sephadex treatment of fraction A, the incubation may be performed in the presence of 0.1 μM AY-9944, which will completely inhibit Δ^7-reductase even in the presence of excess NADPH (cf. Table II).

[8] The exact concentration of the $\Delta^{5,7}$-cholestadienol solution is determined from its absorbance at 281.5 mμ.

Benzene, Thiophene free
Methanol, analytical reagent
Acetone, analytical reagent
2,5-Diphenyloxazole (scintillation grade), 0.3% (w/v) in toluene

Incubation and Extraction of Sterols. One-tenth milliliter of the solution of Δ^7-cholestenol-4-^{14}C (2.4 × 10³ cpm, 0.026 micromole) in propylene glycol is mixed with 2 ml of enzyme system (30 mg protein) in a 20-ml beaker and incubated for 30 minutes at 37° in a Dubnoff shaker under oxygen. Ethanol, 2.1 ml of 95% (v/v), is added to the reaction mixture to terminate the reaction. Sterols (>95% of original radioactivity) are extracted with 4 × 8.2 ml petroleum ether (conveniently performed with a 25 × 150 mm test tube, a Vortex mixer, and a disposable capillary pipette). The extract is filtered through shark skin paper (Schleicher and Schuell), dried under nitrogen, and dissolved in 10 ml of benzene. An aliquot (0.5 ml) is evaporated to dryness in a glass counting vial, dissolved in scintillation fluid, and assayed for radioactivity in a scintillation counter.

Formation of the Epiperoxide Derivative. The remaining benzene solution (9.5 ml) is evaporated to dryness under nitrogen in a Pyrex test tube (25 × 200 mm), and 30 ml (24 mg) of the $\Delta^{5,7}$-cholestadienol and 1.0 ml of the eosin-Y solutions are added. The tube is irradiated for 1.5 hours with the use of a fluorescent tube, 15 watt-daylight type, 1–2 cm from the test tube, and a reflector of aluminum foil. During irradiation, oxygen is continuously bubbled through the reaction mixture. This procedure should reduce the absorption of the reaction mixture at 281.5 mμ to 5% or less of the initial value. The solution is diluted with an equal volume of water and extracted with 3 × 124 ml of petroleum ether. The combined extracts are evaporated to dryness under nitrogen, and the residue is applied in 1.5 ml benzene, to a silicic acid-Super-Cel column[9] (1.5 × 9 cm). Elution with 200 ml of benzene removes unchanged starting material, and 15 ml of methanol removes the epiperoxide which is recovered from the methanol by the addition of 15 ml of water and extraction with 3 × 60 ml of petroleum ether. The extract is evaporated under nitrogen, and the epiperoxide (approximately 20 mg) is crystallized from aqueous acetone, dried, weighed, and assayed for radioactivity. The percentage of the initial mixture of isotopically labeled steroids yielding an epiperoxide derivative (i.e., arising from $\Delta^{5,7}$-cholestadienol) is calculated as follows:[10]

---

[9] The column is prepared from a uniform slurry of 2 parts silicic acid to 1 part Super-Cel in benzene; see I. D. Frantz, Jr., *J. Lipid Res.* **4**, 176 (1963).

[10] This percentage is less than one when isotopically labeled cholesterol, Δ^7-choleste-

Percentage epiperoxide

$$= \left[\frac{\text{cpm epiperoxide}}{\text{cpm sterol mixture}} \right] \left[\frac{\text{mmoles } \Delta^{5,7}\text{-cholestadienol}}{\text{mmoles epiperoxide}} \right] \times 100$$

The Δ^7-sterol Δ^5dehydrogenase activity of the enzyme system is linear with time and protein concentration under the conditions of the incubation procedure just described.

The epiperoxide derivative of $\Delta^{5,7}$-cholestadienol may be hydrogenated to cholestan-3β,5α,8α-triol in the presence of a platinum catalyst.[1,11] This reaction is used in studying the conversion of labeled $\Delta^{7,24}$-cholestadienol to $\Delta^{5,7,24}$-cholestatrienol by Δ^7-sterol Δ^5-dehydrogenase (step 2, Fig. 1); both labeled $\Delta^{5,7,24}$-cholestatrienol and unlabeled $\Delta^{5,7}$-cholestadienol yield the same triol derivative, which may be purified by chromatography and crystallization.[1] If sufficient unlabeled $\Delta^{5,7,24}$-cholestatrienol becomes available for use as carrier in the epiperoxide reaction with labeled $\Delta^{5,7,24}$-cholestatrienol, the reduction step will not be necessary, i.e., the epiperoxide derivatives of labeled and unlabeled $\Delta^{5,7,24}$-cholestatrienol could be purified as just described for the epiperoxides of labeled and unlabeled $\Delta^{5,7}$-cholestadienol.

Spectrophotometric Assay

Reagents

Δ^7-Cholestenol, 0.26 mM (Ikapharm, Ramat-Gan, Israel) in propylene glycol solution

Enzyme system in 0.1 M potassium phosphate buffer, pH 7.3, i.e., fraction B after Sephadex treatment (15 mg protein per milliliter)[7]

Oxygen

Nitrogen

Ethanol, 95% (v/v)

Cyclohexane, spectral grade

Procedure. Unlabeled Δ^7-cholestenol is incubated with the enzyme system as described for the isotopic assay, except that to obtain levels of $\Delta^{5,7}$-cholestadienol sufficient for accurate ultraviolet measurement, larger incubation mixtures are prepared, e.g., 0.3 ml Δ^7-cholestenol (0.078 micromole) and 6 ml of enzyme system (90 mg protein). After inactivation of

---

nol, $\Delta^{5,24}$-cholestadienol, $\Delta^{7,24}$-cholestadienol, 4α-methyl-Δ^7-cholestenol, or 5α-cholestanol are present in the reaction mixture. The expected percentage is obtained with samples containing known levels of labeled $\Delta^{5,7}$-cholestadienol.

[11] D. Dvornik, M. Kraml, and J. F. Bagli, *J. Am. Chem. Soc.* **86**, 2739 (1964).

the incubation mixture with alcohol, the sterols are extracted with spectrograde cyclohexane and the extract is filtered and evaporated. The residue is then dissolved in 3 ml of cyclohexane in which its absorbance at 281.5 mμ is measured. A correction is made by subtraction of the absorbance at 281.5 mμ of material similarly extracted from a parallel incubation without added Δ^7-cholestenol. The amount of $\Delta^{5,7}$-cholestadienol synthesized from Δ^7-cholestenol is then estimated from the corrected absorbance and the molar absorptivity value of $\Delta^{5,7}$-cholestadienol at 281.5 mμ, $11.8 \times 10^3 M^{-1}$ cm^{-1} (see footnotes 2 and 12). The spectrophotometric and isotopic assays give identical estimations of Δ^7-sterol Δ^5-dehydrogenase activity except for measurements of low levels of activity or after short incubation times.

The isotopic or spectrophotometric techniques just described may be used to assay for Δ^8-sterol Δ^7-isomerase activity (steps 1 and 6, Fig. 1) by coupling the isomerase to Δ^7-sterol Δ^5-dehydrogenase.[1] The Δ^8-sterol (e.g., Δ^8-cholestenol or $\Delta^{8,24}$-cholestadienol) is incubated under conditions described for the assay of Δ^7-sterol Δ^5-dehydrogenase (i.e., in the absence of NADPH and under oxygen, thus allowing steps 1 and 2 or 6 and 7 to occur) and the appearance of $\Delta^{5,7}$-cholestadienol or $\Delta^{5,7,24}$-cholestatrienol is measured.

$\Delta^{5,7}$-STEROL Δ^7-REDUCTASE (Reaction Steps 3 and 8, Fig. 1)

This activity may also be assayed by either a specific isotope derivative technique, or spectrophotometrically, using commercially available unlabeled substrate.

Isotopic Assay[1,2]

Labeled $\Delta^{5,7}$-cholestadienol is incubated anaerobically with the enzyme and NADPH, the cofactor required for Δ^7-reductase activity.[2] Labeled Δ^5-cholestenol (cholesterol) arising from labeled $\Delta^{5,7}$-cholestadienol is determined by dilution of the recovered sterols with unlabeled cholesterol, which is then purified via the dibromide derivative, followed by crystallization. It is then assayed for radioactivity.

Reagents

$\Delta^{5,7}$-Cholestadienol-4-^{14}C $(9.4 \times 10^4$ cpm) 0.26 mM, in propylene glycol solution[5,12]

Enzyme system in 0.1 M potassium phosphate buffer, pH 7.3, e.g., fraction B after Sephadex treatment (15 mg protein per milliliter)[13]

---

[12] S. Bernstein, L. J. Binovi, L. Dorfman, K. J. Sax, and Y. SubbaRow, *J. Org. Chem.* **4**, 433 (1949).

[13] A combination (total volume, 2 ml) of fraction A or A$_1$ with either fraction C or

5 mM NADPH, enzymatically reduced (Calbiochem), in 0.1 M potassium phosphate buffer, pH. 7.3

Nitrogen, oxygen-free

Petroleum ether (b.p. 30–60°)

Ethanol, 95% (v/v)

Cholesterol, purified by passage through the dibromide

Benzene, Thiophene-free

Diethyl ether, anhydrous

Bromine, analytical reagent

Glacial acetic acid, analytical reagent

Zinc dust

HCl, 0.7 N

NaOH, 2 N

2,5-Diphenyloxazole (scintillation grade), 0.3% (w/v), in toluene

Procedure. Incubation of labeled Δ⁵,⁷-cholestadienol with the enzyme system is performed as described for the isotopic assay of Δ⁷-sterol Δ⁵-dehydrogenase activity, except that 0.2 ml (1 μmole) of NADPH is added to the mixture and incubation is carried out under oxygen-free nitrogen. The enzyme system is inactivated with alcohol (2.3 ml of 95% (v/v) ethanol), extracted with 4 × 9.2 ml petroleum ether, and the total radioactivity of the extract is determined by assay of a 0.5 ml aliquot of a 10 ml solution as described previously.

To the remaining benzene solution (9.5 ml) containing the labeled sterols, a known weight (120–150 mg) of unlabeled cholesterol is added. The benzene is evaporated under nitrogen and the residual sterols are dissolved in 3 ml of anhydrous diethyl ether. The ether solution is cooled in an ice bath and bromine is added dropwise until a dark yellow color persists. After 1 hour in the ice bath, 1.5 ml of cooled glacial acetic acid is added with stirring. The precipitated cholesterol dibromide is collected on a sintered-glass funnel, washed until colorless with cooled glacial acetic acid, and dissolved in 5 ml of diethyl ether. Zinc dust (100 mg) is added and the suspension is stirred frequently with a Vortex mixer during 30 minutes. Addition of water dissolves the zinc bromide and the suspension is stirred with the Vortex mixer until the ether and water layers separate completely. The ether layer is removed and washed with 1 ml of 0.7 N HCl, twice with 1 ml of water, and with 1 ml of 2 N NaOH. After

---

fraction D_1 or D_2 of the enzyme system (cf. Table I) may be used in place of fraction B. For maximal activation of the reductase by fraction A, the ratio of fraction A protein to that of either fraction C or D_1 or D_2 in the incubation should be 18:1 or greater. To achieve maximum activation of the reductase by fraction A_1, a lower protein ratio is required, i.e., 3:1.

evaporation of the ether, cholesterol (50–60 mg) is crystallized from aqueous acetone, dried, weighed, and assayed for radioactivity. The percentage of labeled $\Delta^{5,7}$-cholestadienol converted to cholesterol by Δ^{7}-reductase activity is calculated as follows:

Percentage conversion of $\Delta^{5,7}$-cholestadienol to cholesterol

$$= \left[\frac{\text{cpm cholesterol (final)}}{\text{cpm sterol mixture}} \right] \left[\frac{\text{mg cholesterol (initial)}}{\text{mg cholesterol (final)}} \right] \times 100$$

The $\Delta^{5,7}$-sterol Δ^{7}-reductase activity of the enzyme system is linear with time and protein concentration under the conditions described.

The isotopic assay just described may also be used to measure Δ^{7}-sterol Δ^{5}-dehydrogenase activity by performing an incubation under oxygen with labeled Δ^{7}-cholestenol as substrate and in the presence of NADPH. The rate of the Δ^{7}-reductase is approximately 40% faster than the Δ^{5}-dehydrogenase; therefore, in the presence of excess NADPH $\Delta^{5,7}$-cholestadienol does not accumulate and the amount of cholesterol synthesized is a measure of Δ^{5}-dehydrogenase activity.[2] This method is the most useful for the measurement of Δ^{7}-sterol Δ^{5}-dehydrogenase activity in crude enzyme preparations containing appreciable amounts of endogenous NADPH (cf. Table I).

Spectrophotometric Assay[14]

Enzymatic conversion of $\Delta^{5,7}$-cholestadienol to cholesterol may be determined from the loss in absorbance at 282 mμ by essentially the same procedure as for spectrophotometric assay of Δ^{7}-sterol Δ^{5}-dehydrogenase activity, except that the incubation is performed with unlabeled $\Delta^{5,7}$-cholestadienol in the presence of 0.5 mM NADPH and under oxygen-free nitrogen.[15]

Sites of Action of Inhibitors.[16-18] Several techniques have been developed which permit definition of the specific sites of action of inhibitors of the enzyme system (Table II). Compounds which block Δ^{7}-sterol Δ^{5}-dehydrogenase activity do not inhibit $\Delta^{5,7}$-sterol Δ^{7}-reductase activity

[14] This procedure is similar to that described by A. A. Kandutsch, J. Biol. Chem. 237, 358 (1962).
[15] It is important that the assay be performed under oxygen-free nitrogen to avoid formation of the epiperoxide derivative, which does not absorb at 282 mμ; if the epiperoxide were formed, a falsely high value for cholesterol synthesis would result.
[16] M. E. Dempsey, Abstr. 6th Intern. Congr. Biochem. IUB Vol. 32, p. 570, VII-39 (1964).
[17] M. E. Dempsey, in "Progress in Biochemical Pharmacology" (D. Kritchevsky, R. Paoletti, and D. Steinberg, eds.), Vol. II, p. 21. Karger, Basel, 1967.
[18] M. E. Dempsey, Ann. N.Y. Acad. Sci. 148, 631 (1968).

and, conversely, Δ^7-reductase inhibitors have no effect on Δ^5-dehydrogenase activity of the enzyme system (compare Table II).

Preparation of Enzyme System

The following procedure for preparation and fractionation of the enzyme system gives highly reproducible results. All operations are performed at 0–4°. Protein concentrations are determined by method of Layne (Vol. III [73]) or Siekevitz.[19]

Preparation and Fractionation of Liver Homogenate. Female albino rats (150–200 g), fed ad libitum, are decapitated and exsanguinated. Livers are rapidly excised, trimmed of connective tissue, and chilled in 0.1 M potassium phosphate buffer, pH 7.3. The livers are then minced and homogenized at moderate speed for 1.5 minutes in fresh buffer (2.5 ml per gram of liver) using a loose-fitting stainless steel pestle (1 mm clearance) and a thick-walled glass test tube (25 × 200 mm).[20]

The sediment obtained after centrifugation of the total homogenate at 500 g for 20 minutes is discarded, and the supernatant fraction is centrifuged for a further 15 minutes at 9000 g. The supernatant from the centrifugation at 9000 g is recentrifuged for 90 minutes at 105,000 g and the floating fat layer is drawn off and discarded. The supernatant is now withdrawn from the microsomal pellet as two equal fractions: an upper fraction (A) and a lower fraction (B).

Treatment of Fraction B with Sephadex. Fraction B, 200 mg, in 10 ml 0.1 M potassium buffer, pH 7.3, is applied to a column of Sephadex G-25 (Pharmacia Fine Chemicals), 2 × 40 cm, which has been equilibrated with the same buffer. The same buffer is also used for elution. Approximately 30 ml of buffer are passed through the column at the rate of 0.6 ml per minute, until the protein (tan color) begins to emerge. The protein is then collected in 2-ml fractions. A total of 10–12 ml of eluate, comprising the main part of the peak, excluding leading and tailing fractions, is collected and contains about 85% of the protein applied to the column. Although this procedure does not lead to appreciable enrichment of Δ^7-sterol Δ^5-dehydrogenase or $\Delta^{5,7}$-sterol Δ^7-reductase activity of the enzyme system, it is useful as a means of removing endogenous cofactors, in particular NADPH required by Δ^7 and Δ^{24}-reductases (cf. Fig. 1). Dialysis of fraction B, to accomplish the same purpose, leads to rapid loss of all the activities of the enzyme system. Fraction B (after Sephadex treatment) catalyzes all the reactions of Fig. 1, and each individual reaction may be defined by choice of the proper substrate and incubation condi-

[19] P. Siekevitz, *J. Biol. Chem.* **195**, 549 (1959).
[20] N. L. R. Bucher, *J. Am. Chem. Soc.* **75**, 498 (1953).

tions.[1] This fraction is the preparation used for the assay procedures described here (e.g., Δ^7-sterol Δ^5-dehydrogenase and $\Delta^{5,7}$-sterol Δ^7-reductase) ; more purified fractions may also be used in the assays as indicated previously.[7,13]

Further Purification of Microsomal Δ^7-Sterol Δ^5-Dehydrogenase and $\Delta^{5,7}$-Sterol Δ^7-Reductase. The microsomal pellet is resuspended by homogenization using a tight-fitting Teflon pestle and glass tube, in a volume of phosphate buffer equal to one-half the total volume of fractions A and B. The suspension is centrifuged for 60 minutes at 105,000 g, the supernatant fraction is decanted, and the pellet is resuspended in phosphate buffer by homogenization as just described. This suspension constitutes fraction C. It is frozen prior to further purification.

A 10-ml aliquot of fraction C is thawed and then mixed for 5 minutes with 32 mg of sodium deoxycholate (special enzyme grade, Mann Research Laboratories). The mixture is centrifuged for 15 minutes at 270 g. One gram of $(NH_4)_2SO_4$ (special enzyme grade, Mann Research Laboratories) is added with stirring to 7 ml of the clear supernatant, and the $(NH_4)_2SO_4$ precipitate is recovered by centrifugation at 10,800 g for 10 minutes. The $(NH_4)_2SO_4$ pellet is suspended by homogenization in 1.5 ml of 0.1 M phosphate buffer, pH 7.3, and then applied to a column of Sephadex G-25, 2×13 cm, which has been equilibrated with the same buffer. The same buffer is used for elution. Approximately 10 ml of buffer is passed through the column at a rate of 1 ml per minute, until a white cloudy solution begins to emerge. This material is collected in one fraction (5 ml) and constitutes fraction D_1. The following 3.5 ml of eluate (light yellow, clear) are collected in one fraction, which constitutes fraction D_2. Fractions C, D_1, and D_2 are assayed for Δ^7-sterol Δ^5-dehydrogenase and $\Delta^{5,7}$-sterol Δ^7-reductase in the presence of excess activator (fraction A) as described previously.[7,13] Fractions A, C, D_1, and D_2 individually exhibit barely detectable Δ^7-sterol Δ^5-dehydrogenase and $\Delta^{5,7}$-sterol Δ^7-reductase activity.[21] A summary of the purification procedure is outlined in Table I.[22]

Purification of the Activator of $\Delta^{5,7}$-Sterol Δ^7-Reductase. Fraction A (12 ml, 204 mg protein) is mixed with 0.6 g $(NH_4)_2SO_4$ and heated at 100° under a stream of nitrogen for 5 minutes. After removal of the pre-

---

[21] M. E. Dempsey, *J. Minn. Acad. Sci.* **34**, 9 (1967).

[22] As may be noted in Table I, the complete procedure for the preparation of fractions D results in a marked activation of Δ^5-dehydrogenase and Δ^7-reductase, i.e., the total activity yield of fraction D_1 plus fraction D_2 exceeds that of fraction C from which fractions D were prepared. This effect is not due solely to activation by deoxycholate because treatment of fraction C with deoxycholate, as described in the purification procedure, does not result in appreciable activation of the enzymes.

TABLE I
SUMMARY OF PURIFICATION PROCEDURE FOR Δ^7-STEROL
Δ^5-DEHYDROGENASE AND $\Delta^{5,7}$-STEROL Δ^7-REDUCTASE

Fraction	Volume (ml)	Protein (mg/ml)	Δ^5-Dehydrogenase[a]		Δ^7-Reductase[b]	
			Specific activity[c]	Yield (%)	Specific activity[c]	Yield (%)
Homogenate[d]	365	48.2	0.1	100	0.2	100
500 g supernatant	248	24.0	0.3	101	0.4	68
Fraction B[e]	89	22.6	0.4	46	0.5	29
Fraction C[f]	89	8.2	1.5[i] (0.4)[l]	62	3.0[j] (0.8)[m]	62
Fraction D$_1$[g]	44	4.0	5.5[i] (1.4)[l]	55[k]	11.0[j] (2.8)[m]	55[k]
Fraction D$_2$[h]	31	1.7	16.0[i] (4.0)[l]	48[k]	30.0[j] (7.5)[m]	45[k]

[a] Incubated under oxygen in the presence of excess NADPH and fraction A (when indicated)[i] and assayed by the dibromide procedure (see text).

[b] Incubated under nitrogen in the presence of excess NADPH and fraction A or A$_1$ (when indicated)[j] and assayed by the dibromide procedure (see text).

[c] Millimicromoles per milligram of protein per 30 minutes.

[d] From 12 rat livers.

[e] Lower half of the 105,000 g supernatant fraction. Treatment of fraction B with Sephadex G-25, as described in the text, does not lead to an appreciable increase in the specific activity of either enzyme; however it does remove endogenous NADPH, allowing the dehydrogenase and reductase to be assayed separately (cf. assay procedures).

[f] 105,000 g sediment, washed with phosphate buffer.

[g] First fraction obtained during gel filtration of fraction C, previously solubilized with sodium deoxycholate and fractionated with (NH$_4$)$_2$SO$_4$.

[h] Second fraction obtained during gel filtration of fraction, previously solubilized with sodium deoxycholate and fractionated with (NH$_4$)$_2$SO$_4$.

[i] Maximally activated by fraction A, i.e., the ratio of fraction A protein to that of either fraction C or D$_1$ or D$_2$ in the incubation was 3:1 or greater. Fraction A alone does not exhibit detectable Δ^5-dehydrogenase activity, and the protein contributed by fraction A is not included in this specific activity figure.

[j] Maximally activated by fraction A or fraction A$_1$, i.e., the ratio of fraction A protein to that of either fraction C or D$_1$ or D$_2$ in the incubation was 18:1 or greater; with A$_1$ it was 3:1 or greater. Fractions A or A$_1$ alone do not exhibit detectable Δ^7-reductase activity, and the protein contributed by these fractions is not included in this specific activity figure.

[k] Preparation of fractions D$_1$ and D$_2$ results in marked activation of these enzymes, i.e., the total activity yield of fraction D$_1$ plus fraction D$_2$ exceeds that of fraction C.

[l] Specific activity including the protein contributed by an optimum level of fraction A, i.e., 3 mg of fraction A per 1 mg of fraction C, D$_1$, or D$_2$.

[m] Specific activity including the protein contributed by an optimum level of fraction A$_1$, i.e., 3 mg fraction A$_1$ per 1 mg of fraction C, D$_1$, or D$_2$.

cipitated protein by filtration, 7.7 g (NH$_4$)$_2$SO$_4$ is added to the filtrate (10 ml) and this mixture is stirred for 1.5 hours at 4°. The resulting precipitate is collected by centrifugation at 13,300 g for 30 minutes; dissolved in 4 ml 0.1 M phosphate buffer, pH 7.4; and applied to a column of Sephadex G-25, 2 $\times$ 13 cm, which has been equilibrated with the same buffer. The same buffer is used for elution. Approximately 10 ml of buffer is passed through the column at a rate of 1 ml per minute until the protein begins to emerge from the column. The protein is collected in one fraction (5 ml, 5 mg protein), which constitutes fraction A_1.

Fraction A_1 may be used in place of fraction A for activation of the $\Delta^{5,7}$-sterol Δ^7-reductase of fraction C, D_1, or D_2.[13] The procedure just described results in an approximately 6-fold purification of the activator of Δ^7-reductase present in fraction A. The use of excess fraction A_1 for activation of Δ^7-reductase does not increase the specific activity of the reductase observed in the presence of excess fraction A and fraction C, D_1, or D_2 (cf. Table I).

Properties of the Enzyme System

Occurrence. Active preparations of the enzyme system (i.e., fractions A, B, and C) have been isolated by the method just described from the livers of dog, cow, mouse, and rabbit.[23] Muscle and kidney homogenates do not have detectable activity (e.g., Δ^7-sterol Δ^5-dehydrogenase and $\Delta^{5,7}$-sterol Δ^7-reductase).[23]

Stability. Fractions A, B, and C may be stored at $-20°$ for at least 4 weeks without appreciable loss of their activities. Fractions D_1 and D_2 are labile to freezing and thawing, but may be stored at $4°$ for approximately 48 hours without appreciable loss of activity. Storage of fraction A_1 at $4°$ for longer than 24 hours results in gradual loss of its ability to activate the $\Delta^{5,7}$-sterol Δ^7-reductase of fractions C, D_1, and D_2. The enzymatic activities of fractions B, C, D_1, and D_2 are completely destroyed by brief exposure to heat (e.g., $100°$, 1–2 minutes). The activator of Δ^7-reductase is heat stable (e.g., $100°$, 10 minutes, and see the procedure for preparation of fraction A_1); it is also nondialyzable and is precipitated by perchloric acid. The material present in fraction A which activates the Δ^7-sterol Δ^5-dehydrogenase of fractions C, D_1, and D_2 is labile to prolonged heat treatments (e.g., longer than 1–2 minutes at $100°$) and is nearly completely lost during preparation of fraction A_1. Properties common to the activators of both Δ^7-reductase and Δ^5-dehydrogenase are that they remain with the protein fractions during gel-filtration on Sephadex G-25 and that they may be extracted into phosphate buffer from an acetone powder preparation of rat liver. In addition, serum albumin will not substitute for fraction A as an activator of Δ^5-dehydrogenase or Δ^7-reductase.

---

[23] J. D. Seaton, Master's Thesis, Univ. of Minnesota, 1963; cf. also the studies of Kandutsch[14] with mouse liver.

TABLE II
INHIBITORS AND THEIR SITES OF ACTION[a]

Inhibitor[b]	Enzymatic steps blocked		
	Δ^5-Dehydrogenase	Δ^7-Reductase	Δ^{24}-Reductase
AY-9944 (40 mμM)[c]	0	+	0
MER-29 (3 μM)[d]	0	+	+
o-Chlorobenzylamine (40 μM)[e]	0	+	+
DBI (0.5 mM)[f]	0	+	0
Chloropropamide (1 mM)[g]	+	0	0
Orinase (0.5 mM)[h]	+	0	0
EDTA (10 mM)[i]	+	0	—
Cysteine (12 mM)[i]	+	0	—
KCN (3 mM)[i]	+	0	—

[a] Fraction B after Sephadex treatment was the source of the enzyme system; incubations were performed as described in the text and in *J. Biol. Chem.* **240**, 4176 (1965); *Abstracts 6th Intern. Cong. Biochem. IUB*, Vol. 32, p. 570, VII-39 (1964); *Federation Proc.* **25**, 221 (1966); *in* "Progress in Biochemical Pharmacology" (D. Kritchevsky, R. Paoletti, and D. Steinberg, eds.), Vol. II, p. 21. Karger, Basel, 1967; and in *Ann. N.Y. Acad. Sci.* **148**, 631 (1968).

[b] The concentrations shown in parentheses are those required for 50% inhibition of the enzymatic step blocked.

[c] AY-9944 (Ayerst Company) is *trans*-1,4-bis-(2-chlorobenzylaminomethyl)cyclohexane dihydrochloride.

[d] MER-29 (Merrell Company) is 1-[p-(β-diethylaminoethoxy)phenyl]-1-(p-tolyl)-2-(p-chlorophenyl)ethanol; it is also called Triparanol. The concentration shown is that required for 50% inhibition of the Δ^7-reductase; approximately 1 μM MER-29 has the same effect on the Δ^{24}-reductase.

[e] A number of other benzylamine derivatives (Ames Laboratories) are also effective inhibitors of Δ^7-reductase; their effect on Δ^{24}-reductase has not been tested. These are: 2,4,-dichlorobenzylamine, o-chloro-N-methylbenzylamine, 3,4,-dichlorobenzylamine, p-chlorobenzylamine, 2,4-dimethylbenzylamine.

[f] DBI (U.S. Vitamin Corporation) is N'-β-phenethylformamidinyliminourea; it is also called Phenformin.

[g] Chlorpropamide (C. Pfizer Company) is N-propyl-N'-p-chlorobenzenesulfonylurea.

[h] Orinase (Upjohn Company) is N-butyl-N'-p-toluensulfonylurea; it is also called Tolbutamide.

[i] The effects of these compounds on Δ^{24}-reductase were not studied.

pH Optimum. The optimum pH for the various reaction steps (Fig. 1) catalyzed by the enzyme system (fraction B or fractions A and C or D) is 7.3. Marked decrease of activity (e.g., Δ^7-sterol Δ^5-dehydrogenase or $\Delta^{5,7}$-sterol Δ^7-reductase) occurs below pH 7.1 and above pH 7.5.

Cofactor Requirements. NADH will not substitute for the NADPH requirement of $\Delta^{5,7}$-sterol Δ^7-reductase activity of fraction B after Sephadex treatment. Δ^7-Sterol Δ^5-dehydrogenase activity is not detected in the absence of oxygen.[2]

Reversibility. Reversal of the reaction steps catalyzed by the enzyme system (Fig. 1) is not detectable.[1]

Specificity. The Δ^5-bond must be present in a sterol before Δ^7-reductase activity (steps 3 or 8, Fig. 1) is detected; the Δ^7-bond must be present in a sterol before Δ^5-dehydrogenase activity (steps 2 or 7, Fig. 1) is detected.[1] The enzyme system does not catalyze reduction of the Δ^5-bond of sterols.[1,2] The Δ^7-sterol Δ^5-dehydrogenase, and $\Delta^{5,7}$-sterol Δ^7-reductase, and Δ^8-sterol Δ^7-isomerase activities of fraction B and fraction C are active with both Δ^{24}- and 24-dihydrosterols. The preparation procedure for fractions D_1 and D_2 results in nearly complete loss of the Δ^{24}-reductase activity of fraction C. The combination of fractions A and C (in the presence of oxygen, NADPH, NADP+, NAD+, nicotinamide, and Mg++) readily catalyzes the conversion of 4,4,14α-trimethyl- Δ^8-cholestenol and 4α-methyl- Δ^7-cholestenol to cholesterol, whereas fraction B (under the same conditions) only poorly catalyzes these steps. Fraction A plus C in the presence of ATP and the cofactors just listed also readily catalyzes the conversion of mevalonic acid to cholesterol, whereas the combination of fractions A and D_1 or D_2 does not.

Kinetics. The apparent K_m for Δ^7-sterol Δ^5-dehydrogenase (fraction B or fraction A plus fraction C) is 20 μM; for $\Delta^{5,7}$-sterol Δ^7-reductase (fraction B or fraction A plus fraction C), it is 12 μM.

Inhibitors. A list of inhibitors and catalytic steps of the enzyme system which each blocks is presented in Table II.[2,16–18] DBI and MER-29 are noncompetitive inhibitors of $\Delta^{5,7}$-sterol Δ^7-reductase (fraction B); the approximate K_i for both drugs is 45 μM.[17,18] The type of inhibition (e.g., competitive, noncompetitive) of the other compounds listed in Table II has not been determined. For additional information on inhibitors of sterol biosynthesis, see D. Steinberg and J. Avigan, this volume [21].

[21] Rat Liver Sterol Δ^{24}-Reductase

By DANIEL STEINBERG and JOEL AVIGAN

This enzyme activity was first extensively investigated in connection with the conversion of desmosterol ($\Delta^{5,24}$-cholestadien-3β-ol) to cholesterol,[1] and has been referred to as "desmosterol reductase." The biological reduction of desmosterol is readily demonstrated in the rat by injecting the labeled sterol intraportally and isolating radioactive cholesterol from

[1] J. Avigan and D. Steinberg, *J. Biol. Chem.* **236**, 2898 (1961).

the liver.[2,3] There is good evidence, discussed below, suggesting that the 24,25-double bond in a variety of sterols can be reduced by the same enzyme, and the more general name "sterol Δ^{24}-reductase" is more appropriate. Until the enzyme is solubilized and purified, however, it should be recognized that the name refers only to activities demonstrable in particulate preparations. Since the substrates are added in suspension, little can be said with respect to kinetics or intimate mechanism of action of the enzyme. Below we discuss its properties, substrate specificity, susceptibility to specific inhibitors, and role in sterol biogenesis.

Assay

The assay described is based on that employed by Steinberg, Avigan, and Goodman for following reduction of desmosterol[1,3] or lanosterol.[4] Because the rate of substrate reduction is so low, radioisotope techniques are needed to provide the required sensitivity. Desmosterol reduction can be measured under either anaerobic or aerobic conditions, since further oxidative metabolism of cholesterol (e.g., to bile acids) is usually slow enough to be neglected. Lanosterol reduction is measured under anaerobic conditions to prevent modifications of the sterol ring structure; 24,25-dihydrolanosterol accumulates as the only product.

Preparation of Radioactive Substrates

Desmosterol-^{14}C can be prepared by the Wittig synthesis from 3β-acetoxy-5-cholen-24-al and ^{14}C-labeled triphenylphosphine isopropylidene phosphorane.[5,6] Another convenient method is that in which radioactive 25-hydroxycholesterol is dehydrated.[7] Dehydration carried out with dioxane and sulfuric acid yields predominantly desmosterol; however, some contaminating Δ^{25}-isomer is always present and should be removed. The Δ^{25}-isomer is also reduced to cholesterol in a cell-free liver preparation, but at rates much lower than desmosterol, and this conversion, like that of desmosterol, is also inhibited by triparanol *in vitro*.[8] The separation of desmosterol from 25-dehydrocholesterol can be carried out by either column chromatography or TLC, as described by Svoboda and Thompson.[9] It was found in the latter study that commercial samples of desmosterol-^{14}C contained up to 40% Δ^{25}-dehydrocholesterol.

[2] W. M. Stokes, F. C. Hickey, and W. A. Fish, *J. Biol. Chem.* **232**, 347 (1958).
[3] D. Steinberg and J. Avigan, *J. Biol. Chem.* **235**, 3127 (1960).
[4] J. Avigan, D. S. Goodman, and D. Steinberg, *J. Biol. Chem.* **238**, 1283 (1963).
[5] V. H. M. Fagerlund and D. R. Idler, *J. Am. Chem. Soc.* **79**, 6473 (1957).
[6] B. Danieli and G. Russo, *J. Labelled Compounds* **1**, 275 (1965).
[7] W. Bergmann and J. P. Dusza, *J. Org. Chem.* **23**, 459 (1958).
[8] J. Avigan and D. Steinberg, unpublished results, 1964.
[9] J. A. Svoboda and M. J. Thompson, *J. Lipid Res.* **8**, 152 (1967).

Lanosterol-^{14}C, like desmosterol-^{14}C, can be prepared by Wittig synthesis from 3β-acetoxy-25,26,27-trisnorlanost-8-en-24-al and the labeled isopropylidene Wittig reagent.[6] The acceptor steroid can be prepared by selective ozonolysis of lanosteryl acetate. Labeled lanosterol can also be prepared biosynthetically, isolating it from livers of rats sacrificed 2 minutes after intravenous administration of mevalonate-2-[14]C.[4] Because of the very small endogenous pool of unlabeled lanosterol, the specific radioactivity of this component is very high, making it particularly suitable for use in the *in vitro* assay.

The instability of the sterol substrates on storage is a matter of concern. Desmosterol can deteriorate rather rapidly and extensively, especially when kept in the solid state and exposed to air.[10] Storage in solution under nitrogen and under refrigeration is always recommended. If it has been stored for any length of time, the substrate should be assayed for radiopurity by TLC or GLC and, if necessary, repurified by chromatography.

Incubation Conditions

Whole rat liver homogenate is prepared in 2.5 ml of Bucher's medium per gram of liver (0.1 M potassium phosphate buffer pH 7.4; 0.03 M nicotinamide, and 0.004 M MgCl$_2$).[11] When subcellular fractions are used, they may be dialyzed against the same medium for assay. To 10 ml of the above, 2 micromoles of NADPH and a suspension of 100 μg of desmosterol-^{14}C are added. The latter may be prepared by dissolving the sterol in a 100-fold amount of Tween-20 (Atlas Powder Company, Wilmington, Delaware) and adding water to make a 10% Tween solution. Alternatively, substrate may be added to the incubation mixture in a small volume of acetone (50–100 μl), while the mixture is held near ice temperature and stirred. Incubations are carried out with shaking under an atmosphere of nitrogen at 37°. At the end of the incubation period, ethanol and KOH are added to final concentrations of 50% and 2%, respectively. In order to achieve optimal recovery of radioactivity, carrier cholesterol and desmosterol in ethanol are added to the sample in amounts depending on the capacity and sensitivity of the fractionation method to be used. After saponification on a steam bath, the nonsaponifiable fraction is extracted three times with petroleum ether or hexane, the combined extracts are filtered, evaporated to dryness, and the sterol substrate and end product are resolved by thin-layer, gas–liquid, or column chromatography. The capacity of 10 ml of homogenate prepared from normal

---

[10] M. J. Thompson, J. N. Kaplanis, and H. E. Vroman, *Steroids* 5, 551 (1965).
[11] N. L. R. Bucher and K. McGarrahan, *J. Biol. Chem.* 222, 1 (1956).

rat liver ranged from 58 to 71 μg of desmosterol reduced in 3 hours out of 100 μg added.[1,3] Similar rates were observed with lanosterol as substrate.[4] However, since activity decreases after as little as 15 minutes of incubation in the system described, assays should be based on short incubations. Almost the entire enzymatic activity of the homogenate is located in particles with only insignificant amounts detectable in the clear 100,000 g supernatant. Microsomes contained 45–62% of the enzyme recovered, and mitochondria 27–50%.[1,4] It is rather likely that all or at least a substantial part of the activity associated with mitochondria was due to their contamination with microsomes. Dempsey et al.[12,13] reported Δ^{24}-reductase activity in the lower zone of "105,000 g supernatant" from rat liver homogenate that probably contained some slowly sedimenting microsomes. The tissue was homogenized in phosphate buffer, rather than in Bucher's medium, and this factor may have possibly influenced the ease of sedimentation of the microsomes.

Thin-Layer Chromatography

Avigan, Goodman, and Steinberg[14] acetylated the sterols with acetic anhydride in the presence of pyridine and separated the sterol acetates on a 40-cm long thin-layer plate of silicic acid, using hexane–benzene 5:1 for development. Under these conditions cholesteryl acetate has a slightly greater R_f than desmosteryl acetate. The fluorescent dye Rhodamine 6G was incorporated into the thin-layer plate, and in its presence the sterol zones were visible under ultraviolet light. According to Claude, the propionyl esters of the two sterols are more readily separated than the acetyl esters when chromatographed on AgNO$_3$-silicic acid thin-layer plates.[15] Bennett and Heftmann[16] describe a very satisfactory separation of the trifluoroacetates on a mixture of equal amounts of kieselguhr G and silica gel G (mobility 0.65 for the desmosterol derivative relative to that of cholesterol). The steryl acetates are separable by TLC on silicic acid impregnated with AgNO$_3$.[14,17] The free sterols are not separated by adsorption chromatography on silicic acid, but incorporation of AgNO$_3$ into the plate makes such a separation possible.[18-20] A reversed-phase

[12] M. E. Dempsey, J. Biol. Chem. 240, 4176 (1965).
[13] M. E. Dempsey, J. D. Seaton, G. J. Schroepfer, Jr., and R. W. Trockman, J. Biol. Chem. 239, 1381 (1964).
[14] J. Avigan, D. S. Goodman, and D. Steinberg, J. Lipid Res. 4, 100 (1963).
[15] J. R. Claude, J. Chromatog. 17, 596 (1965).
[16] R. D. Bennett and E. Heftmann, J. Chromatog. 9, 359 (1962).
[17] H. E. Vroman and C. F. Cohen, J. Lipid Res. 8, 150 (1967).
[18] R. Ikan and M. Cudzinovski, J. Chromatog. 18, 422 (1965).
[19] T. M. Lees, M. J. Lynch, and F. R. Mosher, J. Chromatog. 18, 595 (1965).
[20] J. W. Copius-Peereboom and H. W. Beekes, J. Chromatog. 17, 99 (1965).

system of undecane/acetic acid–acetonitrile[20, 21] may also be useful, although it is not as convenient as adsorption chromatography. Another method for separation of the two sterols, in which paper chromatography is applied, utilizes the difference between the products of bromination of cholesterol and desmosterol, namely dibromo- and tetrabromocholesterol, respectively.[22] After chromatography by any of the methods described above, the two zones are individually eluted with an appropriate solvent.

Gas–Liquid Chromatography

Gas–liquid chromatography of the free sterols has been extensively used for determination of desmosterol in biological samples. The same systems may be applied preparatively with collection of the two components using an appropriate collection device. Cholesterol and desmosterol in free form can be separated in a gas–liquid system on dimethylsiloxane polymer, SE-30,[23, 24] SE-52,[25] or neopentylglycol adipate.[26] Fumagalli, Capella, and VandenHeuvel report that of a number of phases tested by them, best results were obtained (separation factor 1.26) with PhSi 191-43 (a phenylmethylsiloxane polymer, General Electric Corp.).[24] No improvement in separation was achieved when trimethylsilyl ethers were chromatographed instead of the free sterols. However, when a more polar phase, neopentylglycol succinate polymer, was used instead of the silicone polymer, trimethylsilyl ethers gave the advantage of shorter elution time and reduced trailing. Knights has published data on gas chromatography of trimethylsilyl ethers and trifluoroacetyl esters of a number of sterols, including those of desmosterol and cholesterol.[27] The results reveal that both derivatives of the two sterols could be well separated by gas chromatography on neopentylglycol succinate or HiEFF-8B (Applied Science Laboratories, State College, Pennsylvania).

Column Chromatography

A 1-meter long mixed silicic acid-Super-Cel column for the fractionation of unesterified sterol mixtures has been described by Frantz.[28] The

[21] L. Wolfman and B. A. Sachs, *J. Lipid Res.* **5**, 127 (1964).
[22] S. Fabro, *J. Lipid Res.* **3**, 481 (1962).
[23] W. J. A. VandenHeuvel, C. C. Sweeley, and E. C. Horning, *J. Am. Chem. Soc.* **82**, 3481 (1960).
[24] R. Fumagalli, P. Capella, and W. J. A. VandenHeuvel, *Anal. Biochem.* **10**, 377 (1965).
[25] D. Kritchevsky and W. L. Holmes, *Biochem. Biophys. Res. Commun.* **7**, 128 (1962).
[26] M. Kanai, *J. Biochem. (Tokyo)* **56**, 266 (1964).
[27] B. A. Knights, *J. Gas Chromatog.* **2**, 160 (1964).
[28] I. D. Frantz, Jr., *J. Lipid Res.* **4**, 176 (1963).

separation of desmosterol and cholesterol, as published, is satisfactory. The acetates of these two sterols, as well as of lanosterol and dihydrolanosterol, were adequately separated on silica gel columns.[29] It was shown that the separation factors greatly depend on the type of the structure of the silica gel, namely its pore diameter and surface energy.

Calculation

After the two sterols have been separated by one of these methods, their radioactivities are separately determined. The percentage of substrate that underwent enzymatic reduction is calculated from the distribution of radioactivity between unreacted substrate and product. Total recoveries of radioactivity after derivatization and TLC may be as low as 75%, but fractional losses of the sterols (desmosterol, cholesterol; lanosterol, dihydrolanosterol) are generally the same. To check recoveries, a known amount of tritium-labeled cholesterol can be added to the sample before incubation (with desmosterol-^{14}C), and the recovery of cholesterol can be calculated from the recovery of tritium. From this and from the amount of radioactive desmosterol added, the fractional conversion is calculated.

Substrate Specificity and the Actions of Inhibitors

Detailed comparisons of desmosterol reductase and lanosterol reductase activities show that their subcellular distribution, cofactor requirements, and susceptibility to inhibitors—N-ethylmaleimide, p-chloromercuribenzoate, triparanol [1-(p-β-diethylaminoethoxy)phenyl-1-(p-tosyl)-2-(p-chlorophenyl)ethanol], U-18-666A [3β-(2-diethylaminoethoxy)androst-5-en-17-one hydrochloride], and U-5755 (hexestrol bis-β-diethylaminoethyl ether hydrochloride)—are the same.[4] Moreover, addition of unlabeled desmosterol decreased reduction of lanosterol-^{14}C to the same extent as did addition of unlabeled lanosterol itself. Thus, the side-chain reduction in these two sterols, representing the first and the last intermediates with unsaturated side chains, differing so widely in their nuclear structures, appears to be catalyzed by a single enzyme. This strongly supports the interpretation that the same enzyme also catalyzes side-chain reduction of the other Δ^{24}-sterol intermediates, although the evidence is less direct and largely based on studies of the action of inhibitors, mainly triparanol. Evaluation of these results obviously hinges on the specificity of the inhibitors.

In triparanol-treated animals, the rate of incorporation of injected labeled acetate or mevalonate into the total digitonin-precipitable frac-

[29] P. D. Klein and P. A. Szczepanik, *J. Lipid Res.* 3, 460 (1962).

tion of liver is not reduced.[30] If reactions prior to cyclization of squalene are inhibited at all, the inhibition is not sufficient to make them rate-limiting. The nuclear transformations required for the conversion of lathosterol (Δ^7-cholesten-3β-ol) or of zymostenol (Δ^8-cholesten-3β-ol) to cholesterol are not inhibited by pretreatment of the rat with triparanol.[31] Triparanol added to liver homogenates at sufficiently high concentrations, however, inhibits sterol Δ^7-reductase activity,[12, 32] although it is less potent than AY-9944 [trans-1,4-bis(2-dichlorobenzylaminomethyl) cyclohexane dihydrochloride], or the diazasterols (20,25- and 22,25-diazacholesterol). Addition of triparanol to homogenates at concentrations sufficient to inhibit reduction of lanosterol or desmosterol by about 50% did not inhibit conversion of lanosterol-^{14}C to labeled 27-carbon sterols, although the 27-carbon sterols formed were largely those with unsaturated side chains.[1]

If it is assumed that triparanol is not significantly inhibiting nuclear transformations in vivo, the accumulation of various Δ^{24}-sterols could be taken to indicate inhibition of side-chain reduction steps at all levels of the pathway of cholesterol synthesis. It has been found that triparanol does, in fact, cause accumulation of a number of sterols with unsaturated side-chain, e.g., $\Delta^{7, 24}$-cholestadienol, $\Delta^{8, 24}$-cholestadienol, and dehydromethostenol[33, 34] in the skin of rats, C_{28}- or C_{29}-sterols with unsaturated side chain, as well as lanosterol and desmosterol, in rat liver,[35] and $\Delta^{5, 7, 24}$-cholestatrienol in guinea pig intestine.[36] The effect of a number of inhibitors on the reduction of the side chains of labeled sterols have been studied both in vivo[3] and in vitro.[1, 3, 4, 12, 37] Among the compounds found to be active were the experimental drugs mentioned above, triparanol, U-1866A, U-5755, 20,25-diazacholesterol, and 22,25-diazacholesterol. The evidence obtained justifies the tentative conclusion that sterol Δ^{24}-reductase activity is attributable to a single enzyme active with Δ^{24}-sterols of widely differing nuclear structure.

Mechanism of Action

The enzyme is highly sensitive to sulfhydryl reagents.[1, 4] No cofactor requirements have been reported other than that for NADPH, which is

[30] T. R. Blohm and R. D. MacKenzie, Arch. Biochem. Biophys. 85, 245 (1959).
[31] G. J. Schroepfer, J. Biol. Chem. 236, 1668 (1961).
[32] R. Niemiro and R. Fumagalli, Biochim. Biophys. Acta 98, 624 (1965).
[33] L. Horlick and J. Avigan, J. Lipid Res. 4, 160 (1963).
[34] R. B. Clayton, A. N. Nelson, and I. D. Frantz, Jr., J. Lipid Res. 4, 166 (1963).
[35] D. S. Goodman, J. Avigan, and D. Steinberg, J. Biol. Chem. 238, 1287 (1963).
[36] I. D. Frantz, Jr., A. T. Sanghvi, and R. B. Clayton, J. Biol. Chem. 237, 3381 (1962).
[37] T. J. Scallen, Biochem. Biophys. Res. Commun. 21, 149 (1965).

specifically required; NADH is ineffective.[1,4] NADP+ is almost as effective as NADPH under anaerobic conditions, but this can probably be attributed to its reduction to NADPH by endogenous substrates in the homogenate; a total of only about 0.1 micromole of substrate is generally reduced during assay. Direct utilization of NADPH hydrogen was demonstrated in the microsomal system using tritium-labeled reduced nucleotide[1] and desmosterol-^{14}C and it has been shown in a mass spectrometric study that the proton donated to C-25 is derived from water.[38] Attempts to demonstrate reversibility, using as substrates cholesterol,[39] dihydrolanosterol,[4] and zymostenol,[31] have been uniformly negative.

Role in Sterol Metabolism

After cyclization of squalene to lanosterol, reduction of the 24,25-double bond is one of the necessary reactions en route to cholesterol. It is useful to divide the many sterol intermediates into a "Δ^{24}-side-chain series" and a "saturated side-chain series."[3] As discussed above, Δ^{24}-sterol reductase seems to be capable of catalyzing transition from the Δ^{24}-side-chain series to the saturated side-chain series at any of several points during the transformation of lanosterol to cholesterol. The question whether or not there is a "most favored point" for reduction of the side chain has not been answered. Studies in rats indicate that reduction of lanosterol to dihydrolanosterol and desmosterol to cholesterol are relatively unimportant. Reduction after the removal of the methyl groups at carbons 4 and 14, but before the conversion of the Δ^7- to the Δ^5-configuration, seems to be much more significant.[35]

Occurrence of the Enzyme

The Δ^{24}-reductase must presumably be present in all cholesterol-producing tissues. Generally, its activity is high enough relative to that of the enzymes acting on the sterol nucleus to prevent accumulation of any but extremely small amounts of desmosterol and other sterols with a 24,25-unsaturated side chain. There are, however, some important exceptions. The 12–14-day-old chick embryo contains significant amounts of desmosterol (2% of total sterols),[40] as do the developing rat brain[25,41] and also some human brain tumors.[42] The accumulation of desmosterol may simply reflect the rapid sterol synthesis in these tissues, the side-

[38] D. Steinberg, J. Avigan, R. Dexter, and H. Fales, unpublished results, 1964.
[39] R. Dexter, J. Avigan, and D. Steinberg, unpublished results, 1962.
[40] W. M. Stokes, W. A. Fish, and F. C. Hickey, J. Biol. Chem. 220, 415 (1956).
[41] D. Kritchevsky, S. A. Tepper, N. W. DiTullio, and W. L. Holmes, J. Am. Oil Chemists' Soc. 42, 1024 (1965).
[42] R. Fumagalli, E. Grossi, P. Paoletti, and R. Paoletti, J. Neurochem. 11, 561 (1964).

chain reductase activity becoming rate limiting, or it may be that in developing or dedifferentiated tissue the side-chain reductase activity is lower relative to the activity of the other enzymes in sterol biosynthesis. In the case of the barnacle *Balanus glandula,* in which desmosterol is a major sterol component along with cholesterol,[5] there may be an analogous relative deficiency in side-chain reductase activity. Recent findings in the American cockroach, *Periplaneta americana,* and the tobacco hornworm, *Manduca sexta,* are of interest.[43] These insects, typically unable to synthesize cholesterol de novo, are capable of converting desmosterol to cholesterol when fed the former. These organisms, like the German cockroach, *Blatella germanica,*[44] can satisfy their sterol requirements with sitosterol, and the metabolic modification of the latter's side chain has been shown to involve the formation of desmosterol as an intermediate.[43] The Δ^{24}-sterol reductase activity is, therefore, essential in these insects despite the absence of any appreciable *de novo* cholesterol synthesis.

[43] J. A. Svoboda, M. J. Thompson, and W. E. Robbins, *Life Sci.* **6,** 395 (1967).
[44] W. E. Robbins, R. C. Dutky, R. E. Monroe, and J. N. Kaplanis, *Ann. Entomol. Soc. Am.* **54,** 114 (1961).

[22] Hydrolysis and Formation of Cholesterol Esters in Rat Liver

By DeWitt S. Goodman[1]

The liver plays a major role in the overall metabolism of cholesterol in mammals. First of all, under normal conditions, the liver is importantly involved in the biosynthesis of cholesterol. Second, the liver is responsible for the only quantitatively important pathway of cholesterol catabolism, namely the formation of bile acids. Third, the liver is critically involved in cholesterol absorption and participates in an enterohepatic circulation of cholesterol of considerable magnitude. And finally, the liver plays a central role in the regulation of the plasma concentration of cholesterol and of cholesterol ester, via the formation and secretion of plasma lipoprotein molecules containing these moieties.

During the absorption of cholesterol from the gastrointestinal tract, the cholesterol is absorbed as free sterol into the mucosal cells, followed by the esterification of most of the absorbed cholesterol with long-chain fatty acids. This process of esterification shows a relative specificity for

[1] Career Scientist of the Health Research Council of the City of New York under Contract I-399.

the formation of cholesteryl oleate, both in the rat[1a] and in man.[2] The newly formed cholesterol esters are then mainly incorporated into lymph chylomicrons, and transported via the lymphatic route to the vascular compartment. After entering the blood stream the chylomicron cholesterol esters are almost entirely taken up by the liver. This uptake of the cholesterol esters appears to take place without hydrolysis of the esters. After uptake the esters are slowly hydrolyzed in the liver, followed by the reesterification of some of the liberated cholesterol, and by the equilibration of the liberated free cholesterol with the free cholesterol pools of liver, plasma, and subsequently with peripheral tissues.

In this chapter, methods useful in the study of cholesterol ester metabolism will be described, followed by a discussion of the enzymatic hydrolysis and formation of cholesterol esters in rat liver. More comprehensive information about the metabolism of cholesterol esters in the liver and in other organs can be found in a recent review.[3]

Synthetic and Chromatographic Techniques in the Study of Sterol Ester Metabolism

Methods are now available for the synthesis of cholesterol esters on both a macro- and microscale, and for the chromatographic separation of cholesterol esters from other lipids and from each other.

Synthesis

Macroscale. Several different cholesterol esters of high purity are now available from commercial sources. If an ester is not commercially available, it can often be simply and efficiently synthesized by reacting cholesterol with the appropriate fatty acyl chloride. This reaction has been employed by a number of investigators, and the details of the methods employed are described in their publications.[4-6] The reaction releases HCl and is hence customarily conducted in the presence of alkaline organic solvent, such as pyridine, in order to remove HCl and drive the reaction to completion. After synthesis, the ester should be purified by chromatography on columns of alumina[6] or silicic acid.

A method that is equally or more useful for the synthesis of cholesterol esters is the ester interchange method as described by Mahadevan and Lundberg.[7] This method is particularly useful for the synthesis of esters

[1a] A. Karmen, M. Whyte, and D. S. Goodman, *J. Lipid Res.* **4**, 312 (1965).

[2] R. Blomstrand, J. Gürtler, and B. Werner, *J. Clin. Invest.* **44**, 1766 (1965).

[3] D. S. Goodman, *Physiol. Rev.* **45**, 747 (1965).

[4] L. Swell and C. R. Treadwell, *J. Biol. Chem.* **212**, 141 (1955).

[5] W. C. Gray, *J. Chem. Soc.* p. 3733 (1956).

[6] D. Deykin and D. S. Goodman, *J. Biol. Chem.* **237**, 3649 (1962).

[7] V. Mahadevan and W. O. Lundberg, *J. Lipid Res.* **3**, 106 (1962).

of cholesterol with polyunsaturated fatty acids, or with unusual fatty acids, the acyl chlorides of which are not available or are difficult to prepare. As described,[7] the method employs sodium ethylate which is freshly made by dissolving small quantities of clean metallic sodium in absolute ethanol and evaporating to dryness under vacuum. The sodium ethylate (0.05–0.1 g) is mixed with 0.01 mole (4.3 g) of cholesteryl acetate and with 0.01 mole of the methyl ester of the desired fatty acid in a 200-ml round-bottom flask. The flask is connected to a manometer and aspirator through a trap cooled in an acetone–dry ice bath. The flask is flushed several times with nitrogen and then heated under vacuum (20–30 mm Hg) at 80–90° for 1 hour with occasional shaking. The evolution of methyl acetate is observed by effervescence, which subsides toward the end of the reaction. The reaction mixture is cooled, washed with low-boiling petroleum ether or hexane, and filtered to remove insoluble material. The filtrate is evaporated, and the product cholesterol ester is then purified by chromatography on a column of silicic acid.

Microscale. The synthesis of cholesterol esters can also be conducted on a microscale by reacting cholesterol with the appropriate fatty acyl chloride, or by carrying out the ester interchange reaction with small amounts of reactants. Such microsyntheses are particularly required for the preparation of labeled cholesterol esters of high specific radioactivity.

If the acyl chloride of the desired fatty acid is available, milligram or microgram amounts of high specific radioactivity cholesterol ester labeled in the cholesterol moiety can be synthesized by reacting the labeled cholesterol with an excess of acyl chloride and pyridine.[6] This procedure efficiently utilizes the labeled cholesterol. To carry out the reaction, a small amount (e.g., 5 mg) of labeled cholesterol is put in a ground glass-stoppered small centrifuge tube followed by the addition of 20 μl of the appropriate acyl chloride, and 0.3 ml of dry pyridine (distilled from BaO). The mixture is flushed with nitrogen and heated at 50–60° for 3 hours. The mixture is transferred to cold water, and the products are extracted into petroleum ether. The petroleum ether solution is washed with 0.5 N HCl, with 0.1 N NaOH in 50% ethanol (to extract free fatty acids), and with water. After evaporation of the petroleum ether the residues are purified by chromatography on columns of alumina or silicic acid. The yield of labeled cholesterol ester is generally close to 80%.

The fatty acyl chloride employed in this microsynthesis can be obtained commercially, or can be rapidly synthesized by reacting the fatty acid with oxalyl chloride as described by Pinter et al.[8] This method of preparing acyl chlorides is useful for a wide range of fatty acids, includ-

[8] K. G. Pinter, J. G. Hamilton, and J. E. Muldrey, *J. Lipid Res.* **5**, 273 (1964).

ing polyunsaturated acids. To carry out this reaction as described,[8] 1 millimole of fatty acid is placed in a small Florence flask with a side arm, and oxalyl chloride is then added in approximately 3-fold molar excess. The flask is closed with a ground glass stopper and partly evacuated by a water aspirator (30 mm Hg) connected to the side arm. The flask is placed in a water bath at 65°, and the mixture is allowed to boil for 5 minutes to drive off the excess oxalyl chloride. After addition of another 3-fold molar excess of oxalyl chloride and repetition of the procedure, the yield of fatty acyl chloride is usually in the range of 75–85%. The impure reaction product can be used directly to synthesize cholesterol esters.

Cholesterol esters labeled in the fatty acid moiety can be synthesized by reacting free cholesterol with the appropriate labeled fatty acyl chloride. Before conducting the reaction the labeled fatty acyl chloride is synthesized by the method described above, or by the equilibration of the labeled free fatty acid with its unlabeled fatty acyl chloride as described by Borgström and Krabisch.[9] In order to utilize maximally the labeled fatty acid, the reaction should use a 1.5- to 2-fold molar excess of free cholesterol, together with the usual excess of dry pyridine.

Cholesterol esters labeled in either the cholesterol or the fatty acid moiety can be synthesized on a microscale by means of the ester interchange method described above. For example, labeled cholesteryl arachidonate can easily be prepared by mixing together labeled cholesteryl acetate (1 millimole, 0.43 g) with methyl arachidonate (0.33 g), and then adding a few milligrams (e.g., 15 mg) of freshly prepared sodium ethylate. The mixture is flushed with nitrogen and heated under vacuum as described above. At the end of the reaction, the product is extracted and purified as for the macroscale synthesis. The yield, in terms of either the cholesterol or the fatty acid moiety, is close to 80%.

Cholesterol esters can also be prepared enzymatically in high yield with pancreatic extracts. This method has been used for the synthesis of milligram amounts of labeled unsaturated cholesterol esters.[10]

Chromatography

Separation of Cholesterol Esters from Other Lipids. The *en bloc* separation of cholesterol esters from other lipid classes in tissue extracts or in other lipid mixtures can readily be achieved by chromatography on columns of silicic acid or on thin layers of silica gel. Quantitative (more than 98%) recovery of cholesterol esters, virtually uncontaminated (less than 1%) by triglycerides, can be achieved by chromatography on

[9] B. Borgström and L. Krabisch, *J. Lipid Res.* 4, 357 (1963).
[10] L. Swell and C. R. Treadwell, *Anal. Biochem.* 4, 335 (1962).

columns of silicic acid, using mixtures of benzene and n-hexane to elute the different neutral lipids. In order to carry out this separation, the column is packed with silicic acid (Unisil, 100–200 mesh, Clarkson Chemical Co., Williamsport, Pennsylvania) as a slurry in hexane, and the lipid sample is applied to the column in solution in hexane. For a column of 1 cm internal diameter containing 5 g of silicic acid, serial elutions are carried out as follows:[11] 45 ml of 10% benzene in hexane; 90 ml of 21% benzene in hexane; 50 ml benzene followed by 100 ml chloroform; and 100 ml of methanol. The four fractions contain, respectively: hydrocarbons (e.g., squalene) ; cholesterol esters; triglycerides plus partial glycerides plus free cholesterol plus FFA; and phospholipids. The separation should be standardized in each laboratory to ensure the quantitative recovery of cholesterol esters uncontaminated by triglycerides (see, e.g., references cited in footnotes 1a and 11). The optimal concentration of benzene in hexane for elution of cholesterol esters has been found to vary slightly from laboratory to laboratory. The load of total lipid should not exceed 75 mg for 5 g of silicic acid.

Cholesterol esters can readily be separated from other lipid classes by thin-layer chromatography on silica gel G, using mixtures of petroleum ether, ethyl ether, and glacial acetic acid (e.g., in ratios of 80:20:2 or 90:10:1).[12] After visualization under ultraviolet light by means of 2′,7′-dichlorofluorescein or by rhodamine 6G, the spots or streaks comprising the cholesterol esters can be scraped from the plate and eluted with chloroform or ethyl ether.

After isolation of cholesterol esters as a class, the composition of the cholesterol esters can be determined by GLC of the cholesterol ester fatty acids. This procedure has been widely employed during recent years, and considerable information has been acquired on the composition of the cholesterol esters in various tissues, both under normal and abnormal circumstances. Most of the available data on cholesterol ester fatty acid composition of tissues, as determined by GLC, has been summarized in a recent review.[3] It has been pointed out[3] that these data contain a potential uncertainty which derives from the fact that the cholesterol ester fraction isolated during silicic acid column chromatography would also contain other fatty acid monoesters, such as retinyl esters, if these were present in the lipid extract. It has been suggested[3] that studies with GLC on the composition of cholesterol esters in tissues should employ some internal check on the relative amounts of cholesterol and fatty acid in the cholesterol ester fraction.

Separation of Different Cholesterol Esters. Cholesterol esters can be

[11] D. S. Goodman and T. Shiratori, *J. Lipid Res.* **5**, 307 (1964).
[12] H. K. Mangold, *J. Am. Oil Chemists Soc.* **38**, 708 (1961).

separated on the basis of the degree of unsaturation of the fatty acid component by thin-layer chromatography on silica gel impregnated with silver nitrate.[13,14] A useful technique for samples derived from serum and liver[14] employs thin layers prepared from a slurry of 25 g of silica gel plus 55 ml of 0.7% AgNO$_3$ in water. The plates are developed with benzene–hexane, 1:1, as ascending solvent. After development the plates are sprayed with dye and examined under UV light. This system results in the wide separation of saturated cholesterol esters (R_f ca. 0.8), monounsaturated esters (R_f ca. 0.65), diunsaturated esters (R_f ca. 0.4), and more highly unsaturated (e.g., tetraunsaturated) esters (R_f 0.0–0.2). The separated esters can be scraped on to a filter funnel and eluted with 15 ml of chloroform. Polyunsaturated esters containing more than two double bonds in the fatty acid should be eluted further with hot benzene–ethyl ether, 1:1, in order to effect a quantitative elution.

Resolution of a wider range of cholesterol esters can be achieved by an initial chromatography with a polar solvent, so as to separate the highly unsaturated esters from each other, followed by rechromatography of the less unsaturated esters with a less polar solvent mixture. As described,[13] this method employs plates prepared with a slurry of 24 g of silica gel plus 50 ml of 5% AgNO$_3$ in water. Chromatography with pure diethyl ether as ascending solvent results in the resolution of tetraenoic acid esters (R_f ca. 0.5), pentaenoic acid esters (R_f ca. 0.2), and more highly unsaturated esters (which remain at the origin). Saturated, mono- and diunsaturated esters (R_f 0.9–0.95) do not separate in this system, and trienoic acid esters (R_f ca. 0.85) barely separate from the more saturated esters. Resolution of the saturated and the mono-, di-, and triunsaturated esters from each other is achieved by elution of these *en bloc* after the first chromatography followed by rechromatography with ethyl ether–hexane, 1:4, as ascending solvent. The triunsaturated and more saturated esters are eluted from the silica gel with pure ethyl ether, whereas the more unsaturated components are eluted with chloroform–methanol, 2:1. In any laboratory, the elution method should be standardized to ensure the comparable, quantitative elution of the different cholesterol esters after TLC.

Cholesterol esters of the same degree of unsaturation can be separated according to fatty acid chain length by reversed phase paper chromatography. Methods for carrying out such separations have been described by several investigators.[15–17] One effective method consists of reversed-phase

[13] L. J. Morris, *J. Lipid Res.* **4**, 357 (1963).
[14] D. S. Goodman and T. Shiratori, *J. Lipid Res.* **5**, 578 (1964).
[15] J. A. Labarrère, J. R. Chipault, and W. O. Lundberg, *Anal. Chem.* **30**, 1466 (1958).
[16] A. Kuksis and J. M. R. Beveridge, *Can. J. Biochem. Physiol.* **38**, 95 (1960).
[17] Č. Michalec and J. Strašek, *J. Chromatog.* **4**, 254 (1960).

paper chromatography as described in detail elsewhere in this volume for the separation of retinyl esters from each other. Although these paper chromatographic methods are useful and effective, they are relatively time-consuming, and very limited in the amount of material that can be chromatographed at one time. Other methods should, therefore, be considered for achieving these separations. A method that can be employed on a very large scale consists of the reversed-phase chromatographic separation of cholesterol esters by factice column chromatography as described by Hirsch.[18] Another method which has been reported to give rapid separation of cholesterol esters consists of reversed-phase TLC, using thin layers of silicic acid plus gypsum impregnated with paraffin oil, with acetic acid as ascending solvent[19] (see also this volume, pp. 71 and 77). The gas chromatographic separation of intact sterol esters has also been reported,[20] and may become increasingly valuable.

Hydrolysis of Cholesterol Esters in Rat Liver

As already mentioned, hydrolysis within the liver is the immediate metabolic fate of almost all chylomicron cholesterol ester. The liver enzymes responsible for cholesterol ester hydrolysis therefore play a crucial role in the normal enterohepatic circulation of cholesterol and in the metabolism of newly absorbed dietary cholesterol. Since the enterohepatic circulation of cholesterol is of substantial size regardless of the amount of cholesterol in the diet, and since the cholesterol esters in liver are constantly turning over, it is apparent that hydrolysis of cholesterol esters in liver is a quantitatively important reaction in the overall metabolism of cholesterol in the body.

Assay Method

Principle. Radioactive cholesterol ester is cleaved by the enzyme during *in vitro* incubation to form radioactive free cholesterol, which may be separated from the substrate by chromatography on a column of alumina or silicic acid, or on a thin-layer plate of alumina or silica gel. The isolated product may be further characterized by formation of several derivatives.

Incubation. Incubations are carried out in 25-ml Erlenmeyer flasks, at 37°, with shaking, in a metabolic incubator. The incubation mixture contains enzyme protein plus 200 micromoles of potassium phosphate buffer, pH 7.4, in a volume of 2 ml. To the mixture, substrate cholesterol ester (1–2 millimicromoles of labeled cholesteryl palmitate or oleate) is

[18] J. Hirsch, *J. Lipid Res.* **4**, 1 (1963).
[19] Č. Michalec, M. Šulc, and J. Meštan, *Nature* **193**, 63 (1962).
[20] A. Kuksis, *Can. J. Biochem.* **42**, 407 (1964).

added dissolved in 50 μl of acetone, via a 50-μl syringe. Air is used as gas phase. Incubations are usually conducted for 1 hour.

Extraction and Chromatography. Reactions are terminated by transferring the 2-ml incubation mixtures to chloroform–methanol, 2:1 (v/v), and washing with chloroform–methanol to a final volume of 50 ml. The mixture is then separated into two phases by the addition of 8 ml of 0.01 N H_2SO_4, and the entire lower $CHCl_3$ phase, except 0.5 ml, is withdrawn. Carrier cholesterol and cholesteryl palmitate (1 mg each) are added, and the samples are evaporated to dryness with a stream of nitrogen. The free and esterified sterols are then separated by chromatography on a column of alumina (Woelm, neutral, grade II activity). Small (10 mm internal diameter) columns containing 2 g of alumina are charged with the sample dissolved in 1–2 ml of light petroleum ether. The cholesterol ester is eluted with 12 ml benzene-petroleum ether, 1:1. Free sterol is then eluted with 10 ml of acetone-ethyl ether, 1:1. The fractions are eluted directly into counting vials, followed by evaporation of the solvent and assay for radioactivity with a liquid scintillation counter.

Characterization. The radioactivity eluted in the free sterol fraction may be further characterized as free cholesterol by the formation of the digitonide complex, or by addition of a few milligrams of pure unlabeled cholesterol, followed by the formation of the dibromide derivative which can be crystallized serially to constant specific radioactivity.

Enzyme Preparation and Purification[6]

Fed rats weighing approximately 200 g are killed by decapitation. The livers are excised and immediately rinsed in ice-cold 0.1 M potassium phosphate buffer, pH 7.4. The livers are passed through a tissue press, and the pulp homogenized at ice temperature in a Potter-Elvehjem homogenizer with a loose fitting Teflon pestle, using 2.5 ml of the above buffer per gram of liver pulp. After removal of nuclei and cell debris by centrifugation at 2000 g for 30 minutes, the supernatant fraction (S-2) is centrifuged at 10,000 g for 30 minutes to sediment the mitochondria. Microsomes are sedimented from the 10,000 g supernatant (S-10) by centrifugation at 104,000 g for 1 hour. After this latter centrifugation the layer of free-floating fat at the top of the tube is removed and discarded. The clear soluble supernatant fraction (defatted S-104) is then removed, and the microsomes are washed by suspension in a large volume of buffer, followed by centrifugation as before. The microsomes are finally suspended in a volume of buffer equal to one-tenth the volume of the S-10 from which they had been separated. All centrifugations, and subsequent manipulations, are conducted at 0° to 5°.

Assay of the several different subcellular fractions for cholesterol esterase activity revealed that the bulk of the hydrolytic activity was associated with the soluble supernatant fraction, although some activity was detected in both particulate fractions. It has been estimated that the microsomes contribute between 11 and 32% of the total hydrolysis observed with a whole liver homogenate.

Further purification of the soluble cholesterol esterase can be achieved by the addition of solid ammonium sulfate to 30% saturation at 0°C. Thus, 21.2 g of finely ground ammonium sulfate is slowly added per 100 ml of S-104 contained in a beaker packed in ice. After slow stirring for 30 minutes the mixture is centrifuged at 10,000 g for 20 minutes. The precipitate is then dissolved in a few milliliters of water, and the solution passed over a small column of Sephadex G-25 to remove ammonium sulfate. The resulting protein solution, called the F_0^{30}, contains the enzymatic activity and is adjusted to 0.01 M potassium phosphate pH 7.4 before further use.

Additional purification has been achieved by adsorption to and elution from calcium phosphate gel.[6] After dilution of the F_0^{30} to 5 mg protein per milliliter, calcium phosphate gel was added in the amount of 28 mg per 100 mg of protein. The suspension was agitated at ice temperature for 1 minute and allowed to stand for 15 minutes or more. The gel was separated by centrifugation and discarded. The supernatant fluid was acidified to pH 6.5 with 0.1 N acetic acid, and 100 mg of gel per 100 mg of protein added. After centrifugation the sedimented gel was eluted serially with 1 ml per 10 mg of gel of 0.1 and 0.5 M phosphate buffer, pH 7.4, respectively. The latter eluate contained the partially purified cholesterol esterase activity, and was diluted to 0.1 M with respect to phosphate before use.

The results of a typical purification experiment are summarized in

PURIFICATION OF RAT LIVER SOLUBLE CHOLESTEROL ESTERASE

Fraction	Volume (ml)	Protein (mg/ml)	Total protein (mg)	"Relative" enzyme activity[a]	"Total" enzyme activity[b]
S-2	190	40.0	7600	19	146,680
S-10	168	36.5	6132	30	184,573
S-104	137	27.0	3700	160	591,840
F_0^{30}	164	5.0	820	338	277,160
Gel eluate	171	0.3	51	1321	67,371

[a] "Relative" enzyme activity = % hydrolysis $\times$ 5000 cpm per milligram of protein, in a 1 hour incubation, assuming linearity between amount of protein actually incubated and extent of hydrolysis.

[b] "Total" enzyme activity = "relative" enzyme activity times total protein.

the table. The purified fraction after gel elution usually displayed an 8- to 10-fold purification when compared to the S-104. The apparent purification was always very much greater when compared to the S-2 or S-10 fractions. Analysis for free and ester cholesterol revealed that the S-2 and S-10 fractions contained significant amounts of cholesterol ester, whereas the defatted S-104 did not. The estimates of the relative enzyme activity of the S-2 and S-10 fractions were hence in error to the extent that the endogenous substrate in the preparation was available for hydrolysis. This error is apparent in the paradoxically low total enzyme activity of the S-2 and S-10 fractions, and accounts for the much greater apparent purification of the gel eluate when compared to the S-2 or S-10 fractions, rather than to the S-104 fraction.

Properties

Mechanism. No cofactors are required for the reaction. The mechanism of the reaction consists of the hydrolytic cleavage of the ester bond between cholesterol and the fatty acid.

Stability. The soluble cholesterol esterase is quite unstable, and becomes largely inactivated by freezing and thawing, dialysis, or lyophilization. The enzyme can be kept for several days, with only slight loss of activity, by storage at $-20°$ as an ammonium sulfate precipitate (F_0^{30}).

Dispersion of Substrate. Maximum activity was obtained by addition of substrate cholesteryl palmitate to the incubation in solution in 50 μl of acetone. Much less activity was obtained by addition of substrate as a uniform dispersion in saline solutions of 15 mM sodium glycocholate or of 5% Tween 20, or as a uniform dispersion in a mixture of ethanol, acetone, and serum albumin. Addition of Tween 20, sodium glycocholate or taurocholate, human serum albumin, human β-lipoprotein, or rat liver polyglycerolphosphatide to incubations before the addition of substrate in 50 μl of acetone all inhibited ester hydrolysis.

Inhibitors; pH. Both the soluble and the microsomal cholesterol esterases were inhibited by sulfhydryl reagents, including N-ethylmaleimide, iodoacetamide, and p-hydroxymercuribenzoate. These inhibitions were readily overcome by addition of excess GSH. The soluble enzyme was largely inhibited by 10^{-4} M DFP, and was only slightly inhibited by 10^{-4} M eserine; the microsomal enzyme was more strongly inhibited by both of these reagents. The soluble enzyme showed a broad optimal pH range between 6.5 and 7.5, whereas microsomes displayed a sharper pH optimum near 6.1.

Specificity. During purification the soluble enzyme was separated from the major esterase activities involved in the hydrolysis of p-nitrophenylacetate (an ester hydrolyzed by a wide range of esterases) and p-nitrophenylphosphate (a substrate for phosphatases). Most of the activity for hydrolysis of p-nitrophenylacetate was found in the microsomes. Both the soluble enzyme and the microsomes were active in effecting the hydrolysis of ethyl palmitate. The soluble enzyme showed only minimal activity for the hydrolysis of tripalmitin.

Fatty Acid Specificity. The fatty acid specificity of the soluble cholesterol esterase was examined in two ways. In one set of experiments, different cholesterol esters were individually studied at several substrate levels and substrate-concentration curves were determined. Unsaturated cholesterol esters (cholesteryl oleate and linoleate) were hydrolyzed more rapidly than saturated esters (cholesteryl palmitate and stearate). In order to rule out the possibility that these differences resulted from different physical properties of the individual cholesterol ester substrate dispersions, experiments were conducted to measure the hydrolysis of each ester in an equimolar mixture of four long-chain esters. The mixed emulsions so formed minimized any effects dependent on different properties of the individual substrate ester dispersions. Similar results were obtained in both sets of experiments. The order of hydrolysis of cholesterol esters by the soluble enzyme in both experiments was: cholesteryl linoleate $\cong$ oleate $>$ acetate $>$ palmitate $\geqslant$ stearate.

Similar experiments were conducted with microsomes. The microsomes were vastly more active in hydrolyzing cholesteryl acetate than in hydrolyzing long-chain fatty acid esters of cholesterol. The order of specificity for the long-chain fatty acid esters was similar to that seen with the soluble enzyme, with unsaturated cholesterol esters being hydrolyzed more rapidly than saturated esters.

Kinetics. In studies with the soluble enzyme, the rate of hydrolysis was found to be constant for the initial 40 minutes of incubation, and decreased by less than 10% at 90 minutes and less than 20% at 120 minutes. The initial rate of hydrolysis of cholesteryl palmitate was linearly related to enzyme concentration at concentrations of F_0^{30} below 4 mg of protein per milliliter. The rate of hydrolysis was not proportional to protein concentration at concentrations of F_0^{30} above 5 mg of protein per milliliter.

Substrate concentration curves indicated the apparent presence of typical Michaelis kinetics for each of several different cholesterol esters studied. In each instance studied, plotting $1/v$ against $1/s$ in the usual way resulted in a straight line. The extrapolated values for maximal velocities of hydrolysis (expressed as millimicromoles substrate hydrolyzed per 10 mg F_0^{30} per 30 minutes) were: cholesteryl palmitate, 3.1; cholesteryl acetate, 9.1; and cholesteryl oleate and linoleate, 14.3. The approximate substrate concentrations for half-maximal velocity (i.e., the equivalent of the K_m) were: $14.3 \times 10^{-6} M$ for cholesteryl acetate, oleate, and linoleate, and $4.8 \times 10^{-6} M$ for cholesteryl palmitate.

Physiological Implications. The rate of hydrolysis of cholesteryl oleate by the soluble enzyme was in close agreement with the net rate of hydrolysis of chylomicron cholesterol ester observed *in vivo.*

Formation of Cholesterol Esters in Rat Liver

In addition to hydrolytic activity, the liver also contains enzymes that catalyze the formation of fatty acid esters of cholesterol. Although the existence of such activity has been known for some time, it was not until 1958 that Mukherjee et al.[21] obtained evidence that the cofactors ATP and coenzyme A were required for this reaction to occur in liver homogenates. More recent studies[22] have defined most of the details of this reaction. These studies will be summarized here.

Assay

Principle. Cholesterol esterification is studied by means of an assay similar to that used to study cholesterol ester hydrolysis. In this assay, radioactive cholesterol is esterified by the enzyme during an *in vitro* incubation, to form radioactive cholesterol ester, which may be separated from the substrate by chromatography on columns of alumina or silicic acid, or on thin-layer plates of alumina or silica gel. The isolated product may be further characterized by saponification, followed by the isolation and characterization of the liberated labeled free cholesterol.

Incubation. Incubations for the study of cholesterol esterification can be carried out in two ways. In one method, oleyl-CoA is added to effect the esterification of the labeled cholesterol. In the other method the oleyl-CoA is replaced by ATP + CoA + potassium oleate; this method provides an assay for a combination of two reactions, namely the activation of the fatty acid plus the esterification of cholesterol by means of the fatty acyl-CoA thiol ester.

In the first method the 3 ml incubation mixture contains enzyme protein; plus 300 micromoles of potassium phosphate buffer, pH 7.1; plus 3 mg of fatty acid-deficient human serum albumin; plus 75 millimicromoles of oleyl-CoA. In the second method the oleyl-CoA is replaced by 6 micromoles of ATP plus 0.3 micromole of CoA plus 75 millimicromoles of potassium oleate. Radioactive cholesterol (20 millimicromoles) is added in 100 μl of acetone solution, via a 100-μl syringe. The incubations are carried out in 25-ml Erlenmeyer flasks, at 37°, with shaking in a meta-

[21] S. Mukherjee, G. Kunitake, and R. B. Alfin-Slater, *J. Biol. Chem.* **230**, 91 (1958).
[22] D. S. Goodman, D. Deykin, and T. Shiratori, *J. Biol. Chem.* **239**, 1335 (1964).

bolic incubator, using air as the gas phase, for a period of from 30–60 minutes.

Reagents. Fatty acyl-CoA esters are chemically synthesized by the method of Goldman and Vagelos.[23] Free fatty acid-deficient human serum albumin is prepared by extracting dry serum albumin with 5% glacial acetic acid in isooctane.[24] Aqueous solutions of the potassium salts of labeled and unlabeled free fatty acids are prepared by first dissolving the fatty acid in a small volume (e.g., 1 or 2 ml) of ethanol, followed by the addition of exactly that amount of KOH solution required to neutralize the fatty acid. After evaporation of the solvent, the resulting solid potassium salt of the fatty acid is dissolved in hot water, and the solution is diluted to the desired concentration. The ATP and CoA solutions should be neutralized with NaOH to a pH of close to 7 before use.

Extraction and Chromatography are carried out as described above for cholesterol ester hydrolysis.

Enzyme Preparation

Rats weighing approximately 200 g are killed by decapitation, and the livers are perfused *in situ* via the superior vena cava with ice-cold 0.85% NaCl solution. The livers are then excised and homogenized in the manner described above for studies of cholesterol ester hydrolysis, so as to prepare washed mitochondria, washed microsomes, and the soluble supernatant fraction. The enzymes mediating cholesterol esterification in rat liver are exclusively particulate, with mitochondria and microsomes displaying similar activity when prorated according to the amount of homogenate from which they are derived. The soluble fraction displays no activity for cholesterol esterification, and there is no further increase in esterifying activity on addition of the soluble fraction to washed cell particles.

Properties

Cofactor Requirements; Mechanism of Reaction. In experiments with either washed mitochondria or washed microsomes, in which the incubation mixture consisted of enzyme plus phosphate buffer plus labeled cholesterol as substrate, addition of both ATP and coenzyme A were required for cholesterol esterification. There was no divalent cation requirement, and no requirement for added free fatty acid, although the addition of potassium oleate plus human serum albumin usually resulted in an increase in the yield of cholesterol ester.

Omission of serum albumin in the presence of either added free fatty

---

[23] P. Goldman and P. R. Vagelos, *J. Biol. Chem.* **236**, 2620 (1961).
[24] D. S. Goodman, *Science* **125**, 1296 (1957).

acid or fatty acyl-CoA usually resulted in a considerable decrease of cholesterol esterification. This decrease was caused by inhibition of cholesterol esterification by the added free fatty acid or by FFA released from the fatty acyl-CoA by hydrolysis. The serum albumin serves to bind fatty acid anions and thus to prevent them from inhibiting the reaction.

The cofactors ATP and coenzyme A could be completely replaced by addition of preformed fatty acyl-CoA ester (e.g., oleyl-CoA). The mechanism of the reaction hence consists of the transfer of the fatty acyl group from a fatty acyl-CoA ester to the free hydroxyl group of cholesterol. The enzyme involved in cholesterol ester formation in rat liver should be designated a fatty acyl-CoA–cholesterol acyl transferase. Attempts to demonstrate cholesterol esterification by liver preparations with labeled fatty acids derived from lecithin or from free fatty acids in the absence of ATP and CoA have not been successful.

pH; Bile Salts. The effect of pH was studied with two sets of incubations, one with oleyl-CoA and one with ATP + CoA + potassium oleate. The optimal pH was found to lie in a broad range above 6.5, with a suggestion of an optimum close to pH 7.1. Slight differences were observed between the two sets of incubations. The enzyme was strongly inhibited by the addition of conjugated bile salts.

Effect of Added Free Fatty Acid. The esterification of cholesterol with liver particles is remarkably sensitive to the presence of long-chain fatty acid anions. Striking inhibition of cholesterol esterification was observed on adding potassium oleate to a final concentration of 10^{-5} M or greater. This inhibition could be completely overcome by addition of serum albumin to bind the fatty acid anions. When the mole ratio of fatty acid to serum albumin exceeded 7, however, severe inhibition was again observed. This critical mole ratio of 7 corresponds to the saturation of the first two classes of binding sites of albumin for fatty acid anions, beyond which the concentration of unbound fatty acid in solution rapidly rises above 10^{-5} M.[25]

As already mentioned, it was not necessary to add fatty acid to the incubation flasks for cholesterol esterification to occur. Furthermore, addition of free fatty acid (plus serum albumin) resulted in no significant stimulation of esterification of labeled cholesterol, except for the specific addition of oleate. Direct participation of added fatty acid in cholesterol esterification was nevertheless shown by analysis of the composition of the newly synthesized labeled cholesterol esters. Addition of one or another individual free fatty acid (palmitate or oleate or linoleate) to the incubation resulted in the almost exclusive synthesis of the corresponding cholesterol ester. In the absence of added fatty acid, rat liver homogenate

[25] D. S. Goodman, *J. Am. Chem. Soc.* **80**, 3892 (1958).

fractions produced mixtures of labeled cholesterol esters, including saturated, monounsaturated, and diunsaturated esters, with very little cholesteryl arachidonate. Monounsaturated esters predominated in incubations with washed mitochondria or washed microsomes as enzyme. Experiments with ^{3}H-cholesterol and ^{14}C-oleate together also showed the participation of added free fatty acid and suggested that the endogenous fatty acid used for cholesterol esterification was derived from the free fatty acid pool of the particle preparations. It is probable that added fatty acids other than oleate provide little or no stimulation because the endogenous fatty acid pool is of such size and composition as to permit a fairly high rate of cholesterol esterification to occur in the absence of added fatty acid.

Kinetics. The rate of esterification was studied with 3-ml incubations containing 0.4 mg of microsomal protein, potassium phosphate buffer pH 7.4 (300 micromoles), fatty acid-deficient human serum albumin (3 mg), oleyl-CoA (50 millimicromoles), and ^{3}H-cholesterol (20 millimicromoles). The rate was nearly constant for 20 minutes, with less than a 30% decrease in 1 hour. The initial rate was equivalent to the esterification of 16% of the labeled cholesterol in 1 hour. The rate could not be defined more precisely (i.e., in terms of millimicromoles of cholesterol esterified per unit time per unit of enzyme protein) because of the participation of some of the endogenous free cholesterol of the washed cell particles as substrate in the reaction.

The reaction proceeds best when only a small amount of cell particles is used as enzyme. Maximal esterification of labeled cholesterol was observed with less than 1 mg of microsomal protein per incubation. Inhibition of esterification was observed with larger amounts of microsomes.

The rate of esterification was studied under a variety of conditions. In one set of experiments, the percentage esterification of each of two different amounts of ^{3}H-cholesterol (5 and 30 millimicromoles) was studied at varying concentrations (2–50 millimicromoles per incubation) of each of four different fatty acyl-CoA esters. The results showed that considerably more esterification was observed with oleyl-CoA than with any of the other acyl-CoA esters, at each level of added fatty acyl-CoA ester. At both concentrations of added cholesterol the relative rates of esterification were oleyl- > palmityl- > stearyl- > linoleyl-CoA.

Fatty Acid Specificity. The rat liver enzymes responsible for cholesterol ester formation display a strong relative specificity for the formation of cholesteryl oleate, compared to other cholesterol esters. This conclusion derives from several lines of evidence. First, only the addition of oleate, with the concomitant synthesis of predominantly cholesteryl oleate, resulted in a large stimulation of cholesterol esterification. Second, cholesteryl oleate was the predominant ester synthesized from equimolar mixtures of added free fatty acid. Third, the rate of cholesterol esterification with oleyl-CoA was considerably greater than the rate with any of the other fatty acyl-CoA esters studied. This was apparent both when the fatty acyl-CoA esters were used individually (as described above) and when they were added as equimolar mixtures of four CoA esters. The relative rates of cholesterol esterification with the different fatty acyl-CoA esters with microsomes, in both types of experiments, were: oleyl- > palmityl- > stearyl- > linoleyl-CoA. When free fatty acid mixtures plus cholesterol were used, the relative specificity of incorporation of oleate, palmitate, and stearate into cholesterol esters was similar to that observed with the corresponding CoA esters, but the relative incorporation of linoleate was greater than that seen with linoleyl-CoA.

[23] Pancreatic Sterol Ester Hydrolase

By George V. Vahouny and C. R. Treadwell

$$\text{Sterol} + \text{fatty acid} \xrightarrow{\text{bile salt}} \text{sterol ester}$$

Assay Method

Principle. The most convenient methods for measurement of this enzyme are to determine the amount of radioactive sterol ester formed from labeled cholesterol in the presence of excess fatty acid and bile salt, or to determine the amount of labeled free sterol formed by enzymatic hydrolysis of cholesterol-^{14}C ester in the presence of bile salt.[1] During the incubation aliquots of the reaction mixture are added to acetone–ethanol (2:1) and subsequently spotted on micro thin-layer silicic acid chromatoplates for separation of the free and esterified sterol.[2] The silicic acid areas containing these fractions are scraped into scintillation vials, and, after addition of methanol and a scintillation mixture, are counted. The percentage of the total ^{14}C-sterol present in free or esterified form can then be calculated and converted to units of enzyme activity.

Reagents for Sterol Ester Synthesis

Phosphate buffer, 0.154 M, pH 6.2
NH$_4$Cl, 1.0 M

[1] G. V. Vahouny, S. Weersing, and C. R. Treadwell, *Arch. Biochem. Biophys.* **107**, 7 (1964).
[2] G. V. Vahouny, C. R. Borja, and S. Weersing, *Anal. Biochem.* **6**, 555 (1963).

Sodium taurocholate, 0.31 M
Oleic acid, 0.465 M, in petroleum ether
Cholesterol-4-^{14}C, 0.155 M, in petroleum ether
Enzyme: pancreatic juice, 10 mg of protein/ml; or 10% homogenate
of pancreas in glycerol–water (1:1 v/v), physiological saline or
0.154 M phosphate buffer[3, 4]

Reagents for Sterol Ester Hydrolysis

Phosphate buffer, 0.154 M, pH 6.6
Sodium taurocholate, 0.31 M
Lecithin, 30 mg/ml in petroleum ether
Cholesterol-4-^{14}C oleate, 0.155 M, in petroleum ether
Enzyme: See above

Reagents for Isotope Determination

Acetone–ethanol, 1:1 (v/v)
Silica gel G, for thin-layer chromatography
Microscope slides
Developing solvent; hexane–diethyl ether–acetic acid (83:16:1,
v/v)
Methanol, spectroanalyzed
Scintillation mixture; 100 mg of 1,4-di-(2,5-phenyloxazolyl)ben-
zene, 4 g of 2,5-diphenyloxazole per liter of toluene

Preparation of Substrate. The assay mixture for sterol ester synthesis
contains 15.5 micromoles of cholesterol-4-[14]C (specific activity 1×10^5
dpm/mg), 46.5 micromoles of oleic acid, 31 micromoles of sodium taurocholate and 100 micromoles of ammonium chloride in a final volume of
1.5 ml of phosphate buffer, pH 6.2. All components are placed in a Potter-
Elvehjem homogenizing tube, homogenized, and the petroleum ether is
evaporated at 37° under nitrogen. The medium is then rehomogenized
prior to use.

For sterol ester hydrolysis, the medium contains 15.5 micromoles of
cholesterol-4-^{14}C oleate (specific activity 1×10^5 dpm/mg), 31 micromoles of sodium taurocholate, and 18 mg of lecithin in a final volume of
6 ml of phosphate buffer, pH 6.6. The components are homogenized as
above and, after evaporation of the petroleum ether, the assay mixture is
sonicated (15–20 kc/second) for 15 minutes at room temperature.

Procedure. The assay medium is incubated for 15 minutes at 37° prior
to enzyme addition. One-half milliliter of pancreatic juice or pancreatic
extract is added and 20-μl samples are withdrawn initially and at 5-min-

[3] L. Swell and C. R. Treadwell, *J. Biol. Chem.* **182**, 479 (1950).
[4] S. K. Murthy and J. Ganguly, *Biochem. J.* **83**, 460 (1962).

ute intervals with pancreatic juice, or in 15-minute intervals with pancreas, and placed in 20 μl of acetone–ethanol (1:1). Incubation times vary from 5 to 90 minutes, depending on the level of enzyme activity. The thin-layer silicic acid plates are prepared by dipping two microscope slides back to back, into a suspension of silica gel G in chloroform–methanol (2:1). The slides are removed, separated, and placed in an autoclave for 10 minutes at 15 psi. This procedure wets and binds the silicic acid to the glass slide and subsequently dries the thin-layer plates. Slides prepared in this manner and kept in storing racks over $CaCl_2$ do not need additional activation prior to use. The lipid sample in acetone–ethanol is spotted 1.5 cm from one end of a silicic acid slide, the solvent is allowed to evaporate, and the slide is placed in a Coplin staining jar containing 4–5 ml of the developing solvent mixture. When the solvent front is 1 cm from the top (3–5 minutes), the slide is removed and air-dried for 1 minute. Under these conditions, esterified cholesterol is at the solvent front and free cholesterol is approximately 1 cm from the origin as visualized in iodine vapor. The distance of free cholesterol from the origin can be regulated by altering the amount of acetic acid in the developing mixture.

For determination of free and esterified cholesterol-4-^{14}C, the upper half of the silicic acid on the slide is scraped quantitatively, with a razor blade or a second glass slide, into a scintillation vial; the lower half is scraped into a second vial. One milliliter of spectroanalyzed methanol and 10 ml of the scintillation mixture are added, the vial is shaken, and the silicic acid is allowed to settle prior to counting in the liquid scintillation counter.

Other methods reported in the literature involve spectrophotometric,[5] manometric,[6] or nephlometric procedures[7] that involve longer assay times, lengthy analysis, or inaccuracies of assay.

Definition of a Unit. A convenient unit is the amount of enzyme which catalyzes the disappearance or release of 1 micromole of free cholesterol per hour under the conditions described. However, the activity of the enzyme is dependent on the type of suspension used in the assay and the specific fatty acid used to measure esterification.

Preparation of Enzyme

A convenient source for cholesterol ester hydrolase is fresh rat, pig, or beef pancreas, which can either be frozen or from which an acetone powder is prepared. Effective extraction of the enzyme can be accom-

[5] W. M. Sperry and M. Webb, *J. Biol. Chem.* **187**, 97 (1950).
[6] M. Korzenovsky, E. R. Diller, A. C. Marshall, and B. M. Auda, *Biochem. J.* **76**, 238 (1960).
[7] H. H. Hernandez and I. L. Chaikoff, *J. Biol. Chem.* **228**, 447 (1957).

plished by homogenization in 10–20 volumes of glycerol–water (1:1, v/v) or 1.54 M sodium chloride. The homogenate is kept at room temperature for 1 hour with frequent shaking, and centrifuged at 5000 g for 10 minutes. The supernatant fluid contains the enzyme with a specific activity about 10% of that in fresh rat pancreatic juice.

Purification Procedure

Pancreatic Juice. The procedure for obtaining fresh rat pancreatic juice has been described,[8] and cholesterol ester hydrolase from this source has recently been purified about 350-fold to a specific activity of 2400 units per milligram of protein using the following procedure.

Step 1. Cold pancreatic juice is centrifuged at 12,000 g for 10 minutes prior to use to remove insoluble, nonenzyme protein. Ice-cold acetone is slowly added to cold pancreatic juice while stirring to give a final acetone saturation of 35%, and the mixture is kept at 0° for 10 minutes. The precipitate is separated by centrifugation at 23,500 g for 10 minutes in a Servall refrigerated centrifuge, and is redissolved in cold 0.05 M phosphate buffer, pH 6.2. A purification of 15- to 20-fold is usually obtained.

Step 2. DEAE-Cellulose Fractionation. DEAE-cellulose (0.8 meq/g) is equilibrated for 24 hours with 0.05 M phosphate buffer, pH 6.2, and packed in a glass column (30 $\times$ 1 inches) under 2 psi nitrogen pressure. The phosphate buffer solution of the acetone precipitate is added, and the protein is eluted stepwise with 50–70 ml volumes of 0.05 M, 0.10 M, and 0.15 M phosphate buffer, pH 6.2. The enzyme eluted in the 0.15 M buffer fraction is dialyzed against distilled water overnight. This step usually gives an additional 4- to 5-fold purification.

Step 3. Hydroxylapatite Column Fractionation. Hydroxylapatite (Bio-Rad Laboratories) in 0.001 M phosphate buffer, pH 6.5, is packed in a column as described above. The entire 50–70 ml of dialyzed DEAE-cellulose fraction is placed on the column, and after passage of the initial volume, the proteins are eluted with 40-ml volumes of 0.10 M, 0.15 M, and 0.20 M phosphate buffer, pH 6.2, under nitrogen pressure to give a flow rate of 0.25 ml per minute. The enzyme is eluted in the 0.20 M buffer fraction with an additional 4–6-fold purification.

Comments. The purified enzyme after DEAE-cellulose and hydroxylapatite fractionation is labile but can be lyophilized after addition of 25 mg bovine serum albumin, fraction V. The addition of this amount of protein prior to hydroxylapatite chromatography increases stability and recovery of the enzyme during chromatography, and the albumin is eluted with the enzyme in the 0.2 M buffer fraction. The albumin can subsequently be removed by Sephadex G-200 chromatography.

[8] C. R. Borja, G. V. Vahouny, and C. R. Treadwell, *Am. J. Physiol.* **206**, 223 (1964).

TABLE I
SUMMARY OF PURIFICATION PROCEDURE FOR RAT PANCREATIC
JUICE CHOLESTEROL ESTER HYDROLASE

Fraction	Volume (ml)	Units[a]	Protein (mg)	Specific activity (units/mg protein)	Recovery (%)
I Original juice	24	2764	384	7.2	100
II 35% Acetone precipitate	12	1676	14.4	115	60
III DEAE-cellulose, 0.15 M	70	1204	2.4	479	43
IV Hydroxylapatite,[b] 0.20 M	40	168	0.06	2455	6

[a] A unit is 1 micromole of cholesterol esterified per hour under the conditions described in the procedure.

[b] The recovery of enzyme by hydroxylapatite column fractionation is increased from 14 to 50% by stabilization of the enzyme with 25 mg of albumin prior to chromatography.

Pancreas. Hernandez and Chaikoff[7] have reported the following method for 400-fold purification of pork pancreas cholesterol ester hydrolase.

Forty grams of an acetone powder of pork pancreas is extracted five times with 100-ml portions of 0.154 M phosphate buffer, pH 6.2, as described above. Proteins are precipitated by addition of 22 g of ammonium sulfate per 100 ml of extract and the supernatant after centrifugation at 10,000 g for 10 minutes is used for subsequent purification. One hundred grams of solid ammonium sulfate is added to the clear supernatant at room temperature, and the resulting precipitate is separated by centrifugation. The precipitate is dissolved in 100 ml of cold phosphate buffer (pH 6.2) and dialyzed against cold distilled water for 48 hours. The resulting white precipitates from 10 batches carried through the procedure are dissolved in cold phosphate buffer (pH 6.2), combined, and made to 100 ml with buffer. To this solution is added 33 ml of cold 95% ethanol with stirring and the mixture is allowed to stand 10 minutes at 10° during which a flocculent precipitate forms. This precipitate is separated by centrifugation, discarded, and the supernatant is dialyzed against distilled water at 0° for 24 hours. The dialyzate is diluted to 225 ml, the pH is adjusted to 5.0 with 0.1 N HCl, and 25 g of solid ammonium sulfate is added. The resulting precipitate is discarded, and to the supernatant is added 5 g of Whatman cellulose and 5 g of solid ammonium sulfate per 100 ml. This mixture is centrifuged and the protein-cellulose precipitate is suspended in 35% ammonium sulfate solution. This suspension is poured into a glass column and allowed to sediment. For gradient elution of proteins, two reservoirs are set up in series containing 35% ammonium sulfate in the lower reservoir and 20% ammonium sulfate in the upper reservoir. Fifteen-milliliter fractions are collected and assayed for protein and enzyme activity. The enzyme associated with the third protein peak is precipitated by addition of ammonium sulfate (5 g/100 ml) and the protein precipitate is dissolved in 2 ml of distilled water.

TABLE II

SUMMARY OF PURIFICATION PROCEDURE FOR PORK
PANCREAS CHOLESTEROL ESTER HYDROLASE

Fraction	Volume (ml)	Units[a]	Protein (mg)	Specific activity (units/mg protein)	Recovery (%)
I Acetone powder suspension	500	80,000	400,000	0.20	100
II Phosphate extract	500	48,000	96,000	0.50	60
III Ammonium sulfate, 30–60%[b]	100	9,600	1,200	7.8	11
IV 33% Ethanol supernatant	225	8,200	330	25	10
V Second ammonium sulfate supernatant	225	3,700	85	44	4.6
VI Cellulose fractionation 3rd peak	2	550	6	89	0.7

[a] Based on nephlometric assay. One unit is the amount of enzyme producing an increase in turbidity of one Klett unit per minute during the second 15 minutes of a 30-minute incubation.
[b] Ten batches were carried through separately to step IV and combined.

Purification of rat pancreas cholesterol ester hydrolase to a specific activity of 292 units per milligram protein has also been reported[4] and involved primarily aged calcium phosphate gel fractionation of a water extract of rat pancreas acetone powder.

Properties

Specificity. Cholesterol ester hydrolase is highly specific for cholesterol and dihydrocholesterol for esterification but is also active to a lesser extent with β-sitosterol, stigmasterol, and ergosterol. Among the fatty acids employed for esterification, short-chain fatty acids up to C-8 are not esterified; of the long-chain fatty acids, oleic and linoleic acids are equally effective, while the saturated fatty acids have between 25 and 50% of the activity shown by oleic acid. It is possible that the fatty acid specificity is related to the relative solubility of the long-chain fatty acids in the assay system.

For hydrolysis in the micellar substrate systems described here, the enzyme shows no specificity for the fatty acid esterified to cholesterol.

Activators and Inhibitors. Enzyme activity is stimulated 30–40% by $10^{-3}\,M$ ammonium, potassium, and sodium ions, and is inhibited by similar concentrations of mercuric, cupric, and zinc ions. The enzyme is also sensitive to p-chloromercuribenzoate $(10^{-3}\,M)$, atoxyl $(10^{-3}\,M)$, and sodium fluoride and arsenate $(10^{-3}\,M)$. Enzyme activity is completely dependent on the presence of the trihydroxy bile salts, cholic acid or its taurine- and glycine-conjugated derivatives. Other bile salts containing none, one and two hydroxyl groups, and other detergents such as the Tweens, will not support enzyme activity.

pH Optima. The optimal pH for sterol esterification by the pancreatic enzyme is 6.1–6.2, and for sterol ester hydrolysis is 6.6–7.0. At these pH's the reactions go to 80% completion.

[24] Plasma Cholesterol Esterifying Enzyme

(Lecithin: cholesterol acyltransferase)

By John Glomset

The blood of several species contains an enzyme that can catalyze esterification of the cholesterol of plasma lipoproteins. The reaction can easily be demonstrated by incubating fresh plasma at 37° and measuring the decrement in unesterified cholesterol. The total cholesterol remains constant during the incubation; and neither exogenous substrates nor cofactors are required. In native plasma, the rate of the reaction probably depends not only on the esterifying enzyme, but also on the distribution of plasma lipoproteins, since the reaction is faster in the high than in the low density lipoproteins and faster in the smaller high density lipoproteins than in the larger ones.[1] The mechanism of the reaction mainly involves the transfer of fatty acids from the C-2 position of lecithin to cholesterol. The evidence for this is the following: labeled cholesterol esters are formed when plasma is incubated with cholesterol-^{14}C or with lecithin labeled in the C-2 position with linoleic acid-^{14}C but not when plasma is incubated with radioactive free fatty acids;[2] the rates of change

[1] J. Glomset, E. Janssen, R. Kennedy, and J. Dobbins, *J. Lipid Res.* **7**, 639 (1966).
[2] J. Glomset, F. Parker, M. Tjaden, and R. H. Williams, *Biochim. Biophys. Acta* **58**, 398 (1962).

in lecithin and cholesterol are nearly equal;[3] and the fatty acids that are transferred are mainly unsaturated.[3] A critical review of this evidence has recently appeared.[3a]

Assay

General Comments. To date, the activity of the enzyme has not been measured with an artificial substrate in the complete absence of lipoproteins and albumin. Therefore, it is not known whether the presence of these proteins is obligatory nor to what extent they influence the specificity and direction of the reaction. Thus, the fact that lecithin is the major source of fatty acids could be due to lipoprotein composition more than to enzyme specificity. Furthermore, the fact that the reaction strongly favors the formation of cholesterol esters at the expense of lecithin could be due to the subsequent binding of lysolecithin by albumin.[4,5] Under other reaction conditions other phospholipids might serve as fatty acid donors and lyso compounds might be acylated at the expense of cholesterol esters, although this has yet to be demonstrated. In our laboratory, the reaction has always been measured in the direction of cholesterol ester formation, and an attempt has been made to reduce the possible effects of endogenous lipoproteins and albumin by assaying the enzyme in the presence of relatively large amounts of enzyme-free plasma.

Substrate. Human blood plasma (fresh, or 3-week-old, "outdated" plasma can be used) is heated at 56–60°C for 30 minutes and any precipitate that forms is removed by centrifugation for 10^6 g minutes. Then 0.1 volume (or less) of a stable emulsion of carrier-free, labeled cholesterol in a solution of 5% albumin in saline[6] is added to the preheated plasma. Either cholesterol-4-^{14}C or -7-α^3H can be used; and the addition of 10^5 counts per milliliter of preheated plasma is sufficient. The mixture is incubated for 2 hours at 37° to allow equilibration of the labeled cholesterol emulsion with the lipoproteins of the preheated plasma. Then it can either be used directly or stored at $-16°$ for several weeks.

Enzyme. Fresh plasma or a partially purified enzyme can be used. In the latter case, the enzyme is diluted with physiological saline to an activity approximately in the range of fresh plasma.

Incubation. Enzyme, 0.5 ml, is added to 4.5 ml substrate and incubated at 37° in a Dubnoff shaker. Control flasks containing substrate plus physiological saline instead of the enzyme are incubated simultaneously.

---

[3] J. Glomset, *Biochim. Biophys. Acta* **65**, 128 (1962).
[3a] J. Glomset, *J. Lipid Res.* **9**, 155 (1968).
[4] J. Glomset, *Biochim. Biophys. Acta* **70**, 389 (1963).
[5] S. Switzer and H. Eder, *J. Lipid Res.* **6**, 506 (1965).
[6] D. Porte, Jr., and R. Havel, *J. Lipid Res.* **2**, 357 (1961).

Aliquots, 0.5 ml, are taken at the start of the incubation and after 3 and/ or 6 hours.

Analysis. The aliquots are extracted with 10 ml of 1:1 chloroform–methanol, and the labeled cholesterol esters are isolated by chromatography in hexane–ether–acetic acid (90:20:1) on thin-layer plates of silica gel H (Brinkman Instruments, Westbury, New York). The cholesterol esters are visualized by exposure of the plates to iodine vapor, and after the color fades the appropriate areas of silica gel are scraped into counting vials, 10 ml of Liquifluor is added (Pilot Chemicals, Inc., Watertown, Massachusetts), and the radioactivity is determined in a liquid scintillation spectrometer. The enzyme activity can be expressed in terms of the amount of labeled cholesterol ester formed per hour per milliliter of incubation medium. Alternatively, the total amount of cholesterol ester formed can be obtained by dividing the value for the labeled cholesterol ester by the specific activity of the unesterified cholesterol isolated from the thin-layer plates.[1]

Purification

General Comments. The enzyme can be partially purified by one or a combination of methods, including DEAE chromatography, hydroxylapatite chromatography, ammonium sulfate precipitation, gel filtration, zone electrophoresis, and ultracentrifugation.[7] The principal problem is that, while the enzyme in unfractionated plasma is stable for several weeks at 4°, it becomes very labile during purification.[8]

DEAE Chromatography: Plasma is dialyzed for 24 hours against 40 volumes of 0.01 M tris(hydroxymethyl)aminomethane-HCl, pH 7.4, and subsequently applied to a column of DEAE-cellulose (0.1 meq per milliliter of plasma) previously equilibrated with the same buffer. The column is washed with 0.1 N NaCl in Tris buffer until the absorbance of the effluent at 280 mμ is less than 0.1 (usually about 10–20 volumes per original volume of plasma are required). Then the enzyme is eluted with 0.5 N NaCl in Tris buffer. Usually a green band comprised of ceruloplasmin and albumin can be seen to migrate down the column. The effluent corresponding to this band, which also contains the enzyme and other proteins, is collected until the absorbance at 280 mμ is less than 0.1. The total volume collected is usually 2–4 times that of the original plasma, and the enzyme activity is from 6- to 10-fold purified.

Ammonium Sulfate Precipitation. The enzyme-containing fraction from the DEAE column is adjusted to 40% saturation by the addition of

[7] J. Glomset and J. Wright, *Biochim. Biophys. Acta* **89**, 266 (1964).

[8] The addition of 1 M glycerol, glycine, or sucrose does not improve the stability (J. Glomet, unpublished observations, 1966).

solid ammonium sulfate and centrifuged for approximately 8×10^4 g min. The precipitate is resuspended in 100 ml of 40% saturated ammonium sulfate, pH 7.4, centrifuged a second time, and then discarded. The supernatant and the 40% saturated ammonium sulfate wash are combined, adjusted to 66% saturation with respect to ammonium sulfate, and centrifuged as above. The precipitate contains the enzyme. It is suspended in 100 ml of 66% saturated ammonium sulfate, pH 7.4, recentrifuged, and then dissolved in 20–30 ml of 0.07 M potassium phosphate buffer, pH 7.4. (The supernatant solutions from the centrifugations are discarded.) Usually, the ammonium sulfate precipitation step causes a 2- to 3-fold purification.

Hydroxylapatite Chromatography. The enzyme from the ammonium sulfate step, dissolved in 0.07 M potassium phosphate buffer, pH 7.4, is dialyzed for 24 hours against 2 liters of the same buffer and then applied to a column that contains 0.15 volume of hydroxylapatite per volume original plasma (the column should be preequilibrated with the 0.07 M potassium phosphate buffer). The column is washed with 0.07 M potassium phosphate buffer, pH 7.4, the effluent is collected in a fraction collector, and the absorbance of the effluent is determined at 280 mμ. The blue ceruloplasmin should be visible at the top of the column and should not appear in the effluent. The enzyme, however, is present in the first protein peak indicated by the absorbance measurements. The total volume of this peak is usually from 100 ml to 500 ml, depending on the scale of the experiment. It can be reduced by ultrafiltration through dialysis membranes at reduced pressure. Usually the enzyme activity is about 2-fold purified, and the major contaminants are albumin and high density lipoprotein (see below).

PARTIAL PURIFICATION OF LECITHIN:CHOLESTEROL
ACYLTRANSFERASE FROM HUMAN PLASMA

Fraction	Total protein[a]	Total activity[b]	Purification (—fold)	Recovery (%)
Whole plasma	45,458	68,475	—	—
DEAE (0.5 N NaCl)	3,531	35,987	6.8	52.6
Ammonium sulfate (40–66% saturation)	1,213	22,875	12.5	33.4
Hydroxylapatite (0.07 M potassium phosphate, pH 7.4)	300	14,496	32.2	21.2

[a] Absorbance at 280 mμ times volume in milliliters.

[b] Cholesterol ester counts per minute per milliliter of incubation medium per hour of incubation times volume in milliliters.

An experiment in which 825 ml of 3-week-old human plasma was partially purified through the use of the three steps outlined above is shown in the table.

Properties

Distribution. Enzyme activity has been demonstrated in human, monkey, rat, rabbit, dog, cow, and chicken blood. Measurements made in the rat are consistent with the possibility that the lecithin:cholesterol acyltransferase is a "plasma-specific" enzyme, since its concentration in plasma is greater than in liver, lungs, intestinal mucosa, spleen, pancreas, heart, skeletal muscle, skin, fat, brain, or testis.[9]

Specificity. Because of the role of lipoproteins in the reaction, only tentative conclusions can be drawn regarding the specificity of the enzyme. The results of Portman and Sugano[9a] and Goodman[10] suggest that the human plasma enzyme preferentially transfers linoleic acid and oleic acid rather than fatty acids of longer chain length and greater unsaturation. The rat plasma enzyme, however, does not show this preference.[3, 6] These species differences could be true to lipoprotein structure as well as to fatty acid specificity, i.e., the lecithins containing the longer fatty acids might be inaccessible to the enzyme.

The differences in reaction rate among the plasma lipoproteins has already been mentioned. In general, the rate seems to be faster the smaller the lipoprotein and the greater its relative lecithin content,[1] although there may be some exceptions to this rule.

Physical Properties. The enzyme migrates with the α_1-globulins on zone electrophoresis.[7] However, the enzyme normally may exist as a complex with the plasma lipoproteins,[11] and since up to two-thirds of the activity is lost on electrophoresis, it is possible that the only detectable activity may be that of the complex with the α_1-lipoproteins.

pH Optimum. The pH optimum of the cholesterol esterification reaction is between 7.5 and 8.0.

Inhibitors. The enzyme is inactivated by being heated at 56° for 30 minutes, and rapidly loses activity below pH 6.0. It is inhibited by iodacetamide, N-ethylmaleimide, and p-hydroxymercuribenzoate in concentrations of from 0.001–0.1 M. The enzyme is also inhibited by paraoxon $(1 \times 10^{-4}\,M)$[12] and by diisopropylfluorophosphate $(5 \times 10^{-4}\,M)$.[13]

[9] J. Glomset and D. Kaplan, *Biochim. Biophys. Acta* **98**, 41 (1965).
[9a] O. W. Portman and M. Sugano, *Arch. Biochem. Biophys.* **105**, 532 (1964).
[10] D. S. Goodman, *J. Clin. Invest.* **43**, 2026 (1964).
[11] W. Lossow, S. Shah, and I. L. Chaikoff, *Biochim. Biophys. Acta* **116**, 172 (1966).
[12] W. Vogel and E. Bierman, *J. Lipid Res.* **8**, 46 (1967).
[13] R. Rowen and J. Martin, *Biochim. Biophys. Acta* **70**, 396 (1963).

Finally, the reaction is inhibited by calcium and magnesium ions (0.1 M), by urea (4 M), and by bile salts (0.005 M).[7]

Activators. The reaction is *stimulated* by polyvalent anions such as sulfate, phosphate, and citrate.[7] In native plasma, the extent of the reaction is increased by the addition of dimyristoyl lecithin[14] or an extract from streptococci.

[14] A. Wagner, *Circulation Res.* **3**, 165 (1955).

Section IV

Enzyme Systems of Bile Acid Formation

[25] Enzymatic Transformations of the Sterol Nucleus in Bile Acid Biosynthesis

By Olle Berséus, Henry Danielsson, and Kurt Einarsson

The two main bile acids formed from cholesterol in mammalian liver are cholic acid ($3\alpha,7\alpha,12\alpha$-trihydroxy-5β-cholanoic acid) and chenodeoxycholic acid ($3\alpha,7\alpha$-dihydroxy-5β-cholanoic acid). These acids are excreted into the bile as conjugates with taurine or glycine. According to current concepts[1] the conversion of cholesterol to taurocholic acid (or glycocholic acid) and to taurochenodeoxycholic acid (or glycochenodeoxycholic acid) entails fifteen and fourteen steps, respectively. Evidence has been obtained to indicate that in bile acid formation the changes in the sterol nucleus of cholesterol precede the oxidation of the side chain. Thus, 5β-cholestane-$3\alpha,7\alpha$-diol and 5β-cholestane-$3\alpha,7\alpha,12\alpha$-triol appear to be key intermediates in the biosynthesis of chenodeoxycholic acid and cholic acid, respectively.[1,2]

The probable sequence of reactions in the conversion of cholesterol to 5β-cholestane-$3\alpha,7\alpha$-diol and 5β-cholestane-$3\alpha,7\alpha,12\alpha$-triol has been elucidated mainly by means of experiments *in vitro*. Of particular importance in this connection has been the work of Mendelsohn and Staple,[3] who were the first to describe an enzyme system capable of catalyzing the conversion of cholesterol to 5β-cholestane-$3\alpha,7\alpha,12\alpha$-triol. Further work has shown that these reactions are catalyzed by microsomal enzymes except the last two steps, which are catalyzed by soluble enzymes.[1,2,4-6]

In the reactions described below the conditions for those catalyzed by microsomal enzymes have been chosen in such a way that, as far as possible, only one main product is obtained.

Materials

At present, studies of the conversion of cholesterol to 5β-cholestane-$3\alpha,7\alpha,12\alpha$-triol require in almost all instances the use of isotopically labeled compounds. Of the steroids involved in the reactions described below only labeled cholesterol (4-^{14}C-, 26-^{14}C, $1,2$-^{3}H, and uniformly

[1] H. Danielsson and T. T. Tchen, *in* "Metabolic Pathways" (D. M. Greenberg, ed.), Vol. II, p. 117. Academic Press, New York, 1968.
[2] I. Björkhem, H. Danielsson, and K. Einarsson, *European J. Biochem.* **2**, 294 (1967).
[3] D. Mendelsohn and E. Staple, *Biochemistry* **2**, 577 (1963).
[4] H. R. B. Hutton and G. S. Boyd, *Biochim. Biophys. Acta* **116**, 336 (1966).
[5] K. Einarsson, *European J. Biochem.* **5**, 101 (1968).
[6] O. Berséus, H. Danielsson, and A. Kallner, *J. Biol. Chem.* **240**, 2396 (1965).

tritium-labeled cholesterol) is available commercially. All the other steroids can be prepared biosynthetically from labeled cholesterol but the yields are very low. It appears preferable to use labeled cholest-5-ene-3β, 7α-diol for the biosynthetic preparation of the other steroids. Cholest-5-ene-3β,7α-diol is easily labeled with tritium by reduction of 3β-hydroxy-cholest-5-en-7-one 3-benzoate with tritium-labeled sodium borohydride, which is available commercially.

Unlabeled reference compounds are not available commercially. Reference will be made to original articles for the preparation of the different steroids, and only modifications of or additions to published procedures will be described.

Cholest-5-ene-3β,7α-diol-7β-³H. 3β-Hydroxycholest-5-en-7-one 3-benzoate is prepared by oxidation of cholesteryl benzoate with *tert*-butyl chromate according to Heusler and Wettstein.[7] 3β-Hydroxycholest-5-en-7-one 3-benzoate, 100 mg, is dissolved in 50 ml of methanol (slight warming may be necessary) and a solution of 50 mg of tritium-labeled sodium borohydride (prepared by mixing 100 mC of tritium-labeled sodium borohydride with unlabeled sodium borohydride) in 10 ml of methanol is added. The reaction mixture is stirred for 2 hours at room temperature. The reaction mixture is then acidified with 0.1 M hydrochloric acid and extracted twice with ether. The combined ether extracts are washed with 10% solution of sodium carbonate and with water until neutral. The ether is evaporated under reduced pressure and the residue is dissolved in a few milliliters of benzene and applied to a column of 20 g of aluminum oxide, grade II (Woelm, Eschwege, Germany), prepared in benzene. Fractions of 20 ml are collected. Elution is started with benzene which elutes any unchanged starting material. Elution is then continued with 5% ethyl acetate in benzene. This solvent mixture elutes first cholest-5-ene-3β,7β-diol 3-benzoate and then cholest-5-ene-3β,7α-diol 3-benzoate. The separation of these compounds is usually good, and only one or two fractions contain a mixture. A small aliquot of each fraction is analyzed by thin-layer chromatography with benzene–ethyl acetate (95:5) as solvent (see below). The fractions containing predominantly cholest-5-ene-3β,7α-diol 3-benzoate are combined and rechromatographed on a column of 10 g of aluminum oxide, grade II, prepared in a mixture of 5% ethyl acetate in benzene. The column is eluted with the same solvent mixture, and fractions of 10 ml are taken. The cholest-5-ene-3β,7α-diol-7β-³H 3-benzoate thus obtained, 15–20 mg, is hydrolyzed by refluxing with 5 ml of 10% methanolic potassium hydroxide for 1 hour. After dilution with water the hydrolysis mixture is extracted with ether and the ether extract is washed with water until neutral. The residue of the ether extract is crystallized from a methanol-

[7] K. Heusler and A. Wettstein, *Helv. Chim. Acta* **35**, 284 (1952).

water mixture. The yield of cholest-5-ene-3β,7α-diol-7β-³H is about 10 mg with a specific activity of 15–30 μC/mg.

The other steroids, i.e., 7α-hydroxycholest-4-en-3-one, 7α-hydroxy-5β-cholestan-3-one, 5β-cholestane-3α,7α-diol, 7α,12α-dihydroxycholest-4-en-3-one, 7α,12α-dihydroxy-5β-cholestan-3-one, and 5β-cholestane-3α,7α,-12α-triol can then be prepared biosynthetically and in good yields from cholest-5-ene-3β,7α-diol-7β-³H.

Cholest-5-ene-3β,7α-diol is prepared from 3β-hydroxycholest-5-en-7-one 3-benzoate as described above or by bromination of cholesteryl benzoate with N-bromosuccinimide in carbon tetrachloride[8] and treatment of the 7α-bromocompound with silver oxide.[9] The reaction mixture is chromatographed on a column of aluminum oxide, grade II. A 100-fold excess of aluminum oxide is used, and elution is accomplished with benzene and ethyl acetate (5–10%) in benzene. The cholest-5-ene-3β,7α-diol 3-benzoate is crystallized from acetone and hydrolyzed with methanolic potassium hydroxide. The hydrolysis mixture is worked up as described above, and cholest-5-ene-3β,7α-diol is crystallized from a methanol–water mixture.

7α-Hydroxycholest-4-en-3-one can be prepared either as described by Björkhem *et al.*[10] or as described by Naqvi and Boyd as referred to by Hutton and Boyd[4] and by Mendelsohn *et al.*[11]

7α-Hydroxy-5β-cholestan-3-one is prepared by oxidation of 5β-cholestane-3α,7α-diol with aluminum *tert*-butoxide[10] or with aluminum isopropoxide.[11]

5β-Cholestane-3α,7α-diol is prepared by electrolytic coupling of chenodeoxycholic acid and isovaleric acid.[12]

7α,12α-Dihydroxycholest-4-en-3-one is obtained by selenium dioxide dehydrogenation of 7α,12α-dihydroxy-5β-cholestan-3-one.[6]

7α,12α-Dihydroxy-5β-cholestan-3-one is prepared by oxidation of 5β-cholestane-3α,7α,12α-triol with aluminum *tert*-butoxide.[6]

5β-Cholestane-3α,7α,12α-triol is prepared by electrolytic coupling of cholic acid and isovaleric acid.[12]

Methods

Preparation and Fractionation of Homogenates. White male rats weighing 200–250 g are used. Rats of the Sprague-Dawley strain and of the Danish State Serum Institute strain V.S. have been used, and no

[8] H. E. Bide, H. B. Henbest, E. R. H. Jones, R. W. Peevers, and P. A. Wilkinson, *J. Chem. Soc.* p. 1783 (1948).
[9] H. Schaltegger and F. X. Müllner, *Helv. Chim. Acta* 34, 1096 (1951).
[10] I. Björkhem, H. Danielsson, C. Issidorides, and A. Kallner, *Acta Chem. Scand.* 19, 2151 (1965).
[11] D. Mendelsohn, L. Mendelsohn, and E. Staple, *Biochemistry* 5, 1286 (1966).
[12] S. Bergström and L. Krabisch, *Acta Chem. Scand.* 11, 1067 (1957).

major differences between these two strains have been observed. Homogenates, 20% (liver wet weight per volume), are prepared in a modified Bucher medium,[13] pH 7.4, with a Potter-Elvehjem homogenizer equipped with a loosely fitting pestle. The medium consists of 10.8 g of KH_2PO_4, 3.9 g of KOH, 1.0 g of $MgCl_2 \cdot 6 H_2O$, and 3.6 g of nicotinamide in a final volume of 1000 ml; the pH may have to be adjusted with 1 M KOH or 1 M HCl before the solution is brought to final volume. These conditions are used for preparing enzyme solutions for all experiments excluding those concerned with the conversion of 7α-hydroxycholest-4-en-3-one to 5β-cholestane-3α,7α-diol and of 7α,12α-dihydroxycholest-4-en-3-one to 5β-cholestane-3α,7α,12α-triol (see appropriate sections below for description of preparation and fractionation of homogenates in these experiments). Preparation and fractionation of homogenates are carried out at 2–5°. The homogenate is centrifuged at 800 g for 10 minutes. The 800 g supernatant fluid is centrifuged at 20,000 g for 10 minutes. The microsomal fraction is isolated by centrifuging the 20,000 g supernatant fluid for 2 hours at 100,000 g. The microsomal pellet is suspended in the homogenizing medium in a volume corresponding to that of the original 20,000 g supernatant fluid and is homogenized with a loosely fitting pestle for 30 seconds. The suspension is centrifuged for 5 minutes at 800 g, and the supernatant fluid is used as a microsomal preparation.

Incubation Conditions. The substrates are added to the enzyme solutions dissolved in acetone. The amount of acetone used is such as to give a final concentration of acetone in the incubation mixture of 1–2.5%. The amounts of substrate used in the different experiments are given below. Incubations are carried out at 37° with shaking and are terminated by addition of 20 volumes of chloroform–methanol (2:1) or by addition of 3 volumes of 95% ethanol. The latter procedure is used only in the case of incubations of 7α,12α-dihydroxy-5β-cholestan-3-one with partially purified enzyme fractions (see appropriate section below in which the subsequent analytical procedure also will be described).

Extraction of Incubation Mixtures. The chloroform–methanol extract is filtered into a separatory funnel and 0.2 volume of a 0.9% solution of sodium chloride is added. The mixture is shaken well, and the phases are allowed to separate until they are clear (6–12 hours). The chloroform phase (the lower phase) is evaporated to dryness under reduced pressure or blown to dryness with a stream of nitrogen. The residue is dissolved in a small amount (0.1–0.2 ml) of chloroform–methanol (2:1). About 100 μg each of appropriate reference compounds (dissolved in acetone or methanol) are added to the solution, and the mixture (all or part of it) is put as a thin band on a thin-layer plate.

[13] S. Bergström and U. Gloor, *Acta Chem. Scand.* 9, 34 (1955).

Thin-Layer Chromatography. Glass plates, 200×200 mm, are coated with a layer of silica gel G (Merck, Darmstadt, Germany; Brinkmann Instruments, Westbury, New York) about 0.5 mm thick. Prior to use the plate is activated for 1 hour at 110°. Two extracts are put on each plate. Reference compounds are applied as spots on either side of each extract. After development the plates are dried at room temperature and are then exposed to iodine vapor. The plate is put in a chromatographic tank containing some iodine crystals on the bottom. The different zones are marked out and the iodine is allowed to evaporate at room temperature in a hood. The different zones are collected in test tubes by scraping off the layer with a razor blade. (If liquid scintillation counting is used, the layer is scraped directly into the counting vial.)

Radioactivity Assay. Five milliliters of methanol is added to each tube and the mixture is stirred vigorously. The silica gel is allowed to settle for about 1 hour. Aliquots of the supernatant fluid are transferred to aluminum planchettes, the solvent is evaporated on a hot plate and counting is performed with a gas flow counter. If scintillation counting is used, the scintillating solution is added to the silica gel that has been collected directly in the vial. The composition of the scintillating solution should be such as to ensure extraction and solubilization of the steroid. A suitable scintillating solution, for example, is the one described by Kinard.[14]

Identification of Products. If desired, the identity of the products obtained is established by crystallizing part or all of the methanol extract of the appropriate thin layer zone to constant specific activity after addition of 10–25 mg of the appropriate, authentic compound.

Conversion of Cholesterol to Cholest-5-ene-3β,7α-diol

The 20,000 g supernatant fluid, 3 ml, is incubated for 1 hour with 10–20 μg of ^{14}C- or ^{3}H-labeled cholesterol (0.5–1 μC is sufficient) added to the incubation mixture in 30 μl of acetone. The chloroform extract is chromatographed with benzene–ethyl acetate (3:7) as solvent (cf. Table I). The extent of conversion of cholesterol to more polar products is 2–5%. Of the metabolites formed, 30–50% is present in the zones corresponding to 7α,12α-dihydroxycholest-4-en-3-one and cholest-5-ene-3β,7α-diol. Sometimes these compounds do not separate completely with this solvent system. If this is the case the extract of the zone(s) corresponding to these compounds is chromatographed with benzene–ethyl acetate–trimethylpentane (3:7:3) as solvent system. The chromatogram is run for 2 hours giving a complete separation of cholest-5-ene-3β,7α-diol from 7α,12α-dihydroxycholest-4-en-3-one. The usual ratio between the amount

[14] F. E. Kinard, *Rev. Sci. Instr.* **28**, 239 (1957).

TABLE I

APPROXIMATE R_f VALUES IN THIN-LAYER CHROMATOGRAPHY OF SOME C_{27} STEROIDS

	Solvent system	
Compound	Benzene–ethyl acetate 3:7	Benzene–ethyl acetate 1:1
5β-Cholestane-3α,7α,12α-triol	0.18	—
7α,12α-Dihydroxycholest-4-en-3-one	0.33	—
Cholest-5-ene-3β,7α-diol	0.40	0.31
5β-Cholestane-3α,7α-diol	0.61	0.47
7α,12α-Dihydroxy-5β-cholestan-3-one	0.64	—
7α-Hydroxycholest-4-en-3-one	0.75	0.68
Cholesterol	0.87	—
7α-Hydroxy-5β-cholestan-3-one	0.93	0.88

of labeled cholest-5-ene-3β,7α-diol formed and that of labeled 7α,12α-dihydroxycholest-4-en-3-one is 2:1. Another metabolite of cholesterol that can be isolated and identified easily from the first chromatogram is 7α-hydroxycholest-4-en-3-one which accounts for 10–20% of the total amount of cholesterol metabolized.

The extent of metabolism of cholesterol can be increased severalfold by preparing the 20,000 g supernatant fluid from a rat that has had a bile fistula for 3 days. The technique of preparing a bile fistula is simple. Under ether anesthesia a polyethylene tube, PE-50 (Clay-Adams, Inc., New York, New York), is introduced into a small cut in the proximal half of the bile duct and is tied to the duct proximally and distally to the cut. The rat is put in a restraining cage and is given food and 0.9% solution of sodium chloride to drink *ad libitum*. The yield of labeled cholest-5-ene-3β,7α-diol and 7α,12α-dihydroxycholest-4-en-3-one from 10–20 μg of labeled cholesterol in the 20,000 g supernatant fluid of liver homogenate from a bile fistula rat is 5–7% and 2–3%, respectively.

Conversion of Cholest-5-ene-3β,7α-diol to 7α-Hydroxycholest-4-en-3-one

The microsomal fraction, 1 ml (corresponding to 1 ml of 20,000 g supernatant fluid), is diluted with 2 ml of Bucher medium and NAD+, 0.3 μmole in 0.1 ml of Bucher medium, is added. The mixture is incubated for 20 minutes with 40 μg of labeled cholest-5-ene-3β,7α-diol added in 30 μl of acetone. The chloroform extract of the incubation mixture is chromatographed with benzene–ethyl acetate (1:1) as solvent (cf. Table I). About 15–25% of the substrate is metabolized and 7α-hydroxycholest-4-en-3-one accounts for about 60–80% of the metabolites formed.

Conversion of 7α-Hydroxycholest-4-en-3-one to 7α,12α-Dihydroxycholest-4-en-3-one

The microsomal fraction, 3 ml (corresponding to 3 ml of 20,000 g supernatant fluid), is fortified by addition of NADPH, 3 μmoles in 0.1 ml of Bucher medium, and is incubated for 10 minutes with 80 μg of labeled 7α-hydroxycholest-4-en-3-one added in 30 μl of acetone. The chloroform extract is chromatographed with benzene–ethyl acetate (3:7) as solvent (cf. Table I). About 15–25% of the substrate is metabolized, and 7α,12α-dihydroxycholest-4-en-3-one accounts for about 40–50% of the metabolites formed.

Conversion of 7α,12α-Dihydroxycholest-4-en-3-one to 5β-Cholestane-3α,7α,12α-triol

This reaction sequence involves the participation of two soluble enzymes, a 5β-steroid reductase and a 3α-hydroxysteroid dehydrogenase. For practical reasons the two enzymes will be described together in a section on the assay methods and in one on the purification procedures.

Assay Method

The activities of the two enzymes are measured by thin-layer chromatographic analysis of extracts of incubations of tritium-labeled 7α,12α-dihydroxycholest-4-en-3-one and tritium-labeled 7α,12α-dihydroxy-5β-cholestan-3-one, respectively.

Appropriate amount of enzyme solution is diluted to 4 ml with 0.1 M Tris-HCl buffer, pH 7.4, 1 mg of NADPH is added, and the mixture is incubated for 20 minutes with 50 μg of substrate, 7α,12α-dihydroxycholest-4-en-3-one or 7α,12α-dihydroxy-5β-cholestan-3-one, added to the incubation mixture dissolved in 0.1 ml of acetone. The amount of enzyme solution used should be such as to give 15–25% conversion of substrate. Incubations with 7α,12α-dihydroxycholest-4-en-3-one are extracted with chloroform-methanol (2:1) as described above. Incubations with 7α,12α-dihydroxy-5β-cholestan-3-one are terminated by the addition of 3 volumes of 95% ethanol. The mixture is filtered into a separatory funnel. After dilution with water and acidification to pH 1–3 with hydrochloric acid, the mixture is extracted once with 2 volumes of ether. The ether extract is washed with water until neutral, and the ether is evaporated under reduced pressure. This procedure is used as chromatographic artifacts of unknown nature are obtained if extraction is made with chloroform–methanol. The residue of the chloroform extract and that of the ether extract are analyzed by thin-layer chromatography as described above, using benzene–ethyl acetate (3:7) as solvent (cf. Table I).

Protein is determined by reading the absorbance at 280 mμ and by the method of Lowry et al.[15]

Definition of Units. One unit of enzyme, 5β-steroid reductase or 3α-hydroxysteroid dehydrogenase, is defined as that amount which produces 1 μg of product under the conditions of the assay. Specific activity is expressed as units per milligram of protein.

Purification Procedures

Homogenates, 33% (liver wet weight per volume), are prepared from 10 rat livers in a 0.1 M potassium phosphate buffer, pH 7.4, which is 0.125 M with respect to sucrose. A Potter-Elvehjem homogenizer is used. The tissue is first homogenized for 30 seconds with a loosely fitting pestle and then for another 30 seconds with a tightly fitting pestle. Homogenization and subsequent operations are carried out at 2–5°. The homogenate is centrifuged at 800 g for 10 minutes. The 800 g supernatant fluid is centrifuged at 100,000 g for 1 hour.

The 100,000 g supernatant fluid is applied to a column of Sephadex G-25 prepared in 0.01 M Tris-HCl buffer, pH 7.4. The dimensions of the column are 60 mm (diameter) $\times$ 400 mm (height). Elution is carried out with 0.01 M Tris-HCl buffer, pH 7.4. Protein in the effluent is measured by reading the absorbance at 280 mμ. The protein is eluted in a total volume of about 250 ml. A suitable aliquot to assay for enzyme activities is 0.2–0.4 ml.

The protein solution obtained from the Sephadex G-25 column is applied to a column of TEAE-cellulose (triethylaminoethylcellulose, Serva Entwicklungslabor, Heidelberg, Germany) prepared in 0.01 M Tris-HCl buffer, pH 7.4, and with the dimensions 90 mm (diameter) $\times$ 50 mm. The column is eluted with the same buffer until a protein peak has been eluted; 450–500 ml of buffer are required. This protein is practically devoid of 5β-steroid reductase and 3α-hydroxysteroid dehydrogenase activity. The column is then eluted with 0.5 M Tris-HCl buffer, pH 7.4, which elutes all the activity. About 500 ml are required. A suitable aliquot to incubate is 0.3–0.6 ml.

The 0.5 M Tris-HCl eluate is applied to a column of 50 g of hydroxylapatite (Bio-Rad Laboratories, Richmond, California) prepared in 0.01 M Tris-HCl buffer, pH 7.4, and with the dimensions 65 mm (diameter) $\times$ 40 mm. The column is eluted with 0.001 M potassium phosphate buffer, pH 7.4, which elutes inactive protein. About 100–130 ml are required to elute this material. The column is then eluted with 0.03 M potassium phosphate buffer, pH 7.4. This concentration elutes part of the 5β-steroid reductase

[15] O. H. Lowry, N. J. Rosebrough, A. L. Farr, and R. J. Randall, *J. Biol. Chem.* **193**, 265 (1951).

activity, practically free of 3α-hydroxysteroid dehydrogenase activity. About 140–180 ml is required. A suitable aliquot to incubate is 0.5–1.0 ml. The column is then eluted with 0.05 M potassium phosphate buffer, pH 7.4, which elutes a mixture of 5β-steroid reductase activity and 3α-hydroxysteroid dehydrogenase activity. This fraction is discarded. 3α-Hydroxysteroid dehydrogenase, practically free of 5β-steroid reductase activity, is eluted with 0.2 M potassium phosphate buffer, pH 7.4. About 140–180 ml is required. A suitable aliquot to incubate is 0.3–0.6 ml.

5β-Steroid Reductase. This enzyme is further purified by ammonium sulfate fractionation of the 0.03 M phosphate fraction. Neutralized, saturated ammonium sulfate solution is added to 50% saturation. The protein precipitate is collected by centrifugation and is discarded. The solution is then brought to 70% saturation by further addition of neutralized, saturated ammonium sulfate solution. The protein precipitate is collected by centrifugation. The precipitate is dissolved in a small volume of 0.1 M Tris-HCl buffer, pH 7.4, and the solution is applied to a column of Sephadex G-75, prepared in 0.1 M Tris-HCl buffer, pH 7.4, and with the dimensions 30 mm $\times$ 550 mm. The column is eluted with the same buffer and fractions of about 3 ml per 30 minutes are collected. The 5β-steroid reductase activity is eluted mainly in the descending limb of the protein peak. The protein peak is divided into three or four fractions that are analyzed separately. Of the 5β-steroid reductase activity (checked with 7α,12α-dihydroxycholest-4-en-3-one as substrate) originally present in the eluate of the Sephadex G-25 column about 5–10% is recovered from the Sephadex G-75 column in a total volume of about 60 ml. The overall purification is usually about 10-fold (cf. Table II).

Using 7α,12α-dihydroxycholest-4-en-3-one as substrate the 5β-steroid

TABLE II
SUMMARY OF PURIFICATION OF 5β-STEROID REDUCTASE

Step	Volume of solution (ml)	Total units	Specific activity (units/mg protein)	Yield (%)
Sephadex G-25	250	7500	2.8	100
TEAE, 0.5 M Tris-HCl fraction	475	7700	5.6	103
Hydroxylapatite, 0.03 M phosphate fraction	115	1800	5.0	24
Sephadex G-75				
First fraction	30	0	—	—
Second fraction	57	430	25.0	6
Third fraction	65	0	—	—

reductase requires NADPH as cofactor; hardly any activity is observed if NADPH is substituted with NADH. If the rate of reaction with $7\alpha,12\alpha$-dihydroxycholest-4-en-3-one, assayed as described above, is set at 1, the following relative rates with C_{19}, C_{21}, C_{24}, and C_{27} steroids are obtained: androst-4-ene-3,17-dione, 2.9; testosterone, 2.1; progesterone, 1.3; $7\alpha,12\alpha$-dihydroxy-3-ketochol-4-enoic acid, 2.1; 7α-hydroxycholest-4-en-3-one, 1.5; cholest-4-en-3-one, <0.1.

The stability of the 5β-steroid reductase preparation is limited. After storage for 1 week at $-20°$ and subsequent thawing, about half of the activity has been lost.

3α-Hydroxysteroid Dehydrogenase. The dehydrogenase is further purified by ammonium sulfate fractionation of the 0.2 M phosphate fraction from the hydroxylapatite column. This fraction is put in a dialysis bag and is dialyzed for 24 hours against neutralized, saturated ammonium sulfate solution. After this period of time the contents of the dialysis bag are put in a flask and solid ammonium sulfate is added to ensure 100% saturation. After 2 hours of stirring the mixture is centrifuged. The protein precipitate is dissolved in a small volume of 0.1 M Tris-HCl buffer, pH 7.4. The solution is applied to a column of Sephadex G-75, prepared in 0.1 M Tris-HCl buffer, pH 7.4, and with dimensions 30 mm $\times$ 550 mm. Elution is carried out with the same buffer, and fractions of about 3 ml per 30 minutes are collected. The 3α-hydroxysteroid dehydrogenase is eluted mainly in the descending limb of the protein peak. The protein peak is divided into three or four fractions that are analyzed separately. Of the 3α-hydroxysteroid dehydrogenase activity (checked with $7\alpha,12\alpha$-dihydroxy-5β-cholestan-3-one as substrate) originally present in the eluate of the Sephadex G-25 column, about 50–60% is recovered from the Sephadex G-75 column in a total volume of about 60 ml. The overall purification is usually about 70-fold (cf. Table III).

With $7\alpha,12\alpha$-dihydroxy-5β-cholestan-3-one used as substrate, the 3α-hydroxysteroid dehydrogenase requires NADPH as cofactor; hardly any activity is observed under the conditions of the assay, as described above, if NADPH is substituted with NADH. From other experiments it can be estimated that the rate of reaction with NADH is about one-fifteenth of that with NADPH. If the rate of reaction with $7\alpha,12\alpha$-dihydroxy-5β-cholestan-3-one, assayed as described above, is set at 1, the following relative rates with C_{19}, C_{21}, C_{24}, and C_{27} steroids are obtained: 5β-androstan-3,17-dione, 40; 5α-androstan-3,17-dione, 6.0; androst-4-ene-3,17-dione, <0.1; testosterone, <0.1; $11\beta,17\alpha,21$-trihydroxy-5β-pregnane-3,-20-dione, 19; $17\alpha,21$-dihydroxy-5β-pregnane-3,11,20-trione, 80; cortisol, <0.1; cortisone, <0.1; progesterone, <0.1; $7\alpha,12\alpha$-dihydroxy-3-keto-5β-cholanoic acid, 14; 7α-hydroxy-3-keto-5β-cholanoic acid, 8.8; 3-keto-5β-cholanoic acid, 12; 7α-hydroxy-5β-cholestan-3-one, 8.1; 5β-cholestan-3-

TABLE III
SUMMARY OF PURIFICATION OF 3α-HYDROXYSTEROID DEHYDROGENASE

Step	Volume of solution (ml)	Total units	Specific activity (units/mg protein)	Yield (%)
Sephadex G-25	250	6000	2.0	100
TEAE, 0.5 M Tris-HCl fraction	475	6000	4.4	100
Hydroxylapatite, 0.2 M phosphate fraction	180	6300	13.2	105
Sephadex G-75				
First fraction	34	0	—	—
Second fraction	25	1250	27.5	—
Third fraction	60	3750	142.5	62

one, 0.2; 5α-cholestan-3-one, <0.1; 7α,12α-dihydroxycholest-4-en-3-one, <0.1; 7α-hydroxycholest-4-en-3-one, <0.1; cholest-4-en-3-one, <0.1. If NADPH is substituted with NADH, the following relative reaction rates are obtained (the rate 1 being set for 7α,12α-dihydroxy-5β-cholestan-3-one as substrate and NADPH as cofactor): 5β-androstan-3,17-dione, 7.9; 5α-androstan-3,17-dione, 2.5; 11β,17α,21-trihydroxy-5β-pregnane-3,20-dione, 1.5; 17α,21-dihydroxy-5β-pregnane-3,11,20-dione, 9.5.

The stability of the 3α-hydroxysteroid dehydrogenase preparation appears to be good. After storage for 1 month at −20° and subsequent thawing, almost full activity is retained.

Conversion of 7α-Hydroxycholest-4-en-3-one to 7α-Hydroxy-5β-cholestan-3-one

This reaction is catalyzed by the purified 5β-steroid reductase described above.

The 5β-steroid reductase fraction obtained from the Sephadex G-75 column, 1–2 ml, is diluted to 4 ml with 0.1 M Tris-MCl buffer, pH 7.4, 1 mg of NADPH is added and the mixture is incubated for 20 minutes with 50 μg of labeled 7α-hydroxycholest-4-en-3-one, added to the incubation mixture dissolved in 0.1 ml of acetone. The incubation is terminated by addition of 20 volumes of chloroform–methanol 2:1. The subsequent analytical procedure is as described above. The chloroform extract is chromatographed with benzene–ethyl acetate 1:1 as solvent (cf. Table I).

Conversion of 7α-Hydroxy-5β-cholestan-3-one to 5β-Cholestane-3α,7α-diol

This reaction is catalyzed by the purified 3α-hydroxysteroid dehydrogenase described above.

The 3α-hydroxysteroid dehydrogenase fraction obtained from the Sephadex G-75 column, 0.05–0.1 ml, is diluted to 4 ml with 0.1 M Tris-HCl buffer, pH 7.4; 1 mg of NADPH is added and the mixture is incubated for 20 minutes with 50 μg of labeled 7α-hydroxy-5β-cholestan-3-one, added to the incubation mixture dissolved in 0.1 ml of acetone. The chloroform extract of the incubation mixture is chromatographed with benzene–ethyl acetate 1:1 as solvent (cf. Table I).

[26] Enzymatic Degradation of the Cholestane Side Chain in the Biosynthesis of Bile Acids

By EZRA STAPLE[1]

Work on the cleavage of the side chain of cholestanes to form bile acids may be considered to begin with the work of Anfinsen and Horning.[1a] These investigators found that a homogenate of mouse liver and, later, mitochondrial fractions of this homogenate were capable of yielding radioactive carbon dioxide from incubation with cholesterol-26-^{14}C. Later this enzymatic reaction was studied with rat liver mitochondria by Whitehouse, Staple, and Gurin,[2,3] and with liver mitochondria of other species.[4] This reaction has been used to study the influence of various factors on the cleavage process.

Following suggestions first made by Bergström,[5] 5β-cholestane-3α,7α,12α-triol and various derivatives were studied. This work led to the discovery by Danielsson[6] that this triol can be enzymatically hydroxylated in the 26-position by liver mitochondrial preparations. This was confirmed by Suld, Staple, and Gurin.[7] Further, both of these groups of investigators found that 3α,7α,12α-trihydroxy-5β-cholestan-26-oic acid is also produced. Suld et al.[7] also found that production of the 26-oic acid from the 26-hydroxylated compound is dependent on the presence of NAD$^+$. A partial replacement of NAD$^+$ by NADP$^+$ in a crude system is claimed by

---

[1] Deceased.
[1a] C. B. Anfinsen and M. G. Horning, J. Am. Chem. Soc. **75**, 1511 (1963).
[2] M. W. Whitehouse, E. Staple, and S. Gurin, J. Biol. Chem. **234**, 276 (1959).
[3] M. W. Whitehouse, E. Staple, and S. Gurin, J. Biol. Chem. **236**, 68 (1961).
[4] M. W. Whitehouse, M. C. Cottrell, T. Briggs, and E. Staple, Arch. Biochem. Biophys. **98**, 305 (1962).
[5] S. Bergström, Ciba Found. Symp. Biosyn. Terpenes Sterols. Little, Brown, Boston, Massachusetts, 1959.
[6] H. Danielsson, Acta Chem. Scand. **14**, 348 (1960).
[7] H. M. Suld, E. Staple, and S. Gurin, J. Biol. Chem. **237**, 338 (1962).

Dean and Whitehouse[8] for this reaction. This dehydrogenase system has been purified and studied by Herman and Staple[9] and further by Masui, Herman, and Staple[10] to indicate that this is a two-step dehydrogenation reaction, in which the enzyme for the first step can be separated by enzymatic purification techniques. They also have confirmed the obligatory requirement for NAD⁺ as hydrogen acceptor.

As an extension of their work with $3\alpha,7\alpha,12\alpha$-trihydroxy-5β-cholestane-26-oic acid, Suld et al.[7] found that the side chain was cleaved between carbon atoms 24 and 25 of the side chain to yield cholic acid and propionic acid (propionyl-CoA). These authors proposed a thiolytic cleavage similar to that which occurs in the β-oxidation of fatty acids. Consistent with this hypothesis, Masui and Staple[11] have recently reported the formation of $3\alpha,7\alpha,12\alpha,24\xi$-tetrahydroxy-$5\beta$-cholestan-26-oic acid from $3\alpha,7\alpha,12\alpha$-trihydroxy-5β-cholestan-26-oic acid and the conversion of the former compound to cholic acid.

In sum, the enzyme reactions for the cleavage of the cholestane side chain to form the cholanic acids appears to parallel those for ω-oxidation of hydrocarbons and β-oxidation of fatty acids. The enzyme reactions described below give evidence for some of these steps.

I. Preparation of Substrate Materials

A. Cholesterol-26-¹⁴C

This material can be synthesized according to methods previously reported[12-14] and as given below. It may also be obtained commercially from several sources.

Seventy-two milligrams (0.5 millimole) of methyl iodide-¹⁴C (2 mC per millimole) is diluted with 4.5 millimoles of unlabeled methyl iodide and added to 146 mg of magnesium (for Grignard reactions) (6 millimoles) in 4 ml of anhydrous ether. The reaction is initiated by cautious, gentle warming and proceeds for 1 hour. Then 715 mg (1.67 millimoles) of 3β-acetoxy-27-norcholest-5-en-25-one in 5 ml of dry benzene is added slowly with stirring and the mixture is refluxed for 3 hours. It is then allowed to stand overnight. The complex is decomposed by addition of 2 ml of ice water and 4 ml of 50% acetic acid. The mixture is then steam-

[8] P. D. G. Dean and M. W. Whitehouse, *Biochem J.* **98**, 410 (1966).

[9] R. H. Herman and E. Staple, *Federation Proc.* **24**, 661 (1965).

[10] T. Masui, R. Herman, and E. Staple, *Biochim. Biophys. Acta* **117**, 266 (1966).

[11] T. Masui and E. Staple, *J. Biol. Chem.* **241**, 3889 (1966).

[12] A. I. Ryer, W. H. Gebert, and N. M. Murrill, *J. Am. Chem. Soc.* **72**, 4247 (1950).

[13] W. G. Dauben and H. L. Bradlow, *J. Am. Chem. Soc.* **72**, 4248 (1950).

[14] M. G. Horning, D. S. Frederickson, and C. B. Anfinsen, *Arch. Biochem. Biophys.* **71**, 266 (1957).

distilled until no more water-insoluble distillate is obtained. The reaction mixture is extracted with ether, and the ether extract is washed with NaHCO$_3$ solution. The washings are made alkaline and extracted with ether, and the two ether extracts are combined and dried over MgSO$_4$. The MgSO$_4$ is then filtered off and the ether is evaporated, leaving 743 mg of residue of crude cholest-5-ene-3β,25-diol-26-^{14}C.

The residue is dissolved in 5 ml of anhydrous pyridine, and 1.2 ml (12.7 millimole) of acetic anhydride is added. The mixture is refluxed for 7 minutes and then added dropwise to 50 ml of water. The water is extracted with ether, the ether is washed with 2 N H$_2$SO$_4$, and the acid wash is extracted again with ether. The combined ether extracts are dried over MgSO$_4$, after addition of a little dry NaHCO$_3$. After evaporation of the ether, 806 mg of yellowish residue of crude 3β-acetoxy-cholest-5-en-25-ol-26-^{14}C remains.

The crude acetate is dissolved in 20 ml of anhydrous pyridine and 0.85 ml (9.3 millimole) of freshly distilled phosphorus oxychloride is added. A red color, deepening to black, forms on refluxing for 30 minutes. The solution is poured into 50 ml of ice water, diluted with more water, and extracted with ether. The ether extracts are combined and washed with 2 N H$_2$SO$_4$ to remove the pyridine. The extract is then washed with 5% NaHCO$_3$ solution. The acidified washings are extracted with ether, and the ether extract is washed with NaHCO$_3$ solution. The ether extracts are combined and evaporated. The orange residue of mixed 26-^{14}C labeled 24- and 25-dehydrocholesteryl acetates weighs 665 mg. It is hydrolyzed by refluxing for 1 hour in 10 ml of 2 N ethanolic KOH. This solution is diluted with 50 ml of water and extracted several times with ether. The ether extracts are washed repeatedly with water and the water washes are combined and extracted once with ether. The combined ether extracts are dried overnight over MgSO$_4$ and filtered into a 100-ml hydrogenation flask. The solvent is evaporated, leaving 580 mg of orange residue containing the crude 24- and 25-dehydrocholesterols.

The mixture is dissolved in 20 ml of absolute ethanol and hydrogenated at atmospheric pressure using 50 mg of 10% palladium-on-charcoal as catalyst. The process is stopped after uptake of 23 ml (1.0 millimole) of hydrogen per millimole of steroid. The hydrogenation requires about 30 minutes.

The solution of crude cholesterol is filtered, the solvent is evaporated, and the residue is chromatographed on a 30-g column of neutral alumina (Woelm, grade 1). The crude cholesterol is placed on the column, a minimal quantity of 45% (v/v) ether in petroleum ether (b.p. 30–60°) being used; the column is then eluted with 2.3 liters of the same solvent followed by 1.5 liters of 65% ether in petroleum ether. The fractions containing the principal radioactive peak are combined and evaporated. Approximately 509 mg of crude cholesterol-26-¹⁴C is obtained at this point.

The partially purified product is dissolved in 60 ml of 90% aqueous ethanol and to this solution is added a solution containing 1.7 g of digitonin in 200 ml of 90% aqueous ethanol. After it has stood for 1 hour at room temperature, the digitonide is kept at 7° for 60 hours. It is collected by centrifugation and washed with cold 90% aqueous ethanol, cold ether–acetone (2:1), and cold ether, then dried. The cholesterol is recovered by decomposition of the digitonide dissolved in 12 ml of pyridine. Ether, 120 ml, is added and the digitonin is allowed to precipitate several hours at room temperature. The precipitate is centrifuged and washed with ether, and the ether solutions are combined. The precipitated digitonin is redissolved in 85% aqueous ethanol and the above precipitation procedure is repeated; additional cholesterol is obtained in the ether solution. On evaporation of the ether solution, 326 mg (0.84 millimole) cholesterol-26-¹⁴C, specific activity 0.57 μC/mg is obtained. The overall yield from 3β-acetoxy-27-norcholest-5-en-25-one is 51%. On the basis of methyl iodide-¹⁴C the yield is 35%.

B. 5β-Cholestane-3α,7α,12α-triol-26,27-¹⁴C

The synthesis is carried out, as previously reported,[15] as follows:

Triformylcholyl chloride is prepared from cholic acid by the method of Cortese and Bauman.[16] Using benzene-petroleum ether (b.p. 60–80°) as solvent, triformylcholic acid can be crystallized with m.p. 209–211°. The acid is converted to the acid chloride by treatment with oxalyl chloride.

2-Propanol-1,3-¹⁴C (specific activity 1.0 mC per millimole) is diluted 20-fold with unlabeled 2-propanol and then converted to 2-bromopropane-1,3-¹⁴C by reaction with phosphorus tribromide.

A Grignard reagent is prepared from 2.40 g of 2-bromopropane-1,3-¹⁴C (50 μC per millimole) and 600 mg of magnesium (for Grignard reactions) in 25 ml of anhydrous ether. The solution is cooled in an ice bath and stirred. To this 7.5 g of powdered anhydrous cadmium bromide (dried for 3 hours at 120°) is added. The preparation is stirred for 30 minutes, the ice bath is removed, and a solution of 1.50 g of triformylcholyl chloride in 8 ml of benzene is added. Stirring is continued for 30 minutes after this addition, and the reaction mixture is heated to reflux for 1 hour then allowed to stand overnight at room temperature. Ice water and then 3 N HCl solution are added to the reaction mixture. Sufficient acid should be

[15] E. Staple and M. W. Whitehouse, *J. Org. Chem.* **24**, 433 (1959).
[16] F. Cortese and L. Bauman, *J. Am. Chem. Soc.* **57**, 1393 (1935).

added to dissolve the initial precipitate. The benzene–ether layer is separated and washed with water until the washings are neutral. The washed solvent layer is then evaporated to dryness and the residual gum is further dried *in vacuo*. The residue is heated for 1 hour with 10 ml of 5% methanolic KOH. This reaction mixture is then diluted with 100 ml of water and extracted with ether. The ether extract is evaporated to dryness and the residue of crude 24-keto-5β-cholestane-3α,7α,12α-triol-26,27-^{14}C is subjected to silicic acid column chromatography. The column, 2.0 cm in diameter, contains 50 g of silicic acid (Merck) and is prepared with benzene. It is eluted with benzene first and then with ether–benzene (1:2). The ketone is eluted by the latter solvent. After evaporation of this eluate, 698 mg of 24-keto-5β-cholestane-3α,7α,12α-triol-26,27-^{14}C, m.p. 143–145°, is obtained. This product may be recrystallized from acetone if desired.

To 500 mg of 24-keto-5β-cholestane-3α,7α,12α-triol-26,27-^{14}C dissolved in 1 ml of ethanol is added 1 ml of hydrazine hydrate (99%). The mixture is swirled for a few minutes to give a homogeneous solution. Ten milliliters of triethylene glycol and 1 g of KOH are added to this solution, and the mixture is heated under reflux for 30 minutes. Then the reflux condenser is removed and the mixture is heated at 180–200° for 2 hours. At the end of this heating period, the reaction mixture is allowed to cool in a stream of nitrogen gas and it is then poured into 50 ml of water. The precipitated compound is filtered, washed with water, and dried *in vacuo*. This crude product is crystallized from acetone; 269 mg 5β-cholestane-3α,7α,12α-triol-26,27-^{14}C, m.p. 184–185°, is obtained.

C. 3α,7α,12α-Trihydroxy-5β-cholestan-26-oic Acid

The D-isomer of this acid is best obtained from the bile of *Alligator mississippiensis*. Bile from other crocodilia may be used in place of *A. mississippiensis* bile, which is now difficult to obtain. It is necessary, however, thoroughly to identify and characterize the isolated trihydroxy-coprostanic acid in each case. There may be some racemization about carbon atom 25 during the hydrolysis procedure although no definitive information on this has yet been obtained. It may also be obtained synthetically,[17] but with great difficulty and in low yield.

A procedure for isolation of the acid from *A. mississippiensis* bile is given below.

Bile from *A. mississippiensis* is mixed with 2–3 volumes of ethanol to precipitate the mucin and other proteins. The precipitate is removed by centrifugation and again washed with ethanol. The combined alcoholic solutions are evaporated to dryness giving x grams of gummy residual

[17] R. J. Bridgwater, *Biochem. J.* **64**, 593 (1956).

"crude bile salts." This residue is taken up in water, the pH adjusted to 8–9 and 0.5 x g of decolorizing charcoal is added. The mixture is evaporated on a steam bath with occasional stirring. The residue is transferred to a Soxhlet extractor, and extraction with absolute ethanol is continued until fresh extracts are colorless. Evaporation of the alcohol gives a gummy, occasionally brittle, residue, "purified bile salts." In the case where the quantity of "crude bile salts" is small (less than 100 mg) the charcoal purification step can be omitted.

Hydrolysis of the bile salts is carried out in 2.5 N NaOH in an autoclave at a pressure of 20 psi (125° for 16–18 hours).

The alkaline hydrolyzate is diluted with water, saturated with NaCl, and acidified to pH 1 with HCl. The precipitated bile acids are extracted with three portions of ethyl acetate, and the combined extract is washed three times with water. The combined water washings are extracted once with ethyl acetate. The ethyl acetate extracts are combined and evaporated to dryness, and the crude bile acids so obtained are further purified by partition column chromatography.

The chromatographic column system used by Mosbach, Zomzely, and Kendall[18] is recommended. Four milliliters of 70% acetic acid on 5 g of Celite 545 (Johns-Manville Corp.) is the stationary phase, and petroleum ether (b.p. 30–60°) is the first moving phase. The packed column measures 0.9 cm × 18 cm. The sample to be separated is dissolved in 3–4 drops of glacial acetic acid, mixed thoroughly with 0.2–0.3 g of Celite, slurried in petroleum ether, and added to the column. Elution is effected with (1) 45 ml of petroleum ether, (2) 90 ml of 40% isopropyl ether in petroleum ether, and (3) 100 ml of 60% isopropyl ether in petroleum ether. The 3α,7α,12α-trihydroxy-5β-cholestan-26-oic acid is eluted with solvent mixture (2). Material of m.p. 176–178° is readily obtained by isolation from the bile of *A. mississippiensis*.

The acid obtained in this way can be tritiated by the Wilzbach method.[19] Several commercial laboratories offer this tritiation service. The acid which has been through this procedure must be carefully purified by several chromatographic systems.

D. 5β-Cholestane-3α,7α,12α,26-tetraol

This compound is prepared synthetically in high yield from 3α,7α,12α-trihydroxy-5β-cholestan-26-oic acid by reduction with lithium aluminum hydride as follows:

The methyl ester of the trihydroxy acid is prepared by treatment of

[18] E. H. Mosbach, C. Zomzely, and F. E. Kendall, *Arch. Biochem. Biophys.* **48**, 95 (1954).

[19] K. E. Wilzbach, *J. Am. Chem. Soc.* **79**, 1013 (1957).

an ether solution of the acid with an excess of freshly generated diazomethane solution in ether. The mixture is allowed to stand until all the diazomethane decomposes, then the ether is evaporated. The ester in the residue, after removal of the ether, represents a yield of 99+% of the ester.

The ester (50 mg) is dissolved in 10 ml of anhydrous benzene and added slowly to the lithium aluminum hydride (25 mg) dissolved in 15 ml of ether. After the addition, and after the reaction has subsided, the mixture is heated under reflux for 30 minutes. The reaction mixture is then cooled, and ethanol (95%) is then added cautiously, dropwise, to decompose the excess of the hydride. When no further gas evolution occurs after addition of ethanol, the mixture is added to cold water (50 ml) and sufficient 2 N H_2SO_4 is added to dissolve the precipitate. The ether–benzene layer is washed with water until neutral. The original water layer is extracted with ether, and this ether extract is washed with water until neutral. The ether extract is combined with the original ether–benzene layer, and the solvents are removed by evaporation. The residue consists of practically pure 5β-cholestane-3α,7α,12α,26-tetraol, in a yield of 98+%. After crystallization from methanol, the tetraol melts at 203–204°.

E. 3α,7α,12α,24ξ-Tetrahydroxy-5β-cholestan-26-oic Acid-24-[14]C

This compound is prepared from cholic acid-24-^{14}C by a modified Reformatsky reaction[20] as follows. 3α,7α,12α-Trihydroxycholan-24-al-[14]C (0.10 mC, 95 mg) is synthesized from cholic acid-24-[14]C (activity 0.50 mC) (500 mg) by the method of Yashima.[21] This method utilizes the oxidation of homocholane-3α,7α,12α,24,25-pentaol by lead tetraacetate to give cholanyl aldehyde directly. The aldehyde-^{14}C is dissolved in 10 ml of dry toluene and 3 ml of ethyl dl-2-bromopropionate containing granulated zinc and a small amount of both iodine and powdered copper as catalysts. The reaction mixture is then poured into a mixture of crushed ice and excess 10% H_2SO_4. This is then extracted with ethyl acetate. The ethyl acetate extract is washed with dilute $Na_2S_2O_3$ to remove the iodine and then with water. The solvent layer is then evaporated to dryness. The residue is saponified with 5% of methanolic KOH solution for 1 hour on a water bath. After removal of the methanol, the residual aqueous solution is extracted with ethyl acetate to remove 3α,7α,12α-trihydroxycholan-24-al-24-[14]C. This alkaline solution is then acidified with 10% HCl solution and extracted with ethyl acetate. The extract is washed with water

[20] Y. Inai, Y. Tanaka, S. Betsuki, and T. Kazuno, J. Biochem. (Tokyo) 56, 591 (1964).
[21] H. Yashima, J. Biochem. (Tokyo) 54, 47 (1963).

to neutrality and then the solvent is evaporated. A reaction product (activity 0.03 mC) (31 mg) is obtained.

On further purification by thin layer chromatography on silica gel G (Brinkmann Instruments) using benzene–isopropanol–acetic acid 30:10:1 as developing solvent, a mixture of the isomers of $3\alpha,7\alpha,12\alpha,24\xi$-tetrahydroxy-$5\beta$-cholestan-26-oic acid-24-^{14}C, with R_f's of 0.31 and 0.34 and m.p. 98–102° is obtained.

For further separation of the stereoisomers of this acid, see Masui and Staple.[11]

II. Enzymatic Transformation

A. Formation of $^{14}CO_2$ from Cholest-5-en-3β-ol-26-[14]C

Principle. This sequence of enzyme reactions in liver mitochondria involves the cleavage of the terminal side-chain carbon atoms of cholesterol and their oxidation to carbon dioxide. By comparison with 26- or 27-unlabeled cholesterol, the origin of these carbon atoms in the side chain can be identified. In the light of later findings,[7] the formation of carbon dioxide from the terminal fragment is due to the oxidation of propionic acid which is cleaved from the side chain. This represents, therefore, a portion of nonsteroid metabolism in this overall reaction.

Although the cholanoic acids formed in this reaction have been reported by Fredrickson and Ono[22] to represent "unnatural" forms, a recent report differs.[23]

The yield of carbon dioxide from carbon atom 26 of cholesterol-26-^{14}C is variable and is dependent on the preparation of mitochondria and their source. The yield may vary from a few percent to more than 30% in very favorable cases.[3, 24] The yield from other possible steroid intermediates is given by Stevenson and Staple[25] and also by Dean and Whitehouse.[8]

The overall reaction is given by Eq. (1).

$$\longrightarrow \quad C^*O_2 \;+\; \begin{array}{c}\text{Substituted} \\ \text{cholanic acids}\end{array} \qquad (1)$$

[22] D. S. Fredrickson and K. Ono, *Biochim. Biophys. Acta* **22**, 183 (1965).

[23] K. A. Mitropoulos and N. D. Myant, *Biochem. J.* **99**, 51P (1967).

[24] D. Kritchevsky, *in* "Metabolism of Lipids as Related to Atherosclerosis" (F. A. Kummerow, ed.), pp. 106–128. Thomas, Springfield, Illinois, 1965.

[25] E. Stevenson and E. Staple, *Arch. Biochem. Biophys.* **97**, 485 (1962).

Procedure. White, male, Wistar strain rats weighing 150–160 g are sacrificed by decapitation, the livers are excised rapidly and transferred to a beaker containing 10% sucrose (w/v) at 0°. The livers are then minced with a scissors and homogenized in cold aqueous sucrose solution (10%) using 3 volumes of sucrose solution per volume of minced liver. The homogenizer is a loose-fitting (1.5 mm clearance) glass Potter-Elvehjem homogenizer.

The homogenate is then centrifuged at 600 g for 10 minutes for removal of unbroken cells, nuclei, and cellular debris. The resultant, supernatant portion from the previous centrifugation is then centrifuged in an ultracentrifuge (Spinco Model L preparative ultracentrifuge) for 12 minutes at 8500 g to separate the mitochondrial fraction. The mitochondrial suspension obtained from the first centrifugation is resuspended in 10% aqueous sucrose solution and recentrifuged at 8500 g for 12 minutes to wash the mitochondria and remove adhering microsomes. The supernatant fraction from this second centrifugation is discarded. The mitochondria are resuspended in 10% sucrose solution so that a concentration of mitochondria equivalent to 1 g of original wet weight of liver tissue is obtained.

Dispersions of cholesterol-26-^{14}C are most conveniently and consistently prepared by dissolving the cholesterol (usually 0.05 mg) in a small amount of methanol containing 1.0 to 4.0 mg Tween 20 (Atlas Powder Co.) and removal of the methanol in a stream of nitrogen at 40–50°. When the methanol has been entirely removed, the still warm solution of cholesterol in Tween 20 is diluted with the desired quantity of buffer solution at room temperature. This procedure yields a visually transparent dispersion of cholesterol which is consistently reproducible. Other methods of dispersion have been used. These include serum albumin stabilized emulsions (cf. Anfinsen and Horning[1]) and lipoprotein solutions, but the Tween 20 dispersion has been found to be reliably consistent, provided however that quantities of Tween 20 are held sufficiently low so as to have no deleterious effect on the enzymes involved.

Boiled soluble factor (SF), which is added to the incubation, is deproteinized, heat-stable fraction obtained by heating for 5 minutes at 90° the supernatant fraction from the first ultracentrifugation at 8500 g mentioned above. After heating, the solution containing particulate coagulum is filtered or centrifuged at low speed and the resultant supernatant fraction is used in the incubations.

A typical incubation mixture contains 1 ml of mitochondrial suspension (equivalent to 1 g wet weight of liver); 1 ml of a solution containing ATP (25 mg) and AMP (8 mg); NAD$^+$ (5 mg); reduced glutathione (15 mg), sodium citrate monohydrate (30 mg), $Mg(NO_3)\cdot 6\ H_2O$ (10 mg), potassium penicillin G (2000 units), and streptomycin sulfate (1 mg); 5 ml of cholesterol-26-^{14}C dispersion in 0.25 M tris(hydroxymethyl)amino-methane HCl, pH 8.5, and 5 ml of boiled supernatant fluid prepared as described above. The incubations are carried out in 125-ml stoppered Erlenmeyer flasks provided with a center well which contains 2.0 ml of 2.5 N sodium hydroxide solution to trap the carbon dioxide evolved during the incubation.

The flasks are shaken at 37° in air for suitable periods of time (up to 18 hours).

After the incubation periods are completed, 2.5 ml of 25% aqueous trichloroacetic acid are added to each flask, and the flasks are promptly restoppered and shaken for 3 hours to displace [14]CO_2 from the incubation medium. The pH after addition of trichloroacetic acid should not be more than 2.5. This point should be tested and additional acid be added if necessary.

The acidified incubation mixture is shaken, then the contents of the center well are removed with a pipette and added to 2.5 ml of 2 N NH$_4$Cl solution and 1 ml of 1.5 M BaCl$_2$ solution. The precipitated BaCO$_3$ is filtered, washed carefully, dried, weighed, and then counted, either directly in a planchet G-M counter, or dispersed in a thixotropic gel[26] containing appropriate phosphors and counted in a liquid scintillation counter.

B. Formation of 5β-Cholestane-3α,7α,12α,26-tetraol-26,27-^{14}C from 5β-Cholestane-3α,7α,12α-triol-26,27-[14]C

Principle. This reaction involves the enzymatic hydroxylation of 5β-cholestane-3α,7α,12α-triol. Both Danielsson[6] and Suld *et al.*[7] have demonstrated that this hydroxylation can take place in mouse or rat liver mitochondria. Yields of 10–35% have been obtained for the conversion of triol to tetraol using this preparation. The mechanism and stereochemistry of this reaction are yet to be elucidated. It is indicated in Eq. (2):

$$(2)$$

[26] B. L. Funt and A. Hetherington, *Science* **125**, 986 (1957).

Procedure. Washed mitochondria from white, male, Wistar strain rats (150–160 g) are obtained as mentioned in Section II,A. A dispersion of 0.05 mg substrate, 5β-cholestane-$3\alpha,7\alpha,12\alpha$-triol-26,27-^{14}C, is prepared in 5 ml of 0.25 M tris(hydroxymethyl)aminomethane buffer, pH 8.5, with 4 mg Tween 20 as described for cholesterol-26-^{14}C in Section II,A. The dispersion is cooled to room temperature and to it is added 1 ml of a solution containing ATP (25 mg), GSH (15 mg), MgCl$_2$·6 H$_2$O (8 mg), trisodium citrate dihydrate (22 mg). This mixture is added to each incubation flask. A suspension of washed mitochondria in 10% sucrose solution equivalent to 4 g of liver tissue (wet weight) along with 105,000 g supernatant fluid (obtained from the same liver homogenate above) equivalent to 1.5 g liver tissue (wet weight) is added next to each incubation flask. The final volume is made up, as necessary, with 10% sucrose solution, to 12.5 ml.

The incubations are conducted aerobically at 37° for 2.5 hours. At the end of this period, the incubation is terminated by the addition of 40 ml of 95% ethanol. The precipitated proteins and other insoluble substances are removed by centrifugation, and the supernatant ethanolic solution is evaporated to dryness *in vacuo* at 25–30°. The residue, suspended in water (3–4 ml) is acidified with 10 N H$_2$SO$_4$ to pH 1–2, mixed with Celite 535 (Johns-Manville Corp.) which has been previously purified by treatment with water, methanol, and ether and then air dried. The residue–Celite mixture (2 g of Celite per milliliter of residue suspension) is packed firmly into a 2 cm diameter column and eluted with 150 ml of ether. (For isolation of crystalline tetraol the incubation size is increased by 10-fold over that given above and the ether eluates from twenty of these large-scale incubations are combined for paper chromatography.) The ether eluate is evaporated and this residue is then chromatographed on paper using 70% aqueous acetic acid as the stationary phase and isopropyl ether–heptane (60:40) as described by Sjövall[27] and Suld *et al.*[7] Whatman No. 1 paper is used, and the chromatography carried out by the descending technique for periods of 3–10 hours. After development, the paper is dried and may be preliminarily scanned in a suitable scanning radioactivity-detection device or the tetraol spot may be visualized by other techniques.

The radioactive peak corresponding to the 5β-cholestane-$3\alpha,7\alpha,12\alpha,26$-tetraol spot is marked on the paper and cut out, and the material is eluted from the paper with methanol. The tetraol may be further purified by the use of reversed-phase column chromatography.[28] Hydrophobic Hyflo Supercel (Johns-Manville Corp.) is prepared by treatment with

[27] J. Sjövall, *Acta Chem. Scand.* **6**, 1552 (1952).
[28] S. Bergström, R. J. Bridgwater, and U. Gloor, *Acta Chem. Scand.* **11**, 836 (1957).

dimethyldichlorosilane as described by Bergström and Sjövall.[29] The stationary phase is chloroform–heptane (90:10). Four milliliters of this solvent mixture is added to 4.5 g of treated Hyflo Supercel. The moving phase is 50% aqueous methanol. The tetraol is usually eluted by 50–75 ml of the moving phase. Crystalline tetraol, m.p. 203–204° is obtained.

C. Formation of 3α,7α,12α-Trihydroxy-5β-Cholestan-26-oic Acid-26,27-¹⁴C from 5β-Cholestane-3α,7α,12α,26-tetraol-26,27-¹⁴C

Principle. These enzymatic reactions involve the sequence, in which 5β-cholestane-3α,7α,12α,26-tetraol is converted to 3α,7α,12α-trihydroxy-5β-cholestan-26-oic acid. Yields of 20–30% of the acid from enzymatic oxidation of the tetraol by the preparation given here have been reported. A further elaboration of these reactions is given in Section II,D. The reaction is described by Eq. (3).

$$(3)$$

Procedure. The complete system as described in procedure of Section II,B is used here. In addition, NAD⁺ (5 mg) is added to each incubation mixture.

In place of the substrate added above (Section II,B), 0.5 mg of 5β-cholestane-3α,7α,12α,26-tetraol-26,27-¹⁴C is used. The dispersion in Tween 20 is prepared as previously described.

The incubation is carried out at 37°, in air, for 1.5 hours and terminated by addition of 90% ethanol as previously described. The centrifuged ethanolic supernatant is evaporated *in vacuo* and mixed with Celite 535 as described in procedure of Section II,B. The ether eluate from this chromatography is evaporated and the residue is subjected to paper chromatography, as described in Section II,B, for 10 hours. The 3α,7α,12α-trihydroxy-5β-cholestan-26-oic acid-26,27-¹⁴C peak is eluted from the paper with methanol and subjected to partition column chromatography according to the procedure of Mosbach, Zomzely, and Kendall.[18] Celite 545, 4 g, is treated with 5 ml of 70% aqueous acetic acid and packed in a column (1.0 cm × 18 cm). One hundred milliliters of petroleum ether (b.p. 30–60°) is passed through the column. The sample, dis-

[29] S. Bergström and J. Sjövall. *Acta Chem. Scand.* **5,** 1267 (1950).

solved in two drops of glacial acetic acid mixed with 0.2 g Celite 545, is placed on top of the column. The column is eluted with 45 ml of petroleum ether and then with 150 ml petroleum ether–isopropyl ether (60:40) mixture. The $3\alpha,7\alpha,12\alpha$-trihydroxy-5β-cholestan-26-oic acid-26,27-^{14}C can be obtained by evaporation of the solvent of elution. Crystalline acid, m.p. 172–173° can be obtained from ether solution.

D. Formation of $3\alpha,7\alpha,12\alpha$-Trihydroxy-5β-Cholestan-26-oic Acid-^{3}H and $3\alpha,7\alpha,12\alpha$-Trihydroxy-5β-cholestan-26-al-^{3}H from 5β-Cholestane-$3\alpha,7\alpha,12\alpha,26$-tetraol-^{3}H

Principle. The enzyme preparation, which is soluble and partially purified as described, has two component activities: (1) a 26-hydroxyl dehydrogenase which produces the 26-aldehyde, and (2) an aldehyde oxidase which yields the 26-acid. NAD$^+$ is the hydrogen receptor in both these steps although another report indicates a utilization of NADP in an analogous oxidation.[8] Conversions of tetraol to aldehyde in 45% yield and of aldehyde to acid in 5–20% yield have been obtained with the preparation given here. The latter yield of 20% was obtained after addition of ammonium sulfate to the enzyme preparation. The crude system before precipitation with $(NH_4)_2SO_4$ forms the acid from the tetraol in 50% yield. These two activities have been observed in the same preparation. After further purification only the first, or alcohol dehydrogenase, activity is observed. Whether both activities are present in the same enzyme or whether each is present in a separate enzyme remains to be determined. The reactions are indicated in Eq. (4).

$$(4)$$

Procedure. Rat livers (30 g) from male, Wistar strain rats, 150–160 g weight, are homogenized in 10% sucrose solution (75 ml) and then centrifuged at 500 g for 12 minutes. The supernatant layer after removal of nuclei is centrifuged at 105,000 g for 1 hour. $(NH_4)_2SO_4$ is added to the 105,000 g supernatant fraction. The precipitate obtained at between 40 and 65% of the saturation concentration of $(NH_4)_2SO_4$ at pH 7.4 is collected after this mixture has been held at 4° for ½ hour, then centrifuged at 8500 g for 20 minutes. The precipitate (400 mg protein) is dissolved in 20 ml of 0.02 M Tris buffer (pH 8.5). The protein solution is kept at 59–60° for 10 minutes, then cooled rapidly to 4° and centrifuged at 8500 g for 10 minutes. This heat treatment serves to inactivate the 3-hydroxy-steroid dehydrogenase, which interferes with the reaction observed. The precipitate from this treatment is removed and discarded after centrifugation at 8500 g for 10 minutes. The supernatant fraction is treated again with sufficient $(NH_4)_2SO_4$ to attain 40–65% of saturation concentration. The precipitate (120 mg of protein) obtained in this case is dissolved in 0.02 M Tris buffer, pH 8.5 (30 ml) and dialyzed against the same buffer, overnight, at 4°. The resultant dialyzed preparation is the enzyme preparation used in these incubations.

The incubation is carried out with a volume of solution equivalent to 1.2 mg of protein, 0.5 mg substrate, 5β-cholestane-3α,7α,12α,26-tetraol-26-³H in 0.01 ml of methanol, NAD⁺ (2 mg), 0.25 M Tris buffer, pH 9.5, to give a final volume of 3 ml. The resultant mixture is incubated, in air, at 37° for ½ hour. The incubation is terminated by heating on a water bath for 5 minutes. The incubation mixture is then acidified with dilute HCl and extracted with ether.

The resultant extract is chromatographed on thin-layer plates coated with silica gel G using ethyl acetate–acetone (70:30) as developing solvent. The three compounds are separated quite well: the most polar spot is 3α,7α,12α-trihydroxy-5β-cholestan-26-oic acid-³H, $R_f = 0.04$; the intermediate spot corresponds to 5β-cholestane-3α,7α,12α,26-tetraol-³H, $R_f = 0.31$; and the least polar spot is 3α,7α,12α-trihydroxy-5β-cholestan-26-al-³H, $R_f = 0.54$. The aldehyde may be converted to the corresponding acid by an aqueous ethanolic Ag_2O suspension, and the acid identified by comparison with the authentic compound. The spots may also be located by radioactive scanning of the plate, followed by elution of the spot with methanol and determining its radioactivity in a liquid scintillation counter.

This enzyme preparation produces aldehyde predominantly. If a cruder preparation is used[9] the acid is practically the only product observed. The dehydrogenase preparation described here may also be used with NADH in a reverse conversion of the aldehyde to tetraol.[10]

E. Formation of Propionic Acid-1,3-^{14}C from 5β-Cholestane-3α,7α,12α-triol-26,27-[14]C

Principle. The enzymatic transformations carried out here involve the formation of propionic acid-1,3-^{14}C and cholic acid from 5β-cholestane-3α,7α,12α-triol-26,27-[14]C. This involves the formation of the 3α,7α,12α,26-

(5)

tetraol, $3\alpha,7\alpha,12\alpha$-trihydroxy-26-aldehyde, and the $3\alpha,7\alpha,12\alpha$-triol-26-oic acid. This latter acid is believed to be cleaved by a series of reactions, closely resembling the β-oxidation of fatty acids, to yield propionic acid and cholic acid. This reaction has been observed to occur in a mitochondrial fraction of liver homogenate to which a supernatant fraction, centrifuged at 105,000 g for 1 hour, has been added. Yields of propionic acid of 20–34% have been obtained with the preparation given here. The postulated reaction sequence is outlined in Eq. (5).

Procedure. The incubation mixture and additions are the same as for the formation of 5β-cholestane-$3\alpha,7\alpha,12\alpha,26$-tetraol (see procedure of Section II,C) except for the addition of 1.5 mg of potassium propionate to each flask containing 12.5 ml of incubation mixture. The incubation is carried out aerobically, with mechanical shaking at 37° for 1.5 hours. The reaction is terminated by heating for 1–2 minutes on a boiling water bath. The resultant, precipitated proteins are removed by centrifugation and the precipitate is washed thoroughly. The supernatant fluid and washings are treated with 2 mg of potassium propionate and $4 N$ H_2SO_4 to yield a pH of 1–2. The solution is steam-distilled until 150–200 ml of distillate is collected. The steam distillate is neutralized with $0.1 N$ NaOH solution, using phenolphthalein as indicator. An aliquot of the neutralized steam-distillate may be evaporated on a planchet and the radioactivity measured by use of a suitable G-M counter. (Precautions to avoid the loss of radioactive propionic acid when evaporating the propionate solution should be taken by keeping the pH of the evaporating solution at 10 or above.) The remainder of the steam distillate can be chromatographed, as below, to isolate pure propionic acid for further G-M or liquid scintillation counting and derivative formation.

The neutralized steam distillate is evaporated to dryness *in vacuo* and then chromatographed on a Celite 535 column according to the procedure of Swim and Utter.[30] The stationary phase contains 7.5 ml of $0.2 N$ H_2SO_4 mixed with 15 g of Celite 535; the moving phase contains chloroform saturated with $0.2 N$ H_2SO_4. The eluate is collected in 10-ml portions to which 5 ml of CO_2-free water is added. The acidity is then titrated with $0.01 N$ NaOH solution using phenol red as indicator. Aliquots of these fractions can be taken for further counting and the neutralized purified propionic acid can be used also for further solid derivative formation.

F. Formation of Propionyl-1,3-^{14}C-CoA from 5β-Cholestane-3α,7α,12α,26-tetraol-26,27-[14]C

Principle. The formation of propionyl-1,3-^{14}C-CoA from 5β-cholestane-$3\alpha,7\alpha,12\alpha,26$-tetraol-26,27-^{14}C is observed in a microsomal fraction

---

[30] H. E. Swim and M. F. Utter, see Vol. IV, p. 584.

from a rat liver homogenate to which is added a supernatant fraction from this same liver homogenate, spun at 105,000 g for 1 hour. Twenty to thirty percent of the theoretical yield of propionyl-CoA from the cleavage of the side chain of the tetraol has been obtained with this preparation. This reaction sequence is included in Eq. (5).

Procedure. The microsomes as separated by differential centrifugation of the 8500 g supernatant fluid at 105,000 g for 1 hour are used for this incubation. The microsomes equivalent to 1.25 g of rat liver (wet weight) are used for this incubation. The 105,000 g-1 hour supernatant fluid equivalent to 0.5 g of rat liver (wet weight) is also added to the above microsomal preparation. Other rat liver and homogenization specifications are as previously described in procedure of Section II,A.

To microsomes and supernatant fluid in a 50-ml flask is added a dispersion of 0.05 mg of 5β-cholestane-3α,7α,12α,26-tetraol-26,27-^{14}C prepared with 1.5 mg Tween 20 as previously described (Section II,A) in 1.5 ml of 0.25 M phosphate buffer, pH 8.0, containing potassium propionate (0.5 mg), ATP (7.3 mg), NAD$^+$ (1.3 mg), GSH (3.7 mg), nicotinamide (8.8 mg), MgCl$_2\cdot$6 H$_2$O (2.0 mg), trisodium citrate dihydrate (5.5 mg) and Na$_4$P$_2$O$_7$ (11.8 mg). The total volume is 3.5 ml. The mixture is allowed to incubate at 37° for 10 minutes, at which time 5 mg of unlabeled, freshly prepared propionyl-CoA is added to each flask. The mixtures are allowed to incubate at 37° for 10 minutes longer for a total incubation time of 20 minutes.

The incubation mixture can then be worked up for isolation of propionyl hydroxamate or for isolation of propionyl-CoA directly.

The isolation of propionyl hydroxamate is by the method of Stadtman and Barker.[31] After the incubation is terminated, the pH is adjusted to 5.5–6.0 with 1 N HCl. Four milliliters of hydroxylamine reagent, prepared by mixing equal volumes of 3.5 N NaOH solution and 28% hydroxylamine hydrochloride solution, is added to each flask. After 10 minutes at room temperature, 100 ml ethanol is added and the resultant precipitate is removed by centrifugation. The supernatant solution is evaporated *in vacuo* to dryness and the dry residue is extracted with three 2-ml portions of absolute ethanol and the combined ethanolic extracts are evaporated in a stream of nitrogen to a volume of 0.3 ml. The precipitated solid is removed by centrifugation and the supernatant solution is subjected to descending paper chromatography using the octanol–formic acid–water (75:25:75) system[27] for 5–6 hours. The developed chromatograms are dried overnight at room temperature; a narrow lengthwise strip is cut out of the chromatogram and dipped into an acid–ferric chloride

[31] E. R. Stadtman and H. A. Barker, *J. Biol. Chem.* **184,** 769 (1950).

ethanolic solution prepared as described by Thompson.[32] The area of the chromatogram corresponding to the colored area of the indicating strip is then eluted with ethanol and the eluate is evaporated under nitrogen to a volume of 0.3 ml. This solution is in turn rechromatographed on paper using the solvent system n-butanol–acetic acid–water (4:1:5). Again, a narrow lengthwise strip is cut out and stained with ethanolic ferric chloride solution. The area of the chromatogram corresponding to the stained area of the strip is eluted with ethanol, and the propionyl hydroxamate remaining after evaporation of the ethanol is assayed for radioactivity.

For isolation of propionyl-CoA a modification of the procedure of Lynen, Reichert, and Rueff[33] is used. After the incubation is terminated by treatment with 1.2 g of $(NH_4)_2SO_4$, the mixture is extracted with three 1-g portions of phenol. The combined phenol extracts are mixed with an equal volume of ether, and the resultant solution is extracted with five 1-ml portions of water. The aqueous extracts are combined and reextracted with 2 ml of ether. The aqueous layer is evaporated in vacuo (20°) to a volume of 0.3 ml. This solution is spotted on paper (Whatman No. 1) and chromatographed using isopropanol–pyridine–water (1:1:1) as solvent.[34] After development the paper is air-dried. A lengthwise strip is cut from the chromatogram and dipped into nitroprusside reagent[35] made as follows: Dissolve 1.5 g of sodium nitroprusside in 5 ml of 2 N H_2SO_4. Add 95 ml of absolute methanol and then 10 ml of 28% NH_4OH solution. Filter off the white precipitate and use the clear orange filtrate. Then spray the strip with methanolic NaOH solution. The propionyl-CoA area can then be visualized. Cut out the area on the original chromatogram corresponding to this spot and then spray with hydroxylamine reagent solution (previously described); the resulting propionyl hydroxamate is eluted from the paper with ethanol and can be chromatographed by either of the systems mentioned in the preceding paragraph. The eluted propionyl hydroxamate can also be assayed for radioactivity.

G. Formation of 3α,7α,12α,24ξ-Tetrahydroxy-5β-cholestan-26-oic Acid-³H from 3α,7α,12α-Trihydroxy-5β-cholestan-26-oic Acid-³H

Principle. The reaction involved here is the enzymatic formation of the β-hydroxy acid from 3α,7α,12α-trihydroxy-5β-cholestan-26-oic acid in a rat liver mitochondrial preparation to which 105,000 g supernatant

[32] A. R. Thompson, *Australian J. Sci. Res. Ser.* **B4**, 180 (1951).
[33] F. Lynen, E. Reichert, and L. Rueff, *Ann. Chem.* **574**, 1 (1951).
[34] E. R. Stadtman, see Vol. III, p. 940.
[35] G. Toennies and J. J. Kolb, *Anal. Chem.* **23**, 823 (1951).

fraction has been added. The intermediate formation of the 24,25-dehydro acid and subsequent hydration of this dehydro acid by an enzyme similar to the enoyl-CoA hydratase of fatty acid metabolism are probable steps in the conversion. The resultant 24-hydroxy (or β-hydroxy) acid is produced by a stereo-specific enzymatic reaction, the actual stereochemistry of which has not yet been established. It appears likely, by analogy with the report of Masui and Staple[36] on the separation of the stereoisomers of 5β-cholestane-3α,7α,12α,24-tetraols, that the configuration of the hydroxyl group at carbon atom 24 is alpha. A yield of 30% of the tetrahydroxy acid has been obtained with the preparation given here. This sequence is included in Eq. (5).

Procedure. A 40% rat liver homogenate from male Wistar rats (150–160 g) is prepared by homogenizing the tissue in 10% sucrose solution with a loose-fitting (1.5 mm clearance) Potter-Elvehjem glass homogenizer. The homogenate is centrifuged at 600 g for 10 minutes to remove unbroken cells, nuclei, and large cell debris. The mitochondrial portion is isolated by centrifugation of this supernatant fraction at 8500 g for 12 minutes. The supernatant layer from this centrifugation is centrifuged at 105,000 g for 1 hour to give the supernatant fraction used below. The mitochondria are suspended in 10% sucrose to yield a suspension equivalent to 4 g of rat liver (wet weight) in 1 ml. In addition, 2 ml of 105,000 g supernatant fraction prepared as above, is also taken for each incubation.

The incubation medium contains in addition to the above mitochondrial and supernatant fractions, 3α,7α,12α-trihydroxy-5β-cholestane-26-oic acid-^{3}H (0.10 mg, 5×10^4 cpm) in 0.2 ml of methanol; ATP (25 mg), $MgCl_2$ (8 mg), FAD (1 mg), and CoA (2 mg); 0.25 M Tris buffer (5 ml), pH 8.5. The volume is made up to 10 ml with 10% sucrose solution.

The incubations are carried out in air at 37° for 1 hour. At the end of this time the incubation is terminated by the addition of ethanol (20 ml). The precipitated protein is removed by centrifugation and then the precipitate is washed with ethanol. The combined supernatant fractions are evaporated *in vacuo* until most of the ethanol is removed. The aqueous residue is then acidified to pH 1 with diluted HCl and extracted with ethyl acetate. The ethyl acetate solution is then spotted on a thin-layer plate, 0.2 mm silica gel G. A spot of authentic 3α,7α,12α,24ξ-tetrahydroxy-5β-cholestan-26-oic acid is also placed on the same plate. The chromatogram can be developed with various solvent systems: (1) benzene–isopropanol–acetic acid (30:10:1 by volume); (2) benzene–isopropanol–acetic acid (55:25:2 by volume); (3) ether–petroleum ether–methanol–acetic acid (70:30:8:1 by volume); (4) ethyl acetate–acetone (70:30 by

[36] T. Masui and E. Staple, *Steroids* **9**, 443 (1967).

volume). The steroid compounds on the plates can be detected in various ways (see the article by Lisboa[37]). Iodine vapor has the advantage that the spots can be marked after they become visible and then the iodine may be allowed to disappear on standing for some time in air. The marked spots may be scraped off the plate and eluted with methanol and the eluted steroid assayed for radioactivity. Alternatively, the plates, appropriately marked with the positions of substrate and product standard, may be scanned in a radiochromatogram scanner. The enzyme has been found to be stereospecific and if synthetic $3\alpha,7\alpha,12\alpha,24\xi$-tetrahydroxy-5$\beta$-cholestan-26-oic acid is used as standard, radioactivity will be detected in only one (the more polar) of the two spots obtained, using solvent system (4) above.

H. Formation of Cholic Acid-24-^{14}C from $3\alpha,7\alpha,12\alpha,24\xi$-tetrahydroxy-5$\beta$-cholestan-26-oic Acid-24-^{14}C

Principle. The conversion of the 24-hydroxy acid to cholic acid demonstrated here occurs in the supernatant fraction of rat liver homogenate (spun at 105,000 g for 1 hour). It appears to involve a 24-hydroxy-acyl-CoA dehydrogenase and thiolase. The reaction in this crude system has been found to require NAD$^+$ or NADP$^+$, CoA, and ATP. These requirements are consistent with those required for similar conversions of fatty acids. A yield of 15% of cholic acid from the tetrahydroxy acid has been obtained with this preparation. The reaction sequence is shown in the last part of Eq. (5).

Procedure. Two milliliters of 105,000 g supernatant fluid obtained from the 40% rat liver homogenate in 10% sucrose obtained in procedure of Section II,G above is used as the enzyme preparation. To this is added $3\alpha,7\alpha,12\alpha,24\xi$-tetrahydroxy-5$\beta$-cholestan-26-oic acid-24-^{14}C (0.01 mg, 1 $\times$ 10^4 cpm) prepared as described in Section I,E in 0.2 ml of methanol, ATP (28 mg), MgCl$_2$ (8 mg), NAD$^+$ (5 mg), CoA (2 mg), GSH (15 mg), 0.25 M Tris buffer, pH 9.0 (5 ml). Final volume is made up to 10 ml with 10% sucrose solution.

The incubation is carried out in air at 37° for 1 hour. At the end of this time ethanol (20 ml) is added to terminate the incubation. The precipitated protein is removed by centrifugation, washed with ethanol, and recentrifuged. The combined extracts are evaporated *in vacuo* to remove the ethanol. The aqueous residue is acidified with diluted HCl to pH 1, and the steroid acids are extracted with ethyl acetate. This extract is spotted on thin-layer plates (0.2 mm, silica gel G), and the chromatogram is developed with the solvent system, benzene–isopropyl alcohol–acetic acid, 30:10:1. Authentic cholic acid, $R_f = 0.26$, and $3\alpha,7\alpha,12\alpha,24\xi$-tetra-

[37] B. P. Lisboa, this volume [1].

hydroxy-5β-cholestan-26-oic acid, $R_f = 0.54$ are spotted on the plate as standards. The plate is exposed to iodine vapor. The area of the plate containing the radioactive sample and which is opposite the standard spot is scraped from the plate and eluted with methanol; the eluate is counted in a suitable counter to determine its radioactivity. Conversely, the plate may be scanned for radioactivity in a suitable thin-layer plate radio-chromatogram scanner, and the radioactivity opposite the standard spots can be measured from the curve.

Section V

Enzyme Systems of Steroid Hormone Metabolism

[27] The Cholesterol Side-Chain Cleavage Enzyme System of Gonadal Tissue

By Enrico Forchielli

Cleavage of the cholesterol side chain is one of the key reactions leading to the biosynthesis of the steroid hormones in endocrine organs. This complex enzyme system has been studied in some detail in tissue preparations of the various steroid-producing organs.[1-8] The investigations have shown that these enzyme complexes from the various endocrine organs share many properties. The activity seems to be mainly localized in the mitochondrial fraction of the tissue homogenate and availability of TPNH appears to be essential for maximal activity. The sequence from cholesterol to pregnenolone involves at least two hydroxylation steps: on the side chain at C-20 and at C-22 prior to cleavage. The evidence for this is based on the isolation of 20α-hydroxycholesterol from cholesterol incubations with preparations of adrenal tissue,[9] rat testes,[7] and corpus luteum[5] and the isolation of $20\alpha,22R$-dihydroxycholesterol from adrenal[10,12] and corpus luteum tissue incubations[5] with cholesterol-^{14}C. In addition the increasing rate of side-chain cleavage observed going from cholesterol to 20α-hydroxycholesterol and $20\alpha,22R$-dihydroxycholesterol[3] is consistent with the thinking that these two substances are intermediate in the sequence between cholesterol and pregnenolone.

Pregnenolone was established as the immediate steroidal product[11] resulting from the side-chain cleavage, and trapping experiments established the initial C_6 fragment to be isocaproaldehyde.[11,13] Under normal

[1] I. D. K. Halkerston, J. Eichhorn, and O. Hechter, *J. Biol. Chem.* **236**, 374 (1961).

[2] K. Shimizu, R. I. Dorfman, and M. Gut, *J. Biol. Chem.* **235**, PC 25 (1960).

[3] K. Shimizu, M. Hayano, M. Gut, and R. I. Dorfman, *J. Biol. Chem.* **236**, 695 (1961).

[4] G. Constantopoulous and T. T. Tchen, *J. Biol. Chem.* **236**, 65 (1963).

[5] S. Ichii, E. Forchielli, and R. I. Dorfman, *Steroids* **2**, 63 (1963).

[6] D. Toren, K. M. J. Menon, E. Forchielli, and R. I. Dorfman, *Steroids* **3**, 381 (1964).

[7] K. M. J. Menon, M. Drosdowsky, R. I. Dorfman, and E. Forchielli, *Steroids Suppl.* I, 95 (1965).

[8] G. Morrison, R. A. Meigs, and K. J. Ryan, *Steroids Suppl.* **II**, 177 (1965).

[9] S. Solomon, P. Levitan, and S. Lieberman, *Rev. Can. Biol.* **15**, 282 (1956).

[10] K. Shimizu, M. Gut, and R. I. Dorfman, *J. Biol. Chem.* **237**, 699 (1962).

[11] E. Staple, W. S. Lynn, Jr., and S. Gurin, *J. Biol. Chem.* **219**, 845 (1956).

[12] G. Constantopoulos, P. Satoh, and T. T. Tchen, *Biochem. Biophys. Res. Commun.* **8**, 50 (1962).

[13] G. Constantopoulos, A. Carpenter, P. S. Satoh, and T. T. Tchen, *Biochemistry* **5**, 1650 (1966).

incubation conditions only the acid is detected, the aldehyde apparently being oxidized immediately on formation. The entire sequence from cholesterol to pregnenolone may be summarized as follows:

Cholesterol → 20α-hydroxycholesterol
20α-Hydroxycholesterol → 20α,22R-dihydroxycholesterol
20α,22R-Dihydroxycholesterol → pregnenolone + isocaproaldehyde
Isocaproaldehyde → isocaproic acid

Cholesterol → pregnenolone + isocaproic acid

Principle

The enzyme preparation is incubated with cholesterol-26-^{14}C and the rate of cleavage is determined by measuring the amount of steam volatile radioactivity liberated as isocaproic acid-^{14}C.

Reagents

Buffer A: sucrose, 0.07 M; mannitol, 0.21 M; EDTA, 0.0001 M; Tris, 0.1 M at pH 7.2

Buffer B: phosphate buffer, 0.066 M, pH 7.2; prepared by mixing 0.066 M NaH_2PO_4 and 0.066 M Na_2HPO_4

Propylene glycol

$MgCl_2$

TPN

TPNH

H_2SO_4, 3 N

KCl, 0.154 M

0.1 $NaHCO_3$

Anhydrous ether

Ethanolic potassium hydroxide, 0.25 N

Acetone AR, redistilled

Procedure

Preparation of Enzyme from Rat Testis

Young adult male rats (200 g) are sacrificed by decapitation, the testes are removed, and the following operations are carried out in a cold room at 4°. Decapsulated testes are homogenized in 3 volumes of buffer A, and centrifuged at 700 g for 15 minutes at 0° to remove nuclei, unbroken cells, and cell debris. The supernatant fluid is centrifuged at 6500 g for 25 minutes at 0° in a Beckman Model L-2 ultracentrifuge to sediment the mitochondria. The supernatant fluid is discarded, and the residue is resuspended in the same medium and centrifuged at 20,000 g at 0° for 15 minutes. This process is repeated two or more times to obtain the final washed testis mitochondrial fraction.

Incubation Procedure for Rat Testis Preparation

Incubations are carried out in air at 37° with constant shaking for 2 hours with the following additions: cholesterol-26-[14]C (about 15 milli-micromoles) suspended in 0.05 ml of propylene glycol; 20 micromoles of $MgCl_2$; 2.5 micromoles of TPNH, 0.7 ml of phosphate buffer (0.066 M) at pH 7.2; 0.3 ml of distilled water, and 0.8 ml of the enzyme preparation corresponding to 1 g wet weight of the tissue. The final volume of the incubation mixture is 2.0 ml.

Enzyme Preparation from Bovine Corpus Luteum

Bovine corpora lutea are obtained at slaughter and placed on cracked ice until collection is complete. Grossly atrophic corpora lutea and those from pregnant cows are excluded from collection.

The tissues are homogenized with an equal volume of 0.154 M KCl containing 0.1 volume of 0.066 M phosphate buffer at pH 7.4 in a Waring blendor run at half speed. The homogenate is centrifuged at about 700 g for 15 minutes, and the residue containing cell debris, nuclei, and un-broken cells is discarded. The supernatant fraction is used as follows to make the acetone powders.

The cell-free homogenate is slowly added to 10 volumes of acetone precooled to −30°. The precipitate is collected on a Büchner funnel, washed with cold acetone followed by washing with ether also precooled to −30°. The dried precipitated material is pulverized in a glass mortar, placed in a desiccator under vacuum, and stored in a refrigerator at 4°. Prior to incubation the acetone powder is extracted with 1.0 ml of buffer B at pH 7.2 and centrifuged at 105,000 g for 30 minutes at 0° to remove insoluble material.

Incubation Procedure with the Corpus Luteum Preparation

A benzene solution of about 15 millimicromoles of cholesterol-26-^{14}C is placed in 20-ml beakers, 0.05 ml propylene glycol is added, and the benzene is removed by evaporation in a stream of nitrogen. One milliliter of 0.066 M phosphate buffer pH 7.2, 0.1 ml M $MgCl_2$, and 2.5 micromoles TPN dissolved in 1.0 ml phosphate buffer 0.066 M pH 7.2 is added. Finally 1.0 ml of supernatant fluid of the acetone powder extract equivalent to about 30 mg of acetone powder is added and the whole incubated at 37° in air for the length of time desired.

Determination of Rate of Cholesterol Side-Chain Cleavage

Determination of the rate of side-chain cleavage in both the rat testis and bovine corpus luteum preparation is accomplished by measuring the amount of radioactivity liberated from cholesterol-26-^{14}C as isocaproic acid in the following manner.

The incubations are terminated by the addition of 0.5 ml of 3 N H$_2$SO$_4$, and the mixture is subjected to steam distillation until 15 ml of distillate is collected. The distillate is acidified and extracted with 2–20-ml portions of ethyl ether, and the combined ether extracts are partitioned against two 5.0-ml portions of 0.1 M NaHCO$_3$; the combined NaHCO$_3$ extracts are back-extracted once with ether. The alkaline extract is acidified with 3 N H$_2$SO$_4$ and extracted with 20 ml of ether. The ether extract is neutralized with alcoholic KOH and transferred to 20-ml scintillation vials and concentrated to dryness. Two drops of glacial acetic acid are added to the dried extract in the vial, followed by the addition of 10 ml scintillation fluid consisting of 4 g 2,5-diphenyloxazole and 100 mg of 1,4-bis-2-(5-phenyloxazolyl)benzene dissolved in 1 liter of scintillation grade toluene. The radioactivity is measured in a suitable liquid scintillation counter.

Properties

Substrate Specificity

The enzyme preparation of both rat testis and corpus luteum readily cleave, in increasing order of rate, the side chain of cholesterol, 20α-

TABLE I

Comparison of the Rate of Cholesterol-26-^{14}C Side-Chain Cleavage with That of Cholestenone-26-^{14}C by Bovine Corpus Luteum Acetone Powder Preparation[a,b]

	Isocaproic acid liberated (cpm) from	
Cofactor addition	Cholestenone-26-^{14}C	Cholesterol-26-^{14}C
None	60 (0.02)	40 (0.06)
TPNH, 1 mg	1290 (0.32)	25240 (6.30)
TPN, 1 mg	3510 (0.88)	33310 (8.30)
TPNH-generating system	1060 (0.26)	24160 (6.04)

[a] Each incubation flask contained 30 mg of acetone powder suspended in 1.0 ml of 0.066 M phosphate buffer, pH 7.2, 400,000 cpm of each substrate, 0.1 ml of cofactor solution, and 0.1 ml of 0.1 M MgCl$_2$. Incubations were carried out at 37° in air for 1 hour. TPNH-generating system = 1 mg of TPN, 3 mg of glucose 6-phosphate, and 2.0 Kornberg units of glucose-6-phosphate dehydrogenase. Figures in parentheses represent percentage conversion of added precursor.

[b] S. Ichii, R. I. Dorfman, and E. Forchielli, unpublished data.

TABLE II
INHIBITION OF THE CLEAVAGE OF THE SIDE-CHAIN OF CHOLESTEROL-26-[14]C
BY AN ACETONE POWDER OF BOVINE CORPUS LUTEUM BY ADDED
PREGNENOLONE AND PROGESTERONE[a]

Steroid added (μmoles)	Isocaproic acid-^{14}C liberated (cpm)	Percent inhibition
Pregnenolone		
None	28000	—
0.032	10200	60.1
0.096	7140	74.5
0.320	3780	86.5
Progesterone		
0.032	18600	33.6
0.096	14100	49.5
0.320	9700	65.3

[a] Each incubation flask contained 33 mg acetone powder suspended in 2.0 ml 0.066 M phosphate buffer, 500 μg TPNH in 0.1 ml 0.066 M phosphate buffer, 0.1 ml of 0.1 M MgCl₂, 2 × 10⁵ cpm cholesterol-26-^{14}C, and the added inhibitor steroid. Incubation time 30 minutes, 37° in air.

hydroxycholesterol, and $20\alpha,22R$-dihydroxycholesterol. The corpus luteum preparation in addition will also cleave the side chain of cholestenone, but only at about one-tenth the rate of the cholesterol side chain (Table I). The rate of cholesterol side-chain cleavage is markedly inhibited by product. In the case of the corpus luteum preparation addition of 0.1 micromole of pregnenolone to the incubation resulted in more than a 50% inhibition of the cleavage reaction, and similarly progesterone at the same levels resulted in slightly less than a 50% inhibition (Table II). In the case of the testis preparation, about 0.07 micromole of testosterone added to the incubation resulted in better than a 50% inhibition in the rate of the cholesterol side-chain cleavage (Table III).

TABLE III
INHIBITION OF CHOLESTEROL-26-^{14}C SIDE-CHAIN CLEAVAGE BY A RAT
TESTIS MITOCHONDRIAL PREPARATION BY ADDED TESTOSTERONE[a]

Testosterone added (μg)	Isocaproic acid-[14]C (cpm)	Percent inhibition
0	8570	—
20	3220	63.4
40	1700	79.4

[a] Incubation conditions were the same as those described in the text.

Nucleotide Specificity

1. Testis Mitochondrial Preparation. TPNH at a concentration of 0.0025 M was the preferred cofactor for the cleavage reaction to proceed optimally with the testis preparation. DPNH supported the cleavage reaction to a very limited extent. TPN was essentially ineffective, and DPN was inhibitory even in the presence of added TPNH.

2. Bovine Corpus Luteum Preparation. The nucleotide requirement for the bovine corpus luteum preparation was considerably less clear cut than in the case of the testis preparation. This is illustrated in Fig. 1.

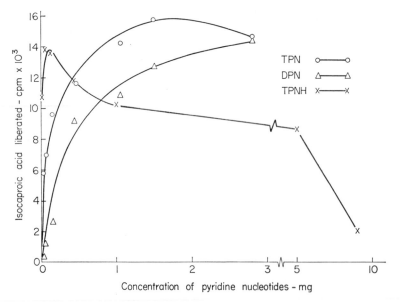

FIG. 1. Influence of varying concentrations (mg/ml) of TPN, DPN, and TPNH, respectively, on rate of cholesterol side-chain cleavage by the bovine corpus luteum enzyme system.

TPNH is quite effective at relatively low concentrations in supporting the cleavage reaction and when the optimal concentration is reached further addition of TPNH becomes inhibitory. Interestingly enough, the cofactor requirement can be met by addition of relatively large amounts of TPN and in fact the optimal rate exceeds that seen at optimal TPNH concentrations. Furthermore TPN does not appear to have the pronounced inhibitory effect of TPNH. It should be noted also that DPN behaves similarly to TPN although in a somewhat less effective manner.

Stability

The rat testis preparation is very unstable and must be prepared fresh for each run. The activity is completely destroyed by sonication, acetone precipitation, freezing and thawing and by placing in hypotonic media.

In contrast, the corpus luteum preparation is highly stable and is readily amenable to acetone precipitation leading to a soluble preparation which can be kept stored in the dry state in a desiccator at −10 to −20° for several months without losing appreciable activity.

[28] A Microassay for the Enzymatic Cleavage of the Cholesterol Side Chain

By CHARLES H. DOERING

Cholesterol → Δ⁵-pregnenolone + isocaproic acid

The method of assay for cholesterol side-chain cleavage activity (see Forchielli, this volume [27]) based on steam distillation and radioassay of labeled isocaproic acid formed from cholesterol-^{14}C-26,[1] which has been used for earlier work on this enzyme system, has the advantage of allowing a specific analysis of the fragment removed from the substrate molecule. This specificity may, however, not always be necessary. A simpler assay has been proposed recently by Kimura, Satoh, and Tchen[2] and has been modified in our laboratory to give greater versatility and reliability.

Principle

An acetone powder prepared from the mitochondrial fraction of tissue homogenate is extracted with buffer. The soluble portion is incubated with cholesterol-7α-^{3}H-26-^{14}C. Samples of the incubation mixture are dried at 120° in the presence of formic acid. The loss of volatile ^{14}C, seen as a change of isotope ratio, is a direct measure of the activity of side-chain cleavage. The activity of different preparations can be compared on the basis of initial rates of reaction. The method was developed particularly for small amounts of tissue such as the adrenal glands of immature rats.

Reagents

Homogenizing medium: 0.25 M sucrose in 0.01 M potassium phosphate buffer, pH 7.4

[1] D. Toren, K. M. J. Menon, E. Forchielli, and R. I. Dorfman, *Steroids* **3**, 381 (1964).
[2] T. Kimura, P. S. Satoh, and T. T. Tchen, *Anal. Biochem.* **16**, 355 (1966).

Acetone, reagent grade, redistilled and anhydrous

Diethyl ether, anhydrous and peroxide-free (freshly distilled from LiAlH$_4$)

Potassium phosphate buffer, 0.10 M, pH 7.4

Solution of cofactors: 20 mM glucose 6-phosphate and 0.5 mM reduced NADP+

Glucose-6-phosphate dehydrogenase, ammonium sulfate suspension, 1 Kornberg unit per 1 μl

Working solution of Tween-80:[3] 100 μg of Tween-80 per 1.0 ml of acetone

Substrate cholesterol: a mixture of 7α-^{3}H-[4] and 26-^{14}C-labeled cholesterol, 10 dpm ^{3}H per dpm ^{14}C, in benzene

Carrier steroid mixture: 2 mg each, of cholesterol, pregnenolone, and progesterone per milliliter of acetone

Formic acid, pure

Dioxane scintillation solution (Bray's solution)[5]

Procedure

Obtaining the Tissue

The following procedure is suggested for obtaining adrenal glands from unstressed and minimally excited animals. The animals to be used should be found and sorted (according to sex, for example) one or more days in advance of the experiment. When the animals are used, an adequate number can be decapitated with a guillotine within 5 minutes of entering the animal colony. The bodies are chilled in ice water until the adrenals can be removed.

Preparing the Mitochondrial Fraction

The whole adrenal glands, kept at 0° on filter paper moistened with homogenizing medium, are freed of fat and excess tissue, blotted, and weighed as a group. Each gland is then halved with a scalpel and the whole batch is homogenized with 5 μl of medium (0.25 M sucrose in 0.01 M potassium phosphate buffer, pH 7.4) per mg of gland in a small glass or quartz tube with a loose-fitting Teflon pestle.[6]

[3] Tween-80 is polyoxyethylene sorbitan monooleate; obtained from Sigma Chemical Company.

[4] Attempts to use 1,2-^{3}H-labeled cholesterol were unsatisfactory, presumably as a result of some loss of tritium from C-2 due to enolization of progesterone formed during incubation.

[5] G. A. Bray, *Anal. Biochem.* 1, 279 (1960).

[6] The Teflon pestles are machined to have a 0.4 mm clearance, i.e., difference in diameters, in 2-ml capacity, quartz certifuge tubes supplied by Beckman, Spinco Division.

The homogenate is then centrifuged in the same tubes in a small swinging-bucket head (SW 39) in a Spinco centrifuge at 0° for 15 minutes at 15,000 g at inside bottom of the tubes (=12,000 rpm). The clear supernatant is discarded; it contains no detectable side-chain cleavage activity.

Preparing the Acetone Powder

The sediment containing the mitochondrial fraction together with cell debris and larger pieces is resuspended with a little homogenizing medium (1 drop per 10 mg whole adrenal weight) and transferred dropwise to 8 ml of acetone at −25 to −30°. The acetone is agitated vigorously during this transfer. After 10 minutes of intermittent agitation, the acetone powder is centrifuged at −20° for 10 min at 1500–2000 g (=3600 rpm) in a swinging-bucket head in a Sorvall centrifuge. The supernatant acetone is discarded, and the sedimented protein is resuspended in 3 ml of peroxide-free, anhydrous ether also at −20° and centrifuged again in the same manner. The supernatant ether is discarded and the sediment is stored over KOH in a partially evacuated desiccator in the dark at −20°.

Reconstituting the Acetone Powder

The acetone-precipitated protein is dissolved at 0° in 5 μl of 0.10 M potassium phosphate buffer (pH 7.4) per 1.0 mg of original whole adrenal fresh weight. It is stirred occasionally with a glass rod for 30 minutes at 0°. The enzyme preparation is freed from insoluble material by centrifugation at 0° for 15 minutes at 1500–2000 g as before. The clear, slightly yellow supernatant is transferred to smaller tubes and stored at 2–3°. The activity remains virtually unchanged for a few days, but vanishes in the course of about 20 days.

The total protein concentration in the reconstituted acetone preparation has been determined by the Warburg-Christian Method[7] and was found to range within 8–12 mg of protein per milliliter.

Incubating with Cholesterol

Incubation flasks (10-ml, conical) are set up as follows in the order indicated and while packed in ice:

x ml 0.10 M potassium phosphate buffer, pH 7.4
0.20 ml of cofactor solution
2 μl of glucose-6-phosphate dehydrogenase
y ml of acetone powder extract
0.55 ml of distilled water
$x + y = 0.20$ ml; the concentration of phosphate in

[7] E. Layne, Vol. III, p. 451.

the final volume of 1.0 ml will thus be not
more than 20 mM.

The volume of reconstituted acetone powder to be used will depend
on the protein concentration, or actually, on the level of activity of the
enzyme preparation. In order to achieve zero-order kinetics (with respect
to substrate) during at least the first 10 minutes of incubation, we have
found it useful to incubate within a range of protein of 300–900 μg per
flask for medium active preparations.

The substrate for each incubation flask is a mixture of 2×10^5 dpm of
cholesterol-26-^{14}C, 1.3 μg ($=3.3$ nanomoles), and 2×10^6 dpm of choles-
terol-7α-^{3}H, about 10 nanograms. A solution of this mixture in benzene
is evaporated to dryness with N_2, the residue is taken up in 0.10 ml of a
solution of 10 μg of Tween-80 in acetone, evaporated to dryness, and
finally the residue is suspended in 50 μl of distilled water.[8]

The flasks are preincubated without substrate for 10 minutes in a
shaking water bath at 30°. Incubation at the same temperature and with
open flasks is started by pipetting 50 μl of an aqueous suspension of
cholesterol and Tween-80 into the reaction mixture, the volume of which
at this point is 1.0 ml. Zero-time samples are withdrawn immediately
after the start of incubation.

To follow the rate of conversion, samples are removed at 0, 3, 6, 9, 12,
and 15 minutes after the start of incubation. Since the method depends
on a change in isotope ratio, accurate pipetting is unnecessary, and to
remove samples in duplicate at such short intervals, a more rapid technique is required. It is convenient to transfer with a disposable pipette
two drops (about 65 μl $\pm$ 5 μl) of the incubation mixture to 20-ml glass
liquid scintillation vials that contain 0.5 mg of unlabeled carrier compounds (cholesterol, pregnenolone, and progesterone) spread over the
bottom of the vial and four drops of formic acid. The vials are heated
in a hood in an aluminum block at 120° for 15 minutes. The formic acid
serves initially to stop the enzymatic reactions and later to permit the
volatilization of the free isocaproic acid. Carrier compounds in the vials
have been found necessary to prevent the evaporation of significant
portions of the very minute amounts of radioactive substrate cholesterol
and reaction products.

Assaying Radioactivity and Calculating the Rate of Enzymatic Activity

The contents of the vials are assayed for radioactivity in 6 ml of
Bray's dioxane scintillation fluid in an automatic 3-channel spectrometer.
In our laboratory, tritium is counted at 11% efficiency (with less than
0.5% overlap in the ^{14}C channel); ^{14}C, at 48% (with 11% of the ^{14}C

---

[8] T. T. Tchen, Vol. VI, p. 512.

CLEAVAGE OF THE CHOLESTEROL SIDE CHAIN BY RAT ADRENAL
PREPARATIONS: DOUBLE ISOTOPE METHOD[a]

Incubation (min)	Ratios: cpm ³H/cpm ¹⁴C		Cholesterol converted per flask		Cholesterol converted per 0.50 mg protein	
	A	B	A (%)	B (%)	A[b] (nmoles)[d]	B[c] (nmoles)
0	1.93	1.96				
3	2.14	2.46	9.7	20.5	0.21	0.22
6	2.37	3.00	18.3	34.8	0.40	0.38
9	2.61	3.37	25.8	41.9	0.57	0.46
12	2.74	3.76	29.4	47.9	0.65	0.53
15	2.93	4.04	34.1	51.5	0.75	0.57

[a] Incubations were carried out as described in the text. In A, 0.75 mg of adrenal protein from 15-day-old Sprague-Dawley rats were incubated with 3.3 nanomoles of cholesterol. In B, 1.50 mg of the same enzyme preparation was incubated in the same way. All values are averages of duplicate determinations.
[b] Plotted as line "a" in Fig. 1.
[c] Plotted as line "b" in Fig. 1.
[d] Unit nmoles = nanomole = 10^{-9} mole.

counts overlapping into the tritium channel). Efficiencies are determined by the internal standard method with calibrated samples of labeled toluene. The background is subtracted automatically, and the counts in the tritium channel are corrected for the ¹⁴C overlap as seen by adding toluene-¹⁴C and recounting. From this information, the ratio of net cpm of ³H/net cpm of ¹⁴C can be calculated for each sample of the incubation mixture. From the changes in isotope ratio *vs* the ratio of zero-time

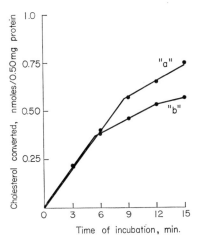

FIG. 1. Cleavage of the cholesterol side chain by rat adrenal preparations. For explanations, see footnotes to the table.

incubation, the amount of substrate cholesterol converted per unit weight of adrenal protein in the incubation mixture is calculated.

An example of the results to be expected from incubation of two different concentrations of enzyme protein is shown in the table and in Fig. 1. The typical curtailment of the linear range of the assay at higher enzyme protein concentrations is illustrated. The cause of this effect has not been defined, although it has been noted by others.[2,9]

[9] P. S. Satoh, G. Constantopoulos, and T. T. Tchen, *Biochemistry* 5, 1646 (1966).

[29] Adrenal Steroid Hydroxylases

By OTTO ROSENTHAL and SHAKUNTHALA NARASIMHULU

I. Introduction[1]

The adrenocortical enzymes that catalyze the monohydroxylations at different positions of the steroid molecule require reduced triphospho-

[1] Terminology and abbreviations: progesterone = pregn-4-en-3,20-dione; cortexone = deoxycorticosterone = 21-hydroxypregn-4-ene-3,20-dione; corticosterone (compound B) = 11β,21-dihydroxypregn-4-ene-3,20-dione; 17-hydroxyprogesterone =

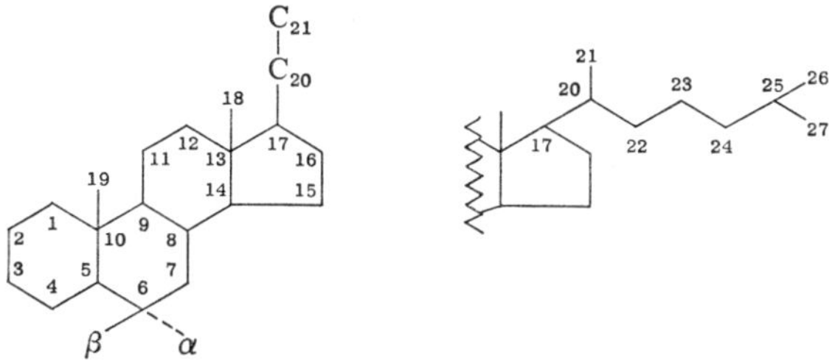

Fig. 1. Steroid nucleus and side chains.

pyridine nucleotide and molecular oxygen as indispensable reactants. These requirements characterize the reactions as mixed function oxidations as defined by Mason[2] and schematically presented in Eq. (1):

$$AH + TPNH + H^+ + O_2 \rightleftharpoons A\text{-}OH + TPN^+ + H_2O \qquad (1)$$

where A stands for a primary, secondary, or tertiary carbon of the steroid.

Incorporation of one atom of $^{18}O_2$ into the steroid molecule has been demonstrated for adrenal 11β-hydroxylations[3,4] and for several fungoid hydroxylations.[5] For adrenal 21-hydroxylations[6] and 11β-hydroxylations[7] it has been shown that one mole of TPNH and one mole of O_2 are utilized for each mole of hydroxylation product formed in accordance with Eq. (1). Although the fate of the second oxygen atom cannot be directly determined, it is most likely that it is reduced to water by TPNH.

After Ryan and Engel's[8] discovery of the light-reversible CO inhibition of steroid 21-hydroxylation, the presence of the carbon monoxide-combining pigment cytochrome P450 was demonstrated in adrenocortical microsomes and mitochondria and its function as the oxygen-activating

---

17α-hydroxypregn-4-ene-3,20-dione; cortexolone (substance S) = $17\alpha,21$-dihydroxy-pregn-4-ene-3,20-dione; cortisol (compound F) = $11\beta,17\alpha,21$-trihydroxypregn-4-ene-3,20-dione; aldosterone = $11\beta,21$-dihydroxy-3,20-dioxopregn-4-en-18-al; pregnenolone = 3β-hydroxypregn-5-en-20-one; TPN and TPNH stand, respectively, for triphosphopyridine nucleotide and its reduced form.

[2] H. S. Mason, *Advan. Enzymol.* **19**, 79 (1957).

[3] M. Hayano, M. G. Lindberg, R. I. Dorfman, J. E. H. Hancock, and W. von E. Doering, *Arch. Biochem. Biophys.* **59**, 529 (1957).

[4] M. L. Sweat, A. Aldrich, C. H. de Bruin, W. L. Fowlks, L. R. Heiselt, and H. S. Mason, *Federation Proc.* **15**, 367 (1956).

[5] M. Hayano, *in* "Oxygenases" (O. Hayaishi, ed.), p. 181. Academic Press, New York, 1962.

[6] D. Y. Cooper, R. W. Estabrook, and O. Rosenthal, *J. Biol. Chem.* **238**, 1320 (1963).

[7] W. Cammer and R. W. Estabrook, *Arch. Biochem. Biophys.* **122**, 721, 735 (1967).

[8] K. J. Ryan and L. L. Engel, *J. Biol. Chem.* **225**, 103 (1957).

component of the steroid 21-,[9, 10] 11-,[11] and 18-aldosterone[12]-hydroxylase systems was established by means of the photochemical action spectrum method.[13] Since there is increasing evidence that cytochrome P450 is the general oxygen activating catalyst of mixed-function oxidations of lipid-soluble organic compounds in tissues of vertebrate animals, it can be assumed that it is also functioning in the various steroid hydroxylations in the adrenals. Up to the present, adrenal hydroxylases for the following positions of the steroid molecule have been partially purified: 2α, 6α, 6β, 11β, 17α, 18, 19, 20, 21, and 22. They are tabulated with references in Volume V of Methods in Enzymology[14] with the exception of the 2α-,[14a] the 20-, and the 22-hydroxylase. The latter two are assumed to be involved in splitting off isocaproic acid in the conversion of cholesterol to pregnenolone.[15] With the exception of the 2α-, 17α-, and 21-hydroxylases, the hydroxylases are localized in the mitochondria. Adrenocortical mitochondria differ from the hepatic and cardiac ones in that they contain a tubular or vesicular type of cristae to which the hydroxylases are apparently attached.[16] The endowment of mitochondria with individual hydroxylases will probably depend on the cortical zone from which the mitochondria are derived.[16a] The complement of enzymes therefore, will vary with the method of preparation and the animal species[16b] employed. The reactions where most progress in enzyme methodology has been made are the 11β-, 21-, and 18-hydroxylations, and only these will be dealt with in detail. The separations of mitochondria and microsomes from the cortex tissue as carried out in different laboratories are largely modifications of the Schneider and Hogeboom procedure.[16c] They have

[9] R. W. Estabrook, D. Y. Cooper, and O. Rosenthal, *Biochem. Z.* **338,** 741 (1963).
[10] D. Y. Cooper, S. Levin, S. Narasimhulu, O. Rosenthal, and R. W. Estabrook, *Science* **147,** 400 (1965).
[11] D. Y. Cooper, B. Novack, O. Foroff, A. Slade, E. Sanders, S. Narasimhulu, and O. Rosenthal, *Federation Proc.* **26,** 341 (1967).
[12] P. Greengard, S. Psychoyos, H. H. Tallan, D. Y. Cooper, O. Rosenthal, and R. W. Estabrook, *Arch. Biochem. Biophys.* **121,** 298 (1967).
[13] O. Rosenthal and D. Y. Copper, see Vol. X, p. 616.
[14] M. Hayano and R. I. Dorfman, see Vol. V, p. 509.
[14a] S. Burstein, B. R. Bhavnani, and M. Gut, *J. Biol. Chem.* **240,** 2845 (1965).
[15] G. Constantopoulos and T. T. Tchen, *J. Biol. Chem.* **236,** 65 (1961).
[16] A. C. Brownie and F. Skelton, *in* "Function of the Adrenal Cortex" (K. McKerns, ed.), Vol. 2, p. 691. Appleton, New York, 1968.
[16a] T. Symington, *Brit. Med. Bull.* **18,** 114 (1962).
[16b] H. W. Deane, "The Anatomy, Chemistry and Physiology of Adrenocortical Tissue." Handb. Exptl. Pharmakol. Ergänzungswerk, Vol. XIV 1. Springer, Berlin, 1962.
[16c] See G. H. Hogeboom, Vol. I, p. 16.

been developed empirically and are not strictly standard methods. The same holds true for the composition of the reaction media, especially with regard to electrolyte concentrations and buffer systems employed.

To avoid repetition as far as possible, we have detailed in the following section the general procedures used in our laboratories for separating mitochondria and microsomes from the adrenal cortex of steers and for preparing a reaction medium for the hydroxylase assays. To this, a brief summary of the most common steroid assays has been added. Modifications of these procedures will be pointed out in the sections on the individual hydroxylase preparations.

II. General Procedures

A. Preparation of Mitochondria and Microsomes from Bovine Adrenal Cortex

1. *Preparation of Cortex Tissue*

Freshly excised glands of steers are collected in ice at the slaughterhouse and brought to the laboratory within 2 hours after slaughter. The following preparative procedures should be done at about 4°.

After adhering fat has been trimmed from the glands, cortical tissue may be obtained either by slicing the cortex with a Stadie-Riggs tissue slicer, care being taken not to enter the medulla,[17] or by bisecting the glands along the major circumference, excising the medullary tissue with curved scissors and scraping the cortical tissue from the capsule with a scalpel blade. The first procedure, more time-consuming and less economical, is recommended if contamination of cortex with medullary components should be kept to a minimum. The second procedure is the method of choice for large-scale preparations as required for isolation of the components of the 11β-hydroxylase system. The slices are briefly rinsed with 0.154 M KCl to remove adhering blood and thereafter with 0.25 M sucrose to remove the KCl solution. Scrapings are collected in a beaker, weighed, and rinsed only once with 0.25 M sucrose to minimize loss of material.

The scissor-minced slices or the scrapings are homogenized in 0.25 M sucrose by means of a Potter-Elvehjem type of homogenizer consisting of a glass grinding vessel and a motor-driven pestle with Teflon grinding head. Approximately 4 parts of 0.25 M sucrose per 1 part of cortex tissue (20% w/v) are used.

[17] D. Y. Cooper and O. Rosenthal, *Arch. Biochem. Biophys.* **96**, 331 (1962).

2. *Initial Fractionation by Centrifugation*[18]

Relative centrifugal forces are computed for the center of the tubes.

Step 2a. The 20% homogenate is centrifuged (horizontal yoke) for 15 minutes at 400 g. The supernatant fraction is pipetted or siphoned off with the exception of some liquid above the sediment containing poorly sedimented material. This step removes most of the unbroken cells, nuclei, erythrocytes and debris.

Step 2b. The supernatant liquid is centrifuged for 15 minutes at 9000 g (angle head), and the supernatant fraction is withdrawn as in step 2a. This fraction is used for preparing the microsomes; the sediment together with the residual liquid phase above it are the source of the mitochondria.

3. *Preparation of Mitochondrial Fraction*

Step 3a. Sediments and residual liquids from step 2b are briefly homogenized, adjusted with 0.25 M sucrose to about 15% of the volume of fraction 2b, and centrifuged for 15 minutes as in step 2a to remove remaining erythrocytes and cell debris. The sediment is discarded.

Step 3b. The supernatant liquid is centrifuged for 15 minutes at 9000 g and the supernatant liquid is decanted and discarded. The sediment, with the exception of a small darkish bottom layer, is transferred to a homogenizer tube and resuspended in an amount of 0.25 M sucrose corresponding to one half the volume of fraction 3b.

Step 3c. The mitochondrial suspension is centrifuged at 10,500 g (15 minutes), and the sediment is resuspended in 0.25 M sucrose as in step 3b.

Step 3d. The mitochondrial suspension is again centrifuged at 12,500 g (15 minutes), and the almost clear supernatant liquid is decanted and discarded. The mitochondrial pellet is the starting material for the resolution of the steroid 11β-hydroxylase system.

For the study of steroid hydroxylation by mitochondrial suspensions where small contaminations with other components are of less concern than the preservation of the intactness of the mitochondria, the following method of Cammer and Estabrook[7] may be used: A 10% homogenate of the scrapings in 0.25 M sucrose solution is centrifuged for 10 minutes at 900 g, and the sediment is discarded. The supernatant fraction is centrifuged for 10 minutes at 9000 g. The supernatant liquid is decanted and discarded, and the sediment is suspended in an amount of 0.25 M sucrose solution corresponding to half the volume of the preceding fraction. The suspension is again centrifuged for 10 minutes at 9000 g. The

[18] D. Y. Cooper, S. Narasimhulu, A. Slade, W. Reich, O. Foroff, and O. Rosenthal, *Life Sci.* **4**, 2109 (1965).

sediment, after suspension in a small volume of 0.25 M sucrose solution to a protein concentration of approximately 30 mg of protein per milliliter is used for the enzyme assays.

4. Preparation of the Microsomal Fraction

Steps 4a and 4b. The supernatant fraction from step 2b is centrifuged for 10 minutes at 10,500 g (4a) and at 12,000 g (4b), and the sediments from each step are discarded.

Step 4c. The supernatant liquid from step 4b is centrifuged for 60 minutes at 78,000 g. The supernatant liquid is decanted and discarded.

Step 4d. The sediment from step 4c is homogenized in 0.15 M KCl to remove soluble proteins from the particles and centrifuged for 30 minutes at 100,000 g. The sediment is suspended in 0.25 M sucrose solution to a concentration of 12–15 mg of protein per milliliter. If stored at −15 to −20° it preserves its enzyme activity for at least one month.

5. Yields

The procedure here described was developed to obtain from the same material within one working day mitochondrial and microsomal preparations with minimal contamination by other tissue components. For this purpose much heterogeneous material has been discarded from which additional amounts of mitochondria and microsomes could have been recovered. Steer adrenals trimmed of adhering fatty tissue vary in weight from about 7 to 11 g. From 250 g of trimmed adrenal glands, the maximal amount that could be conveniently processed in one day, about 80 g of scrapings or 60 g of slices could be obtained. The yield per 100 g of cortex tissue was 1.0–1.5 g of mitochondrial protein and 0.7–1.0 g of microsomal protein.

B. Reaction Medium[19]

A 2% (w/v) aqueous solution of crystalline bovine serum albumin is dialyzed for 24 hours at 4° against 1 mM KCN solution and thereafter twice for 24 hours against double- or triple-distilled water in order to remove contaminating metals, especially copper.

To 100 ml of the dialyzed solution, 0.9 g of NaCl and 0.41 g of glycylglycine are added. The pH is adjusted with 0.1 M NaOH (about 4.5 ml) to pH 7.4 (glass electrode), the volume is brought to 120 ml with distilled water, and 4 ml of 0.154 M KCl and 1 ml of 0.11 M MgCl$_2$ are added.

This mixture is a modified mammalian type of Ringer solution which can also be prepared from the following isotonic stock solutions: 100

[19] S. Narasimhulu, D. Y. Cooper, and O. Rosenthal, *Life Sci.* **4**, 2101 (1965).

parts of 2% albumin-containing $0.154 M$ NaCl adjusted to pH 7.4; 20
parts of $0.154 M$ glycylglycine buffer pH 7.4, 4 parts of $0.154 M$ KCl,
1 part of $0.11 M$ $MgCl_2$. However the first procedure is more convenient
when working with dialyzed albumin solutions. The final concentrations
of the main ingredients are NaCl, 120 mM; glycylglycine, 24 mM; KCl,
4.8 mM; $MgCl_2$, 0.9 mM; albumin, 16 g per liter.

To dissolve steroids in this medium an aliquot of a methanolic solution of the steroid substrate is placed on the bottom of the incubation
flask and the alcohol is driven off with a stream of nitrogen. An appropriate volume of the buffer mixture is added, the flasks are stoppered
and agitated for 30 minutes in a Dubnoff shaker incubator at the temperature selected for the enzyme assay (usually 25°). The flasks are
then removed to be supplemented with the other components of the
particular enzyme assay system.

Alternatively, microliter volumes of concentrated alcoholic solutions
of the steroid substrate can be added to the medium, the albumin pre-
venting precipitation of the steroid in the aqueous medium. If the
presence of albumin is undesirable, the steroid should be added after the
enzyme. Propylene glycol may be substituted for alcohol.

C. Assay of Hydroxylation Products

The enzyme reaction is usually stopped by the addition of the lipid
solvent (e.g., dichloromethane, chloroform, ethyl acetate) which has
been selected for extracting the hydroxylation product from the reaction
medium. If extraction of interfering lipids by a less polar solvent such as
petroleum ether is required prior to the extraction of the hydroxylated
steroid, the reaction is stopped by adding one volume of 5% $HgCl_2$
solution to 4 volumes of the incubation mixture prior to the addition of
the nonpolar solvent.

By means of group-specific reagents quantitation of the hydroxylation products without separating product from precursor is feasible for
hydroxylation at C-21 and C-11. C-21 hydroxylations resulting in the
formation of a dihydroxyacetone side chain are quantitated by means of
the phenylhydrazine reaction of Porter and Silber,[20] while those leading
to the formation of an α-ketol side chain are quantitated by the blue
tetrazolium reaction of Izzo et al.[21] C-11 hydroxylation products can be
estimated by the fluorescence in sulfuric acid–alcohol mixtures described
by Sweat.[22] Versions of this reaction have been discussed by Silber.[23]

[20] C. C. Porter and R. H. Silber, J. Biol. Chem. 185, 201 (1950); R. H. Silber and
C. C. Porter, Methods Biochem. Anal. 4, 139 (1957).
[21] A. J. Izzo, E. H. Keutmann, and R. B. Burton, J. Clin. Endocrinol. Metab. 17,
889 (1957).

After chromatographic separation,[24] methanolic or ethanolic solutions of those hydroxylation products that possess an α,β-unsaturated 3-keto function can be assayed spectrophotometrically by measuring the extinction at 240 mμ. Molar extinction coefficients range from 10,000 to 20,000.[25] Alternatively, the extinction of the thiosemicarbazones at 302 mμ can be measured. The molar extinction coefficients range from 33,000 to 40,000.[26] Thiosemicarbazones of aldehydes and saturated ketones give absorption maxima near 270 mμ and show but negligible absorption at 302 mμ.

Radioisotope tracer techniques in conjunction with chromatography are indispensable for measuring hydroxylase reactions of low activity as well as for many kinetic studies. The hydroxylation product is purified by repeated chromatographies to either constant specific activity or to a constant isotope ratio measured with a liquid scintillation spectrometer if tracer amounts of product labeled with a different isotope (e.g., ^{3}H vs ^{14}C) are used. Thin-layer chromatography,[27] greatly facilitates these procedures. It has recently been reported[28] and confirmed in the writer's laboratory that recoveries on rechromatography are often very poor, particularly with cortisol. The losses are apparently incurred during the evaporation of alcoholic extracts of the spot areas before the material is applied to the next plate. Narasimhulu et al.[29] using commercial silica gel G plates (E. Merck Co., Germany) found that these losses could be avoided and time could be saved by transferring scrapings of the spot area of one plate to a cleared area on the starting line of the next plate. The scrapings can also be transferred to the scintillation vials. The adsorbed steroid dissolves readily in the common scintillation mixtures and the gel settles on the bottom without interfering with the radiometry. A similar procedure has been applied to cut-out areas of filter paper strips.[30]

---

[22] M. L. Sweat, Anal. Chem. 26, 773 (1954).

[23] R. H. Silber, Methods Biochem. Anal. 14, 63 (1966).

[24] R. Neher, "Steroid Chromatography." Elsevier, Amsterdam, 1964.

[25] J. P. Dusza, M. Heller, and S. Bernstein, in "Physical Properties of the Steroid Hormones" (L. L. Engel, ed.), p. 69 ff. Macmillan, New York, 1963.

[26] N. B. Talbot, S. Ulick, A. Kopreianow, and A. Zygmuntowicz, J. Clin. Endocrinol. Metab. 15, 301 (1955). Procedure: Measure absorbance of ethanolic steroid solution at 302 mμ (value A). To 3 ml of this solution in a glass-stoppered test tube add 0.5 ml of 0.1 M thiosemicarbazide (reagent) in 0.1 M HCl. Mix gently. Allow to stand for 90 minutes at room temperature. Read absorbance at 302 mμ (value B). $(B \times 1.17) - A$ = absorbance of thiosemicarbazone, the factor 1.17 correcting for the dilution of the sample by the reagent.

[27] "Thin-Layer Chromatography" (E. Stahl, ed.). Academic Press, New York, 1965.

[28] D. R. Idler, N. R. Kimball, and B. Truscott, Steroids 8, 865 (1966).

[29] S. Narasimhulu, I. Keswani, and G. L. Flickinger, Steroids 12, 1 (1968).

[30] S. Psychoyos, H. H. Tallan, and P. Greengard, J. Biol. Chem. 241, 2949 (1966).

Unlabeled hydroxylation products needed as carriers should be added to the solvent used for the extraction of the enzyme incubation mixture since the hydroxylations are inhibited by high product-to-substrate ratios. Substrate inhibition observed by others[14] has not been encountered by us.

III. Steroid 11β-Hydroxylase (EC 1.14.1.6)

A. Assay of Corticosterone

Although several steroids are converted to their 11β-hydroxylated derivatives by adrenal preparations (cf. Table II, page 509 of Volume V), only the conversion of cortexone (see footnote 1) to corticosterone has been used for the assay of steroid 11β-hydroxylase activity in the preparations dealt with below. The assay depends upon the measurement of the hydroxylated product by fluorometry or other means. Mattingly's version[31] of the fluorometric method of Sweat, which was found to be most convenient for determining corticosterone, will be described in detail. Other methods will be mentioned in conjunction with the different preparations.

Reagents

Dichloromethane: redistill, collecting the fraction distilling between 39 and 40.5°. Spectral grade dichloromethane is used in some laboratories without redistillation.

Fluorescence reagent: Add 3 ml of absolute ethanol to 7 ml of concentrated sulfuric acid (reagent grade) with ice cooling. Unless a color develops, no redistillation of ethanol is required.

Procedure. Dichloromethane (15 ml) is pipetted into 50-ml centrifuge tubes equipped with flat-head ground-glass stoppers. The tubes are placed in an ice bath. At the end of the enzyme incubation period, 2 ml of the reaction mixture is pipetted into the tubes, the contents are mixed for 15 seconds with a Vortex shaker, and the tubes are returned to the ice bath for at least 15 minutes. They are then mounted horizontally on a mechanical shaker and agitated for 4 minutes. The phases are separated by 4 minutes of centrifugation at about 1000 rpm, and the aqueous supernatant layer is removed with an aspirator. A 10-ml aliquot of the organic phase is layered on 3 ml of the fluorescence reagent in a 50-ml conical centrifuge tube fitted with a glass stopper. The tube is vigorously shaken manually for 15 seconds, and the time is recorded. The layers separate promptly, and the dichloromethane layer is removed with an aspirator.

---

[31] D. Mattingly, *J. Clin. Pathol.* **15**, 374 (1962).

The acid layer is transferred to a fluorometer tube and the fluorescence is measured at 20 and 30 minutes after shaking with the reagents. Since fluorescence increases with time—the blank fluorescence somewhat faster than that of corticosterone—no more than 7 or 8 experimental tubes and 3 control tubes should be prepared at one time, so that each tube can be read at exactly the same exposure time to the fluorescence reagent as the control tubes. The latter, consisting of blank and two concentrations of the standard, are carried through all steps of the procedure. They differ from the experimental tubes only in that the electron donor system is omitted and known amounts of corticosterone are added to the standards. Since with this arrangement nonspecific fluorescence is compensated by the controls, the exposure time to the fluorescence reagent can be extended beyond the 13 minutes recommended by Mattingly for corticoid assays in serum, where the blank value is unknown. This extension makes possible the inclusion of up to 11 tubes in one set.

Although light of 470–480 mμ wavelength produces optimal activation, the high intensity 436-mμ emission line is chosen for activation when filter fluorometers equipped with mercury lamps are used. Secondary filters should have maximal transmission at 530–540 mμ and a complete cutoff of light below 510 mμ. From 1.0 to 9.0 μg of corticosterone per 1 ml of fluorescence reagent can be determined with an accuracy of about 5%. Above 9.0 μg per milliliter quenching occurs and smaller aliquots of the dichloromethane extract must be used to extend the range. Recovery of added corticosterone is close to 100%.

The fluorescence of cortexone amounts to about 1/200 of that of corticosterone. The fluorescence of cortisol is 40% of that of corticosterone. The cortisol precursor cortexolone produces a pink color with the fluorescence reagent. Its fluorescence amounts to about 1/80 of that of cortisol. Data on the fluorescence of other 11β-hydroxylated steroids are not available at the present time.

B. Preparations and Properties

1. Intact Mitochondria

The method for the preparation and incubation of bovine adrenocortical mitochondria described is that of Cammer and Estabrook.[7]

Reaction Medium. To 95 parts of 0.25 M sucrose solution are added 2 parts of 1 M KCl; 1.5 parts of 1 M triethanolamine HCl (pH 7.2); 1 part of 1 M potassium phosphate buffer (pH 7.2), and 0.5 part of 1 M MgCl$_2$. The final millimolar concentrations of the components are, respectively, 240, 20, 15, 10, and 5.

Hydroxylase Assay. A suspension of mitochondria (cf. Section II,A,3) is diluted with the reaction medium to a concentration of 1.5 to 3 mg of mitochondrial protein per milliliter. Per milliliter of this mixture, 6.7 μl of a 1 M neutralized solution of Krebs tricarboxylic acid cycle substrates and 5 μl of a 0.48 M ethanolic solution of cortexone are added to start the reaction. The final concentrations of the latter two ingredients are 6.7 and 0.24 mM, respectively. Krebs cycle substrates serve as electron donors (via endogenous TPN) for this mixed function oxidation because exogenous pyridine nucleotides do not diffuse readily into intact mitochondria.

A 9-ml aliquot of the reaction mixture is incubated at 25° in air with magnetic stirring. At intervals of 1 minute, 1-ml aliquots are added to 1 ml of 0.5% $HgCl_2$ followed by the immediate addition of 10 ml of dichloromethane and mixing. After the subsequent addition of 5 ml of dichloromethane, corticosterone is determined by the Mattingly method as outlined in Section III,A. Termination of the reaction may also be done as described in III,A.

Specific Activity. The rates of hydroxylation range from 20 to 29 nanomoles of corticosterone formed per minute per milligram of mitochondrial protein.

Succinate and malate are the most effective electron donors for the reaction. Lower rates, especially with less active preparations, are obtained with isocitrate, fumarate, α-ketoglutarate $+$ malonate, and citrate. In the presence of any of these intermediates of the tricarboxylic acid cycle, cortexone stimulates the rate of oxygen consumption as measured polarographically in aliquots of the incubation mixture. The ratio of corticosterone formation to increase in oxygen consumption is 1.1 with malate; i.e., it is close to the theoretical 1:1 ratio for mixed-function oxidations. With the other Krebs cycle substrates, ratios vary from 1.1 to 1.5. Adenosine diphosphate increases oxygen consumption in the presence of these substrates (transition of the respiratory chain from state 4 to state 3), but does not alter the rate of the cortexone-stimulated oxygen uptake and of corticosterone formation. Pyruvate, β-hydroxybutyrate, and α-glycerophosphate are ineffective as electron donors. Glutamate is also ineffective, indicating the absence of glutamate dehydrogenase.

Inhibitors. With succinate as electron donor, 11β-hydroxylation is inhibited by agents that interrupt the electron flow via the respiratory chain to oxygen (cyanide, Antimycin A) as well as by uncouplers (dinitrophenol, dicumarol). In contrast, malate-supported hydroxylation is but little affected by these agents. This indicates that the electron flow from malate through TPN to cytochrome P450 is mediated, at least in part, by the malic enzyme whereas utilization of succinate as electron donor is mediated by the energy-linked transhydrogenase.

The malate-supported hydroxylation is inhibited by CO, half-inhibition occurring at a CO/O_2 ratio of about 1. Hydroxylation with malate as well as with succinate is inhibited by the 11β-hydroxylase inhibitor Metopirone,[32] which competes with the steroid substrate, and by calcium ions (0.7 mM) which presumably disrupt the structural integrity of the mitochondria and thereby cause loss of the endogenous pyridine nucleotides.

Binding Reactions. The addition of cortexone to the mitochondrial suspension in the absence of added electron donors produces a change in the spectrophotometric difference spectrum characterized by an absorption minimum at 420 mμ and an absorption maximum at about 390 mμ. This reaction will be dealt with in connection with the steroid 21-hydroxylase.

Pigments and Cofactors Participating in 11β-Hydroxylation. The following concentrations in terms of nanomoles per milligram of mitochondrial protein are reported by Cammer and Estabrook[7]: cytochrome P450, 1.5; TPN + TPNH, 2.8; DPN + DPNH, 6.3; NHI (from ESR), 2.6.

Rat Adrenal Mitochondria

The procedure is that of Harding *et al.*[33,34] About 40–50 trimmed adrenal glands of female rats (200–250 g body weight, Holtzman strain) are homogenized in 0.25 M sucrose solution (4 glands/ml) with two passes of a TenBroeck glass homogenizer. Mitochondria are precipitated at 12,000 g (10 minutes) from the supernatant liquid of a 900 g sediment, resuspended in half the previous volume of 0.25 M sucrose, and again precipitated as before. In some experiments[33] the mitochondrial fraction has been separated into a 4000 g sediment (heavy mitochondria) and a 25,000 g sediment (light mitochondria).

The reaction medium contains, besides 0.25 M sucrose, the following: 5.1 mM sodium phosphate buffer, pH 7.3; 3.0 mM MgCl$_2$; 5.1 mM KCl; 3.3 mM ADP; 0.17 mM cortexone; tracer amounts of cortexone-4-^{14}C; 0.1% bovine serum albumin; about 3 mM succinate or malate; and mitochondria corresponding to 0.1–0.3 mg of nitrogen per milliliter of medium. The mixture is incubated at 37° with 5% CO$_2$ in O$_2$ as gas phase. Samples are withdrawn at 4-minute intervals for 20 minutes and quenched by adding 3 volumes of chloroform. Steroids are chromatographed in a Zaffaroni benzene–formamide system and quantitated by means of a radiochromatogram strip counter. The main hydroxylation product is

[32] Metopirone = Ciba SU-4885 = 2-methyl-1,2-bis-(3-pyridyl)-1-propanone.

[33] L. D. Wilson, D. H. Nelson, and B. W. Harding, *Biochim. Biophys. Acta* **99**, 391 (1965).

[34] B. W. Harding, L. D. Wilson, S. H. Wong, and D. H. Nelson, *Steroids,* Suppl. II, 51 (1965).

corticosterone. The minor product is probably 18-hydroxycortexone (18 → 20 hemiketal) according to its mobility.

Properties. 11β-Hydroxylase and apparent 18-hydroxylase activities are restricted to the mitochondrial fraction. Specific 11β-hydroxylase activity of heavy mitochondria is 6.5 μg corticosterone per min per milligram of N or about 3 nanomoles per minute per milligram of protein at 37°. Light mitochondria have about 40% less activity and contain about 30% less cytochrome P450. 18-Hydroxylase activity of the "heavy mitochondria" amounts to about 40% of the 11β-hydroxylase activity. There is no 18-hydroxylase activity in light mitochondria. Effects of Krebs cycle intermediates and of enzyme inhibitors such as CO are similar to those reported for bovine adrenocortical mitochondria.

2. Bovine Submitochondrial Particles

RESOLUTION OF THE 11β-HYDROXYLASE SYSTEM FROM BOVINE ADRENOCORTICAL MITOCHONDRIA

The mitochondrial pellet (Section IIA, step 3d) is homogenized in distilled water (1 ml per 60–80 mg of mitochondrial protein) and is sonicated for 15 minutes (M.S.E. Ultrasonic Disintegrator, 20 kc, 60 watts output) in a Pyrex tube immersed in an ice-salt mixture. Sonication is interrupted at least every 5 minutes to allow cooling of the sonicate and renewal of the ice. The procedure[35] for separating the soluble TPNH-cytochrome P450 reducing system, consisting of the specific flavin-adenine dinucleotide protein Fp and the nonheme iron protein (NHI), from the cytochrome P450-containing particles is outlined in Fig. 2.

The *g* values recorded in the flow scheme are calculated for the center of the centrifuge tubes. The supernatant fraction S-1, which is similar to the "solubilized" 11β-hydroxylase preparation of Sharma *et al.*,[36] displays hydroxylase activity. However, the reaction rates are not proportional to the concentration of mitochondrial protein but decrease precipitously with increasing dilution of the preparation, an indication that the enzyme system is dissociating into several components.[18] Prolonged ultracentrifugation of fraction S-1 results in nearly complete separation of the soluble reducing system from the particulate component. The isolation of Fp and NHI from the supernatant fraction S-2 has been described in detail by Omura *et al.*[37] The precipitate P-2 is freed of most of adhering soluble proteins by sonication in NaCl–phos-

[35] T. Omura, E. Sanders, R. W. Estabrook, D. Y. Cooper, and O. Rosenthal, *Arch. Biochem. Biophys.* **117**, 660 (1966).

[36] D. C. Sharma, E. Forchielli, and R. I. Dorfman, *J. Biol. Chem.* **237**, 1495 (1962).

[37] T. Omura, E. Sanders, D. Y. Cooper, and R. W. Estabrook, see Vol. X, p. 362.

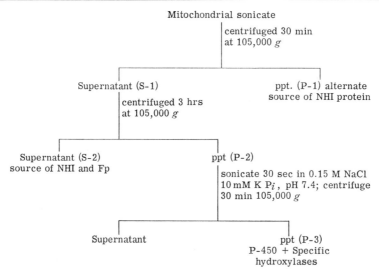

FIG. 2. Resolution of the 11β-hydroxylase system.

phate mixture and reprecipitation by centrifugation. The final particle precipitate, P-3, is suspended by sonication in 10 mM potassium phosphate buffer pH 7.4 to a protein concentration of about 20–30 mg/ml. When used within 2 days the samples are stored on ice. For longer storage periods they are kept in the deep freeze. However some loss in enzyme activity has been noticed in some frozen samples.

The P-3 particles and the purified NHI and Fp samples are standardized on the basis of protein content (biuret method) as well as of the pigment content (spectrophotometrically). For the latter determinations, the following extinction coefficients (cm^{-1} mM^{-1}) have been used:[37, 38] cytochrome P450·CO, difference spectrum (CO + dithionite) − (N$_2$ + dithionite), ϵ(450 mμ − 490 mμ) = 91; NHI, ϵ(Fe 415 mμ) = 5.6; Fp, ϵ(450 mμ) = 11.0.

One hundred grams of cortex scrapings yield approximately the following amounts of the three components of the 11β-hydroxylase system: 130–300 mg of P-3 protein, 0.7–1.5 mg of purified nonheme iron protein, and 2–3 mg of purified flavoprotein. To obtain these yields of NHI and Fp, the pooled S-2 fractions from 1000–1500 g of cortex tissue derived from the adrenal glands of 150–200 steers have to be processed. The S-2 fraction can be stored almost indefinitely in the frozen state (−15 to −20°). Additional amounts of NHI can be recovered from the P-1 precipitate or from extracts of acetone-dried (Suzuki et al.[39]) or

[38] T. Omura and R. Sato, see Vol. X, p. 556.
[39] K. Suzuki and T. Kimura, Biochem. Biophys. Res. Commun. 19, 340 (1965).

freeze-dried (Nakamura et al.[40]) mitochondria. The purification procedure is that described by Omura et al.[37] No additional source for preparing purified Fp from bovine material has as yet been found. The purified Fp in 10 mM potassium phosphate buffer and the purified NHI in 50 mM Tris-HCl buffer can be stored in the frozen state for weeks or months without loss in activity. Freezing of NHI in phosphate buffer causes bleaching and inactivation of the preparation.

RECONSTITUTION OF THE 11β-HYDROXYLASE SYSTEM

Reaction Medium. This is composed of 1.2 ml glycylglycine-albumin medium (Section II,B) containing 200 μg cortexone; 0.4 ml KCl–MgCl$_2$ mixture (100 ml of 0.154 KCl + 3 ml of 0.11 M MgCl$_2$); appropriate aliquots of P-3 particles and Fp (both in 10 mM potassium phosphate buffer, pH 7.4) and of NHI (in 50 mM Tris-HCl buffer, pH 7.4), and sufficient Tris-HCl buffer to adjust the volume to 2.2 ml. The reaction is started by adding 0.3 ml of the KCl–MgCl$_2$ mixture containing 8.4 mM TPN, 35.4 mM glucose 6-phosphate, and 0.5 Kornberg unit of glucose-6-phosphate dehydrogenase.

The millimolar concentrations of the main ingredients of the complete system are: NaCl, 57.5; KCl, 44.4; MgCl$_2$, 1.33; glycylglycine buffer, 11.5; Tris-HCl, 9.1; potassium phosphate buffer, 0.6; glucose 6-phosphate, 4.25; TPN, 1.01; cortexone, 0.24. In addition each milliliter contains 0.2 Kornberg unit of glucose-6-phosphate dehydrogenase and 7.7 mg of bovine serum albumin. Incubation is carried out at 25° either in open Erlenmeyer flasks immersed in a Dubnoff shaker incubator or, with different gas mixtures, in conical Warburg flasks attached to manometers in the Warburg apparatus. In the latter case the TPNH generating system is tipped in from the side arm after equilibration of the contents with the gas mixture. The corticosterone formation is determined with the Mattingly[31] procedure as described.

Saturating Concentrations of Fp and NHI. P-3 particles at a concentration of 1 mg protein per 2.5 ml of incubation medium are adequate for the assay. In the absence of Fp and NHI, hydroxylase activity of the particles is barely detectable (less than 1% of the activity of a saturated system) and is but little increased by adding Fp alone. The addition of NHI alone raises the activity to about 12% of the maximum, an indication that the particles are more completely freed of NHI than of Fp. Maximal activity is reached with 36 μM NHI in combination with 0.7 μM Fp. With 18 μM NHI the hydroxylase system is about 80% saturated. This degree of saturation has proved to be sufficient for assaying the enzyme activity. The reaction rates are proportional to the

[40] Y. Nakamura, H. Otsuka, and B. Tamaoki, *Biochim. Biophys. Acta* **122**, 34 (1966).

amount of P-3 protein over an 8-fold range (0.5–4 mg per 2.5 ml medium), and they are constant for at least 20 minutes. Because of the difficulty of preparing large amounts of NHI and Fp, concentrations of about 20 μM NHI and about 0.6 μM Fp are recommended for studying this system. What necessitates the high NHI concentrations is not understood at the present time.

Specific Activity. With the fully saturated system and fresh particles, rates up to 13.5 nanomoles of corticosterone per minute per milligram of P-3 protein have been observed. Activities of the 80% saturated preparations from fresh or frozen-stored particles range from 5 to 9. There are also variations in the cytochrome P450 content, which ranges from 1 to 1.7 nanomoles per milligram of P-3 protein. Rates per nanomole of cytochrome P450 are more consistent, ranging from 5 to 7. While the particles are virtually free of the pigments of the respiratory chain, they do contain varying amounts of cytochrome P420, an inactive derivative of P450. This may indicate that other components of the hydroxylase system are also inactivated during isolation of the P-3 particles and may explain why the specific hydroxylase activity of even the best reconstituted preparations is lower than that of suspensions of intact mitochondria.

Michaelis Constant. K_m for cortexone is about $2 \times 10^{-5} M$.

CO Inhibition. The degree of inhibition is a function of the CO/O_2 ratio. Warburg's partition constant K,[41] the ratio at which 50% inhibition occurs, is close to unity (range 0.6–1.4). The inhibition is maximally reversed by light of 450 mμ wavelength. The photochemical action spectrum corresponds remarkably well to the spectrophotometric difference spectrum of the CO complex of cytochrome P450. Warburg's light sensitivity factor L,[42] the reciprocal of the quantum energy that doubles the partition constant K, is 9×10^6. These K and L values are typical for hydroxylase systems from tissues of vertebrates.

Other Inhibitors. The following percentage inhibitions were obtained: KCN $(10^{-3} M)$, 28%; epinephrine $(10^{-4} M)$, 33%; diethyldithiocarbamate $(10^{-3} M)$, 26%, and $(10^{-4} M)$, 6%; 8-hydroxyquinoline $(10^{-3} M)$, 20%; bathocuproine $(10^{-4} M)$, 30%, and $(10^{-3} M)$, 97%; Cu^{++} $(10^{-3} M)$, 100%, $(10^{-4} M)$, 80%, and $(10^{-5} M)$, 0%. Inhibition by Metopirone[32] is competitive with cortexone, 50% inhibition occurring at a molar ratio of Metopirone to cortexone of about 6×10^{-3}. Mercurials inactivate P-3 as well as NHI. Ascorbate $(10^{-4} M)$ and glutathione $(5 \times 10^{-3} M)$ have no effect.

Binding Reaction. The addition of cortexone to a P-3 preparation in

---

[41] $K = (n/(1-n)) \times (CO/O_2)$; $n =$ (rate with CO)/(rate without CO).
[42] $L = (1/i) \times (\Delta K/K_d)$; $K_d =$ partition constant in the dark; $\Delta K =$ increase of K by light; $i =$ mole quanta cm^{-1} min^{-1}.

phosphate buffer without TPNH, NHI, and Fp produces the changes in the spectrophotometric absorption spectrum as described for mitochondria, an indication that the particles contain the substrate-specific component of the 11β-hydroxylase system in addition to cytochrome P450, the oxygen activating component.

Other Hydroxylases. The reconstituted system converts cholesterol to pregnenolone. Other hydroxylases have not been studied in this system.

3. *Acetone Powder and Freeze-Dried Preparations of Mitochondria*

The preparation of acetone-dried bovine adrenal mitochondria and the enzymatic properties of 0.154 M KCl extracts of the powders have been described in Volume V.[14,43a] Tomkins found that by extracting the acetone powders successively with H_2O, 0.1 M KCl and 0.5% digitonin, three enzyme fractions were obtained which individually displayed little or no 11β-hydroxylase activity. Recombination of the three fractions usually led to restoration of the enzyme activity. Some reconstituted preparations, however, remained inactive unless they were reinforced with filtrates from boiled adrenals or livers of steers. Nakamura *et al.*[43] observed that the 11β-hydroxylase activity of acetone-dried mitochondria from rat adrenals could be increased 5- to 10-fold by the addition of the filtrate of boiled rat liver. The active component of the filtrates was identified as a heat-stable protein. The purification of this protein and the preparation of freeze-dried mitochondria as described by the authors for the activity assays are outlined below.

PREPARATION OF LIVER FACTOR (LF)

Waring blendor homogenates of rat livers (Wistar strain) in H_2O are heated, and the heat-coagulable protein is removed by filtration. To the filtrate trichloroacetic acid (10% w/v) is added. The precipitate is dissolved in dilute NaOH, placed on a TEAE-cellulose column, which has been equilibrated with 0.05 M acetate buffer pH 5.0, and eluted with increasing concentrations of NaCl in the acetate buffer. The active material is usually found in the 0.4 M NaCl eluate. It is again precipitated with trichloroacetic acid, dissolved in NaOH, and reprecipitated with 5 volumes of ethanol (TEAE preparation).

For further purification, dissolved TEAE material is placed on an acetate buffer-equilibrated DEAE-cellulose column and successively eluted with 0.1 M, 0.12 M, 0.14 M, and 0.15 M NaCl in acetate buffer. Active material is mainly in the 0.14 M and 0.15 M eluates. These eluates are dialyzed, concentrated under reduced pressure, and chromatographed on a Sephadex G-50 column. The active fraction is lyophilized.

[43] Y. Nakamura, H. Otsuka, and B. Tamaoki, *Biochim. Biophys. Acta* **96**, 339 (1965).
[43a] See also this volume, p. 593.

Activity of this preparation is not affected by ribonuclease but is abolished by treatment with pronase or trypsin. The material consists of over 95% of amino acids; it is poor in histidine, tyrosine, and tryptophan, but rich in phenylalanine, which mainly accounts for the ultraviolet absorption between 250 and 280 mμ ($E_{1cm}^{1\%}$ 2.4 at 277 mμ). The factor is not tissue specific. It is present in filtrates of boiled adrenals and blood of rats and of liver and pancreas of pigs. Yeast extracts are ineffective. The factor stimulates 11β- and 18-hydroxylations of steroids but does not influence 21-hydroxylation. It has not been studied by other investigators.

PREPARATION OF FREEZE-DRIED MITOCHONDRIA FROM WHOLE ADRENAL GLANDS[40]

Trimmed adrenal glands of freshly killed Wistar rats are collected and stored at −20°. Ten grams of frozen adrenals are homogenized (as described in Section II,A) in 40 ml of 0.33 M sucrose solution and centrifuged for 10 minutes at 600 g. The sediment, suspended in 20 ml of 0.33 M sucrose, is centrifuged as before. The combined supernatant liquids are centrifuged for 30 minutes at 5000 g. The sediment is suspended in 0.25 M sucrose solution containing 0.1 M Tris-HCl at pH 7.4 and precipitated at 20,000 g. The washing is repeated once more. The final sediment is suspended in a small volume of H_2O and lyophilized. Yield is about 0.5 g from 10 g of frozen adrenals. The supernatant liquids from the 5000 g step and the two washings are combined and used for preparing the NHI protein,[44] and 0.05 M Tris-HCl extracts of lyophilized mitochondria serve as source for preparing Fp. The extracted mitochondrial residue is again freeze-dried and is used for assaying the activity of the three soluble components of the hydroxylase system. Porcine adrenals are processed similarly.

INCUBATION AND ASSAY

A reaction mixture consisting of 1 mg of cortexone, 2 ml of mitochondrial preparation in 0.33 M sucrose solution, 500 mg of TPN, 3 mg of sodium glucose 6-phosphate, 1 Kornberg unit of glucose-6-phosphate dehydrogenase, 200 micromoles of Tris-HCl (pH 7.4), 40 micromoles of MgSO$_4$, and 100 μg of purified LF in 3 ml of H_2O, is incubated for 1 hour at 37° in a Dubnoff shaker incubator in an atmosphere of 5% CO$_2$– 95% O$_2$. Steroids are extracted with 2 × 12 ml of dichloromethane. The combined extracts are dried over anhydrous Na$_2$SO$_4$, evaporated under reduced pressure and chromatographed on thin-layer plates (silica gel

[44] NHI and Fp as defined in Section III,B,2 are referred to by Nakamura *et al.* as SF (soluble-fraction component) and SF-reductase.

GF, Merck) with a benzene–acetone (7:3) solvent system. The ultraviolet quenching areas corresponding to corticosterone and the 18,20-hemiketal of 18-hydroxycortexone are extracted with methanol. The steroids are quantitated by measuring the light absorption of the methanolic extracts at 240 mμ.[45]

Properties of the Extracted Mitochondrial Residue. In the absence of NHI, Fp, and LF the preparation is devoid of hydroxylase activity, nor is it activated by reinforcement with Fp + LF. For complete restoration of activity the presence of all three factors is required.[40] The addition of NHI + LF or NHI + Fp causes, respectively, 18% and 32% restoration, a finding indicative that the extracted mitochondrial residue has not been completely freed of Fp and LF. When supplemented with the three soluble factors, 3 mg of the freeze-dried mitochondrial residue preparations produce 250–300 μg of corticosterone and about equal amounts of 18-hydroxycortexone in 1 hour at 37°. Similar preparations from hog adrenals when reinforced with NHI and Fp from either porcine or rat adrenals as well as with LF produce about 7 times as much corticosterone as 18-hydroxycortexone. The ratio of 11β- to 18-hydroxylation as well as the specific activities of the two mitochondrial hydroxylases are species specific whereas the soluble factors are interchangeable. Absolutely, the 11β-hydroxylase activity per unit weight of the porcine extracted mitochondrial residue amounts to only 6% of that of the rat preparations. While it has now been established that Fp + NHI serve as the TPNH–cytochrome P450 reducing system, the function of the heat-stable liver factor is not known.

IV. Steroid 18-Hydroxylase

A. Species Distribution

Unlike the 11β-hydroxylase, the 18-hydroxylase of mammals is restricted to the zona glomerulosa of the adrenal cortex, beneath the capsule of the organ. Hence, the enzyme activity values reported in the literature depend on the contribution of zona glomerulosa tissue to the adrenal preparation under study. Conversion of cortexone to 18-hydroxycortexone by freeze-dried mitochondrial preparations from rat and hog adrenals (cf. Section III,B,3) demonstrated that 18 hydroxylation requires the same TPNH–cytochrome P450 reducing system as 11β-hydroxylation. Conversion of corticosterone to 18-hydroxycorticosterone

---

[45] Authentic samples of both steroids were available. It is stated that the hydroxylation products were further identified by infrared and sulfuric acid-chromogen spectra and by melting points [Y. Nakamura and B. Tamaoki, *Biochim. Biophys. Acta* **85**, 350 (1964)].

by preparations from fresh adrenal glands of various species has been mainly studied in conjunction with aldosterone formation. Active preparations have been obtained from glands of men, cattle, sheep, and guinea pigs. Adrenal glands of the bullfrog (*Rana catesbeiana*) are by far the richest source. Glands from the leopard frog (*Rana pipiens*) are less active.[30] Frog adrenals do not have the functional zonation of the cortex characteristic of the mammalian gland.

The enzyme assay requires chromatographic separation of the hydroxylation products and quantitation by radioisotope techniques. The procedure of Nicolis and Ulick[46] is best suited for purification and identification of the reaction products and the precise determination of the percentage conversion of steroid precursors to products. It requires prolonged incubation periods and has been as yet applied only to crude preparations. The procedure of Psychoyos et al.[30] though less exacting in regard to purification procedures, can be applied to mitochondria. It measures initial rates under specified conditions and has furnished information on the enzymatic properties of the hydroxylase. Both methods, which are to a certain extent supplementary, will be described.

B. Procedure of Nicolis and Ulick[46]

Tissue Preparations

Beef zona glomerulosa tissue is prepared according to Ayres et al.[47] by passing a scalpel beneath the capsule, keeping the blade parallel with the capsule surface. The capsule is then peeled off with forceps. Bullfrogs weighing 125–450 g are pithed, the kidneys are removed through a ventral or dorsal[30] incision, and the adrenals which are embedded in the ventral surface of the kidneys, are dissected. Weight range of the adrenals is 15–30 mg. The tissues are cut with small scissors into slices 1–2 mm in thickness and are incubated in phosphate or bicarbonate-buffered Krebs-Ringer's solution or in a Tris-buffered medium having the following millimolar concentrations of salts: NaCl, 110; KCl, 5; $CaCl_2$, 2.5; $MgSO_4$, 1.0; KH_2PO_4, 1.0; Tris-HCl buffer (pH 7.4), 33.0. All media contain 2 mg of glucose per milliliter. Tissue concentrations are kept at 20–100 mg per milliliter of medium, and the tritiated steroid precursor (10^6 to 10^7 cpm) is added to the incubation mixture. Incubation is carried out in a Dubnoff shaker incubator in air. Mammalian tissues are incubated for 2 hours at 37°, and frog tissues for 4 hours at 26°.

For the biosynthesis of either unlabeled or 1,2-^{3}H-labeled or ^{14}C-4-

[46] G. L. Nicolis and S. Ulick, *Endocrinology* **76**, 514 (1965).
[47] P. J. Ayres, J. Eichorn, O. Hechter, N. Saba, J. F. Tait, and S. A. S. Tait, *Acta Endocrinol.* **33**, 27 (1960).

labeled 18-hydroxycorticosterone as well as aldosterone-4-^{14}C required for the assay procedure, frog adrenal slices are incubated for 8 hours with corticosterone, corticosterone-1,2-^{3}H, or progesterone-4-^{14}C.

Steroid Assay

The percentage of tritiated precursor steroid converted to 18-hydroxycorticosterone and aldosterone is determined by adding at the end of incubation either the respective unlabelled carrier steroid or a known amount of ^{14}C-labeled product. The hydroxylation products are isolated and purified until either the specific activity or the ^{3}H to ^{14}C ratio after two consecutive chromatographies differ by less than 10%. For specific activity determinations, the amount of carrier is measured with the thiosemicarbazide or the blue tetrazolium reaction. Radioactivity is measured with a Tri-Carb liquid scintillation spectrometer.[48]

Isolation and Purification

The incubation mixture is extracted four times with 0.5 volume of dichloromethane. The extracts are washed with H$_2$O, dried with anhydrous sodium sulfate, evaporated under reduced pressure at 40°, and chromatographed on Whatman No. 1 paper in the ethylenedichloride–ethylene glycol system. Nonvolatile stationary phases are supplied in 25% acetone solution. After location and elution of the areas corresponding to 18-hydroxycorticosterone, aldosterone, and corticosterone (radiochromatogram scanner, 254 mμ light absorption), the steroids are purified by chromatography and formation of derivatives as follows:

18-Hydroxycorticosterone. Chromatography in the toluene–propylene glycol system; oxidation of the eluted steroid in methanol to the etiolactone (0.1 M aqueous periodic acid containing 2% pyridine, 18 hours at room temperature); extraction of the lactone with dichloromethane after addition of H$_2$O; purification by chromatography in the toluene–formamide system; oxidation of the eluted etiolactone to 11-dehydroetiolactone with 1 ml of chromic oxide pyridine complex (=10 mg) overnight at room temperature; purification, by chromatography in the methylcyclohexane:toluene (1:1)–formamide system. When carrier dilution technique is used, the more readily available etiolactone of 18-hydroxycorticosterone is added prior to the treatment with periodic acid. Losses of 18-hydroxycorticosterone before the addition of the carrier are estimated by adding ^{14}C-labeled 18-hydroxycorticosterone to the incubation medium before extraction.

---

[48] S. Ulick, *J. Biol. Chem.* **236**, 680 (1961). Detailed description of the radiometric procedure.

Aldosterone. Eluted material from the first chromatogram is acety-lated with 0.2 ml pyridine $+$ 0.1 ml of acetic anhydride (1 hour at 90° or 18 hours at room temperature) to the 18,21-diacetate, which is purified by chromatographies in the methylcyclohexane:toluene (1:1)–formamide and methylcyclohexane:toluene (4:1)–methanol:water (7:3) systems; the diacetate in 0.2 ml of methanol is hydrolyzed to the 21-monoacetate by shaking it for 10 minutes with 10 ml of 0.1 M HCl, followed by extrac-tion with dichloromethane, washing of the extract successively with 0.1 M NaOH and water, drying, evaporation, and chromatographies of the residue in the toluene–formamide and the toluene–methanol:water (1:1) systems. Usually the monoacetate is radiochemically pure. Otherwise, it is further purified by alkaline hydrolysis to the free steroid and con-version to the etiolactone.

Corticosterone. Chromatography of the free steroid in the toluene:methylcyclohexane (1:1)–propylene glycol system or of the 21-mono-acetate in the same systems as the aldosterone diacetate yields radio-chemically pure material.

The measurement of percentage conversion of radioactive precursors into products does not take into account the dilution of the added pre-cursors by endogenous steroid. The comparison of final specific activities of precursors and products isolated without the addition of carriers eliminates the error due to the dilution by endogenous steroids.

Properties of the Preparations

The ratio of 18-hydroxycorticosterone to aldosterone formed from tritiated corticosterone is approximately 0.5 for sliced bullfrog adrenals; 1.8 for homogenized bullfrog adrenals not supplemented with cofactors; 1.6 for bovine zona glomerulosa slices; and 2.3 for hyperplastic human adrenal tissue. A slight conversion of added tritiated 18-hydroxycorti-costerone to aldosterone is detectable. It amounts to about 2% of the aldosterone yield from corticosterone under identical experimental condi-tions. Endogenous steroid production by 1 g of bullfrog adrenal slices averaged 140 μg of aldosterone, 61 μg of 18-hydroxycorticosterone, and 25 μg of corticosterone (4 hours at 26°). Addition of ACTH (1 unit per gram of tissue) causes tripling of the yield. It does not stimulate the conversion rate of added steroid precursors.

Since the study of Nicolis and Ulick deals only with percentage con-version of steroid substrates at different substrate to tissue ratios, hy-droxylation rates under optimal conditions have not been established. The highest values computable from the conversion data are 16 μg of 18-hydroxycorticosterone by 1 g of bullfrog adrenal slices (4 hours at 26°). For the biosynthesis of authentic 18-hydroxycorticosterne, Ulick and Vetter[49] recommend the use of 2 sliced adrenals, 25 μg of corticosterone, and 2 mg of glucose per milliliter of the Tris medium.

C. Procedure of Psychoyos, Tallan, and Greengard[30]

Tissue Preparations

Freshly excised bullfrog adrenals are collected in ice-cold 0.25 M sucrose solution (previously adjusted to pH 7.25 with 0.1 M KOH) and homogenized in this medium (50 mg of tissue per milliliter) with 5 to 6 strokes of a Potter-Elvehjem glass homogenizer. The homogenate can be stored up to 4 hours at 0° without loss in enzyme activity. An aliquot of the homogenate is centrifuged 10 minutes at 755 g (2500 rpm, Sorvall refrigerated centrifuge, rotor CC-34). The supernatant liquid is centrifuged 3 times for 10 minutes at 13,300 g (10,500 rpm). The sediments and the supernatant liquid are saved. The low-speed sediment is suspended in 0.25 M sucrose solution and again precipitated at 755 g (washed low-speed sediment). The wash is centrifuged at 13,300 g. The supernatant liquid is discarded. The sediment is combined with the 3 other sediments to yield the mitochondrial fraction. The saved supernatant liquid is centrifuged for 1 hour at 105,000 g (Spinco ultracentrifuge, rotor 40) to yield the microsomal sediment and the supernatant (soluble) fraction. The isolated mitochondria show a steady decline in their enzyme activity and should be used as soon as possible.

Cortices with the adhering capsule of bovine adrenals are prepared as in Section II, scissors-minced, suspended in 0.25 M of sucrose solution (250 mg/ml) homogenized first for 30 seconds at high speed in a Vir-Tis homogenizer and thereafter in a Potter-Elvehjem homogenizer. The soluble fraction is prepared by 100,000 g centrifugation of this 25% homogenate, and the particulate fractions are obtained from a 12.5% homogenate by the frog adrenal procedure except that the washing of the low-speed sediment and the third centrifugation of the supernatant liquid are omitted.

Incubation

The incubation mixture contains (in the millimolar concentrations indicated): potassium phosphate buffer (pH 7.25), 20; $MgCl_2$, 4; KCl, 24.6; glucose, 9; fumarate (from free acid and KOH), 8; ATP, 2; TPN, 0.5; corticosterone-1,2-^{3}H (1 μC in 0.1 ml of propylene glycol), 0.0144; sucrose (including sucrose introduced with the tissue preparation), 100. If

[49] S. Ulick and K. K. Vetter, J. Biol. Chem. 237, 3364 (1962).

necessary, constituents are adjusted to pH 7.25 with 0.1 M KOH.[50] Mitochondria derived from 50 mg of frog adrenals are incubated (shaker water bath) in 10 ml of the above mixture for 10 minutes at 37.5° in air. The reaction is stopped by the addition of 15 ml of acetone, and the capped reaction tubes are stored overnight at 4°.

Extraction and Assay of Steroids

The acetone-precipitated protein is packed by centrifugation, suspended in 60% acetone, and again precipitated by centrifugation. The combined supernatant fractions are diluted with water to an acetone concentration of 30%. The mixture is extracted with an equal volume of heptane (previously washed with 30% acetone). The heptane layer is discarded. The watery phase is reduced to about half its volume in a rotary evaporator (temperatures above 37° result in loss of aldosterone) and extracted twice with 2.5 volumes of water-washed dichloromethane. The combined extracts are concentrated (as above) to about 5 ml and transferred to a conical centrifuge tube. After the addition of 30 μg of authentic aldosterone in 0.05 ml of solvent, the extract is concentrated to about 0.1 ml by means of a stream of nitrogen.

Chromatography

The entire material is placed on 1⅜ inches wide Whatman No. 4 paper strips attached to common head and impregnated with the appropriate stationary phase 5 minutes before application of the material (cf. Neher[24]). Aldosterone is separated from 18-hydroxycorticosterone by descending chromatography for 24–48 hours in the formamide–benzene system (impregnation fluid; 30% formamide in acetone; solvent: benzene saturated with formamide). Corticosterone runs off the strip with the solvent. The two major radioactive areas are localized on the air-dried strips by means of a radiochromatogram scanner. The aldosterone area is identified by the ultraviolet quenching of the added authentic aldosterone. The 18-hydroxycorticosterone is tentatively identified by its position on the chromatogram with reference to aldosterone. Methanolic extracts of the two areas are evaporated under nitrogen and taken up in dichloromethane. Aldosterone is rechromatographed as above in the propylene glycol–toluene system (impregnation fluid: 40% propylene glycol in acetone; solvent: toluene saturated with propylene glycol). The apparent 18-hydroxycorticosterone is rechromatographed (descending, 3–

[50] Simplifications of the incubation mixture and the extraction procedure are given below in connection with the CO inhibition of the preparation.

4 hours) in a formamide–chloroform system (impregnation fluid: 35% formamide in acetone; solvent: chloroform saturated with formamide). The ultraviolet quenching of 50 μg of nonradioactive cortisol added to each chromatogram facilitates the localization of the 18-hydroxycorticosterone ($R_{\text{cortisol}} = 0.88$). The areas corresponding to aldosterone and 18-hydroxycorticosterone are cut out from the respective chromatograms and placed into vials to which the scintillation fluid (naphthalene toluene) is added. The quantity of steroid synthesized is calculated from the amount of radioactivity in the reaction products measured as in the method of Nicolis and Ulick.[46] Thus only the conversion of exogenous substrate is determined. Recovery of authentic aldosterone-1,2-^{3}H added to the incubation mixture has ranged from 97 to 103%. In several experiments the aldosterone formed has also been identified by the procedure of Nicolis and Ulick. Because of the unavailability of authentic material, analogous tests have not been performed with 18-hydroxycorticosterone, and the rates of synthesis reported must be taken as minimal values.

Properties

Activity. Homogenates of 50 mg of bullfrog adrenals when added to the standard assay system (10 ml, 37°, 10 minutes) produce on the average 1.1 μg of 18-hydroxycorticosterone and 2.1 μg of aldosterone from tritiated corticosterone. Rates are proportional to the tissue concentration up to 100 mg and are linear for about 30 minutes. Activity resides solely in the mitochondrial fraction. Synthesis of 18-hydroxycorticosterone and aldosterone by the mitochondria at 27° is about half as large as at 37°. The corticosterone concentration producing half-maximal rate of synthesis of both steroids is about $6.8 \times 10^{-6}\,M$ at 37.5°.

Electrolytes. Sodium can be substituted for potassium and Tris buffer for phosphate buffer. Omission of magnesium diminishes synthesis of 18-hydroxycorticosterone and aldosterone by 80–90%. Mn^{++} and Ca^{++} restore 18-hydroxycorticosterone formation by 70 and 90%, respectively. Aldosterone formation is completely restored by Mn^{++} but little enhanced by Ca^{++}.

Electron Donors and Cofactors. TPNH and fumarate (or L-malate) are essential. Omission of glucose, ATP, and TPN does not affect the activity of the mitochondria. DPN and DPNH are without effect. Mitochondria synthesize aldosterone (but not 18-hydroxycorticosterone) at a lower rate than the equivalent amount of adrenal homogenate unless they are fortified with the equivalent amount of soluble fraction prepared from the homogenate. Heating of the soluble fraction for 6 minutes at 100° abolishes its effectiveness. It has recently been found[51] that the

[51] H. H. Tallan, S. Psychoyos, and P. Greengard, *J. Biol. Chem.* **242**, 1912 (1967).

effect is due to the presence of fumarase (EC 4.2.1.2). No soluble fraction is required if L-malate is substituted for fumarate or other intermediates of the citric acid cycle.

Inhibitors. Rotenone (50 μM) inhibits mitochondrial synthesis of the two steroids by 50%. Oligomycin (0.4–2 μg/ml), antimycin A (0.16–2 μg/ml) and 2,4-dinitrophenol (10 μM) have little or no effect.

Carbon monoxide inhibits the biosynthesis of both steroids and the inhibition is reversed by light. The CO/O_2 ratio causing half inhibition and the degree of reversal by monochromatic light of different wavelengths have been determined only for aldosterone synthesis[12] since the formation of 18-hydroxycorticosterone is too low for accurate determination of the degree of inhibition.

The measurements are carried out in Warburg flasks[10] at 25°. Mitochondria derived from 50–60 mg of adrenal homogenate are incubated for 30 minutes in 3 ml of the following medium: KCl 25.7, mM; potassium phosphate buffer (pH 7.25), 20 mM; MgCl$_2$, 4 mM; TPNH, 0.83 mM; L-malate, 10.7 mM; corticosterone-1,2-^{3}H, 0.048 mM (2 μC per vessel); sucrose, 89 mM. The reaction is terminated by transferring 2 ml of the incubation mixture to 15 ml of dichloromethane essentially as described in Section III,A, except that the aqueous phase is extracted once more with dichloromethane. The combined extracts are analyzed as in the standard method.

Rates of aldosterone formation in the incubation mixture equilibrated with 4% O_2 in N_2 average 0.15 nanomole/minute and are constant during the 30 minutes of incubation. Warburg's partition constant K,[41] the CO/O_2 ratio for half-inhibition of the reaction, is 2.0. Maximal reversal of the CO inhibition is effected with light of 450 mμ wavelength. The photochemical action spectrum is in accordance with the difference light absorption spectrum of the CO complex of cytochrome P450. Warburg's light sensitivity factor L[42] averages 1.3 × 10^7. These data are typical for cytochrome P450-catalyzed mixed function oxidations of lipid-soluble organic compounds by vertebrate tissues. They support the conclusion that C-18 hydroxylation of corticosterone is an intermediary step in aldosterone synthesis. Possible reasons for the poor yield of aldosterone from added 18-hydroxycorticosterone have been discussed.[30, 48] The cytochrome P450 content of bullfrog mitochondria ranges from 0.17 to 0.35 nanomole per milligram of protein.

Bovine adrenocortical homogenates produce in the standard assay system (10 minutes at 37°) per unit weight of tissue only one-seventh of the amount of 18-hydroxycorticosterone produced by bullfrog homogenates. There is no detectable aldosterone formation under these conditions. If 250 mg of bovine homogenate is added to 50 mg of bullfrog homogenate, the syntheses of aldosterone and 18-hydroxycorticosterone by the latter are inhibited by about 70%. The inhibitor resides in the mitochondrial fraction of the bovine homogenate and is heat labile. Rat adrenal homogenates are also inhibitory.

D. Relative 18-Hydroxylase Activity of Mammalian Adrenals

Raman *et al.*[52] using radioisotope and chromatographic procedures similar to those described above[53] compared the activity of fresh adrenal cortex preparations from sheep, cow, and guinea pig, as well as of various preparations from an aldosterone-secreting benign human adrenocortical tumor.[54] The incubation system consisted of 2 ml of Krebs-Ringer bicarbonate solution, 4 mg of glucose, TPNH-generating system (see Section II,B) and tissue preparation derived from 1 g of cortex. Incubation was carried out at 37° in air. Adrenocortical homogenates from sheep, cow, and guinea pig when incubated for 90 minutes in the presence of 10 μM cortexone, produced, respectively, 2.1, 0.75, and 0.12 nanomoles of 18-hydroxycorticosterone and 0.56, 0.37, and 0.06 nanomole of aldosterone. Since there was no 18-hydroxycortexone detectable in any of the incubates, Raman *et al.*[52] concluded that 11-hydroxylation must precede 18-hydroxylation. This conclusion was supported by the observation that ovine homogenates 18-hydroxylated corticosterone about twice as fast as cortexone.

The enzymatic properties of various ovine adrenocortical preparations have been studied in greater detail. In the presence of 50 μM corticosterone, the optimal substrate concentration for the hydroxylase system, mitochondria derived from 1 g of cortex produced in 150 minutes 55 nanomoles of 18-hydroxycorticosterone and 7.5 nanomoles of aldosterone.[55] Tris-HCl buffer containing the same Ca^{++} concentration as the Ringer bicarbonate medium could replace the latter.

Formation of 18-hydroxycorticosterone and aldosterone by mito-

[52] P. B. Raman, D. C. Sharma, and R. I. Dorfman, *Biochemistry* 5, 1795 (1966).

[53] Extraction is similar to the standard method of Psychoyos *et al.*, and purification and identification are similar to the procedure of Nicolis and Ulick.

[54] P. B. Raman, D. C. Sharma, R. I. Dorfman, and J. L. Gabrilove, *Biochemistry* 4, 1376 (1965).

[55] If TPNH is substituted for the TPNH-generating system, part of the 18-hydroxycorticosterone formed is converted to the 11-oxo derivative when TPN accumulates. The C-11 dehydrogenation can be mediated by DPN as well as by TPN. The 11-dehydro-18-hydroxycorticosterone has the same mobility as aldosterone in the Bush[24] chromatography systems isooctane–*t*-butanol–water (10:5:9) and benzene–methanol–water (100:50:50).

chondria was inhibited 60–90% by 6–24 μM Metopirone[32] or SU-9055.[56] Added 18-hydroxycorticosterone (4–9 μM) inhibited 18-hydroxylation of corticosterone (14 μM) by about 40%. Varying degrees of inhibition (14–90%) were also observed in the presence of 0.2–30 mM concentrations of the following: diethyldithiocarbamate, p-chloromercuribenzoate, iodoacetate, N-ethylmaleimide, and formamidine acetate. Since several of these agents are inhibitors of glucose-6-phosphate dehydrogenase, TPNH was substituted for the TPNH-generating system.

The fact that inhibition of 18-hydroxycorticosterone synthesis is usually associated with an equal degree of inhibition of aldosterone synthesis lends additional support to the concept that the steroid 18-hydroxylase mediates the biosynthesis of aldosterone. Hence, conversion of corticosterone to the latter product, which can be half as large as the conversion to 18-hydroxycorticosterone, should be taken into account when comparing the relative 18-hydroxylase activity of adrenal tissue from different mammals in order to select the best source for purification and enzymological studies.

A more serious handicap to adequate comparison is the frequent lack of precise data on the specific activity of the preparations under optimal conditions of assay. When expressed as nanomoles of 18-hydroxycorticosterone formed per hour per gram of tissue at 37°, 18-hydroxylase activity (14.4 μM corticosterone, air) is 3 for adrenocortical homogenates of sheep as compared to 40 for the corresponding preparations from steers (Psychoyos et al.[30]). The latter are computed from initial linear reaction rates. The former, however, are derived from the nonlinear average rate during the 90 minutes of incubation required for yielding sufficient material for purification of the hydroxylation products but not optimal for activity measurements. Reaction media and assays also differ. The activity differences between the two species may thus not be real.

When the data of Nakamura et al.[45] are computed in terms analogous to those above, the 18-hydroxylase activity of homogenates of whole rat adrenals (600 μM cortexone, 5% CO_2 in O_2) is of the order of 9000, the hydroxylation product being 18-hydroxycortexone rather than 18-hydroxycorticosterone as in the experiments of Raman et al. The rat preparations are exceptional in that 18-hydroxylase activity at saturating substrate concentrations is about as high as 11-hydroxylase activity. Possibly under these conditions the hydroxylation products formed at one hydroxylase cannot compete with the common cortexone substrate for an additional hydroxylation at the other hydroxylase. Moreover, it is

[56] SU-9055 (Ciba) is 3-(1,2,3,4-tetrahydro-1-oxo-2-naphthyl)pyridine, assumed to be an inhibitor of the steroid-17-hydroxylase.

possible that 18-hydroxylation hinders 11-hydroxylation and, thereby, also aldosterone formation. Clarification of these questions requires further studies.

V. Steroid 21-Hydroxylase (EC 1.14.1.8)

A. Assay Methods

Termination of the enzyme reaction and extraction of the steroids are carried out essentially as described in Section III,A. Two milliliters of the incubation mixture are mixed with 15 ml of redistilled dichloromethane followed by 15 minutes of centrifugation at 1000–1500 rpm; aspiration of the supernatant aqueous layer; brief shaking of the organic phase with 1 ml of 0.1 M NaOH and separation of the phases and removal of the aqueous layer as before. For thin-layer chromatography an additional washing with 1 ml of H_2O is advisable. An occasional slight turbidity remaining after washing can be removed by adding a few crystals of NaCl. Aliquots of this dichloromethane extract are used for quantitation of the reaction products.

1. Determination of 17,21-Dihydroxy-20-ketosteroids[20]

Reagents

    a. Phenylhydrazine·HCl (reagent grade) recrystallized[57] according to Peterson et al.[58] Melting point of the purified salt should be 240–243°.

    b. Dilute sulfuric acid (62%); 62 volumes of concentrated H_2SO_4 (reagent grade) are added to 38 volumes of H_2O with cooling.

    c. Ethanolic sulfuric acid: 100 volumes of 62% sulfuric acid are

---

[57] *Purification of phenylhydrazine:* Add 100 g of the salt to 200 ml of H_2O at 60–70°. Maintain temperature between 50 and 60°. With frequent stirring, it takes up to 3 hours until the salt is dissolved. Heat 1 liter of absolute ethanol to boiling, add to the solution of the salt, and filter the mixture rapidly through Whatman No. 2 filter paper. Leave the filtrate overnight in refrigerator. Collect crystals on sintered-glass (medium) Büchner funnel. Repeat crystallization at least twice or until filtrate is colorless, using smaller amounts of H_2O and lower temperatures (50–25°) each time. The temperature is more critical than the amount of H_2O. Wash the crystals with ice-cold ethanol. Dry thoroughly, first with suction (dust-protected), thereafter in vacuum desiccator. Yield is about 50% of the starting material. Store crystals in glass-stoppered brown bottle in desiccator over $CaCl_2$. Keep in refrigerator. The properly purified and stored salt remains unchanged indefinitely. (Personal communication from Dr. Sidney Levin.)

[58] R. E. Peterson, A. Karrer, and S. L. Guerra, *Anal. Chem.* **29**, 144 (1957).

added to 50 volumes of absolute ethanol. If prepared from purified ethanol[58, 59] the mixture keeps indefinitely.

    d. DKS reagent: 43 mg of purified phenylhydrazine are dissolved
in 100 ml of reagent c or 65 mg phenylhydrazine are added to a
freshly prepared cooled mixture of 100 ml of 62% H_2SO_4 and
50 ml of absolute ethanol. No color should develop in an aliquot
of the DKS reagent that stands for about 20 hours at room
temperature (20–25°). Otherwise purified ethanol should be used.
The reagent can be stored for at most a week in the cold (light-
protected).

    Procedure. A 5–10 ml amount of the dichloromethane extract con-
taining 1–15 μg of dihydroxyketosteroid is added to 1 ml of DKS reagent
in a conical centrifuge tube with ground-glass stopper. The tube is
vigorously shaken for 15 seconds and left at room temperature (light-
protected) for 1 hour before the supernatant solvent layer is aspirated
and discarded. After the preparation has stood for an additional 18–20
hours in the dark, the absorbance at 410 mμ is measured spectrophotometrically in 1-ml microcuvettes.[60] Samples of the enzyme incubation
mixture without TPNH-generating system serve as blanks and for
preparing standards. In an additional sample, the steroid substrate is
also omitted to correct for its contribution to the absorbance. The absorbance of 17-hydroxyprogesterone amounts to about 3.5% of that of
cortexolone. Recovery of the added cortexolone from the incubation mixture described in Section V,B was about 60%. It was not influenced by
the presence or absence of the microsomal preparation.

2. *Determination of 21-Hydroxy-20-ketosteroids*[21]

    Reagents

        a. Absolute ethanol, freshly redistilled.[61] The head and tail frac-
           tions are discarded

        b. Blue tetrazolium:[62] 0.25 g in 100 ml of reagent a. Freshly
           prepared

---

[59] *Purification of ethanol:*[58] Add separately 7 g of silver nitrate and 15 g KOH (each
    dissolved in 100 ml of ethanol) to 4 liters of absolute ethanol and mix. Allow
    mixture to stand overnight and then distill the alcohol off through a Vigreux
    column, discarding the first 700 ml and the last 100 ml.

[60] Pyrocell Company, Westwood, New Jersey.

[61] Izzo *et al.*[21] recommend several methods of purifying alcohol before redistillation.
    However, redistilled absolute ethanol appears to be satisfactory [cf F. M. Kunze
    and J. S. Davis, *J. Pharm. Sci.* **53**, 1259 (1964)].

[62] 3,3'-(3,3'-Dimethoxy-4,4'-biphenylene)bis-(2,5-diphenyl-2*H*-tetrazolium   chloride).
    Dajac Laboratories, The Borden Chemical Corp., Philadelphia, Pennsylvania 19124.

c. Tetramethylammonium hydroxide: 2.5 ml of a commercial 10% aqueous solution diluted to 25.0 ml with reagent a

d. Acid ethanol: 10 ml of concentrated HCl diluted to 100 ml with reagent a

Procedure. An appropriate aliquot of the dichloromethane extract is pipetted into a conical centrifuge tube fitted with a glass stopper and is evaporated to dryness with the aid of a stream of nitrogen. Warming the tube in a water bath at 45° is permissible but not essential. The residue is dissolved in 1.5 ml of the ethanol followed by the successive additions of 0.5 ml of b reagent and 0.5 ml of reagent c, the latter starting the reaction. The tubes are left for 30 minutes at 25° in the dark before the reaction is stopped by adding 0.5 ml of reagent d. The absorbance is measured spectrophotometrically at 510 mμ. Blanks and standards are prepared as in the DKS assay.

With all α-ketolic steroids maximal and equal, color development is reached within 30 minutes. Non-α-ketolic steroids that possess an α,β-unsaturated 3-keto group also reduce blue tetrazolium, but at a slower rate, producing in 30 minutes about 5% of the color of the equivalent amount of α-ketolic steroids.

B. Preparations and Incubation Procedures

The microsomal preparation (Section II,A,4) is thawed and briefly homogenized. Aliquots either of this suspension or of clarified Triton extracts are used for enzyme assays.

Preparation of Triton Extracts[19]

Dilute 4 g of Triton N-101[63] to 100 ml with 0.05 M NaPO$_4$ buffer, pH 7.4, containing 2.5% (w/v) mannitol. To 3 volumes of the microsome suspension add one volume of the 4% Triton solution drop by drop with stirring in order to preclude inactivation of the enzyme by high local concentrations of the detergent. When the Triton concentration reaches 1%, clarification takes place. Allow the mixture to stand for

---

[63] The nonionic Triton surfactants (Rohm and Haas, Philadelphia, Pennsylvania) are either octylphenyl ethers (X-100, X-114) or nonylphenyl ethers (N-101) of polyethylene glycol, the hydrophilic component of the surfactants. There are 9 or 10 ethylene oxide groups in X-100 and N-101 and 7 or 8 groups in X-114. These were the only types of Tritons that yielded enzymatically active microsomal extracts. Specific activity of X-100 extracts is considerably lower than that of N-101 extracts, and the activity of X-114 extracts is considerably higher. However, the latter become cloudy above 10° and, hence, are not suitable for the usual optical measurements at room temperatures.

about 15 minutes at 4° before precipitating insoluble matter by centrifuging for 1 hour at 100,000 g. The supernatant liquid is clear with the
exception of a thin loosely packed surface layer. With some loss of the
subnatant liquid the superficial layer can be cautiously aspirated by
means of a syringe with square-tipped needle. For most purposes, however, this material has not to be removed since it does not interfere with
either the enzyme activity or optical measurements in the visible range.

Yield. With the latter procedure 75% of the mixture, equivalent to
about 40% of the microsomal protein and 40–60% of the hydroxylase
activity, is recovered from microsome suspensions containing about 15 mg
of protein per milliliter. The protein concentration of the extracts ranges
from 3 to 5 mg/ml. Higher concentrations can be obtained by starting
with more concentrated microsome suspensions. However, the percentage
recoveries diminish. With more dilute suspensions—less than 10 mg
protein/ml—percentage recoveries increase, but the lower absolute concentrations of the extracts are inconvenient for analytical procedures.

Storability. Triton N-101 extracts preserve their 21-hydroxylase
activity when stored for 2 days either in the refrigerator (4°) or in the
deep-freeze (−18 to −22°). They lose 50% of their activity during 2
weeks of deep-freeze storage but remain fully active when stored for 2
weeks at −72° (dry ice + cellosolve).

Extracts vs Suspensions

Triton extracts preserve the principal biophysical and biochemical
properties of the microsomes. They are probably not truly solubilized
enzyme preparations, but rather fine dispersions of particulate matter
which aggregate when the surfactant is removed by Sephadex filtration
and which can then be precipitated by high-speed centrifugation. The
sediment is fully active enzymatically but cannot again be clarified by
treatment with Triton.

In comparison with microsome suspensions the Triton extracts have
several advantages. They do not require special instruments for optical
studies. The homogeneity and the absence of sedimentation of particles
make the extracts especially useful for scanning broad spectral regions,
spectrophotometric titrations, and kinetic measurements. The disad-
vantage is the relatively low concentration of microsomal protein in the
Triton extracts, whereas microsome suspensions containing 20–30 mg of
protein can be readily prepared. The latter are thus superior whenever
high reaction rates are desirable. They are also more economical and
are recommended for enzyme assays where clarified preparations are
not essential.

Reaction Medium and Incubation Procedure

The reaction medium and the incubation procedure are essentially as described in Sections II,B and III,B,2 except that 125 μg (0.38 micromole) of 17-hydroxyprogesterone are suspended in 1 ml of the glycylglycine-albumin buffer medium. The total reaction volume after the addition of the enzyme preparation, TPNH-generating system, and 0.154 M KCl is 2.2 ml. The final steroid substrate concentration of 0.17 mM ensures zero-order reaction rates at enzyme protein concentrations of 0.5–3 mg per milliliter of medium for incubation periods up to 30 minutes at 25°.

C. Enzymatic Properties

A more extensive survey is being published elsewhere.[64] Unless specified otherwise, the data presented below have been obtained in the reaction system described either with suspensions (Sp.) or Triton extracts (Tr.) of adrenocortical microsomes at 25°. Data of Ryan and Engel[8] (indicated by R.E.) refer to microsome suspensions isolated from whole steer adrenals and incubated at 37° in Krebs-Ringer phosphate medium (pH 6.8) equilibrated with air. Activities are expressed as nanomoles of hydroxylation product formed per minute per milligram of microsomal protein.

Specific Activity

At saturating concentrations of 17–hydroxyprogesterone, activity (Sp.) ranges from 1.5 to 2.5 at 25° and from 4 to 5 at 37°. Activity of Triton extracts (25°) is about as high as or up to 50% higher than that of the suspension from which they are prepared.

pH Optimum (Sp. 37°)

There is a broad optimum between 6.9 and 7.4. Activity decreases by about 50% toward pH 8 and pH 6.3 (albumin-glycylglycine buffer[7]). For phosphate-buffered assay systems (R.E.) a pH optimum between 6.5 and 7.0 and 50% decreases in activity toward pH 5.5 and 8.0 have been reported.[14]

Substrate Specificity and Michaelis Constants

Most 21-methyl-3,20-oxocorticoids are hydroxylated, although only at much higher concentrations than 17-hydroxyprogesterone. K_m values for the latter steroid are around 10^{-6} M (see Section V,D).

[64] D. Y. Cooper, S. Narasimhulu, O. Rosenthal, and R. W. Estabrook *in* "Function of the Adrenal Cortex" (K. McKerns, ed.), Vol. 2, p. 897. Appleton, New York, 1968.

Carbon Monoxide Inhibition (Sp.[9, 10])

The inhibition is of the competitive type and is reversed by light. Warburg's partition coefficient[41] ranges from 0.6 to about 2.0 and is most frequently close to unity. The light sensitivity factor, L,[42] is 2.7×10^7 (450 mμ, 25°). The photochemical action spectrum is in harmony with the spectrophotometric difference spectrum of cytochrome P450:CO. Triton extracts show also the light-reversible CO inhibition of hydroxylation, but have not been used for determining K and L values.

Other Inhibitors[64]

Unless specified otherwise the percentage inhibitions recorded below have been determined at 37° to make them comparable with the data of Ryan and Engel.[65]

Mercurials[66] and SH-Reagents. HgCl$_2$ (R.E.) 0.1 mM, (Tr. 25°) 0.125 mM, 100%; p-chloromercuribenzoate (Sp. and Tr.) 0.1 mM, (R.E.) 1 mM, 100%, (R.E.) 0.1 mM, 0%; Mersalyl (Tr. 25°) 0.625 mM, (R.E.) 1 mM, 100%;[65] CuSO$_4$ 0.1 mM (Sp.) 100%, (R.E.) 50%; N-ethylmaleimide 4.3 mM (Tr. 25°) 75%. The inhibition by mercurials is due to the conversion of cytochrome P450 to its inactive derivative P420. Iodoacetate, 1 mM (R.E. and Tr.) and o-iodosobenzoate, 1 mM (R.E.) have been ineffective.

"Indirect" Inhibitors and Accelerators. These agents apparently affect steroid 21-hydroxylation by stimulating or suppressing side reactions that compete with the hydroxylase for the electrons of TPNH. The effects of these agents are not as consistent as those of CO and the mercurials.

Cytochrome c. (R.E.) 3 μM 98%, (Sp.) 1 μM 82%, (Tr.) 1 μM 45%. The inhibition is abolished by inhibitors of cytochrome oxidase such as cyanide or azide at 1 mM concentrations.

Ascorbate-Catecholamine Antagonism.[17, 67] At 10 μM concentration ascorbate inhibits steroid 21-hydroxylation by 50% and simultaneously stimulates O$_2$ consumption and TPNH oxidation whereas norepinephrine has the opposite effect.[68] Kitabchi[69] provided experimental evidence sug-

---

[65] The following agents had no significant effect: 10 mM glutathione (R. E., Tr.); 0.1 mM quinacrine (R.E.); 1 mM SKF 525A (diethylaminoethyldiphenylpropylacetic acid, R.E.); 1 μg/ml antimycin A (R.E.); 193 μg/ml catalase (R.E., Sp.); 1 mM H$_2$O$_2$ (Sp.); 50 μg/ml ribonuclease (R.E.).

[66] Mersalyl is the sodium salt of o-(3-hydroxymercuri-2-methoxypropyl)carbamylphenoxyacetic acid.

[67] D. Y. Cooper, O. Rosenthal, V. J. Pileggi, and W. S. Blakemore, *Proc. Soc. Exptl. Biol. Med.* **104**, 52 (1960).

[68] Epinephrine, isopropylepinephrine, and dihydroxyphenylalanine are about equally effective. Dihydroxyphenylethylamine is ineffective. Adrenochrome is ineffective at low concentrations and inhibitory at high concentrations.

gesting that lipid peroxidation is the competing side reaction which is antagonistically affected by ascorbate and catecholamines. Diethyl-dithiocarbamate or cyanide (Sp. 0.1 mM) also stimulates 21-hydroxyla-tion and abolishes the ascorbate inhibition, suggesting the possibility that the latter might be mediated by copper proteins.[70]

It should be noted that the effects of these agents as well as of ascorbate and catecholamines are pronounced only if the cortex is sepa-rated from the medulla as soon as possible after slaughter of the animal because the catecholamines of the medulla start to diffuse into the cortex after death and become adsorbed to the particulate components of the cytoplasm. Hence the effects of the agents are absent in time-consuming large-scale preparations of adrenocortical microsomes or in microsomes isolated from homogenates of whole glands. They are also absent in Triton extracts in which several of the apparent side reactions are diminished or absent.

Various Metal Complexers. Versene (Sp.) 1 mM 20%, (R.E.) 10 mM 0%; azide 1 mM (Sp.) 36%, (Tr.) 0%, (R.E.) 0%; α,α'-dipyridyl (Sp.) 0.1 mM 30%, 1 mM 100%, (R.E.) 1 mM 0%; o-phenanthroline 0.01 mM (Sp.) 60%, (Tr.) 35%, 0.1 mM (Sp.) 96%; 2-thenoxyltrifluoracetone 1 mM (Tr.) 33%.

D. Substrate-Binding Reaction

Method

Observation and quantitation of the spectral change produced by the addition of steroid substrates to microsomal or mitochondrial prepara-tions require difference spectrophotometry in order to compensate for the large background absorption of the material. For the study of Triton extracts any difference recording spectrophotometer with adequate power of resolution can be used. For turbid suspensions of microsomes or other particles the Yang-Chance split-beam recording spectrophotometer[71] or a similar instrument is needed. For kinetic measurements on both types of preparations the Aminco-Chance dual-wavelength spectrophotometer[72] is recommended.

---

[69] A. E. Kitabchi, *Federation Proc.* **24**, 448 (1965); *Nature* **215**, 1385 (1967).

[70] This is the reason for dialyzing the bovine albumin solutions which contained varying amounts of copper (cf. footnote 63).

[71] C. C. Yang and V. Legallais, *Rev. Sci. Instr.* **25**, 801 (1954); B. Chance, see Vol. IV, p. 272 ff. (1957).

[72] American Instrument Company, Silver Spring, Maryland 20910. The Aminco-Chance dual-wavelength/split-beam spectrophotometer, which can be used for both types of measurement, is also now available from this manufacturer.

Characteristics of the Spectral Change

Figure 3 shows the results of the spectrophotometric titration[73] of a microsomal Triton N-101 extract with 17-hydroxyprogesterone in the absence of TPNH. The spectral change is characterized by the formation

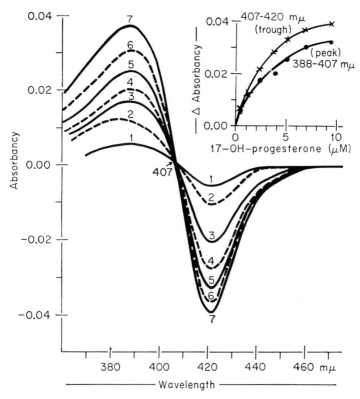

Fɪɢ. 3. The steroid-produced difference spectrum. Spectrophotometric titration of a Triton N-101 extract of adrenocortical microsomes (Cary Model 14 recording spectrophotometer, Cary Instruments, Monrovia, California 91016). The extract (3.7 mg protein/ml) was diluted 1:1 with the standard reaction medium (Section II,B). Aliquots, 0.8 ml each, were placed into two 1-ml optical cuvettes (Pyrocell Company, Westwood, New Jersey) of 1-cm optical path and a baseline of equal light absorption was established. Microliter volumes of a methanolic 17-hydroxyprogesterone solution were transferred to a glass rod at about 2 cm from its tip. The rod was held at a 45-degree angle and rotated during and for 20 seconds after the application to allow spreading of the solution and partial evaporation of the solvent. The semi-dried material was stirred into the experimental cuvette, and the difference spectrum was recorded after each addition. The steroid concentrations corresponding to the successive additions (1–7) are indicated on the abscissa of the inset figure. Reprinted from S. Narasimhulu, D. Y. Cooper, and O. Rosenthal, *Life Sci.* **4,** 2101 (1965).

[73] Carried out with a Cary Model 14 recording spectrophotometer, Cary Instruments, Monrovia, California 91016.

of a sharp absorption minimum at 420 mμ, the simultaneous appearance of a broad absorption maximum at about 390 mμ, and isosbestic points at 407 and 460 mμ. From the inset of Fig. 3 it can be seen that the spectral change becomes evident at a steroid concentration of 0.4 μM, that it approaches a maximum at 10 μM, and that the shape of the saturation curves and the magnitude of the spectral change for peak and trough are quite similar. For split-beam difference spectrophotometry the absorption difference 390–420 mμ has been used as a measure of the magnitude of the spectral change. With turbid solutions, where the resolution around 400 mμ is often unsatisfactory, and with dual wavelength spectrophotometry the absorption difference 460–420 mμ or 450–420 mμ is utilized.

The spectral change persists virtually indefinitely unless the system is supplemented with TPNH or a TPNH-generating system when it gradually disappears. Readdition of the steroid substrate starts a new cycle of appearance and disappearance of the spectral change. The spectral change also disappears after the addition of dithionite to both the experimental and reference cuvettes; but it reappears after the destruction of the reducing agent by aeration.[19] Similarly, the addition of TPNH to both cuvettes in the presence of nitrogen causes disappearance of the spectral change. It rapidly reappears after admission of oxygen to both cuvettes and then gradually disappears owing to 21-hydroxylation of the steroid.[64]

As previously pointed out (Sections III,B,1 and 2), the addition of substrates of the 11β-hydroxylase to mitochondrial preparations also produces the characteristic spectral change. However, the cyclic changes in conjunction with substrate hydroxylation can be most readily demonstrated with the 21-hydroxylase system of adrenocortical microsomes.

Substrate Specificity and Relative Affinity

At saturating concentration of a steroid the magnitude of the spectral change is a linear function of the concentration of microsomal protein.[74] However, the steroid concentration required to produce a maximal spectral change increases with increasing concentration of microsomal protein. The half-saturation concentrations—the apparent substrate dissociation constants, K_s, for the binding reaction—are a nearly linear function of the concentration of microsomal protein within the range studied (0.4–2.2 mg protein per milliliter). The possible reasons for this phenomenon, among them nonspecific substrate binding, have been dis-

[74] D. Y. Cooper, S. Narasimhulu, O. Rosenthal, and R. W. Estabrook, in "Oxidases and Related Redox-Systems" (T. H. King, H. S. Mason, and M. Morrison, eds.), p. 859. Wiley, New York, 1965.

cussed elsewhere.[64] Omission of albumin from the standard buffer medium (Section II,B) does not alter the apparent K_s.

Although this phenomenon, together with solubility limitations, precludes the assignment of definite K_s values to the individual steroids, the average K_s/protein values can serve as rough estimates of affinity to the binding site that mediates the spectral change. The following average K_s values per milligram of protein, calculated from reciprocal Lineweaver-Burk plots, have been obtained with Triton extracts of adrenocortical microsomes: 17-hydroxyprogesterone, 1.9; $11\beta,17\alpha$-dihydroxyprogesterone, 51; 17α-hydroxy-5α-pregnane-3,20-dione, 29; progesterone, 12; 11-oxoprogesterone, 57. 17α-Hydroxy-5β-pregnane-3,20-dione produced turbidity resulting in a distorted spectrum. With all these steroids the spectral change gradually disappears after the addition of TPNH. The steroids stimulate TPNH oxidation; the formation of the C-21 hydroxylation products has been established in every instance. The following steroids fail to produce a spectral change and to be hydroxylated: 5α- and 5β-pregnane-3,20-dione; 20α-hydroxypregn-4-en-3-one; $3\beta,17\alpha$-dihydroxy-5α-pregnan-20-one; pregnenolone (3β-hydroxypregn-5-en-20-one); and 17α-hydroxypregnenolone.

The results indicate that, besides the C-21 methyl group, keto functions at C-3 and C-20 are indispensable for the interaction between steroid and the 21-hydroxylase. Even in the presence of these essential groups, any deviation from the structure of 17-hydroxyprogesterone greatly reduces or abolishes the interaction.

K_s and K_m Values

With 17-hydroxyprogesterone as substrate Michaelis constants have been determined for the stimulation of TPNH oxidation and for hydroxylation. The K_m values do not differ significantly from the K_s values. This has also been the experience with other mixed-function oxidations.[75]

Microsome Suspensions

K_s values have been determined only for 17-hydroxyprogesterone. They also increase with the concentration of microsomal protein. The absolute values are lower, the ratio K_s/protein being approximately 0.5 μM. It is of interest to note that the concentration of cytochrome P450 per milligram of protein in microsomes is lower than in the Triton extracts (see Section E below).

[75] H. Remmer, J. Schenkman, R. W. Estabrook, H. Sasame, J. Gillette, S. Narasimhulu, D. Y. Cooper, and O. Rosenthal, *Mol. Pharmacol.* **2**, 187 (1966).

Specificity of the Spectral Change

While there has been no exception to the rule that C-21 methyl steroids that produce a spectral change are hydroxylated, there are other steroids that produce the characteristic binding reaction without being hydroxylated. The substance most thoroughly studied is the C-19 steroid androst-4-ene-3,17-dione.[19, 64] It produces most of the spectrophotometric changes that 17-hydroxyprogesterone does. It also stimulates TPNH oxidation. However, K_s for the spectral change and K_m for TPNH oxidation—both are about 10 μM—are independent of the concentration of microsomal protein. The spectral change does not disappear when the system is supplemented with a TPNH-generating system in the presence of O_2, and no hydroxylation products can be detected chromatographically. No spectral change has been observed with the 5α-reduced, the 11-oxo, and the 17β-hydroxy derivatives of androstenedione.

Significance of the Spectral Change

There is agreement that the spectral change reflects changes in the light absorption of cytochrome P450. Indeed, the change furnishes the first information on the spectral properties of the functional form of cytochrome P450 of which only the 450 mμ absorption band of the inactive CO derivative of its reduced form is definitely known. The mechanism of interaction between steroid and cytochrome has not been established. The following mechanisms have been proposed: (1) conformational changes of the hemoprotein, (2) changes in the oxidation state, and (3) a combination of (1) and (2). These interpretations have been discussed elsewhere.[64]

E. Hemoprotein Content

The total hemoprotein content in terms of protoheme[76] is, respectively, 0.45 and 0.78 nanomole per milligram of protein for the adreno-

---

[76] Total protoheme content: The heme is converted into the pyridine hemochromogen in the presence of 0.1 M NaOH and 20% pyridine. The absorbance difference 557–575 mμ in the difference spectrum of dithionite-reduced minus oxidized hemochromogen is determined. Readings are continued until ΔA(557–575 mμ) has reached a maximum (usually within 5 minutes). Since the stability of the hemochromogen is limited, readings must start immediately after the addition of the reagents. ϵ(557–575 mμ) = 32.4 cm^{-1}mM^{-1}.

Cytochrome b_5: The absorbance difference 424–409 mμ in the difference spectrum of DPNH-reduced sample minus air-saturated sample is determined. ϵ(424–409 mμ) = 185 cm^{-1}mM^{-1}.

P450·CO: The absorbance difference 450–490 mμ in the difference spectrum of CO-saturated dithionite-reduced sample minus dithionite-reduced sample is determined. ϵ(450–490 mμ) = 91 cm^{-1}mM^{-1}.

cortical microsome preparations and the Triton N-101 extracts prepared therefrom. Cytochrome P450 and cytochrome b_5 account, respectively, for about 50% and 15% of the total. The remaining 35% is assumed to be cytochrome P420, the enzymatically inactive conversion product of P450. The percentage distributions in microsome suspensions and in Triton extracts are identical. The total hemoprotein content of adrenocortical microsomes is about one-fourth and the cytochrome b_5 content about one-third of those in liver microsomes.

The absorbance difference 390–420 mμ in the difference spectrum produced by saturating concentrations of 17-hydroxyprogesterone is about 0.017 and 0.027 cm^{-1} per milligram of protein in microsomes and Triton extracts, respectively; i.e., it is about proportional to the protoheme content. This supports the inference that cytochrome P450 is responsible for the spectral change.

F. Miscellaneous Enzyme Activities of Microsomal Preparations

Contaminating Mitochondrial Enzymes[64]

Enzymes are here considered as mitochondrial contaminants if 50–60% of the activity demonstrable in the supernatant liquid of the mitochondrial fraction (Section II,A, step 2b) is removed during the subsequent fractional centrifugation in steps 4a and 4b (Section II,A) while the specific activity of the intrinsic microsomal enzymes is increasing. Indications of the presence of the following contaminants have been obtained: (1) Pyridine nucleotide transhydrogenase: DPNH supplemented with TPN can serve as electron donor of 21-hydroxylation whereas DPNH alone cannot. (2) Succinate dehydrogenase: Succinate reduces added cytochrome c in the presence of cyanide. (3) Cytochrome oxidase: The cytochrome c inhibition of 21-hydroxylation is abolished by cyanide and azide (Section V,C). These enzyme activities are less pronounced in Triton extracts of microsomes.

Nakamura and Tamaoki[45] reported that microsomes isolated from whole rat adrenals when supplemented with "soluble factor" (Section III,B,3) displayed 11β-hydroxylase and 18-hydroxylase activities. They consider this finding as evidence that the localization of these two steroid hydroxylases is not restricted to the mitochondria. It seems more likely that the activities are due to the carrying over of mitochondrial fragments which have been depleted of soluble components such as NHI and Fp.

Procedures and extinction coefficients are from: T. Omura and R. Sato, *J. Biol. Chem.* **329**, 2370, 2379 (1964).

Intrinsic Microsomal Enzymes

Pyridine Nucleotide Oxidoreductases. DPNH and TPNH reduce added cytochrome *c* in the presence of cyanide. They also reduce, without added cyanide, the microsomal *b*$_5$ which is only slowly reoxidized by oxygen.

DPNH and TPNH are also oxidized in the absence of added cytochrome *c*. TPNH oxidation by Triton extracts has been studied in some detail. It proceeds along two pathways. One is stimulated by steroids and is inhibited by CO while the other is not influenced by these agents. Evidently, the former pathway utilizes cytochrome P450 for the electron transfer to oxygen and functions as the reducing system of 21-hydroxylation. The latter might be involved in lipid oxidation as the TPNH system of liver microsomes described by Orrenius *et al.*[77] Both pathways are competitively inhibited by TPN in the presence of 0.05 *M* nicotinamide, half-inhibition occurring at a TPN/TPNH ratio of two.[64]

Pyridine Nucleosidase. The specific activity (initial rate) in Triton extracts with 2 *µM* TPN is about 1.3 nanomoles per minute per milligram of protein at 25°. The enzyme is inhibited by nicotinamide. Since TPNH is not attacked there is usually no need of supplementing the assay system with nicotinamide when using the standard TPNH generating system. Only for the maintenance of very low levels of TPNH (about 10^{-8} *M*) by the generating system the addition of nicotinamide appears to be required.

Steroid 17-Hydroxylase. Although cortisol is the major corticosteroid in the bovine species and the hydroxylase is supposedly localized in the microsomes,[78] there is little if any 17-hydroxylase activity detectable in the microsomal preparations described above. Whole homogenates of the adrenal cortex of steers do convert added progesterone to cortisol[79] although the rate of conversion is only about one-fourth of the conversion to corticosterone. Because of the hindrance of C-17 hydroxylation by a C-21 hydroxyl group, 17-hydroxylation must precede 21-hydroxylation. Possibly, disintegration of the cellular structure impairs the required reaction sequence. In addition, several investigators[8,78] have emphasized the lability of the 17-hydroxylase. Plager and Samuels,[80] furthermore, have pointed out the low 17-hydroxylase activity of the 20,000 *g* supernatant fraction of adrenocortical homogenates of steers as contrasted to that of heifers.

[77] S. Orrenius, G. Dallner, G. Nordenbrand, and L. Ernster, *Abstr. 6th Intern. Congr. Biochem. New York,* Section IV, 128 (1964).
[78] F. G. Hofmann, *Biochim. Biophys. Acta* 37, 566 (1960).
[79] D. Y. Cooper and O. Rosenthal, *Arch. Biochem. Biophys.* 96, 327 (1962).
[80] J. E. Plager and L. T. Samuels, *J. Biol. Chem.* 211, 21 (1959).

Young, Bryson, and Sweat,[81] using sensitive isotope techniques, found 17-hydroxylase activity exclusively in the 100,000 g supernatant fraction of adrenocortical homogenates of steers and in ammonium sulfate precipitates prepared from this fraction. The authors concluded that the steroid 17-hydroxylase was a soluble cytoplasmic enzyme rather than a component of the endoplasmic reticulum. They did not achieve, however, a separation of the 17-hydroxylase from the 21-hydroxylase. The major hydroxylation product of the soluble preparations was cortexone. The localization of the adrenal steroid 17-hydroxylase and the reasons for the discrepancy between *in vivo* and *in vitro* activities of the gland remain unclear.

Burstein *et al.*[14a] have studied the conversion of cortisol to 2α-hydroxycortisol by adrenal tissue from two strains of guinea pigs genetically distinguished by high (strain 13) and low (Hartley strain) urinary excretion of 2α-hydroxycortisol. The 2α-hydroxylase activity is found to be concentrated in the microsomal fraction separated from homogenates of whole adrenals. The relative 2α-hydroxylase activities of the microsomes from the two strains of guinea pigs corresponds well to the urinary excretion of 2α-hydroxycortisol. A similar correspondence holds true for the cytochrome P450 content of the microsomes.[82] No information on other species with respect to 2α-hydroxylation is available at the present time.

G. Species Distribution of Steroid 21-Hydroxylase

Although the steroid 21-hydroxylase is presumably present in the adrenocortical endoplasmic reticulum of all vertebrates that produce 21-hydroxylated corticoids, there are few data available on the specific activity of the microsomal fraction. Wilson, Nelson, and Harding,[33] using the 100,000 g sediment from the 25,000 g supernatant fraction of whole rat adrenal homogenates and a reaction medium and incubation procedure similar to those employed by them for 11β-hydroxylation (Section III,B,1), report a rate of 20 μg of cortexolone formed per minute per milligram of nitrogen from 17-hydroxyprogesterone at 37° under oxygen. This corresponds to about 10 nanomoles per minute per milligram of protein, a value about twice that of the specific activity of bovine adrenocortical microsomes at 37°. The rates observed by Nakamura and Tamaoki[45] in rat adrenal microsomes also appear quite high but are not expressed in comparable units.

---

[81] R. B. Young, M. J. Bryson, and M. L. Sweat, *Arch. Biochem. Biophys.* **109**, 233 (1965).

[82] S. Burstein, *Biochem. Biophys. Res. Commun.* **26**, 697 (1967).

Matthijssen and Mandel[83] obtained, by ammonium sulfate fractionation of the 15,000 g supernatant liquid from homogenates of whole sheep adrenals, a preparation with a specific activity of about 0.2 nanomoles per minute per milligram of protein (37°, air). In a recent preliminary note[84] they reported that chromatography of this preparation yielded a fraction with about five times higher specific activity. No cytochrome P450 could be detected in this fraction. An evaluation of these findings has to wait until more detailed information is available.

[83] C. Matthijssen and J. E. Mandel, *Acta. Biochem. Biophys.* **82**, 138 (1964).
[84] C. Matthijssen and J. E. Mandel, *Acta Biochem. Biophys.* **146**, 613 (1967).

[30] Purification of Ovarian 20α-Hydroxysteroid Dehydrogenase

By WALTER G. WIEST

Progesterone $+$ TPNH $+$ H$^+$ $\rightleftharpoons$ 20α-hydroxypregn-4-en-3-one $+$ TPN$^+$

Assay Method

Principle. Activity of the enzyme is determined spectrophotometrically by following the rate of increase of the absorbancy at 340 mμ associated with the reduction of TPN to TPNH in the presence of excess 20α-hydroxypregn-4-en-3-one-4-[14]C.

Reagents

TPN$^+$
Tris-HCl buffer, 0.1 M, containing 0.004 M cysteine, pH 8.0
Enzyme. Dilute the enzyme in Tris-cysteine buffer pH 8.0
20α-Hydroxypregn-4-en-3-one-4-^{14}C, 0.1 mg dissolved in 0.1 ml ethanol (specific activity 1000 dpm/μg)

20α-Hydroxypregn-4-en-3-one-4-^{14}C is prepared from progesterone-4-^{14}C by enzymatic reduction. Progesterone-4-^{14}C, approximately 700 μg, is dissolved in 0.7 ml ethanol and incubated in a medium containing 20 ml of pH 6.0 phosphate buffer (0.1 M), 7.0 mg of TPNH, 10.0 mg of glucose 6-phosphate, and 0.2 ml of a crude rat ovary extract (Step 1). After 1 hour at 37° the incubation medium is extracted with ethyl acetate. The extract is dried at 40° under nitrogen, applied to a 14 cm-wide strip of formamide-impregnated Whatman No. 1 filter paper, and chromatographed for 3 hours using hexane as mobile phase. Ultraviolet absorbing bands moving as progesterone and 20α-hydroxypregn-

4-en-3-one are cut out immediately and eluted from the formamide-moist paper with 15–20 ml of hexane. 20α-Hydroxypregn-4-en-3-one-4-^{14}C is diluted to a specific radioactivity of 1000 dpm per microgram with carrier 20α-hydroxypregn-4-en-3-one.

Procedure. 20α-Hydroxypregn-4-en-3-one-4-[14]C (100 μg, 0.316 micromole) dissolved in 100 μl of ethanol, 1.0 micromole of TPN⁺, enzyme equivalent to 0.01–0.02 unit, and 0.1 M Tris-HCl buffer with 0.004 M cysteine, pH 8.0, to make a total volume of 3 ml, are added to cuvettes of 1-cm light path. The reaction is started by the addition of enzyme. The cuvette contents are rapidly mixed, and the rate of change of absorbance at 340 mμ is measured as a function of time. The absorbance change should be linear for at least 1 minute to permit determination of the initial reaction velocity. After 15 minutes the reaction is stopped instantaneously by rapidly mixing ethyl acetate with the reaction mixture. Isolation of the steroid reaction product, progesterone-4-^{14}C, by paper chromatography and quantitation of its ^{14}C content permit a check of the stoichiometry of the reaction.

Definition of Unit and Specific Activity. A unit of enzyme is defined as the amount causing the production of 1 micromole of TPNH per minute at 37°. The specific activity is expressed as units per milligram of protein. Protein is determined by comparing the absorbance at 280 mμ with that of a crystalline bovine serum albumin standard.[1]

Purification Procedure

The following procedure comprises those steps previously reported[2-4] together with modifications appropriate to more recent experience. After removal of ovaries all steps are conducted in a cold room at 4°. Tris-HCl buffer, 0.1 M, containing 0.004 M cysteine, pH 8.0, is used throughout the purification procedure and is prepared fresh daily. Losses of enzyme activity incurred during dialysis and other manipulations are minimized by using freshly prepared buffer.

Step 1. Preparation of Crude Extract. Adult female rats, 200–250 g body weight, are killed by decapitation and bled. The ovaries are immediately removed and transferred to ice cold buffer. After the oviducts and surrounding fat are trimmed away ovaries are blotted dry and weighed. Tissue from 15 animals totaling about 1 g is considered a minimal quantity for the procedure.

The ovaries are homogenized in a glass homogenizer with sufficient

[1] O. Warburg and W. Christian, *Biochem. Z.* **310,** 384 (1941).
[2] W. G. Wiest, *J. Biol. Chem.* **234,** 3115 (1959).
[3] W. G. Wiest and R. B. Wilcox, *J. Biol. Chem.* **236,** 2425 (1961).
[4] R. B. Wilcox and W. G. Wiest, *Steroids* **7,** 395 (1966).

buffer to make a 5% homogenate. This is centrifuged at 25,000 g for 30 minutes, and the supernatant liquid is decanted with care to exclude floating particles of fat. Aliquots containing 0.5–1.0 mg of protein are suitable for enzyme assay.

Step 2. Ammonium Sulfate Precipitation. The supernatant fraction obtained in step 1 is placed in a pretested Visking dialysis bag and dialyzed overnight against that quantity of buffer saturated with ammonium sulfate which will give a concentration of 1.2 M ammonium sulfate at equilibrium. The precipitated protein is removed by centrifuging the contents of the dialysis bag at 600 g for 10 minutes and is discarded. The supernatant liquid is dialyzed a second time against an amount of ammonium sulfate-buffer calculated to give 3.5 M ammonium sulfate at equilibrium. The protein precipitated by this treatment is accumulated by centrifugation at 20,000 g for 30 minutes, and the supernatant liquid is discarded. The precipitated protein is dissolved in a minimum volume of buffer and is freed of ammonium sulfate by dialysis against an excess of buffer; 0.3–0.6 mg of protein is suitable for enzyme assay.

Step 3. DEAE-Cellulose Chromatography. DEAE-cellulose (DEAE or DEAE-SF, Bio-Rad Laboratories, Richmond, California) is washed in deionized water, and the fines are discarded. A column of 1 cm diameter and 10–15 cm length is prepared in water and washed with about 300 ml of buffer. The dialyzed enzyme solution from step 2 is applied to the column. About 10 ml of buffer is placed above the DEAE, and the column is connected for gradient elution. The mixing chamber contains 125 ml buffer, and the reservoir contains 250 ml 0.075 M K_2HPO_4 in buffer. Sixty-five 4-ml fractions are collected and assayed for enzyme activity and protein. Almost instantaneous mixing of the K_2HPO_4 solution should be achieved as it enters the mixing chamber, and its flow rate should equal the flow rate of the column. A considerable quantity of contaminating protein including hemeproteins is not retained by the column and is eluted in tubes 1 through 20. The bulk of 20α-hydroxysteroid dehydrogenase activity appears in tubes 30–50. Enzyme assay requires 0.1–0.2 mg of protein.

Step 4. Second Ammonium Sulfate Precipitation. Tubes from step 3 containing enzyme activity are combined and dialyzed against buffer saturated with ammonium sulfate to give a final salt concentration of 3.5 M. Contents of the dialysis bag are centrifuged at 20,000 g for 30 minutes, and the enzyme pellet is dissolved in 5 ml of buffer. The enzyme solution is freed of ammonium sulfate by dialysis against an excess of buffer. A quantity of enzyme protein from 0.05 to 0.1 mg is suitable for assay.

A summary of the purification procedure used for ovary extracts from 179 rats is given in the table.

PURIFICATION OF 20α-HYDROXYSTEROID DEHYDROGENASE[a]

Step	Total units	Total mg protein	Units per mg	Yield (%)
1. Crude extract	22.20	1355	0.0164	100
2. (NH₄)₂SO₄ precipitate-1	11.51	392	0.0294	52
3. DEAE chromatography	7.46	63	0.118	34
4. (NH₄)₂SO₄ precipitate-2	5.18	23	0.226	23

[a] From 11.7 g of rat ovarian tissue from 179 animals.

Properties[2-4]

Purity. This partial purification of 20α-hydroxysteroid dehydrogenase results in the elimination of reactivity toward DPN⁺ and 17α-hydroxyprogesterone. Fifteenfold purification may be expected.

Stability. Dialyzed precipitates from step 4 were stable for several months when stored at 4°. Inactivation results from 10-minute heating at 42° or from storage of buffered solutions at −15°.

Specificity. The enzyme preparation shows absolute specificity for the 20α-hydroxyl group of certain steroids and for TPN⁺. It does not react with the 20β-hydroxyl, 17α-hydroxyl, 17-ketone, or 21-hydroxyl groups of steroids or with the nonsteroids, secondary butyl alcohol, and ethanol. The enzyme reacts not only with steroids having the Δ⁴-3-keto configuration, but also with steroids saturated in ring A. Saturated steroids with 5β-hydrogen (A-B-ring junction *cis*) are more reactive substrates than those having 5α-hydrogen (A-B *trans*).

pH Optima. Optimum pH range for progesterone reduction is 5.6–6.0. For 20α-hydroxypregn-4-en-3-one oxidation the optimum pH range is 7–9.

Kinetic Properties. K_m (progesterone) 2.60×10^{-5} M; K_m (20α-hydroxypregn-4-en-3-one) 1.81×10^{-5} M.

Activators and Inhibitors. Enzyme activation by cysteine and inactivation by p-chloromercuribenzoate indicate involvement of sulfhydryl groups in the maintenance of enzyme activity.

[31] Δ^5-3-Ketosteroid Isomerase of *Pseudomonas testosteroni*[1-3]

By Rebecca Jarabak, Michael Colvin,
Suresh H. Moolgavkar, and Paul Talalay

Δ^5-Androstene-3,17-dione $\rightarrow$ Δ^4-androstene-3,17-dione

The discovery in 1955 of an enzyme capable of converting $\Delta^{5,6}$- or $\Delta^{5,10}$-3-ketosteroids to the corresponding α,β-unsaturated Δ^4-3-ketosteroids, was followed by the isolation of the induced Δ^5-3-ketosteroid isomerase of *Pseudomonas testosteroni* in crystalline and homogeneous form.[4] The enormous catalytic activity of this protein and the stereospecific intramolecular hydrogen transfer that it catalyzes have stimulated interest in the study of the mechanism of this enzymatic reaction.[5]

Assay Method

The assay method and preparation of the substrate (Δ^5-androstene-3,17-dione), have been described in an earlier volume of this series.[3] Although more modern methods for the preparation of the substrate have been developed[6] we have found the older procedure of Butenandt and Schmidt-Thomé,[7] as modified by Fieser,[8] entirely satisfactory,[3] provided that freshly distilled ether is used for the crystallizations and that full advantage is taken of permitting the crystals to form at $-20°$, thereby avoiding excessive heating at all times. The substrate is stored at $-20°$ over a drying agent and remains perfectly stable for years under these conditions. We are able to prepare Δ^5-androstene-3,17-dione containing no more than 1–2% Δ^4-androstene-3,17-dione without difficulty.

The spectrophotometric assays are carried out at 25° in systems of 3.0 ml containing: 100 micromoles of potassium phosphate buffer (pH 7.0), 0.175 micromole of Δ^5-androstene-3,17-dione in 0.05 ml of methanol and appropriately diluted enzyme to achieve a rate of absorbance change of 0.01–0.08 per minute at 248 mμ when measured against a blank containing all components except the steroid. The enzyme is diluted with

[1] P. Talalay and V. S. Wang, *Biochim. Biophys. Acta* **18**, 300 (1955).
[2] F. S. Kawahara, S.-F. Wang, and P. Talalay, *J. Biol. Chem.* **237**, 1500 (1962).
[3] F. S. Kawahara, Vol. V, p. 527.
[4] F. S. Kawahara and P. Talalay, *J. Biol. Chem.* **235**, PC1 (1960).
[5] P. Talalay, *Ann. Rev. Biochem.* **34**, 347 (1965).
[6] C. Djerassi, R. R. Engle, and A. Bowers, *J. Org. Chem.* **21**, 1547 (1956).
[7] A. Butenandt and J. Schmidt-Thomé, *Chem. Ber.* **69**, 882 (1936).
[8] L. F. Fieser, *J. Am. Chem. Soc.* **75**, 5421 (1953).

crystalline bovine serum albumin solution (10 mg/ml) which has been neutralized with NaOH. One unit of enzyme activity is defined as that amount which under these conditions causes the isomerization of 1 micromole of Δ⁵-androstene-3,17-dione per minute.[9]

Purification Procedure

The isolation of crystalline Δ⁵-3-ketosteroid isomerase from *Pseudomonas testosteroni* was originally reported in 1960,[4] and improvements of the procedure were described subsequently.[2,3] Further refinements of the process have been evolved[10] resulting in substantial simplification of the method and its adaptation to a larger-scale production of the crystalline enzyme.

Large-Scale Growth of Bacteria. Pseudomonas testosteroni (ATCC 11996) is cultured in a medium of the following composition:[11] 1.0 g of $(NH_4)H_2PO_4$, 1.0 g of $(NH_4)_2HPO_4$, 2.0 g of KH_2PO_4, 10 ml of trace element solution,[12] 10 g of Difco yeast extract, and 0.2 ml of Dow Corning Antifoam C, per liter of deionized water. The growth is carried out in a 1000-liter fermentor equipped with facilities for stirring (250 rpm) and aeration (16–18 cu. ft. per minute) and charged with 750 liters of the medium. The temperature is controlled at $30 \pm 0.5°$. The fermentor is inoculated with 7.5 liters (1% by volume) of a starter culture which has been grown for about 16 hours in an aerated flask in the same medium. The growth in the fermentor is monitored by following the absorbance at 650 mμ and the dry cell weight is determined from an appropriate calibration curve which is essentially linear over the absorbance range of 0–0.4. The logarithmic phase of growth terminates in about 6 hours and 150 g of progesterone dissolved in 750 ml of acetone is added to the fermentor at 6–7 hours of growth. The final culture density approaches 2.0 mg dry weight per milliliter. Maximum induction of enzyme activity is observed 2–4 hours after addition of the inducer steroid, at which time, aeration is interrupted and the contents of the fermentor are cooled to 10–12° or lower. The cells are harvested by a single passage at a rate of about 1.5 liters per minute through large-capacity Sharples centrifuges. The use of two or three of these centrifuges is desirable in order to

---

[9] The molar absorbancy coefficient of Δ⁴-androstene-3,17-dione at 248 mμ is assumed to be $16,300 \, M^{-1} \cdot cm^{-1}$. A rate of absorbance change equivalent to 5.43 per minute in the 3.0-ml assay system is equivalent to 1 unit, and hence an absorbance change of 0.001 per minute is produced by 1.84×10^{-4} unit.

[10] P. Talalay and J. Boyer, *Biochim. Biophys. Acta* **105**, 389 (1965).

[11] J. Boyer, D. N. Baron, and P. Talalay, *Biochemistry* **4**, 1825 (1965).

[12] The trace element solution has the following composition: 20 g of $MgSO_4 \cdot 7 \, H_2O$, 1 g of NaCl, 0.5 g of $ZnSO_4 \cdot 7 \, H_2O$, 0.5 g of $MnSO_4 \cdot 3 \, H_2O$, 0.05 g of $CuSO_4 \cdot 5 \, H_2O$, and 10 ml of 0.1 N H_2SO_4 per liter of distilled water.

accomplish the harvesting in a reasonably short period of time (3–4 hours). The packed cell mass is removed from the centrifuge bowls, resuspended in portions with the aid of a slow-running Waring blendor (4 liter capacity) in a total of 25 liters of 0.01 M sodium–potassium phosphate buffer of pH 7.0, containing 0.001 M EDTA. The suspension is then passed quite slowly through a Sharples centrifuge in the cold room (two bowls are required) and washing is completed with about an additional 15 liters of the same buffer. The cell mass is removed, spread thinly and uniformly on 12 rectangular stainless steel trays (20 × 45 cm) and dried in a vacuum from the frozen state (ca. −20°) in a large-capacity lyophilizer. The flaky dried material should be light buff in color. It is ground in a porcelain mortar and stored until used in a vacuum desiccator over P_2O_5 at −20°. The enzymatic activity appears to be completely stable for many months under these conditions. The yield of dried cell powder is 1400–1600 g from a 750-liter fermentation.

Extraction of Powder. The yield of enzyme in the initial extract can be substantially increased by treating the dried cell powder with lysozyme in the presence of EDTA according to the procedure of Repaske.[12a]

All operations are conducted at 2–4°.

Aliquots (200 g) of dried cells are added to 2400 ml of 0.01 M Tris-HCl in 50% ethanol (pH 7.0)[13] in the cooled chamber of a Gifford-Wood mill. The gap is set at 0.005 inch and the mill is operated for 30 minutes at the 40-V mark on the rheostat. The procedure is repeated with a second 200 g portion. The creamy suspension which now measures 5200 ml, receives 4.8 g of thrice crystallized egg lysozyme (Worthington) and 160 ml of 0.05 M EDTA (adjusted to pH 7.0 with NaOH). The suspension is stirred in the cold for 5–7 hours, until there is no further increase in isomerase activity in the supernatant fraction of small centrifuged aliquots. The suspension is centrifuged promptly for 30 minutes at 10,000 rpm in the GSA rotor of the Sorvall centrifuge and the clear chartreuse-colored supernatant fluid is decanted. The sticky and somewhat intractable residue is suspended in 750–1000 ml of the same extraction medium, stirred for 30 minutes, and centrifuged at maximum speed in the GSA rotor for 60 minutes. The second supernatant fluid which is slightly turbid is added to the first extract. The combined extracts from 400 g of dried cells should contain the activity equivalent to about 60 mg of crystalline enzyme. The first extract contains about 85–90% of the total activity found in the combined extracts.

[12a] R. Repaske, *Biochim. Biophys. Acta* **22**, 189 (1956).

[13] Distilled 95% ethanol was used in all operations. All concentrations of ethanol represent the percentage by volume of 95% ethanol, except, of course, where 95% ethanol is specifically designated. The 0.01 M Tris-HCl in 50% ethanol is prepared by mixing 0.02 M Tris-HCl (pH 7.0) with an equal volume of 95% ethanol.

Precipitation of the Enzyme with Magnesium. The pooled centrifuged extract (ca. 5.4 liters) is equally divided between two 12-liter Pyrex bottles, and sufficient distilled 95% ethanol is added to raise the concentration from 50% to 80% (1.5 volumes of ethanol). The solution remains relatively clear, but upon the addition of sufficient 1.0 M MgCl$_2$ to give a final concentration of 5.0 mM, an immediate precipitate forms which contains nearly all the enzyme activity.[14] The precipitate is permitted to settle for 2–3 days, and most of the clear supernatant fluid is carefully siphoned off and discarded. The supernatant fluid contains less than 1% of the total enzyme activity. The residual suspension (ca. 1200 ml) is then centrifuged in the GSA rotor at 5000 rpm for 15 minutes, and the supernatant fluid is discarded. The residue is carefully washed by resuspension in about 300 ml of 80% ethanol–0.005 M MgCl$_2$–0.004 M Tris-HCl (pH 7.0), and again sedimented by centrifugation. The suspension may be stored in the same medium and combined with similar preparations obtained from further 400 g batches of dried cells, or may be processed further by itself. The enzyme activity is stable for long periods at 4° at this stage.

Dialysis and Chromatography. The magnesium precipitate accumulated from 1500 g of dried powder is sedimented by brief centrifugation and the residue is extracted by stirring for 1.5 hours with 200 ml of 0.01 M potassium phosphate at pH 7.0. The insoluble material is removed by centrifuging at 9000 rpm for 30 minutes and the residue subjected to a second extraction for 1 hour with the same buffer. Centrifugation is repeated and the insoluble residue is discarded. The combined centrifuged extract (439 ml) is dialyzed against three 14-liter changes of 0.001 M Tris-phosphate (pH 7.0) during an 18–24 hour period. The dialyzed preparation is applied to a 50 × 420 mm column of DEAE-cellulose (Bio-Rad Cellex D). The exchanger is previously treated successively with 0.5 N NaOH, 0.5 N HCl, and 0.5 N NaOH with intervening and final water washes, and equilibrated thoroughly with 0.001 M Tris-phosphate (pH 7.0). All the enzyme activity is adsorbed. The column is washed with about 2000 ml of 0.001 M Tris-phosphate (pH 7.0) until the absorbance of the effluent is below 0.075 at 280 mμ. The elution is carried out with a convex gradient of Tris-phosphate at pH 7.0 ranging from 0.001 M to 0.4 M with respect to Tris. This gradient is obtained by connecting a constant volume mixing chamber (a three-neck 2000-ml round bottom flask) containing 0.001 M Tris-phosphate buffer to an upper reservoir containing 0.4 M Tris-phosphate. The column is operated at a

---

[14] In earlier procedures it was customary to add calcium phosphate gel to increase the bulk of the precipitate, but this has been found to be unnecessary, since the enzyme precipitates completely (presumably as the magnesium salt) and the gel plays no role in this process.

rate of 250–300 ml per hour and the eluate collected in 20-ml fractions. Most of the enzyme is eluted in a symmetrical band of 14 fractions between 1120 and 1400 ml which corresponds to a Tris concentration of 0.07–0.08 M. The peak of enzyme activity appears just following a yellow band which may be seen moving down the column and provides a useful guide to the position of elution of the enzyme.

The absorbancies at 280 mμ and isomerase activities of the fractions are determined, and the principal enzyme-containing fractions are combined. The enzyme may be stored at 4° for several days or frozen for prolonged periods at this point without loss of activity.

Crystalline isomerase has a specific activity of 62,000 units per milligram of protein.[15] A solution of the pure enzyme containing 1 mg of protein per milliliter has an absorbance at 280 mμ of 0.413.[2] The specific activities of the chromatographic fractions may be calculated on this basis, and should indicate that the enzyme is between 15 and 40% pure. This undoubtedly underestimates the purity of the preparations since the absorbency of isomerase at 280 mμ is low relative to most other proteins.[2]

Dialysis and Calcium Phosphate Gel Chromatography. The combined enzyme fractions (280 ml) are dialyzed against 0.001 M potassium phosphate buffer (pH 7.0) and applied to a column (50 × 120 mm) of calcium phosphate gel prepared according to Anacker and Stoy.[15a] The column is eluted with the same potassium phosphate buffer and fractions of 8–10 ml are collected at a rate of 100 ml/hour. The enzyme activity appears in a broad band. The enzyme is about 50% pure at this stage.

Concentration on DEAE-Cellulose. The dilute enzyme solution is applied directly to a small DEAE-cellulose column (20 × 100 mm) which has been equilibrated with 0.001 M potassium phosphate (pH 7.0). The entire activity is adsorbed and is then eluted in a sharp band (28 ml) with 0.2 M potassium phosphate (pH 7.0). The enzyme retains 50% purity and the protein concentration is around 7 mg/ml.

Crystallization. A saturated ammonium sulfate solution (Mann, enzyme grade; neutralized to pH 7.0 with NH$_4$OH) is added to the concentrated enzyme until incipient turbidity is achieved. Within one to several hours at 4°, the solution becomes very viscous and then

---

[15] This is the most consistent value based on an absorbance of 0.413 at 280 mμ for a solution of 1.0 mg of pure isomerase per milliliter. Earlier reports[3,4] of a specific activity of 169,000 units per milligram of protein were based on an underestimate of the protein concentration.[2]

[15a] W. F. Anacker and V. Stoy, *Biochem. Z.* **330**, 141 (1958). The gel must be extensively washed and decanted until the supernatant fluid rapidly clears on standing, otherwise adequate flow rates are not obtained.

develops a silky sheen on swirling. Microscopic examination confirms the presence of crystals. Long needles and flat-platelets settle out over a period of several days, and more than 90% of the enzyme crystallizes out in 7–14 days. The crystals are collected and recrystallized in the same manner. Usually two crystallizations suffice to bring the specific activity to 62,000 units per milligram of protein.

Uniform large crystals measuring about 1–2 mm in length may be obtained in 7–14 days at 25° with an ammonium sulfate concentration of 26.1% of saturation and at a protein concentration of 4.5 mg/ml. Smaller crystals form if the crystallization proceeds more rapidly.

There is no significant or systematic loss of enzyme activity at any step of the purification, and recoveries at each stage should be 75–100%. The initial extracts derived from 1400–1600 g of lyophilized powder contain the equivalent of 225 mg of pure isomerase, and 50% of this activity can be recovered in the second crystals. The procedure is summarized in Table I.

TABLE I
SUMMARY OF PURIFICATION PROCEDURE[a]

Step	Volume (ml)	Total Activity (units $\times 10^{-7}$)	Specific Activity (units $\times 10^{-3}$/ mg protein)
Initial extract	19,650	1.40	0.20[b]
Combined extracts of magnesium precipitate	439	1.27	1.62[b]
Combined fractions from DEAE-cellulose chromatography after dialysis	355	1.01	7.07[b] 5.58[c]
Calcium phosphate gel eluate	795	0.77	29.5[c]
DEAE-cellulose concentrate	28	0.78	31.1[c]

[a] Lyophilized cells (1500 g) of *Pseudomonas testosteroni* were used for this purification.
[b] Protein determination by the biuret method.
[c] Protein determination by ultraviolet absorbance, assuming $A_{280} = 0.413$ for 1 mg of protein per ml.

Properties

Substrates and Inhibitors.[2, 16, 17] Δ⁵-3-Ketosteroid isomerase converts a variety of $\Delta^{5(6)}$- and $\Delta^{5(10)}$-3-ketosteroids to the corresponding Δ⁴-3-ketosteroids. In the regular assay system at comparable substrate concentrations, the rates of isomerization relative to Δ⁵-androstene-3,17-dione

[16] S.-F. Wang, F. S. Kawahara, and P. Talalay, *J. Biol. Chem.* **238**, 576 (1963).
[17] J. B. Jones and D. C. Wigfield, *J. Am. Chem. Soc.* **89**, 5294 (1967).

(100) are as follows: Δ^5-pregnene-3,20-dione (150), 17β-hydroxy-17α-methyl-Δ^5-androsten-3-one (40), 17β-hydroxy-17α-Δ^5-pregnen-20-yn-3-one (33), 17α-hydroxy-Δ^5-androsten-3-one (30). Δ^5-Androsten-3-one which lacks polar substituents at C-17, is also isomerized by the enzyme, but only in the presence of large concentrations of methanol, suggesting that this substrate does not exist in monodisperse solution in completely aqueous media. Δ^5-Cholesten-3-one is not isomerized at a significant rate. $\Delta^{5(10)}$ Steroids also serve as substrates, but their rates of isomerization relative to Δ^5-androstene-3,17-dione (100) are relatively slow:[2] 17β-hydroxy-$\Delta^{5(10)}$-estren-3-one (0.25, $K_m = 22.3$ μM); $\Delta^{5(10)}$-estrene-3,17-dione (0.27); 17α-methyl-17-hydroxy-$\Delta^{5(10)}$-estren-3-one (0.55); 17α-ethyl-17-hydroxy-$\Delta^{5(10)}$-estren-3-one (0.73); 17α-ethynyl-17-hydroxy-$\Delta^{5(10)}$-estren-3-one (0.65).

A number of steroids are competitive inhibitors of the isomerase when Δ^5-androstene-3,17-dione serves as a substrate. At pH 7.0 under conditions of the standard assay, the following inhibitor constants have been determined: 17β-estradiol ($K_i = 10$ μM); 19-nortestosterone ($K_i = 5.2$ μM) and 17β-dihydroequilenin ($K_i = 6.3$ μM).[2,16]

Effects of Metal Ions and Chelating Agents. When added directly to the assay system, the following chelating agents were without effect on the activity of the enzyme: ethylenediamine tetraacetate (10 mM), α,α-dipyridyl (0.3 mM), o-phenanthroline (80 μM), and 8-hydroxyquinoline (80 μM). The direct addition of MgCl$_2$ (up to 1 mM) did not affect the enzyme assay, and prior incubation of isomerase preparations at 4° with Mg^{++} and Ca^{++} in concentrations up to 1 mM was likewise without effect on the enzyme activity.

Since the enzyme is quantitatively precipitated by 5 mM MgCl$_2$ in the presence of 80% ethanol, the selectivity of this precipitation in crude extracts was investigated with respect to divalent cations and the concentration of ethanol (see Table II). Although 5 mM Ca^{++}, Cd^{++}, and Zn^{++} all caused precipitation of the enzyme at even lower ethanol concentrations than Mg^{++}, the precipitates were much heavier than with Mg^{++}, and the precipitates were less soluble in neutral buffers. It was concluded that within the limits of conditions investigated, Mg^{++} was a more selective precipitant of isomerase from crude extracts and this cation was selected for use in the purification procedure.

Structure. The crystalline isomerase behaved as a homogeneous entity in the ultracentrifuge, and the approach to equilibrium method gave a mean molecular weight of 40,800 $\pm$ 5% in 0.02 M sodium phosphate buffer at pH 6.8 and 20°. The sedimentation coefficient ($S_{20,w}$) was 3.3 S and the frictional ratio (f/f_0) was approximately 1.3.[2] The amino acid composition calculated from analyses obtained at various times of

TABLE II

PRECIPITATION OF ISOMERASE FROM CRUDE EXTRACTS BY VARIOUS
CATIONS IN THE PRESENCE OF ETHANOL[a,b]

Salt, 0.005 M	Concentration of ethanol			
	50%	60%	70%	80%
$MgCl_2$	6.4	12.5	91	98.7
$CdSO_4$	99.8	100	100	100
$CaCl_2$	65.0	99.5	100	100
$Zn(Ac)_2$	99	100	100	100

[a] Values are expressed as percentage of initial activity precipitated.
[b] Five grams of lyophilized steroid-induced *P. testosteroni* were added rapidly to 30 ml of 95% distilled ethanol. After 1 hour of stirring, 30 ml of 0.02 M Tris-HCl (pH 7.0) was added and stirring was continued for 18 hours. The homogeneous suspension was centrifuged at 20,000 *g* for 30 minutes, and the supernatant fraction was used in these experiments. Ethanol was added to achieve the appropriate concentration, and the salts were added as 1.0 M solutions to give a final concentration of 5 m*M*. The solutions were assayed immediately after the addition of the salts, and assays were again carried out on supernatant fluids obtained after 24 hours of storage.

hydrolysis corresponded to the approximate formula:[18] Asx$_{38}$, Thr$_{21}$, Ser$_{16}$, Glx$_{38}$, Pro$_{15}$, Gly$_{29}$, Ala$_{68}$, Val$_{45}$, Met$_8$, Ile$_{12}$, Leu$_{26}$, Tyr$_9$, Phe$_{25}$, Lys$_{13}$, His$_9$, Arg$_{22}$, for a total of 394 residues, where Asx and Glx represent the sums of amidated and unamidated aspartic and glutamic residues, respectively. Tryptophan and cyst(e)ine are absent. Several lines of evidence suggest that the enzyme molecule consists of three identical chains. Hydrolysis of the heat-denatured enzyme with trypsin gives rise to eleven distinct peptides varying in size between 5 and 33 amino acid residues. The N-terminal amino acid residue of each peptide and the N- and C-terminal peptides have been identified. Methionine is the N-terminal residue of each chain.[18]

Ultraviolet Absorption Spectrum.[2] The absorption spectrum of crystalline isomerase at pH 7.0 shows a maximum at 277 mμ with a well-defined shoulder at 282–284 mμ (characteristic of tyrosine), as well as minor peaks at 253, 259, 266, 269 mμ which correspond to the known absorption bands of phenylalanine. There is in addition, a well-defined minimum at 250 mμ. At pH 7.0, the absorbance ratio of $A_{280}:A_{260} = 1.33$, $A_{277}:A_{250} = 1.98$ and $A_{277}:A_{260} = 1.45$. Spectrophotometric titration of the enzyme with base reveals that one-third of the tyrosine residues titrates with a normal pK of about 9.5–10, whereas the remainder become accessible to titration only above pH 12, presumably upon disruption of the native

[18] J. Boyer and P. Talalay, *J. Biol. Chem.* **241**, 180 (1966).

configuration of the protein, and hence the latter residues are presumably buried in the interior of the globular protein.

Mechanism of Isomerization[1, 16, 19]

Detailed studies of the mechanism of this enzymatic reaction have been carried out with isotopic hydrogen. The enzymatic isomerization of Δ^5-androstene-3,17-dione to Δ^4-androstene-3,17-dione in tritium or deuterium-labeled water proceeds without significant incorporation of isotope into the product, in contrast to the acid or base-catalyzed reaction which results in labeling of the product. This leads to the conclusion that enzymatic isomerization proceeds by a direct transfer of a proton from C-4 to C-6. Experiments with substrates specifically labeled at C-4 and C-6 have established that the reaction involves a direct intramolecular transfer of the 4β proton into the 6β position, and hence has *cis* diaxial stereospecificity.

Further insight into the nature of the reaction has been acquired from ultraviolet absorption and fluorescence spectra of steroid-enzyme complexes.[16] Several steroids which are powerful competitive inhibitors of the enzymatic reaction, as for instance, 19-nortestosterone, 17β-estradiol, and 17β-dihydroequilenin undergo marked spectral changes upon interaction with stoichiometric quantities of isomerase. The ultraviolet absorption maximum of 19-nortestosterone is displaced toward longer wavelength by 10 mμ, whereas the spectra of the phenolic steroids at neutral pH resemble those which are observed when the phenolic hydroxyl groups undergo ionization in basic solution. These spectral changes have been ascribed to the formation of structures resembling the enol of the α,β-unsaturated ketone and the phenolate ion of the phenolic steroids, respectively. Spectral titrations indicate the binding of 3 moles of steroid per unit weight of 40,800, with an average dissociation constant of 1.64 μM for 19-nortestosterone. The incubation of 19-nortestosterone, testosterone of Δ^4-androstene-3,17-dione with relatively large quantities of isomerase in a tritium-labeled aqueous medium results in the incorporation of label from the medium into the steroid. The isotope is completely removed under enolizing conditions. These findings support the view that the enzyme-steroid interaction involves the formation of an intermediate resembling the enolic form of the steroid.[16]

Some light has been shed on the nature of the active site by the finding that photooxidation of the enzyme in the presence of methylene blue leads to progressive loss of enzymatic activity which parallels the destruction of histidine. At least one histidine residue appears to be essential for

[19] H. J. Ringold and S. K. Malhotra, *J. Am. Chem. Soc.* **87**, 3228 (1965).

enzymatic activity, and a model has been proposed in which the two nitrogen atoms of the imidazole ring of histidine function alternately as proton donors and acceptors as the isomerization reaction proceeds by way of an intermediate enol.[5]

Acknowledgment

Supported by National Institutes of Health Grants AM 07422 and GM 1183.

[32] 3α-Hydroxysteroid Dehydrogenase of Rat Liver

By SAMUEL S. KOIDE

$$3\alpha\text{-Hydroxysteroid} + \text{DPN}^+(\text{TPN}^+) \rightleftharpoons$$
$$3\text{-ketosteroid} + \text{DPNH}(\text{TPNH}) + \text{H}^+$$

Assay Method

Principle. The dehydrogenase activities are assayed spectrophotometrically by measuring the rate of reduction of DPN or TPN in the presence of a 3α-hydroxysteroid.[1-3] In addition, the enzyme possesses a pyridine nucleotide transhydrogenase activity. This activity is measured in a system containing a TPNH-generating system and DPN. The reduction of DPN is measured optically in the presence of a catalytic amount of 3α-hydroxysteroid or 3-ketosteroid.

Reagents

    Potassium phosphate buffer, 0.1 M, pH 8.0
    Glycine-NaOH buffer, 0.1 M, pH 9.5
    DPN and TPN, 1 micromole/100 μl
    Androsterone in methanol, 0.1 micromole/10 μl
    TPNH-generating system: i.e., glucose-6-phosphate dehydrogenase
      and TPN
    Enzyme: soluble fraction of rat liver

Procedure. DPN-linked dehydrogenase activity is assayed in a total volume of 3 ml containing 200 micromoles of phosphate buffer, pH 8.0, 1 micromole of DPN, and appropriate amounts of enzyme. The TPN-linked dehydrogenase activity is assayed in a system containing 200 micromoles of glycine-NaOH buffer, pH 9.5, 0.5 micromole of TPN, and

---

[1] B. Hurlock and P. Talalay, *J. Biol. Chem.* **233**, 886 (1958); see also Vol. V, p. 512.
[2] S. S. Koide, *Arch. Biochem. Biophys.* **101**, 278 (1963).
[3] S. S. Koide, *Biochim. Biophys. Acta* **110**, 189 (1965).

appropriate amounts of enzyme in a total volume of 3 ml. The reaction is initiated by the addition of 0.1 micromole of androsterone dissolved in 10 μl of methanol to the reaction cuvettes. The control cuvette contains all components except the steroid. An alternative assay for the dehydrogenase activity is based on the determination of the rate of oxidation of DPNH or TPNH in the presence of 3-ketosteroids.[4,5]

The system for the assay of transhydrogenase activity contains 0.6 Kornberg unit of glucose-6-phosphate dehydrogenase, 5 micromoles of glucose 6-phosphate, 5 micromoles of MgCl$_2$, 1 micromole of DPN, 0.02 micromole of TPN, 200 micromoles of phosphate buffer, pH 8.0, and appropriate amounts of enzyme in a total volume of 3 ml. The reaction is initiated by the addition of 0.02 micromole of androsterone dissolved in 10 μl of methanol. The control cuvette contains all components except the steroid. The increase in absorbance at 340 mμ is measured in cuvettes of 1.0 cm light path at 22–23°.

Definition of Units and Specific Activity. One unit of the dehydrogenase and transhydrogenase activities represents the reduction of 1 millimicromole of pyridine nucleotide per minute under the prescribed assay conditions.[3] Specific activity is expressed as units per milligram of protein.

Purification Procedure

Step 1. Preparation of Homogenate. Livers from adult male rats are perfused with 0.9% saline and excised. All subsequent steps are performed at 0–4°. Fifty grams of tissue are homogenized in a Waring blendor at 60 V for 1 minute with 150 ml of 0.1 M potassium phosphate buffer, pH 8.0. The homogenate is centrifuged at 59,000 g for 1 hour.

Step 2. Ammonium Sulfate Fractionation. The volume of the supernatant solution from step 1 is measured. Ammonium sulfate, enzyme grade from Mann Research Lab., New York, is added to the solution in small increments to give 0–0.3, 0.3–0.5, and 0.5–0.7 saturation. The mixture is stirred constantly while the pH at 7.6 is maintained with the addition of dilute ammonium hydroxide solution. After the addition of ammonium sulfate to the appropriate saturation point, the mixture is stirred for 1 hour and centrifuged at 13,000 g for 30 minutes. The precipitate obtained from the 0.5–0.7 ammonium sulfate fraction is dissolved in 20% of the original volume with 30% glycerol (v/v), 0.01 M phosphate buffer, pH 8.0 (solution A) and dialyzed against solution A until the dialyzate gives a negative test for ammonia with Nessler's solution.

Step 3. Sephadex Gel Filtration. Sephadex G-100 is washed ten times

---

[4] G. M. Tomkins, *J. Biol. Chem.* **218**, 437 (1956).

[5] G. M. Tomkins, *Recent Progr. Hormone Res.* **12**, 125 (1956).

with glass-distilled deionized water and three times with solution A. The fine sediment is decanted off after each washing and settling for 2–3 hours. A column of 4.2 cm diameter is prepared and allowed to settle by gravity to a height of 142 cm. The column is further washed with 3 liters of solution A. Ten milliliters of the dialyzed ammonium sulfate preparation are added to the column and eluted with 2 liters of solution A. Fractions of 13–15 ml are collected at a rate of 12 ml/hour (Fig. 1). The tubes containing the enzyme activity are pooled and dialyzed overnight against several changes of solution A.

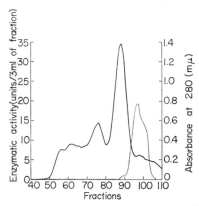

Fig. 1. Sephadex G-100 gel filtration of 3α-hydroxysteroid dehydrogenase preparation from ammonium sulfate precipitation (step 2). In this experiment 10 ml of enzyme preparation containing 300 mg of protein with a total DPN-dehydrogenating activity of 3660 units were placed on the column. The recovery of the activity for this experiment was 15%. Elution of the protein was carried out with solution A. Void volume was 520 ml ...., enzymatic activity (units/3 ml fraction); ———, absorbance at 280 mμ.

Step 4. DEAE-Cellulose Chromatography. Twenty grams of DEAE-cellulose (medium mesh, 0.9 meq/g capacity) is treated successively with 1 N NaOH, 1 N HCl, 1 N NaOH and washed ten times with 1 liter of glass-distilled water and three times with 1 liter of solution A. The DEAE-cellulose is packed to give a column of 2.6 × 26 cm under a pressure of 150–200 mm Hg and washed with 3 liters of solution A. The enzyme preparation from step 3 is added to the column and eluted with 1 liter of solution A, followed by 500 ml of 30% glycerol in 0.1 M phosphate buffer, pH 8.0. The active material is eluted by the latter medium. Fractions of 10–12 ml are collected at a rate of about 60 ml per hour. The tubes containing the enzymatic activity are pooled and dialyzed against several changes of solution A. The pooled samples are concentrated by placing the solution in a Visking casing. The casings are covered with granular sucrose and kept at 0–4° for 48 hours.

A summary of this purification procedure is given in the table. Approximately a hundredfold purification of the enzyme in the 59,000 g supernatant fraction of the homogenate is achieved by the prescribed procedure.[3] Preparations obtained by other procedures have the same degree of purity.[2,6] On subjecting the purified preparation to disc electrophoresis, one major band and three minor bands of protein are observed. The DPN- and TPN-linked dehydrogenase activities are localized only in the major band.[7]

PURIFICATION OF 3α-HYDROXYSTEROID DEHYDROGENASE FROM RAT LIVER

Fraction	DPN-dehydrogenase activities (units)	Specific activity (units/mg protein)			Yield (%)
		Dehydrogenase		Transhydro-genase	
		DPN	TPN		
59,000 g supernatant of rat liver homogenate	4090	3.4	7.3	0.4	100
(NH₄)₂SO₄ fractionation (0.5–0.7 saturation)	3660	12	41	0.6	90
Sephadex G-100 gel filtration	897	184	334	7.8	22
DEAE-cellulose chromatography	257	375	949	9.2	6.3

Distribution

The 3α-hydroxysteroid dehydrogenase activity has been demonstrated in the liver, kidney, testis, and ventral prostate of the rat, rabbit liver, and human prostate gland.[4,5,8] The activity is found in the supernatant and microsomal fractions of rat liver.[1,4,9,10]

Properties

Stability. Thirty percent glycerol prevents denaturation by heating to 57° for 10 minutes or by bubbling oxygen into the enzyme preparation.[11] Various sulfites also protect the enzyme from heat denaturation,

[6] R. Pietruszko and D. N. Baron, *Biochem. J.* **96**, 557 (1965).
[7] E. Doman and S. S. Koide, *Biochim. Biophys. Acta* **128**, 209 (1966).
[8] R. Pietruszko, M. B. R. Gore, and D. N. Baron, *Biochem. J.* **84**, 77P (1962).
[9] B. Hurlock and P. Talalay, *Arch. Biochem. Biophys.* **80**, 468 (1959).
[10] Y. Aoshima, C. D. Kochakian, and D. Jadrijevic, *Endocrinology* **74**, 521 (1964).
[11] S. S. Koide and M. T. Torres, *Biochim. Biophys. Acta* **89**, 150 (1964).

whereas pyridine nucleotides, steroids, and sulfhydryl-containing compounds do not.[11] The purified enzyme preparation is stable for over 6 months at −20° in a medium containing 30% glycerol. In variance with these results, Pietruszko and Baron[6] have reported that mercaptoethanol increases the stability of the enzyme and reactivates inactivated enzyme preparation. Any inhibitory influence of mercaptoethanol was attributed to the use of mercaptoethanol that was not freshly redistilled. Sodium thioglycolate, however, did not protect the enzyme from heat denaturation, whereas all sulfites were protective.[11] Further purification of the enzyme has been achieved with carboxymethyl cellulose and calcium phosphate gel column. In the hands of this author, however, the enzyme preparation was inactivated when eluted from carboxymethyl, phosphorylated, and sulfoethyl cellulose columns. In addition considerable loss in total and specific activities is observed on passage through a calcium phosphate gel column.

Pyridine Nucleotide Requirement. At their optimum pH, the rate of reduction of TPN is greater than of DPN.[4,5] The affinity of the enzyme for TPN and TPNH is greater than for DPN and DPNH.[4,5] The 3-acetylpyridine analog of DPN can act as hydrogen acceptor.[1] Because 3α-hydroxysteroid dehydrogenase can utilize both DPN and TPN as hydrogen acceptors, it was proposed that the transhydrogenase activity was dependent upon the alternate oxidation and reduction of the 3α-hydroxysteroids.[1]

$$3α\text{-Hydroxysteroid} + TPN^+ \rightleftharpoons 3\text{-ketosteroid} + TPNH + H^+$$
$$3\text{-Ketosteroid} + DPNH + H^+ \rightleftharpoons 3α\text{-hydroxysteroid} + DPN^+$$
$$\overline{TPN^+ + DPNH \rightleftharpoons TPNH + DPN^+}$$

Variations in the DPN^+- and TPN^+-linked dehydrogenase and transhydrogenase activities of different fractions in the hands of this author suggest, however, that these might be due to separate enzymatic systems.[3,6]

Effect of pH and Anions. The maximum rates of DPN- and TPN-linked dehydrogenation and transhydrogenation occur at approximately pH 8, 9.5, and 7.5, respectively. Sulfates and phosphates stimulate the DPN-linked dehydrogenase and transhydrogenase activities, but not the TPN-linked dehydrogenase activity.[12] Of the various buffers tested, the most suitable media for the assay of the activities are phosphate buffer for the DPN-linked dehydrogenase and transhydrogenase activities and glycine-NaOH buffer for TPN-linked dehydrogenase activity. An alternative assay depends on the oxidation of DPNH in the presence of 17α,21-

---

[12] D. N. Baron, M. B. R. Gore, R. Pietruszko, and D. C. Williams, *Biochem. J.* **88**, 19 (1963).

dihydroxy-5β-pregnane-3,11,20-trione (dihydrocortisone). The rate of oxidation of DPNH is maximal between pH 7 and 8 with Tris or phosphate buffer.[4,5]

Inhibitors and Metal Requirements. The enzyme is inhibited by Cu^{++}, iodoacetate, *p*-chloromecuribenzoate, urea, α,β-unsaturated 3-ketosteroids, quinones with short side chains, and purine and pyrimidine derivatives.[4,5,12,13] In general, the transhydrogenase activity is more markedly inhibited by these agents than the dehydrogenase activities. None of the metal ions tested stimulated the activity.[12]

Steroid Specificity. The 3α-hydroxysteroids and 3-ketosteroids with saturated ring A of the C_{18}, C_{19}, and C_{21} series are substrates for the enzyme.[1,4,5] Steroids with benzenoid ring A, C_{24}, and C_{27} steroids do not react. α,β-Unsaturated 3-ketosteroids, however, are inhibitors of the reaction. The 5β-series react more rapidly than the 5α-isomers.[1,4,5] The purified enzyme preparation is completely devoid of 3β-hydroxysteroid dehydrogenase activity.[14]

Kinetic Properties. The apparent Michaelis constant for androsterone is in the vicinity of $1.7 \times 10^{-6} M$. The K_m value for dihydrocortisone is $10^{-5} M$ and that for DPNH or TPNH is $1.5 \times 10^{-5} M$.[4,5] The value for DPN is 6.4×10^{-4} at pH 8.0 and for TPN is $1.3 \times 10^{-4} M$ at pH 9.6. The thermodynamic equilibrium constant is defined as

$$K_H = \frac{[\text{ketone}][\text{DPNH}][\text{H}^+]}{[\text{alcohol}][\text{DPN}^+]}$$

The calculated constant is 7.5 mμ M for DPN-linked dehydrogenation and 1.5 mμ M for TPN-linked dehydrogenation.[6] The standard free energy is defined as $\Delta F° = -RT \ln K_H$. The free energy for androsterone with DPN is +11 kcal/mole and that for androsterone and TPN is +12 kcal/mole at 25°.[6]

[13] P. M. Frearson and D. C. Williams, *Biochem. J.* **91**, 76 (1964).
[14] S. S. Koide, *Steroids* **6**, 123 (1965).

[33] The Δ⁴-5β-Steroid Dehydrogenase of *Pseudomonas testosteroni*

By SAMUEL JAMES DAVIDSON

5β-Androstane-3,17-dione + quinonoid cofactor →
         androst-4-ene-3,17-dione + reduced quinonoid cofactor

Numerous microorganisms are able to introduce unsaturation into carbon-carbon bonds within steroid A-rings.[1-3] Several workers have studied the properties of the enzymes involved.[4-6] *Pseudomonas testosteroni* posesses three enzymes of this type.[6] One introduces Δ⁴-double bonds into 5α-steroids, another does the same for 5β-steroids, and the third introduces Δ¹-double bonds into either type of steroid. Detailed studies of these enzymes were facilitated by the development of speedy assays.[6-8] Because it is a soluble enzyme unlike the others, the Δ⁴-5β-steroid dehydrogenase is especially suitable for detailed study.[8]

Assay Method

Principle. Δ⁴-5β-Steroid dehydrogenase transfers hydrogen from substrate to phenazine methosulfate (PMS), which can then reduce cytochrome c.[6,7] This reduction is measured as a change in cytochrome absorbance at 550 mμ.

The Blank Reaction. Solutions of PMS react with cytochrome c in the absence of steroid or enzyme, causing an additional increase in the cytochrome absorbance at 550 mμ, which must be taken into account in the assay.[8] The reaction is biphasic, showing an initial burst followed by a slower reaction linear with time. The rate of reaction is increased by room light and by high pH.

Unsaturated Steroid. In order to be able to measure the Δ⁴-5β-steroid dehydrogenase activity in the presence of Δ¹-steroid dehydrogenase, the assay substrate should be a Δ¹-5β-steroid. We use Δ¹-5β-androstene-3, 17-dione, synthesized as follows after a procedure furnished by Dr. D. H. Peterson of the Upjohn Company, Kalamazoo, Michigan.

Organic solvents used in the synthesis and analyses should be distilled.

Our hydrogenation apparatus consists of a 55-ml flask with side arm and ground-glass neck, provided with a three-way stopcock to permit connection either to a mercury manometer, or to an aspirator, or to the supply of hydrogen gas. The mercury reservoir of the manometer is adjustable in height, allowing some control of pressure. The flask contains a magnetic stirring bar.

A 700 mg amount of 3 or 5% Pd on $CaCO_3$ catalyst is introduced

[1] E. Vischer and A. Wettstein, *Experientia* 9, 371 (1953).
[2] J. Fried, R. W. Thoma, and A. Klingsberg, *J. Am. Chem. Soc.* 75, 5764 (1953).
[3] H. L. Herzog, C. C. Payne, M. T. Hughes, M. J. Gentles, E. B. Hershberg, A. Nobile, W. Charney, C. Federbush, D. Sutter, and P. L. Perlman, *Tetrahedron* 18, 581 (1962).
[4] C. J. Sih and R. E. Bennett, *Biochim. Biophys. Acta* 56, 584 (1962).
[5] H. J. Ringold, M. Hayano, and V. S. Stefanovic, *J. Biol. Chem.* 238, 1960 (1963).
[6] H. R. Levy and P. Talalay, *J. Biol. Chem.* 234, 2014 (1959).
[7] V. Massey, *Biochim. Biophys. Acta* 34, 255 (1959).
[8] S. J. Davidson and P. Talalay, *J. Biol. Chem.* 241, 906 (1966).

into the reaction flask together with 18 ml of pyridine. The flask should be evacuated and filled with hydrogen three times, and the magnetic stirring begun. Hydrogen is taken up in the absence of steroid for about 15 minutes; this rate is about 14–16 ml per hour.

The magnetic stirring is interrupted, and 200 mg of $\Delta^{1,4}$-androstadiene-3,17-dione in 1 ml of pyridine is added through the side arm, then rinsed in with another milliliter of pyridine. Air must be excluded from the flask. Stirring is started, and allowed to proceed until one equivalent of hydrogen is taken up. Hydrogen uptake by steroid is usually calculated as being additional to the rate of blank uptake; with this assumption, reaction takes about 15 minutes. Yields possibly may be improved by ignoring the blank uptake, to minimize the problem caused in purification by the presence of saturated steroid. The reaction is stopped by evacuating the flask and readmitting air.

The catalyst is separated immediately by filtration of the reaction mixture through a sintered-glass filter which is rinsed with pyridine. The pooled filtrates are evaporated under reduced pressure to a greenish sirup that is dissolved in ether and washed successively with $0.5\,M$ HCl, water, $0.5\,M$ NaOH, and water. The solution is dried with anhydrous sodium sulfate, filtered, and evaporated. As a preliminary check on the course of the reaction, a small portion of the resulting sirup may be dissolved in methanol and chromatographed on paper, with standards, as described below. The developed chromatogram includes one spot for isomers of Δ^1-androstene-3,17-dione, as well as spots for isomers of Δ^4-androstene-3,17-dione and for the $\Delta^{1,4}$ starting material. Saturated steroid probably also is present.

For isolation of the desired product, the residue is dissolved in a mixture of methylene chloride-n-hexane (3:17) and applied to a Florisil column containing about 20 g of adsorbant. A height to width ratio of about 12:1 is satisfactory. Acetone–n-hexane (3:97) is used for development, and 10 ml fractions are collected and evaporated. The first material to be eluted is saturated steroid in fractions between about 250 and 380 ml total eluate volume. Δ^1-5β-Androstene-3,17-dione partly overlaps, appearing between about 300 and 500 ml eluate volume. After elution of Δ^1-steroid, about 300 ml of eluate is collected before appearance of Δ^4-product. Paper chromatography may be used to select those fractions containing Δ^1-steroid and little or no saturated material. Poor separation is improved by a second, identical column step, applying only the material from the best fractions. Finally, the selected fractions should be pooled and recrystallized from hexane-acetone until a melting point of 170–171.5° is obtained. The 5α-isomer melts at 140–141.5°, and the addition of a small amount of the 5α-isomer to the 5β-isomer causes a marked melting point depression. Both substances absorb maximally at 230 mμ in ethanol, with $\epsilon = 10,000$ liter mole^{-1} cm^{-1}.

Descending partition chromatography on paper, after Zaffaroni[9] is performed as follows. Whatman No. 43 paper (15 $\times$ 52 cm) is dipped into a methanol-formamide mixture (1:1) and blotted exhaustively between layers of heavy filter paper, using a hand-held rubber roller. The paper is then dried at room temperature for 2–3 hours; preferably in an atmosphere of low humidity. The steroid mixtures are applied in small volumes of ethanol or methanol, and the papers, with samples applied, are equilibrated with solvent vapor in the tanks for 30–60 minutes before the developing solvent is added. Development requires 2–4 hours.

Unsaturated steroids can be detected by fluorescence quenching. Both saturated and unsaturated ketosteroids may be detected by spraying either with 2,4-dinitrophenylhydrazine reagent[10] or with *m*-dinitrobenzene[11,12] by the Zimmerman method. The latter reagent is made up just before use by mixing equal volumes of a 2% solution of pure *m*-dinitrobenzene[12] and 30% (w/v) KOH, each in ethanol.

The chromatographic system cannot separate 5α- from 5β-isomers, but it clearly separates the androstane-3,17-diones, the Δ^1-, the Δ^4-, and $\Delta^{1,4}$-compounds from each other.

Saturated Substrate. In enzyme preparations lacking Δ^1-dehydrogenase activity, it is more convenient to assay for Δ^4-5β enzyme using 5β-androstane-3,17-dione, which is commercially available, as substrate.[13] This steroid is recrystallized from hexane, or from hexane-acetone mixtures.

Phenazine Methosulfate. Stock aqueous PMS solution is stored in the dark, frozen. Dilute solutions may be stored at 4°, in the dark. Solutions containing PMS should be manipulated in subdued light, and removed from light as soon as possible.

Cytochrome c. Stock solutions of commercial cytochrome *c* can be stored frozen, thawed for use, and then refrozen.

Assay System. The enzyme is assayed in a final volume of 3 ml, containing 200 micromoles of Tris-chloride at pH 8.0, 1.24 micromoles of PMS, 0.075 micromole of 5β-androstane-3,17-dione or 0.1 micromoles of 5β-Δ^1-androstenedione in 0.01 ml of 95% ethanol, 2 mg of cytochrome *c*, and enzyme. A control incubation is set up containing all the above com-

[9] A. Zaffaroni, R. B. Burton, and E. H. Keutmann, *Science* 111, 6 (1950).
[10] I. E. Bush, "The Chromatography of Steroids," p. 375. Macmillan (Pergamon), New York, 1961.
[11] I. E. Bush, "The Chromatography of Steroids," p. 377. Macmillan (Pergamon), New York, 1961.
[12] N. H. Callow, R. K. Callow, and C. W. Emmens, *Biochem. J.* 32, 1312 (1938).
[13] See Vol. V [70b].

ponents except steroid. Both the complete system and the control are read against a blank containing only buffer and cytochrome c. The enzymatic reaction rate is the difference in rates between the complete system and the control. The absorbancy in the control increases about 0.007 unit per minute if the enzyme is somewhat purified, but this rate may be about twice as high for crude preparations.

In performing the assay, buffer and cytochrome are introduced into the cuvette. Then PMS is added in subdued light, using a fresh pipette for each addition because traces of photodecomposed PMS added to the assay mixture affect the blank reaction (they increase the extent of the initial burst). The enzyme is added after PMS because enzyme in dilute solution is stabilized by PMS. The reaction is initiated by addition of steroid to the complete system and ethanol to the control. Time-saving simplifications of the assay, for less precise work, have been described.[8]

In this article, the amount of enzyme reducing 1 micromole of 5β-androstane-3,17-dione per minute, in the standard assay, will be described as one unit of enzyme. One unit of enzyme reduces 0.7 micromoles of Δ^1-5β-androstene-3,17-dione per minute, under these conditions.

On a molar basis, the amount of steroid oxidized is half the cytochrome c reduced.[3] For cytochrome c, $a_m^{red} - a_m^{ox}$ is taken as $19,100\ M^{-1}$ cm^{-1} at 550 mμ.[14] Somewhat different values have been reported.[15]

Assay of Δ^1-Steroid Dehydrogenase. It is desirable to test for the presence of this enzyme in the course of purification of the Δ^4-5β-steroid dehydrogenase. The assay is performed exactly as above, except that the substrate is 0.1 micromole of Δ^4-androstene-3,17-dione.[8, 13]

Purification Procedure

Growth of Bacteria. Composition of the medium and technique of culture for *Pseudomonas testosteroni* are described in an earlier volume.[16] Cultures on a smaller scale may be grown by appropriate modifications of the procedure.

A much larger-scale preparation of *P. testosteroni* has also been described.[17] (See this volume, p. 643.)

The enzyme purification scheme described here was developed for acetone powders of *P. testosteroni*. The preparation of bacterial acetone powders is described in these volumes and elsewhere.[17-19]

---

[14] W. W. Umbreit, R. H. Burris, and J. F. Stauffer, "Manometric Techniques," p. 301. Burgess, Minneapolis, Minnesota, 1957.

[15] See Vol. II [122, 133].

[16] See Vol. V [70a].

[17] J. Boyer, D. N. Baron, and P. Talalay, *Biochemistry* 4, 1825 (1965).

[18] See Vol. I [7], p. 55.

[19] F. S. Kawahara, S.-F. Wang, and P. Talalay, *J. Biol. Chem.* 237, 1500 (1962).

Δ⁴-5β-Steroid dehydrogenase may also be liberated from whole cells by sonication. Fifteen minutes of sonication is satisfactory, with 25-ml volumes of suspension, in a 20-kc 60-watt, MSE-Mullard instrument. The preparation should be on ice during sonication. The copious precipitable material is separated by two successive centrifugations for 30 minutes, at 20,000 g, in the cold. The precipitates are discarded.

Fractionation. All steps are performed at ice temperatures. The specified pH's of buffers with glycerol are those measured after the addition of glycerol. Protein concentrations have been measured either by a biuret method, or from ultraviolet absorbance ratios.[20]

TABLE I

PURIFICATION OF Δ⁴-5β-STEROID DEHYDROGENASE OF *Pseudomonas testosteroni*

| | | Δ⁴-5β-Steroid dehydrogenase | | | |
Fraction	Volume (ml)	Total activity (units)	Specific activity (units/mg)	Recovery overall (%)	Total protein (mg)
1. Initial extract	75	128,000	63.3	100	2020
2. Protamine sulfate supernatant	112	134,000	—	105	—
3. Ammonium sulfate precipitate, 45%–60% saturation	4.6	98,200	255	77	386
4. Dialyzed supernatant	5.0	105,000	312	82	337
5. Pooled fractions from chromatography	110	45,000	2560	35	17.6
6. Eluates from calcium phosphate gel					
A. Pilot scale	1.05	2,200	5000	1.7	0.44
B. Large scale	6.9	18,500	6600	14.5	2.8

The course of our purification is summarized in Table I. Because of the absence of Δ¹-dehydrogenase activity in the crude extract, 5β-androstanedione was the substrate for assay of the Δ⁴-5β enzyme throughout the tabulated preparation. The absence of Δ¹ activity in acetone powder extracts is unusual and may reflect the age of the acetone powder used. Δ¹ Activity will commonly be present and testing should be continued until its absence can be demonstrated.

Step 1. Extraction. Acetone powder, 10 g, is added slowly and steadily to about 50 ml of 0.05 M potassium phosphate buffer, pH 7.3, with vigorous stirring. The powder can conveniently be added through a funnel, the stem of which contains a freely rotating, motor-driven glass rod.

[20] See Vol. III [73].

This method prevents the formation of large gummy particles. After all the powder is added, the viscous suspension is aspirated a few times with a large pipette to destroy any aggregates.

Rapid stirring is continued for about an hour after addition of the powder. The suspension is centrifuged for 15 minutes at 20,000 g and the supernatant is collected and recentrifuged in the same way. The combined pellets are reextracted with another 25 ml buffer and this suspension is centrifuged as before. The supernatants from this and the first extraction are combined. The precipitates are discarded.

Step 2. Treatment with Protamine Sulfate. Immediately after the preceding step a 1% aqueous solution of protamine sulfate is added dropwise with stirring to the crude extract, at the rate of 1 mg of protamine sulfate for each 5 mg of protein in the crude extract. Stirring is continued for 20 minutes more. The bulk of the aggregated, sticky white precipitate can be removed with a stirring rod or glass spoon, and the remainder is separated by centrifugation for 15 minutes at 20,000 g. The precipitate is discarded, leaving a clear yellow supernatant. This step, in conjunction with the next one, causes a 2-fold augmentation of the 280 mμ:260 mμ absorbance ratio.

Step 3. Ammonium Sulfate Precipitation. Immediately after removal of the protamine sulfate precipitate, a solution of ammonium sulfate, saturated at 2–5°, and neutralized with NH_4OH, is added dropwise, with stirring, to the enzyme preparation to a final concentration of 45% saturation. The solution is allowed to stand for 45 minutes after the addition of salt, and the precipitate is separated by centrifugation at 20,000 g for 15 minutes. The supernatant is collected and brought to 60% saturation by further additions of ammonium sulfate solution. After the preparation has stood for 45 minutes, the precipitate, containing the bulk of the activity, is collected by centrifugation at 20,000 g for 15 minutes. For transfer to dialysis tubing, the precipitate should be suspended in saturated and neutralized ammonium sulfate solution. If this precaution is not observed, losses of activity may result.

Step 4. Dialysis and Centrifugation. The suspension of the 45–60% ammonium sulfate precipitate is dialyzed against two successive portions, 100 volumes each, of 0.001 M potassium phosphate buffer in 20% glycerol, pH 7.3. It is stored a few hours; if a precipitate forms, this is separated by centrifugation and discarded.

Step 5. Chromatography on DEAE-Cellulose. Bio-Rad Cellex D high-capacity DEAE-cellulose is prepared by successive rinses with 1 N NaOH, 1 N HCl, H_2O, 1 N NaOH, and water again.[21, 22] The cellulose is suspended in several volumes of wash liquid; rinsing may be by decanta-

[21] See Vol. V [1].
[22] E. A. Peterson and H. A. Sober, *J. Am. Chem. Soc.* **78**, 751 (1956).

tion or filtration. A yellow color may appear in the first rinse; all colored material should be eluted before proceding to the equilibration step.

For equilibration, the washed cellulose is suspended in 0.001 M potassium phosphate, 20% glycerol, pH 7.3, and the pH readjusted to pH 7.3 by addition of concentrated phosphoric acid or potassium hydroxide solution. The cellulose is then washed four times in 0.001 M potassium phosphate in 20% glycerol, pH 7.3. The column is packed under gravity flow.

A preparation of the size described in Table I requires a column approximately 2.6 × 24 cm, and a mixing chamber volume of about 1070 ml. Flow rate is about 130 ml/hour.

The dialyzed enzyme solution is applied to the column and rinsed in with 0.001 M potassium phosphate, 20% glycerol, pH 7.3 until nonadsorbed protein has passed through the column and the effluent is essentially free of protein. Gradient elution is then begun. A gently convex gradient is satisfactory; obtained by using a large mixing chamber, which will initially contain the same buffer used for rinsing. The upper chamber contains 0.038 M potassium phosphate buffer in 20% glycerol, pH 7.3. The effluent fractions containing enzyme of high specific activity are pooled. With the column described above, these appear between 425 ml and 530 ml after the gradient is begun and are pale yellow.

The steroid dehydrogenase activity is not adsorbed onto DEAE-cellulose at room temperature. Thus, although no loss of activity occurs in a few hours at room temperature, in glycerol solution, chromatography must be performed in the cold.

The pooled column eluates are dialyzed against 15 volumes of 0.001 M potassium phosphate in 20% glycerol, pH 7.2, and the dialysis is repeated with another fifteen volumes of buffer.

Step 6. Adsorption on Calcium Phosphate Gel. This procedure serves the purposes of purification and concentration. Calcium phosphate gel is prepared according to the method of Tsuboi and Hudson[23, 24] and is suspended in 20% glycerol.

Adsorption by gel should first be performed on a small aliquot of the enzyme preparation, adding small increments of gel to determine the best proportions of gel and enzyme to use with the preparations at hand. After the bulk of the enzyme is adsorbed, further additions of gel take up proportionately less enzyme, and these adsorbates are discarded.

The following is a description of a larger-scale gel purification step,

[23] To prepare gel, add 6 ml of concentrated NH$_4$OH to 200 ml of 0.5 M Na$_2$HPO$_4$; immediately add 1500 ml of 0.1 M CaCl$_2$. When pH falls to 6 (about 30 minutes), wash with deionized water, by centrifugation. Cease washing when supernatants are free of chloride ion.

[24] K. K. Tsuboi and P. B. Hudson, *J. Biol. Chem.* **224**, 879 (1957).

performed following a pilot experiment on a small aliquot. The gel contained 30 mg dry weight per milliliter in 20% glycerol, and the enzyme preparation contained 40,000 units in a 102-ml volume. Gel suspension containing about 20 mg of gel was added to the enzyme solution, and the mixture was stirred for 5 minutes. After centrifugation at 500 g for 10 minutes, the gel, which had adsorbed about 8% of the enzymatic activity, was discarded. Addition of 250 mg (8.3 ml) more of gel to the supernatant, and another 5 minutes of stirring, left only 12% of the initial activity in the supernatant. No further additions of gel were profitable after this stage.

The gel with adsorbed enzyme is washed several times in the dilute phosphate buffer used for dialysis, until little material with absorbance at 280 mμ is removed. Elution is performed with successively smaller volumes of 0.1 M potassium phosphate buffer in 20% glycerol, pH 7.3.

Properties of Δ^4-5β-Steroid Dehydrogenase[8]

Evidence Concerning a Possible Prosthetic Group. When partially purified preparations of this enzyme are precipitated from solution by acid ammonium sulfate almost the entire dehydrogenase activity is lost.[25] A substantial portion of the activity may be restored by the addition of very low concentrations of purified flavin mononucleotide. Flavin adenine dinucleotide, at similar or much higher concentrations, restores no activity. The Michaelis constant for FMN is $3.3 \times 10^{-7}\,M$.

Specificity for Hydrogen Acceptors. A number of artificial and biologically occurring quinones can act as hydrogen acceptors in place of PMS in the standard assay system. The most active acceptors seen under the usual assay conditions were 1,4-naphthoquinone and 1,2-naphthoquinone; each is about 50% again as active as PMS. 1,4-Benzoquinone possesses 58% of the activity of PMS. Within the series of 1,4-naphthoquinones, increases in the number and size of substituents at the 2 and 3 positions decrease activity, in an order roughly correlated with the redox potential. The most active "natural" acceptor tested was coenzyme Q_{10}; a 2,3-methoxy-5-methyl-1,4-benzoquinone, substituted in the 6 position with a chain consisting of ten isoprenoid units. Under the assay conditions, it is one-quarter as active as PMS. Menadione (2-methyl-1,4-naphthoquinone), with 29% of the activity of PMS, shows proportionately much lower blank reaction in the assay and might be a convenient substrate for assay purposes. Among nonquinone hydrogen acceptors, potassium ferricyanide, 2,6-dichlorophenol indophenol, NAD+, NADP+, and the acetylpyridine analog of NAD+ were tested. All are inactive with this enzyme.

[25] O. Warburg and W. Christian, *Biochem. Z.* **298**, 368 (1938).

Substrate Specificity and Substrate Inhibition. The 5β-H configuration and a ketone function at C-3 are essential requirements for substrates of the Δ⁴-5β-dehydrogenase reaction.[8] Various substituents at C-17 may enhance or diminish, but never eliminate, susceptibility to enzymatic dehydrogenation (see Table II).

TABLE II
STEROID SPECIFICITY OF STEROID Δ⁴-5β-DEHYDROGENASE[a]

Substrates	Relative rates of oxidation
5β-Androstane-3,17-dione	100
5α-Androstane-3,17-dione	0
Δ¹-5β-Androstene-3,17-dione[b]	70
Δ¹-Androstene-3,17-dione	0
5β-Androstan-17β-ol-3-one	60
5β-Androstan-3α-ol-17-one	0
5β-Androstan-3β-ol-17-one	0
Androsterone (5α-androstan-3α-ol-17-one)	0
5β-Pregnane-3,20-dione	41
5β-Pregnan-21-ol-3,20-dione	94
Δ⁴-Androstene-3,17-dione	0
Δ¹,⁴-Androstadiene-3,17-dione	0

[a] Assays were carried out at 25° in systems of 3.0 ml final volume containing: 200 micromoles of Tris-chloride buffer, pH 8.0, 0.075 micromole[b] of steroid in 0.01 ml 95% ethanol, 1.24 micromoles of PMS, 2 mg of cytochrome c, and 0.01 ml of enzyme, purified through step 5 (11,000 units/ml; specific activity 1600 units per milligram of protein). Readings were taken at 550 mμ at 15- to 30-second intervals for at least 8 minutes, against a blank containing all components except the steroid.
[b] Twelve hundredths micromole of Δ¹-5β-androstene-3,17-dione was used in the assay with this substrate.

As substrate concentration increases above an optimum value which depends on the steroid, inhibition is seen.

pH Effect on Reaction Rate. With our assay, the rate of enzymatic dehydrogenation increases monotonically and sharply between pH 6.0 and 9.0.[8] Above this level, it was not possible to make measurements, because of the increasing blank reaction.

Steroid and Other Inhibitors. A variety of C_{19} nonsubstrate steroids inhibit at a concentration of $10^{-4} M$ (see Table III).[8] A 3β-hydroxyl group was the only feature found to prevent inhibition; 5β-androstan-3β-ol-17-one does not inhibit, while the 3α-hydroxy isomer causes a 68% inhibition. The 3α-hydroxy, 5α-isomer is less effective, inhibiting 29%.

None of the usual metal-chelating agents is inhibitory.[8] "Sulfhydryl reagents" are strongly inhibitory, although o-iodosobenzoate has no

TABLE III
INHIBITION OF Δ^4-5β-STEROID DEHYDROGENASE REACTIONS BY STEROIDS[a]

Steroids	Inhibition (%)
5β-Androstan-3α-ol-17-one	68
5β-Androstan-3β-ol-17-one	0
5α-Androstan-3α-ol-17-one (androsterone)	29
Δ^4-Androstene-3,17-dione	31
$\Delta^{1,4}$-Androstadiene-3,17-dione	22
Δ^1-5α-Androstene-3,17-dione	19
5α-Androstane-3,17-dione	23

[a] Assays were performed in total volumes of 3 ml at 25°. The complete system contained: 200 micromoles of Tris-chloride of pH 8.0; 1.24 micromoles of PMS; 0.075 micromole of 5β-androstane-3,17-dione; 0.3 micromole of inhibitor; 2 mg of cytochrome c; 0.05 ml of 95% EtOH; and 0.01 ml of enzyme, purified through step 5 of the purification procedure. Readings were made at 550 mμ, against a blank vessel containing all components except steroid and inhibitor.

effect. HgCl$_2$ inhibited completely at $10^{-3}\,M$; Cu^{++} and Zn^{++} inhibit significantly at this concentration.

Stabilization by Glycerol. The partly purified steroid dehydrogenase is most stable if dissolved in 0.05 M potassium phosphate, 20% glycerol, pH 7.3.[8] In such solutions, the half-life of the activity is about 1 month. Stability in these solutions is not improved by freezing. If the enzyme purification has been stopped after the ammonium sulfate precipitation step, the half-life in 0.05 M potassium phosphate, 20% glycerol, pH 7.3 is only about a week at 4°, but then the half-life becomes 3 days if the glycerol is omitted or 1 day if, in addition, the pH is lowered to 6.6. If such an enzyme preparation is stored in 0.05 M Tris-Cl, 20% glycerol, pH 8, the half-life is also 3 days at 4°.

It is remarkable that, when glycerol is first added, enzyme activity is progressively and markedly enhanced over a period of several days.

Stabilization by PMS. Δ^4-5β-Steroid-dehydrogenase is quite unstable in catalytic concentrations, except in the presence of PMS in the concentration used for assay.[8] This is why PMS is added to the assay system before enzyme.

Solubility. Since the Δ^4-5β-steroid dehydrogenase is not sedimented by prolonged centrifugation of sonicates or acetone powder extracts at 105,000 g, it is in the soluble fraction of the cell—or at least, more loosely bound than the other Δ-dehydrogenases of *P. testosteroni*.[6]

[34] 21-Hydroxysteroid Dehydrogenases of Liver and Adrenal

By Carl Monder and Charles S. Furfine

21-Hydroxycorticosteroid $+$ NAD$^+$(NADP$^+$) $\rightleftharpoons$
$$21\text{-dehydrocorticosteroid} + NADH(NADPH) + H^+$$

Pyridine nucleotide-dependent 21-hydroxysteroid dehydrogenase reducing activity has been described in the tissues of the rat,[1,2] steer,[3] calf,[3] pig,[3] sheep,[3,4] and man.[5] In the rat, activity has been detected in the heart, liver, adrenal, kidney, and spleen. In the other species mentioned activity has been primarily detected in the liver. Lamb liver contains three 21-hydroxysteroid dehydrogenases. Two are NAD-dependent, and one is NADP-dependent. An NAD-requiring activity has been isolated from the bovine adrenal gland.[6] Though a greater degree of purification has been attained with the bovine adrenal enzyme, the initial and final specific activities of the major NAD-dependent lamb liver enzyme was higher than that of the adrenal enzyme. The physiological role of these enzymes is as yet unknown.

Assay Method

Principle. The assay of 21-hydroxysteroid dehydrogenases is based on the enzymatic reduction of 21-dehydrocortisol by the appropriate reduced pyridine nucleotide. The rate of decrease in absorbance at 340 mμ is measured spectrophotometrically.

Reagents

Sodium phosphate, 0.1 M, pH 6.9
21-Dehydrocortisol, 0.011 M, in 50% aqueous propylene glycol
NADH or NADPH, 0.005 M, in 0.01 M sodium phosphate pH 8.0
Enzyme, in phosphate buffer diluted if necessary

Procedure. The components are added as follows to a cuvette having a 1-cm light path: 2.7 ml of sodium phosphate buffer, 0.1 ml of 21-dehydrocortisol solution, 0.1 ml of enzyme. After 3 minutes of incubation at

[1] C. Monder and A. White, *Biochim. Biophys. Acta* **46**, 410 (1961).
[2] J. J. Schneider, *J. Am. Chem. Soc.* **75**, 2024 (1953).
[3] C. Monder and A. White, *J. Biol. Chem.* **238**, 767 (1963).
[4] C. Monder and A. White, *J. Biol. Chem.* **240**, 71 (1965).
[5] C. Monder, unpublished observations.
[6] C. Furfine and A. White, *J. Biol. Chem.* **243**, 1190 (1968).

30°, the reaction is initiated by addition of 0.1 ml of reduced pyridine nucleotide. Change in absorbance is followed at 340 mμ using an automatic recording spectrophotometer thermostatically controlled at 30°. Blank correction is made using a system containing 50% aqueous propylene glycol in place of the solution of the steroid. Specific activities are expressed as micromoles of NADH oxidized per minute per milligram of protein. Protein concentrations are estimated by the method of Kalckar[7] or by the biuret procedure.[8]

Preparation of 21-Dehydrocortisol.[9] To 2 g of cortisol, dissolved in 250 ml of methanol in a 1-liter round-bottom flask, is added a solution of 500 mg of cupric acetate in 250 ml of methanol. The mixture is aerated with bubbling for 2 hours. One gram of disodium EDTA suspended in 100 ml of 1% sodium carbonate and adjusted to pH 9, is poured into the steroid solution. Methanol is removed *in vacuo* at a temperature below 45°, and the concentrated solution is diluted to 1 liter with water. Steroid is extracted into ethyl acetate. The organic layer is washed with 2% sodium bicarbonate and with water, then evaporated to dryness *in vacuo*. The residue is dissolved in a minimum of acetone and poured into 500 ml of 0.01 M sodium phosphate, pH 7.5. Needles of 21-dehydrocortisol crystallize out overnight at 4°. Water of hydration may be removed from the product by prolonged drying under vacuum over phosphorus pentoxide. Yield varies from 1.4 to 1.7 g.

λ_{max} (in ethanol), 240 mμ; $\epsilon = 15,200$; m.p. (uncorrected) of steroid hydrate, 175–180° (dec.); of 2-quinoxaline[10] derivative, 265°. Sulfuric acid chromogen:[11] Maxima at 470 (shoulder), 400, 325, 285 mμ; minima at 355, 237 mμ. The preparation may be checked for unconverted cortisol by thin-layer chromatography using the solvent systems[12] chloroform–acetone–acetic acid 10:10:1 (v/v) and chloroform–ethanol 25:1 (v/v).

NAD+-Dependent 21-Hydroxysteroid Dehydrogenase from Lamb Liver

Purification Procedure

The procedure is a modification of one described earlier[3] and results in substantially higher specific activity.

[7] See Vol. III, p. 454.

[8] See Vol. III, p. 450.

[9] This procedure is a modification of M. L. Lewbart, and V. R. Mattox, *J. Org. Chem.* **28**, 2001 (1963). The systematic designation for 21-dehydrocortisol is 3,20-dioxo-11β,17α-dihydroxypregn-4-en-21-al. It has also been called cortisol-21-aldehyde and cortisol glyoxal.

[10] W. J. Leanza, J. P. Conbere, E. F. Rogers, and K. Pfister, *J. Am. Chem. Soc.* **76**, 1691 (1954).

[11] A. Zaffaroni, *J. Am. Chem. Soc.* **72**, 3828 (1950).

[12] C. Monder, *Biochem. J.* **90**, 522 (1964).

Step 1. Preparation of Acetone Powder. Sheep livers are homogenized in a blendor for 2 minutes with 4 volumes of acetone previously cooled to −20°, then collected on a Büchner funnel with suction. The "cake" is treated with a second portion of chilled acetone, powdered, and allowed to dry overnight in air at room temperature.

Step 2. Extraction of Acetone Powder. Fifteen grams of acetone powder is suspended in 150 ml of 0.1 M sodium phosphate buffer, pH 7.0, and stirred at room temperature (24–27°) for 60 minutes. The suspension is cooled to 4° and centrifuged for 15 minutes at 10,000 g.

Step 3. Heat Treatment. The extract (110 ml) is heated at 55° with efficient stirring for 4 minutes and rapidly cooled to 4°. The precipitate of denatured protein is removed by centrifugation at 10,000 g for 15 minutes.

Step 4. Ammonium Sulfate Fractionation. To 90 ml of the supernatant solution from step 4 cooled to 0° is added solid ammonium sulfate to 60% saturation (2.4 M). The salt is added over a period of 5 minutes with constant stirring. Fifteen minutes later, the precipitate is removed by centrifugation and reconstituted to a volume of 22 ml with 0.02 M sodium phosphate buffer, pH 7.0.

Step 5. Gel Filtration. The solution from step 4 is passed in two 10-ml portions through a column of Sephadex G-100, 90 cm high and 5 cm in diameter previously equilibrated with 0.02 M sodium phosphate, pH 7.0. Elution is performed with 0.02 M sodium phosphate, pH 7.0. Fractions (5 ml) are collected at a rate of 8 ml per hour, and those obtained between 60 and 130 ml containing the highest specific activities are combined. Although purification during this step is not great, it serves to separate the major NAD-dependent enzyme activity from the minor NAD-dependent and the NADP-dependent activities. At this stage, the protein solution is free of ammonium sulfate and is in 0.02 M sodium phosphate, pH 7.0.

Step 6. Adsorption and Elution from Alumina C_γ Gel. The combined column effluent from step 5 is adjusted to pH 5.8 with 2 M hydrochloric acid. One milligram of alumina C_γ (calculated on a dry weight basis) is added for every milligram of protein. Stir gently for 10 minutes and centrifuge. The supernatant fluid is discarded. The gel is washed three times with 10-ml portions of 0.05 M sodium phosphate, pH 7.0, followed by 0.1 M sodium phosphate, pH 7.0. Enzyme is eluted from the gel with 10-ml portions of 0.1 M sodium phosphate containing 10% ammonium sulfate. With some preparations of gel, elution of enzyme occurs with 0.1 M sodium phosphate and no added ammonium sulfate.

Step 7. Column Chromatography on DEAE-Cellulose. A column of microgranular DEAE-cellulose (Whatman DE-52, dimensions 45 cm high, 2 cm diameter) is equilibrated with 0.005 M sodium phosphate, pH

7.4. Ten milliliters of enzyme solution containing 12 mg of protein is placed on the column and elution is performed by increasing the molarity of phosphate buffer in a linear manner. This is achieved by adding 1000 ml of 0.3 M phosphate, pH 7.6, at a controlled rate to 1000 ml of 0.005 M phosphate, pH 7.4 and passing the resulting mixture through the column. The enzyme is usually eluted within the second void volume.

Step 8. Second Adsorption and Elution from Alumina C_γ. The combined effluents from step 7 are treated with alumina C_γ in a manner similar to that described in step 6. Elution of enzyme from the gel is performed using 2-ml portions of 0.1 M sodium phosphate containing 10% ammonium sulfate. A summary of the purification procedure is given in Table I.

TABLE I

PURIFICATION OF NAD-DEPENDENT 21-HYDROXYSTEROID
DEHYDROGENASE FROM LAMB LIVER

Step	Volume (ml)	Total protein (mg)	Total activity	Specific activity (μmoles/min/ mg protein)	Yield
Acetone powder extract	110	2057	235	0.114	100
Supernatant fluid from heat step	90	900	218	0.242	92.7
Ammonium sulfate fractionation	20	530	175	0.330	74.5
Sephadex G-100 effluent	17.3	125	49.5	0.396	21.1
Eluate from alumina C_γ adsorption	26	28.5	18.4	0.645	7.8
Effluent from DEAE- cellulose	10.5	2.53	5.24	2.07	2.2
Second alumina C_γ step	12.7	0.79	2.55	3.23	1.1

Properties

Stability. The dehydrogenase, prepared as described, and kept at —15° loses about half its activity in 3 months. Some less purified preparations (through step 5) have been kept in the freezer for over one year with no appreciable loss in potency.

Substrate Specificity. The enzyme reduced 21-dehydro derivatives of the following steroids: cortisol, cortisone, corticosterone, 11-deoxycortisol, Δ^1-cortisone, Δ^1-cortisol, 11-deoxycorticosterone, and 9α-fluorocortisol. Other carbonyl compounds, including acetaldehyde, DL-glyceraldehyde, *p*-tolualdehyde, *o*-anisaldehyde, salicylaldehyde, benzaldehyde, methylglyoxal, and glyoxal were not reduced by the enzyme. Steroids which were found not to be substrates include cortisol, cortisone, 11-deoxy-corticosterone, aldosterone, isoandrosterone, testosterone, and estrone. No activity was obtained with NADPH.

Inhibitors. The enzyme was inhibited in a noncompetitive manner by *p*-chloromercuribenzoate, acrolein, silver, and mercuric ions. Competitive inhibition was observed with a number of steroids, including androsterone, 17β-estradiol, estrone, testosterone, 11-deoxycorticosterone, and cortisol. Cortisol-21-acetate, dihydroxyacetone, methylglyoxal, and glyoxal were not inhibitors.

Effect of pH. The dehydrogenase has a pH optimum of 6.9 in 0.1 M sodium phosphate buffer and an optimum of 6.2 in 0.1 M Tris-maleate buffer. Irreversible inactivation rapidly occurs above pH 10 and below pH 4.

Reversibility. The equilibrium constant of the reaction is of the order of magnitude of 10^{14} and strongly favors the reduction of 21-dehydrocortisol.

Kinetic Constants. The K_m values with respect to NADH depends on pH of measurement. At pH 6.3, a value of $1 \times 10^{-5} M$, and at pH 7.2, a value of $0.23 \times 10^{-5} M$ were obtained. The K_m value for 21-dehydrocortisol was $4 \times 10^{-4} M$ at pH 7.2. Values on the direction of oxidation could not be obtained.

NADPH-Dependent 21-Hydroxysteroid Dehydrogenase from Lamb Liver[4]

Purification Procedure

Step 1. Extraction of Acetone Powder. Lamb liver acetone powder (20 g), prepared as described in the preceding section, is stirred with 10 volumes of 0.1 M sodium phosphate buffer, pH 7.4, for 1 hour at room temperature. The suspension is centrifuged at 1000 g for 15 minutes at 3°, and the supernatant fluid is decanted off.

Step 2. Ammonium Sulfate Precipitation. To the supernatant fluid of step 1 is added ammonium sulfate to 50% saturation (2.0 M). The precipitate which forms is centrifuged down after 15 minutes, and redissolved in 10 ml 0.02 M sodium phosphate, pH 7.4.

Step 3. Gel Filtration. The solution from step 2 is transferred to a column of Sephadex G-100, 90 cm high and 5 cm in diameter. Void volume is 300–350 ml. Filtration is performed at 4° with 0.02 M sodium phosphate buffer, pH 7.4 as eluent. Five-milliliter fractions are collected. Each tube is assayed for activity toward NADH and NADPH. Activity specific for NADH is found during the first 150 ml of filtration. Subsequently, NADPH-dependent activity emerges beginning at approximately 250 ml. This fraction also contains an NADH-dependent enzyme. Separation of NADH activity from the NADPH-dependent enzyme can be achieved at this stage by starch block electrophoresis, but considerable losses in activity are encountered. A summary of the purification of the NADPH-dependent enzyme is presented in Table II.

TABLE II
PURIFICATION OF NADPH-DEPENDENT 21-HYDROXYSTEROID
DEHYDROGENASE FROM LAMB LIVER

Step	Volume (ml)	Total protein (mg)	Total activity (μmole/min)	Specific activity (μmole/min/ mg protein)
Acetone powder extract	210	5250	0.83	0.00016
50% ammonium sulfate precipitate	10	1000	0.48	0.00048
Sephadex G-100 effluent	170	53	0.27	0.0051

Properties

Heat Stability. The NADPH-dependent enzyme is 80% inactivated after heating in a water bath for 5 minutes at 56°.

pH Optimum. The enzyme shows a pH optimum in the range 5.9–6.1 in sodium phosphate, Tris-maleate, or imidazole buffers. The relative activities of the enzyme in these buffers is phosphate > imidazole > Tris-maleate.

Substrate Specificity. All 21-dehydrosteroids are reduced by the NADPH-dependent dehydrogenase. Glyoxal, methylglyoxal, and benzaldehyde are also reduced. However, acetaldehyde, pyruvic acid, cortisone, cortisol, 11-deoxycorticosterone, aldosterone, epiandrosterone, estrone, and testosterone are not substrates.

Inhibitors. The enzyme is irreversibly inhibited by *p*-chloromercuribenzoate, mercuric, and silver ions. Other reagents capable of reacting with sulfhydryl groups are ineffective. No inhibition of activity occurs with cortisol, 11-deoxycorticosterone, corticosterone, androsterone, testosterone, or 17β-estradiol. Reduction of 21-dehydrocortisol is inhibited in a competitive manner by NAD$^+$ and NADP$^+$.

Reversibility. Although evidence for oxidation of cortisol to 21-dehydrocortisol by NADP$^+$-dependent hydroxysteroid dehydrogenase has been obtained, the reaction strongly favors reduction. It has not been possible to determine an equilibrium constant.

Kinetic Constants. K_m values for NADPH are constant from pH 6 to 7.5, ranging from 1.5×10^{-5} to $2.1 \times 10^{-5} M$. K_m values for steroids vary with the substrate used. Values found with 21-dehydrocortisol;

21-dehydrocortisone; 21-dehydro-11-deoxycorticosterone, and 21-dehydrocorticosterone, are 1.77, 4.61, 1.18, and $2.33 \times 10^{-4} M$, respectively.

NADH-Dependent 21-Hydroxysteroid Dehydrogenase from Bovine Adrenals

Purification Procedure

Step 1. Preparation of Acetone Powder. Unless otherwise indicated, all operations are carried out in a cold room at 0–4°. Fresh bovine adrenals are trimmed of adhering fat, cut into small pieces, and homogenized in 250-g batches with 5 volumes of cold acetone (−20°) for 1 minute in a Waring blendor operated at maximum speed. The resulting homogenate is sucked dry on a Büchner funnel, and the process is repeated. The filter cake is scraped from the filter paper, crumbled, and allowed to remain at room temperature until the acetone has evaporated. A fine powder is subsequently obtained after passing the coarse acetone-treated tissue through a strainer.

Step 2. Ammonium Sulfate Fractionation. The powder (63 g) is extracted with 10 volumes of 0.10 M sodium phosphate, pH 7.4, at room temperature for 45 minutes. The extract is cooled in ice and centrifuged at 15,000 g for 15 minutes. Solid ammonium sulfate is added to the supernatant solution to 25% saturation. After standing for 15 minutes, the solution is centrifuged at 15,000 g for 15 minutes and the supernatant fluid is collected. This supernatant solution is now brought to 65% saturation with respect to ammonium sulfate, and the precipitate which forms is separated by centrifugation at 15,000 g for 15 minutes. The precipitate is then dissolved in 200 ml of 0.001 M EDTA, pH 7.4, and dialyzed against 4 liters of this same solution for 24 hours, with two changes of the external buffer solution. Any resulting turbidity is removed by a 10-minute centrifugation at 15,000 g.

Step 3. Adsorption and Elution from Calcium Phosphate Gel. Acetic acid (1 N) is added to the enzyme solution until the pH drops to 5.8, after which calcium phosphate gel is added (1.5 mg of solids per milligram of protein). Further acetic acid is added if the pH rises above 5.8. The solution is stirred with a magnetic stirrer for 10 minutes, then centrifuged for 5 minutes at 15,000 g. If more than 10% of the original activity remains with the supernatant fluid, the process is repeated. The combined solids are then washed with 300 ml of 0.05 M sodium phosphate, pH 7.5, and centrifuged. Enzymatic activity is eluted from the gel by washing it with 600 ml of 0.10 M sodium phosphate, pH 7.5, containing 10% ammonium sulfate. The enzymatic activity is concentrated by dissolving the precipitate which forms between 25 and 65% saturation ammonium sulfate in 60 ml of 0.001 M EDTA, pH 7.4.

Step 4. Removal of Diaphorase Activity. To the solution obtained from step 3, 0.20 volume of saturated ammonium sulfate solution is added, followed by sulfuric acid added dropwise with rapid stirring until the pH falls to 1.2. Immediately thereafter, 2 volumes of saturated ammonium sulfate solution are added with rapid stirring, and the turbid solution is centrifuged at 30,000 g for 15 minutes. The precipitate is suspended in a total volume of 500 ml of 0.10 M sodium phosphate, pH 8.0. Insoluble material is sedimented by centrifugation at 30,000 g for 10 minutes, and the pH of the solution is brought to 8.0 by the addition of ammonium hydroxide (58% diluted 1:1 with 0.001 M EDTA). The enzymatic activity is concentrated by precipitation with ammonium sulfate between 25 and 65% saturation, as described in step 3. Final volume is 65 ml.

Step 5. DEAE-Cellulose Chromatography. An aliquot (20 ml) is dialyzed for 16 hours against 1 liter of 0.01 M sodium phosphate, pH 8.0, with two changes of buffer, then placed on a DEAE-cellulose column (2 × 25 cm, Schleicher and Schuell, Type 40) previously equilibrated with 0.01 M sodium phosphate, pH 8.0. The enzymatic activity is eluted with this same buffer. Four milliliter fractions are collected and tubes 12 through 16, containing the enzymatic activity are combined. The solution is brought to 30% saturation with ammonium sulfate, centrifuged at 30,000 g for 15 minutes, and the precipitate is discarded. The supernatant solution is brought to 60% saturation with ammonium sulfate, centrifuged, and the precipitate is dissolved in 10 ml of 0.01 M phosphate, pH 8.0. This solution must be stored frozen. The enzyme at this stage of purification (280-fold) is very labile, losing 25% of its

TABLE III
PURIFICATION OF NADH-DEPENDENT 21-HYDROXYSTEROID
DEHYDROGENASE FROM BOVINE ADRENALS

Treatment	Total volume (ml)	Total protein (mg)	Total activity[a]	Specific activity (units/mg)
1. Acetone powder extract	500	107,000	324	0.003
2. Ammonium sulfate precipitation (25–65%)	260	83,500	226	0.0272
3. Calcium phosphate gel treatment and concentration	72	17,800	143	0.0805
4. Acid treatment	65	7,250	73.5	0.101
5. DEAE-cellulose chromatography[b]	10	10	8.4	0.84

[a] At 24° instead of 30°.

[b] Data shown were obtained after chromatography of 20 ml of the total 65 ml shown in step 4.

activity in 1 week. A summary of the purification procedure is given in Table III.

Properties

Specificity. In addition to 21-dehydrocortisol, the enzyme will reduce 21-dehydrocortisone, 21-dehydrocorticosterone, 21-dehydrodeoxycorticosterone, and 21-dehydro-11-deoxycortisol. Acetaldehyde, DL-glyceraldehyde, salicylaldehyde, glyoxal, methylglyoxal, hydroxypyruvaldehyde, and phosphohydroxypyruvaldehyde are not reduced at concentrations up to 10^{-3} M. The enzyme is specific for NADH. NADPH and the 3-acetylpyridine analog of NADH are inactive in the most purified preparations.

Inhibitors. Cortisol and NAD inhibit the reaction. Dioxane ($K_i =$ 0.3 M) is a noncompetitive inhibitor. Steroids competitive with respect to 21-dehydrocortisol include progesterone, testosterone, epitestosterone, androsterone, isoandrosterone, dehydroepiandrosterone, 17α- and 17β-estradiol, and Nilevar.

Other Properties. The pH maximum is broad, but maximal at pH 7.4. Using the sucrose density technique of Martin and Ames,[13] the molecular weight is found to be 129,000.

[13] R. G. Martin and B. N. Ames, *J. Biol. Chem.* **236**, 1372 (1961).

[35] Microbial Steroid Esterases

By M. A. RAHIM and C. J. SIH

Hydrolytic cleavage of steroid esters has been observed in both microorganisms and animal tissues. An early example of microbiological hydrolysis of steroidal esters was provided by the work of Mamoli,[1] who demonstrated the combined reduction and hydrolysis of estrone acetate, propionate, and butyrate to estradiol by fermenting yeast. He further observed[2] the hydrolysis and oxidation of 21-acetoxy-3β-hydroxy-pregn-5-en-20-one to deoxycorticosterone by *Corynebacterium mediolanum*. Tamm and Gubler[3] investigated the effects of incubation of a variety of acetylated cardiac glycosides with *Fusarium lini*. The hydrolytic enzyme of this organism removed the acetyl groups from both carbon 3 and carbon 16. On the other hand, interesting selectivity was observed by Okada

[1] L. Mamoli, *Chem. Ber.* **71**, 2696 (1938).
[2] L. Mamoli, *Chem. Ber.* **72**, 1863 (1939).
[3] C. Tamm and A. Gubler, *Helv. Chim. Acta* **42**, 239 (1959).

et al.,[4] with *Gibberella saubinetti,* which efficiently removed acetyl groups from 3-acetyldigitoxigenin and 3-acetylgitoxigenin, but no hydrolysis was noted with 16-acetylgitoxigenin. Further selectivity on steroidal ester hydrolysis has been demonstrated by Charney *et al.*[5] and Noguchi *et al.*[6]

The methods of assay and isolation of two bacterial steroid esterases that have been studied in the authors' laboratory are described. The substrate specificities of these two enzymes, which show important differences, are also indicated.

Assay Method

Principle. Two methods are used to assay these esterases. All assays are performed at 25° in a Beckman DU spectrophotometer, equipped with a Gilford multiple absorbancy recorder model 2000.

p-NITROPHENOL METHOD. *p*-Nitrophenol acetate is hydrolyzed by the esterases to *p*-nitrophenol which has a strong absorption at 400 mμ. The rate of increase in optical density per minute at 400 mμ is a measure of esterase activity.

PAPER CHROMATOGRAPHIC METHOD. The substrate and product in the reaction mixture are separated by paper chromatography and the steroids are eluted from the paper chromatogram. Since Δ^4-3-keto steroids have absorption maxima at 240 mμ, both testosterone and testosterone acetate can be quantitatively estimated by measuring the absorbance at this wavelength.

Reagents

p-Nitrophenyl acetate, dissolved in methanol, 1 mg/ml
Testosterone acetate, dissolved in dimethylformamide, 10 mg/ml
Ethanol, 95%, redistilled
Whatman No. 1 paper, previously washed twice with 95% ethanol and air dried
Sodium phosphate buffer: 0.06 M, pH 7.0, for the assay of *Nocardia restrictus* esterase; 0.1 M, pH 8.0, for the assay of *Cylindrocarpon radicicola* esterase

Procedure. *p*-NITROPHENOL METHOD. The reaction is initiated by adding suitable aliquots of enzyme solution (diluted whenever required)

---

[4] M. Okada, A. Yamada, and M. Ishidate, *Chem. Pharm. Bull. Tokyo* **8,** 530 (1960).
[5] W. Charney, A. Nobile, C. Federbush, D. Sutter, P. L. Perlman, H. L. Herzog, C. C. Payne, M. E. Tully, M. J. Gentles, and E. B. Hershberg, *Tetrahedron* **18,** 597 (1962).
[6] S. Noguchi, K. Morita, and M. Nishikawa, *Chem. Pharm. Bull. Tokyo* **8,** 568 (1960).

to a 3.0-ml cuvette, containing 100 μg of p-nitrophenyl acetate. The total volume of the reaction mixture is brought up to 3.0 ml with either 0.06 M phosphate buffer, pH 7.0, or 0.1 M phosphate buffer, pH 8.0. The increase in optical density at 400 mμ during the initial 30–90-second interval is followed as a measure of reaction velocity.

PAPER CHROMATOGRAPHIC METHOD. The reaction mixture consists of 1 mg of testosterone acetate and a suitable aliquot of enzyme in a final volume of 4.0 ml of phosphate buffer. After 10 minutes of incubation at 25°, the reaction is terminated with 2 N HCl and the preparation is extracted with 1 ml of chloroform. The chloroform layer is separated by centrifugation, and 0.2 ml is spotted on the Whatman No. 1 paper. The chromatogram is then developed in the toluene–propylene glycol system[7] for 3 hours. The reaction product, testosterone, is located on the chromatogram by means of an ultraviolet scanner and is eluted from the paper with 95% ethanol. The absorbance at 240 mμ is taken as a measure of testosterone concentration.

Definition of Unit and Specific Activity. p-NITROPHENOL UNIT. One enzyme unit is that amount which causes an increase of 0.01 optical density units (for esterase of *N. restrictus*) or 0.10 optical density units (for esterase of *C. radicicola*) at 400 mμ during the initial 30–90-second interval.

TESTOSTERONE UNIT. One unit of enzyme is defined as that amount of protein which catalyzes the hydrolysis of 1 micromole of testosterone acetate in 1 hour.

Specific activity denotes units per milligram of protein.

Of the two methods, the p-nitrophenol assay is the more sensitive and is simpler to perform. Although in general it gives good correspondence with the results of the steroid acetate assay, the two activities may diverge under certain conditions (see comments on purification procedure of Table III). It should be noted also that the steroid acetate assay may be subject to inaccuracies due to side reactions in impure enzyme systems.

Steroid Esterase of *Nocardia restrictus*

Growth of Organism

Nocardia restrictus (ATCC 14887) is maintained on Difco nutrient agar slants, supplemented with 1% glucose and 1% yeast extract. The cells are grown in 15-liter batches for approximately 80 hours at 25° in a medium that contains 0.6% corn steep liquor, 0.3% $NH_4H_2PO_4$, 0.35% $CaCO_3$, 0.22% corn oil, 0.25% yeast extract, and 1% glucose. The cells

---

[7] A. Zaffaroni, R. B. Burton, and E. H. Keutman, *Science* **111**, 6 (1950).

are harvested by centrifugation, washed with 0.06 M phosphate buffer pH 7.0, and stored in a deep freeze.

Purification Procedure

All operations are performed at 4°. The cells are suspended in 3 volumes of 0.06 M phosphate buffer pH 7.0 and sonicated in 70-ml aliquots for 20 minutes in a 250-watt, 10-kc Raytheon oscillator. The sonic extract is centrifuged for 20 minutes at 2500 g, and the sediment is discarded (stage 1). To 630 ml of this solution is added 160 ml of CH_2Cl_2 (cooled to −20°), and the two layers are vigorously shaken for 1 minute in a separatory funnel. The resulting emulsion is centrifuged for 7 minutes at 2500 g, and the aqueous layer is recovered (stage 2). An increase of 15–20% in enzyme activity is consistently observed at this stage. To this enzyme solution (600 ml) is slowly added 630 mg of protamine sulfate (1 mg of protamine sulfate per 10 mg of protein) with stirring. The enzyme solution is left standing for 2 hours before it is centrifuged at 2500 g for 10 minutes. The supernatant solution (580 ml) is recovered after centrifugation (stage 3) and sufficient ammonium sulfate is added with constant stirring, to bring the concentration to 20% saturation. To maintain the pH at neutrality, 6 N NH_4OH is added. After 2 hours this mixture is centrifuged for 10 minutes at 2500 g, and the precipitate is discarded. Sufficient ammonium sulfate is now added to increase the concentration to 70% saturation, and after standing for 2 hours, the suspension is again centrifuged for 10 minutes at 2500 g. The residue is taken up in 64 ml of cold 0.06 M phosphate buffer, pH 7.0, and centrifuged again for 10 minutes at 2500 g (stage 4). The supernatant fluid is further centrifuged at 135,000 g for 150 minutes in a Spinco ultracentrifuge, model L. It is then dialyzed against 0.01 M potassium phosphate buffer, pH 7.0, for 16 hours (stage 5). The dialyzed preparation is chromatographed on three DEAE-cellulose columns, 1.4 × 8 cm, that have been equilibrated with 0.01 M phosphate buffer, pH 6.8. Each column is eluted according to a linear gradient elution technique in which the mixing vessel contains 150 ml of a 0.01 M phosphate solution, pH 6.8, and the reservoir vessel, 250 ml of a 0.2 M potassium phosphate buffer solution, pH 6.8. Fractions of 4 ml of eluate are collected in each test tube. The major portion of the esterase activity is collected in tubes 51 through 73 (0.15–0.17 M phosphate buffer). The contents of these tubes are pooled, and the enzyme solution is brought to 70% saturation with ammonium sulfate. After the solution has stood for 30 minutes, the precipitate is removed by centrifugation at 2500 g for 10 minutes. The precipitate is dissolved in 10 ml of 0.06 M phosphate buffer, pH 7.0 (stage 6). A summary of the purification procedure is given in Table I.

TABLE I
Purification Procedure for Steroid Esterase from *Nocardia restrictus*

	p-Nitrophenol method		Testosterone acetate method	
Stage	Total units	Specific activity (units/mg)	Total units	Specific activity (units/mg)
1. Sonic extract	25,200	1.10	2300	0.10
2. CH$_2$Cl$_2$ treatment	29,400	4.65	2800	0.45
3. Protamine sulfate	27,260	30.3	2100	2.34
4. (NH$_4$)$_2$SO$_4$, 20–70%	26,500	35.1	2000	2.75
5. 135,000 g	20,300	52.0	1850	4.75
6. DEAE-cellulose eluate	5,000	78.0	550	8.45

Properties[8]

Stability. Preparations of steroid esterase are generally unstable; they lose their activity in 2–3 weeks when stored at 0–4°. Freezing and thawing rapidly inactivates the enzyme.

Effect of pH. The rates of reaction with p-nitrophenyl acetate and testosterone acetate are maximal at pH 8.0. At pH 7.0, the rates are about one-half of that observed at pH 8.0.

Inhibitors. The following compounds show no inhibition when incubated with this enzyme at a concentration of $10^{-3} M$: NaF, NaASO$_2$, NaN$_3$, KCN, EDTA, iodoacetate, and eserine. SnSO$_4$ and CuSO$_4$ show no inhibition at $10^{-4} M$. However, the enzyme is inhibited by HgCl$_2$ and by p-chloromercuribenzoate. At concentrations of $10^{-3} M$, these compounds inhibit the hydrolysis of p-nitrophenyl acetate to the extents of 50% and 23%, respectively. DFP inactivates the enzyme irreversibly.

Substrate Specificity. The purified enzyme preparation does not hydrolyze succinate, benzoate, and propionate esters of testosterone, whereas acetate and pyruvate esters are readily cleaved by the enzyme. The specificity and relative rates of hydrolysis are shown in Table II.

Steroid Esterase of *Cylindrocarpon radicicola*

Growth of the Organism

Stock cultures of *Cylindrocarpon radicicola* (ATCC 11011) are maintained on Difco nutrient agar slants, supplemented with 1% glucose and 1% yeast extract. This organism is grown in 500-ml Erlenmeyer flasks containing 100 ml of the following medium: 0.6% corn steep liquor,

[8] C. J. Sih, J. Laval, and M. A. Rahim, *J. Biol. Chem.* **238**, 566 (1963).

TABLE II

SUBSTRATE SPECIFICITY OF PURIFIED STEROID ESTERASE FROM *N. restrictus*

Substrate[a]	Percent relative hydrolysis[b]
17α-Acetoxyandrost-4-en-3-one	100
17β-Acetoxyandrost-4-en-3-one	85
17α-Acetoxypregn-4-ene-3,20-dione	0
16α-Acetoxypregn-4-ene-3,20-dione	100
15α-Acetoxypregn-4-ene-3,20-dione	4
15β-Acetoxypregn-4-ene-3,20-dione	0
12β-Acetoxy-17α-hydroxypregn-4-ene-3,20-dione	0
12α-Acetoxy-5β-pregnane-3,20-dione	0
11α-Acetoxypregn-4-ene-3,20-dione	0
11β-Acetoxyandrost-4-ene-3,17-dione	0
7α-Acetoxyandrost-4-ene-3,17-dione	0
7β-Acetoxyandrost-4-ene-3,17-dione	0
6α-Acetoxyandrosta-1,4-diene-3,17-dione	50
6β-Acetoxyandrosta-1,4-diene-3,17-dione	0
6β-Acetoxypregn-4-ene-3,11,20-trione	0
3β-Acetoxy-5α-pregnan-20-one	85
3α-Acetoxy-5α-androstan-17-one	80
3β-Acetoxy-5β-pregnan-20-one	0
3α-Acetoxy-5β-pregnan-20-one	0
3β-Acetoxypregn-5-en-20-one	85
2β,17β-Diacetoxyandrost-4-en-3-one	3[c]
1α-Acetoxyandrost-4-ene-3,17-dione	0
19-Acetoxyandrost-4-ene-3,17-dione	90
3β,20β-Dihydroxy-5α-pregnan 20-acetate	20
17α,21-Dihydroxypregn-4-ene-3,20-dione 21-acetate	15
3-Acetoxyestra-1,3,5(10)-trien-17-one	95[d]

[a] Non-ultraviolet-absorbing steroids were estimated directly on paper chromatographs by visual comparison with a standard after they were sprayed with 20% phosphomolybdic acid in ethanol.

[b] Values are given relative to 17α-acetoxyandrost-4-en-3-one (arbitrarily = 100) determined as described on p. 677.

[c] Refers to the rate of hydrolysis of the 2β-hydroxyl group obtained by measuring the formation of 2β,17β-dihydroxyandrost-4-en-3-one.

[d] Estrone was estimated by measuring the absorbance at 280 mμ.

0.3% $NH_4H_2PO_4$, 0.35% $CaCO_3$, 0.22% soybean oil, 0.25% yeast extract, and 1% glucose. After incubation at 25° on a rotary shaker (250 rpm, 1 inch stroke) for 24 hours, 50 mg of progesterone in 0.8 ml of dimethylformamide is added to each flask to act as an inducer of the esterase, and the incubation is continued for an additional 8 hours. The enzyme level increases 2- to 3-fold over the noninduced level. The mycelia are harvested by filtration through cheesecloth, washed successively with demineralized water and cold 0.1 M phosphate buffer pH 8.0. The cell cake is stored in a deep freeze.

Preparation of the Cell-Free Extract

The frozen mycelium is thawed and suspended in 4 times its weight of 0.1 M phosphate buffer pH 8.0. The suspension is divided into 100-ml portions and sonicated for 15 minutes with a Branson S-75 sonifier at full power. The combined cell extract is centrifuged at 18,400 g for 20 minutes to remove the cell debris and the turbid supernatant (640 ml) is adjusted to pH 7.5 by the gradual addition of 1 N NaOH.

Purification Procedure

Freezing and Thawing of the Crude Extract. The supernatant is stored in a deep freeze ($-20°$) for 48 hours. The preparation is then thawed and centrifuged at 18,000 g for 5 minutes to remove the denatured protein. A recovery of 90–95% enzyme activity with 2- to 3-fold purification is consistently observed at this stage.

Ammonium Sulfate Fractionation. To the 640-ml of enzyme solution from the previous step, solid ammonium sulfate is added slowly with mechanical stirring until a 0.50 saturation is reached. The pH is adjusted to 7.5 by the addition of 3 N NH$_4$OH. The solution is now allowed to stand for 30 minutes, and the precipitate is discarded after centrifugation at 18,000 g for 10 minutes. The supernatant solution is brought to 0.70 saturation by the addition of 114 g of ammonium sulfate, and again the pH is adjusted to 7.5 with ammonium hydroxide. After 30 minutes, the turbid solution is centrifuged as before and the sediment, containing 80% of the initial activity, is dissolved in 55 ml of 0.005 M phosphate buffer, pH 7.2.

DEAE-Sephadex Chromatography. A column (6.5 $\times$ 25 cm) of DEAE-Sephadex A-50 (medium) is made up in 0.005 M phosphate buffer, pH 7.2. To the top of this column is added Sephadex G-50 mixed with 1% cellulose powder to an additional height of 15 cm. The entire column is thoroughly equilibrated with the same buffer. The enzyme solution from the previous step is layered at the top of the column. The column is developed with 0.005 M phosphate buffer, pH 7.2, and 9.5 ml fractions are collected. Esterase activity resides in fractions 29 through 44. A recovery of 5–15% in excess of the enzyme units applied to the column is observed at this stage, probably owing to the removal of an inhibitor. DEAE-cellulose chromatography, under the same elution conditions, consistently gives material of 2-fold purification with a recovery of 50%.

Filtration through Sephadex G-50 (Medium). A 150-ml sample of the enzyme solution is filtered through two Sephadex G-50 columns (6.5 $\times$ 14 cm), equilibrated with 0.005 M phosphate buffer, pH 7.2. The columns are eluted with the same buffer, and the enzyme activity eluted from both is collected (165 ml).

Chromatography on Hydroxylapatite. A hydroxylapatite column (5×55 cm), thoroughly equilibrated with $0.005\,M$ phosphate buffer, is charged with 160 ml of Sephadex G-50 effluents. Elution of the column is effected with potassium phosphate buffer, pH 7.2, in the following order: 50 ml of $0.005\,M$; 50 ml of $0.025\,M$; 200 ml of $0.05\,M$, $0.1\,M$, $0.15\,M$, and $0.2\,M$. The flow rate is 2 ml per minute, and 8.5 ml fractions are collected. Fractions 71–95 (eluted between 0.15 and $0.20\,M$) containing the highest enzyme activity (225 ml) are pooled. A summary of the purification procedure is given in Table III.

TABLE III

PURIFICATION PROCEDURE FOR STEROID ESTERASE FROM *Cylindrocarpon radicicola*

Fractions	Total volume (ml)	Total protein (mg)	p-Nitrophenol method		Testosterone method	
			Total units	Specific activity	Total units	Specific activity
1. Initial extract	674	10,784	350,420	32.5	12,626	1.08
2. Freezing and thawing	640	6,912	318,000	52.0	11,950	1.73
3. Ammonium sulfate (0.50–0.70)	548	904	282,330	313	7,081	7.8
4. DEAE-Sephadex effluents	150	629	300,500	477	7,302	11.9
5. Sephadex G-50	160	198	217,248	1098	4,059	24.5
6. Hydroxylapatite	225	13	115,500	8885	2,685	202

Comments on Purification Procedure. The pH of the crude extract must be adjusted to pH 7.5 before freezing. Freezing and thawing is an essential step in order to obtain a higher recovery in the subsequent step, otherwise a fatty disk is invariably obtained on centrifuging the ammonium sulfate precipitate. While following the purification procedure, the enzyme solution can be stored (at $-20°$ or at $0–6°$) only under ammonium sulfate. Removal of inhibitor on DEAE-Sephadex column can be achieved only by working at near capacity of the column, e.g., equilibration at pH 7, development with phosphate buffer (anionic buffer). Step 5 is to be followed only when the DEAE-Sephadex effluents are associated with ammonium sulfate. Otherwise, step 5 can be avoided and enzyme solution from step 4 can be adsorbed directly on the hydroxylapatite column; recovery of units as high as 50% can be obtained.

The consistent divergence of enzyme activities toward testosterone acetate and p-nitrophenol acetate during purification may be plausibly attributed to inactivation of the catalytic activity toward testosterone acetate, especially in the final two steps. Apparently, the quasisubstrate, *p*-nitrophenol acetate, is not as severely affected. However, the specificity and relative activity of the esterase on different steroid substrates remained unchanged from stage 3 through 6.

Properties[9]

Stability. Purified preparations are unstable. On dialysis for 1 hour against 0.01 M phosphate buffer with or without $10^{-3} M$ EDTA, mer-

TABLE IV

SUBSTRATE SPECIFICITY OF THE PURIFIED STEROID ESTERASE FROM *C. radicicola*

Substrate[a]	Percent relative hydrolysis[b]
17α-Acetoxyandrost-4-en-3-one	100
17β-Acetoxyandrost-4-en-3-one	78
11β-Acetoxyandrost-4-ene-3,20-dione	65[c]
21-Acetoxypregn-4-ene-3,20-dione	40
3-Acetoxyestra-1,3,5(10)-trien-17-one	33[d]
Testosterone pyruvate	55[c]
Testosterone propionate	35
Testosterone succinate	0
Testosterone benzoate	0
11α-Acetoxypregn-4-ene-3,20-dione	0
17α-Acetoxypregn-4-ene-3,20-dione	0
6α-Acetoxyandrost-4-ene-3,17-dione	+[e]
6β-Acetoxyandrost-4-ene-3,17-dione	+
1α-Acetoxyandrost-4-ene-3,17-dione	+
1β-Acetoxyandrost-4-ene-3,17-dione	0
2α-Acetoxyandrost-4-ene-3,17-dione	+
2β-Acetoxy-9α-hydroxyandrost-4-ene-3,17-dione	0
3β-Acetoxy-5β-pregnan-20-one	0
3α-Acetoxy-5α-androstan-17-one	+
3β-Acetoxy-5α-pregnan-20-one	+
3β-Acetoxypregn-5-en-20-one	+
3α-Acetoxy-5β-pregnan-20-one	0
12α-Acetoxy-5β-pregnane-3,20-dione	0
19-Acetoxyandrost-4-ene-3,17-dione	0
16α-Acetoxypregn-4-ene-3,20-dione	+
15α-Acetoxypregn-4-ene-3,20-dione	0
15β-Acetoxypregn-4-ene-3,20-dione	0

[a] See Table II, p. 680.
[b] See Table II, p. 680.
[c] To avoid excessive nonenzymatic hydrolysis at pH 8.0, the reaction was run at pH 7.3 and corrected for nonenzymatic hydrolysis.
[d] See Table II, p. 680.
[e] + =, Significant amount of hydrolysis only on prolonged incubation.

[9] M. A. Rahim and C. J. Sih, to be published.

captoethanol, or dithiothreitol, the enzyme preparation loses 46% of its activity, which cannot be reconstituted by the addition of Mg, Fe^{++}, NH_4, Ca, or Cu^{++} ions. The enzyme activity appears to be stable when stored in ammonium sulfate.

Effect of pH. The rates of hydrolysis of *p*-nitrophenyl acetate and testosterone acetate are maximal at pH 8–9. The rate of hydrolysis of *p*-nitrophenyl acetate at pH 8 is about 3 times faster than at pH 7.0, whereas the hydrolysis of testosterone acetate at pH 8.0 is only 1.5 times faster than at pH 7.0.

Inhibitors. $HgCl_2$ ($5 \times 10^{-4} M$) and *p*-hydroxymercuribenzoate ($1 \times 10^{-3} M$) inhibit the esterase activity to the extent of 80 and 75%, respectively. NaF ($2 \times 10^{-3} M$) and EDTA ($1 \times 10^{-3} M$) cause an inhibition of 63 and 15%, respectively.

Substrate Specificity. The relative rates of hydrolysis of various steroidal esters are listed in Table IV. Acetate, propionate, and pyruvate esters of testosterone are readily cleaved whereas testosterone succinate and benzoate are not attacked. Acetoxy esters, substituted at positions 17α, 17β, 21, and 11β are hydrolyzed rapidly. Acetoxy esters at positions 16α, 6α, 6β, 1α, 2α, and 3—(5α or pregn-5-ene series) are hydrolyzed very slowly.

[36] Hydrolysis of the Sulfuric Acid Esters of Steroid Compounds

By EUGENE C. SANDBERG and R. CLIFTON JENKINS

$$R\text{—O-}\xi\text{-SO}_3^- \xrightarrow{\text{HOH}} ROH + HSO_4^-$$

Interest in the possible role of sulfate esters in the biosynthesis and intermediary metabolism of steroid hormones in mammals is a very recent development in the field of hormone biology. Despite the accumulation of a relatively substantial amount of information on this general subject during the past few years, data concerning the biological existence of the many potential steroid sulfates and the number, properties, and specificities of enzymes involved in steroid sulfate hydrolysis remain limited. Information relative to only one steroidal sulfatase is currently available in any detail, that related to others being purely inferential at present.

While the currently recommended nomenclature for sulfatase enzymes is less than optimal, it will be used here to maintain consistency with published terminology.

Sterol Sulfatase (Sterol-Sulfate Sulfohydrolase) (EC 3.1.6.2)

Assay Method

Principle. METHOD A. An assay method employing radioactive substrate is based on the hydrolysis of [3]H- or ^{14}C-labeled steroid sulfate substrate of known specific activity followed by selective recovery of the product of hydrolysis by solvent partition.[1,2]

METHOD B. An assay method employing nonradioactive substrate is based on the hydrolysis of unlabeled steroid sulfate substrate with recovery and colorimetric quantitation of the unconverted substrate.[3,4]

Reagents

GENERAL

Dehydroisoandrosterone sulfate,[5] radioactive and/or nonradioactive
Tris [2-amino-2-(hydroxymethyl)-1,3-propanediol]-HCl buffer, 0.25 M, pH 8.0

[1] J. C. Warren and A. P. French, *J. Clin. Endocrinol. Metab.* **25**, 278 (1965).
[2] S. Burstein and R. I. Dorfman, *J. Biol. Chem.* **238**, 1656 (1963).
[3] A. B. Roy, *Biochem. J.* **66**, 700 (1957).
[4] A. B. Roy, *Biochem. J.* **62**, 41 (1956).
[5] For methods of preparation and properties of steroid sulfates see article by Jenkins and Sandberg, this volume [10], p. 351.

Sulfuric acid esters of steroids, unlike the unconjugated steroids, are water soluble and have limited solubility in solvents of low polarity, including scintillation counting fluids with a toluene base. Consequently, it is necessary to add ethanol or some other toluene-miscible, steroid sulfate-solubilizing solvent to such samples in order to obtain reproducible results in radioactivity quantitation. For this purpose 0.5–1.0 ml of ethanol per 5 ml of scintillation fluid is generally adequate. The degree of solvent quenching is determined by the use of internal standards.

Preservation of steroid sulfates is best assured by storage of crystalline material in an evacuated desiccator within an unlighted refrigerator or freezer. Solubilized steroid sulfate esters are reasonably stable for periods of months when dissolved in absolute ethanol and kept under refrigeration. Spontaneous hydrolysis of a minor degree (less than 5%) during storage in such a solution is anticipated. Material to be used as substrate should be partitioned between water and ether immediately before use. Sulfurylated substrates of relatively low polarity should be partitioned between water and a solvent of lower polarity than ether in order to prevent an undue transfer of steroid sulfate to the organic phase. In the case of cholesterol sulfate, hexane has been found to be an appropriate solvent.

For Method A

Acetate buffer, 3 *M* (3 *M* sodium acetate and 3 *M* acetic acid), pH 4.0
Diethyl ether
Dry ice–isopropanol bath

For Method B

Absolute ethanol
Methylene blue reagent: 250 mg of methylene blue chloride, 50 g of anhydrous sodium sulfate, and 10 ml of concentrated sulfuric acid brought to 1 liter with water
Chloroform

Procedure. Method A. A solution of dehydroisoandrosterone sulfate of desired concentration and specific activity is made up in absolute ethanol. The quantity appropriate for the particular incubation is transferred to a glass-stoppered tube and evaporated to dryness. Enzyme preparation, 2 ml constituted in 0.25 *M* Tris-HCl buffer, is introduced. The dried material is thoroughly dissolved by shaking, and the mixture is incubated (with agitation) at 37°. Cofactors are not required. At the conclusion of incubation, 2 ml of acetate buffer (or a quantity sufficient to reduce the pH of the mixture to 4.0) is added and the components are thoroughly mixed in order to halt enzymatic activity. Two volumes of diethyl ether are then introduced. The mixture is again shaken thoroughly, the solvent fractions are allowed to separate, and the tube is immersed in a dry ice–isopropanol bath. After total freezing of the aqueous layer, the ether fraction is decanted into a scintillation counting vial and evaporated to dryness; the radioactivity of the residue is quantitated in the usual fashion. Incubations halted at zero time may be used as controls.

The reliability of this assay obviously depends on the partition characteristics of the conjugated substrate and the unconjugated product between aqueous and organic phases. Control studies of the partition characteristics of several Δ^5-3β-hydroxy steroids and their 3β-monosulfate esters have been performed over a broad range of concentration and specific activity. Under the conditions described above, more than 95% of unesterified steroid (representing product) consistently partitions into the ether phase. Conversely, less than 5% of sulfurylated steroid substrate is extracted from the aqueous layer. In the case of cholesterol–cholesterol sulfate, it is necessary to substitute hexane for ether to maintain these extraction characteristics. The applicability of this assay method for other sulfurylated steroids and their unconjugated derivatives has not been ascertained. While it is reasonable to anticipate a broad applicability, an examination of the partition characteristics of each newly used substrate-product combination should be undertaken.

METHOD B. Dehydroisoandrosterone sulfate is incubated with the enzyme preparation in Tris-HCl buffer using a total incubation volume of 1 ml. Following incubation the enzyme and other proteins are precipitated by the addition of 5 ml of absolute ethanol, and the precipitate is removed by centrifugation. A 5-ml sample of supernatant is evaporated to dryness in a glass-stoppered tube, and the residue is taken up in 2 ml of methylene blue reagent. The methylene blue–steroid sulfate complex that is formed is then extracted into 5 ml of chloroform. The aqueous layer is decanted, and the concentration of the unconverted steroid sulfate present in the chloroform phase is quantitated spectrophotometrically at 660 mμ.

The recovery of dehydroisoandrosterone sulfate by this method is in excess of 95% except in the presence of protein. The presence of chlorides or perchlorates will cause an excess transfer of methylene blue into the chloroform phase and will result in erroneously high steroid concentration values.

Recovery of product by the technique employed for product recovery in Method A would appear to be equally applicable, quantitation then being accomplished by any colorimetric technique obeying the Lambert–Beer law.

Purification Method

While it has been determined that sterol sulfatase is located in the microsomal cell fraction of mammalian tissues, it has not yet been possible to bring the enzyme into true solution despite efforts in this direction. A method for obtaining a 50-fold purification of the molluscan enzyme from an acetone powder extract of *Patella vulgata* has been described by Roy.[4] A comparison of the specific activity of the purified acetone powder preparation with that of a microsomal cell fraction has not been reported.

The following method has been employed to obtain a satisfactorily active microsomal cell fraction from human term placental tissue in our laboratory. Fresh tissue is minced, washed several times with 0.25 M sucrose solution, and blotted dry. A 30% tissue homogenate (w/v) is prepared in 0.25 M sucrose solution and centrifuged for 15 minutes at 900 g. The supernatant is centrifuged at 13,500 g for 15 minutes and the supernatant from this centrifugation is then recentrifuged at 105,000 g for 1 hour. The resulting pellet is washed by resuspension in 0.154 M KCl. After repeat centrifugation at 105,000 g for 1 hour, the pellet is resuspended in a minimal quantity of 0.154 M KCl and lyophilized.

Properties

Specificity. Available data[3, 4, 6, 7] suggest that both the mammalian and molluscan enzymes are quite specific for 3β sulfates of the 5α and Δ^5 series of steroids and have little or no activity toward the other isomeric 3-sulfates. This would suggest that the planarity of the molecule is a critical factor in the enzyme-substrate interaction. The nature of the C-17 substituent is apparently influential only in regard to enzyme affinity and rate of hydrolysis. This is seen in the following list of relative rates of hydrolysis of certain Δ^5-3β-sulfoxy compounds by a microsomal cell fraction of human term placenta:

17α-Hydroxypregnenolone sulfate	100
Pregnenolone sulfate	70
Cholesterol sulfate	50
Dehydroisoandrosterone sulfate	40
Androstenediol sulfate (Δ^5-androstene-3β,17β-diol-3-sulfate)	20

While sterol sulfatase preparations from both mammalian and molluscan sources additionally hydrolyze estrone sulfate, there is evidence that a different, but as yet inseparable, microsomal sulfatase (arylsulfate sulfohydrolase) is responsible for this activity. Sterol sulfatase obtained from ox liver, human placenta, and human polycystic ovary has failed to show hydrolytic activity toward 16α-, 17α-, 17β-, 20α-, 20β-, and 21-sulfoxy steroids.[3, 7, 8]

Stability. The enzyme appears to be exceptionally stable. Ethanolic precipitates of rat and cattle liver homogenates which have been lyophilized or dried with ether–ethanol have been reported to be stable for 6 months.[9] Lyophilized human placental microsomal preparations stored in a desiccator at 4° have retained full activity for as long as a year.[7]

Optimum pH. The pH optimum of this enzyme, obtained from human placenta and suspended in 0.25 M Tris-HCl buffer, is 7.8–8.6. The enzyme is completely inactivated at pH 4.5. A slightly lower optimum has been noted for microsomal preparations of rat liver,[2] rat testis,[2] and ox liver[3] suspended in 0.1 and 0.13 M Tris-acetic acid buffer. The optimal pH for molluscan sterol sulfatase activity is 4.5–5.0 in 0.125–0.5 M acetate buffer.[4, 6]

Inhibitors. Both the mammalian and molluscan enzymes are inhibited by PO_4^{3-} and SO_3^{2-} ions at concentrations of 10^{-2} and $10^{-4} M$, respectively.[3, 10] Molluscan enzyme activity is additionally inhibited by F⁻

---

[6] J. Jarrige, J. Yon, and M. F. Jayle, *Bull. Soc. Chim. Biol.* **45**, 783 (1963).

[7] E. C. Sandberg and R. C. Jenkins, unpublished observations.

[8] A. P. French and J. C. Warren, *Steroids* **8**, 79 (1966).

[9] H. Gibian and G. Bratfisch, *Z. Physiol. Chem.* **305**, 265 (1956).

[10] J. Jarriage, *Bull. Soc. Chim. Biol.* **45**, 761 (1963).

$(2 \times 10^{-3} M)$, Cl^- $(5 \times 10^{-2} M)$, BO_3^{3-} $(2 \times 10^{-3} M)$ and SO_4^{2-} $(5 \times 10^{-2} M)$ ions,[4,10] none of which has been reported to substantially influence the mammalian enzyme. Certain cations (Hg^{++}, Zn^{++}, and Ca^{++}, $7.5 \times 10^{-3} M$) have also been shown to inhibit molluscan sterol sulfatase activity.[10]

Reversibility. This enzymatic hydrolysis is essentially irreversible. The biological sulfurylation of steroid compounds requires sulfate-activating enzymes (ATP sulfurylase and APS phosphokinase) in addition to group specific sulfate-transferring enzymes (sulfokinases).[11]

Kinetic Properties. Using a lyophilized microsomal preparation of human term placental tissue at pH 8.0 (0.25 M Tris-HCl buffer), the reaction velocity with Δ^5-3β-sulfoxy steroid substrates is maximal at a concentration of about $10^{-4} M$. The average K_m values for pregnenolone sulfate, 17α-hydroxypregnenolene sulfate, androstenediol sulfate, cholesterol sulfate, and dehydroisoandrosterone sulfate are $5.4 \times 10^{-6} M$, $6.4 \times 10^{-6} M$, $6.8 \times 10^{-6} M$, $8.7 \times 10^{-6} M$, and $1.2 \times 10^{-5} M$, respectively. Each of these compounds competitively inhibits the hydrolysis of dehydroisoandrosterone sulfate giving average K_i values of $2.2 \times 10^{-6} M$ (pregnenolone sulfate), $3.5 \times 10^{-6} M$ (androstenediol sulfate), $4.3 \times 10^{-6} M$ (17α-hydroxypregnenolone sulfate) and $6.5 \times 10^{-6} M$ (cholesterol sulfate).

Sources of the Enzyme

Molluscan sources of the enzyme include *Patella vulgata*, *Helix pomatia*, and *Otala punctata*[4,10,12] while mammalian sources include the liver of cattle, rats, and humans; the testis of rats, guinea pigs, and humans; and also the human myometrium, endometrium, ovary, fetal membranes, and placenta.[1,2,9] The latter is the richest source discovered to date.

Arylsulfatase (Aryl-Sulfate Sulfohydrolase) (EC 3.1.6.1)

Three distinct arylsulfatases have been described. Arylsulfatase A and arylsulfatase B are soluble enzymes located in the mitochondrial cell fraction. Neither is known to cleave phenolic steroid sulfates. Arylsulfatase C is said to hydrolyze estrone sulfate[11] but studies of the kinetic properties of this enzyme have all been performed using a nonsteroidal phenolic sulfate (p-nitrophenol sulfate) as substrate.[13,14]

Arylsulfatase C, like sterol sulfatase, appears to be firmly asso-

[11] A. B. Roy, *in* "Advances in Enzymology" (F. F. Nord, ed.), Vol. 22, p. 205. Wiley (Interscience), New York, 1960; see also this volume, p. 732.

[12] K. Savard, E. Bagnoli, and R. I. Dorfman, *Federation Proc.* 13, 289 (1954).

[13] A. B. Roy, *Biochem. J.* 64, 651 (1956).

[14] K. S. Dodgson, B. Spencer, and C. H. Wynn, *Biochem. J.* 62, 500 (1956).

ciated with the microsomal cell components. It has been generally impossible to solubilize this enzyme although one exception has been reported.[15] While the biological distribution of arylsulfatase C has not been broadly studied, enzyme preparations from the various tissues which have been shown to hydrolyze estrone sulfate (i.e., *Patella, vulgata*,[16] rat liver,[17] and human placenta,[17] endometrium,[7] and ovary[18,19]) have also been shown to hydrolyze Δ^5-3β-sulfoxy steroids. Hydrolytic activity relative to the two groups of compounds has thus far proved to be inseparable.

Preparations of the fungus *Aspergillus oryzae* have been demonstrated to be capable of hydrolyzing estrone sulfate but not dehydroisoandrosterone sulfate,[20] and the same has been reported for human fetal liver.[17] These findings constitute the most salient current evidence supporting the concept that phenolic 3-sulfoxy steroids and Δ^5-3β-sulfoxy steroids are hydrolyzed by separate enzymes.

Hydrolysis of Other Steroid Sulfates

C-21 sulfurylated steroids are hydrolyzed by three species of molluscs[4,6,12,21] and by rat liver,[22] all of which additionally hydrolyze dehydroisoandrosterone sulfate. It is uncertain from the molluscan studies whether sterol sulfatase or a separate sulfatase is responsible for the hydrolysis of C-21 sulfurylated steroids. The inability of ox liver[3] and human term placenta[7] to hydrolyze cortisone 21-sulfate, though both cleave dehydroisoandrosterone sulfate vigorously, would suggest that the specificity of sterol sulfatase does not encompass 21-sulfoxy steroids and that a separate sulfatase is required for the hydrolysis of these compounds.

The enzymatic hydrolysis of steroid sulfates esterified at other positions (e.g., 3α, 16α, 17α, 17β, etc.) has not been reported.

[15] K. S. Dodgson, F. A. Rose, and B. Spencer, *Biochem. J.* **66**, 357 (1957).
[16] S. R. Stitch, I. D. K. Halkerston, and J. Hillman, *Biochem. J.* **63**, 705 (1956).
[17] M. O. Pulkkinen, *Acta Physiol. Scand.* **52**, Suppl. 180 (1961).
[18] E. C. Sandberg and R. C. Jenkins, *Biochim. Biophys. Acta* **113**, 190 (1966).
[19] E. C. Sandberg, R. C. Jenkins, and H. M. Trifon, *Steroids* **8**, 237 (1966).
[20] K. H. Ney and R. Ammon, *Z. Physiol. Chem.* **315**, 145 (1959).
[21] K. S. Dodgson, *Biochem. J.* **78**, 324 (1961).
[22] J. R. Pasqualini and J. Faggett, *J. Endocrinol.* **31**, 85 (1964).

[37] Enzymes of Estrogen Metabolism

By HEINZ BREUER and RUDOLF KNUPPEN

Introduction

The enzymology of the biogenesis and metabolism of estrogens has received increasing attention during recent years. This is primarily due to the fact that in addition to the three classic estrogens—estrone, 17β-estradiol, and estriol—a great number of previously unknown phenolic steroids have been either isolated from human pregnancy urine or identified in human urine after injection of radioactive estrone or 17β-estradiol. The following more recently discovered phenolic steroids are now known to occur in man: 2-hydroxyestrone,[1,2] 2-methoxyestrone,[3-5] 2-methoxy-17β-estradiol,[6] 2-methoxyestriol,[7] 6α-hydroxyestrone,[8] 6-hydroxyestriol,[9] 11-dehydro-17α-estradiol,[10] 15α-hydroxyestrone,[11] 15α-hydroxy-17β-estradiol,[12,13] 15β-hydroxyestrone,[14] 15β-hydroxy-17β-estra-

[1] J. Fishman, R. I. Cox, and T. F. Gallagher, Arch. Biochem. Biophys. 90, 318 (1960).
[2] V. Notchev and B. F. Stimmel, Excerpta Med. 51, 175 (1962).
[3] L. L. Engel, B. Baggett, and P. Carter, Endocrinology 61, 113 (1957).
[4] S. Kraychy and T. F. Gallagher, J. Biol. Chem. 229, 519 (1957).
[5] K. H. Loke and G. F. Marrian, Biochim. Biophys. Acta 27, 213 (1958).
[6] V. A. Frandsen, Acta Endocrinol. 31, 603 (1959).
[7] J. Fishman and T. F. Gallagher, Arch. Biochem. Biophys. 77, 511 (1958).
[8] R. Knuppen, O. Haupt, and H. Breuer, Biochem. J. 101, 397 (1966).
[9] J. Breuer, F. Breuer, H. Breuer, and R. Knuppen, Z. Physiol. Chem. 346, 279 (1966).
[10] T. Luukkainen and H. Adlercreutz, Biochim. Biophys. Acta 107, 579 (1965).
[11] R. Knuppen, O. Haupt, and H. Breuer, Biochem. J. 96, 33c (1965).
[12] B. P. Lisboa, U. Goebelsmann, and E. Diczfalusy, Acta Endocrinol. 54, 467 (1967).
[13] R. Knuppen and H. Breuer, Z. Physiol. Chem. 348, 581 (1967).
[14] R. Knuppen, O. Haupt, and H. Breuer, Steroids 8, 403 (1966).

diol,[14] 16α-hydroxyestrone,[15,16] 16β-hydroxyestrone,[17,18] 16-oxoestrone,[19-21] 16-oxo-17β-estradiol,[17,22]   16-epiestriol,[23]   17-epiestriol,[24,25]   16,17-epiestriol,[24,26] and 18-hydroxyestrone.[27]

The biogenesis and metabolism of these estrogens has been studied *in vitro* and *in vivo* under a variety of experimental conditions (for reviews see footnotes 28–32). The main reactions of the intermediary metabolism of C-2, C-6, C-7, C-11, C-15, and C-16 substituted estrogens are summarized in Figs. 1–7. All reactions have been demonstrated *in vivo* and/or *in vitro*, and with improved modern analytical methods, quantitative kinetic studies of some enzymes related to estrogen metabolism have been made possible. Of the multiplicity of enzymes involved, however, only one, estradiol-17β-oxidoreductase of human placenta, has so far been highly purified.[33-35] Recently this enzyme has been crystallized (see this volume [39]). In addition, a specific estrogen sulfotransferase has been purified 35-fold from bovine adrenal glands. No other enzyme specific to estrogen metabolism has been obtained in a more than slightly purified form. Many of these enzymes (e.g., aromatizing enzyme, hydroxylases, oxidoreductases, demethylase) are microsomal and resist solubilization, and the purification of the soluble cytoplasmic enzymes except the recently described 16α-hydroxysteroid oxidoreductase from

[15] G. F. Marrian, E. J. D. Watson, and M. Panattoni, *Biochem. J.* **65**, 12 (1957).
[16] G. F. Marrian, K. H. Loke, E. J. D. Watson, and M. Panattoni, *Biochem. J.* **66**, 60 (1957).
[17] D. S. Layne and G. F. Marrian, *Biochem. J.* **70**, 244 (1958).
[18] B. T. Brown, J. Fishman, and T. F. Gallagher, *Nature* **182**, 50 (1958).
[19] G. Serchi, *Chimica* **8**, 9 (1953).
[20] W. R. Slaunwhite, Jr. and A. A. Sandberg, *Arch. Biochem. Biophys.* **63**, 478 (1956).
[21] C. J. Migeon, P. E. Wall, and J. J. Bertrand, *J. Clin. Invest.* **38**, 619 (1959).
[22] M. Levitz, J. R. Spitzer, and G. H. Twombly, *J. Biol. Chem.* **222**, 981 (1956).
[23] G. F. Marrian and W. S. Bauld, *Biochem. J.* **59**, 136 (1955).
[24] H. Breuer and G. Pangels, *Z. Physiol. Chem.* **322**, 177 (1960).
[25] H. Breuer, *Nature* **185**, 613 (1960).
[26] H. Breuer and G. Pangels, *Biochim. Biophys. Acta* **36**, 572 (1959).
[27] K. H. Loke, G. F. Marrian, and E. J. D. Watson, *Biochem. J.* **71**, 43 (1959).
[28] E. Diczfalusy and C. Lauritzen, "Oestrogene beim Menschen." Springer, Berlin, 1961.
[29] H. Breuer, *Vitamins Hormones* **20**, 285 (1962).
[30] H. Breuer, *Excerpta Med.* **83**, 1106 (1965).
[31] R. I. Dorfman and F. Ungar, "Metabolism of Steroid Hormones." Academic Press, New York, 1965.
[32] H. Breuer, *Rev. European Endocrinol. Suppl.* **2**, p. 295.
[33] L. J. Langer and L. L. Engel, *J. Biol. Chem.* **233**, 583 (1958).
[34] L. J. Langer, J. A. Alexander, and L. L. Engel, *J. Biol. Chem.* **234**, 2609 (1959).
[35] J. Jarabak, J. A. Adams, H. G. Williams-Ashman, and P. Talalay, *J. Biol. Chem.* **237**, 345 (1962).

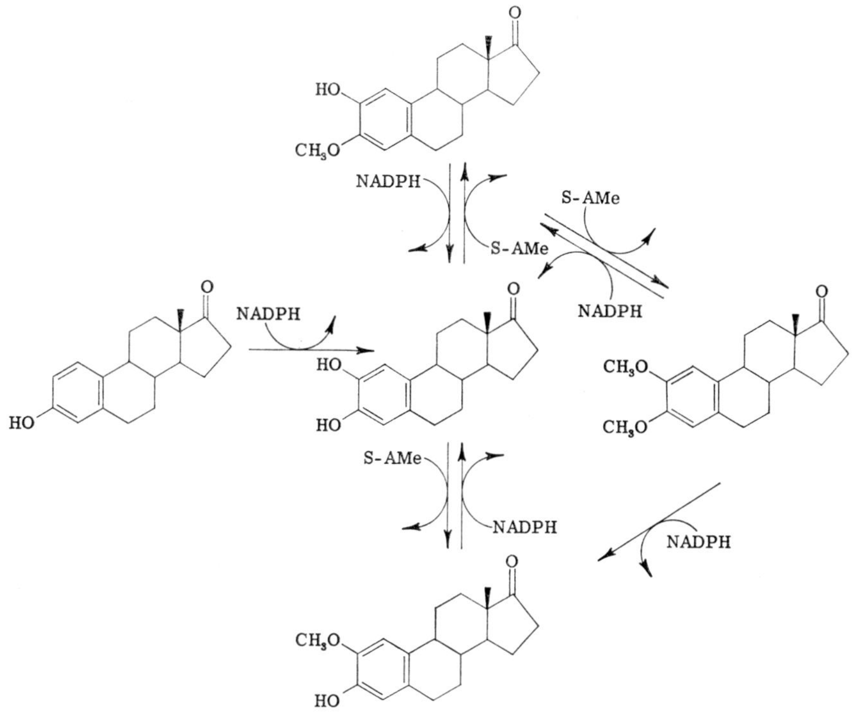

FIG. 1. Biogenesis, methylation, and demethylation of C-2 substituted estrone in liver. S-AMe = S-adenosylmethionine.

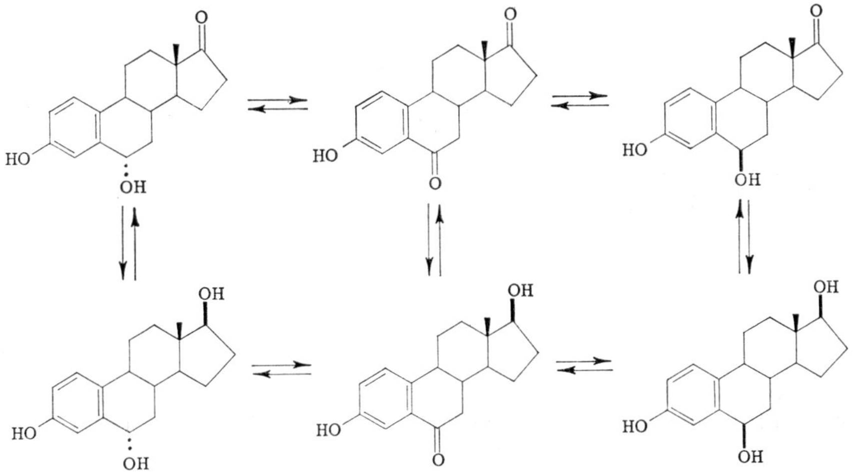

FIG. 2. Intermediary metabolism of C-6 substituted estrone and 17β-estradiol in liver.

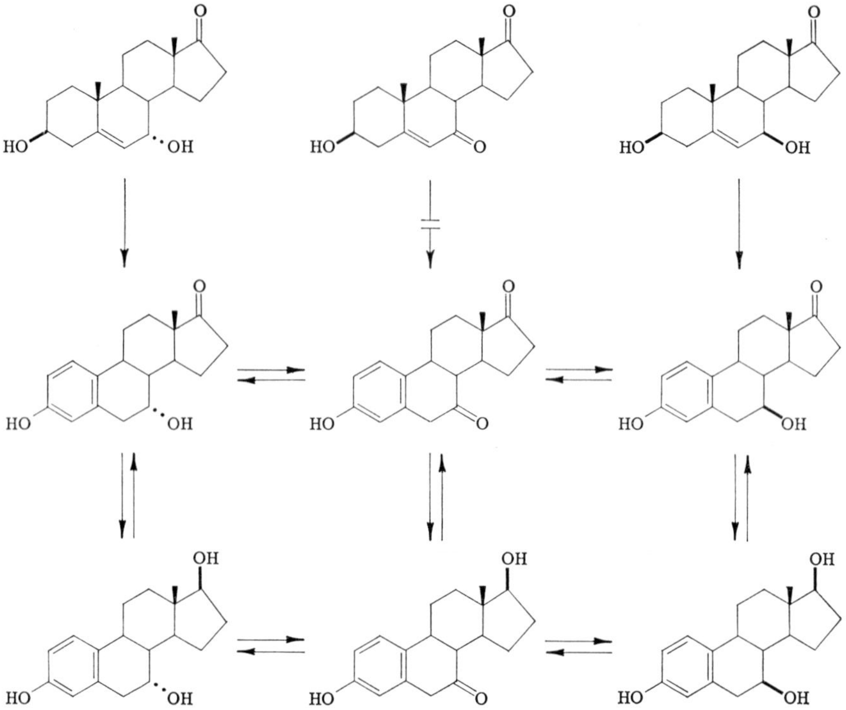

FIG. 3. Biogenesis and metabolism of C-7 substituted estrone and 17β-estradiol.

rat kidney[36] (e.g., oxidoreductases, glucuronyl transferases) has been attempted only in preliminary fashion by precipitation with ammonium sulfate. The only enzyme that has been further purified is the O-methyltransferase; this, however, is by no means a specific enzyme of estrogen metabolism, but reacts also with nonsteroidal substrates. On the basis of the available information, then, the specificity of most of the enzymes involved in the metabolism of estrogens is uncertain.

It is also important to note that the investigation of estrogen metabolism in animal tissue preparations is hampered by the comparatively low activity of the enzymes involved. Hence it is necessary to use relatively long incubation times (usually 60–120 minutes) and the most sensitive chemical and physical methods for analysis of the steroid substrates or their products. For example, most NAD+- and NADP+-linked estrogen oxidoreductases, in their present state of purity, cannot be measured by change in optical density at 340 mμ, but are assayed by paper chromatographic isolation of the reaction products, which are then determined fluorometrically. Moreover, in crude enzyme preparations, the steroid substrate and the primary metabolites may be exposed to many enzymes

[36] R. A. Meigs and K. J. Ryan, J. Biol. Chem. 241, 4011 (1966).

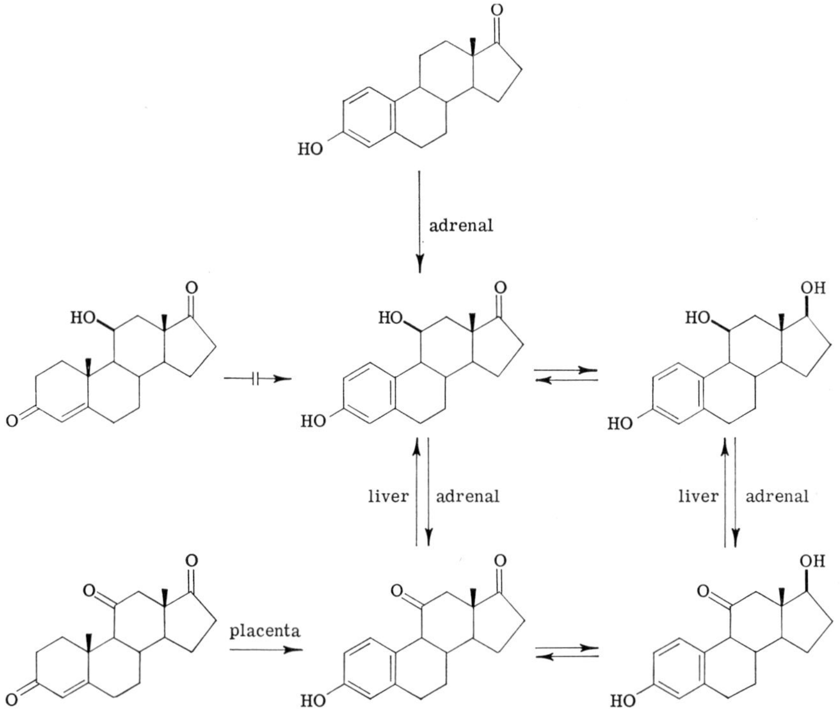

FIG. 4. Biogenesis and metabolism of C-11 substituted estrone and 17β-estradiol.

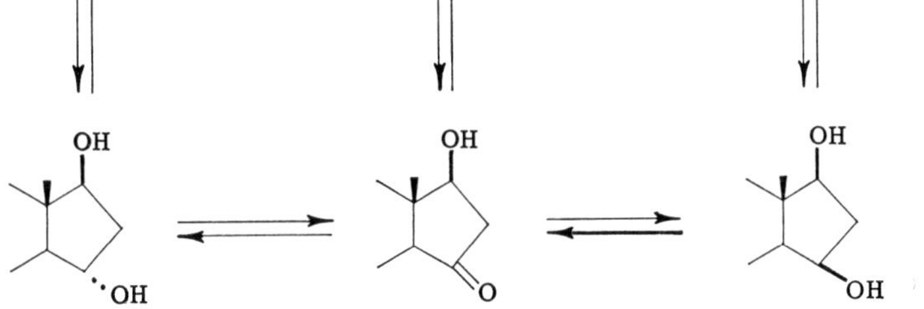

FIG. 5. Intermediary metabolism of C-15 substituted estrone and 17β-estradiol in liver.

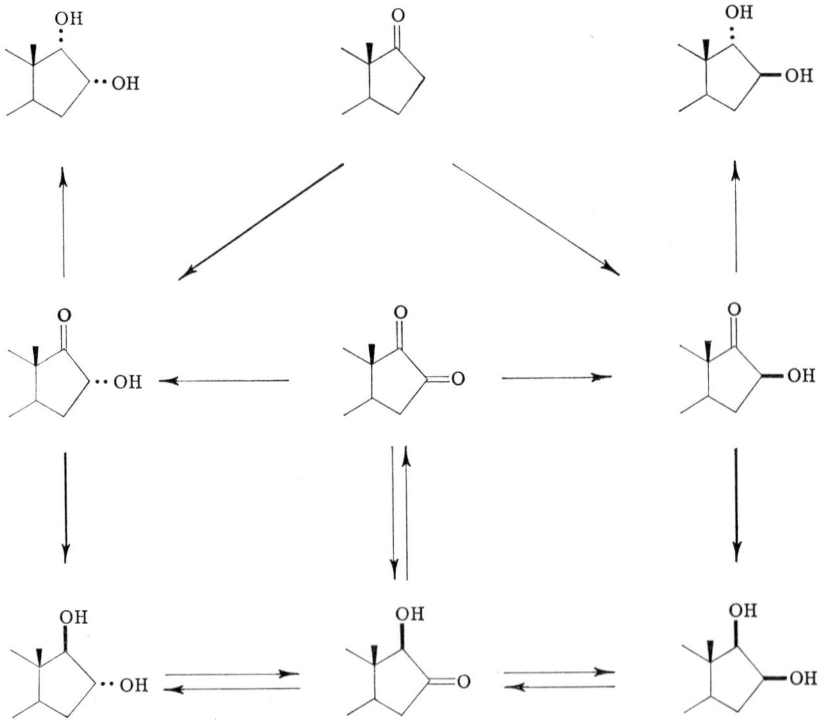

FIG. 6. Intermediary metabolism of C-16 substituted estrogens in liver.

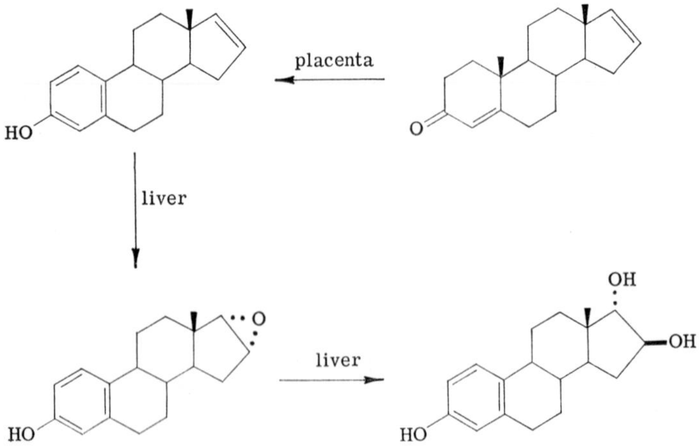

FIG. 7. Biogenesis and metabolism of 16α,17α-epoxyestratrien-3-ol.

simultaneously, and hence further transformations may occur. It is obvious that great care has to be exercised in the interpretation of kinetic data obtained under such conditions.

In the following presentation, an attempt is made to draw together most of the available data relating to estrogen metabolism, bearing in mind the limitations indicated above. Since several analytical methods are referred to recurrently throughout the subsequent discussion, these are first outlined in a separate section.

I. Some Analytical Methods

Preparation of Adrenal Brei

Bovine adrenals are collected as soon after slaughter as possible, freed of adhering fat, and immediately placed in ice. The tissue is taken to the laboratory, and the beef adrenal brei is made by finely grinding the fresh adrenal glands in a meat grinder.

Preparation of the Microsomal and Cytoplasmic Fractions

The tissue is chilled immediately after removal by immersion in $0.25\,M$ sucrose at $0°$, blotted, weighed, and then minced with scissors into pieces. Usually, cold $0.25\,M$ sucrose is added to 1 g of tissue to give a final volume of 5 ml. The tissue is then homogenized for 1 minute in an all-glass homogenizer. During homogenization, the glass tube of the homogenizer is kept in ice water. The homogenate is centrifuged at $+2°$ in a refrigerated ultracentrifuge (Spinco model L-2) at $1000\,g$ for 15 minutes. The sediment is discarded and the supernatant is centrifuged at $20,000\,g$ for 45 minutes. For further fractionation, the $20,000\,g$ supernatant is centrifuged at $100,000\,g$ for 60 minutes. The firmly packed, reddish transparent sediment, containing the microsomal fraction, is redispersed with an all-glass homogenizer in a small volume of $0.25\,M$ sucrose, and this solution again is centrifuged at $100,000\,g$ for 30 minutes; this procedure is repeated twice. The pellet thus obtained is resuspended in $0.25\,M$ sucrose, so that 1 ml of the medium contains the washed microsomal fraction of 1 g of fresh tissue.

The supernatant obtained after the first centrifugation at $100,000\,g$ contains the cytoplasmic fraction. To prepare a further purified cytoplasmic fraction which is free from ribosomes, the $100,000\,g$ supernatant is centrifuged at $150,000\,g$ for 60 minutes. The $150,000\,g$ supernatant is designated as ground plasma.

Solubility and Stability of Estrogens

Because of their relative insolubility in water, it is not possible to add phenolic steroids in aqueous solutions directly to the incubation

TABLE I

Paper Chromatographic Systems, R_f Values, and Mobilities of Phenolic C_{18}-Steroids (Estrogens) Used as Substrates or Formed as Reaction Products in Incubation Experiments with Enzymes of Estrogen Metabolism[a]

Steroid	Paper chromatographic system	R_f value	Mobility	System
Estrone	Monochlorobenzene[b]	0.76	11.1 cm/hour	C
17β-Estradiol	Monochlorobenzene[b]	0.39	5.7 cm/hour	C
Estriol	Chloroform[b]	0.65	6.8 cm/hour	A
	Chloroform[b]	0.04	0.43 cm/hour	A
	Chloroform–ethyl acetate (5:1)	—	25.6 cm/15 hours	D
2-Methoxy-17β-estradiol	Monochlorobenzene[b]	0.83	12.1 cm/hour	C
	Cyclohexane[b]	—	18.2 cm/20 hours	
Equilin	Cyclohexane[c]	—	28 cm/30 hours	G
Equilenin	Cyclohexane[c]	—	20 cm/30 hours	G
6α-Hydroxyestrone	Chloroform[b]	0.18	1.9 cm/hour	A
	Monochlorobenzene–ethyl acetate (3:1)[b]	0.18	34.1 cm/10 hours	B
6β-Hydroxyestrone	Chloroform[b]	0.18	1.9 cm/hour	A
	Monochlorobenzene–ethyl acetate (3:1)[b]	0.17	30.0 cm/10 hours	B
6-Oxoestrone	Chloroform[b]	0.81	8.4 cm/hour	A
	Monochlorobenzene[b]	0.24	3.5 cm/hour	C
6α-Hydroxy-17β-estradiol	Chloroform[b]	0.04	0.43 cm/hour	A
	Chloroform–ethyl acetate (5:1)[b]	—	21.0 cm/15 hours	D
6β-Hydroxy-17β-estradiol	Chloroform[b]	0.04	0.43 cm/hour	A
	Chloroform–ethyl acetate (5:1)[b]	—	17.5 cm/15 hours	D
7α-Hydroxyestrone	Chloroform[b]	0.18	1.9 cm/hour	A
	Monochlorobenzene–ethyl acetate (3:1)[b]	0.17	31.1 cm/10 hours	B
7β-Hydroxyestrone	Chloroform[b]	0.18	1.9 cm/hour	A
	Monochlorobenzene–ethyl acetate (3:1)[b]	0.19	35.2 cm/10 hours	B
7-Oxoestrone	Chloroform[b]	0.81	8.4 cm/hour	A
	Monochlorobenzene[b]	0.24	3.5 cm/hour	C
11β-Hydroxyestrone	Chloroform[b]	0.35	3.6 cm/hour	A
	Monochlorobenzene–ethyl acetate (3:1)[b]	0.29	5.1 cm/hour	B
11-Oxoestrone	Monochlorobenzene[b]	—	17.5 cm/6 hours	C

Compound	Solvent system			
11β-Hydroxy-17β-estradiol	Chloroform[b]	0.06	0.71 cm/hour	A
	Chloroform–ethyl acetate (5:1)[b]	—	32.9 cm/15 hours	D
15α-Hydroxyestrone	Chloroform[b]	0.18	1.9 cm/hour	A
	Monochlorobenzene–ethyl acetate (3:1)[b]	0.26	4.6 cm/hour	B
15β-Hydroxyestrone	Chloroform[b]	0.19	2.0 cm/hour	A
	Monochlorobenzene–ethyl acetate (3:1)[b]	0.33	5.7 cm/hour	B
15-Oxo-17β-estradiol	Monochlorobenzene–ethyl acetate (3:1)[b]	0.29	5.2 cm/hour	B
15α-Hydroxy-17β-estradiol	Chloroform[b]	0.04	0.43 cm/hour	A
	Chloroform–ethyl acetate (5:1)[b]	—	22.9 cm/15 hours	D
	Benzene–methanol–water (100:55:45)	—	12.0 cm/24 hours	E
15β-Hydroxy-17β-estradiol	Chloroform[b]	0.04	0.43 cm/hour	A
	Chloroform–ethyl acetate (5:1)[b]	—	23.2 cm/15 hours	D
	Benzene–methanol–water (100:55:45)	—	14.0 cm/24 hours	E
16α,17α-Epoxyestratrienol	Cyclohexane[c]	—	3.7 cm/6 hours	G
16α-Hydroxyestrone	Chloroform[b]	0.35	3.6 cm/hour	A
	Monochlorobenzene–ethyl acetate (3:1)[b]	0.34	5.9 cm/hour	B
	Light petroleum–benzene–methanol–water (6:4:7:3)	0.25	3.2 cm/hour	F
16β-Hydroxyestrone	Chloroform[b]	0.22	2.3 cm/hour	A
	Monochlorobenzene–ethyl acetate (3:1)[b]	0.26	4.5 cm/hour	B
	Light petroleum–benzene–methanol–water (6:4:7:3)	0.17	2.1 cm/hour	F
16-Oxo-17β-estradiol	Chloroform[b]	0.30	3.1 cm/hour	A
	Monochlorobenzene–ethyl acetate (3:1)[b]	0.32	5.6 cm/hour	B
	Light petroleum–benzene–methanol–water (6:4:7:3)	0.25	3.2 cm/hour	F
16-Oxoestrone	Light petroleum–benzene–methanol–water (6:4:7:3)	0.41	5.1 cm/hour	F
16-Epiestriol	Chloroform[b]	0.15	1.6 cm/hour	A
	Monochlorobenzene–ethyl acetate (3:1)[b]	0.26	4.5 cm/hour	B
	Benzene–methanol–water (100:55:45)	0.38	4.9 cm/hour	E
17-Epiestriol	Chloroform[b]	0.18	1.9 cm/hour	A
	Monochlorobenzene–ethyl acetate (3:1)[b]	0.29	5.0 cm/hour	B
	Benzene–methanol–water (100:55:45)	0.46	5.9 cm/hour	E

[a] For details, see text footnote 37.
[b] On formamide-impregnated paper.
[c] On propylene glycol-impregnated paper.

medium. Therefore, stock solutions are made with methanol, ethanol, or propylene glycol and stored in a refrigerator. Usually, 4 mg of nonradioactive steroids are dissolved in 1.0 ml of organic solvent. Aliquots of these solutions are added to the incubations; the final concentration of the organic solvent in the incubation medium should not exceed 0.5%. Radioactive steroids are stored in benzene–methanol (95:5). Before use, the organic phase is evaporated *in vacuo,* and the dry residue, which must be free of traces of benzene, is redissolved in a small amount (0.02 ml) of propylene glycol. To this propylene glycol solution the incubation mixture is then added. In general, stock solutions of phenolic steroids in highly purified alcohols are stable at +2° for at least 6 months, with the exception of 2-hydroxylated estrogens. These steroids are very unstable and should be dissolved in organic solvents only shortly before use.

Many of the phenolic steroids, particularly those hydroxylated at positions 2, 6α, 6β, 15α, 15β, 16α, 16β, and 18 are sensitive toward alkali and/or acid. For example, in alkaline solutions, 2-hydroxylated estrogens are rapidly oxidized, whereas 16α- and 16β-hydroxyestrone undergo rearrangement to 16-oxo-17β-estradiol; 18-hydroxyestrone looses formaldehyde and is thereby converted to 18-norestrone. In acidic solutions, 6α- and 6β-hydroxyestrogens are interconvertible; under the same conditions, 15α- as well as 15β-hydroxyestrone are converted to 15- and 14-dehydroestrone. It is therefore advisable to avoid alkaline or acidic steps during the working-up procedures.

Paper Chromatography of Estrogens

Extensive work on the paper chromatography of estrogens has been carried out in the past, and this has been reviewed in detail by Knuppen[37] and by Oakey.[38] Strips of Schleicher & Schüll 2043 b Mgl paper (45 cm × 15 cm) are immersed in methanol-formamide (1:1, v/v), blotted with filter paper, and dried in a horizontal position at room temperature for 30 minutes. After this time, the steroids, dissolved in methanol or ethanol, are applied to the paper. All chromatograms are equilibrated for at least 1 hour at 22±2°. The chromatograms are developed with various systems, which are summarized in Table I. After development of the chromatograms, the paper is dried between 70 and 80° for about 3 hours, and the positions of the estrogens are located by Folin-Ciocalteu reagent.

As there is no single paper chromatographic system that is satisfactory for separating all estrogens, and the assumption that a spot corresponds to a particular estrogen is an extremely unreliable interpretation, different systems have to be applied to obtain a preliminary answer

[37] R. Knuppen, *Z. Vitamin-, Hormon- Fermentforsch.* **12,** 355 (1962/63).
[38] R. E. Oakey, *J. Chromat.* **8,** 2 (1962).

as to the identity of an estrogen. This is particularly true for estrogens that are very similar in their mobilities.

Kober Reaction of Estrogens

In 1931, Kober[39] described a color reaction for estrone. This consisted of heating with a mixture of concentrated sulfuric acid and phenol, diluting with water and reheating, when a bright rose pink solution resulted. The Kober reaction, which proved to be highly specific for estrogens, was further studied and modified by Bauld[40] and Brown.[41] Nocke[42] reexamined the conditions of the color reaction in an attempt to improve its sensitivity and reproducibility. He found that, depending on the quality of the sulfuric acid reagent and the time of heating, the color intensities of different estrogens varied considerably. The optimal conditions for the classical and some of the more recently discovered estrogens have been described by Nocke,[42] Breuer,[43] and in the appropriate papers (see text). If necessary, the optimal conditions for estrogens in the Kober reaction can be determined as follows.

Between 10 and 20 Kober tubes, each containing the same amount (5–10 μg) of the steroid to be tested and 4 mg of quinol, are used in each experiment. To determine the optimal sulfuric acid concentration in the second stage of the reaction, increasing amounts of the appropriate Kober reagent[42] are added to the steroid-containing residues; the volume can vary between 0.5 and 3.2 ml of the reagent, depending upon the sulfuric acid concentration required in the second stage of the reaction. The tubes are heated for 20 minutes in a boiling water bath and then cooled for 5 minutes. Water is added to make up the mixture to a final volume of 3.2 ml and, after shaking, the tubes are heated for a further 10 minutes. The extinctions are read at three different wavelengths ($E_1 = 470$–490 mμ; $E_2 = 515$–525 mμ; $E_3 = 550$–560 mμ), and the corrected reading [$E_{corr} = 2 \times E_2 - (E_1 + E_3)$] is plotted against the sulfuric acid concentrations. The optimal sulfuric acid concentration is finally adopted. In the next series of experiments, the heating time in the second stage of the reaction is varied between 0 and 30 minutes, after the first stage of the reaction has been carried out as before. In this way, the optimal heating time for each steroid is estimated. In the third series of experiments, the influence of sulfuric acid concentrations in the first stage of the reaction is tested.

[39] S. Kober, *Biochem. Z.* **239**, 209 (1931).
[40] W. S. Bauld, *Biochem. J.* **56**, 426 (1954).
[41] J. B. Brown, *Biochem. J.* **60**, 185 (1955).
[42] W. Nocke, *Biochem. J.* **78**, 593 (1961).
[43] H. Breuer, *in* "Estrogen Assays in Clinical Medicine," (C. A. Paulsen, ed.), p. 88. Univ. of Washington Press, Seattle, Washington, 1965.

Ten different color reagents with sulfuric acid concentrations of 50–96% are heated with 5 μg of steroid for 20 minutes in the first stage. After cooling, water is added to give the required final optimal concentration of sulfuric acid. The tubes are reheated for the time found optimal in the previous experiment. The volumes of Kober reagents and water are adjusted to give a final volume of 3.2 ml for all reaction mixtures. In the last series of experiments, the heating times in the first stage are varied between 0 and 50 minutes, the remaining reaction conditions being optimal.

II. Aromatization

Phenolic steroids (estrogens), ring B-unsaturated phenolic steroids (equilin and equilenin), and ring B-aromatic C_{18}-steroids (Heard's ketone) arise by aromatization of neutral C_{19}- or C_{18}-steroids (for review see Talalay,[44] Breuer,[29] Ryan[45] and recent detailed discussion of metabolic pathways by Dorfman and Ungar[31]). The highest known activity of the aromatizing enzyme system in mammalian tissue is in human placenta.[46] In the conversion of Δ^4-androstene-3,17-dione and 19-hydroxy-Δ^4-androstene-3,17-dione to estrone by human placental microsomes, formaldehyde is split off as the volatile component.[47] These results support the following course of aromatization: neutral C_{19}-steroid → 19-hydroxy compound → 19-oxo compound → phenolic C_{18}-steroid + formaldehyde. However, Dorfman and his colleagues,[48] in confirming the observation that formaldehyde is liberated from the 19-hydroxy compound, also found that the 19-oxo compound yields formic acid in addition to the estrogen. The enzymatic processes of microbiological aromatization have been studied by a number of investigators (for review see Dorfman and Ungar,[31] Djerassi,[49] Čapek, Hanč, and Tadra[50]). Levy and Talalay[51] have shown that *Pseudomonas testosteroni*, like a number of other microorganisms, can convert the saturated ring A of 3-oxosteroids to the 1,4-dien-3-one grouping. Desaturation also occurs in neutral C_{18}-

[44] P. Talalay, *Physiol. Rev.* 37, 362 (1957).
[45] K. J. Ryan, *Proc. 5th Intern. Congr. Biochem. Moscow, 1961.* Vol. 7, p. 381.
[46] K. J. Ryan, *J. Biol. Chem.* 234, 268 (1959).
[47] H. Breuer and P. Grill, *Z. Physiol. Chem.* 324, 254 (1961).
[48] T. Morato, M. Hayano, R. I. Dorfman, and L. R. Axelrod, *Biochem. Biophys. Res. Commun.* 6, 334 (1961).
[49] C. Djerassi, "Steroid Reactions, An Outline for Organic Chemists." Holden-Day, San Francisco, California, 1963.
[50] A. Čapek, O. Hanč, and M. Tadra, "Microbial Transformations of Steroids." Academia, Prague, 1966.
[51] H. R. Levy and P. Talalay, *J. Biol. Chem.* 234, 2009 (1959).

steroids (19-norsteroids), which then undergo energetically favorable tautomerization with aromatization of ring A.

A. AROMATIZATION OF TESTOSTERONE TO 17β-ESTRADIOL BY HUMAN PLACENTAL MICROSOMES

Assay Method[46, 52]

Δ^4-Androstene-3,17-dione (0.7 micromole) is incubated with the microsomal fraction of human placenta (corresponding to 12.5 g wet weight of fresh tissue), NAD+ (2.5 micromoles) and ATP (10 micromoles) in phosphate buffer (5 ml; 0.05 M, pH 7.0) for 60 minutes at 37° under air. The incubation mixture is extracted with chloroform (3 × 5 ml); the pooled chloroform solutions are washed with distilled water (10 ml) and evaporated to dryness. The residue is dissolved in pentane (20 ml), and the pentane solution is extracted with aqueous methanol (90%, v/v) (3 × 20 ml). The pooled methanolic extracts are evaporated, and the residue is dissolved in toluene (50 ml). The toluene is extracted with sodium hydroxide (1 N) (4 × 15 ml) and water (2 × 10 ml). The combined alkali and water extracts are adjusted to pH 8.5 and extracted with ether (3 × 50 ml). The residue is chromatographed in system C. Estrone, R_f 0.76, and 17β-estradiol, R_f 0.39, are eluted and assayed by the Kober reaction.[42] The yield of estrone and 17β-estradiol from Δ^4-androstene-3,17-dione varies from 10 to 30%.

A lyophilized enzyme preparation from human placenta ("Aromatase")[52a] which converts testosterone into 17β-estradiol is suitable for the specific and sensitive measurement of microgram quantities of testosterone in body fluids. The method follows the original procedure as described by Finkelstein, Forchielli, and Dorfman.[53] Testosterone (3.5 nanomoles) is incubated with the enzyme preparation [2 mg of "Aromatase" suspended in phosphate buffer (0.5 ml, 0.1 M, pH 7.2)], NADP+ (0.4 micromoles), glucose 6-phosphate (5.4 micromoles), and glucose-6-phosphate dehydrogenase (0.5 Kornberg unit) in potassium chloride (0.6 ml, 0.154 M) for 60 minutes at 37°. At the end of the incubation, ethanol (96%, v/v) (5 ml) is added to the mixture. The precipitate is centrifuged out and washed with ethanol (2 × 2 ml). To the pooled ethanol supernatant, sodium borohydride (3 mg) is added to convert estrone into 17β-estradiol. After 15 minutes, the solution is neutralized with acetic

---

[52] H. Breuer and R. Ortlepp, *Acta Endocrinol.* **35**, 508 (1960).
[52a] Commercially available from Mann Research Laboratories, Inc., 136 Liberty Street, New York, New York.
[53] M. Finkelstein, E. Forchielli, and R. I. Dorfman, *J. Clin. Endocrinol.* **21**, 98 (1961).

acid (0.05 ml, 1 N). The ethanolic solution is evaporated almost to dryness, the residue is taken up in water (20 ml), and the aqueous phase is extracted with ether (3 × 10 ml); the ether phase is washed with sodium bicarbonate (2 × 2 ml, 1 N) and water (2 × 2 ml). After drying over sodium sulfate, the ether is evaporated to dryness and the residue is subjected to paper chromatography in system F. After elution, 17β-estradiol (R_f 0.54) is determined by fluorometry.[54]

Properties

Distribution. The aromatizing enzyme system of human placenta is localized in the microsomal fraction and requires both NADPH and oxygen.[46] Aromatization of neutral C_{19}-steroids has been found in human placenta (testosterone,[46] 19-hydroxy-Δ^4-androstene-3,17-dione[55] and numerous substituted C_{19}-steroids[31]), equine placenta (testosterone, 19-hydroxytestosterone, 19-oxo-Δ^4-androstene-3,17-dione[56]), bovine placenta (19-hydroxy-Δ^4-androstene-3,17-dione[55]), human ovary (testosterone,[57,58] Δ^4-androstene-3,17-dione[59]), rat ovary (testosterone[60]), mouse ovary (testosterone[61]), dog ovary (testosterone[62]), frog ovary (testosterone[63]), stallion testis (testosterone[64]), human adrenal carcinoma (testosterone[64,65]), cow adrenal (19-hydroxy-Δ^4-androstene-3,17-dione[55]) and human fetal liver (testosterone, Δ^4-androstene-3,17-dione[66]).

Specificity. Evaluation of the substrate specificity of the aromatizing enzyme system has shown that it is restricted to C_{19}-steroids.[45,67] However, a number of substituted C_{19}-steroids and of neutral C_{18}-steroids are aromatized by the placental enzyme to the corresponding phenolic

---

[54] S. Landany and M. Finkelstein, *Steroids* **2**, 297 (1963).

[55] A. S. Meyer, *Biochim. Biophys. Acta* **17**, 441 (1955).

[56] L. Stárka, J. Breuer, and H. Breuer, *Naturwissenschaften* **52**, 540 (1965).

[57] B. Baggett, L. L. Engel, K. Savard, and R. I. Dorfman, *J. Biol. Chem.* **221**, 931 (1956).

[58] H. H. Wotiz, J. W. Davis, H. M. Lemon, and M. Gut, *J. Biol. Chem.* **222**, 487 (1956).

[59] O. W. Smith and K. J. Ryan, *Endocrinology* **69**, 869 (1961).

[60] S. R. Stitch, R. E. Oakey, and S. S. Eccles, *Biochem. J.* **88**, 70 (1963).

[61] N. Hollander and V. P. Hollander, *Cancer Res.* **19**, 290 (1959).

[62] N. Hollander and V. P. Hollander, *J. Biol. Chem.* **233**, 1097 (1958).

[63] R. Ozon and H. Breuer, *Z. Physiol. Chem.* **337**, 61 (1964).

[64] B. Baggett, L. L. Engel, L. Balderas, G. Lanman, K. Savard, and R. I. Dorfman, *Endocrinology* **64**, 600 (1959).

[65] L. L. Engel, *Cancer* **10**, 711 (1957).

[66] S. Mancuso, S. Dell'Acqua, G. Eriksson, N. Wiqvist, and E. Diczfalusy, *Steroids* **5**, 183 (1965).

[67] C. Gual, T. Morato, M. Hayano, M. Gut, and R. I. Dorfman, *Endocrinology* **71**, 920 (1962).

steroids (see Table II and also footnote 67). The Δ^1-3-oxo structure is not transformed; $\Delta^{1,4}$-dienones and 19-nor (10β) compounds are aromatized slowly. The presence of an 11β-hydroxyl group interferes with ring A aromatization[67,68]; however, aromatization takes place readily in the presence of an 11α-hydroxyl group, as it does also in structures with substituents at the 9α position.[67]

Stability. Active preparations of the microsomal aromatizing enzyme system of human placenta can be stored in the freezer for weeks without loss of activity.[46] The lyophilized microsomal enzyme is stable for months.

TABLE II

RELATIVE SUBSTRATE ACTIVITY IN STEROID AROMATIZATION
BY HUMAN PLACENTA[a]

Substrate	Product	Percent activity relative to Δ^4-androstene-3,17-dione
Δ^4-Androstene-3,17-dione	Estrone	100
Testosterone	17β-Estradiol	100
19-Hydroxy-Δ^4-androstene-3,17-dione	Estrone	184
17α-Methyltestosterone	17α-Methyl-17β-estradiol	44
1β-Hydroxytestosterone	1-Hydroxy-17β-estradiol	30
17β-Hydroxy-$\Delta^{1,4}$-androstadien-3-one	17β-Estradiol	25
$\Delta^{1,4}$-Androstadiene-3,17-dione	Estrone	22
19-Nor-Δ^4-androstene-3,17-dione	Estrone	21
6β-Hydroxy-Δ^4-androstene-3,17-dione	6β-Hydroxyestrone	21
Epitestosterone	17α-Estradiol	20
19-Nortestosterone	17β-Estradiol	20
2β-Hydroxytestosterone	2-Hydroxy-17β-estradiol	13
11α-Hydroxytestosterone	11α-Hydroxy-17β-estradiol	12
11β-Hydroxytestosterone	—	0
17β-Hydroxyandrostan-3-one	—	0
Δ^1-Androstene-3,17-dione	—	0
2α-Hydroxy-Δ^4-androstene-3,17-dione	—	0

[a] K. J. Ryan, *Proc. 5th Intern. Congr. Biochem., Moscow, 1961*, Vol. 7, p. 381.

Inhibition. Aromatization is inhibited in the absence of oxygen. Disodium ethylenediaminetetraacetate, Cu^{++}, Hg^{++}, cyanide, and iodoacetate at 10^{-3} M concentration have slight or no inhibitory effects.[46] A 50% decrease in the aromatizing activity of placental preparations is obtained with 4.42 micromoles 2-methyl-1,2-bis-(3-pyridyl)-1-propanone

[68] H. Breuer, *In* "Symposion über Krebsprobleme" (K. G. Ober, H. M. Rauen, J. Schoenmackers, and J. Zander, eds.), p. 62. Springer, Berlin, 1961.

(SU 4885) ; 0.44 micromole has little effect, whereas 22.1 micromoles have an effect similar to that produced by 4.42 micromoles.[69]

pH Effect. The optimum pH for the microsomal aromatizing enzyme is at 7.0.

Specific Activity. No unit of enzymatic activity has been defined. Results are reported in micrograms or micromoles of product formed.

B. Aromatization of 19-Nor-Δ^4-androstene-3,17-dione to Estrone by Microorganisms

Aromatization of C_{19}-norsteroids (19-nortestosterone, 19-nor-Δ^4-androstene-3,17-dione) to estrogens is accomplished by a number of microorganisms (for review, see footnote 70). These include *P. testosteroni* (19-nortestosterone[71, 72]), 19-nor-Δ^4-androstene-3,17-dione[50, 71, 72]), *Corynebacterium simplex* (19-norprogesterone,[73, 74] 11-oxo-19-norprogesterone,[74] 19-nor-$\Delta^{4, 7}$-androstadiene-3,17-dione,[75] 19-nortestosterone acetate[76]), and *Septomyxa affinis* (2α-methyl-19-nortestosterone,[77] 4-methyl-19-nortestosterone,[77] 19-nor-Δ^4-androstene-3,17-dione[71]). 19-Hydroxy-Δ^4-androstene-3,17-dione is aromatized to estrone by *Pseudomonas* sp.[78] It should be mentioned that Δ^4-androstene-3,17-dione is converted to 3-hydroxy-9,10-seco-$\Delta^{1,3,5(10)}$-androstatriene-9,17-dione by incubation with species of *Pseudomonas* and *Arthrobacter*,[79] and with *Mycobacterium smegmatis*;[80] according to Dodson and Muir,[79] this microbial aromatization of Δ^4-androstene-3,17-dione resembles, in many respects, the sequences for the conversion of androgenic steroids to estrogens in mammals.

The aromatization of 19-nor-Δ^4-androstene-3,17-dione to estrone by a Δ^1-dehydrogenase from *P. testosteroni* has been studied in detail by Levy and Talalay.[50, 72] This reaction has been described in an earlier volume in this series (cf. H. R. Levy, Vol. V, p. 533).

[69] C. Giles and K. Griffiths, *J. Endocrinol.* **28**, 343 (1964).
[70] C. Tamm, *Angew. Chem. Intern. Ed.* **1**, 178 (1962).
[71] H. R. Levy and P. Talalay, *J. Am. Chem. Soc.* **79**, 2568 (1957).
[72] H. R. Levy and P. Talalay, *J. Biol. Chem.* **234**, 2014 (1959).
[73] A. Bowers, C. Casas-Campillo, and C. Djerassi, *Tetrahedron* **2**, 165 (1958).
[74] A. Bowers, J. S. Milz, C. Casas-Campillo, and C. Djerassi, *J. Org. Chem.* **27**, 361 (1962).
[75] J. A. Zderic, A. Bowers, H. Carpio, and C. Djerassi, *J. Am. Chem. Soc.* **80**, 2596 (1958).
[76] S. Kushinsky, *J. Biol. Chem.* **230**, 31 (1958).
[77] D. H. Peterson, L. M. Reineke, H. C. Murray, and O. K. Sebek, *Chem. J. Ind.* p. 1301 (1960).
[78] R. M. Dodson and R. D. Muir, *J. Am. Chem. Soc.* **83**, 4631 (1961).
[79] R. M. Dodson and R. D. Muir, *J. Am. Chem. Soc.* **83**, 4627 (1961).
[80] K. Schubert, K. H. Boehme, and C. Hoerhold, *Z. Naturforsch.* **15b**, 584 (1960).

C. Biogenesis of Phenolic Unsaturated Ring-B and Nonphenolic Aromatic Ring-B C_{18}-Steroids from Androstenolone

Aromatization plays an important role in the biogenesis of phenolic unsaturated ring-B C_{18}-steroids[81] (equilin, equilenin) and of nonphenolic aromatic ring-B C_{18}-steroids[82] (Heard's ketone = 3β-hydroxy-$\Delta^{5,7,9}$-estratrien-17-one) from 3β-hydroxy-Δ^5-androsten-17-one (dehydroepiandrosterone). The formation of equilin follows this sequence of intermediates:[83] androstenolone → 7α-hydroxyandrostenolone → 3β-hydroxy-$\Delta^{5,7}$-androstadien-17-one → $\Delta^{4,7}$-androstadiene-3,17-dione → equilin (3-hydroxy-$\Delta^{1,3,5(10),7}$-estratetraen-17-one) → equilenin (3-hydroxy-$\Delta^{1,3,5(10),6,8}$-pentaen-17-one). Nonphenolic aromatic ring-B C_{18}-steroids are produced from 3β-hydroxy-$\Delta^{5,7}$-androstadien-17-one. 3β-Hydroxy-$\Delta^{5,7}$-androstadien-17-one is formed by loss of water from 7α-hydroxyandrostenolone in the 23,000 g supernatant of horse liver or placenta, or during perfusion through human placenta; the oxidation to $\Delta^{4,7}$-androstadiene-3,17-dione takes place in the cytoplasmic fraction of liver or placenta.[83] Aromatization of $\Delta^{4,7}$-androstadiene-3,17-dione to equilin and of 3β-hydroxy-$\Delta^{5,7}$-androstadien-17-one to Heard's ketone occurs in the microsomal fraction of human or horse placenta.

Assay Method[83]

3β-Hydroxy-$\Delta^{5,7}$-androstadien-17-one (0.7 micromole) is incubated with the microsomal fraction (1 ml) of horse placenta (corresponding to 1 g wet weight of fresh tissue), NADP+ (2 micromoles), glucose 6-phosphate (16 micromoles), and glucose-6-phosphate dehydrogenase (2.0 Kornberg units) in Tris-maleate buffer (4 ml, 0.2 M, pH 7.4) for 60 minutes at 38° under oxygen. At the end of incubation, ethanol (5 ml, 90%, v/v) is added and the precipitated protein removed by centrifugation. After evaporation of the ethanol, the aqueous phase is diluted with water (final volume 10 ml). The aqueous phase is then extracted with ether (3 $\times$ 10 ml). The combined ether extracts are washed with a saturated sodium bicarbonate solution (1 $\times$ 10 ml) and with water (2 $\times$ 10 ml). After evaporation of the ether *in vacuo,* the residue is dissolved in methanol (20 ml, 90%) and the methanolic solution extracted with *n*-hexane (2 $\times$ 20 ml). After reextraction of *n*-hexane with methanol (90%), the methanolic solutions are combined and evaporated *in vacuo.* The residue is dissolved in ether (20 ml) and extracted with sodium hydroxide (2 $\times$ 10 ml, 1 N). The remaining ether-soluble material is the

[81] L. Stárka, H. Breuer, and L. Cedard, *J. Endocrinol.* 34, 447 (1966).

[82] L. Stárka and H. Breuer, *Biochim. Biophys. Acta* 115, 306 (1966).

[83] L. Stárka and H. Breuer, *Z. Physiol. Chem.* 344, 124 (1966).

neutral fraction. The sodium hydroxide extracts are neutralized to pH 6–7 by the addition of hydrochloric acid (5 N), and the neutralized solution is extracted with ether (2 × 20 ml). The ether extracts, containing the *phenolic fraction*, are evaporated to dryness.

The residue of the *neutral fraction* is subjected to thin-layer chromatography on silica gel in the system chloroform–ethanol 9:1 (R_f values of 3β-hydroxy-$\Delta^{5,7,9}$-estratrien-17-one, 0.57 and of $\Delta^{5,7,9}$-estratriene-3β, 17β-diol, 0.42). After elution, the nonphenolic aromatic ring-B C_{18}-steroids are determined by measuring the ultraviolet absorption (267–269 mμ) in ethanolic solution (yield 15–25%).[83]

The residue of the *phenolic fraction* is chromatographed on paper in system C (R_f values of 3-hydroxy-$\Delta^{1,3,5(10),7}$-estratetraen-17-one [I], 0.73; 3-hydroxy-$\Delta^{1,3,5(10),6,8}$-estrapentaen-17-one [II], 0.64; $\Delta^{1,3,5(10),7}$-estratetraene-3,17β-diol [III], 0.24; and $\Delta^{1,3,5(10),6,8}$-estrapentaene-3-17β-diol [IV], 0.17).[37] For further purification, substances I–IV are eluted and rechromatographed on paper in the system benzene–methanol–water 2:1:1 (R_f values of I, 0.89; II, 0.87; III, 0.77; and IV, 0.75). After elution, the phenolic ring-B unsaturated C_{18}-steroids are determined by the Kober reaction[42] (yield 12%).[83]

Properties

The enzyme system which converts $\Delta^{4,7}$-androstadiene-3,17-dione to equilin and 3β-hydroxy-$\Delta^{5,7}$-androstadien-17-one to Heard's ketone is localized in the microsomal fraction of placental tissue and requires both NADPH and oxygen; so far, it has been found in human[81, 82, 84] and horse[83] placenta. The enzyme system is most probably identical with the microsomal aromatizing enzyme system.

III. Hydroxylation

Hydroxylations by mammalian and nonmammalian tissue preparations have been noted in positions 2, 4, 6α, 6β, 7α, 10, 11β, 14α, 15α, 16α, 16β, and 18 of the estrogen molecule. Microbiological hydroxylations of estrogens have been shown to occur in positions 6β, 7α, 15α, and 16α.

Like C_{21}- and C_{19}-hydroxylases, the estrogen hydroxylases require NADPH and molecular oxygen, and similar mechanisms probably operate in all cases. No estrogen hydroxylase has yet been solubilized; therefore, the properties of these enzymes have been studied only in a limited way. Thus, it remains to be shown whether the same, or different enzymes are involved in the introduction of a hydroxyl group into a given position of C_{21}-, C_{19}-, and C_{18}-steroids. The mechanism and general aspects of

[84] L. Stárka, J. Janata, H. Breuer, and R. Hampl, *European J. Steroids* **1**, 37 (1966).

steroid hydroxylation are discussed in detail by Dorfman and Ungar[31] and by Abraham *et al.*[85] (See also this volume, p. 596.)

The following estrogen hydroxylases will be discussed (see Table III): 6α-, 6β-, 7α-, 11β-, 14α-, 15α-, 16α-, 16β-, and 18-hydroxylase. The mammalian hydroxylations at carbons 2, 4, and 10 are dealt with in this volume [38].

TABLE III

PARTIALLY PURIFIED ESTROGEN HYDROXYLASES IN ANIMAL TISSUES

Hydroxylase	Tissue preparation	Steroid substrate	Reference
6α	Rat liver microsomes	17β-Estradiol	Breuer *et al.*[a]
6β-	Rat liver microsomes	17β-Estradiol	Breuer *et al.*[a]
7α-	Beef adrenal brei	Estrone	Knuppen *et al.*[b]
11β-	Beef adrenal brei	Estrone	Knuppen and Breuer[c]
14α-	Beef adrenal brei	Estrone	Knuppen *et al.*[d]
15α-	Beef adrenal brei	Estrone	Knuppen and Breuer[e]
16α-	Rat liver microsomes	17β-Estradiol	Pangels and Breuer[f]
16β-	Chicken liver microsomes	17β-Estradiol	Ozon and Breuer[g]
18-	Ox adrenal homogenate	Estrone	Loke *et al.*[h]

[a] H. Breuer, R. Knuppen, and G. Pangels, *Biochim. Biophys. Acta* **65**, 1 (1962).
[b] R. Knuppen, O. Haupt, and H. Breuer, *Steroids* **3**, 123 (1964).
[c] R. Knuppen and H. Breuer, *Biochim. Biophys. Acta* **58**, 147 (1962).
[d] R. Knuppen, O. Haupt, and H. Breuer, *Biochem. J.* **105**, 971 (1967).
[e] R. Knuppen and H. Breuer, *Z. Physiol. Chem.* **337**, 159 (1964).
[f] G. Pangels and H. Breuer, *Naturwissenschaften* **49**, 106 (1962).
[g] R. Ozon and H. Breuer, *Z. Physiol. Chem.* **341**, 239 (1965).
[h] K. H. Loke, G. F. Marrian, and E. J. D. Watson, *Biochem. J.* **71**, 43 (1959).

A. 6α-HYDROXYLASE

$$\text{Estrone} + \text{NADPH} + \text{H}^+ + \text{O}_2 \rightarrow 6\alpha\text{-hydroxyestrone} + \text{NADP}^+ + \text{H}_2\text{O}$$

Assay Method[86]

Estrone (0.37 micromole) is incubated with the microsomal fraction (2 ml) of rat liver (corresponding to 2 g wet weight of fresh tissue) and NADPH (4 micromoles) in Krebs phosphate buffer (2 ml, pH 7.4) for 60 minutes at 37° under oxygen. The incubation mixture is extracted with ethyl acetate (3 × 5 ml); the extracts are combined, dried over sodium sulfate, and evaporated to dryness. The residue is chromato-

[85] R. Abraham, E. Balke, K. Krisch, S. Leonhäuser, K. Leybold, K. A. Sack, and H. Staudinger, *in* "Handbuch der Physiologisch- und Pathologisch-Chemischen Analyse" (K. Lang and E. Lehnartz, eds.), 10th ed., Vol. VI, part A, p. 917. Springer, Berlin, 1964.
[86] H. Breuer, R. Knuppen, and G. Pangels, *Biochim. Biophys. Acta* **65**, 1 (1962).

graphed in system A for 8 hours. The fraction containing 6α-hydroxy-estrone (R_f 0.18, 1.9 cm/hour) is eluted and rechromatographed for 10 hours in system B in which 6α-hydroxyestrone (mobility 34.1 cm/10 hours) separates from other hydroxyestrones. After elution, 6α-hydroxy-estrone is determined by the Kober reaction[42, 43] (yield 5%).

Properties

The 6α-hydroxylase is localized in the microsomal fraction and re-quires both NADPH and oxygen.[86]

6α-Hydroxylation of estrogens is found in liver of mouse (17β-estradiol[87]), rat (estrone[88, 89], 17β-estradiol,[86, 88, 90] estriol[91]), human fetus (17β-estradiol[92]), man (estrone,[93] 17β-estradiol[93]), and amphibians (17β-estradiol[94]), in human ovary (17β-estradiol[95]) and placenta (estrone,[96] 17β-estradiol[96]), and in beef adrenal (estrone[97]).

B. 6β-HYDROXYLASE

$$\text{Estrone} + \text{NADPH} + \text{H}^+ + \text{O}_2 \rightarrow \text{6β-hydroxyestrone} + \text{NADP}^+ + \text{H}_2\text{O}$$

Assay Method[86]

The procedure is identical with that described for the 6α-hydroxylase. In the final chromatography in system B, the mobility of 6β-hydroxy-estrone is 30 cm/10 hours. After elution, 6β-hydroxyestrone is deter-mined by the Kober reaction[42, 43] (yield 1%).

Properties

The 6β-hydroxylase is localized in the microsomal fraction and re-quires both NADPH and oxygen.[86]

So far, 6β-hydroxylation of estrogens has been found in rat liver (estrone,[89] 17β-estradiol[86]), and in the microorganisms *Fusarium monili-*

[87] G. C. Mueller and G. Rumney, *J. Am. Chem. Soc.* **79**, 1004 (1957).
[88] H. Breuer, L. Nocke, and R. Knuppen, *Z. Physiol. Chem.* **315**, 72 (1959).
[89] E. Hecker and F. Marks, *Biochem. Z.* **343**, 211 (1965).
[90] H. Breuer and R. Knuppen, *Biochim. Biophys. Acta* **39**, 408 (1960).
[91] H. Breuer, R. Knuppen, and H. Schriefers, *Z. Physiol. Chem.* **319**, 136 (1960).
[92] H. Breuer, R. Knuppen, R. Ortlepp, G. Pangels, and A. Puck, *Biochim. Biophys. Acta* **40**, 560 (1960).
[93] H. Breuer, R. Knuppen, and M. Haupt, *Nature* **212**, 76 (1966).
[94] R. Ozon and H. Breuer, *Z. Physiol. Chem.* **333**, 282 (1963).
[95] H. Breuer, R. Knuppen, and G. Pangels, *Z. Physiol. Chem.* **321**, 57 (1960).
[96] L. Cedard and R. Knuppen, *Steroids* **6**, 307 (1965).
[97] R. Knuppen, M. Behm, and H. Breuer, *Z. Physiol. Chem.* **337**, 145 (1964).

forme (17β-estradiol 3-methyl ether[98]) and *Mortierella alpina* (estrone,[99] 17β-estradiol[99]).

C. 7α-HYDROXYLASE

Estrone → 7α-hydroxyestrone

Assay Method[99a]

Estrone-4-^{14}C (1 μC; specific activity 33.7 mC/millimole) is incubated with beef adrenal brei (10 g) in a medium (40 ml), containing sodium chloride (6.15 millimoles), potassium chloride (0.25 millimoles), magnesium sulfate (0.06 millimoles), sodium phosphate dibasic (0.4 millimole), hydrochloric acid (0.08 millimoles), and sodium citrate (0.90 millimoles). Incubation is carried out with stirring under a continuous flow of oxygen for 120 minutes at 37°. At the end of incubation, methanol is added to give a final concentration of 33% (v/v). The mixture is then dialyzed (dialyzing tubing of Kalle, Wiesbaden, Germany) against 40% aqueous methanol (v/v), and the outside phase is changed three times during the following 3 days. The outside solvents are combined, the methanol is removed under reduced pressure, and the remaining aqueous phase is extracted three times with the same volume of ether. The ether extract is washed with water, dried over sodium sulfate, and evaporated to dryness. The residue is chromatographed for 10 hours in system A (mobility of 7α-hydroxyestrone, 1.9 cm/hour) to separate 7α-hydroxyestrone from other hydroxylated estrones. The fraction containing 7α-hydroxyestrone is rechromatographed for 10 hours in system B. 7α-Hydroxyestrone (mobility 31.1 cm/10 hours) is eluted and purified further by chromatography for 10 hours in system F (mobility of 7α-hydroxyestrone 8.5 cm/10 hours). 7α-Hydroxyestrone is determined either by counting directly the radioactivity from the paper chromatogram (using a paper strip scanner) or, after elution from the paper, by measuring the radioactivity in a liquid scintillation counter (yield 0.5–1%).

Properties

The 7α-hydroxylase is localized in the mitochondrial fraction and requires both NADPH and oxygen.[99b] 7α-Hydroxylation of estrone has

[98] P. Crabbé and C. Casas-Campillo, *J. Org. Chem.* **29**, 2731 (1964).

[99] A. I. Laskin, P. Grabowich, B. Junta, C. de Lisle Meyers, and J. Fried, *J. Org. Chem.* **29**, 1333 (1964).

[99a] Modification of the original preparative method[100] by R. Knuppen and W. Hoffmann.

[99b] W. Hoffmann, R. Knuppen, and H. Breuer, *Z. Physiol. Chem.*, in press.

been shown to occur in beef adrenal,[100] in the liver of human fetus[101] and man,[93] and in the microorganism *Glomerella fusarioides*.[99]

D. 11β-HYDROXYLASE

Estrone → 11β-hydroxyestrone

Assay Method[101a]

The conditions of incubation and method of working up the reaction products into ether are as described for the 7α-hydroxylase. The residue of the ether extract is chromatographed in system A (mobility of 11β-hydroxyestrone, 3.6 cm/hour[102]). To remove 16α-hydroxyestrone, the 11β-hydroxyestrone fraction is rechromatographed in system B (mobility of 11β-hydroxyestrone 5.1 cm/hour[102]). 11β-Hydroxyestrone is determined either by direct radioassay (using a paper strip scanner), or by assay of the eluted radioactivity in a liquid scintillation counter (yield 0.5%).

Properties

The 11β-hydroxylase is localized in the mitochondrial fraction and requires both NADPH and oxygen.[99b] 11β-Hydroxylation of estrone has been shown to occur in beef adrenal.[102] 11β-Hydroxyestrone has also been isolated from a human feminizing adrenal carcinoma by Mahesh and Herrmann.[103]

E. 14α-HYDROXYLASE

estrone → 14α-hydroxyestrone

Assay Method[103a]

The conditions of incubation and extraction of the reaction products into ether are as described for the 7α-hydroxylase. The residue of the ether extract is chromatographed in system B (mobility of 14α-hydroxyestrone, 5.2 cm/hour[103b]). To separate 14α-hydroxyestrone from other phenolic steroids, the 14α-hydroxyestrone fraction is rechromatographed in system A. 14α-Hydroxyestrone (mobility, 2.3 cm/hour[103b]) is deter-

---

[100] R. Knuppen, O. Haupt, and H. Breuer, *Steroids* **3**, 123 (1964).

[101] R. Knuppen, H. Breuer, and E. Diczfalusy, *Excerpta Med.* **111**, 171 (1966).

[101a] Modification of the original preparative method[102] by R. Knuppen and W. Hoffmann.

[102] R. Knuppen and H. Breuer, *Biochim. Biophys. Acta* **58**, 147 (1962).

[103] V. B. Mahesh and W. Herrmann, *Steroids* **1**, 51 (1963).

[103a] Modification of the original preparative method[103b] by R. Knuppen and W. Hoffmann.

[103b] R. Knuppen, O. Haupt, and H. Breuer, *Biochem. J.* **105**, 971 (1967).

mined either by direct radioassay, using a paper strip scanner, or by elution and radioassay in a liquid scintillation counter (yield 6–8%).

Properties

The 14α-hydroxylase is localized in the mitochondrial fraction and requires both NADPH and oxygen.[99b] 14α-Hydroxylation of estrone was found in beef adrenal.[103b]

F. 15α-Hydroxylase

Estrone → 15α-hydroxyestrone

Assay Method[103c]

The conditions of incubation and extraction of the reaction products into ether are as described for the 7α-hydroxylase. The residue of the ether extract is chromatographed in system A (mobility of 15α-hydroxy-estrone, 1.9 cm/hour[104]). To separate 15α-hydroxyestrone from 6α- and 7α-hydroxyestrone, the 15α-hydroxyestrone fraction is rechromatographed in system B. 15α-Hydroxyestrone (mobility 4.6 cm/hour[104]) is determined either by direct radioassay, using a paper strip scanner, or by elution and radioassay in a liquid scintillation counter (yield 5–6%).

Properties

The 15α-hydroxylase is localized in the mitochondrial fraction and requires both NADPH and oxygen.[99b] 15α-Hydroxylation of estrogens is found in liver of human fetus (estrone,[101] 17β-estradiol[101]) and man (17β-estradiol[93]), in human (17β-estradiol[105]) and beef adrenal (estrone,[104] 17β-estradiol[106]), and in the microorganisms *Glomerella fusarioides* (estrone[99]), *Glomerella glycines* (estrone[99]), *Aspergillus carneus* (17β-estradiol[99]), and *Fusarium moniliforme* (estrone,[98] 17β-estradiol[98]).

G. 16α-Hydroxylase

$$17\beta\text{-Estradiol} + \text{NADPH} + \text{H}^+ + \text{O}_2 \rightarrow \text{estriol} + \text{NADP}^+ + \text{H}_2\text{O}$$

Assay Method[107]

17β-Estradiol (0.37 micromole) is incubated with the microsomal fraction (1 ml) of rat liver (corresponding to 1 g wet weight of fresh

---

[103c]Modification of the original preparative method[104] by R. Knuppen and W. Hoffmann.
[104] R. Knuppen and H. Breuer, *Z. Physiol. Chem.* **337**, 159 (1964).
[105] R. Knuppen, M. Haupt, and H. Breuer, *J. Endocrinol.* **33**, 529 (1965).
[106] H. Levy, B. Hood, C. H. Cha, and J. J. Carlo, *Steroids* **5**, 677 (1965).
[107] G. Pangels and H. Breuer, *Naturwissenschaften* **49**, 106 (1962).

tissue) and NADPH (4 micromoles) in Krebs phosphate buffer (3 ml, pH 7.4) for 30 minutes at 37° under oxygen. The incubation mixture is extracted with ether–chloroform (3:1, v/v) (3 × 3 ml); the extracts are combined and evaporated to dryness. The residue is then chromatographed in system A (mobility of estriol 0.43 cm/hour). The estriol fraction is then rechromatographed in system D (mobility of estriol 25.6 cm/15 hours), and finally in system E (mobility of estriol 0.57 cm/hour). After elution, estriol is determined by the Kober reaction[42] (yield 5%).[107]

Properties

The 16α-hydroxylase is localized in the microsomal fraction and requires both NADPH and oxygen.[89, 107, 108]

16α-Hydroxylation of estrogens is found in liver of mouse (17β-estradiol[108]), rat (estrone,[88, 89, 109] 17β-estradiol,[90, 109, 110] 6α-hydroxy-17β-estradiol,[91] 2-methoxy-17β-estradiol[91]), human fetus (17β-estradiol[92, 111, 112]), man (estrone,[93] 17β-estradiol[93]), chicken (17β-estradiol[113]), and amphibians (17β-estradiol[94]); it has also been found in various human organs (17β-estradiol[114]), in beef adrenal (estrone,[102] 17β-estradiol[106]) and in the microorganisms *Streptomyces halstedii* (estrone,[115] 17β-estradiol[115]), *Streptomyces mediocidicus* (estrone,[115] 17β-estradiol[115]), *Streptomyces griseus* (17α-estradiol,[116] 17β-estradiol[116]), *Streptomyces* sp. (17α-estradiol,[116] 17β-estradiol[116]), and *Streptomyces bikiniensis* (estrone[116]).

H. 16β-HYDROXYLASE

$$17\beta\text{-Estradiol} + \text{NADPH} + \text{H}^+ + \text{O}_2 \rightarrow 16\text{-epiestriol} + \text{NADP}^+ + \text{H}_2\text{O}$$

Assay Method[113]

17β-Estradiol (0.37 micromole) is incubated with the microsomal fraction (1 ml) of chicken liver (corresponding to 2 g wet weight of fresh tissue) and NADPH (1.5 micromoles) in phosphate buffer (3 ml, 0.1 M, pH 7.4) for 60 minutes at 38° under oxygen. The incubation mixture is extracted with ether–chloroform (3:1, v/v) (3 × 10 ml); the

[108] G. Rumney, *Federation Proc.* **15**, 343 (1956).
[109] H. Breuer, L. Nocke, and G. Pangels, *Acta Endocrinol.* **34**, 359 (1960).
[110] M. Hagopian and L. K. Levy, *Biochim. Biophys. Acta* **30**, 641 (1958).
[111] L. L. Engel, B. Baggett, and M. Halla, *Biochim. Biophys. Acta* **30**, 435 (1958).
[112] L. L. Engel, B. Baggett, and M. Halla, *Endocrinology* **70**, 907 (1962).
[113] R. Ozon and H. Breuer, *Z. Physiol. Chem.* **341**, 239 (1965).
[114] R. M. Dowben and J. L. Rabinowitz, *Nature* **178**, 696 (1956).
[115] D. A. Kita, J. L. Sardinas, and G. M. Shull, *Nature* **190**, 627 (1961).
[116] B. F. Stimmel, T. E. Bucknell, and V. Notchev, *Federation Proc.* **19**, 115 (1960).

extracts are combined and evaporated to dryness. The residue is chromatographed in system A (mobility of 16-epiestriol 1.6 cm/hour). After elution, 16-epiestriol is methylated with methyl iodide and the 3-methyl ether subjected to column chromatography on alumina with ethanol–benzene.[117] 16-Epiestriol 3-methyl ether is determined by the Kober reaction[42,43] (yield 5%).

Properties

The 16β-hydroxylase is localized in the microsomal fraction and requires both NADPH and oxygen.[89,113]

16β-Hydroxylation of estrogens was found in liver of rat (estrone[89]), chicken (17β-estradiol[113]), human fetus (estrone,[101] 17β-estradiol[101,112]), and man (estrone[93]), and also in bovine adrenal (17β-estradiol[106]).

I. 18-HYDROXYLASE

Estrone → 18-hydroxyestrone

Assay Method[105]

Estrone-4-^{14}C; (1 μC; specific activity 33.7 mC/millimole) is incubated with human adrenal tissue slices (400 mg) in Krebs phosphate buffer (5 ml, pH 7.4) containing glucose (20 mM), for 60 minutes at 37° under air. The incubation mixture is extracted with ether–chloroform (3:1, v/v) (3 × 5 ml); the extracts are combined and evaporated to dryness. The residue is first chromatographed in system A. The fraction containing 18-hydroxyestrone (mobility 1.4 cm/hour) is eluted and rechromatographed on formamide-impregnated paper in system B. 18-Hydroxyestrone (mobility 2.5 cm/hour) is determined by counting directly the radioactivity from the paper chromatogram, using a paper strip scanner (yield 1%).

The intracellular localization of the 18-hydroxylase has not been studied, and the enzyme has not been further characterized. So far, 18-hydroxylation of estrone has been found only in a bovine homogenate[27] and in human adrenal tissue.[105]

IV. Oxidoreduction

Hydroxysteroid + NAD(P)$^+$ $\rightleftharpoons$ oxosteroid + NAD(P)H + H$^+$

Reversible interconversions of hydroxyl and ketone functions of phenolic steroids (estrogens) are catalyzed by nicotine adenine dinucleotide-dependent hydroxysteroid oxidoreductases. With the exception of the

[117] W. Nocke, *Clin. Chim. Acta* **6**, 449 (1961).

placental estradiol-17β oxidoreductases (EC 1.1.1.62), which is described elsewhere (this volume [39]), the preparations of estrogen oxidoreductases are relatively crude. Oxidoreductions by mammalian and nonmammalian tissue preparations have been found in positions 6α, 6β, 7α, 7β, 11β, 15α, 15β, 16α, 16β, 17α, and 17β of the estrogen molecule. In this presentation the enzymes are named according to the position of the hydroxyl group of the estrogens used as substrates. However, since most of the enzyme preparations have not been adequately purified, the number of enzymes involved and their substrate specificities remain doubtful.

The following estrogen oxidoreductases will be described: 6α, 6β, 7α, 7β, 11β, 15α, 15β, 16α, 16β, and 17α. Most of these enzymes are localized in the microsomal fraction of rat liver; 16β-hydroxyestrogen oxidoreductases are present in both the microsomal and the cytoplasmic fractions, whereas the 16α- and the 17α-hydroxyestrogen oxidoreductases are to be found in the cytoplasm.[118]

A. Microsomal Enzymes

6α-Hydroxyestrone + NAD(P)$^+$ ⇌ 6-Oxoestrone + NAD(P)H + H$^+$

6β-Hydroxyestrone + NAD(P)$^+$ ⇌ 6-Oxoestrone + NAD(P)H + H$^+$

7α-Hydroxyestrone + NAD(P)$^+$ ⇌ 7-Oxoestrone + NAD(P)H + H$^+$

7β-Hydroxyestrone + NAD(P)$^+$ ⇌ 7-Oxoestrone + NAD(P)H + H$^+$

11β-Hydroxyestrone + NAD(P)$^+$ ⇌ 11-Oxoestrone + NAD(P)H + H$^+$

15α-Hydroxy-17β-estradiol + NAD(P)$^+$ ⇌
$$15\text{-Oxo-}17\beta\text{-estradiol} + \text{NAD(P)H} + \text{H}^+$$

15β-Hydroxy-17β-estradiol + NAD(P)$^+$ ⇌
$$15\text{-Oxo-}17\beta\text{-estradiol} + \text{NAD(P)H} + \text{H}^+$$

16-Epiestriol + NAD(P)$^+$ ⇌
$$16\text{-Oxo-}17\beta\text{-estradiol} + \text{NAD(P)H} + \text{H}^+$$

Assay Method[118]

When the oxidation of a hydroxysteroid is measured, the reaction system (final volume 4 ml) contains the appropriate hydroxysteroid (0.37 micromole), the microsomal fraction (1 ml) of rat liver (corresponding to 1.5 g wet weight of fresh tissue), NAD$^+$ or NADP$^+$ (4 micromoles), and phosphate buffer (3 ml, 0.15 M, pH 8.3). The reduction of an oxosteroid is followed in a system which consists of the appropriate oxosteroid (0.37 micromole), the microsomal fraction of rat liver (corresponding to 1.5 g wet weight), NADH or NADPH (4 micromoles), and phosphate buffer (3 ml, 0.15 M, pH 6.5). Incubation is carried out for 30 minutes at 37°.

[118] H. Breuer and M. Lindlau, unpublished data, 1966.

The incubation mixture is extracted with ether–chloroform (3:1, v/v) (3 × 5 ml); the extracts are then combined and evaporated to dryness. The residue is subjected to paper chromatography in a suitable system and, if necessary, rechromatographed in a second paper chromatographic system (for R_f values and mobilities of the various estrogens see Table I). The estrogens are determined either directly on paper by the Folin-Ciocalteu reaction,[119, 120] using a densitometer[121] (Spinco Analytrol model RA), or, after elution, by the Kober reaction,[14, 42, 43] No unit of enzymatic activity has been defined.

Properties

Distribution. So far, the following oxidoreductions of estrogens have been found: 6α-oxidoreduction in the liver of rat (6α-hydroxy-17β-estra-diol[86]), man (6-oxo-17β-estradiol[122]), triton (*Pleurodeles waltlii* Michah) and frog (*Rana temporaria*) (6-oxo-17β-estradiol[123]); 6β-oxidoreduction in rat liver (6β-hydroxy-17β-estradiol,[86] 6-oxo-17β-estradiol[122]); 7α-oxidoreduction in rat liver (7α-hydroxyestrone,[124] 7-oxoestrone[124]); 7β-oxidoreduction in rat liver (7β-hydroxyestrone,[118] 7-oxoestrone[118]) and in human placenta (7β-hydroxyestrone[125]); 11β-oxidoreduction in liver of rat (11β-hydroxyestrone,[118] 11-oxoestrone[118, 126]), cattle (11-oxoestrone[126]), pig (11-oxoestrone[126]) and in bovine adrenal (11β-hydroxyestrone[126]); 15α- and 15β-oxidoreduction in rat liver (15α-hydroxy-17β-estradiol,[118] 15β-hydroxy-17β-estradiol[118]) and human liver (15-oxo-17β-estradiol[127]); 16β-oxidoreduction in human liver (16-oxo-17β-estradiol,[128] 16-oxoestrone[129]), human ovary (16-oxoestrone[129]), human placenta (16-oxo-17β-estradiol,[130] 16-oxoestrone[130]), rabbit liver (16-oxo-17β-estradiol,[131] 16-oxoestrone[131]), and rat liver (16-oxo-17β-estradiol,[118] 16-oxoestrone[118]).

Nicotinamide Nucleotide Specificity. The microsomal hydroxyestrogen

[119] O. Folin and V. Ciocalteu, *J. Biol. Chem.* **73**, 627 (1927).
[120] F. L. Mitchell and R. E. Davies, *Biochem. J.* **56**, 690 (1954).
[121] H. Breuer, I. Petershof, and R. Knuppen, *Z. Physiol. Chem.* **334**, 259 (1963).
[122] H. Breuer, R. Knuppen, and G. Pangels, *Biochem. J.* **79**, 32P (1961).
[123] H. Breuer, R. Ozon, and C. Mittermayer, *Z. Physiol. Chem.* **333**, 272 (1963).
[124] B. P. Lisboa, R. Knuppen, and H. Breuer, *Biochim. Biophys. Acta* **97**, 557 (1965).
[125] L. Cedard, B. Fillmann, R. Knuppen, B. P. Lisboa, and H. Breuer, *Z. Physiol. Chem.* **338**, 89 (1964).
[126] R. Knuppen and H. Breuer, *Bull. Soc. Chim. Biol. (France)* **46**, 192 (1964).
[127] R. Knuppen and H. Breuer, *Naturwissenschaften* **53**, 506 (1966).
[128] H. Breuer, L. Nocke, and R. Knuppen, *Z. Physiol. Chem.* **311**, 275 (1958).
[129] H. Breuer, R. Knuppen, and G. Pangels, *Acta Endocrinol.* **30**, 247 (1959).
[130] W. D. Lehmann and H. Breuer, *Acta Endocrinol.* **58**, 215 (1968).
[131] H. Breuer, R. Knuppen, and G. Pangels, *Z. Physiol. Chem.* **317**, 248 (1959).

oxidoreductases have essentially equal reaction rates with NAD⁺ and NADP⁺.

Kinetic Properties. The reactions catalyzed by the microsomal hydroxyestrogen oxidoreductases are freely reversible near neutral pH. The equilibria are displaced toward oxidation of the hydroxylated estrogens at high pH and favor the reduction of the oxoestrogens at low pH.

B. Cytoplasmic Enzymes

1. *16α-Hydroxyestrogen Oxidoreductase*

16-Oxo-17β-estradiol + NAD(P)H + H⁺ ⇌ estriol + NAD(P)⁺

The partial purification of this enzyme from rat kidney has been reported recently by Meigs and Ryan[36] and is described elsewhere in this volume [41].

Distribution. 16α-Oxidoreduction of estrogens has been found in liver of trout (*Salmo iridens*), triton (*Pleurodeles waltlii* Michah), frog (*Rana temporaria*) (16-oxo-17β-estradiol[123]), chicken liver (16-oxo-17β-estradiol[113]), rat kidney (estriol,[132, 36] 16-oxo-17β-estradiol,[36] 16-oxo-17α-estradiol,[36] 16-oxoestrone,[36] 16-estrone,[36] 17-epiestriol,[36] 16α-estradiol[36]), rat liver (16-oxo-17β-estradiol[118]), human liver (16-oxo-17β-estradiol,[128] 16-oxoestrone[129]), human ovary (16-oxoestrone[129]), rabbit liver (16-oxo-17β-estradiol,[131] 16-oxoestrone[131]), human placenta (16-oxoestrone[130, 133]), human erythrocytes (16-oxoestrone[134]), and cat blood (16-oxo-17β-estradiol[135]).

2. *16β-Hydroxyestrogen Oxidoreductase*

16-Oxo-17β-estradiol + NAD(P)H + H⁺ ⇌ 16-epiestriol + NAD(P)⁺

Assay Method[123]

16-Oxo-17β-estradiol (0.75 micromole) is incubated with the enzyme preparation (5 ml, containing 1.5 mg protein) and NADPH (5 micromoles) in Teorell-Stenhagen buffer[136] (2 ml, pH 6.4) for 60 minutes at 37° under air. The incubation is extracted with ether–chloroform (3:1, v/v) (2 × 10 ml); the extracts are combined and evaporated to dryness. The residue is then chromatographed in system A; 16-epiestriol (mobility 1.6 cm/hour) is eluted and determined by the Kober reaction.[42]

[132] R. J. B. King, *Biochem. J.* **76**, 7P (1960).
[133] K. J. Ryan, *Endocrinology* **66**, 491 (1960).
[134] H. Breuer, *Arzneimittel-Forsch.* **9**, 667 (1959).
[135] D. Trachewsky and R. Hobkirk, *Biochim. Biophys. Acta* **71**, 748 (1963).
[136] T. Teorell and E. Stenhagen, *Biochem. Z.* **299**, 416 (1938).

Purification Procedure

All procedures are carried out at 0–4°. Liver tissue of frog (*Rana temporaria*) is homogenized in 0.25 M in an all-glass homogenizer. The homogenate is first centrifuged in an ultracentrifuge at 23,000 g for 45 minutes. The supernatant is then centrifuged at 100,000 g for 120 minutes. The supernatant is now treated with finely powdered ammonium sulfate until 80% saturation is reached; the precipitate formed after 24 hours is removed by centrifugation at 23,000 g. The supernatant is used as an enzyme preparation and stored at +2°. The specific activity (micrograms of steroid formed × 10 per milligram of protein per hour) of the enzyme is 68.5.

Properties

The soluble 16β-hydroxyestrogen oxidoreductase of frog liver requires specifically NADPH and does not react with NADH.

Distribution. 16β-Oxidoreduction of estrogens has been found in liver of trout (*Salmo iridens*), triton (*Pleurodeles waltlii* Michah), frog (*Rana temporaria*) (16-oxo-17β-estradiol[123]), chicken liver (16-oxo-17β-estradiol[113]), rat kidney (estriol[132]), rat liver (16-oxo-17β-estradiol[118]), human liver (16-oxo-17β-estradiol,[128] 16-oxoestrone[129]), human ovary (16-oxoestrone[129]), human placenta (16-oxo-17β-estradiol,[130] 16-oxoestrone[130]), human erythrocytes (16-oxoestrone[134]), and rat blood (16-oxo-17β-estradiol[135]).

3. *17α-Hydroxyestrogen Oxidoreductase*

$$\text{Estrone} + \text{NAD(P)H} + \text{H}^+ \rightleftharpoons 17\alpha\text{-estradiol} + \text{NAD(P)}^+$$

Assay Method[137]

Estrone-4-[14]C (0.05 μC; specific activity 1 μC/milligram) is incubated with the enzyme preparation (0.3 ml, containing 10 mg of protein) and NADH (3.2 micromoles) in phosphate buffer (3 ml, 0.3 M, pH 7.0) for 60 minutes at 37° under air. The incubation mixture is extracted with ether–chloroform (3:1, v/v) (3 × 5 ml); the extracts are combined, dried over sodium sulfate, and evaporated to dryness. The residue is subjected to chromatography on formamide-impregnated paper with monochlorobenzene (R_f value of 17α-estradiol, 0.46; of 17β-estradiol, 0.39). 17α-Estradiol is determined either by counting directly the radioactivity from the paper chromatogram (using a paper strip scanner) or, after elution from the paper, by measuring the radioactivity in a liquid scintillation counter.

[137] E. Döllefeld and H. Breuer, *Biochim. Biophys. Acta* **124**, 187 (1966).

Purification Procedure

All procedures are carried out at 0–4°. Horse placental tissue[137] or chicken liver[113] is homogenized in sucrose (0.25 M) in an all-glass homogenizer. The homogenate (30%) is first centrifuged at 23,000 g for 45 minutes. The 23,000 g supernatant is then centrifuged at 100,000 g for 120 minutes. The 100,000 g supernatant is brought to 20% saturation by addition of finely powdered ammonium sulfate, and the precipitate formed after 24 hours is removed by centrifugation at 23,000 g. The supernatant solution is then brought to 40% saturation with ammonium sulfate. After 24 hours, the precipitate is removed by centrifugation and dissolved in a small volume of medium, containing 0.01 M potassium phosphate, glycerol (50%), 0.005 M EDTA and 0.007 M β-mercaptoethanol (medium B[35]). The specific activity (nanomoles, of steroid formed per milligram of protein per hour) of the 17α-hydroxyestrogen oxidoreductase of horse placenta varies between 5 and 6, whereas the specific activity of the chicken enzyme is about 0.8. Both enzyme preparations contain also 17β-hydroxyestrogen oxidoreductases.

Properties

Distribution. 17α-Oxidoreduction of estrogens has been found in liver of man (16α-hydroxyestrone,[128] 16β-hydroxyestrone,[138] 16-oxoestrone[139]), rabbit (estrone,[140] 16α-hydroxyestrone,[131] 16β-hydroxyestrone[131]), chicken (estrone,[113] 16α-hydroxyestrone[113]), and triton (*Pleurodeles waltlii* Michah) (estrone[141]), in placenta of horse (estrone[137]) and in erythrocytes of various species (estrone[142]).

Nicotinamide Nucleotide Specificity. The 17α-hydroxyestrogen oxidoreductase of chicken liver has somewhat higher reaction rates with NADP+ than with NAD+.[113]

pH Effect. The oxidation of 17α-estradiol to estrone by the chicken enzyme shows a maximum at pH 8.0.[113]

V. Equilin Dehydrogenase

3-Hydroxy-$\Delta^{1,3,5(10),7}$-estratetraen-17-one (equilin) $\rightarrow$

3-hydroxy-$\Delta^{1,3,5(10),6,8}$-estrapentaen-17-one (equilenin)

Assay Method[143]

Equilin (0.73 micromole) is incubated with the microsomal fraction (1 ml) of rat liver (corresponding to 1 g wet weight of fresh tissue),

[138] H. Breuer and L. Nocke, *Biochim. Biophys. Acta* **36**, 271 (1959).
[139] K. Dahm, M. Lindlau, and H. Breuer, unpublished data, 1966.
[140] H. Breuer and G. Pangels, *Acta Endocrinol.* **33**, 532 (1960).
[141] R. Ozon and H. Breuer, *Gen. Comp. Endocrinol.* **6**, 295 (1966).
[142] H. J. Portius and K. Repke, *Naturwissenschaften* **47**, 43 (1960).
[143] C. Mittermayer and H. Breuer, *Biochim. Biophys. Acta* **77**, 191 (1963).

NAD$^+$ (5 micromoles), and NADP$^+$ (5 micromoles) in Teorell-Stenhagen buffer (2 ml, pH 8.2)[136] and Krebs-Ringer solution (0.5 ml, containing 20 millimoles of glucose per liter) for 60 minutes at 37° under oxygen. The incubation mixture is extracted with ether–chloroform (3:1, v/v) (2 × 5 ml); the extracts are combined, dried over sodium sulfate, and evaporated to dryness. The residue is chromatographed in system G (mobility of equilin 28 cm/30 hours and of equilenin 20 cm/30 hours). Equilenin is determined either directly on paper by the Folin-Ciocalteu reaction,[119, 120] using a densitometer[121, 144] (Spinco Analytrol model RA) or, after elution, by the Kober reaction (yield 10%).

Properties[143]

Distribution. The equilin dehydrogenase is localized in the microsomal fraction; either NAD$^+$ or NADP$^+$ can be utilized as cofactors, although the latter is more efficient at higher concentrations. Enzymatic dehydrogenation of equilin occurs under aerobic as well as under anaerobic conditions.

Inhibition. Equilin dehydrogenase activity is inhibited 100% by $10^{-3} M$ p-chloromercuribenzoate, 70% by $10^{-3} M$ zinc chloride and $10^{-3} M$ ethylenediamine tetraacetate, 30% by $10^{-3} M$ iodoacetamide and $10^{-3} M$ o-iodosobenzoate; it is unaffected by $10^{-5} M$ cyanide and $10^{-5} M$ o-iodosobenzoate. 10^{-5} and $10^{-3} M$ iodoacetate, and $10^{-5} M$ zinc chloride have an activating effect.

pH Effect. The dehydrogenation of equilin to equilenin shows two maxima at pH 5.0 and 8.4.

Kinetic Properties. At pH 5.0, the Michaelis-Menten constant is $5.5 \times 10^{-5} M$ for equilin.

VI. O-Methyl Transfer

2-Hydroxy-17β-estradiol + S-adenosylmethionine →

$$2\text{-methoxy-}17\beta\text{-estradiol}$$

Assay Method[145]

2-Hydroxy-17β-estradiol (0.5 micromole) is incubated with the enzyme preparation (1 ml, containing 9 mg of protein), S-adenosylmethionine (0.5 micromole) and magnesium chloride (1 ml, 0.05 M) in phosphate buffer (1 ml, 0.5 M, pH 7.8) for 60 minutes at 37° under air or nitrogen. The incubation mixture is extracted with ether–chloroform (3:1, v/v) (2 × 5 ml); the extracts are combined, dried over sodium sulfate, and evaporated to dryness. To separate 2-methoxy-17β-estradiol from the isomeric 2-hydroxy-17β-estradiol 3-methyl ether,[146] the residue

[144] H. Breuer and C. Mittermayer, *Biochem. J.* **86**, 12P (1963).
[145] H. Breuer, W. Vogel, and R. Knuppen, *Z. Physiol. Chem.* **327**, 217 (1962).
[146] R. Knuppen and H. Breuer, *Z. Physiol. Chem.* **346**, 114 (1966).

is chromatographed with cyclohexane on formamide-impregnated paper (mobilities of 2-methoxy-17β-estradiol 18.2 cm/20 hours and of 2-hydroxy-17β-estradiol 3-methyl ether 14.0 cm/20 hours). For further purification, 2-methoxy-17β-estradiol is eluted from the paper and the residue is subjected to column chromatography on alumina (3 g), using first benzene (0–60 ml) and then benzene with 0.5% ethanol (60–120 ml).[146] 2-Methoxy-17β-estradiol is eluted between 60 and 120 ml and then determined by the Kober reaction.[146]

Purification Procedure[145, 147]

The procedure is based on the method for the preparation of the catechol O-methyltransferase from rat liver as described by Axelrod and Tomchick.[147] Ten adult rats are killed by a blow on the back of the neck and exsanguinated; the livers are immediately removed and chilled. All further purification procedures are carried out at 0–5°. Livers (100 g) are homogenized with isotonic potassium chloride (400 ml) and centrifuged at 78,000 g for 30 minutes. The supernatant fraction (320 ml) is adjusted to pH 5.0 with acetic acid (1 M), allowed to stand for 10 minutes, and centrifuged at 3000 g. To the supernatant fraction (300 ml) is added ammonium sulfate (52 g; 0–30% saturation). The precipitate is then discarded and ammonium sulfate (34 g; 30–50% saturation) is added to the supernatant. After centrifugation, the precipitate is dissolved in water (50 ml) and dialyzed for 6 hours against phosphate buffer (0.001 M; pH 7.0). The dialyzed solution is diluted to contain 9 mg of protein per milliliter in acetate buffer (0.02 M, pH 5.0). To this solution (100 ml) calcium phosphate gel (45 ml; 18 mg solids/ml) is added. After 15 minutes, the sample is centrifuged and the gel is eluted with phosphate buffer (2 × 25 ml, 0.02 M, pH 6.9). The preparation obtained by this procedure is purified about 25–30-fold with a 15–30% yield.

Properties

Distribution. The S-adenosylmethionine:o-diphenol-O-methyltransferase (EC 2.1.1.6) is localized in the soluble fraction of rat liver and requires S-adenosylmethionine as cofactor. Under certain experimental conditions, S-adenosylmethionine can be replaced by ATP and methionine.[148] Methylation of 2-hydroxyestrogens to 2-methoxyestrogens has been found in liver of rat (2-hydroxy-17β-estradiol,[145, 149] 2-hydroxyes-

[147] J. Axelrod and R. Tomchick, *J. Biol. Chem.* 233, 702 (1958) ; J. Axelrod, see Vol. V [101b].
[148] H. Breuer, G. Pangels, and R. Knuppen, *J. Clin. Endocrinol.* 21, 1331 (1961).
[149] R. Knuppen, H. Breuer, and G. Pangels, *Z. Physiol. Chem.* 324, 108 (1961).

triol[150]) and man (2-hydroxy-17β-estradiol[148,151]), in human fetal membranes, leiomyoma, myometrium, endometrium, functioning and menopausal ovary (2-hydroxyestrone,[152] 2-hydroxy-17β-estradiol[152]), and in human placenta (2-hydroxy-17β-estradiol[149,152]); it has also been shown to occur in kidney and spleen of rat (2-hydroxy-17β-estradiol[149]), guinea pig (2-hydroxy-17β-estradiol[153]), and man (2-hydroxy-17β-estradiol[149]).

Specificity. It has been shown that the conversion of 2-hydroxyestrogens to 2-methoxyestrogens is catalyzed by the same S-adenosylmethionine:o-diphenol-O-methyltransferase which also transfers the methyl group of S-adenosylmethionine to the 3-hydroxy group of epinephrine and other catechols. This enzyme was first described and purified by Axelrod and Tomchick.[147] Since the methyltransferase catalyzes the methylation of catecholamines as well as that of 2-hydroxyestrogens, its specificity seems to be low. It should be mentioned that, after incubation of 2-hydroxy-17β-estradiol with the cytoplasmic fraction of rat liver, not only 2-methoxy-17β-estradiol, but also 2-hydroxy-17β-estradiol 3-methyl ether is formed as a metabolite.[146]

Stability. The purified enzyme preparation is stable at $-10°$ for at least 3 months[147] or at $-5°$ for at least 3 weeks.[145]

Inhibition. The enzyme (catechol O-methyltransferase) is inhibited by sulfhydryl binding agents.[147] The presence of catecholamines interferes with the methylation of 2-hydroxy-17β-estradiol.[152]

pH Effect. The methylation of 2-hydroxy-17β-estradiol to 2-methoxy-17β-estradiol has a pH optimum at 7.8.[145]

Kinetic Properties. The Michaelis-Menten constant for 2-hydroxy-17β-estradiol is $8.7 \times 10^{-4}\ M$;[145] for comparison, the K_m for epinephrine is $1.2 \times 10^{-4}\ M$.[147]

VII. Demethylation

2-Methoxy-17β-estradiol + NADPH + H$^+$ + O$_2$ →
$$\text{2-hydroxy-17}\beta\text{-estradiol} + \text{NADP}^+ + \text{CH}_2\text{O} + \text{H}_2\text{O}$$
17β-Estradiol 3-methyl ether + NADPH + H$^+$ + O$_2$ →
$$\text{17}\beta\text{-estradiol} + \text{NADP}^+ + \text{CH}_2\text{O} + \text{H}_2\text{O}$$

Assay Method[154]

Since demethylation of 2-methoxyestrogens leads to the very unstable 2-hydroxyestrogens, it is advisable to determine the activity of the

[150] R. J. B. King, *Biochem. J.* 79, 361 (1961).
[151] H. Breuer and R. Knuppen, *Naturwissenschaften* 47, 280 (1960).
[152] O. J. Lucis, *Steroids* 5, 163 (1965).
[153] L. R. Axelrod, *Arch. Biochem. Biophys.* 91, 152 (1960).

ether-cleaving enzyme system with 17β-estradiol 3-methyl ether as substrate.

17β-Estradiol 3-methyl ether (0.34 micromole) is incubated with the lyophilized microsomal fraction (90 mg, containing 1.2 mg protein/10 mg) of rat liver and NADPH (5 micromoles) in Tris-maleate buffer (3 ml, 0.2 M, pH 7.4) containing 50 micromoles of nicotinamide and 25 micromoles of magnesium chloride, for 60 minutes at 37° under oxygen. The incubation mixture is then extracted with ether–chloroform (3:1, v/v) (2 × 10 ml); the extracts are combined, dried over sodium sulfate, and evaporated to dryness. For further purification, the residue is dissolved in a small amount of methanol (0.1 ml) to which benzene is then added (10 ml). The organic solution is extracted with aqueous sodium hydroxide (3 × 5 ml, 1 N); the combined aqueous extracts are adjusted to pH 5–6 by the addition of hydrochloric acid and reextracted three times with the same volume of ether. The ether extracts are combined, washed with water, and evaporated to dryness. The residue is chromatographed in system C. After elution, estrone (R_f 0.75) and 17β-estradiol (R_f 0.39) are determined by the Kober reaction[42] (yield 20–25%).

Properties[154]

Distribution. The estrogen ether-cleaving enzyme (dealkylase) is localized in the microsomal fraction and requires both NADPH and oxygen; no activity of the enzyme was found in the mitochondrial or cytoplasmic fractions. So far, demethylation of 2- and 3-methoxy estrogens has been studied only in rat liver.

Specificity. Apart from methoxyestrogens, the enzyme can cleave a great number of aromatic ethers, yielding phenols and aldehydes.[155,156]

Stability. The lyophilized enzyme preparations are stable at +1° for at least 3 months. Lyophilization causes a slight rise in activity.

Inhibition. The demethylation of 17β-estradiol 3-methyl ether is markedly inhibited by β-diethylaminoethyl diphenylpropylacetate (SKF 525-A; final concentration 0.5 × 10⁻⁴ M); the type of inhibition seems to be noncompetitive.

pH Effect. The optimum pH for the demethylation of 17β-estradiol 3-methyl ether is at 7.4.

Kinetic Properties. The Michaelis-Menten constant for 17β-estradiol 3-methyl ether is 2.0 × 10⁻⁴ M.

---

[154] H. Breuer, M. Knuppen, D. Gross, and C. Mittermayer, *Acta Endocrinol.* **46**, 361 (1964).

[155] J. Axelrod, *Biochem. J.* **63**, 634 (1956).

[156] J. W. Daly, J. Axelrod, and B. Witkop, *J. Biol. Chem.* **235**, 1155 (1960).

VIII. Epoxidation

Estra-1,3,5(10),16-tetraen-3-ol $+ 1/2\,O_2 \rightarrow$
$$16\alpha,17\alpha\text{-epoxyestra-1,3,5(10)-trien-3-ol}$$

Assay Method[157]

Estra-1,3,5(10),16-tetraen-3-ol (0.38 micromole) is incubated with rat liver slices (200 mg) in Krebs phosphate buffer (5 ml, pH 7.4) containing glucose (20 mM) for 30 minutes at 37° under air. The incubation mixture is extracted with ether–chloroform (3:1, v/v) (3 × 5 ml). The extracts are then combined, evaporated to dryness, and chromatographed in system G. The 16α,17α-epoxyestratrienol (mobility 3.7 cm/6 hours) is eluted and determined by the Kober reaction.[42, 43]

Properties

The intracellular localization of the steroid 16α,17α-epoxidase has not been studied. So far, epoxidation of unsaturated estrogens has been found only in rat liver.[157]

IX. 16α,17α-Epoxide Cleavage

16α,17α-Epoxyestra-1,3,5(10)-trien-3-ol $\rightarrow$
$$\text{estra-1,3,5(10)-triene-3,16}\beta,17\alpha\text{-triol}$$

Assay Method[121]

16α,17α-Epoxyestratrienol (0.37 micromole) is incubated with the lyophilized microsomal or mitochondrial fraction (10 mg, containing 1.2–1.7 mg protein) of rat liver in Krebs phosphate buffer (3 ml, pH 7.4) for 60 minutes at 37° under air. The incubation mixture is then extracted with ether–chloroform (3:1, v/v) (2 × 5 ml); the extracts are combined and evaporated to dryness. The residue is chromatographed in system D [mobility of estra-1,3,5(10)-triene-3,16β,17α-triol (16,17-epiestriol) 33.1 cm/15 hours]. 16,17-Epiestriol is determined directly on paper by the Folin-Ciocalteu reaction,[119, 120] by means of a densitometer[121] (Spinco Analytrol model RA).

Properties[121]

Distribution. The splitting of 16α,17α-epoxyestratrienol to 16,17-epiestriol has been observed in the following rat organs: kidney, testis, liver, spleen, uterus, ovary, adrenal, and diaphragm. The highest activity of the epoxyestratrienol-lyase was found to occur in kidney, and smaller amounts were shown to be present in testis and liver. Both the mitochon-

[157] H. Breuer and R. Knuppen, *Biochim. Biophys. Acta* **49**, 620 (1961).

drial and the microsomal fractions of rat liver contain an epoxyestra-trienol-lyase. The specific activity (micromoles of substrate transformed per minute per milligram of protein) of the microsomal enzyme (27×10^{-4}) is significantly higher than that of the mitochondrial enzyme (11×10^{-4}).

Stability.[121] The lyophilized enzyme preparations are stable at $+1°$ for at least 3 months. Lyophilization causes a slight rise in activity.

Inhibitors. The mitochondrial and microsomal enzymes are not in-hibited (final concentrations in parentheses) by p-chloromercuribenzoate ($10^{-5} M$), o-iodosobenzoate ($10^{-5} M$), iodoacetate (10^{-3} to $10^{-5} M$), or potassium cyanide (10^{-3} to $10^{-5} M$).

pH Effect. The optimum pH for the microsomal enzyme is at 7.0, and for the mitochondrial enzyme at 7.4.

Kinetic Properties. The Michaelis-Menten constants for $16\alpha,17\alpha$-epoxyestratrienol are $2.0 \times 10^{-4} M$ (microsomal enzyme) and $2.9 \times 10^{-4} M$ (mitochondrial enzyme).

X. Glucuronyl Transfer

Since the estrogen molecule contains phenolic as well as alcoholic hydroxyl groups, two different types of glucuronides are known: (1) glucuronides in which the glucuronic acid is conjugated with a phenolic hydroxyl group (phenolic glucuronides); (2) glucuronides in which the glucuronic acid is conjugated with an alcoholic hydroxyl group (alcoholic glucuronides). The enzymes involved in the formation of the two types of estrogen glucuronides have been partially purified and will be dealt with separately.

A. FORMATION OF ESTROGEN 3-MONOGLUCURONIDES

Estrone + uridine-5′-diphosphoglucuronic acid →
estrone 3-monoglucuronide + uridine 5′-diphosphate

Assay Method[158]

Estrone (0.1 micromole) is incubated with the microsomal fraction (1 ml) of rabbit liver (corresponding to 1 g wet weight of fresh tissue) and uridine-5′-diphosphoglucuronic acid (0.4 micromole) in phosphate buffer (1.7 ml, $0.15 M$, pH 8.2) for 60 minutes at 37° under air. The incubation mixture is extracted with ether (3×8 ml) to remove the unconjugated estrone and then treated with ethanol (12 ml). The pre-cipitated protein is removed by centrifuging at 500 g for 10 minutes and the ethanolic solution is evaporated *in vacuo* to dryness. The residue is

[158] E. R. Smith and H. Breuer, *Biochem. J.* 88, 168 (1963).

taken up in ice-cold hydrochloric acid (4 ml, 2 N) and the solution is extracted with ethyl acetate (3 × 4 ml) in the cold. The recovery of estrone 3-monoglucuronide by this extraction was found to be 90%.[158] The combined ethyl acetate extracts are washed with hydrochloric acid (4 ml, 2 N) at 4° and evaporated *in vacuo* to dryness. The residue, which contains the conjugated estrone, is hydrolyzed with hydrochloric acid (4 ml, 1.6 N) for 60 minutes at 100°. The cooled aqueous solution is extracted with benzene (2 × 5 ml), and the benzene is then extracted with sodium hydroxide (2 × 5 ml, 2 N). Sodium hydrogen carbonate (40 ml, 1 M) is added to the alkaline solution, and the estrone is extracted with ether (2 × 40 ml). The ether is dried over sodium sulfate and evaporated to dryness. Estrone is then determined by the Kober reaction.[42] From recovery experiments, the overall yield is calculated to be approximately 80%.[158]

Properties

Distribution. The enzyme conjugating the phenolic hydroxyl group of estrogens with glucuronic acid is localized in the microsomal fraction of liver and requires uridine-5′-diphosphoglucuronic acid as cofactor. Glucuronidation of the phenolic hydroxyl group of estrogens has been found in the liver of rat (estrone,[159] 17β-estradiol,[159] estriol[139, 160]), guinea pig (16α-hydroxyestrone,[139] estriol[161]), and rabbit (estrone,[158] 17β-estradiol[162]), in human intestine (17β-estradiol,[163, 164] estriol[164]) and in amphibians (estrone,[141] 17α-estradiol,[141] 17β-estradiol[141]). No further purification of the microsomal enzyme has been described.

Specificity. The microsomal UDPglucuronate glucuronyltransferase (EC 2.4.1.17), which was first described by Dutton and Storey[165] (for review see Dutton and Storey, Vol. V, p. 159) seems to catalyze glucuronic acid transfer from UDP-glucuronic acid to a wide variety of acceptors; these include many phenolic and alcoholic hydroxy compounds, such as o-aminophenol, p-nitrophenol, phenolphthalein, (−)-menthol, thyroxine, neutral and phenolic steroids. For example, the microsomal enzyme of human intestine forms not only 17β-estradiol 3-monoglucuronide and estriol 3-monoglucuronide, but also, *inter alia*, testosterone 17β-monoglucuronide and androsterone 3α-monoglucuronide.[164]

[159] K. Dahm, M. Lindlau, and H. Breuer, *Acta Endocrinol.* **56**, 403 (1967).
[160] K. Dahm and H. Breuer, *Z. Klin. Chem.* **4**, 153 (1966).
[161] U. Goebelsmann, E. Diczfalusy, J. Katz, and M. Levitz, *Steroids* **6**, 859 (1965).
[162] H. Breuer and D. Wessendorf, *Z. Physiol. Chem.* **345**, 1 (1966).
[163] E. Diczfalusy, C. Franksson, B. P. Lisboa, and B. Martinsen, *Acta Endocrinol.* **40**, 537 (1962).
[164] K. Dahm, H. Breuer, and M. Lindlau, *Z. Physiol. Chem.* **345**, 139 (1966).
[165] G. J. Dutton and I. D. E. Storey, *Biochem. J.* **57**, 275 (1954).

Stability. The microsomal enzyme of rabbit liver is stable at 2–4° for at least 4 days.[158]

Inhibition. The formation of 17β-estradiol 3-monoglucuronide is inhibited by $10^{-5} M$ p-chloromercuribenzoate.[162]

pH Effect. A fairly broad pH optimum exists for the 3-glucuronidation of estrone (8.0–8.2)[158] and 17β-estradiol (8.5–8.7)[162] by the microsomal enzyme of rabbit liver.

Kinetic Properties. The Michaelis-Menten constants are $9.7 \times 10^{-5} M$ for estrone[158] and $2.8 \times 10^{-4} M$ for 17β-estradiol.[162] With 17β-estradiol as substrate, the activation energy was found to be $+5.5$ kcal/mole, whereas the inactivation energy amounted to -8.5 kcal/mole.

B. Formation of Estrogen 16α-Monoglucuronides

Estriol + uridine-5'-diphosphoglucuronic acid →
$$\text{estriol 16}\alpha\text{-monoglucuronide} + \text{uridine 5'-diphosphate}$$

The formation of alcoholic glucuronides of the estrogen series, of which estriol 16α-monoglucuronide is quantitatively the most important[166, 167] can be catalyzed by the microsomal enzyme system as well as by a soluble enzyme which has recently been partially purified from human intestine.[168, 169]

1. *Microsomal Enzyme*

Assay Method[170]

Estriol-16-^{14}C (0.025 μC) and estriol (0.175 micromole) are incubated with the microsomal fraction (1–2 ml) of human liver (corresponding to 0.1–0.2 g wet weight of fresh tissue) and uridine-5'-diphosphoglucuronic acid (0.18–0.36 micromoles) in phosphate buffer (2–3 ml, 0.05 M, pH 7.4) for 60 minutes at 37° under air. The reaction is terminated by the addition of ethanol (95%, 4–5 volumes). After the mixture has stood in the cold over night, the precipitated protein is centrifuged down and washed once with alcohol. After 1 ml of the pooled alcoholic extract has been taken for assay of radioactivity, the ethanol is removed at 40° in a stream of air. The residue, taken up in water (15 ml), is extracted with ethyl acetate (2 × 45 ml). The latter is pooled, evaporated, and counted.

---

[166] J. G. D. Carpenter and A. E. Kellie, *Biochem. J.* **84**, 303 (1962).
[167] Y. Hashimoto and M. Neeman, *J. Biol. Chem.* **238**, 1273 (1963).
[168] K. Dahm and H. Breuer, *Biochim. Biophys. Acta* **113**, 404 (1966).
[169] K. Dahm and H. Breuer, *Biochim. Biophys. Acta* **128**, 306 (1966).
[170] W. R. Slaunwhite, Jr., M. A. Lichtman, and A. A. Sandberg, *J. Clin. Endocrinol.* **24**, 638 (1964).

All radioactivity is assayed in a gas-flow counter. Percentage of conjugation is calculated by the relation

$$\left[1 - \frac{\text{counts in ethyl acetate}}{\text{counts added}} \right] \times 100$$

The recovery of radioactivity in the alcoholic extracts ranges from 75 to 90%. One unit of activity is defined as that amount of enzyme required to conjugate 1 nanomole of estriol per minute under standardized conditions.[170]

Properties

Distribution. The microsomal enzyme conjugating the 16α-hydroxyl group of estriol with glucuronic acid has been found in human liver[170, 171] and human intestine.[164, 172] The formation of 16α-hydroxyestrone 16α-monoglucuronide occurs in rat liver.[171]

Specificity. The microsomal glucuronyl transferase of *human liver* seems to possess a high specificity toward the 16α-hydroxy group of estriol; neither estriol 3-monoglucuronide nor estriol 17β-monoglucuronide was found after incubation of estriol with the microsomal fraction of human liver.[173]

Stability. Freezing the human liver tissue preserves the glucuronyl transferase activity well. After storage for 38 days, the enzyme can conjugate three-quarters of a tracer amount of estriol, and after 8 months, 36% conjugation is observed.[170]

pH Effect. The 16α-glucuronidation of estriol (by human liver homogenate) has a pH optimum at 7.5.

2. Soluble Enzyme

Assay Method[169]

Estriol-16-^{14}C (0.1 μC) and estriol (0.35 micromole) are incubated with the enzyme preparation (3 ml, containing 10 mg of protein) and uridine-5'-diphosphoglucuronic acid (5 micromoles) in phosphate buffer (3 ml, 0.15 M, pH 7.4) for 60 minutes at 37° under air. The incubation mixture is extracted with ether–chloroform (3:1, v/v) (3 × 5 ml) to remove the unconjugated estriol. The protein is precipitated with ethanol (10 ml), and the supernatant is evaporated *in vacuo* to a third of its original volume. This solution is then extracted with butanol (2 × 10

[171] K. Dahm and H. Breuer, *Acta Endocrinol.* **52,** 43 (1966).

[172] E. Diczfalusy, C. Franksson, and B. Martinsen, *Acta Endocrinol.* **38,** 59 (1961).

[173] D. A. Boon, N. J. Wagner, and W. R. Slaunwhite, Jr., *Abstr. 47th Meeting Endocrine Soc. New York, 1965,* p. 57.

ml), and the residue of the butanolic extracts is subjected to paper chromatography in the system acetic acid–water–*tert*-butanol–dichloroethane (6:14:5:15, by volume) (mobility of estriol 16α-monoglucuronide, 23.5 cm/20 hours). Estriol 16α-monoglucuronide is determined either by counting directly the radioactivity from the paper chromatogram, using a paper strip scanner, or, after elution from the paper, by measuring the radioactivity in a liquid scintillation counter.

Purification Procedure

All procedures are carried out at 0–4°. The mucosa of human small intestine is carefully separated from the submucosa, and the sheet of mucosa thus obtained is cut into small pieces; these are suspended in 0.25 M sucrose and homogenized in an all-glass homogenizer. The homogenate is first centrifuged in an ultracentrifuge at 1000 g for 15 minutes, and then at 20,000 g for 45 minutes. The 20,000 g fraction is further centrifuged at 105,000 g for 60 minutes and the microsomal pellets are removed. The 105,000 g supernatant is subjected to centrifugation at 150,000 g for 120 minutes in order to remove the small microsomal particles and the ribosomes; the supernatant after 120 minutes of centrifugation at 150,000 g is considered to be free from particles. Finely powdered ammonium sulfate (35.2 g) is added to this supernatant (200 ml) within 2 hours. The precipitate formed after 24 hours is removed by centrifugation at 105,000 g for 60 minutes and dissolved in phosphate buffer (28 ml, 0.15 M, pH 7.4) (0–30% ammonium sulfate fraction). The enzyme preparation thus obtained is stored at 0°. The specific activity (nanomoles of estriol 16α-monoglucuronide formed per milligram of protein per hour) of the purified enzyme is 1.8. A 3-fold purification is achieved, as compared to the 150,000 g supernatant.

Properties

Distribution. The soluble enzyme conjugating the 16α-hydroxyl group of estriol with glucuronic acid has been found in human liver[174] and human intestine.[169]

C. FORMATION OF ESTROGEN 17β-MONOGLUCURONIDES

Estriol + uridine-5′-diphosphoglucuronic acid →
$$\text{estriol } 17\beta\text{-monoglucuronide} + \text{uridine 5′-diphosphate}$$

The human intestine contains a glucuronyl transferase which can be separated from other glucuronyl transferases and partially purified by treatment of the 150,000 g supernatant with ammonium sulfate (60–80%

[174] K. Dahm and H. Breuer, unpublished data, 1966.

saturation). This enzyme catalyzes exclusively the glucuronidation of the 17β-hydroxy groups of 17β-estradiol and estriol.[169]

Assay Method[169]

Estriol-16-[14]C (0.1 μC) and estriol (0.35 micromole) are incubated with the enzyme preparation (3 ml, containing 22.5 mg protein) and uridine-5'-diphosphoglucuronic acid (5 micromoles) in phosphate buffer (3 ml, 0.15 M, pH 7.4) for 60 minutes at 37° under air. The incubation mixture is extracted with ether–chloroform (3:1, v/v) (3 × 5 ml) to remove the unconjugated estriol. The protein is precipitated with ethanol (10 ml), and the supernatant is evaporated *in vacuo* to a third of its original volume. This solution is then extracted with butanol (2 × 10 ml), and the residue of the butanolic extracts is subjected to paper chromatography in the system acetic acid–water–*tert*-butanol–dichloroethane (6:14:5:15 by volume) (mobility of estriol 17β-monoglucuronide, 32 cm/20 hours). Estriol 17β-monoglucuronide is determined either by counting directly the radioactivity from the paper chromatogram (using a paper strip scanner) or, after elution from the paper, by measuring the radioactivity in a liquid scintillation counter.

Purification Procedure

The 150,000 g supernatant of human intestine (for preparation see preceding section) serves as source for the enzyme. All procedures are carried out at 0–4°. Finely powdered ammonium sulfate (70.4 g) is added to ground plasma (400 ml) within 2 hours. The solution is allowed to stand overnight and the precipitate (0–30% ammonium sulfate fraction) is removed by centrifugation at 150,000 g for 60 minutes. To the supernatant solution is added ammonium sulfate (79.2 g), and the precipitate is removed by centrifugation as described above (30–60% ammonium sulfate fraction). The supernatant solution is further treated with ammonium sulfate (57.2 g), the precipitated protein is collected by centrifugation and dissolved in Sørensen phosphate buffer (0.15 M, pH 7.4)[175] (60–80% ammonium sulfate fraction). The enzyme preparation is stored at 0°. The specific activity (nanomoles of estriol 17β-monoglucuronide formed per milligram of protein per hour) of the preparation is 1.4.

Properties

Distribution. The soluble enzyme conjugating the 17β-hydroxyl groups of 17β-estradiol and estriol with glucuronic acid has been found in human intestine.[169]

[175] S. P. L. Sørensen, *Biochem. Z.* **22**, 352 (1909).

Specificity. The partially purified 17β-hydroxysteroid glucuronyl transferase of human intestine is free of any activity toward the phenolic 3-hydroxyl and the alcoholic 16α-hydroxyl groups of estriol. The enzyme acts specifically on the 17β-hydroxyl groups of 17β-estradiol and estriol.

Stability. The enzyme preparation is stable at 0° for at least 6 weeks.

pH Effect. The optimum pH for the 17β-glucuronidation of estriol is at 6.8.

Kinetic Properties. The Michaelis-Menten constant for estriol is $3.4 \times 10^{-4} M$.[169] The activation energy was found to be 12.2 kcal/mole; Q_{10} amounted to 2.1 between 10 and 20°, and to 1.4 between 25–35°.[169]

XI. Sulfate Transfer

Sulfurylation of estrogens seems to be confined mainly to the phenolic hydroxyl group at position 3 of the molecule, although a few reports have appeared in the literature which indicate that sulfurylation of alcoholic hydroxyl groups (17α,17β) may also occur to a small extent. The enzyme catalyzing the transfer of sulfate from 3'-phosphoadenosine 5'-phosphosulfate (PAPS) to the phenolic hydroxyl group of estrone has been partially purified and separated from an androstenolone (dehydroepiandrosterone) sulfotransferase by Nose and Lipmann.[176] Recently, Banerjee and Roy[177] have carried out similar studies.

$$\text{PAPS} + \text{estrone} \rightarrow \text{PAP} + \text{estrone 3-sulfate}$$

Assay Method[176]

Estrone (0.1 micromole) is incubated with the rabbit liver enzyme (0.2 ml), a yeast sulfate-activating enzyme mixture (see Vol. V [129]), phosphate buffer (100 micromoles; pH 7.0), magnesium sulfate (12.5 micromoles), neutralized cysteine hydrochloric acid (5 micromoles), and potassium ATP (10 micromoles) in a total volume of 1 ml for 60 minutes at 37°. The production of estrone 3-sulfate is measured by the method of Roy,[178] which depends on the extraction of the methylene blue complex of the product into chloroform (see Vol. V, p. 980).

Purification Procedure

The enzyme preparation, containing estrone sulfotransferase activity, is obtained from rabbit liver according to the procedure of Nose and Lipmann[176] (see Vol. V, p. 982). The estrone sulfotransferase can be completely separated from the androstenolone sulfotransferase by resin electrophoresis.

[176] Y. Nose and F. Lipmann, *J. Biol. Chem.* **233**, 1348 (1958).
[177] R. K. Banerjee and A. B. Roy, *Mol. Pharmacol.* **2**, 56 (1966).
[178] A. B. Roy, *Biochem. J.* **62**, 41 (1956).

Properties

Distribution. Sulfurylation of the phenolic hydroxyl group of estrogens has been found in liver of rat (estrone,[176, 179–182] 17α-estradiol,[183] 17β-estradiol,[181–184] estriol[185]), rabbit (estrone,[176, 181–184, 186] 17β-estradiol,[181, 186] estriol,[186] and equilin[186]), lamb (estrone[176]), ox (estrone,[181] 17β-estradiol[181]), and man (estriol[187]); in adrenal of ox (estrone,[188] 17β-estradiol[189]) and man (estrone,[190–192] estriol[192]); in bovine ovary (17β-estradiol[189]) and corpus luteum (17β-estradiol[189]); and in rat testis (17β-estradiol[189]). For further details see footnote 193.

Specificity. Since the estrone sulfotransferase also sulfurylates other phenols, such as *p*-nitrophenol and 2-naphthylamine, statements on specificity have little meaning.

pH Effect. Sulfurylation of estrone by human liver shows a pH optimum between 7 and 8.[194]

Addendum

Recently, Adams and Poulos[195] have succeeded in isolating an enzyme from bovine adrenal glands, the estrogen sulfotransferase (3′-phosphoadenylylsulfate:estrone sulfotransferase, EC 2.8.2.4), which is free of various other types of sulfotransferases. Using DEAE-cellulose chromatography and a concave phosphate gradient, two separate enzymes, fraction A and fraction B enzymes, are obtained. Both enzymes fractions catalyze the formation of estrone sulfate from estrone and [³⁵S] PAPS. The enzyme is specific for natural estrogens. Only the phenolic 3-hy-

---

[179] H. L. Segal, *J. Biol. Chem.* **213**, 161 (1955).
[180] A. B. Roy, *Biochem. J.* **63**, 294 (1956).
[181] R. H. DeMeio, C. Lewycka, M. Wizerkaniuk, and O. Salciunas, *Biochem. J.* **68**, 1 (1958).
[182] B. Spencer, *Biochem. J.* **77**, 294 (1960).
[183] B. Wengle and H. Boström, *Acta Chem. Scand.* **17**, 1203 (1963).
[184] A. H. Payne and M. Mason, *Biochim. Biophys. Acta* **71**, 719 (1963).
[185] B. Wengle, *Acta Chem. Scand.* **18**, 65 (1964).
[186] J. J. Schneider and M. L. Lewbart, *J. Biol. Chem.* **222**, 787 (1956).
[187] H. Boström and B. Wengle, *Acta Soc. Med. Upsalien.* **69**, 41 (1964).
[188] A. Sneddon and G. F. Marrian, *Biochem. J.* **86**, 385 (1963).
[189] A. H. Payne and M. Mason, *Steroids* **5**, 21 (1965).
[190] J. B. Adams, *Biochim. Biophys. Acta* **71**, 243 (1963).
[191] J. B. Adams, *Biochim. Biophys. Acta* **82**, 572 (1964).
[192] H. Boström, C. Franksson, and B. Wengle, *Acta Endocrinol.* **47**, 633 (1964).
[193] E. Döllefeld and H. Breuer, *Z. Vitamin-, Hormon- Fermentforsch.* **14**, 193 (1966).
[194] H. Boström, *Abstr. Xth Meeting Scand. Soc. Clin. Chem. Clin. Physiol., Jönköping, 1965*, p. 33.
[195] J. B. Adams and A. Poulos, *Biochim. Biophys. Acta* **146**, 493 (1967).

droxyl group of estrogens appears to enter the reaction, since the 3-methyl ether gives no sulfoconjugated product.

In another communication, Adams and Chulavatnatol[196] report the properties of the two estrogen sulfotransferases. Molecular weight determinations show that the B enzyme possesses an associated structure. The rapid loss of activity of the B enzyme on standing is attributed to its conversion to the A form of the enzyme. The kinetic curve of the A form in presence of cysteine is observed to be complex and similar to that of the B form. The associated form of the enzyme—that is the form obtained by isolation with mercaptoethanol—exhibited homotropic effects, thereby indicating that this enzyme could belong to the allosteric class.

The increased rate of sulfoconjugation of estrone at high substrate concentrations, when compared to the lower rate with 17β-estradiol and estriol, suggests a specific role for estrone. It has subsequently been shown[197] that estrone is bound to the purified form A sulfotransferase, suggesting estrone to be a true substrate of the enzyme.

Assay Method[195]

Estrone (0.8 millimole; 5 μl of a stock solution in propylene glycol), [^{35}S] PAPS (0.25 millimole; 10^5 counts/minute) and $MgCl_2$ (20 millimoles) are incubated in Tris-HCl buffer (0.1 M, pH 8.1) with 5–200 μg of the enzyme, in a total volume of 0.135 ml for 20 minutes at 37°. The reaction is stopped by placing the tubes in boiling water for 1 minute. Estimation of the steroid sulfate is as follows: 1 ml of 0.1 M $Ba(OH)_2$ is added to each tube followed by 1 ml of 0.05 M H_2SO_4. After mixing, the heavy precipitate, containing inorganic sulfate and PAPS, is centrifuged and the supernatant decanted into 10 ml centrifuge tube. Carbon dioxide is bubbled into the solution to remove excess Ba^{2+}; the precipitate is centrifuged and 1 ml of the supernatant is used for measurement of [^{35}S] steroid sulfate by liquid scintillation counting. Propylene glycol 5 μl in place of estrogen is employed as a blank.

Purification Procedure

After removing the adhering fat the bovine adrenal glands are homogenized in the cold with 2 volumes of phosphate buffer (0.01 M) in saline (0.155 M, pH 7.4) in a Waring blendor. The homogenate is then centrifuged at 100,000 g for 60 minutes at 4° and the supernatant containing about 20 mg of protein/ml is used for further purification. The enzyme is concentrated by adding a saturated solution of ammonium sulfate. The fraction precipitating between 0.55 and 0.8 saturation is

[196] J. B. Adams and M. Chulavatnatol, *Biochim. Biophys. Acta* **146**, 509 (1967).
[197] J. B. Adams, *Biochim. Biophys. Acta* **146**, 522 (1967).

collected, dissolved in sodium phosphate (45 ml, 0.1 M, pH 7.5), and dialyzed overnight against 5 liters of the same buffer. After repeating the dialysis against fresh buffer, the protein solution (106 ml, containing 17.3 mg protein/ml) is placed on a column (35 cm $\times$ 4 cm) of DEAE-cellulose which was equilibrated against sodium phosphate buffer (5 mM, pH 7.5). The column is eluted with increasing concentrations of phosphate buffer delivered as a concave gradient, achieved with phosphate buffer (500 ml, 5 mM, pH 7.5) in a large-diameter mixing vessel and phosphate buffer (500 ml, 20 mM, pH 6.5) in a small diameter reservoir. Three fraction pools are collected, dialyzed against distilled water, and concentrated with the aid of Carbowax 20 M. An overall purification of 35-fold is obtained through the above procedure.

Properties

Distribution. The estrogen sulfotransferases are found to occur in bovine adrenal glands.

Specificity. Fraction A enzyme fails to sulfoconjugate dehydro-epiandrosterone, phenol, p-nitrophenol, α-naphthol, β-naphthol, β-naphthylamine, and 17β-estradiol 3-methyl ether. Stilbestrol and hexestrol give low yields of the sulfoconjugate. At substrate concentrations of 0.1 mM, the rates of sulfoconjugation compared to estriol are: estriol 1.0, 17β-estradiol 0.6, estrone 0.53. Equilenin is sulfoconjugated at a rate of 0.42 times less than that of estriol, indicating that a saturated ring B is not essential for sulfotransferase activity.

Stability. Fraction A enzyme is more stable, losing only a small percentage of its activity on standing at 0° for a period of 2 weeks. The B form loses 50% of its initial activity during this period.

pH Effect. The enzyme has an optimum in the vicinity of 8.0.

Kinetic Properties. Fraction A enzyme shows normal Michaelis-Menten kinetics. The apparent K_m for 17β-estradiol was $1.4 \times 10^{-5} M$ and is unaffected by the concentration of PAPS present; this indicates a sequential rather than a ping-pong mechanism of action. The apparent K_m for PAPS in the presence of saturating levels of 17β-estradiol was $7.0 \times 10^{-5} M$. ADP, a noncompetitive inhibitor, has a K_i value of $1.1 \times 10^{-3} M$.

[38] Aromatic Hydroxylation and Protein Binding
of Estrogens *in Vitro*

By Friedrich Marks and Erich Hecker

Hydroxylation of estrogens at C-2, which is catalyzed by the microsomal hydroxylase systems of liver[1,2] and of other tissues[3] is quantitatively one of the most important reactions of estrogen metabolism.[2-6] The two other types of aromatic *o,p*-hydroxylation 4- and 10β-hydroxylation seem to be of minor importance.[2,7,8] Besides hydroxylated estrogens, considerable quantities of water-soluble and protein-bound metabolites are formed[2,7,9,10] on incubation of estrogenic steroids with liver microsomes and NADPH. The chemical nature of these metabolites is largely unknown.

Survey of the Methods of Isolation and Determination
of 2-Hydroxy Estrogens

Up to the present time, 2-hydroxylation has been studied almost exclusively by using radioactively labeled estrogens. 2-Hydroxyestradiol-17β, 2-hydroxyestrone, and 2-hydroxyestriol together with other less polar estrogen metabolites can be extracted from hydrolyzed urine or incubation mixtures by organic solvents. Purification of crude extracts from accompanying substances is achieved by paper chromatography,[2,3,6,10-12]

[1] A. H. Conney and A. Klutch, *J. Biol. Chem.* **238**, 1611 (1963); R. Kuntzman, M. Jacobson, K. Schneidman, and A. H. Conney, *J. Pharmacol. Exptl. Therap.* **146**, 280 (1964); R. Kuntzman, D. Lawrence, and A. H. Conney, *Mol. Pharmacol.* **1**, 163 (1965).

[2] E. Hecker and F. Marks, *Biochem. Z.* **343**, 211 (1965); F. Marks and E. Hecker, *Z. Physiol. Chem.* **345**, 22 (1966).

[3] H. F. Acevedo and S. C. Beering, *Steroids* **6**, 531 (1965); H. F. Acevedo and J. W. Goldzieher, *Biochim. Biophys. Acta* **97**, 571 (1965).

[4] J. Fishman, R. I. Cox, and T. F. Gallagher, *Arch. Biochem. Biophys.* **90**, 318 (1960).

[5] J. Fishman, L. Hellman, B. Zumoff, and T. F. Gallagher, *J. Clin. Endocrinol. Metab.* **25**, 365 (1965).

[6] R. J. B. King, *Biochem. J.* **79**, 361 (1961).

[7] E. Hecker and S. M. A. D. Zayed, *Z. Physiol. Chem.* **325**, 209 (1961).

[8] E. Hecker and F. Marks, *Z. Physiol. Chem.* **340**, 229 (1965).

[9] J. L. Riegel and G. C. Mueller, *J. Biol. Chem.* **210**, 249 (1954); P. H. Jellinck, C. Lazier, and M. L. Copp, *Can. J. Biochem. Physiol.* **43**, 1774 (1965). E. Hecker, G. Walter, and F. Marks, *Biochim. Biophys. Acta* **111**, 546 (1965).

[10] E. Hecker and G. Nowoczek, *Z. Physiol. Chem.* **337**, 257 (1964).

[11] P. H. Jellinck and I. Lucieer, *J. Endocrinol.* **32**, 91 (1965).

[12] H. Breuer, R. Knuppen, and M. Haupt, *Nature* **212**, 76 (1966).

thin-layer chromatography,[13] and column chromatography on silica gel,[4] or, after acetylation, on alumina.[4] Furthermore, 2-hydroxyestrone may be isolated by Girard's reagent.[4] Radioactive estrogen may be localized on paper chromatograms by means of radioautography,[11] or by the use of an automatic scanner.[8] The Folin-Ciocalteu reagent for detection of unlabeled estrogens, in conjunction with radioactivity measurements, provides a means of determining specific activity.

Recrystallization of the free steroids[3, 4, 6, 11] or their acetates[4, 10, 12] has been carried out in the course of characterization of the 2-hydroxyestrogens, and the 2,4-dinitrophenylhydrazones of 2- and 4-hydroxyestrone have also been used.

Detection of the 2- and 4-Hydroxylations in Rat-Liver Microsomes

General. As yet all experiments aiming at solubilization and isolation of the microsomal steroid hydroxylase from liver have resulted in partial or complete loss of enzyme activity. Therefore, for *in vitro* studies the microsomal fraction is used. From such incubation mixtures 2-hydroxyestrogens can be isolated in good yields (10–30%) only under certain well-established conditions.[2]

1. The choice of incubation medium: Tris-HCl buffer, pH 7.4, has proved to be favorable; phosphate buffer, on the contrary, seems to activate secondary reactions that consume 2-hydroxyestrogen faster than it is formed, so that only very small quantities of this metabolite are detectable; because Fe^{2+} ions facilitate these secondary reactions, the water used for buffers has to be freed from heavy metals by double distillation in a quartz apparatus; the consumption of 2-hydroxyestrogen seems to be inhibited by 100,000 g supernatant or tetrahydrofolic acid.[6]

2. The choice of incubation time: Even in Tris-buffered systems 2-hydroxyestrogen is further metabolized and may be no longer detectable after too extended incubation. The optimal reaction time depends on the proportion of microsomes to substrate; under the conditions described below maximal yields can be expected with male animals after 10–30 minutes and with females after 20–60 minutes (Fig. 1).

3. NADPH-concentration: The cofactor essential for hydroxylation has to be added in great excess, since it is consumed by competing reactions (microsomal aerobic NADPH-oxidation); it can be regenerated also *in situ;* glucose-6-phosphate, glucose-6-phosphate dehydrogenase,[6] and glucose 6-phosphate/100,000 g-supernatant[11] have been used as regenerating systems.

Even under these conditions the yield of 4-hydroxyestrone remains

[12] F. Marks and E. Hecker, *Z. Physiol. Chem.* **349**, 523 (1968).

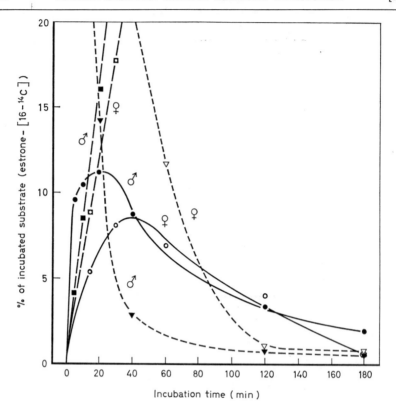

Fɪɢ. 1. Consumption of estrone (- - - ▼ - - - for ♂, - - - ▽ - - - for ♀), formation of protein-bound and water-soluble metabolites (- - ■ - - for ♂, - - □ - - for ♀) and formation of 2-hydroxyestrone (- ● - for ♂, - ○ - for ♀) by rat-liver microsome NADPH as functions of incubation time. The conditions are described in the text.

below 1%. Preincubation of microsomes for 15 minutes has no influence on 2- or 4-hydroxylation.

Preparation of Microsomes. Wistar rats, approximately 3–5 months old, after a starvation period of 18–24 hours are killed by a blow in the neck and bled by decapitation. The livers are transferred as quickly as possible into ice-cold isotonic KCl solution. A 1:10 homogenate in 0.25 M sucrose is prepared using a Potter-Elvehjem homogenizer. After removal of cell debris, nuclei, and mitochondria by centrifugation for 15 minutes at 15,000 g, the supernatant (with as little as possible of the "fluffy layer") is centrifuged for 45 minutes at 100,000 g. The microsomal pellet is thoroughly rinsed with isotonic KCl and resuspended in isotonic KCl so that 1 ml of the suspension contains the microsomes of 1 g of liver (fresh weight). Washing by resuspension and further centrifugation at 100,000 g results in a loss of microsomes of 20–30% measured as dry weight after precipitation with perchloric acid. Enzymatic activity, however, does not seem to be reduced. The microsomal suspension may be stored for 1 day at $-30°$ without loss of hydroxylase activity.

If it is desirable to incubate with 15,000 g supernatant, a 1:5 homogenate in isotonic KCl may be used to produce the same microsomal concentration in the incubation mixture without exceeding its volume.

All these operations are carried out at $0°–4°$.

Incubation. The incubation mixtures are made up at $0°$ in the following order:

> Microsomes from 200 mg of liver in isotonic KCl (0.20 ml) (or 0.75 ml of 15,000 g supernatant of 20% homogenate in isotonic KCl)
> Tris-HCl, 0.15 millimole, pH 7.4
> Nicotinamide, 0.13 millimoles
> NADPH 2.00 micromoles
> Estrone-^{14}C or 17β-estradiol-^{14}C, 0.03 micromole (0.75 ml) (i.e., 3 to 7 $\times$ 10^4 cpm)

The final volume is 2.0 ml. NADPH$_2$ and other additions (inhibitors, etc.) are first dissolved in buffer. The labeled steroid is dissolved in hot ethanol (approximately 1 mg/ml) and diluted with buffer to a final concentration of 10.0 μg of estrogen per milliliter. Frequently this procedure results in a turbid solution which becomes clear by heating briefly on a boiling water bath. The solution, cooled to $0°$, has to be used soon, since turbidity will appear again after some hours. The specific radioactivity of an aliquot of the solution should be determined for each series of experiments.

The mixtures are incubated at $38°$ in open 25-ml Erlenmeyer flasks by shaking in a water bath. To stop the reaction at the end of the incubation time the flasks are cooled quickly and 1 ml of 1 N HCl is added.

Separation of the Steroid Metabolites. The acidified incubation mixture is extracted three times with 3 ml of peroxide-free ether or ethyl acetate, and the combined extracts are dried with anhydrous sodium sulfate. Unlabeled 2- or 4-hydroxyestrone[2] and estrone are added as carrier, the quantities depending on the separation method used (described below). The solvent is evaporated *in vacuo* or under nitrogen. If the residue has to be stored, it is kept under nitrogen at $-20°$.

Isolation of Estrone and 2- and 4-Hydroxyestrone by Thin-Layer Chromatography. The three steroids can be separated from each other and from 6α- and 6β-hydroxyestrone (but not from 17β-estradiol and 16-hydroxyestrone) by thin-layer chromatography on layers of silical gel with a thickness of 1.5 mm (silica gel PF$_{254}$ of E. Merck, A. G., Darmstadt, Germany). The residue of the ethereal extract (containing

5 mg of hydroxyestrone or 15 mg of estrone as carrier, respectively) is dissolved in a small volume of methanol–chloroform 1:4 and applied to the plate as a streak, 10–15 cm long and 0.5–0.8 cm wide. The chromatogram is developed with acetic acid–cyclohexane–chloroform 1:2:2 in an equilibrated chamber. Under such conditions the R_f values are about 0.3 for 2-hydroxyestrone, 0.4 for 4-hydroxyestrone, and 0.5 for estrone as well as 0.45 for estradiol-17β, 0.3 for 16-hydroxyestrones, and 0.2 for 6-hydroxyestrones. The ultraviolet light absorbing zones are scraped from the plate and eluted with acetone–chloroform 1:2; the solvent is removed in centrifuge tubes with a stream of nitrogen. Estrone-^{14}C is recrystallized from methanol to constant specific radioactivity (5 times). The 2- and 4-hydroxyestrones are each dissolved in 0.7 ml ethanol and 0.7 ml of 2,4-dinitrophenylhydrazine reagent[14] is added. The mixture is warmed to 50–60° for 1 minute and allowed to cool, giving the yellow crystalline 2,4-dinitrophenylhydrazone (DNPH). After a few hours they are centrifuged and washed three times with 20% ethanol. The DNPH of 2-hydroxyestrone is radiochemically almost pure. It is freed from nonradioactive impurities by recrystallization from ethanol by dropping water into the hot saturated ethanolic solution until the turbidity produced does not disappear even if it is boiled. This procedure is repeated twice. Then the crystalline yellow powder is dried *in vacuo* and recrystallized to constant specific radioactivity from nitromethane–benzene 1:1. The same result may be achieved by thin-layer chromatography of the DNPH of 2-hydroxyestrone (1.5 mm layer of silica gel, acetic acid–chloroform–cyclohexane 1:2:2, saturated atmosphere, R_f value approximately 0.4) followed by recrystallization from nitromethane–benzene (1:1).

The 2,4-dinitrophenylhydrazone of 4-hydroxyestrone is recrystallized three times from *n*-propanol and then to constant specific activity from nitromethane–benzene 1:1 (five times).

The radioactivity of the derivatives is measured in a flow counter.[15] For this purpose the samples are dissolved in cold acetone; the solution is distributed over aluminum planchets and evaporated to constant weight by means of an infrared lamp. An even distribution of microcrystalline material is obtained by rubbing with a thin glass rod while heating with the lamp. The data are corrected for self-absorption using a protein standard curve.

[14] R. L. Shriner, R. C. Fuson, and D. Y. Curtin, "The Systematic Identification of Organic Compounds," 5th Ed., p. 126, Wiley, New York, and Chapman & Hall, London, 1964.
[15] Because of the strong quenching effect to be expected by the yellow DNPH-derivatives, liquid scintillation has not been applied in our laboratory.

Isolation of Estrone and 2- and 4-Hydroxyestrone by Paper Chromatography. Estrone as well as 2- and 4-hydroxyestrone can be separated from each other and from 6- and 16-hydroxyestrones[2] by descending paper chromatography in system I. For this system and the systems described below a proper atmosphere is generated by filter papers covering the walls of the chamber and dipping into the stationary phase. The systems are ready for use after 12 hours. If a second phase separates in the alcohol–water mixture, the systems have to be replaced. The R_f values may vary with the distance traveled by the liquid front. Therefore, reference substances must be applied for identification of spots. Average R_f values are shown in the table. Before application to the paper, 0.15 mg

AVERAGE R_f VALUES FOR PAPER CHROMATOGRAPHY
OF ESTROGEN METABOLITES

Compound	R_f value in system[a]			
	I	II	III	IV
Estrone	0.7	0.8	0.9	0.5
2-Hydroxyestrone	0.2	0.6	0.7	0.25
4-Hydroxyestrone	0.3	0.7	0.8	—
6α- and 6β-Hydroxyestrone	<0.1	0.2	0.4	—
16α- and 16β-Hydroxyestrone	0.1	0.4	0.65	—
17-Oxoestra-p-quinol-(10β)	0.25	—	0.8	0.3
17β-Estradiol	0.35	—	0.8	0.4
2-Hydroxyestradiol-(17β)	—	—	0.4	0.1
17β-Hydroxyestra-p-quinol-(10β)	—	—	0.5	0.1

[a] Composition of systems:
  I. Mobile phase: cyclohexane–benzene 4:1; stationary phase: ethanol–water 2:2 (Bush-type system).
  II. Mobile phase: cyclohexane–benzene 6:6; stationary phase: ethanol–water 5:5.
  III. Mobile phase: benzene, 12.5 volumes; stationary phase: methanol–water 5:5.
  IV. Mobile phase: isooctane–n-hexane 4:1; stationary phase: n-propanol–water 2:2.

of each compound is added as carrier and the dry residue of the organic extract is dissolved in a little methanol–chloroform (1:4). This solution is applied as 15–20 cm long and 1 cm wide strip (paper 2043 of Schleicher & Schuell Corporation). On both sides of the sample, spots of about 10 μg of reference substance will help to localize the compounds. After termination of chromatography, the reference strips are cut off and stained with Folin-Ciocalteu phenol reagent: 2- and 4-hydroxyestrone appear at once as blue spots, estrone only after treatment with ammonia vapor. For the isolation of 6- and 16-hydroxyestrones chromatography in system II is used (see the table).

The paper zones corresponding to the reference spots are cut off, clamped between microscope slides, and eluted with methanol overnight. The effluents are mixed with 15 mg (estrone) or 5 mg (hydroxyestrone) as carrier and concentrated in a stream of nitrogen. Estrone is recrystallized as such, 2- or 4-hydroxyestrone after precipitation as 2,4-dinitrophenylhydrazone (see p. 740) to constant specific radioactivity.

Isolation of 2-Hydroxyestradiol-17β.[10] After addition of 3.0 mg of carrier, the ethereal extract of the incubation mixture is applied on two paper sheets as stripes 20–25 cm long and 1 cm wide (2043 b from Schleicher & Schuell Corporation). The 2-hydroxysteroid is separated by chromatography in system III (see the table) and rechromatography in system IV (see the table) by descending technique. Identification of the compound is accomplished by reference spots on either side of the chromatogram stained with Folin-Ciocalteu reagent. The corresponding zone is eluted with ethanol overnight.

Then another 5–8 mg of carrier is added, and the sample is treated with a mixture of 1 ml of pyridine and 1 ml of acetic anhydride. After 48 hours, excess acetic anhydride–pyridine is removed *in vacuo* and the residue (2,3,17β-triacetoxy-$\Delta^{1,3,5(10)}$-estratriene) dried over P_2O_5 and KOH. The derivative is recrystallized to constant specific activity from methanol.

Separation of Estrone and Estradiol-17β.[16] In many cases it may be useful to know the yields of 2- or 4-hydroxyestrogens as percentage of all the metabolites formed during incubation. For this purpose and for measurement of the rate of estrogen inactivation, it is necessary to determine the amount of hormone not metabolized. After addition of 15 mg of each carrier to the ether or ethyl acetate extracts of the incubation mixtures, the two estrogens are separated by thick-layer chromatography on silica gel (20 × 20 cm, 1.5 mm thick) with carbon tetrachloride–ethyl acetate 3:1. The R_f values are 0.8 for estrone and 0.5 for 17β-estradiol. After elution of the ultraviolet light absorbing zones with acetone–chloroform 1:2, estrone is recrystallized from ethanol up to constant specific radioactivity (5 times). Estradiol is recrystallized 4 times from ethanol–water 8:1 and once or twice from methanol.

Demonstration of 10β-Hydroxylation with Rat Liver Microsomes

General. 10β-hydroxylation of estrogens producing estra-*p*-quinols-(10β) is catalyzed by an enzyme system in rat liver microsomes, which is probably not identical with the 2- or 4-hydroxylase. The reaction runs much quicker than 2-hydroxylation—also at low temperatures—but, for

[16] F. Marks, R. Süss, H. P. Morris, and E. Hecker, *Z. Krebsforsch.* **69**, 361 (1967).

reasons not yet clarified, it stops after a few minutes. Like other microsomal hydroxylations it requires oxygen; but its demand for NADPH is unclear; at least it needs much less NADPH than 2-hydroxylation and protein binding.[8] 10β-Hydroxylation is inhibited by p-chloromercuribenzoic acid, 2,4-dichlorophenol, tetrahydrofolic acid, 2-hydroxyestrone and cyanide and activated by peroxidase. Heat-inactivated microsomes, too, are able to catalyze the reaction occasionally.[8]

For the formation of the p-quinols a homolytic as well as a heterolytic mechanism has been discussed.[7,17] For example, first a phenoxyl radical may be generated by microsomal redox factors. This radical possibly reacts with free or enzymatically activated oxygen in the p-position. The resulting peroxide may be cleaved hydrolytically or reductively to give the p-quinol. An analogous mechanism has been demonstrated for the nonenzymatic autoxidation of 2,4,6-trialkyl phenols.[18] As a model for an ionic mechanism the oxidation of p-alkyl phenols with lead tetraacetate may be considered.[19]

As yet 10β-hydroxylation has been studied exclusively *in vitro* by using radioactively labeled estrogens. For isolation and purification of estra-p-quinols-(10β) the ethereal extracts of incubation mixtures are mixed with a little carrier followed by paper chromatography. After further addition of carrier the p-quinols are transformed into (2,4-dinitrobenzene)-<1-azo-3>-$\Delta^{1,3,5(10)}$-estratriene derivatives (see formulas I and II below) by reaction with 2,4-dinitrophenylhydrazine.[19] These compounds are excellently suited for recrystallization in small amounts and column chromatography on alumina.

(I)

[17] E. Hecker and G. Nowoczek, *Z. Naturforsch.* **21b**, 153 (1966).

[18] H. Musso, *Angew. Chem.* **75**, 974 (1963); M. S. Kharasch and B. Joshi, *J. Org. Chem.* **22**, 1439 (1957); K. Ley, *Angew. Chem.* **70**, 74 (1958); H. R. Gersmann and A. F. Bickel, *J. Chem. Soc.* p. 2711 (1959); *ibid.*, p. 2356 (1962).

[19] E. Hecker and E. Walk, *Chem. Ber.* **93**, 2928 (1960); E. Hecker, *ibid.* **92**, 3198 (1959).

(II)

Carrier estra-p-quinols-(10β) are prepared according to Hecker *et al.*;[19] the 10β position of the hydroxyl group has been demonstrated by measurements of optical rotatory dispersion.[20]

Preparation of Microsomes and Incubation. Preparations of microsomes and of incubation mixtures are in general similar to those described for 2-hydroxylation. The incubations are carried out in phosphate buffer (Sörensen) of pH 7.4. A quarter of the quantity of NADPH required for 2-hydroxylation suffices if 2-hydroxylation and protein binding are not to be determined in the same incubation mixture. But in view of the small yields of estra-p-quinols (0.5–5%), it is suitable to double or quadruple all components and the volume of the incubation mixture as compared to 2-hydroxylation. After incubation at 38° (usually 15 minutes), the reaction is stopped by cooling to 0° and adding 1 N HCl. For isolation of the ether-soluble metabolites, the mixture is extracted three times with peroxide-free ether.

Detection of 17′-Oxoestra-p-quinol-10β.[8] The ether extract is mixed with 1.5 mg of carrier[19] and applied to chromatography paper (2043 b of Schleicher & Schuell Corporation) as a streak 15–20 cm long and about 1 cm wide. Descending chromatography is carried out in system I. Estra-p-quinols absorb ultraviolet light penetrating through paper; the R_f value is about 0.25. The corresponding paper zone is cut out, clamped between slides, and eluted with methanol overnight. Then the eluate is mixed with another 5–8 mg of carrier and evaporated under nitrogen. The residue is dissolved in 1–2 ml acetic acid and 1 ml of a saturated solution of 2,4-dinitrophenylhydrazine in glycol monomethyl ether (about 20 mg/ml) is added at 50°. After 24 hours (room temperature), the red precipitate of 2,4-dinitrophenylhydrazone of (2,4-dinitrobenzene)-<1-azo-3>-$\Delta^{1,3,5(10)}$-estratrien-17-one (I) is centrifuged and washed with a small quantity of nitromethane–benzene (1:1). The derivative is recrystallized three times from the same solvent mixture and then chromatographed on alumina (Woelm, neutral, activity 2) with chloro-

[20] E. Hecker, R. Lattrell, and E. Meyer, *Chem. Ber.* 95, 985 (1962).

form. After this procedure it is recrystallized from nitromethane–benzene 1:1 to constant specific radioactivity (once or twice). For plating, 1.5–2 mg of the derivative is suspended in 0.1–0.2 ml of warm tetrahydrofuran; the suspension is spread evenly on aluminum planchets and dried to constant weight by an infrared lamp. Radioactivity of the sample is measured in a flow counter and corrected for self-absorption (protein standard). Because of the intense color of the derivative liquid-scintillation counting has not been applied.

Detection of 17β-Hydroxyestra-p-quinol-(10β).[7, 21] Carrier (1.5 mg) is added to the ethereal extract of the incubation mixture. The solution is applied to paper 2043 b (Schleicher & Schuell) as a streak 20 cm long and 1 cm wide. *p*-Quinol is separated by descending paper chromatography first with system I, then rechromatographed in system III. After chromatography, the zone which absorbs penetrating ultraviolet light[22] is cut out, eluted with methanol, and mixed with an additional 5 mg of carrier. The solution is concentrated under nitrogen; after addition of 1.5 ml of acetic acid and 1.0 ml of a saturated solution of 2,4-dinitrophenylhydrazine in glycolmonomethylether, the residue is heated on a boiling water bath for 1 hour. This procedure converts the *p*-quinol to (2,4-dinitrobenzene)-<1-azo-3>-$\Delta^{1,3,5(10)}$-estratrien-17β-ol (II). After 12 hours (room temperature) the reaction mixture is evaporated to dryness and chromatographed with benzene on alumina (Woelm, anionotrop, activity 3). The red band finally appearing is collected, and the dry residue is recrystallized from acetone to constant specific radioactivity.

Formation of Protein-Bound and Water-Soluble Metabolites in Rat Liver Microsomes

General. The formation of protein-bound and water-soluble metabolites presumably takes place by similar mechanisms. Reactive conversion products of 2-hydroxyestrogens are probably bound to proteins, peptides, and amino acids. There is good evidence that the binding sites are SH-groups.[9] The reaction requires oxygen and NADPH. The yield of protein-bound metabolites is maximal with microsomes only; additions of glutathione, 100,000 *g* supernatant, or thioglycolic acid reduce protein-binding and raise the yield of water-soluble metabolites.

Determination of protein-binding may be used as a rapid test for the *in vitro* activity of the estrogen hydroxylases in microsomes.[16, 23]

Preparation of Microsomes and Incubation. The same incubation

[21] E. Hecker and D. Betz, *Z. Physiol. Chem.* 338, 260 (1964).

[22] Because of their low concentration on paper, the ultraviolet absorptions of the other hydroxyestrogens are not detectable.

[23] P. H. Jellinck, *Proc. 6th Can. Cancer Conf., Honey Harbour, Ontario, 1964.* Macmillan (Pergamon), New York, 1965.

mixtures already described for 2-hydroxylation also yield protein-bound and water-soluble estrogen metabolites. Under these conditions maximal protein binding will be reached after an incubation time of 2 hours.

Determination of Protein-Bound and Water-Soluble Metabolites. After extraction of incubation mixtures with ether or ethyl acetate, 3 ml of 0.6 N perchloric acid is added to the aqueous phase. The protein precipitated is centrifuged and washed once with 5 ml of 0.6 N perchloric acid, three times with 5 ml of ethanol, and three times with 5 ml of ether. The dry white protein powder which contains *ca.* 20% RNA, is weighed (3–5 mg per 200 mg of fresh liver) and dissolved in 80% formic acid (about 10 mg/ml). Of this solution, 0.1–0.2 ml is evenly distributed on aluminum planchets and dried to constant weight by means of an infrared lamp. Radioactivity is measured in a flow counter and corrected for self-absorption by means of a protein standard.

In order to determine the water-soluble radioactivity (which does not precipitate with perchloric acid), the supernatant of the protein precipitation is neutralized at 0° with 6 N KOH against methyl red, decanted from precipitated $KClO_4$, and filled up to 10.0 ml with water. For measurements of radioactivity, 0.5-ml aliquots are evaporated by an infrared lamp on aluminum planchets. In order to correct for the considerable self-absorption it is necessary to determine the dry weight of the samples.

Protein-bound and water-soluble radioactivity may be measured also in the liquid-scintillation counter.[11] For this purpose the protein precipitate is warmed up together with 1–2 ml of formamide at 180° for 2 hours. A mixture of 10 ml of toluene–ethanol 3:2 and 1 ml of hyamine hydroxide serves as solvent.

[39] Soluble 17β-Hydroxysteroid Dehydrogenase of Human Placenta[1,1a]

By JOSEPH JARABAK

$$17\beta\text{-Estradiol} + NAD^+(NADP^+) \rightleftarrows \text{estrone} + NADH(NADPH) + H^+$$

The hydroxysteroid dehydrogenases are a group of pyridine nucleotide-linked enzymes that catalyze the interconversion of steroid hydroxyl

[1] These studies were supported by a grant from the United States Public Health Service (AM 07422).
[1a] J. Jarabak, J. A. Adams, H. G. Williams-Ashman, and P. Talalay, *J. Biol. Chem.* **237**, 345 (1962).

and carbonyl groups.[2] The initial description of one of these, a soluble 17β-hydroxysteroid dehydrogenase obtained from human placenta and possessing a dual nucleotide specificity, was given by Langer and Engel.[3] Shortly thereafter it was recognized that this enzyme could also catalyze transhydrogenation between pyridine nucleotides in the presence of *catalytic* amounts of steroid substrate:[4, 5]

$$17β\text{-Estradiol} + NAD^+ \rightleftarrows estrone + NADH + H^+$$
$$\underline{Estrone + NADPH + H^+ \rightleftarrows 17β\text{-estradiol} + NADP^+}$$
$$Net:\quad NADPH + NAD^+ \rightleftarrows NADP^+ + NADH$$

Methods for isolation and assay of this enzyme are described in Vol. V [69]. The method described below achieves a more effective stabilization of the enzyme than the previous procedure and also gives a 30-fold greater purification.

Assay Method

Dehydrogenation. The enzyme is assayed by spectrophotometric measurement of the reduction of the pyridine nucleotide. The assay system contains, in a final volume of 3.0 ml: 440 micromoles of sodium pyrophosphate buffer at pH 10.2, 25 mg of crystalline bovine serum albumin, 0.3 micromole of 17β-estradiol (added in 0.04 ml of 95% ethanol), 1.35 micromoles of NAD+ (or 1.1 micromoles of NADP), and appropriate quantities of enzyme. The final pH is between 9.2 and 9.3. The reaction is initiated by the addition of enzyme, and the rate of formation of the reduced nucleotide is followed at 340 mμ against a blank cuvette containing all ingredients except steroid. Reaction velocities are calculated from the initial linear slopes. While the assay using NAD+ or NADP+ is quite satisfactory, the linearity may be improved and the sensitivity increased by a factor of 6 if 3-acetylpyridine adenine dinucleotide (1.9 micromoles) is used as the cofactor. The reduction of 3-acetylpyridine adenine dinucleotide is measured at 363 mμ.

Transhydrogenation. The transhydrogenation assay system contains, in a final volume of 3.0 ml: 300 micromoles of Tris-hydrochloride buffer at pH 7.4, 10 micromoles of sodium glucose 6-phosphate, 2.5 μg of purified yeast glucose-6-phosphate dehydrogenase (obtained from the Boerhinger Mannheim Corporation and added in 0.01 ml of 20% glycerol), 0.01

[2] P. Talalay, in "The Enzymes" (P. D. Boyer, H. Lardy, and K. Myrbäck, eds.), 2nd ed., Vol. VII, p. 177. Academic Press, New York, 1963.
[3] L. J. Langer and L. L. Engel, *J. Biol. Chem.* **233**, 583 (1958).
[4] P. Talalay and H. G. Williams-Ashman, *Proc. Natl. Acad. Sci. U.S.* **44**, 15 (1958).
[5] P. Talalay, B. Hurlock, and H. G. Williams-Ashman, *Proc. Natl. Acad. Sci. U.S.* **44**, 862 (1958).

micromole of NADP+, 1.35 micromoles of NAD+, 0.015 micromole of 17β-estradiol (added in 0.02 ml of 95% ethanol), and suitable quantities of the placental enzyme. The final pH is between 7.2 and 7.3. The rate of the reduction of NAD+ is followed spectrophotometrically at 340 mμ against a blank containing all ingredients except the steroid. Generation of NADPH is completed within a period of several minutes, after which there is no detectable change in absorbance at 340 mμ until the placental enzyme is added.

One unit of enzyme (in either the dehydrogenation or transhydrogenation assay) is defined as the amount of enzyme reducing 1 micromole of pyridine nucleotide per minute in a cuvette of 1.0-cm light path at 25°. Initial rates obtained from assays are converted to enzyme units by dividing the change in absorbance per minute by 2.07 for the systems containing NAD+ or NADP+ or by 3.03 for the systems containing 3-acetylpyridine adenine dinucleotide. In earlier publications[1a, 4, 5] one unit of enzyme was defined as causing a change in absorbance of 0.001 per minute at 340 or 363 mμ.

Purification Procedure

The procedure to be described is a modification of that given by Jarabak et al.[1a] Unless otherwise specified all operations are performed at 2–5°. The activities and yields are summarized in the table. Because a single normal term placenta contains between 10 and 20 mg of the 17β-hydroxysteroid dehydrogenase, it is advisable to use more than one placenta if the purification is to be carried to completion. Steps 1 and 2 are carried out at once with each placenta and material is accumulated after step 2.

Step 1. Homogenization. Term placenta are obtained immediately after delivery[6] and chilled on ice during transport from the birthrooms. The villi are rapidly dissected away from major vessels, connective tissue, and fetal membranes (200–450 g of tissue are obtained per placenta). Seventy-five grams of the villous tissue are added to 150 ml of Medium A (20% glycerol,[7] 5 mM potassium phosphate, 1 mM EDTA, at a final pH

---

[6] It is often difficult to obtain a large number of term placentas immediately after their delivery. Frozen placentas are not a good source of the enzyme since freezing is accompanied by a loss of more than 80% of the 17β-hydroxysteroid dehydrogenase activity. If term placentas are stored in a 20% glycerol solution (as described in step 1) and then homogenized in the same solution (10–20% of the 17β-hydroxysteroid dehydrogenase is found in the storage solution after 24 hours), the yield of enzyme per placenta is about the same as for the term placenta homogenized immediately after delivery. All of the placentas used in the purification outlined in the table were stored in glycerol prior to homogenization.

[7] Spectroscopic grade glycerol is used in all steps of the purification. Glycerol concentrations are expressed by volume.

of 7.0) and homogenized for 15 sec at full speed in a 1-liter Waring blendor container. The homogenate is centrifuged at 10,000 g for 30 minutes. The supernatant solution is decanted, and the precipitate is discarded. It is advisable to perform the next step without delay.

Step 2. Ammonium Sulfate Precipitation. Solid ammonium sulfate (Mann, enzyme grade) is added to the supernatant solution slowly and with continuous stirring to achieve 50% saturation.[8] At the same time, the pH is maintained at about 7.0 by addition of 1 N ammonium hydroxide. The precipitate is allowed to accumulate for 2 hours and then is centrifuged for 1 hour at 10,000 g. The supernatant solution is discarded and the precipitate is dissolved in a minimum amount (about 50–75 ml) of Medium B (50% glycerol, 5 mM potassium phosphate, 1 mM EDTA, at a final pH of 7.0).[9]

Step 3. Heat Treatment. Dialysis tubing (flat width, 1.3 inch) is filled with 75-ml portions of the enzyme solution from the preceding step and placed in a 4-liter Erlenmeyer flask containing Medium B preheated to 67–68°. (Generally the precipitates from three to five placentas are heated at once.) The temperature is held at 67–68° and the buffer is stirred magnetically for 3 hours. During this time the color of the solution becomes darker and a copious precipitate forms. After the heating is completed, the dialysis bags are transferred to a 4-liter Erlenmeyer flask containing Medium A at 2–5°. The dialysis is continued at 2–5° for 16–24 hours with several changes of buffer; then the contents of the bags are centrifuged for 1 hour at 39,000 g. The precipitate is discarded.

Step 4. Second Ammonium Sulfate Precipitation. The enzyme is precipitated at pH 7.0 by adding solid ammonium sulfate to 50% saturation as described in step 2. The precipitate is allowed to accumulate for 2 hours and then is centrifuged for 1 hour at 10,000 g. The supernatant solution is discarded and the precipitate is dissolved in a minimum amount of Medium B.

Step 5. First DEAE-Cellulose Chromatography.[10] Material combined from twenty to twenty-five placentas is dialyzed against several changes of Medium A and then applied to a 25 × 500 mm DEAE-cellulose[11] (Bio-Rad) column which has been thoroughly equilibrated with Medium A. After the enzyme is applied, the elution gradient is started. The column is connected to a constant-volume 1000-ml mixing vessel which is stirred magnetically and initially is filled with Medium A. The mixing vessel

---

[8] The quantities of ammonium sulfate added are calculated and expressed as though the solutions were completely aqueous.

[9] This is a convenient stage to accumulate material for a large-scale purification since the enzyme may be stored for extended periods without significant loss of activity.

[10] Performed at room temperature.

[11] See Vol. V [1] for the method of preparation of the anion-exchange cellulose.

is connected to a reservoir which contains 20% glycerol, $0.4\,M$ K_2HPO_4, 1 mM EDTA. The gradient formed in this manner is convex.[12] The rate of elution is usually 45–60 ml/hour. The elution of the enzyme starts at a phosphate concentration of $0.08\,M$ and is essentially completed by a concentration of $0.16\,M$.

Step 6. Hydroxylapatite Chromatography.[10] Fractions from step 5 having the highest specific activity are combined and dialyzed at 2–5° against several changes of Medium C (20% glycerol, 5 mM potassium phosphate, at a final pH of 7.0). The resulting solution is then applied to a 25×450 mm hydroxylapatite[13] column which has been thoroughly equilibrated with Medium C. After the enzyme is applied, the column is

PURIFICATION OF 17β-HYDROXYSTEROID DEHYDROGENASE FROM HUMAN PLACENTA

Step	Fraction	Volume (ml)	Total activity[b] (units)	Specific activity (units/mg protein[c])	Yield (%)
1	Homogenization	16,185[a]	834	0.00178	100
2	First ammonium sulfate precipitate	1,990	584	0.00830	70
3	Supernatant from heat treatment	1,650	460	0.0286	55
4	Second ammonium sulfate precipitate	201	362	0.0440	43
5	First DEAE-cellulose column effluent	200	270	0.314	32
6	Hydroxylapatite column effluent	270	203	0.725	24
7	Second DEAE-cellulose column effluent	65	157	2.05	19

[a] Volume of the supernatant solution after centrifugation.

[b] Activities were measured by the rate of NAD reduction.

[c] Protein concentrations were determined by the method of O. H. Lowry, N. J. Rosebrough, A. L. Farr, and R. J. Randall, *J. Biol. Chem.* **193**, 265 (1951). The specific activity of the enzyme from the second DEAE-cellulose chromatography is 4.55 units per milligram of protein if the protein concentration is determined by the method of O. Warburg and W. Christian, *Biochem. Z.* **310**, 384 (1941), using the formula: $1.50 \times A_{280} - 0.75 \times A_{260}$ = protein concentration in milligrams per milliliter.

[12] R. M. Bock and N.-S. Ling, *Anal. Chem.* **26**, 1543 (1954).

[13] W. F. Anacker and V. Stoy, *Biochem. Z.* **330**, 141 (1958). If vigorous stirring is avoided during the preparation of the hydroxylapatite and the fine particles of the hydroxylapatite are thoroughly removed by decantation, a column of the size described in the text will have a flow rate of at least 100 ml/hour even after it is equilibrated with 20% glycerol.

washed with 200–300 ml of Medium C and then a linear elution gradient[12] is started. The column is connected to a mixing vessel which is stirred magnetically and contains 1000 ml of Medium C. The mixing vessel is joined to a vessel of identical cross section containing 1000 ml of 20% glycerol, 0.1 M potassium phosphate, at a final pH of 7.0. The rate of elution is usually 45–60 ml per hour. The enzyme is eluted between phosphate concentrations of 0.02 and 0.03 M.

Step 7. Second DEAE-Cellulose Chromatography.[10] The fractions from step 6 containing the highest specific activity are combined and applied without dialysis to a 16 × 100 mm DEAE-cellulose column which has been equilibrated with Medium A. Elution is with a linear gradient, the mixing vessel containing 1000 ml of Medium A and the other vessel containing 1000 ml of 20% glycerol, 0.2 M K_2HPO_4, 1 mM EDTA. The fractions having the highest specific activity are combined and dialyzed against Medium B. The purified enzyme then may be stored at 2–5° for over a year with little loss of activity.

Discontinuous polyacrylamide gel electrophoresis[14] of this material, performed on samples containing 40 μg of protein, reveals a single band of protein. Histochemical staining[15] of the polyacrylamide gels indicates that this band contains the 17β-hydroxysteroid dehydrogenase activity.

Properties

Stability.[15] If highly purified solutions of the enzyme, which have been diluted so that their glycerol concentration is reduced to 1%, are cooled to 0° by storage in crushed ice, they rapidly lose enzymatic activity. The loss of enzymatic activity is much slower when the 1% glycerol solutions are stored at 25°. Solutions which have been stored at 0° for as long as several days may be partially reactivated by warming them to 25°, but more prolonged cooling leads to complete denaturation. The inactivation appears to involve both a change in the shape of the enzyme and the formation of a series of polymeric species. Both placental homogenates and highly purified enzyme solutions lose over 90% of the 17β-hydroxysteroid dehydrogenase activity if they are stored for 24 hours at 0° without any protective agent. This loss may be greatly slowed by the addition of NAD+, NADP+, and certain other pyridine nucleotides, 17β-estradiol,[3] high concentrations of polyvalent anions (phosphate and sulfate), and certain alcohols.[1a, 3, 16] Glycerol in concentrations of 20% prevents the loss of enzyme activity for several weeks

[14] B. J. Davis, *Ann. N.Y. Acad. Sci.* **121**, 404 (1964).
[15] J. Jarabak, A. E. Seeds, Jr., and P. Talalay, *Biochemistry* **5**, 1269 (1966).
[16] P. Talalay, *in* "On Cancer and Hormones" (E. Boyland *et al.*, eds.), p. 271. Univ. of Chicago Press, Chicago, Illinois, 1962.

while concentrations of 50% completely protect the enzyme for many months.

Inhibitors. Crude preparations of the enzyme are inactivated by certain heavy metal ions (Hg^{++}, Cu^{++}, and Fe^{3+}) but are unaffected by others (Fe^{++}, Mn^{++}, Mg^{++}, and Co^{++}).[3] Certain monovalent anions (chloride and nitrate) both inhibit the enzyme and decrease its stability at $0°$.[15, 17] Inactivation of highly purified samples of the enzyme by either *p*-(hydroxy)mercuribenzoate, urea, or sodium dodecyl sulfate is accompanied, as during cold inactivation, by the formation of a series of polymeric species.

Pyridine Nucleotide Specificity. The enzyme catalyzes the reduction of NAD^+, $NADP^+$, 3-acetylpyridine adenine dinucleotide, 3-acetylpyridine adenine dinucleotide phosphate, and 3-pyridinealdehyde adenine dinucleotide. There is little change in the rates of reduction of NAD^+, $NADP^+$, and 3-acetylpyridine adenine dinucleotide phosphate between pH 7 and 9, while there is a greater than 10-fold increase in the rate of reduction of 3-acetylpyridine adenine dinucleotide over the same pH range.[18] The Michaelis constants have been determined at pH 7.4 for some of these nucleotides: 3-acetylpyridine adenine dinucleotide, 0.5 mM; NAD^+, 80 μM; NADH, 30 μM; $NADP^+$ and NADPH, less than 1 μM.[3]

Steroid Specificity.[19-21] The steroids most readily oxidized or reduced by the enzyme possess both a phenolic A ring and a 17β-hydroxyl or a 17-ketone group. If the hydroxyl group is absent in the A ring (estra-1,3,5(10)-trien-17β-ol) the rate of oxidation of the 17β-hydroxyl group is slightly reduced, while the presence of a ketone at C-3 and a partial saturation of the A ring (19-nortestosterone) reduce the rate of oxidation to about 3% that of 17β-estradiol. Although 17α-estradiol and 16β-estradiol are unreactive, the specificity for the 17β-hydroxyl or 17-ketone group is not absolute. Thus, the reduction of the 20-ketone group of progesterone to a 20α-hydroxyl group is also catalyzed by this enzyme, but at a rate one-fiftieth that of the reduction of estrone.[22]

Equilibrium Constant. $K_H = $ [estrone][NADH][H^+]/[estradiol]-[NAD^+] $= 1.8 \times 10^{-8}$ at $25°$.[3]

[17] J. C. Warren and S. G. Cheatum, *Biochemistry* **5**, 1702 (1966).
[18] P. Talalay and H. G. Williams-Ashman, *Recent Progr. Hormone Res.* **16**, 1 (1960).
[19] V. P. Hollander, H. Nolan, and N. Hollander, *J. Biol. Chem.* **233**, 580 (1958).
[20] L. J. Langer, J. A. Alexander, and L. L. Engel, *J. Biol. Chem.* **234**, 2609 (1959).
[21] J. A. Adams, J. Jarabak, and P. Talalay, *J. Biol. Chem.* **237**, 3069 (1962).
[22] R. H. Purdy, M. Halla, and B. Little, *Biochim. Biophys. Acta* **89**, 557 (1964).

[40] 16α-Hydroxysteroid Dehydrogenase from Rat Kidney[1]

By Robert A. Meigs and Kenneth J. Ryan

Estriol + NAD⁺(NADP⁺) $\rightleftarrows$
16-keto-17β-estradiol[2] + NADH(NADPH) + H⁺

16α-Hydroxysteroid dehydrogenase from rat kidney is an adaptive enzyme regulated by sex hormones.[3] Significant activity is found only in mature females, but the enzyme can be induced in castrated males or immature animals by estrogen administration. Testosterone inhibits this induction. Cat, chicken, and pigeon kidneys have lesser activity, and no striking sex difference is evident in these species. Mouse, hamster, guinea pig, rabbit, dog, pig, sheep, cow, and human kidneys show little or no activity.[4]

Assay Method[5]

Principle. The rate of disappearance of NADH due to the reduction of 16-keto-17β-estradiol is measured with a spectrophotometer at 340 mμ.

Reagents

Quartz-distilled water, for all solutions
Purified enzyme preparation in 20 mM potassium phosphate, 20% glycerol, pH 7.0
KH_2PO_4, 50 mM
16-Keto-17β-estradiol, 2.5 mg dissolved in 1 ml of redistilled 95% ethanol
NADH, 4 mM, freshly prepared
Plumpers (Calbiochem)

Procedure. A 0.5 ml amount of 50 mM KH_2PO_4 is added slowly, with mixing, to 0.4 ml of purified enzyme preparation in a 1-cm microcuvette. The mixture is warmed to 37°. Then 0.02 ml of substrate solution (0.175 micromole steroid) is added and mixed. Addition of 0.05 ml of NADH solution (0.2 micromole) with subsequent rapid mixing results in a total volume of 0.97 ml and a final pH of 6.2, and initiates the reaction. The absorbance at 340 mμ is measured immediately and at 15-second intervals

[1] R. J. B. King, *Biochem. J.* **76**, 7P (1960).
[2] 3,17β-Dihydroxyestra-1,3,5(10)-trien-16-one.
[3] K. J. Ryan, R. A. Meigs, Z. Petro, and G. Morrison, *Science* **142**, 243 (1963).
[4] C. A. Bush and K. J. Ryan, unpublished observations, 1965.
[5] R. A. Meigs and K. J. Ryan, *J. Biol. Chem.* **241**, 4011 (1966).

for 5 minutes in a spectrophotometer with cuvette chamber maintained at 37°. Correction for nucleotide metabolism unrelated to 16α-hydroxysteroid dehydrogenase is made with a control mixture lacking only steroid, and the initial, steroid-dependent reaction velocity, constant for at least 2 minutes, is calculated. Protein is determined by the spectrophotometric method of Warburg and Christian.[6]

Assay of Crude Preparations. The spectrophotometric assay is suitable only for purified enzyme preparations having low endogenous nucleotide metabolism and in which 16β-hydroxysteroid dehydrogenase activity is lacking or low. 16α-Hydroxysteroid dehydrogenase can be assayed in crude tissue extracts by extraction and measurement of the 16-keto-17β-estradiol produced in incubations with estriol and a NAD+-regenerating system.[5]

Purification Procedure[5]

Kidneys from 6–8 sexually mature *female* rats are removed and chilled in crushed ice. All subsequent processing is carried out at 0–5°. Deionized water, doubly distilled from a quartz apparatus is used for all solutions. Twelve to fifteen grams of kidneys are minced with scissors and homogenized together with 33 ml of 1 mM potassium phosphate, 50% glycerol, pH 7.0 in a glass vessel with a motor-driven Teflon pestle. Use of a blender will reduce the activity of the homogenate. Centrifugation at 31,000 g for 10 minutes and then at 105,000 g for 1 hour produces a particle-free supernatant solution containing all activity.

This crude soluble fraction is equilibrated with 1 mM potassium phosphate, 50% glycerol, pH 7.0 by filtration, under pressure, through a 2.2 × 18 cm column of Sephadex G-25 medium grade (Pharmacia) (approximately 20 g) prepared in that buffer. The total protein fraction (20–25 ml) of the effluent, detected by its color, is collected and diluted with 1.5 volumes of 1 mM potassium phosphate, pH 7.0, to achieve a final glycerol concentration of 20%. (The Sephadex column, after re-equilibration with fresh buffer, can be used repeatedly for over a year.)

Purification and concentration of the enzyme is accomplished with a carboxymethyl cellulose (CM-cellulose) column prepared in the following manner: CM-cellulose, 0.77 meq/g (Schleicher & Schuell), is purified as described by Peterson and Sober[7] and stored moist. It is washed, as needed, with 1 mM potassium phosphate, 20% glycerol, pH 7.0, until a pH of 7.1 is achieved, then packed under pressure in this buffer, at room temperature, to form a column 1 × 6.5 cm with a column volume of 3

---

[6] O. Warburg and W. Christian, *Biochem. Z.* **310**, 384 (1942); Vol. III [73].
[7] E. A. Peterson and H. A. Sober, Vol. V [1].

ml. The column is cooled to 4° overnight in the cold room and recompressed just prior to addition of the enzyme solution.

The crude enzyme solution is passed through the column under pressure, and the column is washed with 6 ml of each of the following buffers: 1 mM potassium phosphate, 20% glycerol, pH 7.0; 20 mM potassium phosphate, 20% glycerol, pH 7.0; and 50 mM potassium phosphate, 20% glycerol, pH 7.0.

After the yellow unretained protein fraction has left the column, an additional 6 ml of effluent (the 1 mM phosphate, 20% glycerol wash) is discarded. Collection of the enzyme fraction (20 mM phosphate, 20% glycerol) is begun and continued until the dark red band eluted by 50 mM phosphate, 20% glycerol is just about to leave the column. Six to 6.5 ml of enzyme solution is obtained. The pH is adjusted to between 6.95 and 7.00 at 0° and the solution is centrifuged at 16,000 rpm for 5 minutes to remove any precipitate. The clear supernatant solution is kept at 0° and used immediately.

The CM-cellulose fractionation step results in up to 18-fold purification and yields a preparation with low endogenous nucleotide-metabolizing activity. 16β-Hydroxysteroid dehydrogenase and 17β-hydroxysteroid dehydrogenase activities are removed or reduced to insignificant levels. Typical preparations catalyze the steroid-dependent oxidation of 30 nanomoles of NADH per minute per milliliter of enzyme or 9 nanomoles per minute per milligram of protein in the spectrophotometric assay.

Properties

Stability. The enzyme is unstable and is only partially protected by glycerol. The purified preparation retains full activity for at least 8 hours after processing.

Cofactor Requirement. The enzyme has been characterized using NADH as cofactor. A K_m of $0.83 \times 10^{-5} M$ was found. NADPH is also utilized with a K_m of $4.1 \times 10^{-5} M$. There is no evidence, at the present state of purity, for multiple nucleotide-specific activities in the preparation.

Substrate Specificity. The enzyme is specific for the 16α-hydroxyl (16β-hydrogen) configuration of a variety of steroids. 16-Keto-17β-estradiol (K_m $8.9 \times 10^{-5} M$), 3,17α-dihydroxyestra-1,3,5(10)-trien-16-one (K_m $7.0 \times 10^{-5} M$), 3-hydroxyestra-1,3,5(10)-triene-16,17-dione, 3-hydroxyestra-1,3,5(10)-trien-16-one, 17β-hydroxyandrost-4-ene-3,16-dione (K_m $28.0 \times 10^{-5} M$), and 3β,17β-dihydroxyandrost-5-en-16-one are all reduced to 16α-hydroxysteroids.

Estriol, 3-methoxyestra-1,3,5(10)-triene-16α,17β-diol, estra-1,3,5(10)-triene-3,16α,17α-triol, estra-1,3,5(10)-triene-3,16α-diol, 16α,17β-dihydroxyandrost-4-en-3-one, and androst-5-ene-3β,16α,17β-triol, but not 16α-hydroxyandrost-4-ene-3,17-dione or 3β,16α-dihydroxyandrost-5-en-17-one, are oxidized to 16-ketosteroids by the purified enzyme preparation plus NAD$^+$.

pH Optimum. The pH optimum under the conditions of the spectrophotometric assay is 6.1–6.2.

Sulfhydryl Requirement. The enzyme is inhibited 90–100% by low concentrations (10^{-4} to 10^{-5} M) of 2-iodoacetamide, o-iodosobenzoate, N-ethylmaleimide, p-(hydroxymercuri)benzoate, Ag$^+$, and Cu^{++}.

Metal Requirement. The enzyme is inhibited 90–100% by 8-hydroxy-5-quinolinesulfonate or 2,2'-bipyridine at $1 \times 10^{-3}\,M$, by 1,10-phenanthroline at $1 \times 10^{-4}\,M$, and by EDTA at $1 \times 10^{-5}\,M$. No activation by Zn^{++} is found.

Other Sex-Dependent Enzymes of Steroid Metabolism

Δ^4-5α-Reductase activities are 3 to 10 times greater in the livers of female rats than in those of males,[8-11] while male rat liver has higher levels of 3β-hydroxysteroid dehydrogenase, 11β-hydroxysteroid dehydrogenase, and 20-keto reductase activities.[11-13] Rat liver hydroxylase activities against both androgens and estrogens show a similar sex-dependence, being more active in the male than in the female.[10, 14-16] With all these activities, androgens are the primary endocrine regulators although a uniform mechanism of action is not involved. In species other than the rat, these sex differences may be absent or reversed.

Purification. All of these sex-dependent rat liver enzymes are localized in the microsomal fraction of differentially centrifuged homogenates. Solubilization attempts have been unsuccessful and most studies of these enzymes have utilized, at best, crude, washed particulate preparations. Talalay has described a partial purification of male rat liver microsomes applicable to the 11β-hydroxysteroid dehydrogenase.[17]

Assay. The incubation of multiactive microsome particles with multi-

[8] F. E. Yates, A. L. Herbst, and J. Urquhart, *Endocrinology* 63, 887 (1958).
[9] E. Forchielli, K. Brown-Grant, and R. I. Dorfman, *Proc. Soc. Exptl. Biol. Med.* 99, 594 (1958).
[10] K. Leybold and H. Staudinger, *Biochem. Z.* 331, 389, 399 (1959).
[11] A. A. Hagen and R. C. Troop, *Endocrinology* 67, 194 (1960).
[12] B. L. Rubin, H. J. Strecker, and E. B. Koff, *Endocrinology* 72, 764 (1963).
[13] H. J. Hübener and D. Amelung, *Z. Physiol. Chem.* 293, 137 (1953).
[14] A. Colás, *Biochem. J.* 82, 390 (1962).
[15] R. Kuntzman, M. Jacobson, K. Schneidman, and A. H. Conney, *J. Pharmacol. Exptl. Therap.* 146, 280 (1964).
[16] P. H. Jellinck and I. Lucieer, *J. Endocrinol.* 32, 91 (1965).
[17] P. Talalay, Vol. V [69].

functional steroids results in complex mixtures of conversion products. Assay of a particular enzyme component in terms of the rate of product formation usually entails laborious steroid fractionations prior to any quantitative measurement. For assaying Δ^4-reductases, the loss of the characteristic absorption at 240 mμ of Δ^4-3-ketosteroid substrates can be more readily monitored, but is valid only if 3-hydroxysteroid dehydrogenase catalyzed reduction to allylic alcohols is negligible. Tomkins has stated that the method described for the Δ^4-5β-reductase can also be adapted for use with the 5α-enzymes.[18] Spectrophotometric measurement of changes in pyridine nucleotide cofactor concentration has been used for the assay of reductases, dehydrogenases, and hydroxylases when the multiplicity of microsomal activities could be restricted by proper selection of the sex of the microsome source, by the limitation of reactive groups of the substrate, and by selective enzyme inactivations.[10, 17, 19]

[18] G. M. Tomkins, Vol. V [67].
[19] I. E. Bush, S. A. Hunter, and R. A. Meigs, *Biochem. J.* **107**, 239 (1968).

Section VI

Corticosteroid-Binding Globulin and
Other Steroid-Binding Serum Proteins

[41] Assay and Properties of Corticosteroid-Binding Globulin and Other Steroid-Binding Serum Proteins

By ULRICH WESTPHAL

Introduction

Steroid hormones and other steroids interact with certain serum proteins to form dissociable complexes. Much information on these interactions has been obtained with serum albumin of human and bovine origin.[1-8] This protein has a remarkable ability to combine with a great variety of different types of substances, anions and cations as well as neutral molecules.[9] The binding of steroid hormones to serum albumin, therefore, cannot be considered a particularly specific type of interaction, and the binding affinities are of moderate strength. A binding system of considerable specificity and high combining affinity was recognized when the corticosteroid-binding globulin (CBG) or transcortin was discovered.[8, 10-13] A third distinct serum protein, the α_1-acid glycoprotein[14, 15] or orosomucoid,[16] has been found to interact with progesterone and other

[1] S. Roberts and C. M. Szego, *Ann. Rev. Biochem.* **24**, 543 (1955).

[2] A. A. Sandberg, W. R. Slaunwhite, Jr., and H. N. Antoniades, *Recent Progr. Hormone Res.* **13**, 209 (1957).

[3] W. H. Daughaday, *Physiol. Rev.* **39**, 885 (1959).

[4] H. N. Antoniades (ed.), "Hormones in Human Plasma." Little, Brown, Boston, Massachusetts, 1960.

[5] C. H. Gray and A. L. Bacharach (eds.), "Hormones in Blood." Academic Press, New York, 1961.

[6] U. Westphal, *in* "Mechanism of Action of Steroid Hormones" (C. A. Villee and L. L. Engel, eds.), p. 33. Macmillan (Pergamon), New York, 1961.

[7] W. K. Brunkhorst and E. L. Hess, *Arch. Biochem. Biophys.* **111**, 54 (1965).

[8] A. A. Sandberg, H. E. Rosenthal, S. L. Schneider, and W. R. Slaunwhite, Jr., *in* "Steroid Dynamics" (G. Pincus, T. Nakao, and J. F. Tait, eds.), p. 1. Academic Press, New York, 1966.

[9] J. T. Edsall and J. Wyman, "Biophysical Chemistry," Chapt. 11. Academic Press, New York, 1958.

[10] W. H. Daughaday, *J. Lab. Clin. Med.* **48**, 799 (1956).

[11] I. E. Bush, *Ciba Found. Colloq. Endocrinol.* **11**, 263 (1957).

[12] W. R. Slaunwhite, Jr., and A. A. Sandberg, *J. Clin. Invest.* **38**, 384 (1959).

[13] U. S. Seal and R. P. Doe, *in* "Steroid Dynamics" (G. Pincus, T. Nakao, and J. F. Tait, eds.), p. 63. Academic Press, New York, 1966.

[14] K. Schmid, *J. Am. Chem. Soc.* **75**, 60 (1953).

[15] R. W. Jeanloz, *in* "Glycoproteins" (A. Gottschalk, ed.), p. 362. Elsevier, New York, 1966.

[16] R. J. Winzler, *in* "The Plasma Proteins" (F. W. Putnam, ed.), Vol. I, pp. 309, 317. Academic Press, New York, 1960.

steroid hormones with a relatively high affinity.[17-19] The cholesterol-containing lipoproteins[20] will not be discussed here.

Whereas early considerations suggested that the interactions of the steroid hormones with serum proteins served a transport function, it has been recognized in recent years that the formation of the dissociable complexes has a regulatory effect on the physiological function of the hormones. The steroid hormones are inactivated when bound to the proteins; only the unbound portion is considered biologically active. A change in the concentration of the binding protein or in its binding affinity, therefore, results in a quantitative change of the hormonal activity. In a similar way, distribution of steroid hormones in the body, as well as their metabolism and excretion, is influenced by association with serum proteins.[21] These various consequences of protein interaction are particularly important for the corticosteroid hormones and for progesterone, due to their high binding affinity to CBG.

Interaction between steroid hormones and protein structures has attained new significance through recent observations on the binding of active hormones to nuclear receptor sites of target tissues. Aldosterone, prior to exerting its hormonal action, has been found to be bound with high affinity to nuclear receptors, present in rat kidney, which are specific for mineralocorticoids.[22] Estradiol becomes firmly attached to estrogen-specific nuclear receptors in the uterus.[23, 24] In both cases, the receptors appear to be of protein nature. Their role in the overall mechanism of action of these hormones has not yet been elucidated. Similarly, the important question of interaction of steroid hormones with genetic repressor substances, resulting in derepression of the genes,[25] awaits conclusive evidence. Repressors of gene activity may be proteins, possibly belonging to the class of histones.[26]

[17] U. Westphal, B. D. Ashley, and G. L. Selden, *Arch. Biochem. Biophys.* **92**, 441 (1961).
[18] U. Westphal, *J. Am. Oil Chemists' Soc.* **41**, 481 (1964).
[19] M. Ganguly, R. H. Carnighan, and U. Westphal, *Biochemistry* **6**, 2803 (1967).
[20] F. T. Lindgren and A. V. Nichols, *in* "The Plasma Proteins" (F. W. Putnam, ed.), Vol. 2, p. 2. Academic Press, New York, 1960.
[21] J. F. Tait and S. Burstein, *in* "The Hormones" (G. Pincus, K. V. Thimann, and E. B. Astwood, eds.), Vol. V, p. 441. Academic Press, New York, 1964.
[22] D. D. Fanestil and I. S. Edelman, *Proc. Natl. Acad. Sci. U.S.* **56**, 872 (1966).
[23] E. V. Jensen, H. I. Jacobson, J. W. Flesher, N. N. Saha, G. N. Gupta, S. Smith, V. Colucci, D. Shiplacoff, H. G. Neumann, E. R. DeSombre, and P. W. Jungbluth, *in* "Steroid Dynamics" (G. Pincus, T. Nakao, and J. F. Tait, eds.), p. 133. Academic Press, New York, 1966.
[24] D. Toft and J. Gorski, *Proc. Natl. Acad. Sci. U.S.* **55**, 1574 (1966).
[25] P. Karlson, *Deut. Med. Wochschr.* **86**, 668 (1961).
[26] J. Bonner, *J. Cell. Comp. Physiol.* **66** (Suppl. 1), 77 (1965).

A number of different principles have been applied to demonstrate steroid–protein interaction.[6] Most of them are based on a physical separation of the bound and unbound portion of the ligand; some techniques (solubility increase, spectrophotometry) do not rely on such separation. Quantitative assays of steroid-binding proteins are essentially concerned with four objectives:

1. Determination of binding activity, or affinity, for a given steroid hormone in blood serum or in other body fluids, expressed either as percent of total or as micrograms per 100 ml bound and unbound. The method of choice is considered to be equilibrium dialysis.

2. Determination of binding capacity, of a specific binding protein in blood serum or plasma, for a steroid hormone, expressed as concentration of binding sites of the binding protein (=concentration of steroid maximally bound) in moles/liter. If the molecular weight and the number of binding sites of the protein are known, its concentration may be expressed as grams per liter. Equilibrium dialysis can be used for assay of binding capacity only by a series of experiments at different concentrations; this procedure is used for steroid-protein complexes of high as well as of low affinity. For comparative studies under certain limiting conditions, gel filtration may be the preferable procedure for proteins of high binding affinity, such as CBG.

3. Determination of the number of binding sites in a protein.

4. Determination of the association constant(s) of a steroid–protein complex.

The two last-mentioned problems can be solved with accuracy by equilibrium dialysis if the binding protein is available in pure form, as will be outlined below. The seemingly simpler case of the quantitative determination of the binding activity of a protein in serum, as for example CBG, has inherent difficulties. There is, in the author's opinion, no simple method available which measures the CBG activity or its concentration in a serum or plasma to complete satisfaction. Among the reasons for the difficulties are the very low CBG concentration in the blood, possible interference with other binding proteins, and competitive as well as noncompetitive inhibition of binding by other steroid hormones which may not be known as to quantity and/or structure.

The methods described on the following pages are used to determine the various binding parameters involved in the interaction of steroids with serum proteins. Their usefulness is not limited to these particular two components of a dissociable complex. The techniques are directly applicable to solutions of other nondialyzing macromolecular compounds, and to any low-molecular ligand for which sufficiently sensitive methods for quantitative determination are available. This would include studies on binding of hormones, hormone metabolites, drugs or other biologically important compounds to serum proteins, interaction of various substances with enzyme proteins or with other types of macromolecules such as soluble tissue proteins, nucleic acids or synthetic polymers. The equilibrium dialysis procedures can be readily modified for the investigation of suspended material;[27, 28] the interpretation and quantitative evaluation of such systems, however, requires careful consideration of the specific properties of the components involved.

Determination of CBG Activity by Equilibrium Dialysis

The method which can be considered most lucid and firm in its theoretical basis, and which is applied most widely in studies on protein–ligand interaction, is that of equilibrium dialysis,[29] introduced into the field of steroid-protein binding more than 10 years ago.[30] For the assay of CBG, a trace amount of radiolabeled corticosteroid is added to a dialysis system in which diluted serum is equilibrated against a buffer. At equilibrium, the bound and unbound portions of the total, i.e., endogenous plus added corticoid, are determined; they are used to obtain the convenient expression, C, the combining affinity, which is defined[31] as:

$$C = \frac{S_{bound}}{S_{unbound} \cdot P} \tag{1}$$

where S and P indicate concentrations of steroid and protein, respectively, in grams per liter. The term C is closely related[6] to the apparent association constant, k; it serves to calculate the bound and unbound portion in the undiluted serum. The equilibrium dialysis technique is used in the author's laboratory either as single or as multiple[17] equilibrium dialysis procedure.

The combining affinity, C, is a useful term for convenient assessment of binding affinity in serum or other protein mixtures. In this application, it serves as a relative measure and does not have the precision of the association constant, k, which is defined as

$$k = \frac{[SP]}{[S] \cdot [P]} \tag{2}$$

where [SP] and [S] are the concentrations of bound and unbound steroid, respectively, and [P] denotes the concentration of *unbound* protein in

[27] F. DeVenuto, P. C. Kelleher, and U. Westphal, *Biochim. Biophys. Acta* **63**, 434 (1962).
[28] E. T. Davidson, F. DeVenuto, and U. Westphal, *Endocrinology* **71**, 893 (1962).
[29] I. M. Klotz, F. M. Walker, and R. B. Pivan, *J. Am. Chem. Soc.* **68**, 1486 (1946).
[30] J. A. Schellman, R. Lumry, and L. T. Samuels, *J. Am. Chem. Soc.* **76**, 2808 (1954).
[31] W. H. Daughaday, *J. Clin. Invest.* **37**, 511 (1958).

moles/liter. The C value differs from k mainly by the use of *total* protein concentration, P, expressed in grams per liter. It therefore gives an accurate measurement of binding affinity only if the total protein concentration is large in relation to the concentration of bound protein, so that $P_{unbound} \cong P_{total}$. In the practical determination of the C value as an indicator of the CBG binding activity of a serum, a compromise is made between measuring binding affinity and binding capacity; the concentration of the endogenous plus added radiolabeled corticosteroid is small enough to be well below saturation of the binding sites, and high enough to reveal differences in concentration of binding sites in different sera. This necessary compromise is the main reason why the use of a single C value is not considered completely satisfactory as a measure of binding activity. This is also the reason why the application of the multiple equilibrium dialysis technique is preferred for a critical comparison of different sera, e.g., in animal experimentation. The results obtained with these methods over many years have provided ample evidence for their validity.

Single Equilibrium Dialysis

Reagents and Materials

Deionized doubly distilled water

Sodium phosphate buffer, 0.05 M, pH 7.4

Sulfosalicylic acid, 10% in water

Cortisol-4-^{14}C, specific activity about 45 mC/millimole, or corticosterone-4-^{14}C, specific activity about 56 mC/millimole, stock solution containing 0.2 μg/ml in methanol-benzene $(1 + 9$ v/v). The radiochemical purity of the steroids has to be ascertained, usually by paper or thin-layer chromatography[32]

Streptomycin sulfate, USP; stock solution 5.0 mg/ml in water

Buffered potassium penicillin G; stock solution 200,000 units/ml in water

Dialysis tubing (Visking Company, Chicago, Illinois), size 20, 24 mm flat diameter, washed with distilled water, then with a solution containing 0.002 M sodium EDTA and 0.002 M ascorbic

---

[32] Check of radiochemical purity is particularly important for radiolabeled cortisol which readily decomposes with cleavage of the side chain and other molecular alterations [U. Westphal, G. J. Chader, and G. B. Harding, *Steroids* **10**, 155 (1967)]. The stability of unlabeled cortisol is also limited; loss of the 17-dihydroxyacetone function has been described [K. J. Kripalani and D. L. Sorby, *J. Pharm. Sci.* **56**, 687 (1967)]. These observations caution of possible chemical changes during prolonged dialyses at elevated temperature; in experiments involving cortisol or other steroid structures of limited stability, an examination of the steroid material by paper or thin-layer chromatography may therefore be necessary.

acid in the 0.05 M phosphate buffer of pH 7.4 for 24 hours on a shaking machine, and finally rinsed with four changes of double-distilled, deionized water over a period of 18–24 hours

Procedure

Serum is obtained by allowing the blood[33] to clot overnight at 4°. The protein concentration is determined by a biuret technique[35] using for calibration on analyzed crystalline bovine albumin (Protein Standard Solution, Armour Pharmaceutical Co., Kankakee, Illinois). The cortisol concentration in human serum, and in other sera in which cortisol is the predominant corticosteroid, is determined by a colorimetric procedure;[36] the corticosterone level in serum of rat and other species is assayed by a fluorometric technique.[37, 38] The serum is diluted 1 + 9 with the phosphate buffer, and 5.0 ml[39] of the 10-fold diluted serum is pipetted into a dialysis bag. The bag is placed in a glass-stoppered 25-ml Erlenmeyer, or other glass-stoppered vessel of 20 to 25 ml volume, which contains 10.0 ml of the "outside solution." This dialysis system is set up in duplicate or triplicate.

For the preparation of the outside solution, 0.08 μg of radiolabeled cortisol (for assay of human serum) or corticosterone (for assay of rat serum) is pipetted from the stock solution into a 125-ml Erlenmeyer flask, and the solvent is evaporated at room temperature under a stream of cotton-filtered nitrogen. The film of radioactive corticosteroid is dissolved in 40 ml of phosphate buffer at 40° with gentle shaking for approximately 2–3 hours, and the resulting outside solution is filtered through a medium porosity fritted-glass filter. Part of the filtered solution, containing 0.002 μg of radiolabeled steroid per milliliter, is stored at 4° for subsequent radiocounting. For dialysis experiments at 37° or at room temperature, streptomycin and penicillin are added to the outside solution at the level of 30 μg and 750 units/ml, respectively; after equi-

---

[33] No diurnal variation of CBG activity has been observed in human plasma.[34] However, the time of day of blood withdrawal should be controlled because of the varying corticosteroid level.

[34] P. DeMoor, K. Heirwegh, J. F. Heremans, and M. Declerck-Raskin, *J. Clin. Invest.* **41**, 816 (1962).

[35] A. G. Gornall, C. J. Bardawill, and M. M. David, *J. Biol. Chem.* **177**, 751 (1949).

[36] R. E. Peterson, A. Karrer, and S. L. Guerra, *Anal. Chem.* **29**, 144 (1957).

[37] R. H. Silber, R. D. Busch, and R. Oslapas, *Clin. Chem.* **4**, 278 (1958).

[38] If the volume of serum or plasma is limited it may be advantageous to use a scaled-down modification of the fluorometric method, such as that described by D. Glick, D. von Redlich, and S. Levine [*Endocrinology* **74**, 653 (1964)] for the determination of corticosterone and cortisol in 0.02–0.05 ml of plasma.

[39] Diluted serum, 4.0 ml, may also be used, with proportional changes in the other components of the dialysis system.

libration, two-thirds of these concentrations will be present throughout the dialysis system. The antibiotics in these quantities prevent bacterial growth and do not measurably interfere with the binding activity in the sera studied; however, the influence of the antibiotics on the binding system must be ascertained for each case, especially with pure protein solutions.

The stoppers of the dialysis vessels are secured by tape and Parafilm, and the vessels are completely wrapped in two sheets of Parafilm or in plastic "Baggies." They are then mounted on a slowly rotating disk with horizontal axle, submerged in a 37° water bath. After a period of 48 hours, the dialysis bags are removed and briefly dried with tissue; the inside volumes are measured. There is ordinarily no change in volume. The outside solutions are tested for absence of protein using sulfosalicylic acid. Triplicate 1-ml samples of each inside solution are then plated on lens paper-covered aluminum planchets and air-dried for solid counting of radioactivity in a Nuclear-Chicago gas-flow counter; all necessary corrections are applied to the counting data.[40]

The radioactivity may preferably be determined in a Packard Tri-Carb liquid scintillation spectrometer. Since many samples of aqueous solutions have to be assayed, the efficiency of several liquid scintillation systems, capable of dissolving aqueous samples, have been tested. The phosphor-containing solvent mixture described by DeMoor and Steeno[41] is found well suitable. It consists of

 Naphthalene, 40 g
 2,5-Diphenyloxazole (PPO), 4 g
 1,4-Bis-2-(4-methyl-5-phenyloxazolyl)benzene (dimethyl POPOP),
  0.10 g

These reagents are dissolved in 400 ml of reagent-grade p-dioxane. To 10 ml of this scintillator solution, 1 ml of the aqueous sample is added. A counting efficiency of 7.5% and 55% is obtained for tritium and carbon-14, respectively. The comparable counting efficiencies for these two isotopes in a water-free toluene system[42] are 28% and 74%, respectively. A minimum volume of 5 ml of the phosphor solution in dioxane plus 0.5 ml aqueous sample is found to be necessary for the counting to be independent of volume. A very useful procedure for monitoring aqueous fractions from chromatographic columns for radioactivity (see below)

---

[40] E. M. Pearce, F. DeVenuto, W. M. Fitch, H. E. Firschein, and U. Westphal, *Anal. Chem.* **28**, 1762 (1956).

[41] P. DeMoor and O. Steeno, *J. Endocrinol.* **26**, 301 (1963).

[42] Solution described by F. N. Hayes: 5.0 g of PPO and 0.30 g of dimethyl POPOP per liter of toluene (Packard Instrument Co., Tech. Bull. No. 1, January, 1962).

is the addition of 0.2 ml of the aqueous sample to 5 ml of the above scintillator solution in dioxane. It is found that the counting efficiency for ^{14}C and ^{3}H is not reduced when up to 20 mg of protein is contained per milliliter of aqueous sample that is added to 10 ml of the phosphor solution.[43]

The number of counts taken (at least 6000 cpm for each sample) permits an error of 2.4% or less at a 95% confidence level. The outside solutions at the start and end of the experiment are counted for radio-activity, in at least quadruplicate 1-ml samples. The total radioactivity used in each dialysis bottle, calculated from the cpm value of the outside solution at the start of the experiment, is balanced against the sum of total inside and total outside radioactivity at the end. The recovery of radioactivity should not be below 90%.

The cpm value of the outside solution after 48 hours at 37° is given by the unbound corticosteroid; the value for the bound portion is obtained by subtracting the cpm unbound value from the cpm value of the inside solution which corresponds to the total, i.e., bound plus unbound steroid. Since complete equilibration between endogenous and added corticosteroid is achieved under the conditions of the dialysis at 37°, the cpm_bound and $\text{cpm}_\text{unbound}$ values may be used directly for the calculation of the C value:

$$C = \frac{\text{cpm}_\text{bound}}{\text{cpm}_\text{unbound} \cdot P} \tag{3}$$

It has been ascertained[44] in determinations of cortisol-binding affinity with sera of man, rabbit, and guinea pig, that the C values obtained with the 10-fold diluted sera are essentially the same as those of the undiluted sera (Table I). The C values, therefore, can be applied to a calculation of the bound and unbound portion of the corticosteroid in the undiluted serum. When x equals the percent unbound steroid and $100 - x$ the percentage bound, then

$$C = \frac{100 - x}{x \cdot P} \tag{4}$$

[43] Unpublished experiments with J. Kerkay. Under comparable conditions, two other dioxane-containing phosphor solutions which could also hold 1 ml of aqueous sample per 10 ml, were found to have an efficiency for tritium counting of 4.4% [phosphor solution described by R. V. Quincey and C. H. Gray, *J. Endocrinol.* **26**, 509 (1963)] and 2.9% [phosphor solution of G. A. Bray, *Anal. Biochem.* **1**, 279 (1960)]. A third system [R. C. Meace and R. A. Stiglitz, *Intern. J. Appl. Radiation and Isotopes* **13**, 11 (1962)], containing the quaternary ammonium chloride Hyamine-10x, had a tritium counting efficiency of 18.6%; however, only 0.5 ml of an aqueous solution could be taken up by 19.5 ml of this scintillator mixture.

[44] R. G. Allen, G. B. Harding, and N. Rust, unpublished experiments.

TABLE I
INFLUENCE OF SERUM DILUTION ON CORTISOL-BINDING AFFINITY[a]

Species	Serum diluted 1:10	Serum undiluted[b]
Human	0.29	0.28
Rabbit	0.24	0.22
Guinea pig	0.06	0.06

[a] Average C values of duplicate comparative determinations with several sera for each species; in 0.05 M phosphate, pH 7.4, 37°. Unpublished experiments with G. B. Harding and R. G. Allen.
[b] Concentration of added radiocortisol was 10-times that used with the diluted serum.

and x is obtained by using the protein concentration (mg/ml) of the undiluted serum. The concentrations of bound and unbound corticosteroid are obtained in micrograms per 100 ml by multiplying $(100 - x)$ and x, respectively, by the total corticoid concentration in micrograms per 100 ml, divided by 100.

Multiple Equilibrium Dialysis

In this technique, the above procedure is modified by including in each of duplicate glass-stoppered vessels a number of bags, up to 20 or 30, containing 10-fold diluted samples of different sera. In general, 4.0 ml is used in each bag. The volume of the outside solution is twice the sum of all inside solutions. Since one outside solution serves for the analysis of many sera, not less than six 1-ml samples of the final, equilibrated outside solution are counted for high accuracy of the $cpm_{unbound}$ value. A normal serum of the species investigated, taken from the same large pool, is included in each multiple equilibrium dialysis as a standard in order to check the reliability of the experiment.

Comments

The equilibrium dialysis procedure, in the author's opinion, gives the most satisfactory answer to the question of what percentage of a given steroid hormone in a serum is bound and what portion is unbound.[45] One condition for the validity of the C values as a predominant expression of CBG activity is that the CBG capacity, i.e., the concentration of high-affinity binding sites for corticosteroids, is not exceeded by the addition of the radioactive test corticosteroid. When increasing amounts of radio-corticosteroid are added to a given serum in an equilibrium dialysis

[45] The term "unbound" is preferred to "free" to avoid confusion with the "free" form of a steroid as opposed to its conjugated form.

experiment, the C values remain almost constant until the CBG binding sites for corticoids are occupied. With further addition of radiocorticosteroid, the C values decline and reach the level characteristic of binding to serum albumin.[31] In contrast to the high-affinity, low-capacity CBG, albumin has a low affinity for corticosteroids, but the high albumin concentration provides a practically unlimited capacity. It follows from these considerations that, in the determination of the C value as a measure of CBG activity, the sum of endogenous and added corticosteroid must be well within the limit of the concentration of high-affinity (CBG) binding sites. Even then, the binding value obtained includes a contribution from binding protein other than CBG, i.e., mainly albumin; it is usually of the order of 8–10% of the total binding value.

In the case of high endogenous corticoid levels, for example, in serum from a stressed organism, a very high portion of the total binding sites may be occupied by endogenous steroid so that the addition of the radioactive corticoid would lead to relatively low C values. This would make a comparison of C values as a measure of CBG activity uncertain whenever the endogenous corticosteroid levels in the different sera vary markedly. The application of the multiple equilibrium dialysis procedure at 37° overcomes the difficulty. In this technique at the elevated temperature, the endogenous steroid hormones of all the sera present are equilibrated[46] throughout the dialysis system, so that all sera are tested under identical conditions of endogenous steroid level. The C values are therefore well comparable with each other. Experience[47] has shown that the same standard serum included in a number of multiple equilibrium dialysis experiments yields approximately the same C value in the different experiments, indicating that the use of a larger number of sera in combination tends to average the individual endogenous steroid concentrations to a relatively uniform level.

Another way to eliminate the interference of endogenous steroids with the assay of CBG activity is the complete removal of the endogenous steroids from the serum. A technique for this will be described below; it is considered the method of choice for determination of accurate binding parameters in special cases as, for example, of the association constants and concentration of CBG binding sites in mammalian sera listed in Table VIII. However, the procedure is not practical in the clinical assay of CBG activity or in animal experimentation involving many samples.

The interference by endogenous steroids is not limited to corticosteroid hormones. In confirmation of earlier observations,[48] progesterone

[46] J. C. Warren and H. A. Salhanick, *Proc. Soc. Exptl. Biol. Med.* **105**, 624 (1960).
[47] R. R. Gala and U. Westphal, *Acta Endocrinol.* **55**, 47 (1967). This paper refers to five preceding publications in which the above procedures were used.
[48] W. H. Daughaday and I. Kozak, *J. Clin. Invest.* **37**, 519 (1958).

has been found to form a strong complex with CBG;[8, 13] the association constant is three times that of the CBG-cortisol complex in human serum (see Table VIII). The association apparently occurs at the same binding sites with which cortisol and corticosterone interact. The presence of progesterone, especially in pregnancy sera, has to be considered, therefore, in the interpretation of experimental data on CBG activity.

Other Methods for Determination of CBG Activity

Equilibrium dialysis against albumin solution[2] at 37° has been used[49] in conjunction with a graphic procedure to determine CBG concentration and affinity constants in human serum. A technique has been developed to distinguish between binding to CBG and to albumin by selective thermal inactivation of CBG at 60°.[50] Rate and degree of this inactivation, however, are dependent on the concentration of cortisol present.[51] In another procedure,[52] the amount of cortisol required to saturate the CBG binding sites is determined by a series of six equilibrium dialyses at increasing cortisol concentrations at 9°, and the data are used to obtain the CBG capacity in micrograms per 100 ml. Most recently, an elaborate procedure has been developed in Yates' laboratory[53] to determine the concentration of binding sites and their association constants for CBG and albumin in undiluted plasma or serum of the rat by ultrafiltration at 37°. Eight different concentrations of corticosteroid hormone, including tritiated tracer, are equilibrated with the plasma pool; the binding data obtained are subjected to computer analysis using multiple mass action theory. A graphic method based on similar equations of multiple binding equilibria is used[54] by Sandberg and Slaunwhite's group[8] to obtain the physicochemical parameters of a binding system which consists of one ligand and one or more binding macromolecules.

Determination of CBG Capacity by Gel Filtration

A method for the assay of CBG capacity[34, 51] is widely used and, due to its technical simplicity, is gaining in popularity, especially in clinical laboratories. It is based on the principle of molecular sieving[55] in which

[49] M. Booth, P. F. Dixon, C. H. Gray, J. M. Greenaway, and M. J. Holness, *J. Endocrinol.* **23**, 25 (1961).
[50] W. H. Daughaday, R. E. Adler, I. K. Mariz, and D. C. Rasinski, *J. Clin. Endocrinol.* **22**, 704 (1962).
[51] R. P. Doe, R. Fernandez, and U. S. Seal, *J. Clin. Endocrinol.* **24**, 1029 (1964).
[52] B. P. Murphy and C. J. Pattee, *J. Clin. Endocrinol.* **23**, 459 (1963).
[53] N. Keller, L. R. Sendelbeck, U. I. Richardson, C. Moore, and F. E. Yates, *Endocrinology* **79**, 884 (1966).
[54] H. E. Rosenthal, *Anal. Biochem.* **20**, 525 (1967).
[55] J. Porath and P. Flodin, *Nature* **183**, 1657 (1959).

the unbound corticosteroid is separated from the protein–corticosteroid complex by gel filtration. The unbound ligand penetrates the network ("pores") of Sephadex, a cross-linked dextran, and is thus retained on a filtration column, whereas the protein including the bound ligand is eluted without retention (molecular exclusion) and appears in the eluate immediately following the void volume. By adding corticosteroid to the test serum in excess of its binding capacity, the serum to be eluted is saturated with the steroid, and the steroid analysis yields the binding capacity directly in micrograms per 100 ml serum. The association constant of the corticosteroid-albumin complex is low enough, and the equilibration rate high enough, to allow dissociation so that the eluted albumin is free of corticosteroid. The gel filtration technique described below follows published procedures[34, 51] with only minor modifications.

Materials

Serum[56] analyzed for endogenous cortisol[36] or corticosterone[37, 38]
Sephadex G-50, coarse (Pharmacia Fine Chemicals, Inc., Piscataway, New Market, New Jersey)
Unlabeled cortisol and corticosterone, recrystallized
Other materials as described for the equilibrium dialysis procedure

Procedure

The Sephadex powder is thoroughly washed and decanted several times using approximately 20 times its weight of $0.05 M$ phosphate, pH 7.4, to remove possible fines. The column (14 mm inner diameter, 380 mm length, sealed-in-glass frit) is packed at room temperature to a height of 330 mm, and a filter paper disk is placed on top of the Sephadex. The column is then equilibrated in the cold room at 4° and the buffer flow rate is adjusted to 1 ml/minute. In a 10-ml Erlenmeyer flask, a mixture of 1.0 μg cortisol-4-^{14}C (corticosterone-4-^{14}C) in methanol–benzene 1:9 (v/v) plus 1.0 μg unlabeled cortisol (corticosterone) in methanol is evaporated to dryness under a stream of cotton-filtered nitrogen. To the remaining film, 2.0 ml of the serum to be tested is added; the flask is then incubated at 37° for 90 minutes[57] on a water bath shaker to dissolve and equilibrate the corticosteroid with the endogenous steroid.[58]

---

[56] Plasma may be used with equivalent results.[34, 51]
[57] Equilibrium at 37° is completed within minutes.[51]
[58] In confirmation of published procedure,[51] it has also been found practical to admix the 1.0 μg of radioactive plus 1.0 μg of unlabeled corticosteroid in 0.4 ml of aqueous buffer to the 2.0 ml of serum and to equilibrate for 60 minutes at 37°. The volume applied to the column then was 0.6 ml corresponding to 0.5 ml of serum.

After cooling to 4°, 0.5 ml of the serum[59] containing a total of 100 μg/100 ml plus the original endogenous concentration of corticosteroid hormone, in a homogeneous mixture with the radiosteroid, is allowed to drain into the column. By adding the 0.5 ml serum in small portions, the height of the serum layer on top of the Sephadex is kept to a minimum. By use of a disposable pipette, the serum is followed by 5 buffer rinses without exceeding the height of the original serum layer. The elution buffer (phosphate) is then placed on top of the column in a separatory funnel and the flow rate is maintained at 1 ml/minute. The transition from void volume to the appearance of protein in the eluate is checked by dripping single drops of eluate into 10% sulfosalicylic acid on a glazed black spot plate; the protein fraction is collected in a 5-ml graduated cylinder. The end of protein elution is again determined by sulfosalicylic acid. Complete separation of protein-bound and unbound corticosteroid should be ascertained by collecting additional 0.5-ml fractions and demonstrating an interval practically free of radiocarbon, before the unbound corticoid appears in the eluate. The protein is contained in a volume < 5.0 ml; it is brought to 5.0 ml with buffer. Quadruplicate 1-ml samples corresponding to 0.1 ml original serum, are plated on lens paper-covered aluminum planchets for solid counting of ^{14}C in a gas-flow counter or are counted in a liquid scintillation system as described above. The gel filtration is performed in duplicate. After each run, the column is washed with at least 300 ml of buffer in order to remove the unbound corticosteroid.

The remainder of the 2 ml of test serum is used to plate quadruplicate 0.2-ml samples in order to determine the cpm per milliliter of serum, i.e., per total (added plus endogenous) corticosteroid contained in one milliliter. The result, in micrograms of bound corticosteroid per 100 ml is obtained by the formula

$$\frac{cpm_{bound}}{cpm_{total}} \cdot S \tag{5}$$

where cpm_{bound} is the radioactivity count of the protein filtrate, calculated per milliliter of undiluted serum, cpm_{total} is the ^{14}C count for 1 ml of test serum before gel filtration, and S is the total corticosteroid concentration (added plus endogenous) in micrograms per 100 ml. Assuming a molecular weight of 52,000 and one cortisol-binding site for human CBG (see below, Table XII), the value micrograms of bound cortisol per 100 ml can readily be converted to weight concentration of CBG in human serum:

[59] The column described can also be used for 1.0-ml serum samples. The bed volume to sample volume would then be about 50:1, vs 100:1 with 0.5 ml of serum. Ratios of 30:1 (footnote 34) to 88:1 (footnote 51) are being applied in other laboratories.

micrograms of bound cortisol per 100 ml multiplied by 1.44 will yield milligrams of CBG per liter. This is a minimum value as will be discussed below.

Comments

The determination of CBG capacity by gel filtration is a convenient method for the routine analysis of clinical serum or plasma samples. It gives reproducible results that are comparable within one species if the same experimental conditions are maintained from one experiment to the other. The column can be regenerated by washing and can be reused indefinite numbers of times.

The gel filtration procedure has certain limitations, however, which have to be recognized in the interpretation of the results. Most of these restrictions originate from the fact that the method is not based on attainment of an equilibrium between unbound and protein-bound steroid but rather on a separation of the entities in a continuously changing system. The component parts of the CBG–corticosteroid complex are held together by noncovalent bonds; protein and ligand, therefore, are in a binding equilibrium subject to the law of mass action.[9] As a consequence, the complex undergoes continuous dissociation as it moves down the Sephadex column. Due to a relatively high association constant and a slow rate of equilibration, dissociation is small at 4°; at higher temperatures, the affinity decreases and the equilibration rate becomes greater, so that at 46° complete dissociation has taken place when the CBG emerges from the column.[51] This is a temperature at which the equilibrium dialysis technique still shows definite binding, although at a fraction of the affinity observed at 4°. Another consequence of the dissociation is the inability of the gel filtration column to demonstrate albumin binding of corticosteroid hormones, a deficiency turned to advantage in the determination of CBG capacity.

The following observations[44] provide experimental evidence for the limitations of the gel filtration method. It has been reported[34] that stepwise changes of the elution rate between 0.2 and 2.5 ml/min resulted in a slight increase of binding capacity, i.e., from 22.1 to 26.1 μg of cortisol bound per 100 ml of normal human plasma at room temperature. Greater differences in CBG binding capacity with changing elution rates were observed[60] at 37°. The shape of the column appears to be without measurable influence on the results as demonstrated by experiments in which the length of the column was varied in the approximate ratio of 1:2:4 at equal volume (Table II).

[60] R. V. Quincey and C. H. Gray, *J. Endocrinol.* **26,** 509 (1963).

TABLE II

FILTRATION OF CORTICOSTEROID-LOADED SERA OVER SEPHADEX COLUMNS OF DIFFERENT LENGTHS BUT EQUAL VOLUMES[a]

Species	Endogenous corticosteroid (µg/100 ml)	Radiocorticoid added	Columns used[b]		
			S	M	L
Human[c]	11.0 F	F	15.9 (15.0, 16.7)	15.6 (15.6, 15.5)	15.5 (15.6, 15.4)
Guinea pig	81.8 F	F	18.5 (19.6, 17.4)	24.5 (27.3, 21.7)	20.3 (18.2, 22.3)
Rat	59.0 B	B	66.4 (63.9, 68.9)	69.5 (70.2, 68.7)	70.7 (72.9, 68.5)
		F	—	48.7[d]	—
Rabbit	11.2 B	B	28.8 (28.3, 29.3)	31.5 (30.8, 32.1)	26.3 (24.7, 27.9)
		F		28.3[d]	—

[a] Unpublished experiments with R. G. Allen, using 0.5 ml of serum containing 100 µg/100 ml added radioactive cortisol (F) or corticosterone (B). Figures indicate µg/100 ml bound corticosteroid. Average values from two experiments (in parentheses). Temperature, 4°.

[b] Inner diameter, height, volume for S = 20 mm, 170 mm, 53 ml; M = 14 mm, 330 mm, 50 ml; L = 10 mm, 670 mm, 53 ml.

[c] Diluted serum, 3.85% protein.

[d] Average value from 4 experiments.

The above considerations of the CBG-corticosteroid complex undergoing continuous dissociation during transport over the Sephadex were verified by experiments recorded in Table III. The height (volume) of the column was changed in the ratio 1:2:3 at constant diameter. It is evident that the value for micrograms of cortisol bound per 100 ml serum obtained with the single volume is about 60% higher than that given by the triple volume. Although such drastic differences in experimental design are unlikely to occur in practice, the results show the limitation of the method; the binding capacities obtained even under technically faultless conditions must be considered minimum values and interpreted with caution. An approximation to the correct values can be obtained by gel filtration at three different column heights (volumes) such as those shown in Table III, and extrapolation to height (volume) zero. Proportionality between height (volume) and amount of Sephadex should be verified in this extrapolation procedure.

TABLE III

FILTRATION OF CORTISOL-LOADED HUMAN SERUM OVER SEPHADEX
COLUMNS OF DIFFERENT LENGTHS AND VOLUMES[a]

Column		Cortisol bound (μg/100 ml)
Height (mm)	Volume (ml)	
103	72	14.7 (14.4, 15.0)
205	145	11.5 (11.1, 11.8)
310	219	9.2 (8.5, 9.9)

[a] Unpublished experiments with R. G. Allen, using 0.5 ml of diluted (3.85% protein) serum containing 50 μg/100 ml added cortisol-4-[14]C. Temperature, 4°.

Another experimental variable which influences the result is the quantity of corticosteroid added to "overload" the serum. Table IV shows that even at a ratio of bed volume:sample volume of about 100:1, the differences in CBG capacity values obtained are considerable. This is evident for the human serum tested; it is also marked with the guinea pig serum. In the latter case, the relatively low association constant (see below, Table VIII) may be responsible for the high dissociation during gel filtration. It should be noted that the CBG concentration and the serum cortisol level in the pregnant guinea pig are extremely high.[13, 47, 61]

The equilibrium dialysis method measures the portion of bound and unbound corticosteroid at binding equilibrium, and is responsive to the

[61] N. Rust, unpublished studies.

TABLE IV
SEPHADEX FILTRATION[a] OF SERUM CONTAINING
DIFFERENT AMOUNTS OF ADDED CORTISOL

	Cortisol bound in	
Radiocortisol added[b] (μg/100 ml)	Human serum[c] (μg/100 ml)	Serum of pregnant guinea pig[d] (μg/100 ml)
50	14.7 (14.4, 14.9)	381 (378, 384)
100	15.8 (avg. of 6 exp.)	414 (422, 405)
200	18.7 (17.7, 19.6)	402 (409, 394)
400	28.2 (27.1, 29.3)	470 (483, 456)
600	—	527 (530, 524)
800	—	539 (540, 538)

[a] Unpublished experiments with R. G. Allen and N. Rust. Column 14 mm i.d., 330 mm high; temperature, 4°.

[b] Total of labeled plus unlabeled cortisol.

[c] 0.5 ml dilute serum (3.85 g of protein, 11.0 μg of cortisol per 100 ml), containing different amounts of added mixtures of cortisol-4-^{14}C and unlabeled cortisol, applied to column.

[d] 0.6 ml, composed of 0.5 ml of serum (4.03 g of protein, 432 μg of "cortisol" per 100 ml) and 0.1 ml aqueous solutions of mixtures of cortisol-4-^{14}C and unlabeled cortisol, applied to column.

quantity of binding protein (concentration of binding sites) and to the binding affinity (association constant) of the protein for the steroid. This includes the small contribution of proteins interacting at relatively low affinity, such as albumin. It gives the result as percentage of corticosteroid actually bound and unbound, which can be readily expressed as micrograms per 100 ml.

In contrast, the gel filtration technique provides a submaximal value for CBG binding capacity, expressed as micrograms of corticosteroid bound per 100 ml of serum or plasma. The value, not originating from a system at equilibrium, is critically dependent on the given experimental conditions. The method is not responsive to finely graded differences in binding affinity, but resembles more an "all or nothing" response to the steroid-protein complex; for example, most of the cortisol-CBG complex can be eluted whereas the cortisol-albumin complex dissociates completely during the filtration. The gel filtration procedure in its simple form is unable to yield thermodynamically valid data which can be used for the determination of number of binding sites and association constants; it is not possible to measure the equilibrium concentration of the unbound corticosteroid. These limitations do not necessarily detract from the usefulness of the Sephadex technique for a comparison of CBG capacities in a series of clinical sera, as long as the experimental conditions are kept strictly constant, and as long as the interpretation of the results is in accord with the theoretical limits.[62]

In Table V, a comparison is made of the CBG activity values of sera of four species determined by equilibrium dialysis and by gel filtration. A fair proportionality is seen between C values at 37° and binding capacity values obtained at 4° for human, rat, and rabbit serum. For the guinea pig serum, however, the C value is much smaller in relationship to the binding capacity value. This is presumably caused by the fact that

TABLE V

CBG ACTIVITY OF SERA DETERMINED BY EQUILIBRIUM DIALYSIS AND BY SEPHADEX FILTRATION[a]

| | | | Equilibrium dialysis | | Gel filtration binding capacity (μg/100 ml at 4°) |
Species	Endogenous corticosteroid[a,b] (μg/100 ml)	Radioactive corticosteroid added for assay[b]	C value[c] at 37°	Corticosteroid actually bound (μg/100 ml)[d]	
Human[e]	18.6 F	F	0.16	17.0	26.7
Rat	59.0 B	B	0.46	57.2	69.5
Rabbit	11.2 B	B	0.14	10.1	31.5
Guinea pig	81.8 F	F	0.05	51.0	25.7

[a] Average values from duplicate experiments.
[b] F = cortisol; B = corticosterone.
[c] Multiple equilibrium dialysis for the two sera having the same endogenous corticosteroid.
[d] At 37°.
[e] Diluted serum; values corrected for normal concentration of 6.5% protein.

the association constant of guinea pig CBG is only about one-third to one-tenth that of the CBG's of the other species (see Table VIII); it has been pointed out that the C value provides an integrative measure for concentration of binding sites as well as association constant.

Removal of Endogenous Steroid from Serum by Gel Filtration at 45°

Sephadex G-25, medium, or Sephadex G-50, coarse, is washed as described above and packed into a jacketed column of 22 mm inner diameter and 400 mm length to a height of about 370 mm; a filter paper disk is placed on top. Phosphate buffer, 0.05 M, pH 7.4, is applied as suspending

[62] Sephadex gel has been used [W. H. Pearlman and O. Crépy, *J. Biol. Chem.* **242**, 182 (1967)] in an equilibrium binding system to determine the binding affinity of testosterone to a specific serum protein. This method avoids some of the disadvantages of the gel filtration; however, it has other limitations mainly due to adsorption of steroids on the large surface of the dextran gel.

and eluting medium; it is briefly boiled before use to eliminate dissolved gas and is maintained at 45° during the experiment. The temperature of the column is equilibrated by a Brinkman-Haake Ultra Thermostat, Type F, set at 45° ± 0.01°. The void volume of the column is approximately 50 ml. The flow rate is adjusted to 1.0 ± 0.1 ml/minute. Beginning and end of protein appearing in the eluate are tested by 10% sulfosalicylic acid as described above.

A total of 5 ml serum is applied to the column in small portions with the aid of a propipette, and "washed in" with buffer as outlined before. After the void volume has emerged, the protein is collected in approximately 20 ml. The collection, possibly combined from several experiments, is either frozen and kept at −85°, or lyophilized and stored in the dry state at −85°. The column is regenerated by washing with at least 300 ml of buffer at 45°. The removal of corticosteroids and progesterone from human serum by this procedure has been found to be 99–99.5% complete.[63] It has been ascertained that the binding affinity of CBG in serum is not affected by the gel filtration at 45°.[63]

Determination by Equilibrium Dialysis of Number of Progesterone-Binding Sites and Association Constant in Serum Albumin

Materials

Deionized distilled water
Sodium phosphate buffer, 0.05 M, pH 7.4
Crystalline human serum albumin
Progesterone-4-^{14}C, 46 mC/millimole; stock solution containing 0.2 μg/ml in methanol–benzene 1:9 (v/v)
Progesterone, unlabeled, recrystallized, m.p. 130°; stock solutions of 1, 10, and 100 μg/ml in redistilled methanol
Dialysis tubing, streptomycin sulfate, and penicillin G as described above

Procedure

The albumin is dissolved in buffer to a concentration of 1 mg/ml calculated on the basis of water-free (105°) protein. Eight outside solutions are prepared containing 2.0, 2.5, 3.0, 4.0, 6.0, 8.0, 10.0 and 12.0 μg of progesterone per milliliter of buffer, respectively; 0.02 μg of each of these quantities consist of progesterone-4-^{14}C. For the dissolution of the progesterone, the appropriate volumes of the stock solutions of labeled and unlabeled steroid are mixed, evaporated to dryness under a stream of cotton-filtered nitrogen; the residue is dissolved in 0.15 ml of absolute

---

[63] U. Westphal, *Arch. Biochem. Biophys.* **118**, 556 (1967).

ethanol,[64] and 25 ml of "sterile" (short boiling) buffer at 55° added rapidly. The solution is filtered through a medium-porosity fritted-glass filter. In addition, the outside solutions contain 30 μg of streptomycin sulfate and 750 units of penicillin G per milliliter.[65] Eight duplicate sets of glass-stoppered vessels are prepared, each containing 4 ml of the albumin solution in a dialysis bag plus 8 ml of one of the eight outside solutions. The equilibrium dialysis is performed for 72 hours at 4° or for 48 hours at 37°. The duplicate values obtained for the concentrations of bound and unbound progesterone are averaged. Other details of the equilibrium dialysis procedure have been described above.

General Calculations[9]

Under the simplifying assumption of one class of equivalent and independent binding sites, the number of binding sites for the steroid on each protein molecule, n, and the association constant for each binding site, k, may be obtained from the simple equation[29]

$$\frac{1}{\bar{\nu}} = \frac{1}{nk[S]} + \frac{1}{n} \tag{6}$$

where $\bar{\nu}$ equals the moles of steroid bound per mole of total protein, and [S] the molar concentration of unbound steroid. The quotient, $1/\bar{\nu}$, is plotted as ordinate vs $1/[S]$ as abscissa, and the experimental points are fitted to the optimal straight line by the method of least squares. The intersection of this line with the ordinate, $(1/[S]) = 0$, gives $1/n$; the slope of the line is given by $1/nk$. This method of reciprocal plots determines both n and k directly if the binding sites are indeed equivalent and independent.

In practical cases, however, the situation often is more complex, and the assumption of only one class of equivalent binding sites may not be justified. The graphical solution of equation (6) by reciprocal plots does not necessarily reveal the discrepancy, since the accuracy in determining the intersection of the curve with the ordinate at high values of [S] (where $1/[S]$ values on the abscissa are close to 0) is limited. It is then preferable to use another presentation of the data, in the form of the equation of Scatchard[66]

$$\frac{\bar{\nu}}{[S]} = k \cdot (n - \bar{\nu}). \tag{7}$$

---

[64] The small concentration of ethanol does not measurably influence the binding in this case and in serum samples; however, this point must be checked for each binding system studied.

[65] The antibiotics are ordinarily omitted in experiments at 4°.

[66] G. Scatchard, *Ann. N.Y. Acad. Sci.* **51**, 660 (1949).

If the binding groups are equivalent and independent, a straight line is again obtained by plotting $\bar{v}/[S]$ as ordinate against $\bar{v}$ as abscissa. The intercept on the abscissa gives n (for $\bar{v}/[S] = 0, \bar{v}$ equals n). The intercept on the ordinate, $\bar{v} = 0$, gives kn. The results obtained by the two procedures are identical in the example described.

If there is more than one set of equivalent binding sites, the Scatchard plot reveals a break in the line, straight portions of which may be capable of extrapolation to the abscissa and to the ordinate, indicating different values for n and k. In this case the binding equilibrium would be represented by more complex equations. Equations (6) and (7) are derived from the general equation

$$\bar{v} = \frac{nk[S]}{1 + k[S]} \tag{8}$$

Assuming a protein containing m different sets of binding sites, each having an intrinsic association constant, Eq. (8) would be extended to

$$\bar{v} = \sum_{i=1}^{m} \frac{n_i k_i[S]}{1 + k_i[S]} \tag{9}$$

Of the numerous possible sets of binding sites and k values, a simple example for the use of equation (9) is given by the assumption of two sets of binding sites with two corresponding association constants:

$$\bar{v} = \frac{n_1 k_1[S]}{1 + k_1[S]} + \frac{n_2 k_2[S]}{1 + k_2[S]} \tag{10}$$

To test the validity of the binding parameters obtained graphically from the Scatchard plot, these values, i.e., n_1, k_1, n_2, k_2, are used in Eq. (10) with the different values of [S], and the resulting values for $\bar{v}$ are used for comparison with the experimental points. The values for n_1, n_2, and k_1, k_2 may have to be readjusted until the theoretical and experimental curve, Eq. (7), coincide. A computer is very useful for these calculations.

It should be noted that all the above equations for binding equilibria are based on the assumption of independent sites, i.e., no account is made of interaction between the sites. For further discussion and evaluation of binding data see footnote 9.

The calculation of the association constant for a steroid complex with a pure protein by Eqs. (7)–(10) requires determination of binding data at a number of steroid concentrations. An abbreviated procedure can be applied under certain conditions to find the affinity constant, utilizing equilibrium dialysis at a single concentration.[19] The association constant is defined as

$$k = \frac{[S_{bd}]}{[S][P]} \tag{11}$$

where $[S_{bd}]$ is the molar concentration of bound steroid. Equation (11) resembles Eq. (1), except for the use of molar concentration of unbound protein, $[P]$, instead of total protein concentration in grams per liter. If the total protein concentration applied, $[P_{total}]$, is very large compared to $[S_{bd}]$, the concentration of unbound protein becomes practically equal to that of total protein, $[P] \cong [P_{total}]$. Equation (11) then becomes

$$k = \frac{[S_{bd}]}{[S][P]} \cong \frac{[S_{bd}]}{[S][P_{total}]} = \frac{\bar{v}}{[S]} \tag{12}$$

Values for $\bar{v}/[S]$ are valid expressions for the binding affinity of a steroid–protein complex, if they are determined with homogeneous proteins at molar steroid concentrations 1% or less of the molar protein concentration.[19] They are in agreement with k or nk for proteins with one or n steroid-binding sites, respectively. It should be realized that the overall experimental error in the measurement of the binding parameters, including radioactivity counting, is generally greater than that caused by equating Eqs. (11) and (12).

Distribution of a Steroid between Two Binding Proteins

If the concentrations of binding sites for the steroid and the corresponding association constants are known for the two proteins, the binding distribution can be calculated. On the basis of the law of mass action

$$\frac{[SP]}{[S][P]} = k; [SP] = k[S][P] \tag{13, 14}$$

where $[SP]$ denotes the concentration of the steroid–protein complex, and $[S]$ and $[P]$ the concentrations of unbound steroid and protein, respectively; k is an association constant characteristic of the particular steroid and protein involved and is valid for a given temperature, pH, solvent, and other conditions. All concentrations are given in molarities. The equations hold for any binding protein, e.g., for CBG = transcortin $= P_T$, or for serum albumin $= P_A$:

$$[SP_T] = k_T[S][P_T]; [SP_A] = k_A[S][P_A] \tag{15, 16}$$

Dividing one equation by the other results in

$$\frac{[SP_T]}{[SP_A]} = \frac{k_T[P_T]}{k_A[P_A]} \tag{17}$$

substituting convenient terms and rearranging[67] the equation leads to the quadratic equation

[67] F. E. Yates and J. Urquhart, *Physiol. Rev.* **42**, 359 (1962).

$$\left(1 - \frac{k_T}{k_A}\right)[SP_A]^2 + \left([S] - [\Sigma S] - [\Sigma P_A] - \frac{k_T}{k_A}[\Sigma P_T] + \frac{k_T}{k_A}[\Sigma S]\right.$$

$$\left. - \frac{k_T}{k_A}[S]\right)[SP_A] + [\Sigma P_A]([\Sigma S] - [S]) = 0 \quad (18)$$

where $[\Sigma S]$ = total concentration of steroid; k_T and k_A = association constants (for each binding site) of steroid complex with CBG (transcortin) and albumin, respectively; $[\Sigma P_T]$ and $[\Sigma P_A]$ = total concentration of binding sites on CBG and albumin, respectively. The concentration of unbound steroid, $[S]$, is obtained experimentally by equilibrium dialysis.

Equation (18) is solved for $[SP_A]$, the concentration of albumin-bound steroid, by use of the quadratic formula. The concentration of the CBG-bound steroid can be obtained by deducting $[SP_A]$ from the total protein-bound steroid:

$$[SP_T] = [\Sigma S] - [S] - [SP_A] \quad (19)$$

The calculation can be made for the binding of cortisol, corticosterone, or progesterone to the same binding serum proteins; examples have been given.[68] Since the above equations refer to an unreal system involving only one steroid in the presence of two binding proteins, additional formulas have been developed to assess the binding of two or more steroids, present simultaneously, to serum proteins.[68] These formulas are valid under the condition that the steroids compete for the same binding sites on the protein, as is the case, for example, for the corticosteroid hormones and progesterone with respect to CBG.

Properties of the CBG in Serum

Influence of Temperature

The chemical properties of serum albumin[69] and α_1-acid glycoprotein[15, 16] will not be discussed here; their role in the binding of steroids has been reviewed.[2, 4-6, 8, 18, 19, 68] Many characteristics of the CBG have been recognized before the isolation of a purified protein. A strong temperature dependency of the binding affinity of the cortisol-CBG complex has been observed in the serum; the dissociation increases considerably with increasing temperature.[8, 12, 13, 34, 51, 70, 71] The CBG molecule is more labile to heat than serum albumin; at 60° and above, CBG is irreversibly denatured with loss of binding affinity.[70] This effect has been utilized[50]

[68] U. Westphal, Z. Physiol. Chem. 346, 243 (1966).
[69] J. F. Foster, in "The Plasma Proteins" (F. W. Putnam, ed.), Vol. 1, p. 179. Academic Press, New York, 1960.
[70] W. H. Daughaday and I. K. Mariz, Metabolism 10, 936 (1961).
[71] U. S. Seal and R. P. Doe, J. Biol. Chem. 237, 3136 (1962).

for a fairly selective thermal inactivation of CBG in serum at 60°, a temperature at which cortisol-binding ability of albumin is not altered significantly. However, for application of this effect in quantitative determination of CBG in serum, it must be considered that the rate and degree of CBG inactivation are dependent on the concentration of corticosteroid present.[51]

Table VIA shows the corticosterone-binding affinity in human, rat, and rabbit serum measured at different temperatures; the decrease with rising temperature is very marked.[63, 68] This temperature effect resembles

TABLE VI
EFFECT OF TEMPERATURE ON CORTICOSTERONE-BINDING
AFFINITY IN HUMAN, RAT, AND RABBIT SERUM[a]

A. *Affinity values at different temperatures*

Species	4°	25°	37°	45°	50°
Human	1.17	0.69	0.24	0.18	0.06
Rat	4.06	1.80	0.43	0.18	0.06
Rabbit	0.62	0.17	0.06	0.06	0.04

B. *Affinity values at 4°, at 50°, and at 4° after exposure to 50°*

Species	4°	50°	4°[b]	Percent of original activity lost by heat exposure
Human	2.62	0.03	0.19	93
Rat	7.92	0.04	0.15	98
Rabbit	0.60	0.04	0.12	80

[a] Average C values from duplicate multiple equilibrium dialysis experiments in 0.05 M phosphate, pH 7.4. The sera from different pools were cleared of endogenous steroids by gel filtration at 45°. See also U. Westphal, *Z. Physiol. Chem.* **346**, 243 (1966); *Arch. Biochem. Biophys.* **118**, 556 (1967).

[b] Equilibrium dialysis at 50° for 48 hours, then for an additional period of 48 hours at 4°. Unpublished experiments with G. B. Harding.

that on the association constant of the progesterone-orosomucoid ($=\alpha_1$-acid glycoprotein) complex.[18, 19] The heat inactivation experiments summarized in Table VIB indicate that the low corticosteroid-binding activity observed at 50° is to a major extent irreversible, confirming the heat lability of the CBG molecules for several species. This is in contrast to the thermostability of the α_1-acid glycoprotein which can be heated as a pure protein for 24 hours at 60° at pH 7.4 in phosphate buffer without significant loss of progesterone-binding affinity.[19] Under similar conditions, human serum albumin does not show any loss of progesterone-binding affinity after exposure to 45° for at least 24 hours.[19]

Influence of pH

The binding affinity between the corticosteroid hormones and CBG is dependent on the hydrogen ion concentration; a maximal value has been observed[72, 73] at approximately pH 8. At pH 5 and below, irreversible denaturation appears to take place with loss of the binding affinity.[70, 71] Table VII shows the influence of pH on the CBG activity in

TABLE VII

Corticosterone-Binding Affinity of Human and
Rat Serum at Different pH Values[a]

pH	Buffer[b]	C Values for serum of	
		Human	Rat
2.8	CP	0.11	0.09
3.5	CP	0.13	0.13
4.7	CP	0.36	0.08[c]
5.6	CP	1.05	2.68
6.2	CP	2.55	5.68
6.9	CP	4.14	7.51
7.3	CP	6.05	8.72
7.9	CP	9.66	12.8
7.9	A	9.08	9.92
8.8	A	8.21	8.38
9.8	A	5.10	7.15
$2.8 \rightarrow 7.35$[d]	CP	0.46	1.63
$9.8 \rightarrow 7.35$[d]	A	6.52	9.35

[a] Unpublished experiments with G. B. Harding. Average C values from duplicate multiple equilibrium dialysis experiments for 48 hours at 4°. Sera cleared of endogenous steroids by gel filtration at 45°.

[b] CP = 0.05 M citrate-phosphate; A = 0.05 M 2-amino-2-methyl-1,3-propanediol (ammediol).

[c] Reason for low value unknown.

[d] Samples dialyzed for 48 hours, at pH 2.8 or 9.8, at 4°, then equilibrated for 24 hours at pH 7.35, 4°; C values determined at this pH.

human and rat serum, measured as corticosterone binding.[74] It is evident that some irreversible inactivation takes place at pH 2.8. However, the very low binding affinity at this pH is not entirely explained by this inactivation; the CBG molecule appears to be somewhat protected while in the serum. The decrease of binding affinity from the maximum at pH about 8 to that at pH 9.8 is completely reversible.

[72] I. H. Mills, Recent Progr. Hormone Res. 15, 261 (1959).
[73] P. DeMoor, O. Steeno, and R. Deckx, Acta Endocrinol. 44, 107 (1963).
[74] U. Westphal and G. B. Harding, unpublished.

Concentration of Binding Sites and Association Constants

Table VIII shows the concentration of corticosteroid-binding sites on CBG, for five species, and the association constants for the steroid-CBG complexes at 4° and 37°. These parameters have been determined[63] with pooled normal sera which had been cleared of endogenous steroids by gel filtration at 45°. The equilibrium dialysis technique has been used with five different concentrations of radiolabeled steroid for each value. The concentration of binding sites is determined graphically by plotting the concentrations of unbound steroid against the corresponding total steroid concentration.[75] Since the steroid concentrations applied cover

TABLE VIII
CONCENTRATION OF BINDING SITES AND ASSOCIATION CONSTANTS
OF STEROID-CBG COMPLEXES IN MAMMALIAN SERA[a]

Serum	[CBG] $10^{-7} M$	k for cortisol		k for corticosterone		k for progesterone	
		4° $10^8 M^{-1}$	37° $10^8 M^{-1}$	4° $10^8 M^{-1}$	37° $10^8 M^{-1}$	4° $10^8 M^{-1}$	37° $10^8 M^{-1}$
Human	7.2	6	0.3	10	0.3	7	0.9
Monkey	9.3	3	0.3	—	1.4	—	—
Rat	11.3	3	0.1	5	0.3	3	—
Rabbit	3.4	10	0.4	8	0.2	4	—
Guinea pig	5.7	0.5	0.04	**1.1**	0.14	48	3.9

[a] U. Westphal, *Arch. Biochem. Biophys.* **118**, 556 (1967). Sera cleared of endogenous steroids by gel filtration at 45°. Each association constant, k, from duplicate multiple equilibrium dialysis experiments at 5 concentrations of radiolabeled steroid. Concentration of binding sites, [CBG], average of all values obtained for the serum.

several orders of magnitude, the plot describes high-affinity binding to CBG at the low concentrations of total steroid, and low-affinity binding to serum albumin at the high steroid levels.[67] The intersection of the two parts of the curve indicates the concentration of high affinity binding sites, n[CBG], where n is the apparent number of binding sites per mole CBG, and [CBG] is the concentration of CBG. Since n has been found to equal one for the isolated human CBG (see below), it may be omitted. The value [CBG] is expressed in moles per liter. This graphic determination of [CBG] is approximate since it disregards the binding to albumin; however, the procedure appears justified for the present purpose in view of the 10^3 to 10^5 times larger association constants of the CBG–steroid complexes compared with those of the albumin complexes.

The association constants were determined by the method of recip-

[75] R. E. Peterson, G. Nokes, P. S. Chen, Jr., and R. L. Black, *J. Clin. Endocrinol. Metab.* **20**, 495 (1960).

rocal plots,[9] in a similar way as done by Booth *et al.*[49] The following formulas indicate the calculation of k, the apparent association constant for the complex between steroid hormone and the high-affinity protein, CBG.

$$\frac{1}{[S_{bd}]} = \frac{1}{kn[CBG]} \cdot \frac{1}{[S]} + \frac{1}{n[CBG]} \tag{20}$$

$$y = m \cdot x + a \tag{21}$$

where $[S_{bd}]$ and $[S]$ denote the molar concentrations of bound and unbound steroid, respectively. With the values obtained for the lowest steroid concentrations applied, the reciprocal of the concentration of bound steroid, y, was plotted against the reciprocal of the unbound concentration, x. The apparent association constant, k, was then found from the slope m according to

$$m = \frac{1}{kn[CBG]}; k = \frac{1}{mn[CBG]} \tag{22}$$

The values of $[CBG]$ used in this equation were obtained from the graphic determination described above. The results are shown in Table VIII.

As far as comparative data are available, this procedure[63] yields values that are in agreement with those obtained with highly purified CBG.[13,76,77] Table IX gives the corresponding association constants for human serum albumin and α_1-acid glycoprotein (orosomucoid) calculated for each of two or one binding sites, respectively.

TABLE IX

ASSOCIATION CONSTANTS[a] FOR STEROID-SERUM PROTEIN COMPLEXES

Protein	n^a	k for cortisol		k for corticosterone		k for progesterone	
		4° $10^3 M^{-1}$	37° $10^3 M^{-1}$	4° $10^3 M^{-1}$	37° $10^3 M^{-1}$	4° $10^5 M^{-1}$	37° $10^5 M^{-1}$
Human serum albumin	2^b	3.5	3	3	2.5	1.0^c	0.6^c
Human orosomucoid^d	1	29	18	92	37	10.8	3.7

[a] Intrinsic association constants, k, calculated for each of the number of binding sites, n.

[b] $n = 2$ used for the three steroids.

[c] Values obtained with delipidated crystalline albumin [U. Westphal, *Z. Physiol. Chem.* **346**, 243 (1966)].

[d] Unpublished values obtained by M. Ganguly.

Preparation of a Homogeneous CBG from Human Blood

The purification of human CBG,[76] and subsequently that of rabbit CBG,[77] are given as examples for the isolation from serum of a trace protein that can be recognized only by its high affinity for a radioactive ligand. The procedures described, and appropriate modifications thereof, may be applicable to the isolation of other proteins with analogous properties. Additional details and documenting illustrations have been given elsewhere.[76,77]

Reagents and Materials

Glass-redistilled water of specific resistivity, 2×10^5 ohm cm
Human ACD blood-bank blood, outdated (containing acid-citrate-dextrose)
$CaCl_2$, 1.0 M, reagent grade
Ammonium sulfate, special enzyme grade (Mann Research Laboratories, Inc.); or reagent grade, recrystallized from 1.5×10^{-3} M disodium EDTA
H_2SO_4, 0.1 N
DEAE-cellulose (Eastman-Kodak Co.)
Cortisol-4-^{14}C, specific activity approximately 40 mC/millimole
Tris-phosphate, prepared from tris(hydroxymethyl)aminomethane, primary standard grade; or reagent grade, recrystallized from 95% ethanol
Phosphoric acid, reagent grade
Na_2HPO_4, NaH_2PO_4, reagent grade
Hydroxylapatite (Bio-Rad Laboratories, Inc.)
Sephadex G-25; Sephadex G-200 (Pharmacia Fine Chemicals, Inc.)
Ascorbic acid, reagent grade

Procedure

The procedure[76] is a modification of that employed by Seal and Doe.[71] The final preparation is homogeneous by several rigorous criteria and the method is reproducible. Glass-redistilled water of low conductivity is used throughout. All manipulations are performed at 4° unless otherwise specified.

Step 1. Either normal human plasma or serum is suitable as a source of CBG. Plasma is obtained from ACD blood by gentle aspiration and pooled. It is dialyzed for 24 hours against two changes of 4 volumes of water; the euglobulin precipitate is removed by centrifugation at 8000 g

---

[76] T. G. Muldoon and U. Westphal, *J. Biol. Chem.* **242**, 5636 (1967).
[77] G. J. Chader and U. Westphal, *J. Biol. Chem.* **243**, 928 (1968).

for 20 minutes and discarded. Serum may then be prepared from the plasma by recalcification. A solution of 1.0 M $CaCl_2$ is added to the plasma to a final concentration of 11.5 mg Ca^{++} per 100 ml. Fibrin precipitation is allowed to proceed with gentle stirring for 1 hour at 25°, followed by 8 hours at 4°. The material is then centrifuged for 20 minutes at 8000 g, and the serum is decanted.

Step 2. This step is advantageous only when more than 1 liter of plasma or serum is used; smaller volumes are more conveniently and efficiently handled by direct application of DEAE-cellulose chromatography, as described below. The plasma sample is brought to 40% saturation with a saturated solution of ammonium sulfate. After gentle stirring for 8 hours, the material is centrifuged at 10,000 g for 20 minutes, and the precipitate is discarded. The supernatant is adjusted to pH 6.4 with 0.1 N H_2SO_4 and brought to 63% saturation with saturated ammonium sulfate solution. After 8 hours, this fraction is centrifuged for 20 minutes at 10,000 g and the supernatant is discarded. The precipitate is dissolved in a minimum of water and dialyzed against water until the conductivity of the dialyzate is less than 2 parts per million. The dialyzed protein solution is then adjusted to a concentration of 7 g/100 ml in water. This fraction contains 85% of the total CBG activity and 17% of the total protein applied.

Step 3. DEAE-cellulose is prepared for use as described by Peterson and Sober.[78] The adsorbent is equilibrated with water, and columns of 48 × 460 mm are poured. Under the conditions used, columns of this size are suitable for fractionation of protein quantities of 75 g or less. The sample (plasma or ammonium sulfate fraction) is equilibrated for 24 hours with cortisol-4-^{14}C at a concentration of 2 μg/100 ml. The total, i.e., endogenous plus added cortisol level must exceed the binding capacity of the CBG present, as seen from the appearance of unbound cortisol in the DEAE-cellulose eluate immediately following the void volume. This has to be considered with sera of abnormally high CBG activity such as pregnancy serum. The serum is then applied to the DEAE-cellulose column and allowed to penetrate under gravity at a flow rate of 1 ml/min. Elution with 800 ml of water removes 85–90% of the protein applied to the column; the fraction does not contain CBG and is discarded. At this point a nonlinear Tris phosphate gradient is applied to the column using a nine-chambered Buchler Varigrad (No. 3-6034). The gradient is prepared using 0.005 M Tris-phosphate, pH 8.0, as starting buffer and 0.5 M Tris-phosphate, pH 5.5, as limit buffer (see tabulation).

The gradient elution is performed under a hydrostatic head of 300–

---

[78] E. A. Peterson and H. A. Sober, Vol. V, p. 3.

Chamber No.	Starting buffer (ml)	Limit buffer (ml)	Water (ml)
1	0	0	440
2	220	0	220
3	192.5	27.5	220
4	165	55	220
5	137.5	82.5	220
6	110	110	220
7	82.5	137.5	220
8	55	385	0
9	0	440	0

350 mm at a constant flow rate of 1 ml/min. Fractions of 20 ml are collected. Each fraction is analyzed for radioactivity either by gas flow or by liquid scintillation counting (see above). Protein is determined by measurement of absorbance at 279 mμ. At least 90% of the radioactivity applied is recovered in a single protein peak. Fractions of high specific activity (counts per minute per milligram of protein) are combined, dialyzed against water, and lyophilized. The dry protein fractions are stored at $-85°$ between consecutive steps of the purification procedure. Prolonged exposure to the low temperature does not affect the CBG activity at any stage of purity.

Step 4. The dry protein is dissolved in 0.001 M phosphate buffer, pH 6.8, at a concentration of 20 mg/ml and subjected to a single hydroxyl-apatite gel batch separation as described by Seal and Doe.[71] The gel used may be either a commercial preparation or may be prepared according to the method of Levin.[79] The hydroxylapatite is equilibrated to the 0.001 M sodium phosphate buffer, pH 6.8, and lightly packed by centrifugation for 5 minutes at 1000 g. The protein solution is then mixed with an equal volume of the lightly packed hydroxylapatite gel and allowed to stand for 1 hour at 4°. The mixture is centrifuged at 2000 g for 10 minutes. The precipitate is washed twice with volumes of buffer each equivalent to half the original volume of protein solution, and the three supernatants are combined and analyzed for protein and radioactivity. This material is then dialyzed against water, lyophilized, and stored at $-85°$.

Step 5. Hydroxylapatite is equilibrated with 0.001 M phosphate buffer, pH 6.8, and poured in a thin slurry into columns to a final size of 25 $\times$ 250 mm. After settling of the columns for 2 hours at maximal flow rate, 5 ml of a solution containing 50–60 mg protein per milliliter of 0.001 M

[79] O. Levin, Vol. V, p. 27.

phosphate, pH 6.8, are applied. Elution is performed with the same buffer at a flow rate of about 0.2 ml/minute. Fractions of 3 ml are collected and analyzed. Active protein fractions are pooled, desalted by gel filtration with Sephadex G-25, and lyophilized. Hydroxylapatite column chromatography is repeated under identical conditions until a product of constant specific activity is obtained; this requires up to five chromatographies, depending on the nature of the material.

Step 6. Protein which is homogeneous by hydroxylapatite chromatography is dissolved at a concentration of 10 mg/ml in 0.05 M sodium phosphate buffer, pH 7.4, containing 5×10^{-4} M ascorbate. Three milliliters of this solution is applied to a 14×210 mm column of Sephadex G-200 and the protein is eluted with the same buffer. Fractions of 3 ml are collected and analyzed for radioactivity and protein content. The filtration procedure is repeated until protein and radioactivity are eluted in a single symmetrical peak of constant specific activity.

Pure human CBG (12–13 mg per liter of plasma) has been isolated by this procedure with an overall yield of about 50% and with a purification factor of approximately 4000. Table X summarizes the results obtained by this method. The final preparation is found to be

TABLE X

CBG Isolation Procedure from 600 ml of ACD Plasma[a]

Purification step	Protein (mg)	Specific activity (cpm/mg protein)	Purification[b]	Recovery Total (%)	Recovery Utilized[c] (%)
DEAE-cellulose column eluate	3600	24	11.3	92	73
Hydroxylapatite batch fractions	530	141	6.0	87	87
Hydroxylapatite column eluate	10.7	5460	42.2	87	85
Sephadex G-200 column eluate	7.2	7530	1.38	98	91
Final CBG preparation	7.2	7530	3950	68	49

[a] Preliminary ammonium sulfate fractionation (step 2) was not performed. With plasma samples of 2–3 liters, initial salt fractionation gives a 5-fold purification with an 85% yield.

[b] Purification is calculated on the basis of increase in specific activity at each step in the isolation procedure, after correction for losses of radioactivity upon dialysis.

[c] The percentage actually introduced into the next fractionation step. For instance, 100,000 cpm in starting plasma would give 92,000 cpm total recovery in DEAE-cellulose eluate; 73,000 cpm, constituting the fractions of highest specific activity, were subjected to hydroxylapatite batch fractionation. The remaining active fractions (corresponding to 19,000 cpm) were combined from several experiments and rechromatographed. This improves the overall yield to a value greater than 49% and not higher than 68%. However, this additional yield is not included in the 7.2-mg final preparation listed in this table.

homogeneous by chromatography, by free boundary, paper, and agar electrophoresis, immunoelectrophoresis, sedimentation velocity and diffusion. One binding site for cortisol is found by equilibrium dialysis at 4° and at 37°; the association constants at these temperatures are $5.2 \times 10^8 M^{-1}$ and $2.4 \times 10^7 M^{-1}$, respectively. Complete removal of sialic acid does not significantly affect the cortisol-binding affinity. Pure CBG is inactivated by gel filtration at 45° or by dialysis against glass-redistilled water. The physicochemical properties of human CBG are given in Table XII.

Preparation of a Homogeneous CBG from Rabbit Serum

A brief description of this procedure is given since it differs in several ways from the above preparation of human CBG. Full details and illustrations have been published.[77]

Reagents and Materials

Rabbit serum, from female animals (Pel-Freez Biologicals, Inc., Rogers, Arkansas)
DEAE-Sephadex (Pharmacia Fine Chemicals, Inc.)
Other reagents as specified for the preparation of human CBG

Procedure

One liter of serum is dialyzed for 24 hours against 2 to 3 changes of doubly distilled water; it is not found necessary to remove the small quantities of euglobulin precipitated. All experiments are conducted at 4° unless otherwise noted. The serum is then added to a container in which a methanolic solution of 2.5 μg cortisol-4-^{14}C has previously been evaporated to a dry film. The mixture is stirred at room temperature for 1–2 hours and at 4°C for 2–3 hours; it is then applied to a 45×570 mm DEAE-Sephadex column. A 450 ml sample of $0.0025 M$ Tris-phosphate, pH 7.5, is passed down the column followed by 1.5 liter of $0.06 M$ Tris phosphate, pH 7.0. Eluted protein is discarded. A gradient is applied (see accompanying tabulation) by means of a Buchler Varigrad apparatus with a total of 440 ml in each of seven chambers. The starting buffer is $0.0075 M$ Tris-phosphate, pH 8.0, and the limit buffer is $0.4 M$ Tris-phosphate, pH 5.5. Flow rate is maintained at 1.0 ml per minute. Protein and radioactivity are determined; fractions of high specific activity are combined, dialyzed, and lyophilized.

The lyophilized protein is dissolved at a concentration of 75–100 mg/ml in $0.05 M$ phosphate buffer, pH 7.4, and applied to a 45×550

mm column of Sephadex G-200 equilibrated with the phosphate buffer. After elution with the same buffer, fractions of high specific activity are combined, partially dialyzed and lyophilized. This pooled material is then rechromatographed on a third Sephadex G-200 column in a similar manner.

The combined dialyzed and lyophilized fractions from the last Sephadex chromatography are subjected to successive chromatographies on small hydroxylapatite columns of about 25×100 to 150 mm equilibrated with $0.005\,M$ phosphate buffer of pH 6.8. Protein is applied to these columns at a concentration of about 60–80 mg/ml in the $0.005\,M$ buffer in amounts not greater than 750–1000 mg per column. Stepwise elution is performed with $0.005\,M$, $0.02\,M$, $0.05\,M$, and $0.2\,M$ phosphate buffers of pH 6.8; the next higher buffer concentration is used only when the protein concentration eluted by a particular buffer falls markedly

Chamber	Starting buffer (ml)	Limit buffer (ml)	Water (ml)
1	192.5	27.5	220
2	165	55	220
3	137.5	82.5	220
4	110	110	220
5	82.5	137.5	220
6	55	275	110
7	0	440	0

and approaches zero. The protein fractions of high specific activity are combined, equilibrated with $0.005\,M$ phosphate buffer, pH 6.8, by passage over Sephadex G-25, and rechromatographed until the eluted CBG protein peak exhibits constant specific radioactivity in symmetrical fashion on both the ascending and descending limbs. Three or four chromatographies are usually necessary before the eluted CBG is homogeneous by the criteria given below. The final product is desalted on Sephadex G-25.

Hydroxylapatite used in these experiments was obtained from Bio-Rad Laboratories, Inc. It was found to have considerably higher adsorptive affinity than batches prepared in this laboratory according to Tiselius et al.[79] It was thus necessary to use buffer strengths about 5-fold higher with commercial apatite preparations to elute the CBG fractions. In both cases it became increasingly difficult to elute the CBG fractions with the lower buffer strengths as the preparation became more pure.

Since the amount of CBG in rabbit serum is relatively low (12–14 mg/liter), it is advantageous to start with more than 1 liter of serum in this procedure. The purification steps summarized in Table XI show the procedure starting with 2 liters of serum. The final homogeneous preparation obtained was 13.5 mg of CBG, at a yield of about 50% and an overall 6200-fold purification. In a number of experiments, the yield varied from 50% to about 75%. The final CBG preparation has been found homogeneous by chromatography, free boundary and paper electrophoresis, immunoelectrophoresis, sedimentation velocity at different concentrations, and diffusion.

TABLE XI

PURIFICATION PROCEDURE FOR RABBIT CBG[a]

Purification step	Protein applied (g)	Label applied (cpm)	Purified fraction (g)	Recovery (cpm)	Percent cumulative recovery	Specific activity (cpm/mg protein)[b]
DEAE-Sephadex	156	593,556	8.73	544,975	92	68
Sephadex G-200						
1st	8.73	544,975	3.18	524,370	88	169
2nd	3.18	524,370	2.17	473,200	80	218
Hydroxylapatite						
1st	2.17	473,200	0.890	367,455	62	412
2nd	0.89	367,055	0.471	356,880	60	758
3rd	0.471	357,280	0.102	340,095	57	3,330
4th	0.102	340,095	0.0135	317,720	53	23,500

[a] The procedure outlined starts with 2 liters of rabbit serum; this is divided in half and applied to two DEAE-Sephadex columns. The total resulting CBG fraction (8.73 g) is then applied to a Sephadex G-200 column. The CBG fractions are combined and rechromatographed on Sephadex G-200. The CBG fraction (2.17 g) is again divided in half and applied to two small hydroxylapatite columns; the eluted CBG fractions are combined and subjected to three further hydroxylapatite chromatographies.

[b] Specific activity of starting material 3.8 cpm per milligram of protein.

The pure rabbit CBG, isolated under conditions of saturation with endogenous corticosterone and added cortisol-4-^{14}C, contains 1 mole of corticosteroid per 40,700 g of CBG; the number of high-affinity binding sites has been found to be $n = 1$. The association constants at 4° and 37° are $k = 9.0 \times 10^8 \, M^{-1}$ and $4.7 \times 10^7 \, M^{-1}$, respectively. Thermodynamic calculations show negative enthalpy change and negative entropy change for the interaction. Removal of corticosteroid from the isolated pure CBG by gel filtration at 45° results in loss of binding activity.

Physicochemical Properties of CBG Preparations

The CBG preparations isolated in different laboratories from human plasma[71, 76, 80] and from rabbit serum[77] have been reported to be homogeneous by several criteria. They are glycoproteins and belong to the

TABLE XII
Physicochemical Properties of CBG and Orosomucoid

Property	Human CBG[a]	Human CBG[b]	Rabbit CBG[c]	Human orosomucoid[d]
Sedimentation coefficient, $S_{20,w}$	3.79S^e	3.0	3.55[e]	3.2[e]
Diffusion coefficient, $D_{20,w}{}^f$	6.15[e]	—	7.02[e]	6.54
Partial specific volume (ml/g)	0.708	0.718	0.695	0.689[g]
Molecular weight	51,700	52,000[h]	40,700	41,600[g]
Electrophoretic mobility[i]	4.9	—	5.1	5.4
Extinction coefficient $E_{1cm}^{1\%}$ at 279 mμ	6.45	7.4	8.4	9.3
Carbohydrate content, %	26.1	14.1	29.2	40.5
Hexose, %	11.5	5.4	10.4	14.6
Hexosamine, %	9.0	4.7	9.5	12.7
Fucose, %	1.5	0.8	0.8	1.2
Sialic acid, %	4.1	3.2	8.5	12.0
Number of binding sites[j]	1	1	1	1

[a] T. G. Muldoon and U. Westphal, *J. Biol. Chem.* **242,** 5636 (1967).

[b] U. S. Seal and R. P. Doe, *J. Biol. Chem.* **237,** 3136 (1962).

[c] G. J. Chader and U. Westphal, *J. Biol. Chem.* **243,** 928 (1968).

[d] Unpublished studies with J. Kerkay. The carbohydrate values were averaged with data obtained by M. Ganguly, R. H. Carnighan, and U. Westphal, *Biochemistry* **6,** 2803 (1967).

[e] Extrapolated to infinite dilution.

[f] $D_{20,w}$ is the diffusion coefficient given in Fick units (10^{-7} cm²/sec) and reduced to water at 20°.

[g] Bezkorovainy, *Biochim. Biophys. Acta* **101,** 336 (1965).

[h] U. S. Seal and R. P. Doe, *in* "Steroid Dynamics" (G. Pincus, T. Nakao, and J. F. Tait, eds.), p. 63. Academic Press, New York, 1966.

[i] Determined by the Tiselius method in Veronal buffer pH 8.6, and given in -10^{-5} cm²/volt/sec.

[j] High-affinity binding sites for cortisol (CBG) and progesterone (orosomucoid).

electrophoretic group of α_1-globulins. The high-affinity corticosteroid binding proteins in the serum of other species, as far as examined, are also α-globulins.[81] Some physicochemical properties of human and rabbit CBG are listed in Table XII, in comparison with the corresponding data

[80] W. R. Slaunwhite, Jr., S. Schneider, F. C. Wissler, and A. A. Sandberg, *Biochemistry* **5,** 3527 (1966).

[81] U. Westphal and F. DeVenuto, *Biochim. Biophys. Acta* **115,** 187 (1966).

for human α_1-acid glycoprotein (orosomucoid). The reason for the discrepancies in the carbohydrate content of human CBG prepared in two laboratories[71, 76] is not known. The amino acid compositions of human CBG,[13, 76, 80, 82] rabbit CBG,[77] and human α_1-acid glycoprotein[15, 83, 84, 85] have been reported.

[82] U. S. Seal and R. P. Doe, *Proc. 2nd Intern. Congr. Endocrinol., London,* 1964. (*Excerpta Med.* **83,** p. 325.)

[83] A. Bezkorovainy and D. G. Doherty, *Nature* **195,** 1003 (1962).

[84] W. E. Marshall and J. Porath, *J. Biol. Chem.* **240,** 209 (1965).

[85] K. Schmid, T. Okuyama, and H. Kaufmann, *Biochim. Biophys. Acta* **154,** 565 (1968).

Author Index

Numbers in parentheses are reference numbers and indicate that an author's work is referred to although his name is not cited in the text.

A

Abelson, D., 21
Abraham, R., 39, 709
Acevedo, H. F., 736, 737(3)
Acker, L., 44
Adamec, O., 39, 150
Adams, J. A., 692, 720(35), 746, 748(1a), 751(1a), 752
Adams, J. B., 733, 734(195)
Adler, R. E., 771, 783(50)
Adlercreutz, H., 49, 691
Aexel, R., 498
Ageta, H., 201, 202(10), 213, 214, 216, 217
Agnello, E. J., 334
Agranoff, B. W., 425, 426, 437(40)
Aguilera, P. A., 37, 155(214), 158(214)
Ahrens, E. H., Jr., 238, 246(4), 272, 278 (64), 288
Ahmad, N., 174
Akahori, A., 31
Akhrem, A. A., 3
Aldrich, A., 597
Alexander, J. A., 692, 752
Alfin-Slater, R. B., 533
Ali, S. S., 279
Alkemeyer, M., 30
Allen, R. G., 768, 769, 774(44), 775, 776, 777
Amelung, D., 756
Ames, B. N., 675
Ammon, R., 690
Anacker, W. F., 646, 750
Anders, D. A., 201, 218(5)
Anderson, C. A., 55
Anderson, D. G., 456, 459
Anfinsen, C. B., 562, 563
Angliker, E., 22, 42(95), 43(95)
Anker, L., 141
Anthony, W. L., 34
Antoniades, H. N., 761, 771(2), 783(2, 4)
Antonucci, R., 342
Aoki, H., 467

Aoshima, Y., 654
Aplin, R. T., 216, 268
Applegate, H. E., 312
Aquilera, P. A., 255
Archer, B. L., 426, 428(39), 476, 477, 480
Arthur, H. R., 216
Ashley, B. D., 762, 764(17)
Attal, J., 157
Auda, B. M., 539
Audley, B. G., 476
Aurich, O., 63, 108, 113, 118
Averill, W., 159(5), 160
Avigan, J., 514, 515(1), 516(4), 517(1, 3, 4), 519(4), 520(1, 3, 4), 521(1, 4, 35), 63, 67, 71
Axelrod, J., 722, 723(147), 724
Axelrod, L. R., 44, 321, 702, 723
Axen, V., 124
Ayres, P. J., 615
Azarnoff, D. L., 52

B

Baba, S., 344
Bacharach, A. L., 761, 783(5)
Baggett, B., 691, 704, 714
Bagli, J. F., 505
Bagnoli, E., 689, 690(12)
Bailey, E., 179
Baisted, D. J., 498
Baitsholts, A. D., 6
Bakinouskii, L. V., 109
Balderas, L., 704
Balke, E., 709
Ball, S., 468
Balle, G., 48
Bandi, L., 355, 356(c), 357(20, c), 358
Banerjee, R. K., 732
Banes, D., 310
Barbier, M., 26, 35(134), 38(134), 54, 85
Bardawill, C. J., 766
Barker, H. A., 578
Barnard, D., 426, 428(39), 480

Gray, C. H., 761, 768, 771, 774, 783(5), 787(49)
Gray, W. C., 523
Green, S., 154, 156(351)
Greengard, P., 598, 603, 615(30), 618, 620, 621(12, 30), 623(30)
Greenspan, G., 133, 135
Greenway, J. M., 771, 787(49)
Gregg, J. A., 76, 81, 250, 251
Greve, H., 44
Griffiths, K., 706
Grill, P., 702
Groen, D., 189
Gross, D., 723(154), 724
Grossi, E., 521
Groot, K., 152, 355
Grove, J. F., 481(2), 482, 490
Grower, D. B., 199
Grubb, R., 271
Grundy, S. M., 272, 278(64), 288
Gual, C., 704, 705(67)
Gubler, A., 675
Guerra, S. L., 624, 625(58), 766, 772(36)
Guerra-Garcia, R., 185
Gürtler, J., 523
Guimarães, C. V., 35, 36(201), 46(201)
Gupta, G. N., 762
Gurin, S., 562, 563(7), 569(3, 7), 571(7), 572(7), 585
Gustafsson, J.-Å., 47
Gut, M., 308, 321, 325, 327, 328(31), 330, 337, 342, 343, 344, 585, 598, 637(14a), 704, 705(67)
Guther, A., 22, 37(98), 61
Gutzwiller, J., 324

H

Haahti, E. O. A., 164
Haaki, E., 33
Haenny, E. O., 310
Hänsel, R., 30
Hagen, A. A., 31, 756
Hagopian, M., 714
Halkerston, I. D. K., 585, 690
Hall, N. F., 5
Hall, P. F., 199
Halla, M., 714, 752
Halpaap, H., 8
Halpern, E., 281
Hamilton, J. G., 251, 256, 524

Hamilton, P. B., 220
Hamilton, R. J., 200
Hamlet, J. C., 290
Hammarsten, O., 27
Hampl, R., 708
Hanaineh, L., 166
Hanč, O., 702, 706(50)
Hancock, J. E. H., 597
Handloser, J. H., 46
Hanford, W. E., 361
Hanson, J. R., 481(2), 482, 490
Hanze, A. R., 289
Hara, S., 76, 79, 80, 87, 95, 108
Harashima, K., 467
Harding, B. W., 607, 637
Harding, G. B., 765, 768, 769, 774(44), 785
Harel, S., 61, 63
Harris, F., 154, 156(351)
Hartman, I. S., 164, 166, 168, 174
Hartman, J. A., 308, 310
Hashimoto, Y., 728
Haslewood, G. A. D., 31, 199, 237, 249, 253(1)
Hasunuma, M., 92, 93
Haupt, M., 736, 737(12)
Haupt, O., 691, 692(14), 709, 711(100), 712, 713(93), 714(93), 715(93, 105), 717(14)
Havel, R., 544, 547(6)
Hayaishi, O., 462, 470(12), 471(12), 474 (12)
Hayano, M., 325, 327, 344, 585, 597, 598, 604(14), 612(14), 628(14), 657, 702, 704, 705(67)
Hayes, F. N., 767
Heard, R. D. H., 27
Hechter, O., 585, 615
Hecker, E., 710, 714(89), 715(89), 736, 737(2, 10), 739(2), 741(2), 742(10), 743(7, 8), 744(8), 745(7, 16)
Heftmann, E., 3, 4, 16, 17, 22, 45, 48, 49 (267), 61, 67(7), 71(77), 94, 95(77), 103(77), 108, 109(312), 150, 498, 517
Heilbron, I. M., 345
Heirwegh, K., 766, 771(34), 772(34), 773 (34), 774(34), 783(34)
Heiselt, L. R., 597
Heitzman, R., 200
Held, G., 111

Subject Index

Abbreviations used in the index: CC, column chromatography; GLC, gas-liquid chromatography; IEC, ion-exchange chromatography; PC, paper chromatography; TLC, thin-layer chromatography; T, table; UV, ultraviolet. T followed by a number (both in parentheses) means that the entry is the systematic name of a compound numbered as indicated and listed in a table under a trivial name. Methyl esters of bile acids are listed under Methyl and the name of the acid.

Aromatase, 703
Aromatic aldehyde, in reagents for
    visualization in TLC, 28–32
Aromatic hydroxylation, and protein
    binding of estrogens, 736–746
Aromatic steroids, and GLC, 169
Aromatization, 702–708
    general intermediates of, 702
    in microorganisms, 706
    of C_{19}-norsteroids, 706
    in placental microsomes, 703–706
    of ring B, 707–708
        assay method, 704
    of steroids, by human placenta, 705
Aromatization system
    distribution of, 704
    placental
        assay method, 703
        properties of, 704–706
Arsenomolybdic acid, in TLC, 26
Arundoin, relative retention times on
    GLC, 215
Arylsulfatases, 689–690
Aryl-sulfate sulfohydrolases, 689–690
Ascending TLC, development, 13–14
Ascorbate inhibition, 630
Asiatic acid, mobilities on TLC, 127
Asiatic acid lactone, mobilities on TLC,
    129
Association constants, of corticosteroid-
    binding globulin, 786–787
ATP:mevalonate 5-phosphotransferase,
    402–410
ATP:5-phosphomevalonate
    phosphotransferase, 413–417
ATP:5-pyrophosphomevalonate carboxy-
    lyase, 419–425
Atiserene, mobilities on TLC, 484
Autoradiograms, 46
Avenacoside, mobilities on TLC, 110
AY-9944, 503, 513, 520

B

Bacterial enzymes, induced, 280
Barringtogenol C, TLC of, 125
Bassiac acid, mobilities on TLC, 127
Bauerenol methyl ether, relative retention
    times on GLC, 215
Bayogenin
    mobilities on TLC, 126
    TLC of, 119

Bayogenin lactone, mobilities on TLC, 129
Benzaldehyde, in TLC, 31, 32
Benzoate formation, for TLC, 49
Benzoyl chloride, 49
Betulafolianetriol tritrimethylsilyl ether,
    relative retention time on GLC, 205
Betulic acid, mobilities on TLC, 127
Betulin
    mobilities on TLC, 121, 123
    relative retention time on GLC, 210
Betulin diacetate
    mobilities on TLC, 121, 123
    relative retention time on GLC, 210
Betulin trimethylsilyl ether, relative
    retention time on GLC, 210
Betulinic acid, mobilities on TLC, 127
Bidimensional TLC, multiple runs, 17
Bile, 270
    bile acids in, 270–272
Bile acid(s), see also individual systematic
    names
    absorption chromatography of 238–242
    analytical separation of, 246–280
    in bile, 270–272
    biosynthesis
        assay method, 557
        enzymatic degradation of side chain,
            562–582
        enzyme systems, 551–562
            purification procedures, 558–561
        extraction of incubation mixtures, 554
        incubation conditions, 554
        preparation of enzyme systems,
            553–554
        steam distillation of propionic acid,
            577
    in blood, quantitative analysis, 272–274
    characterization of, in TLC, 20
    chromatographic identification, in
        biological materials, 268–270
    conjugated
        defined, 238
        hydrolysis of, 271
        retention volumes in reversed phase
            CC, 245
        TLC of, 81, 250–251
    detection in TLC, 22, 23, 26, 27, 29, 32,
        34, 46, 253
    dimethylhydrazone formation of, 260
    with elongated side chains, 253

Diterpenes
  cyclic, 481
  enzymatic synthesis of, 481–490
  TLC of, 483
DNPH, see 2,4-Dinitrophenylhydrazine
Dodecyl sulfate, as detergent for enzyme
    substrate, 472
Double bonds
  behavior on GLC, 216
  ditertiary, detection in TLC, 44
  isolated, characterization of, in TLC, 20
  GLC and, 166
  reactions for, in TLC, 44
Dowex 50W-X4
  preparation of column, 221
    of resin, 220
  separation of estrogens, 232
Doysinolic acid, mobilities on TLC, 131
Dragendorff's reagent, 33
DS-5, Camag, 4

E

Echinocystic acid methyl ester, mobilities
    on TLC, 129
EDTA, 513
Ekkert, reaction of, 28–29
Elatography, 53
  TLC of, 54
Electrolytic coupling of carboxylic acids,
    553
Electron capture in GLC, 160–161
Electron capture detectors, contamination
    of, 161
Eluents for ion-exchange chromatography,
    222
Eluotropic series of solvents, 10
Elution
  of compounds from adsorbents, 155–158
    choice of solvents, 156–157
  on IEC, of 17-hydroxycorticosteroids,
    226–228
    of 17-ketosteroids, 226–230
Emmolic acid, mobilities on TLC, 127
Emmolic acid dimethyl ester, mobilities on
    TLC, 129
Emulsification of substrate by detergents,
    472
Enolization, and preparation of
    halogenated esters, 164, 189

Enoyl-CoA hydratase, and bile acid
    biosynthesis, 580
Enzymatic assay of bile acids and related
    3α-hydroxysteroids, 280–288
14α-Epiadynerigenin acetate, TLC of, 18
Epi-β-amyrin, mobilities on TLC, 120, 122
Epiandrosterone, see also 3β-Hydroxy-5α-
    androstan-17-one
  detection in TLC, 33
  elution on IEC, 226, 230
  IEC of, 229
  relative elution volume on IEC, 224
  TLC of, 133
Epiandrosterone sulfate, TLC of, 153
3-Epicanarigenin, TLC of, 89
Epicholestanol
  mobilities on reversed-phase TLC, 64
    on TLC, 55, 56, 58
      with silver nitrate, 68
Epicholestanol dinitrobenzoate, TLC of,
    55
Epicholesterol
  mobilities on reversed-phase TLC, 64
    on TLC, 56, 58
Epicholesterol acetate, mobilities on
    reversed-phase TLC, 64
Epicoprostanol
  mobilities on reversed-phase TLC, 64
    on TLC, 55, 56, 58
      with silver nitrate, 68
11α-Epicorticosterone, detection in TLC,
    37
3-Epidigoxigenin, mobilities on TLC,
    90–91
16-Epiestriol, 125, see also Estra-1,3,5(10)-
    triene-3,16β,17β-triol
  elution on IEC, 232–234
  formation of, 714
  IEC of, 235
  mobility on PC, 699
  PC of, 715
  retention times in GLC, 171
  TLC of, 51
  urinary, analysis of, 176
16-Epiestriol acetate, retention times in
    GLC, 171
16-Epiestriol 3-methyl ether, PC of, 715
16-Epiestriol trimethylsilyl ether,
    retention times in GLC, 171

I

Identification
  of bile acids, 268–270
  of compounds by GLC, 173
Ilinski-von Knorre reaction, 43
18,20-Imino-5α-pregnane, mobilities on
    TLC, 115, 117
Impregnation of plates in TLC, 6–8
  with boric acid, 8
  with ethylene glycol, TLC and, 6
  with formamide, TLC and, 7
  with silver nitrate, 8
  with sodium borate, 8
  with undecane, 7
Inhibitors of cholesterol biosynthesis,
    513
Inouye, reaction of, 29–30
Iodine, in TLC, 37
Iodine reagent, visualization in TLC, 37
Iodine vapor, in TLC, 37
Ion-exchange chromatography
  analysis of urinary steroids, 231
  of bile acids, 273
  composition of eluents, 222
  flow rate, 230
  preparation of column, 221–223
    of stationary phase, 220
  recovery of steroids from column, 225
  separation of steroids, 219–237
  of steroids, initial separation, 225
Ishidate reaction, 41
Isoandrosterone 3β-sulfate, physical
    properties of, 356
Isoarborinol methyl ether, relative
    retention times on GLC, 215
Isoatiserene, mobilities on TLC, 484
Isochiapugenin, mobilities on TLC, 105,
    107
Isoconessimine, mobilities on TLC, 115,
    117
Isofernene, relative retention times on
    GLC, 211
Isokaurene
  mobilities on TLC, 484
  TLC of, 483
Isomasticadienonic acid, mobilities on
    TLC, 63
Isomiliacin, relative retention times on
    GLC, 215
Isomorphism, 307

Isonicotinic acid hydrazide, in TLC, 38
Isonicotinic hydrazide, 50
Isonicotinic hydrazone formation, for
    TLC, 50
Isopentenol, 382, 383
  GLC of, 432
  phosphorylation of, 385
Isopentenol-1-^{14}C, 382
Isopentenyl phosphate, dilithium,
    synthesis of, 383
Isopentenyl phosphoromorpholidate,
    synthesis and application, 383, 384
Isopentenyl pyrophosphate, 431, 433, 454,
    459, 476, 494
  analysis of, 384
  chromatography of, 385
  electrophoresis of, 450
  enzymatic preparation from mevalonate,
    450
  incorporation into rubber, 476
  isomerization reaction mechanism, 425,
    426
  synthesis of, 382–385
  trilithium, synthesis of, 383, 384
Isopentenyl-1-^{14}C pyrophosphate, 385,
    490–492
Isopentenyl-4-^{14}C pyrophosphate, 455
Isopentenyl pyrophosphate isomerase,
    425–438, see also Prenyl isomerase
  assay for, 456–457
  distribution in ammonium sulfate
    fractions, 429
  purification steps (T), 430
Isophyllocladene, mobilities on TLC, 484
Isopimaradiene, mobilities on TLC, 484
Isoprenoid hydrocarbons, mobilities on
    TLC, 484
Isopropyl triphenyl phosphonium iodide,
    348
Isopyrocalciferol, mobilities on TLC, 57,
    59
Isoramanon, TLC of, 151
Isoreductodehydrocholic acid, mobilities
    on TLC, 73, 75
Isorhodeasapogenin, mobilities on TLC,
    105, 107
Isosawamilletin, relative retention times
    on GLC, 215
Isoserratene, relative retention times on
    GLC, 214

Linoleic acid, as detergent for enzyme
  substrate, 472
Lipid, peroxidation, 630
Lipophilic substances, detection in TLC,
  46
Liquid phases for GLC of triterpenes, 200
Liquid scintillation counting of aqueous
  samples, 767
Lithium borodeuteride reduction, 373
Lithium borotritide reductions, 374–375
Lithocholic acid, *see also* 3α-Hydroxy-
  5β-cholanoic acid
  CC of, 245
  detection in TLC, 32
  detection of, 281
  loss on aluminum oxide
    chromatography, 274
  mobilities on TLC, 72, 74
  from serum, 288
Liver enzymes, *see* specific enzymes
Liver factor, in steroid hydroxylations,
  preparation and properties of, 612–613
Longispinogenin triacetate, relative
  retention time on GLC, 208
Longispinogenin trimethylsilyl ether
  GLC of, 201
  relative retention time on GLC, 208
Lophenol, mobilities on TLC, 62
Low-temperature TLC, 19
Lumisterol, mobilities on reversed-phase
  TLC, 65
Lupane compounds, relative retention time
  on GLC, 210
Lupane derivatives, correlation of
  retention times, 216
α-Lupene, relative retention times on
  GLC, 210, 213
α-Lupene terpenes and derivatives,
  mobilities on TLC, 121, 123
Lupeol
  GLC of, 203
  mobilities on TLC, 121, 123
    with silver nitrate, 69
  relative retention time on GLC, 210
  in silkworm blood, isolation by GLC,
    219
Lupeol acetate, relative retention time on
  GLC, 210
Luvigenin, mobilities on TLC, 104, 106
Lycopersane, GLC of, 456
Lycopersene, mobilities on TLC, 484

Lysolecithin, 544
  as detergent for enzyme substrate, 472

M

Machaeric acid, mobilities on TLC, 127
Magnesium oxide, TLC and, 5
Majaloside, mobilities on TLC, 97, 99
Mallogenin, mobilities on TLC, 90–91
Malloside, mobilities on TLC, 97, 99
Marinobufogenin, mobilities on TLC, 93
Mass spectrometry
  combined with GLC, 267–268
  of 4-D.-labeled mevalonic lactones, 374
Masticadienolic acid, mobilities on TLC,
  63
Matthews, reaction of, 30
MER-29, *see* Triparanol
Medicagenic acid, mobilities on TLC, 127
Mercury arc lamp, in visualization of
  compounds on TLC, 22
Mesityl oxide formation from acetone on
  aluminum oxide, 241
Metagenin, mobilities on TLC, 105, 107
Metal salts, for visualization in TLC,
  35–37
Meteogenin, mobilities on TLC, 104, 106
Methanol, in TLC, 12
Methanolic nitroprusside paste reagent, 41
β-Methasone, recovery after TLC, 155
Methostenol, mobilities on TLC, 62
p-Methoxybenzaldehyde, in TLC, 28
2α-Methoxy-4,4-dimethylcholestan-3-one,
  R_f in TLC, 54
2β-Methoxy-4,4-dimethylcholestan-3-one,
  R_f in TLC, 54
3-Methoxyestra-2,5(10)-dien-17β-ol-6,7-
  [3]H, synthesis of, 339
3-Methoxyestra-2,5(10)-dien-17-one-6,7-
  [3]H, synthesis of, 339
2-Methoxy-17β-estradiol
  formation of, 721
  16α-hydroxylation of, 714
  mobility on PC, 698
  occurrence in man, 691
  PC of, 722
2-Methoxyestradiol-17β, demethylation of,
  723
3-Methoxyestradiol-17β, demethylation of,
  723
2-Methoxyestra-1,3,5(10)-triene-3,17β-
  diol, mobilities on TLC, 131

Mevalonate, 5R-5-^{3}H$_1$-labeled, preparation of, 398

Mevalonate, 5R-5-D$_1$-labeled, preparation of, 398

Mevalonate kinase, 400, 402, 410, 411, 445
assay of, 402–407
from pig liver, purification procedures (T), 408
preparation of, 444
properties of, 410
purification of, 407

Mevalonate: NAD oxidoreductases, 394

Mevalonate: NADP oxidoreductases, 394

Mevalonic acid, see Mevalonate, Mevalonolactone

Mevalonic lactone, synthesis of, 377

Mevalonolactone, see also Mevalonate
labeled, synthesis of, 361
synthesis of, 360, 361, 367, 377

Mevalonolactone, 4-D$_1$-labeled, mass spectrometry of, 374

4-^{14}C-Mevalonolactone, synthesis of, 365–367

4-T$_2$-Mevalonolactone, synthesis of, 365

5-^{14}C-Mevalonolactone, synthesis of, 365

5-D$_2$-Mevalonolactone, synthesis of, 365–367

R-Mevalonolactone, synthesis of, 362–365

R-(−)-Mevalonolactone, properties of, 364

Mevalonolactone-3′,4-^{13}C$_2$, synthesis of, 362

Microbial steroid esterases, 675–684

Microchemical reaction on layers of TLC, 54

Microcircular TLC, 10

Microorganisms, and aromatization, 706

Microsomal enzymes, intrinsic, 636

Microsomal fraction, preparation of, 697

Microsomal suspension, clarification of, 626

Microsomes
adrenocortical, preparation of, 599–601
liver, preparation of, 452

Migration, order of, in TLC, of some sterols, 55

Miliacin, relative retention times on GLC, 215

Milius reaction, in TLC, 37

Mirtillogenic acid, mobilities on TLC, 126

Mitochondria
acetone powders and freeze-dried preparations, 612–614
adrenocortical, 598
preparation of, 599–601
incubation of, 605

Mobilities
on PC, see individual compounds
on reversed phase TLC, see individual compounds
on TLC, see individual compounds

Mobility in chromatography, see specific compound or group name of compounds

Molecular ion peak, 270

Molecular shape
GLC retention times, 217
retention time in GLC, 165–166

Molecular weight, and retention time in GLC, 166

Molecular weight determination of bile acids, 270

Monoacetylcratalgolic acid methyl ester, mobilities on TLC, 129

Monochloroacetates in GLC, 164

Monochloroacetic anhydride, 186

Monochloroacetylation, 188

Monochlorodifluoroacetic anhydride, 182, 184

Monomethyl 3-hydroxy-3-methyl-glutarate, synthesis of, 377

Monoolein, as detergent for enzyme substrate, 472

Monosubstituted compounds, and GLC, 166–168

Moretane, relative retention times on GLC, 211

Morin, visualization in TLC, 21

Morolic acid, mobilities on TLC, 126

Morpholine, 382

Multiflorenol methyl ether, relative retention times on GLC, 215

Multiple bidimensional TLC, 17

Multiple one-dimensional TLC, 14–16

N

NADH-4R-4-D$_1$, preparation of, 399

β-Naphthol, in TLC, 34

β-Naphthol-sulfuric acid reagent, 34

Naphthoquinone-perchloric acid reagent, 35